Fungal Biology

Series Editors

Vijai Kumar Gupta, Biorefining and Advanced Materials Research Center, Scotland's Rural College (SRUC), SRUC Barony Campus, Parkgate, Dumfries, Scotland, UK

Maria G. Tuohy, School of Natural Sciences, National University of Ireland Galway, Galway, Ireland

Fungal biology has an integral role to play in the development of the biotechnology and biomedical sectors. It has become a subject of increasing importance as new fungi and their associated biomolecules are identified. The interaction between fungi and their environment is central to many natural processes that occur in the biosphere. The hosts and habitats of these eukaryotic microorganisms are very diverse; fungi are present in every ecosystem on Earth. The fungal kingdom is equally diverse, consisting of seven different known phyla. Yet detailed knowledge is limited to relatively few species. The relationship between fungi and humans has been characterized by the juxtaposed viewpoints of fungi as infectious agents of much dread and their exploitation as highly versatile systems for a range of economically important biotechnological applications. Understanding the biology of different fungi in diverse ecosystems as well as their interactions with living and non-living is essential to underpin effective and innovative technological developments. This series will provide a detailed compendium of methods and information used to investigate different aspects of mycology, including fungal biology and biochemistry, genetics, phylogenetics, genomics, proteomics, molecular enzymology, and biotechnological applications in a manner that reflects the many recent developments of relevance to researchers and scientists investigating the Kingdom Fungi. Rapid screening techniques based on screening specific regions in the DNA of fungi have been used in species comparison and identification, and are now being extended across fungal phyla. The majorities of fungi are multicellular eukaryotic systems and therefore may be excellent model systems by which to answer fundamental biological questions. A greater understanding of the cell biology of these versatile eukaryotes will underpin efforts to engineer certain fungal species to provide novel cell factories for production of proteins for pharmaceutical applications. Renewed interest in all aspects of the biology and biotechnology of fungi may also enable the development of "one pot" microbial cell factories to meet consumer energy needs in the 21st century. To realize this potential and to truly understand the diversity and biology of these eukaryotes, continued development of scientific tools and techniques is essential. As a professional reference, this series will be very helpful to all people who work with fungi and should be useful both to academic institutions and research teams, as well as to teachers, and graduate and postgraduate students with its information on the continuous developments in fungal biology with the publication of each volume.

More information about this series at https://link.springer.com/bookseries/11224

N. Amaresan • A. Sankaranarayanan
Mitesh Kumar Dwivedi • Irina S. Druzhinina
Editors

Advances in *Trichoderma* Biology for Agricultural Applications

Springer

Editors
N. Amaresan
C.G. Bhakta Institute of Biotechnology
Uka Tarsadia University
Surat, Gujarat, India

A. Sankaranarayanan
School of Life Sciences, Sri Sathya Sai
University for Human Excellence
Kalaburagi, Karnataka, India

Mitesh Kumar Dwivedi
C.G. Bhakta Institute of Biotechnology
Uka Tarsadia University
Surat, Gujarat, India

Irina S. Druzhinina
Nanjing Agricultural University
Nanjing, China

ISSN 2198-7777 ISSN 2198-7785 (electronic)
Fungal Biology
ISBN 978-3-030-91649-7 ISBN 978-3-030-91650-3 (eBook)
https://doi.org/10.1007/978-3-030-91650-3

This Springer imprint is published by the registered company Springer Nature Switzerland AG
The registered company address is: Gewerbestrasse 11, 6330 Cham, Switzerland

Foreword

At the beginning of my mycological career, I fell in love with *Trichoderma* because of its looks. The colorful stromata and usually ornamented ascospores, and the complex branching patterns and green conidia of the asexual states, were eye-catching. Now, after more than 20 years of working with this genus, I have come to know it for more than its appearance, and I still love it. The book *Advances in Trichoderma Biology for Agricultural Applications* presents an overview of all those other amazing aspects of *Trichoderma* that cannot be seen by the naked eye: from its complex taxonomy and species identification to its known applications in biological control, growth promotion, and stress alleviation. The book also includes chapters on mechanisms involved in plant protection and the genomic and metabolomic tools used to understand them. In addition, even though the book's title relates to the agricultural applications, it also contains chapters on industrial uses, such as biosynthesis of nanoparticles, bioremediation, medicine, and wine and beer production. However, as it is mentioned in the preface, *Trichoderma* also has an "evil" side that we, *Trichoderma*-lovers, sometimes want to evade. For example, potent mycotoxin producers, agents causing disease in immunocompromised patients, or mycoparasites of commercial edible mushrooms.

I believe there are many facets of *Trichoderma* biology that need further studies, and hopefully this book will provide motivation to delve deeper into them. For example, some interrogations may be: How many species of *Trichoderma* are there and how ubiquitous are they in natural ecosystems? Does host-specificity of mycoparasitic species exist, which would then have implications in the efficacy of the biofungicides? Can we introduce additional microorganisms in a biofungicide/biostimulant consortium that will not be outcompeted by *Trichoderma*? What are the roles, in the plant, of truly endophytic species? Will species used in bioremediation convert pesticides into more toxic by-products? These, and many more questions, will keep us busy for many more years to come.

Priscila Chaverri
Visiting Scientist – ORISE Fellow
U.S. Department of Agriculture, Agricultural Research Service
Mycology and Nematology Genetic Diversity and Biology Laboratory
Beltsville, MD, USA

Associate Professor
School of Biology, Natural Products Research Centre (CIPRONA)
University of Costa Rica
San Pedro, San José, Costa Rica

Adjunct Associate Professor
Department of Plant Science and Landscape Architecture
University of Maryland
College Park, MD, USA

Preface

Some names shed light on the biology and ecology of organisms. For example, the Chinese name of the common filamentous green mold, *Trichoderma* (meaning a hairy thin skin in Latin), is mùméi (木霉) or, literally, the wood mold. Indeed, recent ecological surveys confirm that deadwood colonized by other fungi is the most frequent habitat for *Trichoderma*. However, decoding the higher taxonomic names related to these fungi appears even more meaningful. *Trichoderma* is the currently accepted generic name for fungi that were previously known as *Hypocrea* (sexual stage, teleomorph, and holomorph) and "imperfect" *Trichoderma* (asexual stage, anamorph). Even though the name *Hypocrea* is now outdated, the genus is still considered as the type for the family Hypocreaceae, which, in turn, is the type family for the order Hypocreales. But what is the meaning of this taxonomic name that is so tightly linked to *Trichoderma*? Interestingly, the root of these Latin names comes from the ancient Greek *ὑπόκρῐσῐς* (hupókrisis) or hypocrisis, which can be roughly translated as something hypocritical or insincere. Apparently, this linguistic nuance is not random.

The contemporary scientific results presented in this book show that the hypocrealean filamentous fungal genus *Trichoderma* (Ascomycota) has at least two faces, similar to Janus, the ancient Roman god of duality. Indeed, the genus contains numerous environmentally opportunistic species that have a multitude of beneficial properties and therefore can be used in industry and agriculture; this characteristic is *Trichoderma*'s "God" face. However, some of these fungi are so powerful that their domestication and rigorous use in industry and farming may become risky. Avoiding the other, "evil," face of *Trichoderma* requires cautious and science-based applications. Thus, recent findings on *Trichoderma* genomics explain the hypocritical features of its biology that were predicted by the first taxonomists who proposed the name *Hypocrea*.

It is not surprising that we mainly know *Trichoderma* spp. for their "God" face: besides the industrial cellulase producer *T. reesei*, several dozen *Trichoderma* species are used as unspecific plant-beneficial *bio*effectors with outstandingly high efficiency and a broad spectrum of applications in agriculture. In these cases, the innate mycoparasitism of *Trichoderma* is very useful in the development of *bio*fungicides for environmentally friendly crop protection known as *bio*control. Furthermore, being unable to attack plants and having low phytotoxicity, *Trichoderma* spp. are efficient stimulants of plant growth and immunity. This property is rigorously exploited in the development of modern *bio*fertilizers. The environmentally opportunistic species of *Trichoderma* are also highly resistant to chemical fungicides, which makes them suitable *bio*effectors for product applications together with chemical pesticides and integrated pest management.

However, the success of *Trichoderma*-plant interactions should not allow us overlooking its "evil" face: many *Trichoderma* spp. are causative agents of devastating green mold diseases on mushroom farms (*e.g.*, champignons, oyster mushrooms, and shiitake) and cause destruction to mushroom crops and severe worldwide economic damage. Furthermore, the same *Trichoderma* spp. that impact crops (plants and mushrooms) are also known as opportunistic pathogens of mainly, but not exclusively, immunocompromised humans and are capable of causing severe, frequently life-threatening mycoses.

In this book, we paid attention to the dual nature of *Trichoderma*. We first focused on the beneficial facets of *Trichoderma* and reviewed not only the classical aspects of its biology and applications in agriculture and industry but also outlined several innovative developments based on these outstanding microorganisms. Nevertheless, we found it necessary to also highlight the hazardous features of the fungus. Therefore, we also invited the specialists addressing the potentially harmful sides of *Trichoderma* biology, hoping that their knowledge will help our readers to develop new generation *bio*products that will be safe for the total environment and valuable for people.

Reading this book should kindle further discussions among researchers working in fungal biotechnology, microbiology, agriculture, environmental science, forestry, and other allied subjects and thus lead to a broader scope of *Trichoderma*-based products and technologies. The knowledge shared in this book should also provide a warning on the potential risks associated with *Trichoderma*. The editors are highly obliged to each author or a team of the authors who found time to write a comprehensive chapter on their expertise in *Trichoderma* science. We also hope that the slight unavoidable overlaps in the contents of independent chapters will help the students and the young *Trichoderma* scientists to retrieve the topics that attract most of research attention and therefore will progress into the future.

Surat, Gujarat, India — N. Amaresan
Kalaburagi, Karnataka, India — A. Sankaranarayanan
Surat, Gujarat, India — Mitesh Kumar Dwivedi
Nanjing, China — Irina S. Druzhinina
August 2021

Contents

About the Editors

N. Amaresan is an assistant professor at C.G. Bhakta Institute of Biotechnology, Uka Tarsadia University, Gujarat. He has over 15 years of experience in teaching and research in various allied fields of microbiology, mainly plant-microbe interactions, bioremediation, and plant pathology. He has been awarded young scientist awards by the Association of Microbiologists of India and National Academy of Biological Sciences. He also been awarded visiting scientist fellowship from the National Academy of India to learn advanced techniques. He has published more than 100 research articles, book chapters, and books of national and international repute. He also deposited over 550 16S rDNA, 28S rDNA, and ITS rDNA sequences in the Genbank (NCBI, EMBL & DDBJ) and also preserved over 150 microbial germplasm in various culture collection centers of India. Dr. Amaresan has successfully completed research projects by national funding agencies such as SERB-DST, GUJCOST, UTU, & GEMI and guided students for their doctoral and master's degrees.

A. Sankaranarayanan is Associate Professor of Life Sciences, Sri Sathya Sai University for Human Excellence, Kalaburagi, Karnataka, India, since June 2021. His current research focus is on fermented food products. He has published 30 book chapters and 60 research articles in international and national journals of repute, guided 5 PhD and 16 MPhil scholars, and operated 5 minor funded projects in microbiology. From 2002 to 2015, he worked as an assistant professor and head of the Department of Microbiology at K.S.R. College of Arts & Science, Tiruchengode, Tamil Nadu, and from August 2015 to May 2021, he was associated with Uka Tarsadia University, Surat, Gujarat, India. Dr. Sankaranarayanan has awarded the Indian Academy of Sciences (IASc), the National Academy of Sciences (NAS), and the National Academy of Sciences (TNAS) sponsored summer research fellowship for young teachers consecutively for 3 years, and his name is included as a mentor in DST-Mentors/Resource persons for summer/winter camps and other INSPIRE initiatives, Department of Science & Technology, Govt. of India, New Delhi. He is a grant reviewer with the British Society of Antimicrobial Chemotherapy (BSAC), UK.

Mitesh Kumar Dwivedi is Assistant Professor of Microbiology at C. G. Bhakta Institute of Biotechnology, Uka Tarsadia University. He has published 51 research papers in reputed journals, written 14 book chapters, and is editor of 6 books. He has h-index of 21 with 1521 citations for his research papers. He has more than 13 years of experience in research and teaching in various allied fields of microbiology and life sciences including host-microbe interaction, probiotics, herbal medicine, and immunobiology. He has been serving as an editorial board member and reviewer of many international journals. He has been honored with international as well as national awards for excellent research performance [Best Researcher Award (2020), INSA Visiting Scientist Award (2019), DST-SERB Early Career Research Award (2018), Young Scientist Awards (2011, 2013, 2018)]. He secured all-India rank "32" in CSIR-NET national examination (2011; Life Sciences). He has successfully completed research projects from national funding agencies such as SERB-DST, GUJCOST, UTU, and Neosciences & Research Solutions Pvt. Ltd. and guided students for their doctoral and master's degrees.

Irina S. Druzhinina is Professor of Microbiology in the College of Resources and Environmental Sciences at Nanjing Agricultural University (China); head of the International Committee on *Trichoderma* Taxonomy; member of the International Committee of Taxonomy of Fungi, IUMS; and an editor of the *Applied and Environmental Microbiology* (AEM) journal (ASM). Before moving to China in 2019, for many years, Irina ran the group Microbiology and Applied Genomics at TU Wien (Vienna, Austria), where she studied fungal DNA barcoding and molecular evolution, genomics, and ecophysiology of *Trichoderma* and other hypocrealean fungi. The newly established Fungal Genomics Laboratory (FungiG) in Nanjing continues the research on the genus *Trichoderma*, focusing on its ecological genomics and fitness. In particular, the group is interested in the development of a *Trichoderma*-based model for systems biology investigation of filamentous fungi. Furthermore, the group studies the function and production of fungal surface-active proteins (hydrophobins, cerato-platanins) and their role in the fungal lifestyle. The applied research of Irina and her colleagues targets the improvement of the biological degradation of synthetic polymers and the development of biological products for plant protection and growth promotion.

Part I
Diversity of *Trichoderma*

The Current State of *Trichoderma* Taxonomy and Species Identification

Feng Cai, Kai Dou, Ping Wang, Komal Chenthamara, Jie Chen, and Irina S. Druzhinina

Contents

1 Introduction

Molds from the genus *Trichoderma* (*Hypocreales*, *Ascomycota*) are among of the most common fungi; they are easy to isolate and handle in a pure culture (Migheli et al. 2009; Zachow et al. 2009; Chen et al. 2021). Consequently, the taxonomy of

F. Cai (✉) · I. S. Druzhinina
School of Ecology, Sun Yat-sen University, Shenzhen, China

Fungal Genomics Laboratory (FungiG), Nanjing Agricultural University, Nanjing, China

Institute of Chemical, Environmental, and Bioscience Engineering (ICEBE), TU Wien, Vienna, Austria
e-mail: caif8@mail.sysu.edu.cn

K. Dou · J. Chen (✉)
School of Agriculture and Biology, Shanghai Jiaotong University, Shanghai, China
e-mail: jiechen59@sjtu.edu.cn

P. Wang
Ausliva International, Suzhou, China

K. Chenthamara
Institute of Chemical, Environmental, and Bioscience Engineering (ICEBE), TU Wien, Vienna, Austria

N. Amaresan et al. (eds.), *Advances in* Trichoderma *Biology for Agricultural Applications*, Fungal Biology, https://doi.org/10.1007/978-3-030-91650-3_1

Trichoderma started with the beginning of the modern fungal taxonomy in the eighteenth century (Persoon 1794). Similar to other fungi, it was in the descriptive stage for two centuries and before entering a period of turbulence caused by molecular methods (Bissett 1984; Bissett 1991a, b, c; Kuhls et al. 1997; Kindermann et al. 1998; Kullnig et al. 2000). Ideally, taxonomy should reflect the nature of the organism and help its investigation. The biology of *Trichoderma* offers a convenient example to illustrate this relationship. Many *Trichoderma* strains have properties of environmental opportunism meaning that they are capable of fast colonization of a great variety of natural and artificial substrates, are highly competitive in microbial communities, are resistant to xenobiotics including chemical fungicides, and are potent producers of various metabolites such as enzymes, secondary metabolites, or surface-active proteins (Druzhinina et al. 2011; Sun et al. 2019; Gao et al. 2020; Druzhinina and Kubicek 2017; Pang et al. 2020). Some *Trichoderma* species can survive in soil and colonize rhizosphere possessing almost no harm to plants but stimulating their growth and development (Druzhinina et al. 2011; Harman et al. 2004; Marra et al. 2019; Rivera-Méndez et al. 2020). Being mycoparasitic, a growing number of *Trichoderma* species are proposed as biofungicides for plant protection in agriculture (Ding et al. 2020; Wu et al. 2018). However, the same property also makes *Trichoderma* species causative agents of the green mold disease on mushroom farms (Komoń-Zelazowska et al. 2007; Kredics et al. 2010) (see Kredics et al. in this book). Finally, some *Trichoderma* strains also have clinical significance as causative agents of nosocomial mycoses in immunocompromised humans (Chouaki et al. 2002; Myoken et al. 2002; Kredics et al. 2003). These versatile, largely beneficial, but also harmful properties of *Trichoderma* make the taxonomy of this genus a high priority task because the correct identification of a species can predict its properties and thus facilitate applications. The taxonomy of *Trichoderma* has been intensively studied over the last two decades resulting in a hundred-fold increase in the species number from a few "species aggregates" of Rifai (1969) to several hundred molecularly defined species enumerated in several recent reviews (Druzhinina et al. 2006; Atanasova et al. 2013; Bissett et al. 2015; Cai and Druzhinina 2021). Thus, today *Trichoderma* comprises the genus of very common fungi with most species that have been characterized using modern molecular techniques.

The large number of species in *Trichoderma* appears to be reasonable: Whole genomic investigations of this genus and other hypocrealean fungi have estimated the origin of the genus at the edge of Cretaceous-Paleogene mass extinction event 66–67 million years ago (Kubicek et al. 2019). The most recent phylogenomic tree (Kubicek et al. 2019) indicates that the formation of the major infrageneric clades such as Sections *Trichoderma* and *Longibrachiatum* recognized by John Bissett in the 1990s or the *Harzianum* Clade (Bissett 1984; Chaverri et al. 2003) was formed somewhat 20–25 million years ago, while some closely related species such as *T. reesei* and *T. parareesei* shared a common ancestor 4–8 million years ago. This vast evolutionary time and the relatively high evolutionary rates (compared to, e.g., vertebrates) offer the genus *Trichoderma* tremendous possibilities for the adaptation to the environmental conditions and speciation. However, similar to other fungi, many evolutionary different strains of *Trichoderma* still share remarkable morphological and ecophysiological similarities. It appears that many traits suitable and accessible for direct examination by taxonomists are homoplasious and appeared due to convergent evolution. Thus, the most difficult task of modern taxonomy of *Trichoderma* is to retrieve the traits that would allow one to distinguish a great number of species.

The general fungal taxonomy is regulated by the Code, i.e., CN International Code of Nomenclature for algae, fungi, and plants (Turland et al. 2018), that now contains an advanced section for fungi in Chapter F, San Juan Chapter F (May et al. 2019). Even though the Code strictly regulates nomenclatural acts, it assumes a heterogeneity of approaches to define species (Turland et al. 2018). This can be explained by the complexity of lineage-dependent evolutionary processes (Steenkamp et al. 2018; Inderbitzin et al. 2020) or numerous pragmatic criteria used by the taxonomists for the classification of particular fungal groups. Lücking et al. (2020) found that the best practice depends on the group in question and the required level of precision. Some fungi can be grouped based on phenotype characteristics; however, most fungi, especially asexual forms such as *Trichoderma*, require time-consuming and labor-intensive methods that include culturing, DNA barcoding, and phylogenetic analysis as well as discipline- or taxon-specific approaches such as physiological profiling (Lücking et al. 2020). Therefore, it is common for species concepts determined by the taxonomy providers to vary even within one genus. However, taxonomy users expect that the identification of species should be precise and accurate. For *Trichoderma*, this collision of possibly vague species delimitation and the need for the exact species identification was recently addressed in Cai and Druzhinina (2021). This topic requires a thoughtful discussion that will also be presented in this chapter and continued elsewhere.

The biology of *Trichoderma* offers a number of exclusive opportunities to the taxonomists. Fungi from this genus are ubiquitous and relatively simple to recognize and collect in natural and human-made habitats. They are easy to isolate directly from specimens and from a broad range of substrates based on the characteristic genus-specific features. Most strains have fast growth in vitro on all common laboratory media and do not require demanding cultivation conditions such as temperature, illumination, or humidity. Importantly, and as it will be described in most chapters of this book, many *Trichoderma* spp. have highly valuable properties for industry and agriculture. Respectively, *Trichoderma* has attracted the attention of classical mycologists and people focusing on applied microbiology and developmental applications. Therefore, all collections of microorganisms have numerous *Trichoderma* isolates. Public depositories of gene sequences contain thousands of *Trichoderma* DNA barcodes, and the number of the whole genome sequences has grown exponentially. However, the identification of *Trichoderma* is also considered to be extremely difficult. Fungal taxonomists including experts working with this genus for many years now frequently fail to determine the species (Cai and Druzhinina 2021).

In this chapter, we investigate the theoretical background of these collisions in *Trichoderma* research aiming for a concise review of the taxonomic state of the genus. We present a brief synopsis of *Trichoderma* taxonomy through January 2021, list all *Trichoderma* species names, and explain the latest identification protocol for *Trichoderma* species.

2 The Numerical State of *Trichoderma* Taxonomy and Species Identification

After the implementation of the "One fungus – One name" concept of fungal nomenclature (Taylor 2011)—and based on the voting organized by the International Commission on *Trichoderma* Taxonomy (ICTT) (formerly www.isth.info, now www.trichoderma.info) of the International Commission on the Taxonomy of Fungi (ICTF, www.fungaltaxonomy.org)—*Trichoderma* was selected as a single generic name that should be used for all stages such as holo-, ana-, and teleomorphs. Consequently, the taxonomy of the genus *Trichoderma* was updated to include the species names previously attributed to teleomorphs from such genera as *Hypocrea*, *Sarawakus*, and *Protocrea* (Jaklitsch 2009a; Jaklitsch et al. 2014). The formal transfer of a few species of *Hypocrea* to *Trichoderma* is still pending (Cai and Druzhinina 2021); nevertheless, these species are valid names of the genus (Table 1).

As of January 2021, the genus *Trichoderma* contains 468 species epithets, among which 379 names are currently in use, while 89 names (19%) are synonyms of different categories (abandoned names, orthographic variants, synonyms) (Cai and Druzhinina 2021) updated with materials from Gu et al. (2020). Forty names were introduced before the twentieth century. Of these, only five are currently in use including such important species as *T. viride* and *T. atroviride*. Sixty species were introduced in the twentieth century based on their morphology, (sometimes) ecophysiological properties, and biogeography (Rifai 1969; Bissett 1984, 1991a, b, 1992). The end of the century coincided with the introduction of molecular methods in *Trichoderma* taxonomy and the proposal of the genealogical concordance phylogenetic species recognition concept (GCPSR) as the most powerful approach to distinguish fungal taxa (Taylor et al. 2000; Lücking et al. 2020). These changes resulted in a rapid increase in the number of taxa adding the majority of modern *Trichoderma* species names (364, 78%) delineated in the first two decades of the twenty-first century. Consequently, only 14 (4%) currently valid *Trichoderma* species have not been characterized by molecular markers (Cai and Druzhinina 2021), while 365 species (96%) have been DNA barcoded. This makes the genus *Trichoderma* a suitable model for DNA barcoding and molecular evolutionary studies in fungi.

The largest database of *Trichoderma* names is available in MycoBank (http://www.mycobank.org/) followed by Index Fungorum (http://www.indexfungorum.org). Most species names are recorded in both taxonomic depositories, but MycoBank still has 14 and Index Fungorum has 8 unique records. Therefore, none of the official depositories of fungal taxonomy has the full list of *Trichoderma* species names (Fig. 1). To date, the most complete list of *Trichoderma* species can be found in Table 1 (sorted alphabetically for convenience). Alternatively, the newly

Table 1 The alphabetic list of all species names deposited for *Trichoderma* in Index Fungorum (http://www.indexfungorum.org/), MycoBank (https://www.mycobank.org/), NCBI Taxonomy Browser (https://www.ncbi.nlm.nih.gov/Taxonomy/Browser/wwwtax.cgi), and scientific literature as of February 2021

Species name	Author(s)	Year	Reference strain
Trichoderma acremonioides	Zhang & Zhuang	2018	HMAS 279611
Trichoderma adaptatum	Chen & Zhuang	2017	HMAS 248800
Trichoderma aeroaquaticum	Yamag., Tsurumi, Chuaseehar. & Nakagiri	2012	NBRC 108034
Trichoderma aerugineum	Jaklitsch	2009	CBS 120541
Trichoderma aeruginosum	Link	1816	not in use
Trichoderma aestuarinum	Gonçalves & Alves	2019	MUM H-19.05
Trichoderma aethiopicum	Mulaw, Kubicek & Samuels	2012	CBS 130628
Trichoderma afarasin	Chaverri & Rocha	2015	CBS 130755
Trichoderma afroharzianum	Chaverri, Rocha, Degenkolb & Druzhin.	2015	CBS 124620
Trichoderma aggregatum	Chen & Zhuang	2017	HMAS 248863
Trichoderma aggressivum	Samuels & Gams	2002	DAOM 222156
Trichoderma albocorneum	(Doi) Jaklitsch & Voglmayr	2014	G.J.S. 97-28
Trichoderma albofulvopsis	Qin & Zhuang	2016	HMAS 273760
Trichoderma albofulvum	(Berk. & Broome) Jaklitsch & Voglmayr	2014	CBS 114787
Trichoderma albolutescens	Jaklitsch	2011	CBS 119286
Trichoderma alboviride	Chen & Zhuang	2017	HMAS 247224
Trichoderma album	Preuss	1851	not in use
Trichoderma alcalifuscescens	(Overton) Jaklitsch & Voglmayr	2014	CBS 122303
Trichoderma alni	Jaklitsch	2008	CBS 120633
Trichoderma alpinum	Chen & Zhuang	2017	HMAS 248821
Trichoderma alutaceum	Jaklitsch	2011	CBS 120535
Trichoderma amazonicum	Chaverri & Gazis	2011	CBS 126898
Trichoderma americanum	(Canham) Jaklitsch & Voglmayr	2014	CBS 976.69
Hypocrea ampulliformis	Doi & Yamat.	1989	JCM 11982
Trichoderma andinense	(Samuels & Petrini) Samuels, Jaklitsch & Voglmayr	2014	CBS 345.97
Trichoderma angustum	Qin & Zhuang	2017	HMAS 273784

(continued)

Table 1 (continued)

Species name	Author(s)	Year	Reference strain
Trichoderma appalachiense	Samuels & Jaklitsch	2013	CBS 133558
Trichoderma applanatum	Zhu & Zhuang	2015	HMAS 245081
Trichoderma arachnoidea	Kuritzina & Sizova	1967	not in use
Trichoderma arachnoideum	Kuritzina & Sizova	1967	not in use
Trichoderma arenarium	Cai, Ding & Druzhin.	2020	CGMCC 19611
Trichoderma arundinaceum	Zafari, Gräfenhan & Samuels	2008	CBS 119575
Trichoderma asperelloides	Samuels	2010	CBS 125938
Trichoderma asperellum	Samuels, Lieckf. & Nirenberg	1999	CBS 433.97
Trichoderma asterineum	Qin & Zhuang	2016	HMAS 271353
Trichoderma atlanticum	Jaklitsch	2011	CBS 120632
Trichoderma atrobrunneum	Rocha, Chaverri & Jaklitsch	2015	CBS 548.92
Trichoderma atrogelatinosum	(Dingley) Jaklitsch & Voglmayr	2014	CBS 237.63
Trichoderma atroviride	Bissett	1984	not in use
Trichoderma atroviride	Karst.	1892	IMI 206040
Trichoderma attinorum	Montoya, Meirelles, Chaverri & Rodrigues	2016	CBS 139783
Trichoderma auranteffusum	Jaklitsch	2011	not in use
Trichoderma aurantioeffusum	Jaklitsch	2011	CBS 119284
Trichoderma aureoviride	Rifai	1969	CBS 120536
Trichoderma aureum	Pers.	1796	not in use
Trichoderma austriacum	Jaklitsch	2011	CBS 122494
Trichoderma austrokoningii	Samuels & Druzhin.	2006	CBS 119092
Trichoderma avellaneum	(Rogerson & Carey) Jaklitsch & Voglmayr	2014	CBS 121667
Trichoderma azevedoi	Valadares-Inglis & Inglis	2020	CEN 1422
Trichoderma balearicum	Jaklitsch & Voglmayr	2015	CBS 133222
Trichoderma bannaense	Chen & Zhuang	2017	CGMCC 3.18394
Trichoderma barbatum	Samuels	2012	CBS 125733
Trichoderma bavaricum	Jaklitsch	2011	WU 29196a
Trichoderma beijingense	Chen & Zhuang	2017	HMAS 248804
Trichoderma beinartii	du Plessis, Druzhin., Atan., Yarden & Jacobs	2018	PPRI 19281
Trichoderma bifurcatum	Chen & Zhuang	2017	HMAS 248795
Trichoderma bissettii	Sand.-Den. & Guarro	2014	CBS 137447
Trichoderma bomiense	Zhang & Zhuang	2019	W.Z. 2018a
Trichoderma brassicae	Schumach.	1803	not in use

(continued)

Table 1 (continued)

Species name	Author(s)	Year	Reference strain
Trichoderma breve	Chen & Zhuang	2017	CGMCC 3.18398
Trichoderma brevicompactum	Kraus, Kubicek & Gams	2004	CBS 109720
Trichoderma brevicrassum	Chen & Zhuang	2017	CGMCC 3.18407
*Trichoderma brevipes**	(Mont.) Samuels	2015	CBS 139044
Trichoderma britannicum	(Rifai & Webster) Jaklitsch & Voglmayr	2014	CBS 253.62
Trichoderma britdaniae	(Jaklitsch & Voglmayr) Jaklitsch & Voglmayr	2014	WU 31610
Trichoderma brunneoviride	Jaklitsch	2008	CBS 121130
Trichoderma byssinum	Chen & Zhuang	2017	CGMCC 3.18393
Trichoderma caeruleimontis	du Plessis & Jacobs	2018	PPRI 23903
Trichoderma caerulescens	(Jaklitsch & Voglmayr) Jaklitsch & Voglmayr	2014	CBS 130011
Trichoderma caesareum	Samuels	2012	CBS 124369
Trichoderma caesium	Pers.	1794	not in use
Trichoderma calamagrostidis	Jaklitsch	2011	WU 29198a
Trichoderma camerunense	Chaverri & Samuels	2015	CBS 138272
Trichoderma candidum	Chaverri & Samuels	2003	not in use
Trichoderma candidum	Alb. & Schwein.	1805	not in use
Trichoderma capillare	Samuels & Kubicek	2012	CBS 130629
Trichoderma caribbaeum	Samuels & Schroers	2006	CBS 119093
Trichoderma carneum	Schumach.	1803	not in use
Trichoderma catoptron	Chaverri & Samuels	2003	CBS 114232
Trichoderma ceciliae	Jaklitsch & Voglmayr	2015	CBS 130010
Trichoderma centrosinicum	Qin & Zhuang	2016	HMAS 252910
Trichoderma ceraceum	Chaverri & Samuels	2003	BPI 843654
Trichoderma ceramicum	Chaverri & Samuels	2003	CBS 114576
Trichoderma ceratophylli	Yu	2019	YMF 1.04621
Trichoderma cerebriforme	(Berk.) Samuels	2015	G.J.S. 85-245
Trichoderma cerinum	Bissett, Kubicek & Szakács	2003	DAOM 230012
Trichoderma changbaiense	Chen & Zhuang	2017	HMAS 247198
Trichoderma chetii	du Plessis, Druzhin., Atan., Yarden & Jacobs	2018	PPRI 19363
Trichoderma chlamydosporicum	Chen & Zhuang	2017	HMAS 248850
Trichoderma chlamydosporum	Chen & Zhuang	2017	not in use

(continued)

Table 1 (continued)

Species name	Author(s)	Year	Reference strain
Trichoderma chlorosporum	Chaverri & Samuels	2003	CBS 114231
Trichoderma christiani	Jaklitsch & Voglmayr	2015	CBS 132572
Trichoderma christianii	Jaklitsch & Voglmayr	2015	not in use
Trichoderma chromospermum	Chaverri & Samuels	2003	CBS 114577
Trichoderma cinnabarinum	Wallr.	1833	not in use
Trichoderma cinnamomeum	Chaverri & Samuels	2003	G.J.S. 97-237
Trichoderma citrinella	(Ellis) Zhuang & Zeng	2017	
Trichoderma citrinoviride	Bissett	1984	CBS 258.85
Trichoderma citrinum	(Pers.) Jaklitsch, Gams & Voglmayr	2014	CBS 894.85
Trichoderma collae	(Schwein.) Sacc.	1886	not in use
Trichoderma compactum	Yu & Zhang	2007	CBS 121218
Trichoderma composticola	Samuels & Jaklitsch	2013	CBS 133497
Trichoderma concentricum	Chen & Zhuang	2017	HMAS 248833
Trichoderma confertum	Chen & Zhuang	2017	HMAS 248896
Trichoderma confluens	Qin & Zhuang	2016	HMAS 244993
Hypocrea coprosmae	Dingley	1952	PDD 10453
Trichoderma cordobense	Speg.	1926	not in use
Trichoderma corfecianum	Sacc.	1911	not in use
Trichoderma corneum	(Pat.) Jaklitsch & Voglmayr	2014	CBS 100541
Trichoderma cornu-damae	(Pat.) Zhu & Zhuang	2014	G.J.S. 06-03
Trichoderma corrugatum	(Doi, Liu & Tamura) Liu, Zhu & Zhuang	2014	not in use
Trichoderma costaricense	(Chaverri & Samuels) Chaverri, Jaklitsch & Voglmayr	2014	P.C. 21
Trichoderma crassum	Bissett	1992	CBS 336.93
Trichoderma cremeoides	Jaklitsch & Voglmayr	2015	S112
Trichoderma cremeum	Chaverri & Samuels	2003	CBS 111146
Trichoderma croceum	Bissett	1992	not in use
Trichoderma crystalligenum	Qin & Zhuang	2017	not in use
Trichoderma crystalligenum	Jaklitsch	2006	CBS 118980
Trichoderma cuenisporum	Chaverri & Samuels	2003	not in use
Trichoderma cuneisporum	Chaverri & Samuels	2003	not in use
Trichoderma cyanodichotomus	Li & Chen	2018	not in use
Trichoderma dacrymycellum	Jaklitsch	2009	WU 29042a
Trichoderma danicum	(Jaklitsch) Jaklitsch & Voglmayr	2014	CBS 121273
Trichoderma decipiens	(Jaklitsch, Põldmaa & Samuels) Jaklitsch & Voglmayr	2014	G.J.S. 97-207

(continued)

Table 1 (continued)

Species name	Author(s)	Year	Reference strain
Trichoderma delicatulum	Jaklitsch	2011	CBS 120631
Trichoderma deliquescens	(Sopp) Jaklitsch	2011	CBS 121131
Trichoderma densum	Qin & Zhuang	2016	HMAS 273758
Trichoderma desrochii	Sartory & Bainier	1913	
Hypocrea dichromospora	Doi	1968	CBS 337.69
Trichoderma dimorphum	Chen & Zhuang	2017	HMAS 247199
Trichoderma dingleyae	Samuels & Dodd	2006	CBS 119056
Trichoderma dorotheae	Samuels & Dodd	2006	CBS 119089
Trichoderma dorothopsis	Tomah & Zhang	2020	HMAS 248251
Trichoderma dubium	Pers.	1801	not in use
Trichoderma dubium	Alb. & Schwein.	1805	not in use
Trichoderma effusum	Bissett, Kubicek & Szakács	2003	DAOM 230007
Trichoderma eijii	Kim & Maek.	2013	CBS 133190
Trichoderma endophyticum	(Jaklitsch, Põldmaa & Samuels) Jaklitsch & Voglmayr	2015	CBS 130729
Trichoderma epimyces	Jaklitsch	2008	CBS 120534
Trichoderma erinaceum	Bissett, Kubicek & Szakács	2003	DAOM 230018
Trichoderma estonicum	Chaverri & Samuels	2003	CBS 111147
Trichoderma eucorticioides	(Overton) Jaklitsch & Voglmayr	2014	G.J.S. 99-61
Trichoderma europaeum	Jaklitsch & Voglmayr	2015	CBS 121276
Trichoderma euskadiense	Jaklitsch & Voglmayr	2015	CBS 130013
Trichoderma evansii	Samuels	2009	CBS 123079
Trichoderma fasciculatum	Bissett	1992	not in use
Trichoderma fassatiae	Nováková, Kubátová, Valinová, Hubka & Kolařík	2015	PRM 933821
Trichoderma fertile	Bissett	1992	CBS 339.93
Trichoderma flagellatum	Mulaw, Kubicek & Samuels	2012	CBS 130626
Trichoderma flavescens	Zhu, Zhuang & Li	2017	HMJAU 34730
Trichoderma flaviconidium	(Chaverri, Druzhin. & Samuels) Jaklitsch & Voglmayr	2014	CBS 130688
Trichoderma flavipes	(Peck) Seifert, Jaklitsch & Voglmayr	2014	CBS 123070
Trichoderma flavofuscum	(Mill., Giddens & Foster) Bissett	1992	not in use
Trichoderma flavum	Sommerf.	1826	not in use
Trichoderma floccosum	Samuels	2011	CBS 124372
Trichoderma foliicola	(Jaklitsch & Voglmayr) Jaklitsch & Voglmayr	2014	CBS 130008

(continued)

Table 1 (continued)

Species name	Author(s)	Year	Reference strain
Trichoderma fomiticola	Jaklitsch	2009	CBS 121136
Trichoderma fomitopsis	(Liu & Doi) Liu, Zhu & Zhuang	2014	not in use
Trichoderma fragile	(Doi) Jaklitsch & Voglmayr	2014	
Trichoderma fujianense	Zhu, Zhuang & Li	2017	HMJAU 34830
Trichoderma fuliginoides	Pers.	1801	not in use
Trichoderma fuscum	Schumach.	1803	not in use
Trichoderma gamsii	Samuels & Druzhin.	2006	CBS 120075
Trichoderma ganodermatis	Chen & Zhuang	2017	HMAS 248856
Trichoderma gelatinosum	Chaverri & Samuels	2003	CBS 114246
Trichoderma ghanense	Doi, Abe & Sugiy.	1987	ATCC 208858
Trichoderma gillesii	Samuels	2012	CBS 130435
Trichoderma glaucum	Abbott	1927	not in use
Trichoderma gliocladium	Jaklitsch & Voglmayr	2015	CBS 130009
Trichoderma globoides	Qin & Zhuang	2017	HMAS 248747
Trichoderma globosum	Schwein.	1822	not in use
Trichoderma gracile	Samuels & Szakács	2012	CBS 130714
Trichoderma grande	Qin & Zhuang	2016	HMAS 248749
Trichoderma granulosum	Fuckel	1870	not in use
Trichoderma gregarium	Chen & Zhuang	2017	HMAS 248887
Trichoderma guizhouense	Li, McKenzie & Wang	2012	CBS 131803
Trichoderma guttatum	Alb. & Schwein.	1805	not in use
Trichoderma hainanense	Chen & Zhuang	2017	HMAS 248837
Trichoderma hamatum	(Bonord.) Bainier	1906	CBS 102160
Trichoderma harzianum	Rifai	1969	CBS 226.95
Trichoderma hausknechtii	Jaklitsch & Voglmayr	2015	CBS 133493
Trichoderma hebeiense	Chen & Zhuang	2017	HMAS 248743
Trichoderma helicolixii	Jaklitsch & Voglmayr	2015	CBS 133499
Trichoderma helicum	Bissett, Kubicek & Szakács	2003	DAOM 230022
Trichoderma henanense	Qin & Zhuang	2016	HMAS 252891
Trichoderma hengshanicum	Chen & Zhuang	2017	HMAS 248852
Trichoderma hexasporum	(Boedijn) Jaklitsch & Voglmayr	2014	
Trichoderma hirsutum	Chen & Zhuang	2017	HMAS 248834

(continued)

Table 1 (continued)

Species name	Author(s)	Year	Reference strain
Trichoderma hispanicum	(Jaklitsch & Voglmayr) Jaklitsch & Voglmayr	2014	CBS 130540
Trichoderma hongkongensis	(Zhu & Zhuang) Zeng & Zhuang	2017	HMAS 75530
Trichoderma hubeiense	Qin & Zhuang	2016	HMAS 252888
Trichoderma hunanense	Chen & Zhuang	2017	HMAS 248841
Trichoderma hunua	(Dingley) Jaklitsch & Voglmayr	2014	CBS 238.63
Trichoderma hypoxylon	Sun, Liu & Hyde	2016	CGMCC 3.17906
Trichoderma ingratum	Chen & Zhuang	2017	HMAS 248822
Trichoderma inhamatum	Veerkamp & Gams	1983	CBS 273.78
Trichoderma intricatum	Samuels & Dodd	2006	CBS 119059
Trichoderma istrianum	Jaklitsch & Voglmayr	2015	CBS 130539
Trichoderma italicum	Jaklitsch & Voglmayr	2015	CBS 132567
Trichoderma ivoriense	Samuels	2012	CBS 125734
Trichoderma izawae	(Doi) Jaklitsch & Voglmayr	2014	
Trichoderma junci	Jaklitsch	2011	WU 29229a
Trichoderma konilangbra	Samuels, Petrini & Kubicek	1998	CBS 100808
Trichoderma koningii	Oudem.	1902	G.J.S. 96-117
Trichoderma koningiopsis	Samuels, Carm. Suárez & Evans	2006	CBS 119075
Trichoderma koreanum	Oh, Park & Lim	2019	SFC 20131005-S066
Trichoderma kunigamense	Yabuki & Okuda	2014	TNS-F 38436
Trichoderma kunmingense	Yu & Li	2018	YMF 1.02659
Trichoderma lacteum	Bissett	1992	not in use
Trichoderma lacuwombatense	(Lu, Druzhin. & Samuels) Jaklitsch & Voglmayr	2014	CBS 122668
Trichoderma laeve	Pers.	1796	not in use
Trichoderma laeve	Schumach.	1803	not in use
Trichoderma laevisporum	Qin & Zhuang	2016	not in use
Trichoderma lanuginosum	Samuels	2012	CBS 125718
Trichoderma lateritio-roseum	Lib. ex Cooke	1880	not in use
Trichoderma latizonatum	(Peck) Samuels	2015	
Trichoderma leguminosarum	Jaklitsch & Voglmayr	2015	CBS 130014
Trichoderma lentiforme	(Rehm) Chaverri, Samuels & Rocha	2015	CBS 100542
Trichoderma lentinulae	Sun & Liu	2020	HMAS 248256
Trichoderma leucopus	Jaklitsch	2011	CBS 122499
Trichoderma liberatum	Chen & Zhuang	2017	HMAS 248831

(continued)

Table 1 (continued)

Species name	Author(s)	Year	Reference strain
Trichoderma lieckfeldtiae	Samuels	2009	CBS 123049
Trichoderma lignorum	(Tode) Harz	1872	not in use
Trichoderma limonium	Qin & Zhuang	2016	HMAS 248751
Trichoderma linzhiense	Chen & Zhuang	2017	HMAS 248846
Trichoderma lixii	(Pat.) Chaverri	2015	CBS 110080
Trichoderma longibrachiatum	Rifai	1969	CBS 816.68
Trichoderma longifialidicum	Montoya, Meirelles, Chaverri & Rodrigues	2016	CBS 139785
Trichoderma longipile	Bissett	1991	CBS 120953
Trichoderma longipilis	Bissett	1992	not in use
Trichoderma longipilum	Bissett	1992	not in use
Trichoderma longisporum	Chen & Zhuang	2017	HMAS 248843
Trichoderma luteffusum	Jaklitsch	2011	not in use
Trichoderma luteocrystallinum	Jaklitsch	2011	CBS 123828
Trichoderma luteoeffusum	Jaklitsch	2011	CBS 120537
Trichoderma lycogaloides	(Berk. & Broome) Jaklitsch, Lechat & Voglmayr	2014	CBS 123493
Trichoderma mangshanicum	Chen & Zhuang	2017	HMAS 248810
Trichoderma margaretense	Jaklitsch	2011	CBS 120540
Trichoderma martiale	Samuels	2008	CBS 123052
Trichoderma matsushimae	(Webster) Yamag., Tsurumi, Chuaseehar. & Nakagiri	2012	IMI 266915
Trichoderma mediterraneum	Jaklitsch & Voglmayr	2015	CBS 136469
Trichoderma medusae	Samuels	2012	CBS 125719
Trichoderma megalocitrinum	(Doi) Jaklitsch & Voglmayr	2014	B.E.O. 00-09
Trichoderma melanomagnum	Chaverri & Samuels	2003	G.J.S. 99-153
Trichoderma microcitrinum	(Doi) Jaklitsch & Voglmayr	2014	G.J.S. 91-61
Trichoderma mienum	Kim, Nakagiri & Maek.	2012	CBS 132690
Hypocrea mikurajimensis	Doi	2001	JCM 12018
Trichoderma minima	(Speg.) Gunth. Müll.	1965	not in use
Trichoderma minimum	(Speg.) Gunth. Müll.	1965	not in use
Trichoderma minutisporum	Bissett	1992	CBS 341.93
Trichoderma minutum	Bainier	1906	not in use
Trichoderma moravicum	Jaklitsch	2011	CBS 120539

(continued)

Table 1 (continued)

Species name	Author(s)	Year	Reference strain
Hypocrea muroiana	Hino & Katum.	1958	NBRC 31293
Trichoderma mycophilum	(Pers.) Schwein.	1822	not in use
Trichoderma narcissi	(Tochinai & Shimada) Tochinai & Shimada	1931	not in use
Trichoderma neocrassum	Samuels	2015	CBS 114230
Trichoderma neokoningii	Samuels & Soberanis	2006	CBS 120070
Trichoderma neorufoides	Jaklitsch	2011	CBS 119506
Trichoderma neorufum	(Samuels, Dodd & Lieckf.) Jaklitsch & Voglmayr	2014	CBS 111144
Trichoderma neosinense	Samuels & Jaklitsch	2013	CBS 134884
Trichoderma neotropicale	Chaverri & Rocha	2015	CBS 130633
Trichoderma nigrescens	Pers.	1794	not in use
Trichoderma nigrovirens	Goddard	1913	not in use
Trichoderma nigrovirens	Chaverri & Samuels	2001	not in use
Trichoderma nigrovirens	Chaverri & Samuels	2003	not in use
Trichoderma nothescens	Samuels & Jaklitsch	2013	CBS 134882
Trichoderma novae-zelandiae	(Samuels & Petrini) Jaklitsch & Voglmayr	2014	CBS 639.92
Trichoderma nunbergii	Svilv.	1932	not in use
Trichoderma nybergianum	(Ulvinen & Chamb.) Jaklitsch & Voglmayr	2014	CBS 122500
Trichoderma oblongisporum	Bissett	1992	CBS 343.93
Trichoderma ochroleucum	(Berk. & Ravenel) Jaklitsch & Voglmayr	2014	CBS 119502
Trichoderma odoratum	Qin & Zhuang	2016	HMAS 271354
Trichoderma oligosporum	Zhu & Zhuang	2015	HMAS 252870
Trichoderma olivascens	Jaklitsch, Samuels & Voglmayr	2013	CBS 132574
Trichoderma orientale	(Samuels & Petrini) Jaklitsch & Samuels	2014	CBS 130428
Trichoderma ovalisporum	Samuels & Schroers	2004	CBS 113299
Hypocrea pachybasioides	Doi	1972	not in use
Trichoderma pachypallidum	Jaklitsch	2011	CBS 122126
Trichoderma panacis	Liu, Zhang, Yu & Zhang	2020	CGMCC 3.18297
Trichoderma paraceramosum	Bissett	1992	not in use
Trichoderma parapiluliferum	(Lu, Druzhin. & Samuels) Jaklitsch & Voglmayr	2014	CBS 112771
Trichoderma parareesei	Atan., Jaklitsch, Komoń-Zel., Kubicek & Druzhin.	2010	CBS 125925
Trichoderma pararogersonii	Jaklitsch & Voglmayr	2015	CBS 133496

(continued)

Table 1 (continued)

Species name	Author(s)	Year	Reference strain
Trichoderma paratroviride	Jaklitsch & Voglmayr	2015	CBS 136489
Trichoderma paraviridescens	Jaklitsch, Samuels & Voglmayr	2013	CBS 119321
Trichoderma parceramosum	Bissett	1992	not in use
Trichoderma parepimyces	Jaklitsch	2009	CBS 122769
Trichoderma parestonicum	Jaklitsch	2009	CBS 120636
Trichoderma parmastoi	(Overton) Jaklitsch & Voglmayr	2014	TFC 97-143
Trichoderma patella	(Cooke & Peck) Jaklitsch & Voglmayr	2014	CBS 110081
Trichoderma patellotropicum	Samuels	2015	CBS 110084
Trichoderma paucisporum	Samuels, Carm. Suárez & Solis	2006	CBS 118645
Trichoderma peberdyi	Valadares-Inglis & Inglis	2020	CEN 1426
Trichoderma pedunculatum	Schumach.	1803	not in use
Trichoderma peltatum	(Berk.) Samuels, Jaklitsch & Voglmayr	2014	G.J.S. 08-207
Trichoderma penicillatum	Wallr.	1833	not in use
Trichoderma perviride	Qin & Zhuang	2017	HMAS 273786
Trichoderma petersenii	Samuels, Dodd & Schroers	2006	G.J.S. 91-99
Trichoderma pezizoides	(Berk. & Broome) Samuels, Jaklitsch & Voglmayr	2014	G.J.S. 01-257
Trichoderma pezizoideum	Wallr.	1833	not in use
Trichoderma phellinicola	Jaklitsch	2011	CBS 119283
Trichoderma phyllostachydis	Chaverri & Samuels	2003	CBS 114071
Trichoderma piluliferum	Webster & Rifai	1969	CBS 120927
Trichoderma pinicola	Oh, Park, & Lim	2019	KACC 48486
Trichoderma pinnatum	Samuels	2012	CBS 131292
Trichoderma placentula	Jaklitsch	2011	CBS 120924
Trichoderma pleuroti	Yu & Park	2006	CBS 124387
Trichoderma pleuroticola	Yu & Park	2006	CBS 124383
Trichoderma pleurotum	Yu & Park	2006	not in use
Trichoderma pollinicola	Liu & Cai	2018	CGMCC 3.18781
Trichoderma polyalthiae	Nuankaew & Boonlue	2018	TBRC 8737
Trichoderma polypori	Chen & Zhuang	2017	HMAS 248855
Trichoderma polysporum	(Link) Rifai	1969	CBS 820.68
Trichoderma poronioideum	(Möller) Samuels	2015	CBS 139046
Trichoderma priscilae	Jaklitsch & Voglmayr	2015	CBS 131487
Trichoderma protopulvinatum	(Doi) Jaklitsch & Voglmayr	2014	CBS 739.83

(continued)

Table 1 (continued)

Species name	Author(s)	Year	Reference strain
Trichoderma protrudens	Samuels & Chaverri	2008	CBS 121320
Trichoderma pruinosum	Chen & Zhuang	2017	HMAS 247217
Trichoderma pseudobritdaniae	Qin & Zhuang	2016	HMAS 271355
Trichoderma pseudocandidum	Minnis, Samuels & Chaverri	2009	BPI 843652
Trichoderma pseudodensum	Chen & Zhuang	2017	HMAS 248828
Trichoderma pseudogelatinosa	(Komatsu & Doi) Kim	2012	not in use
Trichoderma pseudogelatinosum	(Komatsu & Doi) Kim	2017	TUFC 60186
Trichoderma pseudokoningii	Rifai	1969	CBS 408.91
Trichoderma pseudolacteum	Kim & Maek.	2013	CBS 133191
Trichoderma pseudonigrovirens	Minnis, Samuels & Chaverri	2009	G.J.S. 99-64
Trichoderma pseudostraminea	(Doi) Kim	2012	not in use
Trichoderma pseudostramineum	(Doi) Kim	2012	TUFC 60104
Trichoderma psychrophilum	Jaklitsch	2011	CBS 119129
Trichoderma pubescens	Bissett	1992	CBS 345.93
Trichoderma pulvinatum	(Fuckel) Jaklitsch & Voglmayr	2014	CBS 121279
Trichoderma purpureum	Qin & Zhuang	2017	HMAS 273787
Trichoderma pyramidale	Jaklitsch & Chaverri	2015	CBS 135574
Trichoderma pyrenium	Pers.	1801	not in use
Trichoderma pyrenium	Schumach.	1803	not in use
Trichoderma racemosum	McAlpine	1902	not in use
Trichoderma reesei	Simmons	1977	CBS 383.78
Trichoderma restrictum	du Plessis & Jacobs	2018	PPRI 19367
Trichoderma rhododendri	(Jaklitsch) Jaklitsch & Voglmayr	2014	CBS 119288
Trichoderma rifaii	Rocha, Chaverri & Samuels	2015	CBS 130746
Trichoderma rodmanii	(Samuels & Chaverri) Jaklitsch & Voglmayr	2014	CBS 120895
Trichoderma rogersonii	Samuels	2006	G.J.S. 94-115
Trichoderma rosellum	Jaklitsch & Voglmayr	2014	
Trichoderma roseum	Pers.	1794	not in use
Trichoderma rossicum	Bissett, Kubicek & Szakács	2003	ATCC MYA-4839
Trichoderma rosulatum	Zhu & Zhuang	2015	HMAS 244906

(continued)

Table 1 (continued)

Species name	Author(s)	Year	Reference strain
Trichoderma rubi	Jaklitsch & Voglmayr	2015	CBS 127380
Trichoderma rubropallens	Schwein.	1832	
Trichoderma rufobrunneum	Zhu & Zhuang	2015	HMAS 252547
Trichoderma rugosum	Zhang & Zhuang	2018	not in use
Trichoderma rugulosum	Park, Oh & Lim	2019	SFC 20180301-001
Trichoderma sambuci	(Jaklitsch & Voglmayr) Jaklitsch & Voglmayr	2014	WU 29467
Trichoderma samuelsii	Jaklitsch & Voglmayr	2012	CBS 130537
Trichoderma saturnisporopsis	Samuels & Jaklitsch	2012	CBS 128829
Trichoderma saturnisporum	Hammill	1970	CBS 330.7
Trichoderma scalesiae	Samuels & Evans	2006	CBS 120069
Trichoderma semiorbis	(Berk.) Jaklitsch & Voglmayr	2014	CBS 130716
Trichoderma sempervirentis	Jaklitsch & Voglmayr	2013	CBS 133498
Trichoderma seppoi	Jaklitsch	2008	CBS 122498
Trichoderma shaoguanicum	Chen & Zhuang	2017	HMAS 248809
Trichoderma shennongjianum	Chen & Zhuang	2016	HMAS 245009
Trichoderma sichuanense	Chen & Zhuang	2017	HMAS 248737
Trichoderma silvae-virgineae	Jaklitsch	2011	CBS 120922
Trichoderma simmonsii	Chaverri, Rocha, Samuels, Degenkolb & Jaklitsch	2015	CBS 130431
Trichoderma simplex	Chen & Zhuang	2017	HMAS 248842
Trichoderma sinense	Bissett, Kubicek & Szakács	2003	DAOM 230004
Trichoderma sinensis	Bissett, Kubicek & Szakács	2003	not in use
Trichoderma sinoaustrale	Zhu & Zhuang	2014	HMAS 23403
Trichoderma sinokoningii	Qin & Zhuang	2016	HMAS 271397
Trichoderma sinoluteum	Zhu & Zhuang	2015	HMAS 252868
Trichoderma sinuosum	Chaverri & Samuels	2003	CBS 114247
Trichoderma solani	Samuels	2012	CBS 130506
Trichoderma solum	Chen & Zhuang	2017	HMAS 248848
Trichoderma songyi	Park, Seung Oh & Lim	2014	CBS 138099
Trichoderma sordidum	(Doi) Jaklitsch & Voglmayr	2014	

(continued)

Table 1 (continued)

Species name	Author(s)	Year	Reference strain
Trichoderma spadiceum	Schwein.	1822	not in use
Trichoderma sparsum	Qin & Zhuang	2016	HMAS 273759
Trichoderma speciosum	Yu & Du	2018	CGMCC 3.19079
Trichoderma sphaerosporum	Qin & Zhuang	2016	HMAS 273763
Trichoderma spinulosum	(Fuckel) Jaklitsch & Voglmayr	2014	CBS 311.5
Trichoderma spirale	Bissett	1992	CBS 346.93
Hypocrea splendens	Phillips & Plowr.	1885	CBS 336.69
Trichoderma sporulosum	(Link) Hughes	1958	not in use
Trichoderma stellatum	(Lu, Druzhin. & Samuels) Jaklitsch & Voglmayr	2014	not in use
Trichoderma stercorarium	(Barrasa, Martínez & Moreno) Jaklitsch & Voglmayr	2015	CBS 148.85
Trichoderma stilbohypoxyli	Samuels & Schroers	2006	CBS 992.97
Trichoderma stipitatum	Zhu & Zhuang	2015	HMAS 266613
Trichoderma stramineum	Chaverri & Samuels	2003	BPI 843667
Trichoderma strictipile	Bissett	1992	CBS 347.93
Trichoderma strictipilis	Bissett	1992	not in use
Trichoderma strigosellum	López-Quint., Gams, Boekhout & Druzhin.	2013	CBS 102817
Trichoderma strigosum	Bissett	1992	CBS 348.93
Trichoderma stromaticum	Samuels & Pardo-Schulth.	2000	CBS 101875
Trichoderma subalni	Zhang & Zhuang	2018	not in use
Trichoderma subalpinum	Jaklitsch	2011	CBS 119128
Hypocrea subcitrina	Kalchbr. & Cooke	1880	J.A.C. 14420
Trichoderma subeffusum	Jaklitsch	2011	W.M.J. 2009-17
Trichoderma subiculoides	Zeng & Zhuang	2019	not in use
Trichoderma subsulphureum	(Syd. & Syd.) Jaklitsch & Voglmayr	2014	not in use
Trichoderma subtrachycarpum	(Doi) Jaklitsch & Voglmayr	2014	
Trichoderma subviride	Qin & Zhuang	2016	HMAS 273761
Trichoderma succisum	(Rifai) Jaklitsch & Voglmayr	2014	
Trichoderma sulawesense	(Doi) Jaklitsch & Voglmayr	2014	GJS 85-228
Trichoderma sulphureum	(Schwein.) Jaklitsch & Voglmayr	2014	CBS 119929
Trichoderma surrotundum	Chaverri & Samuels	2003	BPI 843668
Trichoderma sympodianum	Kulik	1960	not in use
Trichoderma taiwanense	Samuels & Wu	2006	CBS 119058
Trichoderma tardum	Chen & Zhuang	2017	HMAS 248798

(continued)

Table 1 (continued)

Species name	Author(s)	Year	Reference strain
Trichoderma tawa	Chaverri & Samuels	2003	CBS 114233
Trichoderma taxi	Zhang, Lin & Kubicek	2007	CGMCC 1672
Trichoderma tenue	Qin & Zhuang	2017	HMAS 273785
Trichoderma texanum	Montoya, Meirelles, Chaverri & Rodrigues	2016	CBS 139784
Trichoderma thailandicum	Chaverri & Samuels	2003	CBS 114234
Trichoderma thelephoricola	Chaverri & Samuels	2003	CBS 114237
Trichoderma theobromicola	Samuels & Evans	2006	CBS 119120
Trichoderma thermophilum	Qin & Zhuang	2016	HMAS 252912
Trichoderma tiantangzhaiense	Zhu & Zhuang	2015	HMAS 252872
Trichoderma tibetense	Chen & Zhuang	2016	HMAS 245010
Trichoderma todica	Sokoloff & Toda	1967	not in use
Trichoderma tomentosum	Bissett	1992	CBS 349.93
Trichoderma trachycarpum	(Syd.) Jaklitsch & Voglmayr	2014	
Trichoderma tremelloides	Jaklitsch	2011	CBS 121140
Trichoderma trixiae	Samuels & Jaklitsch	2013	CBS 134702
Trichoderma tropicosinense	(Liu) Zhu & Zhuang	2015	HMAS 252546
Trichoderma tsugarense	Yabuki & Okuda	2014	NBRC 109641
Trichoderma tuberculatum	Pers.	1795	not in use
Trichoderma turrialbense	Samuels, Degenkolb, Nielsen & Gräfenhan	2008	CBS 112445
Trichoderma undatipile	Chen & Zhuang	2017	not in use
Trichoderma undatipilosum	Chen & Zhuang	2017	not in use
Trichoderma undulatum	du Plessis & Jacobs	2018	PPRI 19365
Trichoderma valdunense	Jaklitsch	2011	CBS 120923
Trichoderma varians	Sartory & Bainier	1912	not in use
Trichoderma varium	Ehrenb.	1818	not in use
Trichoderma velutinum	Bissett, Kubicek & Szakács	2003	DAOM 230013
Trichoderma vermifimicola	Sun & Liu	2020	HMAS 248255
Trichoderma vermipilum	Samuels	2012	CBS 127103
Trichoderma verticillatum	Chen & Zhuang	2017	HMAS 248740
Trichoderma victoriense	(Overton) Jaklitsch & Voglmayr	2014	CBS 140064
Trichoderma vinosum	Samuels	2006	CBS 119087
Trichoderma violaceum	Oudem.	1904	not in use

(continued)

Table 1 (continued)

Species name	Author(s)	Year	Reference strain
Trichoderma virens	(Mill., Giddens & Foster) Arx	1987	CBS 249.59
Trichoderma virescentiflavum	(Speg.) Jaklitsch & Voglmayr	2014	P.C. 278
Trichoderma virgatum	Cserjesi & Johnson	1972	not in use
Trichoderma viridarium	Jaklitsch, Samuels & Voglmayr	2013	CBS 132568
Trichoderma viride	Schumach.	1803	not in use
Trichoderma viride	Pers.	1794	not in use
*Trichoderma viride***	Pers.	1832	CBS 119325
Trichoderma viridescens	(Horne & Will.) Jaklitsch & Samuels	2006	CBS 433.34
Trichoderma viridialbum	Jaklitsch, Samuels & Voglmayr	2013	CBS 133495
Trichoderma viridicollare	Zhang & Zhuang	2019	W.Z. 2018b
Trichoderma viridiflavum	Zhu & Zhuang	2014	HMAS 252549
Trichoderma viridulum	Qin & Zhuang	2017	HMAS 273865
Trichoderma virilente	Jaklitsch & Voglmayr	2013	CBS 132569
Trichoderma voglmayrii	Jaklitsch	2006	CBS 117711
Trichoderma vulgatum	Chen & Zhuang	2017	HMAS 248796
Trichoderma vulpinum	Fuckel	1874	not in use
Trichoderma xanthum	Chen & Zhuang	2017	HMAS 247202
Trichoderma xixiacum	Sun & Liu	2020	HMAS 248253
Trichoderma yui	Zhu & Zhuang	2015	HMAS 266633
Trichoderma yunnanense	Yu & Zhang	2007	CBS 121219
Trichoderma zayuense	Chen & Zhuang	2017	HMAS 248835
Trichoderma zelobreve	Sun & Liu	2020	HMAS 248254
Trichoderma zeloharzianum	Yu & Du	2018	CGMCC 3.19082
Trichoderma zonatum	Zhu, Zhuang & Li	2017	CGMCC 3.18758

* *T. brevipes* was transferred from *Cordyceps* (Hypocreales) to *Trichoderma* (Bissett et al. 2015). No DNA barcoding information is available for this species.

** The name of *Trichoderma viride* is presented diferently in the three databases, namely the NCBI Taxonomy Browser contains *T. viride* Pers. 1832, while MycoBank and Index Fungorum refer to *T. viride* Pers. 1794.

re-established website of the ICTT (www.trichoderma.info) contains the other copy of the complete list of species and is designed to be regularly updated. The interactive, updated, and searchable version of the complete list of *Trichoderma* species is available as a supplementary tool in the species identification protocol

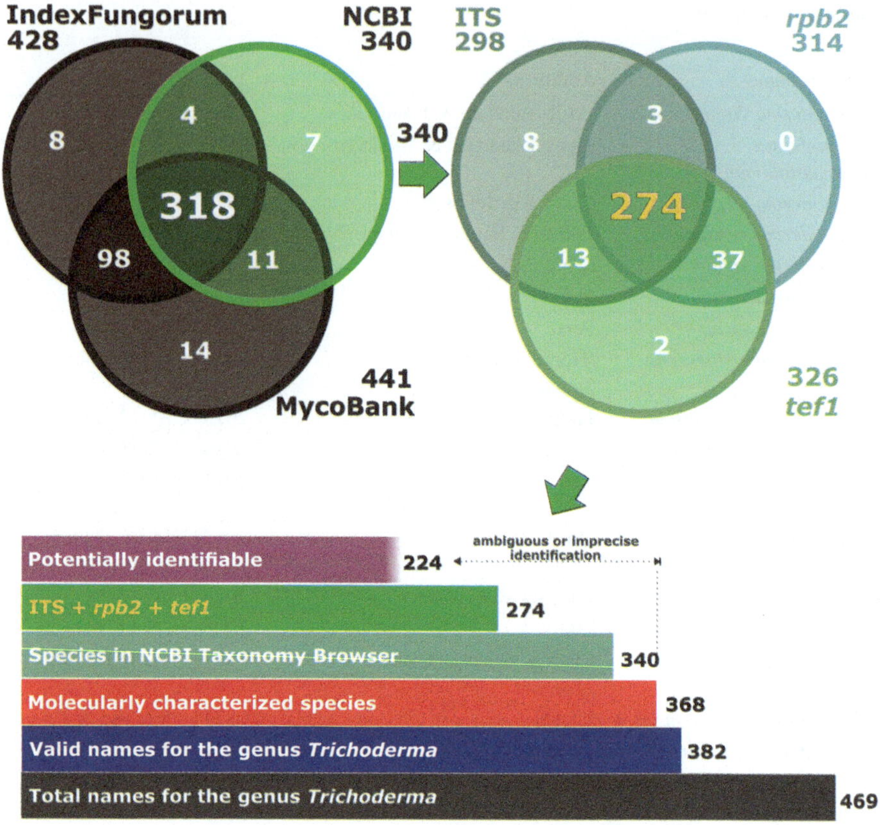

Fig. 1 The numerical representation of *Trichoderma* taxonomy. The left Venn diagram shows the number of *Trichoderma* species deposited in the major depositories of fungal taxonomy such as Index Fungorum (http://www.indexfungorum.org/), MycoBank (https://www.mycobank.org/), and NCBI Taxonomy Browser (https://www.ncbi.nlm.nih.gov/Taxonomy/Browser/wwwtax.cgi). The right Venn diagram shows the numbers of species that have one or several of the three DNA barcode sequences required for the molecular identification of *Trichoderma*. The bar plot illustrates the alarming situation related to identifiability of *Trichoderma* species. Numbers near the bars show the numbers of species (based on the estimates updated from Cai and Druzhinina 2021, www.trichokey.com and www.trichoderma.info)

(www.trichokey.com) (Cai and Druzhinina 2021). However, as the number of species grows rapidly (Cai and Druzhinina 2021), it has been suggested to screen the most recent taxonomic literature and compare it to the data on recent website updates.

The introduction of molecular methods in *Trichoderma* taxonomy not only resulted in the rapid growth of the species number but it also ended the morphological identification of *Trichoderma* (Kullnig-Gradinger et al. 2002; Druzhinina and Kubicek 2005; Druzhinina et al. 2005). Regardless of the experience and training of the taxonomist, the analysis of many morphological features cannot lead to unambiguous diagnosis of *Trichoderma* taxa even at the level of clades or sections. Thus, identification can only be achieved via analysis of DNA barcodes.

Even though 96% of *Trichoderma* species are characterized molecularly and the sequences are preserved in public databases, the Taxonomy Browser of NCBI (https://www.ncbi.nlm.nih.gov/taxonomy) contains only 340 species names (89% from all and 93% from molecularly characterized) meaning that sequence records for at least several dozen described species were not updated; however, these are still deposited as taxonomically undefined records (i.e., *Trichoderma* sp. strain ID). Consequently, these species will not appear in the results of the sequence similarity search using NCBI BLAST. The vouchered sequences can be retrieved based on sequence accession numbers provided in the publications.

Due to the high number of cryptic and closely related species, the accurate molecular identification of *Trichoderma* species requires analysis of at least three DNA barcodes (Cai and Druzhinina 2021) (see below). Considering the updated records for early 2021, the largest number of species have been DNA barcoded for *tef1* (86%) followed by *rpb2* (82%) and ITS (78%); only 270 (71%) have all 3 DNA barcodes (Fig. 1). Other commonly provided DNA barcodes (*chi18-5=ech42*, *cal1*, *act*, *acl1*, 18S rRNA=SSU, and 28S rRNA=LSU) are sequenced for less than one-half of the species; therefore, they currently have limited or no suitability for molecular identification regardless of their properties.

We notice that the number of species suitable for accurate species identification based on molecular markers is even lower than the estimate provided above (71%, Fig. 1). Our analysis showed that the identification of at least 50 recently described species is compromised by either incomplete reference sequences or sequences indistinguishable from the sister species (Cai and Druzhinina 2021). Thus, we counted only 224 (60%) of *Trichoderma* species that can be potentially identified based on available DNA barcodes (ITS, *tef1*, and *rpb2*). Still, this number appears to be an overestimate because the individual analysis of species frequently reveals further taxonomic collisions and leads to ambiguous results.

Thus, we conclude that while the taxonomy of *Trichoderma* attracted considerable attention over the last two decades, the taxonomic situation in the genus is alarming and requires urgent improvements (Fig. 1). The reasons for this unfortunate state of *Trichoderma* taxonomy and possible measures that can be taken for its improvement will be discussed below.

3 Three Stages of *Trichoderma* DNA Barcoding

The development of DNA barcoding of *Trichoderma* went through three pronounced stages: First, the species could be identified based on the combination of diagnostic oligonucleotide sequences in specific areas of ITS sequences of the rRNA gene cluster when the total diversity of the genus did not exceed 100 taxa (Druzhinina et al. 2005). This method was implemented in the web-based tool *TrichO*KEY and was supported by the public database of the reference sequences. At least for a decade, the *TrichO*KEY tool was appreciated by users of *Trichoderma* taxonomy because of its simplicity. For most species recognized at that time, a pasting of an ITS sequence in the web form provided an unambiguous and final identification result that did not require further analyses (reviewed at Druzhinina et al. (2006)). The identification could be performed by people having no experience in fungal taxonomy or molecular phylogeny. However, there were already several pairs of species that shared the same phylotypes of ITS and therefore were not distinguishable. Upon subsequent introduction of more and more new species, insufficient variability of ITS was demonstrated for many infrageneric groups especially for the clades within Section *Trichoderma* and Section *Longibrachiatum* as well as the *Harzianum* Clade. Therefore, ITS started to lose its reputation as the diagnostic marker for *Trichoderma* species (Druzhinina et al. 2012; Atanasova et al. 2010).

A new effort was focused on a search for the so-called "secondary" DNA barcode loci that would aid in unambiguous species identification. At that stage, the suitability of various loci was tested based either on the random use of recently cloned and characterized genes (e.g., *ech42* = *chi18-5*) or more commonly following the practices used for the large DNA barcoding initiatives such as the Fungal Tree of Life project (Lutzoni et al. 2004). Thus, *rpb2* (Liu et al. 1999), *cal1* (Carbone and Kohn 1999), *act* (Carbone and Kohn 1999), 18S rRNA=SSU (White et al. 1990), and 28S rRNA=LSU were sequenced for a broad range of species, but only *tef1* locus received broad support by the community (Cai and Druzhinina 2021). Therefore, the second phase of *Trichoderma* DNA barcoding was associated with the use of the large intron of *tef1* gene (Kopchinskiy et al. 2005) for sequence similarity search. The sequences of *tef1* were sufficiently polymorphic and allowed species identification with quite high precision versus the curated database of vouchered sequences using such tools as *Tricho*BLAST or (with more caution) NCBI BLAST. At that stage, we estimated that intraspecific variability of *tef1* large (4th) intron could be as high as 4–5% meaning there was a 95% similarity threshold for most of the species in BLAST.

Rahimi et al. (2021) recently offered a way to identify *T. reesei* strains by searching for the long (400 bp) sequence of *tef1* fragment that they postulated to be diagnostic for this species. However, no such hallmarks were reported for other *Trichoderma* spp. This "*tef1*" stage ended with the so-called species boom that occurred in *Trichoderma* in 2014–2015 when more than 100 new species were added mainly due to the taxonomic studies in Europe and China (reviewed in Cai and Druzhinina 2021). Dou et al. (2020) were the first group to realize that the single secondary barcode—the partial *tef1* sequence—was no longer sensitive enough for the identification of *Trichoderma* species. For this purpose, they programmed MIST (The Multiloci Identification System for *Trichoderma* (http://mmit.china-cctc.org/)) that relied on the gradual application of sequence similarity search for the three loci: ITS, *tef1*, and *rpb2*. This started the third stage of *Trichoderma* DNA barcoding. This program offered a reasonable replacement to *TrichO*KEY that was consequently shut down (Cai and Druzhinina 2021). The strength of MIST was the most complete database of the reference sequences for *Trichoderma* and included the tree DNA barcoding loci for many type strains; it also contained numerous unverified records and thus could not result in highly accurate or precise identification. Interestingly, the two secondary DNA barcodes (the partial sequences of *tef1* and *rpb2*) have unequal levels of polymorphism. Therefore, no single value of the similarity threshold could be used for either markers. To overcome this issue, we recently collected all DNA barcoding records for all contemporary valid *Trichoderma* species and proposed the species identification protocol (Cai and Druzhinina 2021). There, we reviewed the interspecific polymorphism of ITS, *tef1*, and *rpb2* sequences of closely related *Trichoderma* species to find the most reasonable sequence similarity values for each of the three DNA barcoding loci. This allowed us to formulate the sequence similarity standard:

$$\textit{Trichoderma}\ \left[\mathrm{ITS}_{76}\right] \sim \mathrm{sp}\exists!\ \left(\textit{rpb}2_{99} \cong \textit{tef}1_{97}\right).$$

Here, "Trichoderma" means the genus *Trichoderma*, "sp" means a species, "~" indicates an agreement between ITS and other loci, "≅" refers to the concordance between "*rpb2*" and "*tef1*," and "∃!" indicates the uniqueness of the condition (only one species can be identified). Subscripts show that the similarity per locus is sufficient for identification based on the assumptions of the protocol. This standard was then implemented in the molecular identification protocol (Cai and Druzhinina 2021) that required a manual analysis of every set of sequences per individual strain. Still, due to the high number or poorly characterized reference taxa, this protocol would also result in some ambiguous identifications. Moreover, the application of the identification procedure requires training in sequence analysis and can be difficult for inexperienced people. However, no "easy" solution appears to be feasible at this phase of *Trichoderma* taxonomy.

The current (third) stage of DNA barcoding of *Trichoderma* is based on the three DNA loci that are considered to be the most reliable. Still the identification process remains complex. Even though Cai and Druzhinina (2021) argue that all three loci are required for the accurate and precise species identification, ITS can only be used to identify *Trichoderma* at the generic level. Most species recognition comes from the diagnostic fragments of *tef1* and *rpb2* gene sequences. The choice of these loci is not determined by their particular suitability for the purpose but rather by their availability in public databases for most species (Fig. 1).

The advantage of *tef1* is the high polymorphism of its large (4th) intron sequence that is 250–300 base pairs long. We determined that individual strains within most of the contemporary species share >97% similarity of this fragment meaning that the polymorphism can reach up to 3% or 20–25 single mutations. This "identification window" is small versus that during the second stage of DNA barcoding, but it still offers a reasonable resolution and may potentially lead to unambiguous identification of strains having *tef1* phylotypes highly similar to that of the type strain for a given species. However, the disadvantage of *tef1* is also linked to its high polymorphism because it prevents combining strains from different infrageneric clades on a single alignment (Jaklitsch 2009a, 2011). Consequently, many *Trichoderma* taxonomy providers keep sequencing *tef1* for newly described species but have largely abandoned the polymorphic fragment and shifted toward the 3′ end of the gene to the highly conserved fragment of the last (6th) exon (Jaklitsch 2009b, 2011). Consequently, the taxonomic value of this version of the *tef1* DNA barcode locus is neglectable. This shift coincided with the "species boom" and resulted in the description of the large number of species that cannot be distinguished based on existing DNA barcodes (Cai and Druzhinina 2021).

The properties of *rpb2* are the reverse versus *tef1*: The DNA barcoding fragment of this gene covers an area of relatively highly conserved exon sequence. Contrary to *tef1*, these sequences are easily aligned genus-wide and therefore are suitable for the construction of whole genus phylograms (Atanasova et al. 2013; Cai and Druzhinina 2021). Consequently, the polymorphism of *rpb2* is essentially lower than *tef1*, and such well-defined pairs of sister species such as *T. asperellum* and *T. asperelloides, T. reesei* and *T. parareesei*, and *T. harzianum* and *T. afroharzianum* differ by only 1% or a few single mutations of *rpb2* (usually less than eight). Unfortunately, we have detected numerous recently described species that share identical or highly similar (>99%) sequences of *rpb2* (Cai and Druzhinina 2021). The consideration of above-described limitations of *tef1* and *rpb2* DNA barcodes is the main but not the only source of identification complexity.

The other issue causing the identification ambiguity is related to the cases of unconcordant similarities of the three DNA barcoding loci. For example, Cai and Druzhinina (2021) pointed to the ambiguous taxonomic position of their model whole genome sequenced strain NJAU 4742 (Zhang et al. 2016, 2019; Pang et al. 2020; Cai et al. 2020; Gao et al. 2020; Druzhinina et al. 2018; Kubicek et al. 2019; Jiang et al. 2019; Zhao et al. 2021). This strain has the *tef1* DNA barcode identical to the type strain of *T. guizhouense*. Therefore, it was attributed to this species at the second stage of DNA barcoding of *Trichoderma*. However, the *rpb2* sequence of this strain is less than 95% similar to that of the type strain of *T. guizhouense* and has most affinity to *T. pyramidale* (97.8%, which is still below the identification threshold). Interestingly, we came across several other strains with the same haplotype of *tef1* and *rpb2* as NJAU 4742. These data suggest the existence of a putative new species (*T. shenii* nom. prov., Cai and Druzhinina 2021). This and numerous other cases of incongruent similarities point to the need for phylogenetic analyses of *tef1* and *rpb2* alignments along with the consideration of the similarities. In turn, these data explain why any attempts at automated identification of sequences such as *TrichO*KEY and MIST do not appear feasible.

4 Notes on the Identification of *Trichoderma* Species

The protocol for molecular identification of a single *Trichoderma* strain is detailed in Cai and Druzhinina (2021). That work also contains several dozen practical examples that provide an overview of various situations related to the implementation of this protocol. In this chapter, we do not repeat the description of the protocol but rather comment on it and highlight a few aspects that appear critical for its understanding and correct use (Fig. 2).

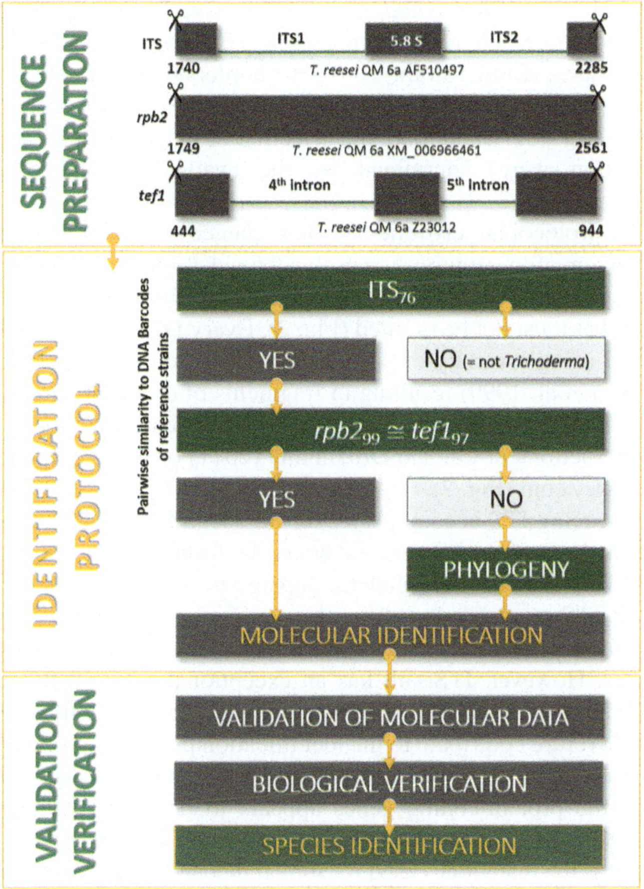

Fig. 2 The summary of the current molecular identification protocol for *Trichoderma* species (Cai and Druzhinina 2021)

First, it is important to bear in mind that neither the choice of DNA barcode markers nor the sequence similarity threshold values were selected based on their properties or particular suitability for the species recognition in *Trichoderma*. The decision to use these loci was merely pragmatic because these were the only three DNA barcoding markers that were available in public databases for the majority of species (Fig. 1). Accordingly, the similarity values were picked such that they could distinguish most of the contemporary species (Cai and Druzhinina 2021). We admit that the whole genome sequences for *Trichoderma* (Druzhinina et al. 2018; Kubicek et al. 2019) could be used for the detection of essentially more powerful DNA barcoding loci in a hypothetical situation of a taxonomic revision of the entire genus. However, it is important to understand that no such revision appears to be envisioned in the near future for nonscientific reasons. The comparison of closely related *Trichoderma* strains is impeded by the strain exchange barriers between countries.

For instance, at least 100 *Trichoderma* species have been recently described in China, and this number will likely keep growing (Cai and Druzhinina 2021). Due to the quarantine rules, sending strains across the borders between some specific countries for examination in other laboratories appears to be difficult. Thus, at this stage of DNA barcoding of *Trichoderma*, the selection of diagnostic loci and criteria for the identification were determined by the availability and other practical considerations.

Second, the protocol largely relies on the sequence similarity values, and its successful implementation requires precisely defined sequence fragments per each locus. Consequently, preparation of the protocol by trimming the sequences is an essential step that must not be omitted (Fig. 2). Every DNA barcoding locus can be PCR amplified using a variety of primer pairs (Jaklitsch et al. 2005; Carbone and Kohn 1999; Liu et al. 1999) resulting in fragments of different lengths. Therefore, the base pairs flanking the diagnostic regions must be removed either manually following the instructions in Cai and Druzhinina (2021) or using online support such as www.trichokey.com (Fig. 2).

Third, sequencing ITS is compulsory for the identification of *Trichoderma* species and the analysis of infrageneric diversity. Unfortunately, to date, the database of vouchered ITS sequences is smaller compared to *tef1* and *rpb2* (Fig. 1) because sequencing of ITS was abandoned by some providers of *Trichoderma* taxonomy after this locus lost its power in distinguishing many pairs or groups of closely related species. However, ITS still has an exceptional value in fungal taxonomy (Schoch et al. 2012). Even in *Trichoderma*, many species have unique phylotypes of ITS and can therefore contribute to the identification precision. More critically, ITS is highly diagnostic at the generic border of *Trichoderma* where the limited polymorphism of the protein-coding genes appears to be less informative (Cai and Druzhinina 2021). It is also necessary to determine ITS sequences for all new fungal taxa because it is the main locus used for fungal metagenomic studies and has a vast database of environmental records (reviewed in Lücking et al. (2020)).

Fourth, it is important to specify that the protocol allows one to identify some species through the analysis of sequence similarity values with no need to run phylogenies. For example, it might be common when a certain strain has the trimmed ITS and *rpb2* phylotypes identical to that of *T. asperelloides* CBS 125938 (type) and the trimmed *tef1* phylotype having one or two SNPs different from that of the above strain. In this case, the application of the *Trichoderma* $[\mathrm{ITS}_{76}]\sim\mathrm{sp}\exists!(rpb2_{99} \cong tef1_{97})$ standard is unambiguous and leads to the molecular identification of the query strain as *T. asperelloides*. Many other cases require phylogenetic analysis. This is in particular necessary when *tef1* and *rpb2* are not concordant or the reference DNA barcoding material is incomplete. The quality of phylogenetic analysis is also strongly influenced by the taxonomic completeness of the reference materials. The dataset suitable for phylogeny should have no gaps, i.e., it should include all species reported for this infrageneric group. The protocol of Cai and Druzhinina (2021) offers a list of *Trichoderma* species and reference strains sorted based on their phylogenetic relation (PhyloOrder in Table 2 there and on www.trichokey.com). This should assist people searching for a taxonomically complete set of sequences required for their analysis.

The fifth note on the implementation of the molecular identification protocol for *Trichoderma* species refers to the validation and verification steps (Fig. 2). These steps were not considered important at the first and second stages of *Trichoderma* DNA barcoding but now appear critical.

In Cai and Druzhinina (2021), validation refers to the quality control step in the reference materials for DNA barcoding. The most common issue leading to ambiguous identifications is the deposition of the reference *tef1* sequences that contain only a portion of the last large intron (Jaklitsch 2009a) that is diagnostic for *Trichoderma* DNA barcoding. One or another end of this sequence is the mission (more frequently the 5′ end of the intron sequence). The taxonomically relevant map and the structure of the *tef1* gene were provided in Rahimi et al. (2021). As mentioned above, many taxonomists sequence the 3′ end of the *tef1* gene spanning over the last large exon that can be aligned for across the genus, but it has limited or no suitability for DNA barcoding. This refers to numerous new species introduced from Europe and China in prior and over the recent "species boom" in 2009–2015. The missing diagnostic *tef1* DNA barcodes should be provided on the first instance because with the current high number of taxa, even a single incomplete reference sequence per species will result in ambiguous identification.

This situation is less frequently noticed for *rpb2* sequences. However, *rpb2* can sometimes contain sequences of poor quality that are also not suitable for references. For the cases when the DNA barcoding sequences for the reference strains are either incomplete or of poor quality, the protocol of Cai and Druzhinina (2021) suggests using the *T.* cf. [species name] construct. The users of taxonomy (researchers that perform the identification) are advised to seek or request the completion of reference materials from their respective taxonomy providers. Alternatively (and as it was practiced at early stages of *Trichoderma* DNA barcoding), the reference strains can be obtained from the respective strain collections and sequenced.

The validation step can also fail when several species share the same phylotype of one or several DNA barcodes. Unfortunately, this is also a common situation in *Trichoderma* taxonomy (Cai and Druzhinina 2021). For example, *T. afarasin* and *T. endophyticum* share a highly similar *tef1* phylotype (>99% similarity); *T. yunnanense* and *T. kunmingense* share highly similar phylotypes of *rpb2* with each other and with *T. asperellum* (>99%). In this case, the ambiguity of the final identification can be recorded as *T.* aff. *asperellum* if the query strain was isolated from Europe (for instance). If sampling was performed in the Chinese province Yunnan, then the strains can be identified as *T.* aff. *yunnanense* or *T.* aff. *kunmingense*, depending on other properties.

After the results of molecular identification become validated through the quality control of reference materials, the next important step is the biological verification of the identification result. Biological verification requires critical evaluation of such criteria as morphology, ecophysiology, biogeography, habitat, and occurrence. At this stage, the consideration of micromorphological features appears to be reasonable. For example, the three sister species *T. pleuroti*, *T. amazonicum*, and

T. pleuroticola have numerous common and sharply different morphological and ecophysiological features verifying their distinct taxonomic statuses. Cai and Druzhinina (2021) provide a detailed explanation of the verification stage of their protocol.

Finally, the "new species hypothesis" can be an unambiguous, accurate, and precise result of molecular identification. This case ultimately requires validation of reference materials, phylogenetic analysis, and biological verification. In this chapter, we avoid discussing the criteria applicable for the delineation of species in *Trichoderma* as Cai and Druzhinina (2021) had presented a comprehensive discussion of this topic. However, we would like to stress that the correct implementation of the genealogical concordance phylogenetic species recognition concept (Taylor et al. 2000) requires the analysis of single gene topologies. The common use of the single tree based on a combined multilocus alignment is insufficient for the new species proposal.

5 Conclusions

The identification of *Trichoderma* species is an intricate and laborious task that requires a background in mycology, molecular biological skills, training in molecular evolution, and in-depth knowledge of taxonomic literature (Cai and Druzhinina 2021). The contemporary diversity of *Trichoderma* spp. cannot be identified by automated sequence similarity searches (such as NCBI BLAST or MIST BLAST) or oligonucleotide DNA barcodes. All molecular identification results require in silico validation and biological verification. Similarly, *Trichoderma* spp. cannot be identified by phylogenetic analysis without considering the sequence similarity values relative to the complete set of closely related species. The complexity of the identification process points to the need for close interactions between *Trichoderma* taxonomy experts.

In this chapter, we used *Trichoderma* to address the modern taxonomic collision that can also occur in many other genera of common and well-investigated fungi. The taxonomy of these fungi was visited and revisited many times and seemingly progressed with the introduction of new species. The delineation of the cryptic species is considered to be a useful practice because it increases the accuracy and precision of property prediction. However, many of newly recognized species appear to be difficult to identify. Ultimately, the failure to identify species leads to ambiguity but, more dangerously, to the description of more new species that further complicate the identification. This loop has been already reported before and noticed that every single fungal species has been named 2.5 times on average (Hawksworth and Lucking 2017). The good taxonomic practice should include the verification of species identifiability. Even though this process appears to be implemented as a reverse operation to the species recognition, it is frequently obscured by the application of vague species criteria. In an unfortunate case, a species can be recognized based on a comparison with a taxonomically incomplete set of references or based on species criteria that do not correspond to the state of the art in this genus. Even now, the Code will allow the application of the morphological species concept or a description of a *Trichoderma* species based on the morphological characters and the analysis of any single locus, i.e., ITS.

In this chapter, we tried to emphasize that such cases will result in a valid species name, but this species will not be possible to identify because most sister species were delineated based on advanced molecular species criteria such as GCPSR or even an integrated polyphasic approach. The example above is an exaggeration, but the taxonomic reality of *Trichoderma* is highly ambiguous. We assume that this turbulent state was caused by the recent introduction of highly powerful molecular techniques in fungal taxonomy, and the situation will get its rational solution. However, we set a further warning related to the introduction of the whole genus genomic data in *Trichoderma* taxonomy. The whole genome sequences have a still unexplored inter- and intraspecific polymorphism and thus offer essentially more options for taxonomic splitting: Species within the genus may share only 75% similarity genome-wide (Kubicek et al. 2019) and genomes of the two strains of the same clonal species *T. harzianum* have up to 1000 unique genes each. Therefore, the discussion of the unified species concept suitable for such fungi as *Trichoderma* is an urgent task for *Trichoderma* researchers and fungal taxonomists.

Acknowledgments The authors wish to thank Prof. Wenying Zhuang from the Institute of Microbiology, Chinese Academy of Sciences, for her useful discussion on *Trichoderma* taxonomy. This work was supported by grants from the National Natural Science Foundation of China (31801939) to F.C. and the National Special Project of Basic Work Project for Science and Technology (2014FY120900) to J.C.

References

Atanasova L, Jaklitsch WM, Komoń-Zelazowska M, Kubicek CP, Druzhinina IS (2010) Clonal species *Trichoderma parareesei* sp. nov. likely resembles the ancestor of the cellulase producer *Hypocrea jecorina/T. reesei*. Appl Environ Microbiol 76(21):7259–7267. https://doi.org/10.1128/AEM.01184-10

Atanasova L, Druzhinina IS, Jaklitsch WM (2013) Two hundred *Trichoderma* species recognized on the basis of molecular phylogeny. CABI, Wallingford, pp 10–42. https://doi.org/10.1079/9781780642475.0010

Bissett J (1984) A revision of the genus *Trichoderma*. I. Section *Longibrachiatum* sect. nov. Can J Bot 62(5):924–931. https://doi.org/10.1139/b84-131

Bissett J (1991a) A revision of the genus *Trichoderma*. II. Infrageneric classification. Can J Bot 69(11):2357–2372. https://doi.org/10.1139/b91-297

Bissett J (1991b) A revision of the genus *Trichoderma*. III. Section Pachybasium. Can J Bot 69(11):2373–2417. https://doi.org/10.1139/b91-298

Bissett J (1991c) A revision of the genus *Trichoderma*. IV. Additional notes on section *Longibrachiatum*. Can J Bot 69(11):2418–2420. https://doi.org/10.1139/b91-299

Bissett J (1992) Trichoderma atroviride. Can J Bot 70(3):639–641

Bissett J, Gams W, Jaklitsch W, Samuels GJ (2015) Accepted *Trichoderma* names in the year 2015. IMA Fungus 6(2):263–295. https://doi.org/10.5598/imafungus.2015.06.02.02

Cai F, Druzhinina IS (2021) In honor of John Bissett: authoritative guidelines on molecular identification of *Trichoderma*. Fungal Divers. https://doi.org/10.1007/s13225-020-00464-4

Cai F, Gao R, Zhao Z, Ding M, Jiang S, Yagtu C, Zhu H, Zhang J, Ebner T, Mayrhofer-Reinhartshuber M, Kainz P, Chenthamara K, Akcapinar GB, Shen Q, Druzhinina IS (2020) Evolutionary compromises in fungal fitness: hydrophobins can hinder the adverse dispersal of conidiospores and challenge their survival. ISME J 14(10):2610–2624. https://doi.org/10.1038/s41396-020-0709-0

Carbone I, Kohn LM (1999) A method for designing primer sets for speciation studies in filamentous ascomycetes. Mycologia 91(3):553–556. https://doi.org/10.1080/00275514.1999.12061051

Chaverri P, Castlebury LA, Samuels GJ, Geiser DM (2003) Multilocus phylogenetic structure within the *Trichoderma harzianum/Hypocrea lixii* complex. Mol Phylogenet Evol 27(2):302–313. https://doi.org/10.1016/S1055-7903(02)00400-1

Chen P, Pang G, Cai F, Druzhinina IS (2021) Strain improvement and genetic engineering of *Trichoderma* for industrial applications. In: Zaragoza O, Casadevall A (eds) Encyclopedia of mycology. Elsevier, pp 505–517. https://doi.org/10.1016/B978-0-12-819990-9.00029-9

Chouaki T, Lavarde V, Lachaud L, Raccurt CP, Hennequin C (2002) Invasive infections due to *Trichoderma* species: report of 2 cases, findings of *in vitro* susceptibility testing, and review of the literature. Clin Infect Dis 35(11):1360–1367. https://doi.org/10.1086/344270

Ding MY, Chen W, Ma XC, Lv BW, Jiang SQ, Yu YN, Rahimi MJ, Gao RW, Zhao Z, Cai F, Druzhinina IS (2020) Emerging salt marshes as a source of *Trichoderma arenarium* sp. nov. and other fungal bioeffectors for biosaline agriculture. J Appl Microbiol n/a(n/a). https://doi.org/10.1111/jam.14751

Dou K, Lu Z, Wu Q, Ni M, Yu C, Wang M, Li Y, Wang X, Xie H, Chen J, Zhang C (2020) MIST: a multilocus identification system for *Trichoderma*. Appl Environ Microbiol 86(18):e01532–e01520. https://doi.org/10.1128/AEM.01532-20

Druzhinina I, Kubicek CP (2005) Species concepts and biodiversity in *Trichoderma* and *Hypocrea*: from aggregate species to species clusters? J Zhejiang Univ Sci B 6(2):100–112. https://doi.org/10.1631/jzus.2005.B0100

Druzhinina IS, Kubicek CP (2017) Genetic engineering of *Trichoderma reesei* cellulases and their production. Microb Biotechnol 10(6):1485–1499. https://doi.org/10.1111/1751-7915.12726

Druzhinina IS, Kopchinskiy AG, Komoń M, Bissett J, Szakacs G, Kubicek CP (2005) An oligonucleotide barcode for species identification in *Trichoderma* and *Hypocrea*. Fungal Genet Biol 42(10):813–828. https://doi.org/10.1016/j.fgb.2005.06.007

Druzhinina IS, Kopchinskiy AG, Kubicek CP (2006) The first 100 *Trichoderma* species characterized by molecular data. Mycoscience 47(2):55–64. https://doi.org/10.1007/S10267-006-0279-7

Druzhinina IS, Seidl-Seiboth V, Herrera-Estrella A, Horwitz BA, Kenerley CM, Monte E, Mukherjee PK, Zeilinger S, Grigoriev IV, Kubicek CP (2011) *Trichoderma*: the genomics of opportunistic success. Nat Rev Microbiol 9(10):749–759. https://doi.org/10.1038/nrmicro2637

Druzhinina IS, Komoń-Zelazowska M, Ismaiel A, Jaklitsch W, Mullaw T, Samuels GJ, Kubicek CP (2012) Molecular phylogeny and species delimitation in the section *Longibrachiatum* of *Trichoderma*. Fungal Genet Biol 49(5):358–368. https://doi.org/10.1016/j.fgb.2012.02.004

Druzhinina IS, Chenthamara K, Zhang J, Atanasova L, Yang D, Miao Y, Rahimi MJ, Grujic M, Cai F, Pourmehdi S, Salim KA, Pretzer C, Kopchinskiy AG, Henrissat B, Kuo A, Hundley H, Wang M, Aerts A, Salamov A, Lipzen A, LaButti K, Barry K, Grigoriev IV, Shen Q, Kubicek CP (2018) Massive lateral transfer of genes encoding plant cell wall-degrading enzymes to the mycoparasitic fungus *Trichoderma* from its plant-associated hosts. PLoS Genet 14(4):e1007322. https://doi.org/10.1371/journal.pgen.1007322

Gao R, Ding M, Jiang S, Zhao Z, Chenthamara K, Shen Q, Cai F, Druzhinina IS, Drake HL (2020) The evolutionary and functional paradox of cerato-platanins in fungi. Appl Environ Microbiol 86(13):e00696–e00620. https://doi.org/10.1128/AEM.00696-20

Gu X, Wang R, Sun Q, Wu B, Sun JZ (2020) Four new species of *Trichoderma* in the *Harzianum* clade from northern China. MycoKeys 73:109–132. https://doi.org/10.3897/mycokeys.73.51424

Harman GE, Howell CR, Viterbo A, Chet I, Lorito M (2004) *Trichoderma* species-opportunistic, avirulent plant symbionts. Nat Rev Microbiol 2(1):43–56. https://doi.org/10.1038/nrmicro797

Hawksworth DL, Lucking R (2017) Fungal diversity revisited: 2.2 to 3.8 million species. Microbiol Spectr 5(4). https://doi.org/10.1128/microbiolspec.FUNK-0052-2016

Inderbitzin P, Robbertse B, Schoch CL (2020) Species identification in plant-associated prokaryotes and fungi using DNA. Phytobiomes J 4(2):103–114. https://doi.org/10.1094/pbiomes-12-19-0067-rvw

Jaklitsch WM (2009a) European species of *Hypocrea* Part I. The green-spored species. Stud Mycol 63(0):1–91. https://doi.org/10.3114/sim.2009.63.01

Jaklitsch WM (2009b) European species of *Hypocrea* Part I. The green-spored species. Stud Mycol 63:1–91. https://doi.org/10.3114/sim.2009.63.01

Jaklitsch WM (2011) European species of *Hypocrea* part II: species with hyaline ascospores. Fungal Divers 48(1):1–250. https://doi.org/10.1007/s13225-011-0088-y

Jaklitsch WM, Komon M, Kubicek CP, Druzhinina IS (2005) *Hypocrea voglmayrii* sp. nov. from the Austrian Alps represents a new phylogenetic clade in *Hypocrea/Trichoderma*. Mycologia 97(6):1365–1378. https://doi.org/10.1080/15572536.2006.11832743

Jaklitsch WM, Lechat C, Voglmayr H (2014) The rise and fall of *Sarawakus* (Hypocreaceae, Ascomycota). Mycologia 106(1):133–144. https://doi.org/10.3852/13-117

Jiang S-Q, Yu Y-N, Gao R-W, Wang H, Zhang J, Li R, Long X-H, Shen Q-R, Chen W, Cai F (2019) High-throughput absolute quantification sequencing reveals the effect of different fertilizer applications on bacterial community in a tomato cultivated coastal saline soil. Sci Total Environ 687:601–609. https://doi.org/10.1016/j.scitotenv.2019.06.105

Kindermann J, El-Ayouti Y, Samuels GJ, Kubicek CP (1998) Phylogeny of the genus *Trichoderma* based on sequence analysis of the Internal Transcribed Spacer region 1 of the rDNA cluster. Fungal Genet Biol 24(3):298–309. https://doi.org/10.1006/fgbi.1998.1049

Komoń-Zelazowska M, Bissett J, Zafari D, Hatvani L, Manczinger L, Woo S, Lorito M, Kredics L, Kubicek CP, Druzhinina IS (2007) Genetically closely related but phenotypically divergent *Trichoderma* species cause green mold disease in Oyster mushroom farms worldwide. Appl Environ Microbiol 73(22):7415–7426. https://doi.org/10.1128/AEM.01059-07

Kopchinskiy A, Komon M, Kubicek CP, Druzhinina IS (2005) *Tricho*BLAST: a multilocus database for *Trichoderma* and *Hypocrea* identifications. Mycol Res 109(Pt 6):658–660. https://doi.org/10.1017/S0953756205233397

Kredics L, Antal Z, Dóczi I, Manczinger L, Kevei F, Nagy E (2003) Clinical importance of the genus *Trichoderma*. Acta Microbiol Immunol Hung 50(2–3):105–117. https://doi.org/10.1556/amicr.50.2003.2-3.1

Kredics L, Garcia Jimenez L, Naeimi S, Czifra D, Urbán P, Manczinger L, Vágvölgyi C, Hatvani L (2010) A challenge to mushroom growers: the green mould disease of cultivated champignons. In: Technology and education topics in applied microbiology and microbial biotechnology, pp 295–305

Kubicek CP, Steindorff AS, Chenthamara K, Manganiello G, Henrissat B, Zhang J, Cai F, Kopchinskiy AG, Kubicek EM, Kuo A, Baroncelli R, Sarrocco S, Noronha EF, Vannacci G, Shen Q, Grigoriev IV, Druzhinina IS (2019) Evolution and comparative genomics of the most common *Trichoderma* species. BMC Genomics 20(1):485. https://doi.org/10.1186/s12864-019-5680-7

Kuhls K, Lieckfeldt E, Samuels GJ, Meyer W, Kubicek CP, Börner T (1997) Revision of *Trichoderma* sect. *Longibrachiatum* including related teleomorphs based on analysis of ribosomal DNA internal transcribed spacer sequences. Mycologia 89(3):442–460. https://doi.org/10.1080/00275514.1997.12026803

Kullnig C, Szakacs G, Kubicek CP (2000) Molecular identification of *Trichoderma* species from Russia, Siberia and the Himalaya. Mycol Res 104(9):1117–1125. https://doi.org/10.1017/S0953756200002604

Kullnig-Gradinger CM, Szakacs G, Kubicek CP (2002) Phylogeny and evolution of the genus *Trichoderma*: a multigene approach. Mycol Res 106(7):757–767. https://doi.org/10.1017/S0953756202006172

Liu YJ, Whelen S, Hall BD (1999) Phylogenetic relationships among ascomycetes: evidence from an RNA polymerase II subunit. Mol Biol Evol 16(12):1799–1808. https://doi.org/10.1093/oxfordjournals.molbev.a026092

Lücking R, Aime MC, Robbertse B, Miller AN, Ariyawansa HA, Aoki T, Cardinali G, Crous PW, Druzhinina IS, Geiser DM, Hawksworth DL, Hyde KD, Irinyi L, Jeewon R, Johnston PR, Kirk PM, Malosso E, May TW, Meyer W, Öpik M, Robert V, Stadler M, Thines M, Vu D, Yurkov AM, Zhang N, Schoch CL (2020) Unambiguous identification of fungi: where do we stand and how accurate and precise is fungal DNA barcoding? IMA Fungus 11(1):14. https://doi.org/10.1186/s43008-020-00033-z

Lutzoni F, Kauff F, Cox C, McLaughlin D, Celio G, Dentinger B, Padamsee M, Hibbett D, James T, Baloch E (2004) Where are we in assembling the fungal tree of life, classifying the fungi, and understanding the evolution of their subcellular traits. Am J Bot 91:1446–1480

Marra R, Lombardi N, d'Errico G, Troisi J, Scala G, Vinale F, Woo SL, Bonanomi G, Lorito M (2019) Application of *Trichoderma* strains and metabolites enhances soybean productivity and nutrient content. J Agric Food Chem 67(7):1814–1822. https://doi.org/10.1021/acs.jafc.8b06503

May TW, Redhead SA, Bensch K, Hawksworth DL, Lendemer J, Lombard L, Turland NJ (2019) Chapter F of the International Code of Nomenclature for algae, fungi, and plants as approved by the 11th International Mycological Congress, San Juan, Puerto Rico, July 2018. IMA Fungus 10(1):21. https://doi.org/10.1186/s43008-019-0019-1

Migheli Q, Balmas V, Komoñ-Zelazowska M, Scherm B, Fiori S, Kopchinskiy AG, Kubicek CP, Druzhinina IS (2009) Soils of a Mediterranean hot spot of biodiversity and endemism (Sardinia, Tyrrhenian Islands) are inhabited by pan-European, invasive species of *Hypocrea/Trichoderma*. Environ Microbiol 11(1):35–46. https://doi.org/10.1111/j.1462-2920.2008.01736.x

Myoken Y, Sugata T, Fujita Y, Asaoku H, Fujihara M, Mikami Y (2002) Fatal necrotizing stomatitis due to *Trichoderma longibrachiatum* in a neutropenic patient with malignant lymphoma: a case report. Int J Oral Maxillofac Surg 31(6):688–691. https://doi.org/10.1054/ijom.2001.0211

Pang G, Sun T, Yu Z, Yuan T, Liu W, Zhu H, Gao Q, Yang D, Kubicek CP, Zhang J, Shen Q (2020) Azaphilones biosynthesis complements the defence mechanism of *Trichoderma guizhouense* against oxidative stress. Environ Microbiol 22(11):4808–4824. https://doi.org/10.1111/1462-2920.15246

Persoon CH (1794) Disposita methodical fungorum. Romers Neues Mag Bot 1:81–128

Rahimi MJ, Cai F, Grujic M, Chenthamara K, Druzhinina IS (2021) Molecular identification of *Trichoderma reesei*. In: Mach-Aigner AR, Martzy R (eds) *Trichoderma reesei*: methods and protocols. Springer US, New York, pp 157–175. https://doi.org/10.1007/978-1-0716-1048-0_14

Rifai MA (1969) A revision of the genus *Trichoderma*. Mycol Pap 116:1–56

Rivera-Méndez W, Obregón M, Morán-Diez ME, Hermosa R, Monte E (2020) *Trichoderma asperellum* biocontrol activity and induction of systemic defenses against *Sclerotium cepivorum* in onion plants under tropical climate conditions. Biol Control 141:104145. https://doi.org/10.1016/j.biocontrol.2019.104145

Schoch CL, Seifert KA, Huhndorf S, Robert V, Spouge JL, Levesque CA, Chen W, List FBCA (2012) Nuclear ribosomal internal transcribed spacer (ITS) region as a universal DNA barcode marker for *fungi*. Proc Natl Acad Sci U S A 109(16):6241–6246. https://doi.org/10.1073/pnas.1117018109

Steenkamp ET, Wingfield MJ, McTaggart AR, Wingfield BD (2018) Fungal species and their boundaries matter – definitions, mechanisms and practical implications. Fungal Biol Rev 32(2):104–116. https://doi.org/10.1016/j.fbr.2017.11.002

Sun J, Yuan X, Li Y, Wang X, Chen J (2019) The pathway of 2,2-dichlorovinyl dimethyl phosphate (DDVP) degradation by *Trichoderma atroviride* strain T23 and characterization of a paraoxonase-like enzyme. Appl Microbiol Biotechnol 103(21):8947–8962. https://doi.org/10.1007/s00253-019-10136-2

Taylor JW (2011) One Fungus = One Name: DNA and fungal nomenclature twenty years after PCR. IMA Fungus 2(2):113–120. https://doi.org/10.5598/imafungus.2011.02.02.01

Taylor JW, Jacobson DJ, Kroken S, Kasuga T, Geiser DM, Hibbett DS, Fisher MC (2000) Phylogenetic species recognition and species concepts in fungi. Fungal Genet Biol 31(1):21–32. https://doi.org/10.1006/fgbi.2000.1228

Turland N, Wiersema J, Barrie F, Greuter W, Hawksworth D, Herendeen P, Knapp S, Kusber W-H, Li D-Z, Marhold K, May T, McNeill J, Monro A, Prado J, Price M, Smith G (2018) International Code of Nomenclature for algae, fungi, and plants (Shenzhen Code) adopted by the Nineteenth International Botanical Congress Shenzhen, China, July 2017. Regnum Vegetabile Volume 159. Koeltz Botanical Books. doi:https://doi.org/10.12705/Code.2018

White TJ, Bruns T, Lee S, Taylor J (1990) 38 - Amplification and direct sequencing of fungal ribosomal rna genes for phylogenetics. In: Innis MA, Gelfand DH, Sninsky JJ, White TJ (eds) PCR protocols. Academic Press, San Diego, pp 315–322. https://doi.org/10.1016/B978-0-12-372180-8.50042-1

Wu Q, Ni M, Dou K, Tang J, Ren J, Yu C, Chen J (2018) Co-culture of *Bacillus amyloliquefaciens* ACCC11060 and *Trichoderma asperellum* GDFS1009 enhanced pathogen-inhibition and amino acid yield. Microb Cell Factories 17(1):155. https://doi.org/10.1186/s12934-018-1004-x

Zachow C, Berg C, Müller H, Meincke R, Komon-Zelazowska M, Druzhinina IS, Kubicek CP, Berg G (2009) Fungal diversity in the rhizosphere of endemic plant species of Tenerife (Canary Islands): relationship to vegetation zones and environmental factors. ISME J 3(1):79–92. https://doi.org/10.1038/ismej.2008.87

Zhang J, Bayram Akcapinar G, Atanasova L, Rahimi MJ, Przylucka A, Yang D, Kubicek CP, Zhang R, Shen Q, Druzhinina IS (2016) The neutral metallopeptidase NMP1 of *Trichoderma guizhouense* is required for mycotrophy and self-defence. Environ Microbiol 18(2):580–597. https://doi.org/10.1111/1462-2920.12966

Zhang J, Miao Y, Rahimi MJ, Zhu H, Steindorff A, Schiessler S, Cai F, Pang G, Chenthamara K, Xu Y, Kubicek CP, Shen Q, Druzhinina IS (2019) Guttation capsules containing hydrogen peroxide: an evolutionarily conserved NADPH oxidase gains a role in wars between related fungi. Environ Microbiol 21(8):2644–2658. https://doi.org/10.1111/1462-2920.14575

Zhao Z, Cai F, Gao R, Ding M, Jiang S, Chen P, Pang G, Chenthamara K, Shen Q, Bayram Akcapinar G, Druzhinina IS (2021) At least three families of hyphosphere small secreted cysteine-rich proteins can optimize surface properties to a moderately hydrophilic state suitable for fungal attachment. Environ Microbiol. https://doi.org/10.1111/1462-2920.15413

Part II
Trichoderma and Plant-Pathogenic Fungi

Functional Genetics of *Trichoderma* Mycoparasitism

Kai Dou, Guan Pang, Feng Cai, Komal Chenthamara, Jian Zhang, Hongyi Liu, Irina S. Druzhinina, and Jie Chen

Contents

K. Dou · H. Liu · J. Chen (✉)
School of Agriculture and Biology, Shanghai Jiaotong University, Shanghai, China
e-mail: jiechen59@sjtu.edu.cn

G. Pang · F. Cai
Key Laboratory of Plant Immunity, Nanjing Agricultural University, Nanjing, China

Fungal Genomics Laboratory (FungiG), Nanjing Agricultural University, Nanjing, China

K. Chenthamara
Institute of Chemical, Environmental, and Bioscience Engineering (ICEBE), TU Wien, Vienna, Austria

J. Zhang
Key Laboratory of Plant Immunity, Nanjing Agricultural University, Nanjing, China

I. S. Druzhinina
Key Laboratory of Plant Immunity, Nanjing Agricultural University, Nanjing, China

Fungal Genomics Laboratory (FungiG), Nanjing Agricultural University, Nanjing, China

Institute of Chemical, Environmental, and Bioscience Engineering (ICEBE), TU Wien, Vienna, Austria

N. Amaresan et al. (eds.), *Advances in* Trichoderma *Biology for Agricultural Applications*, Fungal Biology, https://doi.org/10.1007/978-3-030-91650-3_2

1 Current Trends in *Trichoderma* Mycoparasitism Research

Sustainable and green agriculture requires the development of ecologically friendly bioeffectors that can act either instead of chemical pesticides and fertilizers or can be applied along with them in an integrated manner (Yadav et al. 2020). Over the last decade, the governments of many countries and political unions have issued regulations restricting the anthropogenic load on the environment from extensive agriculture (Huang and Yang 2017; Gołaś et al. 2020; Nguyen et al. 2020; Priyadarshini and Abhilash 2020). Thus, the environment-friendly products of biological control of pests, in which indigenous plant-beneficial microorganisms or communities of such microbes act as bioeffectors (suppressors of plant pathogens), are currently in high demand on the market worldwide (Fraceto et al. 2018; Ruiu 2018). Among bioeffectors used against fungal diseases, species of the mycoparasitic and cellulolytic fungi from the genus *Trichoderma* (*Hypocreales*, *Ascomycota*) are by far the most efficient, convenient, and, therefore, most commonly used (De Rezende et al. 2020; Ding et al. 2020b). In China, at least four types of *Trichoderma*-based formulations have been developed, including wettable powder, granules, water-dispersible granules, and seed dressing agents (Chen et al. 2014). Among those formulations, six kinds of wettable powder and one kind of water-dispersible granule were registered as biofungicides to protect tomato (*Solanum lycopersicum*), ornamental lily (*Lilium* spp.), and cucumber (*Cucumis sativus*) against seedling damping-off (*Pythium* spp.), root rot (*Pythium* spp., *Fusarium* spp.), gray mold (*Botrytis cinerea*), and downy mildew (*Pseudoperonospora cubensis*) (Chen et al. 2016). Besides China, *Trichoderma*-based biofungicides are widely used in numerous other countries with developed agriculture, for example, Belgium, Germany, Spain, the United States, Brazil, and India (Woo et al. 2014; Fraceto et al. 2018).

Biofungicides are based on interfungal interactions that are most frequently adverse. As heterotrophic organisms, fungi usually compete for nutrients with other fungi in their environment or just take resources directly from them through parasitism or predation. This means that there are usually "fungal wars" during the interfungal interactions (Hiscox et al. 2018). The fungal war could happen in the manner of mutual inhibition and space and nutrient competition (Hiscox et al. 2018; Ujor et al. 2018) or result in induced necroses due to the attacks of facultative mycoparasites (Jeffries and Young 1994). With the profound ability to detoxify toxins secreted by hostile fungi and the secretion of fungal cell wall-lysing extracellular enzymes or antifungal secondary metabolites, *Trichoderma* spp. usually have the upper hand in fungal wars and are considered to be prominent and versatile fighters (Komoń-Zelazowska et al. 2007; Druzhinina et al. 2011).

The discovery of the suitability of *Trichoderma* mycoparasitism for plant protection happened only a few decades ago. Although this research area is young, it is attracting increasing attention because of the rapidly growing availability of "Big Data" for *Trichoderma* and other fungi. It is not surprising, therefore, that, until now, only a few *Trichoderma* species have been tested for their biocontrol potential and used for commercial applications. However, even these studies reveal that the ability of *Trichoderma* species to efficiently suppress various phytopathogenic fungi should be investigated in the ecological and evolutionary concept. For example, the mycoparasitic lifestyle is also reported for the members of the order *Hypocreales* like *Escovopsis weberi* (De Man et al. 2016), many *Hypomyces* spp. (Zeng and Zhuang 2019; Lakkireddy and Khonsuntia 2020; Yu et al. 2020), and *Tolypocladium* spp., but only species from the genus *Clonostachys* have also been proposed for plant protection (Da Silva et al. 2021). *C. rosea* has been used to control plant diseases, including *Fusarium graminearum* (Hue et al. 2009) and *Sclerotinia sclerotiorum* (*Helotiales*, *Ascomycota*) (Rodríguez et al. 2011). *C. chloroleuca* also exhibits potential for biocontrol of various plant pathogens, including *S. sclerotiorum* (Sun et al. 2018).

Studies show that *Trichoderma* and *Clonostachys* species employ versatile strategies to suppress other fungi and may rely on one or several "weapons," such as secretion of hydrolytic enzymes, secondary metabolites, or both, and meanwhile being tolerant to toxic metabolites produced by other fungi. For instance, it has recently been shown that *T. guizhouense* [*T.* sp. NJAU 4742 sensu in Cai and Druzhinina (2021)] can use protease (Zhang et al. 2016) and reactive oxygen species (Zhang et al. 2019) for the antagonism of *F. odoratissimum* (formerly known as FOC4) and protect itself with antioxidant azaphilones (Pang et al. 2020) and a short-chain dehydrogenase (Zhu et al. 2021). The interaction between these two fungi is complex and depends on multiple factors, such as abiotic factors (illumination, availability of the nutrients) and developmental stage. It is remarkable as it indeed resembles a war. When the two fungi contact each other on the surface of the Petri dish, *F. odoratissimum* produces abundant water-soluble toxins that arrest the entire protein synthesis by *T. guizhouense*, as seen in the transcriptomic profile (Zhang et al. 2019). The early images of the contact zone indicate antibiosis and a "deadlock" reaction when the growth of both fungi stops (Fig. 1). However, with time, *T. guizhouense* develops aerial hyphae capable of overgrowing the *F. odoratissimum* aerial hyphae. It has been shown that the direct interaction between the aerial hyphae of both fungi is aided by the formation of special guttation drops that contain proteolytic enzymes, are enriched in hydrogen peroxide, and contain numerous other metabolites (Zhang et al. 2019). Although it was not experimentally verified, the transcriptomic results suggest that the surface-active small secreted cysteine-rich proteins hydrophobins contribute to the stability of such "hunting bags" by assembling in the water-air interface and forming a water-proof membrane (Fig. 1). This interaction allows *T. guizhouense* to overgrow *F. odoratissimum* and form abundant conidiation on its aerial mycelium. As *T. guizhouense* is unable to touch the medium beneath *F. odoratissimum* due to toxins, it is assumed that the nutrients required to support conidiation are taken from the aerial hyphae of the host. The transcriptional profiling of genes involved in this interaction with *F. odoratissimum* suggests their involvement in the parasitism of several other fungi, such as *Alternaria alternata* (*Pleosporales*, *Ascomycota*), *Botrytis cinerea* (*Helotiales*, *Ascomycota*), and others (Zhang et al. 2016, 2019; Pang et al. 2020). However, a certain level of host specificity was also documented because none of the mechanisms mentioned above were exploited by this strain in interactions with *Rhizoctonia solani* (Fig. 1) (Zhang et al.

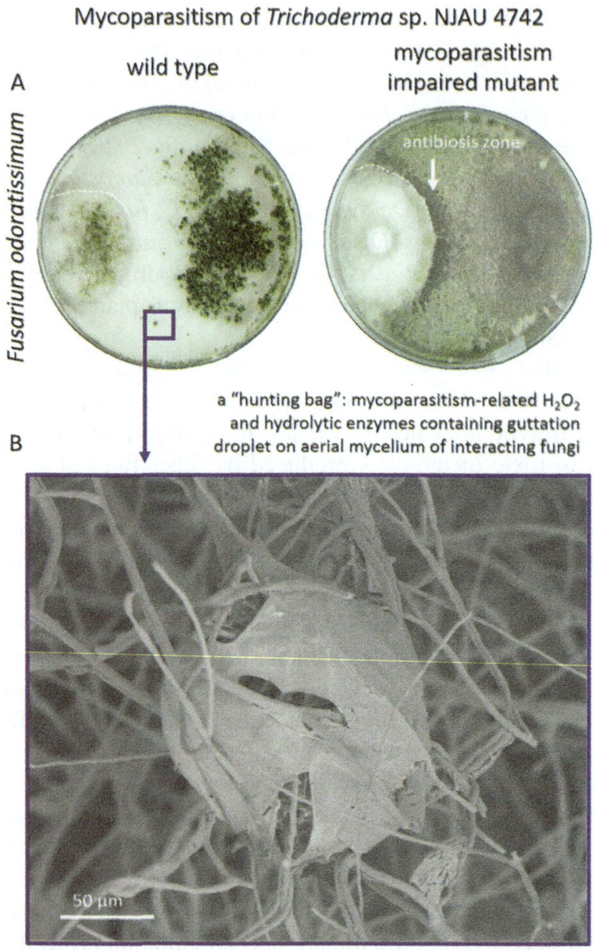

Fig. 1 A unique mycoparasitic strategy of *Trichoderma* sp. NJAU 4742 (formerly known as *T. guizhouense* NJAU 4742) on toxin-producing strain of *Fusarium odoratissimum* (formerly known as FOC4)

At the initial interaction stage, *F. odoratissimum* produces water-soluble toxic metabolites that arrest protein synthesis of *T.* sp. NJAU 4742 (**a**). The resulting antibiosis zone is clearly seen in the case of the mycoparasitism-impaired mutant of this *Trichoderma*. The wild-type strain, however, is able to overgrow and parasitize aerial hyphae of *F. odoratissimum* by producing "hunting bags" (**b**) – the guttation drops putatively coated by hydrophobins and containing ROS and hydrolytic enzymes required for mycoparasitism (Zhang et al. 2019). (Images taken after 13 days of incubation on PDA at 25 °C in darkness)

2016, 2019). Similar to *Trichoderma*, fungal cell wall-degrading enzymes (Chatterton and Punja 2009) and peptaibiotic metabolite production (Rodríguez et al. 2011) are involved in the antagonism of *Clonostachys* (*Hypocreales*, *Ascomycota*) against species of *Fusarium* and *Sclerotinia*. ATP-binding cassette transporters are essential for the xenobiotic tolerance of *Clonostachys* (Dubey et al. 2014).

The rigorous way to verify the functional genetics involved in the mycoparasitism of *Trichoderma* spp. and *Clonostachys* spp. is by comparing the phenotypes of wild-type strains, gene disruption mutants, and reverse complementation mutants. If a gene is related to mycoparasitism, the mycoparasitism ability of gene disruption mutants should be reduced compared with wild-type strains. For example, the *abcG5* disruption mutants of *C. rosea* exhibited reduced antagonism toward *F. graminearum* (Dubey et al. 2014), indicating that *abcG5* plays a role in the antagonism process. The gene reverse complementation mutants, which showed a similar phenotype with the wild-type strains, were used to illustrate that the phenotypes of gene disruption mutants are not a false negative. For instance, while the *CrSsd1* mutants of *C. rosea* showed a weakened mycoparasitism rate against *S. sclerotiorum*, the complementary mutants recovered the lost ability and were similar to the wild-type strains in mycoparasitism (Lv et al. 2020). Although not essential, the overexpression and fluorescent labeling methods can also help explore the role of functional genetics (Pang et al. 2020). Overexpressing of protease in *T. virens* showed increased protection of cotton seedlings against *R. solani* (Pozo et al. 2004). Red fluorescent protein-labeled *T. guizhouense* was used to investigate the interactions with *F. odoratissimum* (Zhang et al. 2019).

To date, the functional genetic studies of *Trichoderma* mycoparasitism have always been determined by the applications, i.e., based on plant pathogenic fungi (Guzmán-Guzmán et al. 2017; Rubio et al. 2017; Estrada-Rivera et al. 2020). However, the interactions between *Trichoderma* and these fungi in nature were not documented. In fact, the most large-scoped ecological surveys of *Trichoderma* diversity performed by W. Jaklitsch in Europe (e.g., Jaklitsch et al. 2008; Jaklitsch 2009, 2011; Jaklitsch and Voglmayr 2012, 2015) demonstrate that the most frequent natural hosts of *Trichoderma* are saprotrophic *Basidiomycota*, such as *Fomes fomentarius*, *Steccherinum ochraceum*, and *Fomitopsis pinicola*, or some plant-beneficial members of this group (Jaklitsch 2009, 2011).

However, the hidden interactions with plant-associated *Ascomycota* have been revealed in the recent study of *Trichoderma* genomic architecture and the evolution of *Trichoderma* cellulolytic ability. Multiple cases of the lateral gene transfers (LGTs) of genes encoding the plant cell wall-degrading enzymes to *Trichoderma* from various filamentous *Ascomycota* have been described (Druzhinina et al. 2018), suggesting abundant and close interactions with the fungi in the evolutionary history of the genus. Thus, the ancestor of genus *Trichoderma* obtained numerous hemicellulases from hemicellulase-enriched species of the black yeast *Aureobasidium* spp. (*Dothideomycetes*). Also, some putatively mycorrhizal fungi, such as *Oidiodendron maius* (*Leotiomycetes*), donated cellulases, hemicellulases, and pectinases, while many plant pathogenic and endophytic fungi served as donors of cellulases and hemicellulases. Interestingly, multiple cases of DNA exchanges were also recorded for nutritionally versatile *Eurotiomycetes*, such as *Aspergillus* and *Penicillium* (Druzhinina et al. 2018). Consequently, the evolutionary and ecological studies revealed the broad range of hosts for *Trichoderma* mycoparasitism.

We would like to stress that although phytopathogenic fungi were usually studied as hosts for mycoparasitism research in vitro, functional genetic studies should address the ecology of the fungus and its natural partners. The improved understanding of molecular pathways will open a way to develop the next generation of bioeffectors for plant protection.

Functional genetic studies of *Trichoderma* mycoparasitism require the availability of the molecular toolkit for efficient genetic transformation. Due to the wide application in industrial cellulase production, *T. reesei* has been the focus of gene engineers for several decades (Toyama et al. 2002; Druzhinina and Kubicek 2017; Cai et al. 2021). The initial assumption that this species is saprotrophic and, therefore, unsuitable for the mycoparasitic studies [see, e.g., Kubicek et al. (2011)] was not confirmed. The recent studies demonstrated the high mycoparasitic potential of this species (Druzhinina et al. 2010, 2018; Atanasova et al. 2013). This opens many new possibilities for mycoparasitic studies (Druzhinina and Kubicek 2017; Druzhinina et al. 2016). Currently, different kinds of genetic transformation methods have been successfully conducted in *T. reesei*, including polyethylene glycol (PEG)-mediated transformation, *Agrobacterium*-mediated transformation, biolistic transformation, and electroporation (Cai et al. 2021; Martzy and Mach-Aigner 2021). Genetic modifications in the genomic or transcriptional level based on homologous recombination, transcription activator-like effector nucleases (TALENs), CRISPR/Cas9, and RNA interference systems have also been adopted in the genetic engineering of *T. reesei* (Chen et al. 2021; Martzy and Mach-Aigner 2021). Within the same genus, the successful protocol of genetic engineering for *T. reesei* could be expanded to other species of *Trichoderma*.

2 Ecological Terms and Definitions Describing *Trichoderma* Mycoparasitism

The development of next-generation biocontrol products requires the standardization of the scientific terms and definitions used to describe the relationship between bioeffectors and phytopathogenic fungi. For instance, such terms as mycoparasitism, mycotrophy, and hyperparasitism need to be explicitly defined. Moreover, as fungi were initially being attributed to plants, many mycological terms originate from botany. However, as fungi form a distinct monophyletic group that shares the last common ancestor with animals at least 1 billion years ago (Lücking et al. 2009; Bengtson et al. 2017), the unique mycological terminology used for the description of the ecological model of interactions and feeding types of fungi, including *Trichoderma*, needs to be refined and used correctly. It will help to avoid confusion by mixing terms used for the description of botanical and zoological interactions. Therefore, we propose the scheme of Druzhinina et al. (2018) that was based on Getz (Getz 2012). It offers the distinction between feeding on live or dead biomass and feeding on animals, plants, or fungi (Table 1) (see also www.fungig.org). For

Table 1 Terms describing feeding types of filamentous *Ascomycota* fungi

State	Resource	Term indicating nutrition type	Comments	References for *Trichoderma*
Live	Parasite			
Live	*Insects sensu lato*	Entomoparasite	Here a colloquial meaning of insects is used: insect may apply to any small arthropod similar to an insect including spiders, centipedes, millipedes, etc.	Moths, aphids (Evidente et al. 2009), bed bugs (Ab Majid et al. 2015), corn borer (Li et al. 2013)
Live	*Fungi*	Mycoparasite	May also include necrotrophic parasites of fungi	Broad spectrum (Druzhinina et al. 2011, Carsolio et al. 1994, El-Katatny et al. 2000, Rocha-Ramirez et al. 2002, Mukherjee et al. 2014, Chenthamara and Druzhinina 2016, Karlsson et al. 2017)
Live	*Plants*	Phytoparasite	The term may also include plant pathogenic organisms, croppers, and also endophytes as symptomless parasites of plants. Note: In this article, we do not use the meaning of this term sensu *Merriam Webster Medical Dictionary* (https://www.merriam-webster.com/medical/phytoparasite [Oct. 2017]: "a parasitic plant")	Mainly endophyte (Bae et al. 2009, Bailey et al. 2009, Bae et al. 2011, Rosmana et al. 2015, Rosmana et al. 2016, Chen et al. 2016), rarely plant pathogen (Li Destri Nicosia et al. 2015)
Live	*Plants and/or fungi and/or animals*	Parasite	Feeding on live biomass of any type, biotrophy	Immunocompromised and immunocompetent humans (Gautheret et al. 1995, Furukawa et al. 1998), nematodes (Sharon et al. 2001, De Souza Maia Filho et al. 2017, Zhang et al. 2017)
Dead	Greek: *phagos* = eat			
Dead	*Insects sensu lato*	Sarcophage, necrophage	May include necrotrophic parasitism	Moths (Ghosh and Pal 2015), aphids (Pacheco et al. 2017), bed bugs (Zahran et al. 2017), corn borer (Li et al. 2012)
Dead	*Fungi*	Mycophage	Also includes necrotrophic mycoparasites	Druzhinina et al. (2011)

(continued)

Table 1 (continued)

State	Resource	Term indicating nutrition type	Comments	References for *Trichoderma*
Dead	*Plants*	Phytophage, saprophytophage	In this article, mainly organisms feeding on non-wooden biomass were studied; xylophagous fungi capable to degrade lignin were not considered	Dead wood and herbaceous biomass (Druzhinina et al. 2011, Cianchetta et al. 2012, Bischof et al. 2016, Keshavarz and Khalesi 2016)
Dead	*Plants and/or fungi and/or animals*	Polyphage	Saprotrophic nutrition	Druzhinina et al. (2011)
Live/dead	*Fungi*	Mycotrophy	All kinds of feeding on fungi and fungal biomass	
Live/dead	*Animal, fungi, and plants*	Nutritional versatility, generalism	If at least two types of resources may be equally well consumed	

the precise description of the ecology of different *Trichoderma* species, it is essential to distinguish between generalists (capable of feeding on any of the listed resources) and specialists (feeding on a particular food source). Because the terms indicating feeding types of gatherers are based on the Latin "vorus" (to swallow), the use of such zoological terms as "carnivorous," "fungivorous," and "herbivorous" seems to be inappropriate for *Trichoderma* or other sessile fungi. However, we should note that the terms mentioned above are frequently applied in fungal biology (Schmidt et al. 2007; Sung et al. 2008; Liu et al. 2016; Yang et al. 2012).

The term parasite refers to any organism that feeds on live biomass of any type. Mechanisms of interactions, interactions types, and benefits and losses for individual partners are not considered here. Consequently, frequently used terms referring to pathogenicity (entomopathogen, plant pathogen, etc.) are not informative for the description of *Trichoderma* interactions with other fungi unless the disease caused by this is being referred to (like in the case of mushroom farms where *Trichoderma* spp. cause the green mold disease). Feeding on dead biomass is described using terms based on the Greek word "phagos" (to eat), while the terms based on the Greek word "trophe" (food, nourishment) are used to describe the act of feeding on live or dead biomass. We also note that the term "environmental opportunist," which has been recently attributed to some *Trichoderma* species (Druzhinina et al. 2011), does not specifically mean nutritional versatility. It also includes the ability to grow rapidly and resist environmental stresses. Consequently, the term "generalist" is used for nutritional versatility on dead or live biomass, while "parasite" and "polyphags" describe each of the latter two types of biomasses, respectively.

Mycoparasitism is a specified definition of a parasitic relationship between two organisms, parasite and host, with a taxonomic limitation to the fungal kingdom. Depending on whether the hosts are dead or alive, the mycoparasitism could be further divided into two types, necrotrophic mycoparasitism and biotrophic mycoparasitism (Chenthamara and Druzhinina 2016). The necrotrophic mycoparasitism occurs when the parasite fungus kills the host fungus and takes nutrients from the dead host, while the biotrophic mycoparasitism occurs when the host fungus benefits by taking nutrients from the living host fungus and brought harmful effects to the host. The necrotrophic mycoparasitism could also be included in the term "mycophage" as one form, while this term highlights the dead state of fungi as the nutrition resource and is irrelevant to the mycoparasitic relationship. Mycoparasitism is different from the term mycotrophy (being synonymous with mycophagy), which describes a broader nutrient relationship – the use of fungi for food. Therefore, for mycotrophy, there must be a fungus as the nutrient resource, while the other organism (could be any kind of creatures such as fungi, bacteria, plants, vertebrates, and invertebrates) and nutrient relationship (could be any kind of saprotrophic and biotrophic) are not limited.

The term hyperparasitism is frequently confused with mycoparasitism (Akrami et al. 2011; Mendoza et al. 2015; Chenthamara and Druzhinina 2016). The hyperparasitic relationship is defined as "parasitism on a parasite" with no taxonomic limitation, and at least three organisms should be involved in – one organism is the primary host, and one organism acts as a parasite of the primary host and, in the meanwhile, serves as a host for the third organism which parasitizes on it and is therefore named as hyperparasite (Chenthamara and Druzhinina 2016). The hyperparasitic relationship is more complicated than the mycoparasitism as more organisms are involved and usually studied among plant, phytopathogen, and biocontrol agents for agriculture practices (Inayati et al. 2020).

This chapter aims to sum up what is known about the genetic mechanisms employed for the fungal wars (mycoparasitism) between *Trichoderma* and other hypocrealean fungi and their mostly plant pathogenic hosts.

3 Taxonomic Diversity of Fungi Used in Functional Genetic Studies of Mycoparasitism

Currently, there are at least 382 valid species names in the genus *Trichoderma* (see Chap. 1). However, only eight species were studied for their mycoparasitism against hosts based on functional genetics (Fig. 2). Among them, the most studied species are *T. virens* and *T. atroviride*, and in most cases, strain *T. virens* Gv29-8 and *T. atroviride* IMI 3206040 or P1 were used to reveal the mycoparasitism mechanisms adopted by *Trichoderma*. *T. virens* and *T. atroviride* also exhibit the broadest scope of hosts, ranging from plant pathogenic fungi and fungi-like Oomycota to saprotrophic fungi and even insects. *T. harzianum* sensu lato (including *T.* sp. NJAU 4742 = *T. guizhouense* NJAU 4742) were proved to be mycoparasites of plant pathogenic fungi and fungi-like Oomycota, showing the comparable host scope with industrial species *T. reesei*, but could parasite more individual hosts. Thus, these

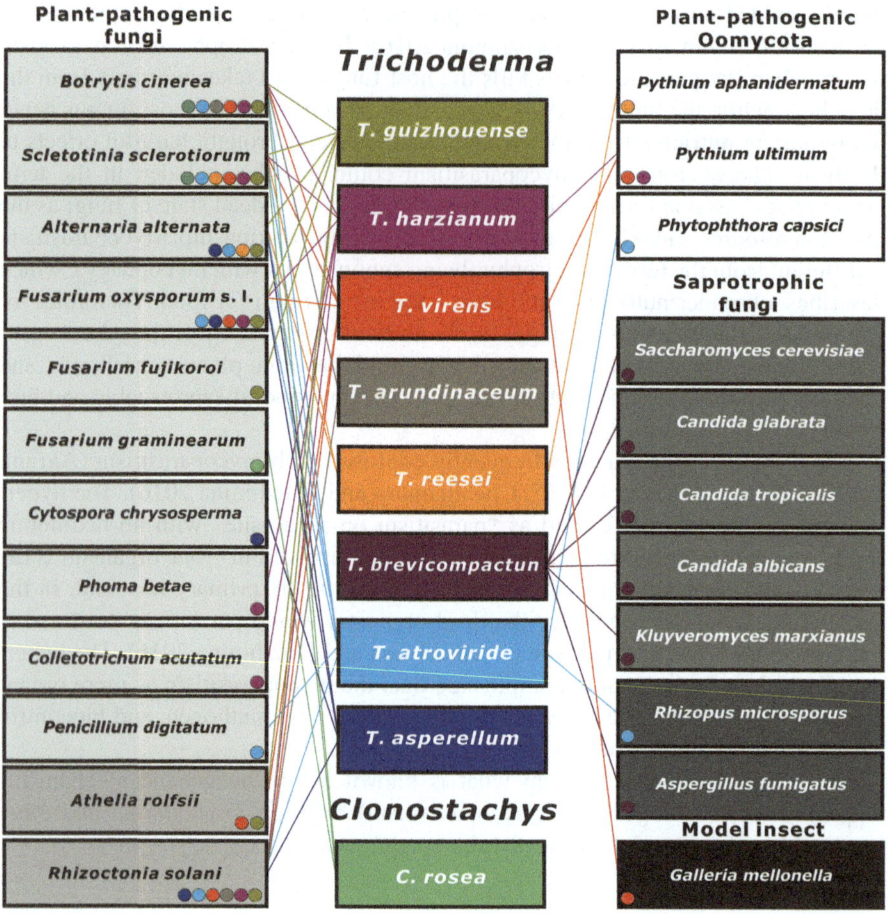

Fig. 2 The diversity of fungi studied in functional genetic research of *Trichoderma* mycoparasitism
The parasitism relationships of hosts and parasites that have been studied based on functional genetics were shown by lines with the same color of the parasites. The middle column represents the parasites, while the two side columns represent the hosts

three species make the golden standard for the genetic research of *Trichoderma* mycoparasitism.

Regarding the host of *Trichoderma* spp. and *Clonostachys* spp., plant pathogenic fungi are attracting the interest of researchers. *Botrytis cinerea*, *S. sclerotiorum*, and *R. solani* [teleomorph, *Thanatephorus cucumeris* (*Cantharellales*, *Basidiomycota*)] are the most studied hosts and were studied being parasitized by at least six species of *Trichoderma* and *Clonostachys* (Fig. 2). The host status of fungi-like Oomycota and saprotrophic fungi are only confirmed by one parasite except for *P. ultimum* that could be parasitized by *T. virens* and *T. harzianum*.

Considering that the mycoparasitism is specified and one *Trichoderma* species can use different mechanisms against different hosts (Zhang et al. 2016, 2019; Pang et al. 2020), further studies related to *Trichoderma* mycoparasitism require standardization and inclusion of ecologically justified hosts to reflect biological meanings.

4 *Trichoderma* Genes Involved in the Attack on Other Fungi

To launch a fungal war, the mycoparasitic hypocrealean fungi should be well-armed. The arsenal of molecular tools used by *Trichoderma* spp. and *Clonostachys* spp. against other fungi comprises three main categories (Table 2). These are the fungal cell wall-degrading enzymes (FCWDEs), antifungal secondary metabolites, and accessory proteins. Additionally, the genes involved in sensing and signal transduction are critical for the successful combat of some fungi (Mukherjee et al. 2003; Yang 2017; Moreno-Ruiz et al. 2021). The success of *Trichoderma* spp. and *Clonostachys* spp. in the interactions with their hosts (i.e., the conditions leading to the improved fitness of *Trichoderma*) is believed to be either a single tool activity or a combined action of all those weapons.

The genes involved in mycoparasitism of the most studied *Trichoderma* species and verified by genetic modification mutants are shown in Fig. 3. These genes are classified into four groups: cell wall-degrading and remodeling, regulatory genes, secondary metabolites, and signal transduction. In *T. virens*, the regulatory genes were most studied, while in *T. atroviride*, the studied genes were distributed more evenly into the four groups.

In total, 13 individual hosts were used for the gene function analysis of *Trichoderma* mycoparasitism. Most of the studied genes were targeted on soilborne basidiomycete *R. solani*. Besides the broad scope of plant hosts, including potato (*Solanum tuberosum*) (Tsror 2010), rice (*Oryza* spp.), and maize (*Zea mays*) (González-Vera et al. 2010), the popularity of *R. solani* in *Trichoderma* mycoparasitism studies may be due to the clear observation of *R. solani* as host of *Trichoderma* spp. at an early period (Elad et al. 1983, 1987) and the fact that the mycelium of *R. solani* is easily distinguished from that of *Trichoderma* spp. based on the pigmentation and diameter size.

Genera *Fusarium* and *Trichoderma* are both located at the order of *Hypocreales* and share the same ancestor approximately 200 million years ago (Kubicek et al. 2019). Initially, *Fusarium* spp. were not considered a proper host for *Trichoderma* due to their ability to produce toxins (Vogelgsang et al. 2008). As described above, many studies show that the direct confrontation with some *Fusarium* spp. results in a deadlock reaction (Zhang et al. 2016, 2019), but the diversity studies revealed that some species could also combat this toxic fungus (Zhang et al. 2019). As cases of LGTs from *Fusarium* to *Trichoderma* have also been documented (Druzhinina et al. 2018) and because not all species of *Trichoderma* can attack them, these fungi offer a reasonable model for the functional genetic research on mycoparasitism.

Table 2 Arsenal genes of *Trichoderma* spp. and *Clonostachys* spp.

					Phenotype			
Process	Gene	General function	Aggressor	Defender	Gene deletion	Gene complementation or overexpression	Comment	References
Cell wall degradation	*bgn3*	β-1,6-Glucanase, secreted	*T. virens*	*Rhizopus oryzae* (*Mucorales*, *Mucoromycotina*), *R. solani*, *P. ultimum*	No effect on growth and development; reduced ability to inhibit growth of *P. ultimum*	Improved antagonism against *P. ultimum*, *R. oryzae*, and *R. solani*.	n.a.	Djonović et al. (2006)
	ech42 (chi18–5)	Endochitinase, secreted	*T. atroviride*	*P. digitatum*	n.a.	Germination of *P. digitatum* spores was highly inhibited by the secreted enzymes of the gene overexpression mutant of *T. atroviride*	n.a.	Deng et al. (2007)
	chi67–1	Endochitinase	*C. rosea*	*S. sclerotiorum*	Increased parasitic rates against *S. sclerotiorum* and higher control efficiencies to soybean *Sclerotinia* stem rot	Increased parasitic rates against *S. sclerotiorum*	n.a.	Sun et al. (2017)

Process	Gene	General function	Aggressor	Defender	Phenotype		Comment	References
					Gene deletion	Gene complementation or overexpression		
Protease	*nmp1*	Neutral deuterolysin metallopeptidase, secreted	*T.* sp. NJAU 4742 (syn. *T. guizhouense* NJAU 4742)	*A. rolfsii*, *R. solani*, *B. cinereal*, *S. sclerotiorum*, *A. alternata*, *F. odoratissimum*, *F. fujikuroi*	Reduced mycoparasitism; no coiling around hyphae of *F. odoratissimum*; reduced ability to produce antifungal secondary metabolites; reduced ability to defend against other fungi	Increased mycoparasitism; self-toxicity	n.a.	Zhang et al. (2016)
Secondary metabolites	*nox2*	NADPH oxidase	*T. atroviride*	*R. solani*, *S. sclerotiorum*	Reduced the VOC-dependent inhibitory effects on *R. solani* and *S. sclerotiorum*	Recovered the VOC-dependent effects on *R. solani* and *S. sclerotiorum*	n.a.	Cruz-Magalh et al. (2019)

(continued)

Table 2 (continued)

Process	Gene	General function	Aggressor	Defender	Phenotype		Comment	References
					Gene deletion	Gene complementation or overexpression		
	nox1	NADPH oxidase	*T. harzianum*, *T. atroviride*, *T.* sp. NJAU 4742 (syn. *T. guizhouense* NJAU 4742)	*P. ultimum*, *R. solani*, *S. sclerotiorum*, *F. odoratissimum*	In *T. atroviride*: (1) lowered the inhibition against *R. solani* and *S. sclerotiorum* on the dual growth plate; (2) enhanced the VOC-dependent inhibitory effects on *R. solani* and *S. sclerotiorum*	In *T. harzianum*: resulted in improved biocontrol potential against *P. ultimum*	In *T. harzianum*: increased ROS production, protease, cellulase, and chitinase activities when confronted with *P. ultimum*	Montero-Barrientos et al. (2011); Zhang et al. (2019); Cruz-Magalh et al. 2019)

Process	Gene	General function	Aggressor	Defender	Phenotype		Comment	References
					Gene deletion	Gene complementation or overexpression		
					In *T.* sp. NJAU 4742 (syn. *T. guizhouense* NJAU 4742): the ability to produce H_2O_2 was diminished and the ability to efficiently overgrow *F. odoratissimum* was lost	In *T.* sp. NJAU 4742 (syn. *T. guizhouense* NJAU 4742): the parental phenotype was recovered by the complementation of *nox1*, and *nox1* overexpression strain formed structures similar to the guttation capsules of wild-type strain, but their surface was damaged	In *T.* sp. NJAU 4742 (syn. *T. guizhouense* NJAU 4742): when confronted with *F. odoratissimum*, *Trichoderma* formed bright yellow macroscopic guttation droplets with content exhibiting activity of chitinolytic and proteolytic enzymes and high concentration of H_2O2	

(continued)

Table 2 (continued)

Process	Gene	General function	Aggressor	Defender	Phenotype		Comment	References
					Gene deletion	Gene complementation or overexpression		
							In *F. odoratissimum*: the genes related to toxicity were upregulated when confronted with *T.* sp. NJAU 4742 (syn. *T. guizhouense* NJAU 4742)	
	tal67	Transaldolase	*C. rosea*	*B. cinerea*, *S. sclerotiorum*	Decreased in growth rate, antagonistic activity against *B. cinerea*, parasitic rate against *S. sclerotiorum*	Recovered the growth rate and fungicidal activity of *C. rosea*	n.a.	Liu et al. (2016)
	zhd101	Zearalenone hydrolase	*C. rosea*	*F. graminearum*	Lowered in vitro ability to inhibit growth of the ZEA-producing *F. graminearum*	n.a.	n.a.	Kosawang et al. (2014)

Process	Gene	General function	Aggressor	Defender	Phenotype		Comment	References
					Gene deletion	Gene complementation or overexpression		
	pks22	Polyketide synthase involved in the biosynthesis of clonorosein	*C. rosea*	*B. cinerea*, *F. graminearum*	A significant reduction in mycelial growth compared with wild type. Lost the ability to produce clonorosein A and B that showed strong antifungal activity against *B. cinerea* and *F. graminearum*	n.a.	n.a.	Fatema et al. (2018)
	pks29	Polyketide synthase	*C. rosea*	*B. cinerea*, *F. graminearum*	Filtrates obtained from Δ*pks29* strains cannot decrease the biomass of *B. cinerea* compared with the wild-type strain	n.a.	n.a.	Fatema et al. (2018)

(continued)

Table 2 (continued)

Process	Gene	General function	Aggressor	Defender	Phenotype		Comment	References
					Gene deletion	Gene complementation or overexpression		
	nps1	Non-ribosomal peptide synthetase	*C. rosea*	*B. cinerea*, *F. graminearum*	The growth and conidiation rate of deletion strains were increased. The growth rate of *B. cinerea* was reduced during interactions with deletion strains compared with the wild-type strain	n.a.	n.a.	Iqbal et al. (2019)

					Phenotype			
Process	Gene	General function	Aggressor	Defender	Gene deletion	Gene complementation or overexpression	Comment	References
	gliP	Involved in the production of gliotoxin	*T. virens*	*S. sclerotiorum*, *R. solani*, *P. ultimum*, *Galleria mellonella* (Lepidoptera, Arthropoda)	Abolition of gliotoxin production; reduced growth; dispersed and less dense mycelium; less branched hyphae; increased sensitivity to oxidative stress (H_2O_2, 10 mM); ineffective as mycoparasites against *P. ultimum* and *S. sclerotiorum*, but retained mycoparasitic ability against *R. solani*	n.a.	n.a.	Vargas et al. (2014)

(continued)

Table 2 (continued)

Process	Gene	General function	Aggressor	Defender	Phenotype		Comment	References
					Gene deletion	Gene complementation or overexpression		
	pks4	Polyketide synthase	*T. reesei*	*R. solani*, *S. sclerotiorum*, *A. alternata*	Loss of green pigmentation in their conidia; reduced resistance to UV; reduced stability of the conidial wall and the antagonistic abilities against *R. solani*, *S. sclerotiorum*, and *A. alternata*; reduced formation of water-soluble antifungal metabolites; altered expression of other PKS-encoding genes	n.a.	Most of the pks genes in Δpks4 mutants were downregulated during the confrontation against *R. solani*	Atanasova et al. (2013)

Process	Gene	General function	Aggressor	Defender	Phenotype		Comment	References
					Gene deletion	Gene complementation or overexpression		
	tri3, *tri11*	Within a *tri* cluster responsible for the biosynthesis of trichodermin	*T. brevicompactum*	n.a.	Δ*tri3* mutant exhibited a sharp decline in the production of trichodermin and trichodermol, while Δ*tri11* did not produce trichodermin	n.a.	n.a.	Shentu et al. (2018)
	tri4	Cytochrome P450 monooxygenase that oxygenates trichodiene to give rise to isotrichodiol	*T. arundinaceum*, *T. brevicompactum*	*B. cinerea*, *R. solani*	In *T. arundinaceum*: reduced antifungal activity against *B. cinerea* and *R. solani* In *T. brevicompactum*: did not produce trichodermin	n.a.	n.a.	Malmierca et al. (2012); Shentu et al. (2018)

(continued)

Table 2 (continued)

Process	Gene	General function	Aggressor	Defender	Phenotype		Comment	References
					Gene deletion	Gene complementation or overexpression		
	tri5	Terpene synthase	*T. brevicompactum*, *T. arundinaceum*	*Saccharomycetales*: *Saccharomyces cerevisiae*, *Kluyveromyces marxianus*, *Candida albicans*, *C. glabrata*, *C. tropicalis*	In *T. arundinaceum*: no production of HA (a non-phytotoxic trichothecene), which has a role in antagonistic activity against fungal plant pathogens and induction of plant genes involved in defense responses; altered the expression of other tri genes involved in HA biosynthesis; altered the expression of hmgR, dpp1, erg9, erg1, and erg7, all genes involved in terpene biosynthetic pathways	In *T. brevicompactum*: increase of the trichodermin; increase in the antibiotic activity against a large panel of yeasts		Tijerino et al. (2011); Malmierca et al. (2013)
				A. fumigatus, *B. cinerea*, *R. solani*				

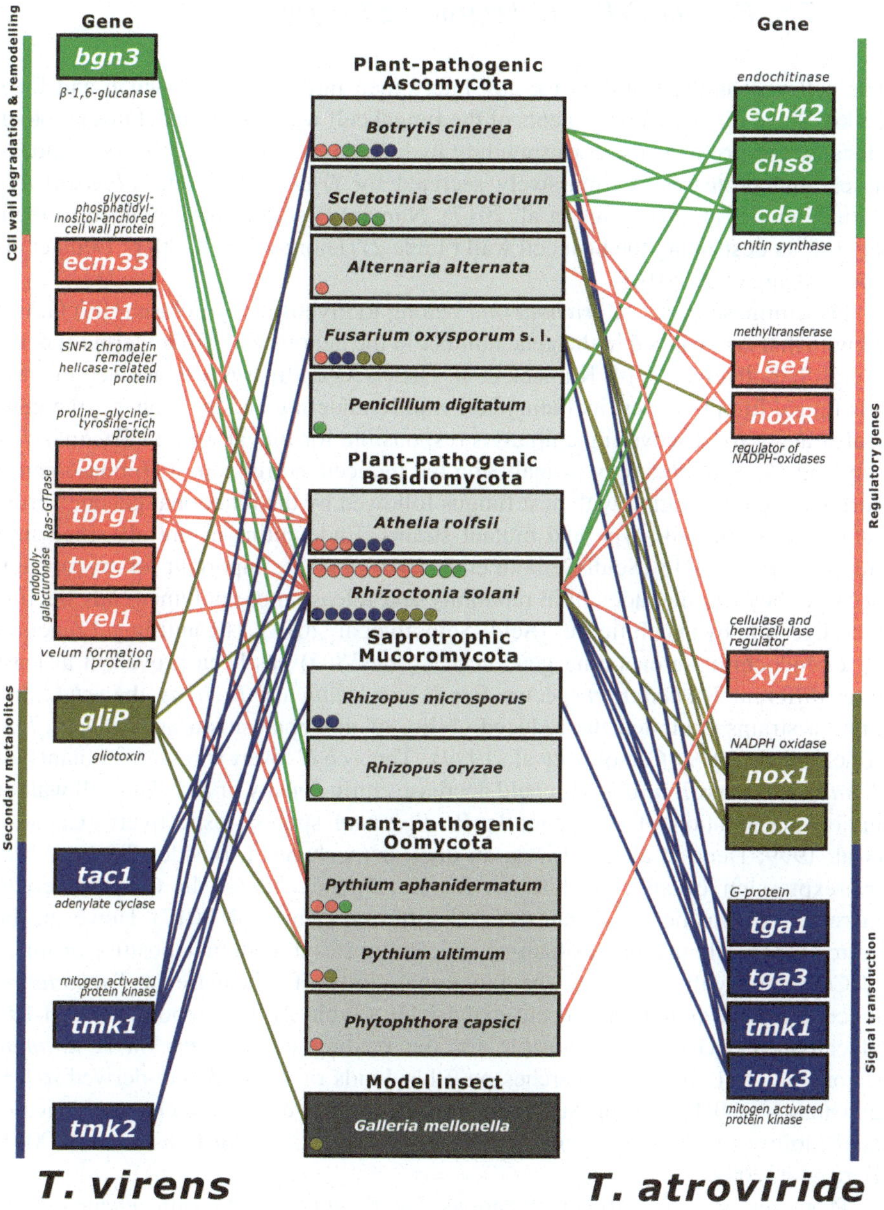

Fig. 3 Genes studied for their involvement in the mycoparasitism of *T. virens* and *T. atroviride* The genes employed by *T. virens* and *T. atroviride* for the mycoparasitism of hosts were classified into four groups and exhibited based on different colors. The genes and targeted hosts were linked by lines with the same color as the corresponding group. The middle column represents the host, while the two side columns represent the related genes

4.1 The Fungal Cell Wall-Degrading Enzymes

The cell wall is the frontline for any interfungal interactions (Bowman and Free 2006). Since the main components of the fungal cell wall are chitin, chitosan, other glucans, and proteins, the corresponding hydrolytic enzymes are chitinases, glucanases, and proteases, all massively secreted by *Trichoderma* and *Clonostachys* (Druzhinina et al. 2011; Sun et al. 2017). Numerous studies have pointed to their key role in destroying the host cell wall (Table 2) (Djonović et al. 2006; Deng et al. 2007; Zhang et al. 2016).

The chitinase genes of *Trichoderma* belong to glycoside hydrolase (GH) family 18 with that is expanded in the gene number in the genomes of *Trichoderma* species (Seidl-Seiboth et al. 2014; Kubicek et al. 2019). As chitinases are involved in various fungal life processes, including cell wall remodeling and mycophagy, the general strategy for discovering chitinases responsible for fungal wars in *Trichoderma* spp. and *Clonostachys* spp. is analyzing the induced expression of chitinase genes during the confrontation with host fungus followed by the functional genetic verification using the wild-type and mutant strains. Endochitinases are classified into subgroup B of GH18 (Seidl-Seiboth et al. 2014) and are important for mycophagy because they can degrade chitin randomly and release more polymer ends for further digestion by exochitinases (Seidl-Seiboth et al. 2014). The antifungal function of an endochitinase-encoding gene *ech42* (=*chi18–5*) has been proved in at least three different *Trichoderma* species. For *T. harzianum* and *T. virens*, the *ech42* disruption strains exhibited the reduced ability of mycoparasitism against *R. solani* (Baek et al. 1999; Carsolio et al. 1999). The *ech42* overexpression mutants of *T. harzianum* and *T. atroviride* could decrease chitin density of *R. solani* cell wall or inhibit germination of stramenopile *P. digitatum* spores, respectively (Carsolio et al. 1999; Deng et al. 2007), via an unclear mechanism. Besides *Trichoderma*, overexpression of an endochitinase-encoding gene *chi67–1* in *C. rosea* could increase its parasitic rates against *S. sclerotiorum* (Sun et al. 2017). However, the search for strictly mycoparasitism-specific chitinases did not give positive results.

Compared with chitinase, the functional study of glucanase in *Trichoderma* mycoparasitism is not carried out extensively (Table 2). It is reported that β-1,6-glucanase TvBGN3 is responsible for the feeding of *T. virens* on *P. ultimum* (Djonović et al. 2006). Researches on other kinds of *Trichoderma*-derived α-1,3-glucanases or β-1,3-glucanases mainly focus on their enzymatic characteristics or applicability for the development of biocontrol products (Ait-Lahsen et al. 2001; Bara et al. 2003).

Proteases are also efficient weapons for *Trichoderma* to gain advantages in mycoparasitism on a broad range of host fungi (Elad and Kapat 1999; Szekeres et al. 2004). The genomic studies reveal that *Trichoderma* shares a common ancestor with hypocrealean parasites on insects, such as species of *Metarhizium*, *Cordyceps*, and others (Kubicek et al. 2019). The inventory of proteolytic genes revealed that most *Trichoderma* species maintained a nearly complete set of these enzymes, likely due to their critical role in fungal wars (Druzhinina et al. 2010;

Kubicek et al. 2019). It is supposed that the secreted proteases of *Trichoderma* could digest the host-derived compounds, and oligopeptides released from the host trigger further mycoparasitic responses of *Trichoderma* (Druzhinina et al. 2011). A metalloprotease NMP1 was proven essential for the success of mycoparasitism of *T. guizhouense* against various host fungi (Fig. 3). The *nmp1*-deficient mutant showed reduced ability for mycoparasitism, while the combination uses of *nmp1*-deficient mutant and purified NMP1 could partially recover the antagonistic strength of the wild-type strain (Zhang et al. 2016). Thus, the function of metalloprotease that was first discovered by the group of Enrique Monte for characteristic analysis (Suarez et al. 2004) was confirmed to be related to mycoparasitism based on a complete functional genetic study by Zhang et al. (2016).

4.1.1 Antifungal Bioactive Compounds

Trichoderma spp. secrete a variety of secondary metabolites, including polyketides, non-ribosomal peptides, and terpenoids, which is consistent with the abundance of secondary metabolite synthesis gene clusters within their genomes (Kubicek et al. 2019). It is now believed that most secondary metabolites are secreted by *Trichoderma* spp. to facilitate the necrotrophic mycoparasitism against host fungi (Kubicek et al. 2011). For example, *pks4* is a polyketide synthase-encoding gene involved in producing green pigment in *T. reesei*. The *pks4* disruption mutant could not produce green pigmentation in their conidia and showed reduced ability against *R. solani*, *S. sclerotiorum*, and *Alternaria alternata* (Atanasova et al. 2013). Gliotoxin is a member of epipolythiodioxopiperazines and was discovered in *T. virens* with strong antimicrobial activity. Gene cluster *gliP* within the genome of *T. virens* was verified to be responsible for gliotoxin biosynthesis. The *gliP* disruption mutant of *T. virens* exhibits abolition of gliotoxin production, reduced colony growth, dispersed and less dense mycelium and branched hyphae, and increased sensitivity to oxidative stress (H_2O_2, 10 mM) and is ineffective in interactions with non-fungal hosts, such as *P. ultimum*, and fungal host such as *S. sclerotiorum* (Vargas et al. 2014). Trichothecenes are sesquiterpenoid mycotoxins and were studied in *T. brevicompactum* and *T. arundinaceum* (Proctor et al. 2018). They were synthesized in two different types, named trichodermin and harzianum A. Disruption of *tri4* and *tri5* genes in *T. arundinaceum* resulted in mutants with no production of harzianum A and reduced antifungal activity against *B. cinerea* and *R. solani* (Malmierca et al. 2012, 2013). Overexpression of *tri5* in *T. brevicompactum* could increase the production of trichodermin and the antibiotic activity against a large panel of yeasts (Tijerino et al. 2011). Not only in *Trichoderma*, but secondary metabolites also function in the mycoparasitism of *C. rosea* against host fungi. The disruption of polyketide synthase-encoding genes *pks22* and *pks29* in *C. rosea* could result in the reduced antifungal ability against *B. cinerea* and *F. graminearum* (Fatema et al. 2018). In rare studied cases, the secondary metabolite biosynthesis gene is negatively related to the antifungal ability. Disruption of a non-ribosomal peptide synthetase-encoding gene *nps1* in *C. rosea* could increase the growth and conidiation rate of the mutant and reduce the growth rate of *B. cinerea* at the third day postinoculation when confronted with the mutant compared with the wild-type strain (Iqbal et al. 2019).

Different species of *Trichoderma* produce numerous other antifungal metabolites such as peptaibols that are pending the comprehensive functional genetic investigation. Purified metabolites secreted by *T. harzianum* sensu lato could inhibit *Gaeumannomyces graminis* (*Magnaporthales*), *R. solani*, and *P. ultimum* at low doses (T22azaphilone and harzianopyridone) or with high concentrations (T39butenolide and harzianolide) (Vinale et al. 2006, 2009). The function of azaphilone in *Trichoderma* mycoparasitism was further verified based on gene deletion mutants by Pang et al. (2020) that revealed their putative role in defense against the oxidative stress (vide infra).

4.2 Genetics for Trichoderma *Self-Defense*

The success of mycoparasitism requires a strong parasitic fungus defense system. The molecular basis that has been revealed in the defense system of *Trichoderma* includes small secreted cysteine-rich proteins (SSCPs), secondary metabolites, and cell wall remodeling enzymes (Table 3).

Studies have shown that the cerato-platanin EPL1 (=SM1) in *T. harzianum* putatively plays a role in self-recognition during the mycoparasitic attacks. When the *epl1* disruption *T. harzianum* mutant was confronted with the wild-type strains, an interaction line separating the two strains and debris from the hypha degradation was observed. This self-hostile phenomenon could be eliminated by reintroducing *epl1* to the disruption mutant, indicating the role of EPL1 to protect *T. harzianum* by avoiding accidental injury by itself (Gomes et al. 2015).

Production of reactive oxygen species (ROS) is a conserved mechanism in eukaryotes and was speculated to be adopted by *Trichoderma* and host fungi as a strategy for mycoparasitic attack and defense, respectively (Zhang et al. 2019; Kappel et al. 2020; Pang et al. 2020). Although the aerial hyphae of *T. guizhouense* could produce ROS for mycoparasitic attacking against *F. odoratissimum* (Pang et al. 2020), excessive production of ROS is also harmful to *T. guizhouense*. As a defense response, azaphilones, a structurally diverse class of fungal polyketides with the characterization of pigment and antioxidant property, was biosynthesized in *T. guizhouense* to eliminate the hydrogen peroxide (Pang et al. 2020). Functional analysis based on gene disruption mutants verified that a transcriptional activator (*aza5*), O-acetyltransferase (*zaz10*), and two PKS-encoding genes (*aza1* and *aza2*), within a gene cluster, are responsible for the biosynthesis of azaphilones (Pang et al. 2020). Thus, the gene cluster and azaphilones are considered as a defense mechanism in *Trichoderma*. However, whether this defense mechanism is also effective against the immune response of host fungus remains unknown.

Table 3 Self-defense genes of *Trichoderma* spp. and *Clonostachys* spp.

Process	Gene	General function	Aggressor	Defender	Phenotype		Comment	References
					Gene deletion	Gene complementation or overexpression		
Accessory proteins	*epl1 (sm1)*	Cerato-platanin, small secreted cysteine-rich protein, surface active	*T. harzianum*	*S. sclerotiorum*, *R. solani*	Putatively impaired self-recognition; no hyphal coiling against *S. sclerotiorum* and *R. solani*	Self-recognition to wild-type strain and hyphal coiling around *S. sclerotiorum* and *R. solani* were reestablished	n.a.	Gomes et al. (2015)
Cell wall remodeling	*chs8*	Chitin synthases	*T. atroviride*	*S. sclerotiorum*, *B. cinerea*, *R. solani*	*chs8* knockout mutants were severely compromised in mycoparasitism	n.a.	n.a.	Kappel et al. (2020)
	cda1	Chitin deacetylases	*T. atroviride*	*S. sclerotiorum*, *B. cinerea*, *R. solani*	*chs8* knockout mutants were severely compromised in mycoparasitism	n.a.	n.a.	Kappel et al. (2020)
Cell wall biogenesis	*crssd1*	Cell wall biogenesis protein phosphatase	*C. rosea*	*S. sclerotiorum*	Δ*crssd1* mutants had flatter and thinner mycelia, lost almost all ability to undergo conidiation, and been more sensitive to osmotic and cell wall stresses; hyphal extension ability was decreased; weakened mycoparasitism rate against *S. sclerotiorum*	Recovered the lost ability and exhibited similar to wild-type strain	n.a.	Lv et al. (2020)

(continued)

Table 3 (continued)

Process	Gene	General function	Aggressor	Defender	Phenotype		Comment	References
					Gene deletion	Gene complementation or overexpression		
Secondary metabolites	*aza1*, *aza2*	PKS-encoding genes within the gene cluster for the biosynthesis of trigazaphilones	*T.* sp. NJAU 4742 (syn. *T. guizhouense* NJAU 4742)	*F. odoratissimum*	Lost the ability of biosynthesis of trigazaphilones	n.a.	n.a.	Pang et al. (2020)
	aza10	O-Acetyltransferase within the gene cluster for the biosynthesis of trigazaphilones	*T.* sp. NJAU 4742 (syn. *T. guizhouense* NJAU 4742)	*F. odoratissimum*	Formed harziphilone but failed to form any of the other trigazaphilones	n.a.	n.a.	Pang et al. (2020)
	aza5	Transcriptional activator of the aza gene cluster responsible for the biosynthesis of trigazaphilones	*T.* sp. NJAU 4742 (syn. *T. guizhouense* NJAU 4742)	*F. odoratissimum*	Lost the ability of biosynthesis of trigazaphilones	Expression of genes within gene cluster responsible for the biosynthesis of trigazaphilones was significantly enhanced	n.a.	Pang et al. (2020)

It is reported that cell wall remodeling is necessary for the mycoparasitic activity of *T. atroviride* against *S. sclerotiorum*, *B. cinerea*, and *R. solani*. The genes involved in the cell wall remodeling of *T. atroviride* were identified as chitin synthases, chitin deacetylases, and chitinolytic enzymes. It is supposed that during the mycoparasitism of *T. atroviride* against phytopathogen fungi, two high responder genes, *chs8* encoding chitin synthases and *cda1* encoding chitin deacetylases, could strength the cell wall of *T. atroviride* by chitin synthesis and deacetylation, thus enhancing the defense to hostile chitinases or immune responses of the host (Kappel et al. 2020). Therefore, cell wall remodeling acts as a defense system for the success of mycoparasitism of *Trichoderma*.

5 The Supporting Role of Regulatory Genes

The supporting system in *Trichoderma* functions by transducing signals or switching the arsenal and defense systems on or off (Table 4). Genes such as *nor1* and *noxR* encode regulators of NADPH oxidases in *T. guizhouense* and *T. atroviride*, respectively (Hernández-Oñate et al. 2012; Zhang et al. 2019). The *nor1* knockout mutant of *T. guizhouense* exhibited a reduced ability to produce H_2O_2 and overgrow *F. oxysporum* (Zhang et al. 2019).

Methyltransferase LAE1 is involved in multiple biological processes in *Trichoderma* (Seiboth et al. 2012). In *T. atroviride*, LAE1 contributes to sporulation, defense, and attack ability during mycoparasitism against phytopathogenic fungi (Aghcheh et al. 2013). The *lae1* knockout mutant of *T. atroviride* (Δ*lae1*) showed less sensitivity to light and injury in the sporulation process with a decrease of 50% conidiation, while the conidiation ability was increased in the overexpression mutant. During the mycoparasitic process, Δ*lae1* exhibits decreased defense ability to oxidative stress, and offensive related genes (proteases, GH16 ß-glucanases, and polyketide synthases) could not be induced as the wild-type strains. Therefore, LAE1 is considered the global regulatory protein in *T. atroviride* for mycoparasitism (Aghcheh et al. 2013).

The mitogen-activated protein kinases (MAPKs) play important roles in intracellular signal transduction pathways by phosphorylation (Schaeffer and Weber 1999). The MAPKs could be further classified into three pathways according to the regulated processes (Moreno-Ruiz et al. 2021). The first pathway is mainly responsible for mycoparasitism activities in *Trichoderma*. The gene [homolog genes with unified name *tmk1* (=*tmkA*, *tvk1*, *task1*) in Table 4] disruption mutants in the first MAPK pathway led to the reduced ability of *T. virens* IMI 304061 (Mukherjee et al. 2003), *T. atroviride* P1 (Reithner et al. 2007), and *T. asperellum* T4 (Yang 2017) to parasitize hosts. In contrast, the *tmk1* disruption mutant of *T. virens* Gv29–8 showed improved mycoparasitism against *R. solani* (Mendoza-Mendoza et al. 2003). The contrast observation was supposed to be the interaction specificity between *Trichoderma* strains and hosts and may also be due to the nutrition conditions during the mycoparasitism (Mendoza-Mendoza et al. 2003). The second MAPK

Table 4 Regulatory genes of *Trichoderma* spp. and *Clonostachys* spp.

Process	Gene	General function	Aggressor	Defender	Phenotype		Comment	References
					Gene deletion	Gene complementation or overexpression		
Regulator	*xyr1*	Regulatory protein for cellulase and hemicellulase gene expression	*T. atroviride*	*P. capsici*, *B. cinerea*, *R. solani*	Reduced transcript levels of *axe1* and *swo1*, which encode accessory cell wall-degrading enzymes; upregulation of *prb1* expression; overall enhanced competition with studied plant pathogens probably due to overexpression of *prb1*	Self-recognition to wild-type strain and hyphal coiling around *S. sclerotiorum* and *R. solani* were reestablished	n.a.	Reithner et al. (2014)
	pgy1	Encoding a proline-glycine-tyrosine-rich protein	*T. virens*	*S. rolfsii*, *R. solani*, *P. aphanidermatum*	Negatively regulates conidiation, cell wall integrity, antagonism of *S. rolfsii* and *R. solani*, inhibition of *P. aphanidermatum* by culture filtrates and production levels of viridin	n.a.	n.a.	Bansal et al. (2019)
	ecm33	Encoding a glycosylphosphatidylinositol-anchored cell wall protein	*T. virens*	*S. rolfsii*, *R. solani*, *P. aphanidermatum*	More sensitive to digestion with cell wall-lysing enzyme. Negatively regulates conidiation, cell wall integrity, antagonism of *S. rolfsii* and *R. solani*, inhibition of *P. aphanidermatum* by culture filtrates and production levels of viridin	n.a.	n.a.	Bansal et al. (2019)

Process	Gene	General function	Aggressor	Defender	Phenotype		Comment	References
					Gene deletion	Gene complementation or overexpression		
	ipa1	Chromatin remodeler/helicase-related proteins, SNF2 family	*T. virens*	*R. solani*	Mycelium-free culture filtrates of Δ*ipa1* showed diminished inhibition of *R. solani* radial growth compared with the filtrates from the wild type	n.a.	n.a.	Estrada-Rivera et al. (2020)
	lae1	Methyltransferase	*T. atroviride*	*A. alternata*, *R. solani*, *B. cinerea*	Decrease in conidiation by 50% in light; no conidiation in darkness; abolishment of sporulation in response to injury; increased sensitivity to oxidative stress; affected expression of genes encoding several proteases, GH16 ß-glucanases, PKSes, and SSCPs; decrease in antagonism against *A. alternata*, *R. solani*, and *B. cinerea*; decrease in production of known antifungal metabolites	Increased conidiation by 30–50% in light; enhanced mycoparasitic vigor; resistance to oxidative stress; increased production of 6-pentyl-2H-pyran-2-one	n.a.	Aghcheh et al. (2013)
	nor1	Regulator of NADPH oxidase *nox1*	*T.* sp. NJAU 4742 (syn. *T. guizhouense* NJAU 4742), *T. atroviride*	*F. odoratissimum*, *R. solani*, *S. sclerotiorum*	In *T.* sp. NJAU 4742 (syn. *T. guizhouense* NJAU 4742): the ability to produce H_2O_2 was diminished, and the ability to efficiently overgrow *F. odoratissimum* was lost	In *T. atroviride*: recovered the VOC-dependent effects on *R. solani* and *S. sclerotiorum*	n.a.	Zhang et al. (2019); Cruz-Magalh et al. (2019)
					In *T. atroviride*: (1) lowered the inhibition against *R. solani* and *S. sclerotiorum* on the dual growth plate; (2) enhanced the VOC-dependent inhibitory effects on *R. solani* and *S. sclerotiorum*			

(continued)

Table 4 (continued)

Process	Gene	General function	Aggressor	Defender	Phenotype		Comment	References
					Gene deletion	Gene complementation or overexpression		
	tbrg1	A new family member of Ras-GTPases	*T. virens*	*R. solani*, *Sclerotium rolfsii*, *F. odoratissimum*	Deregulated conidiation-related genes and produced less conidia, as well as a delayed conidiophore development; upregulated the expression of mycoparasitism-related gene and increased the antagonistic capability against *R. solani*, *S. rolfsii*, and *F. odoratissimum*; upregulated non-ribosomal peptide synthetase gene *gliP*, and polyketide synthetase gene *pk2*, and produced higher amounts of gliotoxin compared to the wild type; the antibiosis effect of mycelium-free culture filtrates against *R. solani* was considerably enhanced	n.a.	n.a.	Dautt-Castro et al. (2020)
	thmbf1	Transcriptional coactivator	*T. harzianum*	*F. odoratissimum*, *B. cinerea*	n.a.	Reduced antifungal activity against *F. odoratissimum* and *B. cinerea* in confrontation assays on discontinuous medium; increased susceptibility of tomato to *F. odoratissimum* and *B. cinerea*	n.a.	Rubio et al. (2017)

Process	Gene	General function	Aggressor	Defender	Phenotype		Comment	References
					Gene deletion	Gene complementation or overexpression		
	tvpg2	Endopolygalacturonase	*T. virens*	*B. cinerea*	Failed to transcribe an inducible endopolygalacturonase gene *tvpg1*	n.a.	n.a.	Sarrocco et al. (2017)
	vel1	Velum formation protein 1	*T. virens*	*R. solani*, *P. ultimum*	Highly hydrophilic; defective gliotoxin production; no conidiation; early chlamydospore formation under nutrient stress conditions; delayed or eliminated chlamydospore formation in nutrient-rich media; absence of mycelial and extracellular pigments; defects in the regulation of many other secondary metabolism-related genes; decrease in mycoparasitism against *R. solani* and *P. ultimum*	Highly hydrophobic as the wild-type strain	n.a.	Mukherjee and Kenerley (2010)
	crtf	Member of tubby transcription factor family	*C. chloroleuca*	*S. sclerotiorum*	The ability to parasitize sclerotia was markedly diminished	Recovered as the wild type	n.a.	Sun et al. (2018)
Signal transduction	*tac1*	Adenylate cyclase	*T. virens*	*Athelia rolfsii (Sclerothium rolfsii)*, *R. solani*, *Pythium* sp.	Retarded morphology; retained only 5–6% of the wild-type growth rate on agar; lowered intracellular cAMP levels to below the detection limit; loss of virulence against *A. rolfsii*, *R. solani*, and *Pythium* sp.; negatively affected production of secondary metabolites	n.a.	n.a.	Mukherjee et al. (2007)

(continued)

Table 4 (continued)

Process	Gene	General function	Aggressor	Defender	Phenotype: Gene deletion	Phenotype: Gene complementation or overexpression	Comment	References
	tga1	G-protein α-subunit	*T. atroviride*	*R. solani*	Light-independent hypersporulation; retarded mycoparasitic-related coiling against *R. solani*; loss of GTPase activity, which is stimulated by peptide toxin	An increase of coiling; inhibited sporulation; increased mycoparasitism on *R. solani*	n.a.	Reithner et al. (2005)
	tga3	G-protein α-subunit	*T. atroviride*, *T. harzianum*	*R. solani*, *B. cinerea*	In *T. atroviride*: reduced growth; defect in chitinase secretion; loss of infection structure formation; avirulence against *R. solani* and *B. cinerea*	In *T. harzianum*: no significant differences to wild type	n.a.	Zeilinger et al. (2005); Ding et al. (2020a)
					In *T. harzianum*: decreasing in hyphal growth, conidia yield, chitinase activity, antagonistic and mycoparasitism abilities against *R. solani*, and transcript abundance of the hydrophobin gene tha_09745 and hyphal surface hydrophobicity			
	tmk3	Mitogen-activated protein kinase	*T. harzianum*, *T. atroviride*	*Phoma betae* (*Pleosporales*, *Ascomycota*), *Colletotrichum acutatum* (*Glomerellales*, *Ascomycota*), *B. cinerea*, *F. odoratissimum*, *R. microsporus*	In *T. harzianum*: reduced osmotic and oxidative stress tolerance; reduced antagonistic activity against the tested plant pathogens. In *T. atroviride*: affected tip polarization, chemotropic growth, contact-induced morphogenesis, and makes the establishment of mycoparasitism highly inefficient to impossible	n.a.	n.a.	Delgado-Jarana et al. (2006); Moreno-Ruiz et al. (2021)

Process	Gene	General function	Aggressor	Defender	Phenotype		Comment	References
					Gene deletion	Gene complementation or overexpression		
	tmk1	Mitogen-activated protein kinase	*T. virens*, *T. atroviride*, *T. asperellum*	*A. rolfsii (S. rolfsii)*, *R. solani*, *B. cinereal*, *F. odoratissimum*, *R. microspores*, *A. alternata*, *Cytospora chrysosperma*	In *T. virens* IMI 304061: reduced ability to parasitize *R. solani* and *S. rolfsii*	In *T. atroviride*: the same mycoparasitism ability as the wild-type strain	n.a.	Mukherjee et al. (2003); Mendoza-Mendoza et al. (2003); Reithner et al. (2007); Moreno-Ruiz et al. (2021); Yang (2017)
					In *T. virens* Gv29–8: reduction in the rate of colony growth and development of aerial hyphae on solid media; production of conidia is lower on solid media but higher in liquid media compared with the wild-type strain; improved mycoparasitism against *R. solani*			
					In *T. atroviride*: reduced mycoparasitic abilities against *R. solani* and *B. cinerea* but showed higher antifungal activity caused by low molecular weight substances compared with wild-type strain; expression of chitinase gene and N-acetylglucosaminidase gene was reduced when confronted with *R. solani* and *B. cinerea*; affected tip polarization, chemotropic growth, contact-induced morphogenesis, and makes the establishment of mycoparasitism highly inefficient to impossible			
					In *T. asperellum*: less aerial mycelium but enhanced conidia production; rough surface of the hyphal wall; reduced antagonistic activity against *F. odoratissimum*, *A. alternata*, and *C. chrysosperma*; increased activity of N-acetylglucosaminidase, antibiotic activity of extracted metabolites including 6-PP			

(continued)

Table 4 (continued)

Process	Gene	General function	Aggressor	Defender	Phenotype		Comment	References
					Gene deletion	Gene complementation or overexpression		
	tmk2	Mitogen-activated protein kinase signaling pathway gene	*T. virens*	*A. rolfsii* (*S. rolfsii*), *R. solani*, *Pythium* sp.	Reduced radial growth; constitutive conidiation in darkness; defects in cell wall integrity; attenuated ability to overgrow the plant pathogen *A. rolfsii*	n.a.	n.a.	Kumar et al. (2010)
	abcG5	ATP-binding cassette (ABC) transporter	*C. rosea*	*F. graminearum*	Reduced antagonism toward *F. graminearum*; decreased tolerance to ZEN and xenobiotics secreted by *F. graminearum*; iprodione- and mefenoxam-based fungicides	n.a.	n.a.	Dubey et al. (2014)
	mfs464	Major facilitator superfamily (MFS) transporter	*C. rosea*	*F. graminearum*	Increased growth inhibitory activity against *F. graminearum* when co-inoculated in liquid medium	n.a.	n.a.	Nygren et al. (2018)

pathway was considered to be related to cell wall integrity. The defects in cell wall integrity of *tmk2* disruption mutant of *T. virens* were presented by autolysis of the mycelia and increased sensitivity to cell wall-degrading enzymes (Kumar et al. 2010). The third MAPK pathway was involved in stress tolerance. In *T. harzianum*, the *tmk3* null mutant exhibited reduced tolerance to osmotic and oxidative stress (Delgado-Jarana et al. 2006).

The regulatory gene *xyr1* was verified to be a negative regulator for the antagonistic behavior of *T. atroviride* IMI206040. The *xyr1* deletion mutant showed enhanced competition with *Phytophthora capsici* (Peronosporales, Oomycota), *B. cinereal*, and *R. solani*, which may be due to overexpression of the protease gene *prb1* (Reithner et al. 2014).

6 Outlook

Mycoparasitism is one of the strategies used in fungal wars, and the outcome depends on the comparative advantage of arsenal and defense systems of the parasite and host fungi. However, few battle details (i.e., physiological responses) were revealed from the mycoparasitism studies of *Trichoderma* spp. and *Clonostachys* spp. against other fungi, including plant pathogens. Since battle strategies likely differ depending on the mycoparasitic *Trichoderma* spp. and *Clonostachys* spp. and host fungal species, the battle details might be more spectacular than expected. The physiological responses of *Trichoderma* spp. and *Clonostachys* spp. and host fungi in mycoparasitism could provide more information and help us improve the application strategy for biocontrol.

Moreover, it is difficult to evaluate whether the condition established for observation of mycoparasitism is biased to *Trichoderma* spp. and *Clonostachys* spp. In that case, the biocontrol application of *Trichoderma* spp. and *Clonostachys* spp. into various environmental conditions may put *Trichoderma* spp. and *Clonostachys* spp. in danger and decrease the efficiency of biocontrol products based on these species. Therefore, lab experiments should be combined with *Trichoderma* spp. and *Clonostachys* spp. diversity studies to provide distribution data and describe the environmental preferences of *Trichoderma* spp. and *Clonostachys* spp. or strains. This combined approach will help form a multidimensional view of the biocontrol practices.

Acknowledgments The authors are thankful to Miss Sabine Schiessler (Vienna, Austria) for her assistance in the experiments shown in Fig. 1. This work was supported by grants from the National Special Project of Basic Work Project for Science and Technology (2014FY120900) to J.C. and National Natural Science Foundation of China (31801939) to F.C.

References

Ab Majid AH, Zahran Z, Abd Rahim AH, Ismail NA, Abdul Rahman W, Mohammad Zubairi KS, Dieng H, Satho T (2015) Morphological and molecular characterization of fungus isolated from tropical bed bugs in Northern Peninsular Malaysia, Cimex hemipterus (Hemiptera: Cimicidae). Asian Pac J Trop Biomed 5:707–713

Aghcheh RK, Druzhinina IS, Kubicek CP (2013) The putative protein methyltransferase LAE1 of *Trichoderma atroviride* is a key regulator of asexual development and mycoparasitism. PLoS One 8:e67144

Ait-Lahsen H, Soler A, Rey M, Cruz JDL, Monte E, Llobell A (2001) An antifungal Exo-α-1,3-Glucanase (AGN13.1) from the biocontrol fungus *Trichoderma harzianum*. Appl Environ Microbiol 67:5833–5839

Akrami M, Golzary H, Ahmadzadeh M (2011) Evaluation of different combinations of Trichoderma species for controlling *Fusarium* rot of lentil. Afr J Biotechnol 10:2653–2658

Atanasova L, Knox BP, Kubicek CP, Druzhinina IS, Baker SE (2013) The polyketide synthase gene pks4 of *Trichoderma reesei* provides pigmentation and stress resistance. Eukaryot Cell 12:1499–1508

Bae H, Sicher RC, Kim MS, Kim S-H, Strem MD, Melnick RL, Bailey BA (2009) The beneficial endophyte *Trichoderma hamatum* isolate DIS 219b promotes growth and delays the onset of the drought response in *Theobroma cacao*. J Exp Bot 60:3279–3295

Bae H, Roberts DP, Lim H-S, Strem MD, Park S-C, Ryu C-M, Melnick RL, Bailey BA (2011) Endophytic *Trichoderma* isolates from tropical environments delay disease onset and induce resistance against *Phytophthora capsici* in hot pepper using multiple mechanisms. Mol Plant-Microbe Interact 24:336–351

Baek J-M, Howell CR, Kenerley CM (1999) The role of an extracellular chitinase from *Trichoderma virens* Gv29-8 in the biocontrol of Rhizoctonia solani. Curr Genet 35:41–50

Bailey BA, Strem MD, Wood D (2009) *Trichoderma* species form endophytic associations within *Theobroma cacao* trichomes. Mycol Res 113:1365–1376

Bansal R, Mukherjee M, Horwitz BA, Mukherjee PK (2019) Regulation of conidiation and antagonistic properties of the soil-borne plant beneficial fungus *Trichoderma virens* by a novel proline-, glycine-, tyrosine-rich protein and a GPI-anchored cell wall protein. Curr Genet 65:953–964

Bara MTF, Lima AL, Ulhoa CJ (2003) Purification and characterization of an exo-β-1,3-glucanase produced by *Trichoderma asperellum*. FEMS Microbiol Lett 219:81–85

Bengtson S, Rasmussen B, Ivarsson M, Muhling J, Broman C, Marone F, Stampanoni M, Bekker A (2017) Fungus-like mycelial fossils in 2.4-billion-year-old vesicular basalt. Nat Ecol Evol 1:0141

Bischof RH, Ramoni J, Seiboth B (2016) Cellulases and beyond: the first 70 years of the enzyme producer *Trichoderma reesei*. Microb Cell Factories 15:106

Bowman SM, Free SJ (2006) The structure and synthesis of the fungal cell wall. BioEssays 28:799–808

Cai F, Druzhinina IS (2021) In honor of John Bissett: authoritative guidelines on molecular identification of *Trichoderma*. Fungal Divers

Cai F, Kubicek CP, Druzhinina IS (2021) Genetic transformation of *Trichoderma* spp. In: Basu C (ed) Biofuels and Biodiesel. Springer US, New York

Carsolio C, Gutiérrez A, Jiménez B, Van Montagu M, Herrera-Estrella A (1994) Characterization of ech-42, a *Trichoderma harzianum* endochitinase gene expressed during mycoparasitism. Proc Natl Acad Sci 91:10903–10907

Carsolio C, Benhamou N, Haran S, Cort SC, Guti Rrez A, Chet I, Herrera-Estrella A (1999) Role of the *Trichoderma harzianum* endochitinase gene, *ech42*, in mycoparasitism. Appl Environ Microbiol 65:929–935

Chatterton S, Punja ZKPK (2009) Chitinase and β-1,3-glucanase enzyme production by the mycoparasite *Clonostachys rosea* f. *catenulata* against fungal plant pathogens. Can J Microbiol 55:356–367

Chen J, Dou K, Gao Y, Li Y (2014) Mechanism and application of *Trichoderma* spp. in biological control of corn diseases. Mygosystema 33:1154–1167

Chen J-L, Sun S-Z, Miao C-P, Wu K, Chen Y-W, Xu L-H, Guan H-L, Zhao L-X (2016) Endophytic *Trichoderma gamsii* YIM PH30019: a promising biocontrol agent with hyperosmolar, mycoparasitism, and antagonistic activities of induced volatile organic compounds on root-rot pathogenic fungi of Panax notoginseng. J Ginseng Res 40:315–324

Chen PJ, Pang G, Cai F, Druzhinina IS (2021) Strain improvement and genetic engineering of *Trichoderma* for industrial applications

Chenthamara K, Druzhinina IS (2016) 12 ecological genomics of mycotrophic fungi. In: Druzhinina IS, Kubicek CP (eds) Environmental and microbial relationships. Springer International Publishing, Cham

Cianchetta S, Galletti S, Burzi PL, Cerato C (2012) Hydrolytic potential of *Trichoderma* sp. strains evaluated by microplate-based screening followed by switchgrass saccharification. Enzym Microb Technol 50:304–310

Cruz-Magalh V, Nieto-Jacobo MF, Van Zijll De Jong E, Rostás M, Padilla-Arizmendi F, Kandula D, Kandula J, Hampton J, Herrera-Estrella A, Steyaert JM, Stewart A, Loguercio LL, Mendoza-Mendoza A (2019) The NADPH oxidases Nox1 and Nox2 differentially regulate volatile organic compounds, fungistatic activity, plant growth promotion and nutrient assimilation in *Trichoderma atroviride*. Front Microbiol 9

Da Silva HAO, Teixeira WD, Borges ÁV, Silva Junior AL, Alves KS, Rodrigues Junior OM, De Abreu LM (2021) Biocontrol of potato early blight and suppression of *Alternaria grandis* sporulation by *Clonostachys* spp. Plant Pathol n/a

Dautt-Castro M, Estrada-Rivera M, Olguin-Mart Nez I, Rocha-Medina MDC, Islas-osuna MA, Casas-Flores S (2020) TBRG-1 a Ras-like protein in *Trichoderma virens* involved in conidiation, development, secondary metabolism, mycoparasitism, and biocontrol unveils a new family of Ras-GTPases. Fungal Genet Biol 136:103292

De Man TJB, Stajich JE, Kubicek CP, Teiling C, Chenthamara K, Atanasova L, Druzhinina IS, Levenkova N, Birnbaum SSL, Barribeau SM, Bozick BA, Suen G, Currie CR, Gerardo NM (2016) Small genome of the fungus *Escovopsis weberi*, a specialized disease agent of ant agriculture. Proc Natl Acad Sci 113:3567–3572

De Rezende LC, De Andrade Carvalho AL, Costa LB, De Almeida Halfeld-Vieira B, Silva LG, Pinto ZV, Morandi MAB, De Medeiros FHV, Mascarin GM, Bettiol W (2020) Optimizing mass production of *Trichoderma asperelloides* by submerged liquid fermentation and its antagonism against *Sclerotinia sclerotiorum*. World J Microbiol Biotechnol 36:113

De Souza Maia Filho F, Da Silva Fonseca AO, Persici BM, De Souza Silveira J, Braga CQ, Pötter L, De Avila Botton S, Brayer Pereira DI (2017) *Trichoderma virens* as a biocontrol of Toxocara canis: in vivo evaluation. Rev Iberoam Micol 34:32–35

Delgado-Jarana J, Sousa S, Gonz Lez F, Rey M, Llobell A (2006) ThHog1 controls the hyperosmotic stress response in *Trichoderma harzianum*. Microbiology 152:1687–1700

Deng S, Lorito M, Penttil M, Harman GE (2007) Overexpression of an endochitinase gene (*ThEn-42*) in *Trichoderma atroviride* for increased production of antifungal enzymes and enhanced antagonist action against pathogenic fungi. Appl Biochem Biotechnol 142:81

Ding J, Mei J, Huang P, Tian Y, Liang Y, Jiang X, Li M (2020a) Gα3 subunit Thga3 positively regulates conidiation, mycoparasitism, chitinase activity, and hydrophobicity of *Trichoderma harzianum*. AMB Express 10:221

Ding MY, Chen W, Ma XC, Lv BW, Jiang SQ, Yu YN, Rahimi MJ, Gao RW, Zhao Z, Cai F, Druzhinina IS (2020b) Emerging salt marshes as a source of *Trichoderma arenarium* sp. nov. and other fungal bioeffectors for biosaline agriculture. J Appl Microbiol n/a

Djonović S, Pozo MJ, Kenerley CM (2006) Tvbgn3, a β-1,6-Glucanase from the biocontrol fungus *Trichoderma virens*, is involved in mycoparasitism and control of *Pythium ultimum*. Appl Environ Microbiol 72:7661–7670

Druzhinina IS, Kubicek CP (2017) Genetic engineering of *Trichoderma reesei* cellulases and their production. Microb Biotechnol 10:1485–1499

Druzhinina IS, Komoń-Zelazowska M, Atanasova L, Seidl V, Kubicek CP (2010) Evolution and Ecophysiology of the industrial producer *Hypocrea jecorina* (Anamorph *Trichoderma reesei*) and a new sympatric agamospecies related to it. PLoS One 5:e9191

Druzhinina IS, Seidl-Seiboth V, Herrera-Estrella A, Horwitz BA, Kenerley CM, Monte E, Mukherjee PK, Zeilinger S, Grigoriev IV, Kubicek CP (2011) *Trichoderma*: the genomics of opportunistic success. Nat Rev Microbiol 9:749–759

Druzhinina IS, Kopchinskiy AG, Kubicek EM, Kubicek CP (2016) A complete annotation of the chromosomes of the cellulase producer *Trichoderma reesei* provides insights in gene clusters, their expression and reveals genes required for fitness. Biotechnol Biofuels 9:75

Druzhinina IS, Chenthamara K, Zhang J, Atanasova L, Yang D, Miao Y, Rahimi MJ, Grujic M, Cai F, Pourmehdi S, Salim KA, Pretzer C, Kopchinskiy AG, Henrissat B, Kuo A, Hundley H, Wang M, Aerts A, Salamov A, Lipzen A, Labutti K, Barry K, Grigoriev IV, Shen Q, Kubicek CP (2018) Massive lateral transfer of genes encoding plant cell wall-degrading enzymes to the mycoparasitic fungus *Trichoderma* from its plant-associated hosts. PLoS Genet 14:e1007322

Dubey MK, Jensen DF, Karlsson M (2014) An ATP-binding cassette pleiotropic drug transporter protein is required for xenobiotic tolerance and antagonism in the fungal biocontrol agent *Clonostachys rosea*. Mol Plant-Microbe Interact 27:725–732

Elad Y, Kapat A (1999) The role of *Trichoderma harzianum* protease in the biocontrol of Botrytis cinerea. Eur J Plant Pathol 105:177–189

Elad Y, Chet I, Boyle P, Henis Y (1983) Parasitism of *Trichoderma* spp. on *Rhizoctonia solani* and *Sclerotium rolfsii*-scanning electron microscopy and fluorescence microscopy. Phytopathology 73:85–88

Elad Y, Sadowsky Z, Chet I (1987) Scanning electron microscopical observations of early stages of interaction of *Trichoderma harzianum* and *Rhizoctonia solani*. Trans Br Mycol Soc 88:259–263

El-Katatny M, Somitsch W, Robra K-H, El-Katatny M, Guebitz G (2000) Production of chitinase and β-1,3-glucanase by *Trichoderma harzianum* for control of the phytopathogenic fungus Sclerotium rolfsii. Food Technol Biotechnol 38:173–180

Estrada-Rivera M, Hernández-Oñate MÁ, Dautt-Castro M, Gallardo-Negrete JDJ, Rebolledo-Prudencio OG, Uresti-Rivera EE, Arenas-Huertero C, Herrera-Estrella A, Casas-Flores S (2020) IPA-1 a putative chromatin remodeler/helicase-related protein of Trichoderma virens plays important roles in antibiosis against Rhizoctonia solani and induction of Arabidopsis systemic disease resistance. Mol Plant-Microbe Interact 33:808–824

Evidente A, Andolfi A, Cimmino A, Ganassi S, Altomare C, Favilla M, De Cristofaro A, Vitagliano S, Agnese Sabatini M (2009) Bisorbicillinoids produced by the fungus Trichoderma citrinoviride affect feeding preference of the aphid Schizaphis graminum. J Chem Ecol 35:533–541

Fatema U, Broberg A, Jensen DF, Karlsson M, Dubey M (2018) Functional analysis of polyketide synthase genes in the biocontrol fungus Clonostachys rosea. Sci Rep 8:15009

Fraceto LF, Maruyama CR, Guilger M, Mishra S, Keswani C, Singh HB, De Lima R (2018) Trichoderma harzianum-based novel formulations: potential applications for management of Next-Gen agricultural challenges. J Chem Technol Biotechnol 93:2056–2063

Furukawa H, Kusne S, Sutton DA, Manez R, Carrau R, Nichols L, Abu-Elmagd K, Skedros D, Todo S, Rinaldi MG (1998) Acute invasive sinusitis due to Trichoderma longibrachiatum in a liver and small bowel transplant recipient. Clin Infect Dis 26:487–489

Gautheret A, Dromer F, Bourhis J-H, Andremont A (1995) Trichoderma pseudokoningii as a cause of fatal infection in a bone marrow transplant recipient. Clin Infect Dis 20:1063–1064

Getz WM (2012) A biomass flow approach to population models and food webs. Nat Resour Model 25:93–121

Ghosh SK, Pal S (2015) Entomopathogenic potential of Trichoderma longibrachiatum and its comparative evaluation with malathion against the insect pest Leucinodes orbonalis. Environ Monit Assess 188:37

Gołaś M, Sulewski P, Wąs A, Kłoczko-Gajewska A, Pogodzińska K (2020) On the way to sustainable agriculture—eco-efficiency of polish commercial farms. Agriculture 10:438

Gomes EV, Costa MDN, De Paula RG, Ricci De Azevedo R, Da Silva FL, Noronha EF, Jos Ulhoa C, Neves Monteiro V, Elena Cardoza R, Guti Rrez S, Nascimento Silva R (2015) The Cerato-Platanin protein Epl-1 from Trichoderma harzianum is involved in mycoparasitism, plant resistance induction and self cell wall protection. Sci Rep 5:17998

González-Vera AD, Bernardes-De-Assis J, Zala M, Mcdonald BA, Correa-Victoria F, Graterol-Matute EJ, Ceresini PC (2010) Divergence between sympatric rice- and maize-infecting populations of Rhizoctonia solani AG-1 IA from Latin America. Phytopathology 100:172–182

Guzmán-Guzmán P, Alemán-Duarte MI, Delaye L, Herrera-Estrella A, Olmedo-Monfil V (2017) Identification of effector-like proteins in Trichoderma spp. and role of a hydrophobin in the plant-fungus interaction and mycoparasitism. BMC Genet 18:16

Hernández-Oñate MA, Esquivel-Naranjo EU, Mendoza-Mendoza A, Stewart A, Herrera-Estrella AH (2012) An injury-response mechanism conserved across kingdoms determines entry of the fungus *Trichoderma atroviride* into development. Proc Natl Acad Sci 109:14918–14923

Hiscox J, O'leary J, Boddy L (2018) Fungus wars: basidiomycete battles in wood decay. Stud Mycol 89:117–124

Huang J, Yang G (2017) Understanding recent challenges and new food policy in China. Glob Food Sec 12:119–126

Hue AG, Voldeng HD, Savard ME, Fedak G, Tian X, Hsiang T (2009) Biological control of fusarium head blight of wheat with Clonostachys rosea strain ACM941. Can J Plant Pathol 31:169–179

Inayati A, Sulistyowati L, Aini LQ, Yusnawan E (2020) Mycoparasitic activity of indigenous Trichoderma virens strains against mungbean soil borne pathogen Rhizoctonia solani: Hyperparasite and hydrolytic enzyme production. AGRIVITA, J Agric Sci 42:229–242

Iqbal M, Dubey M, Broberg A, Viketoft M, Jensen DF, Karlsson M (2019) Deletion of the nonribosomal peptide synthetase gene nps1 in the fungus Clonostachys rosea attenuates antagonism and biocontrol of plant pathogenic Fusarium and nematodes. Phytopathology 109:1698–1709

Jaklitsch WM (2009) European species of *Hypocrea* part I. The green-spored species. Stud Mycol 63:1–91

Jaklitsch WM (2011) European species of *Hypocrea* part II: species with hyaline ascospores. Fungal Divers 48:1–250

Jaklitsch WM, Voglmayr H (2012) Hypocrea britdaniae and H. foliicola: two remarkable new European species. Mycologia 104:1213–1221

Jaklitsch WM, Voglmayr H (2015) Biodiversity of Trichoderma (Hypocreaceae) in Southern Europe and Macaronesia. Stud Mycol:1–87

Jaklitsch WM, Kubicek CP, Druzhinina IS (2008) Three European species of Hypocrea with reddish brown stromata and green ascospores. Mycologia 100:796–815

Jeffries P, Young TW (1994) Interfungal parasitic relationships. Cab International

Kappel L, Münsterkötter M, Sipos G, Escobar Rodriguez C, Gruber S (2020) Chitin and chitosan remodeling defines vegetative development and Trichoderma biocontrol. PLoS Pathog 16:e1008320

Karlsson M, Atanasova L, Jensen DF, Zeilinger S, Heitman J, James TY, Crous PW (2017) Necrotrophic mycoparasites and their genomes. Microbiol Spectrum 5:5.2.08

Keshavarz B, Khalesi M (2016) *Trichoderma reesei*, a superior cellulase source for industrial applications. Biofuels 7:713–721

Komoń-Zelazowska M, Bissett J, Zafari D, Hatvani L, Manczinger L, Woo S, Lorito M, Kredics L, Kubicek CP, Druzhinina IS (2007) Genetically closely related but phenotypically divergent *Trichoderma* species cause green mold disease in oyster mushroom farms worldwide. Appl Environ Microbiol 73:7415–7426

Kosawang C, Karlsson M, Vélëz H, Rasmussen PH, Collinge DB, Jensen B, Jensen DF (2014) Zearalenone detoxification by zearalenone hydrolase is important for the antagonistic ability of *Clonostachys rosea* against mycotoxigenic *Fusarium graminearum*. Fungal Biol 118:364–373

Kubicek CP, Herrera-Estrella A, Seidl-Seiboth V, Martinez DA, Druzhinina IS, Thon M, Zeilinger S, Casas-Flores S, Horwitz BA, Mukherjee PK, Mukherjee M, Kredics L, Alcaraz LD, Aerts A, Antal Z, Atanasova L, Cervantes-Badillo MG, Challacombe J, Chertkov O, Mccluskey K, Coulpier F, Deshpande N, Hren VD, Ebbole DJ, Esquivel-Naranjo EU, Fekete E, Flipphi M, Glaser F, Gómez-Rodríguez EY, Gruber S, Han C, Henrissat B, Hermosa R, Hernández-Oñate M, Karaffa L, Kosti I, Le Crom S, Lindquist E, Lucas S, Lübeck M, Lübeck PS, Margeot A, Metz B, Misra M, Nevalainen H, Omann M, Packer N, Perrone G, Uresti-Rivera EE, Salamov A, Schmoll M, Seiboth B, Shapiro H, Sukno S, Tamayo-Ramos JA, Tisch D, Wiest A, Wilkinson HH, Zhang M, Coutinho PM, Kenerley CM, Monte E, Baker SE, Grigoriev IV (2011) Comparative genome sequence analysis underscores mycoparasitism as the ancestral life style of *Trichoderma*. Genome Biol 12:R40

Kubicek CP, Steindorff AS, Chenthamara K, Manganiello G, Henrissat B, Zhang J, Cai F, Kopchinskiy AG, Kubicek EM, Kuo A, Baroncelli R, Sarrocco S, Noronha EF, Vannacci G, Shen Q, Grigoriev IV, Druzhinina IS (2019) Evolution and comparative genomics of the most common *Trichoderma* species. BMC Genomics 20:485

Kumar A, Scher K, Mukherjee M, Pardovitz-Kedmi E, Sible GV, Singh US, Kale SP, Mukherjee PK, Horwitz BA (2010) Overlapping and distinct functions of two *Trichoderma virens* MAP kinases in cell-wall integrity, antagonistic properties and repression of conidiation. Biochem Biophys Res Commun 398:765–770

Lakkireddy K, Khonsuntia W, Kües U (2020) Mycoparasite *Hypomyces odoratus* infests *Agaricus xanthodermus* fruiting bodies in nature. AMB Express 10:141

Li Destri Nicosia MG, Mosca S, Mercurio R, Schena L (2015) Dieback of Pinus nigra seedlings caused by a strain of *Trichoderma viride*. Plant Dis 99:44–49

Li YY, Tang J, Fu KH, Gao SG, Wu Q, Chen J (2012) Construction of transgenic *Trichoderma koningi* with chit42 of *Metarhizium anisopliae* and analysis of its activity against the Asian corn borer. J Environ Sci Health B 47:622–630

Li Y, Fu K, Gao S, Wu Q, Fan L, Li Y, Chen J (2013) Impact on bacterial community in midguts of the Asian corn borer larvae by transgenic *Trichoderma* strain overexpressing a heterologous chit42 gene with chitin-binding domain. PLoS One 8:e55555

Liu K, Tian J, Xiang M, Liu X (2016) How carnivorous fungi use three-celled constricting rings to trap nematodes. Protein Cell 3:325–328

Liu J-Y, Li S-D, Sun M-H, (2016) Transaldolase gene *Tal67* enhances the biocontrol activity of *Clonostachys rosea* 67-1 against *Sclerotinia sclerotiorum*. Biochem Bioph Res Co 474(3) 503–508 10.1016/j.bbrc.2016.04.133

Lücking R, Huhndorf S, Pfister DH, Plata ER, Lumbsch HT (2009) Fungi evolved right on track. Mycologia 101:810–822

Lv B, Jiang N, Hasan R, Chen Y, Sun M, Li S (2020) Cell wall biogenesis protein phosphatase CrSsd1 is required for conidiation, cell wall integrity, and mycoparasitism in Clonostachys rosea. Front Microbiol 11

Malmierca MG, Cardoza RE, Alexander NJ, Mccormick SP, Hermosa R, Monte E, Guti Rrez S (2012) Involvement of *Trichoderma* trichothecenes in the biocontrol activity and induction of plant Defense-related genes. Appl Environ Microbiol 78:4856–4868

Malmierca MG, Cardoza RE, Alexander NJ, Mccormick SP, Collado IG, Hermosa R, Monte E, Guti Rrez S (2013) Relevance of trichothecenes in fungal physiology: disruption of tri5 in *Trichoderma arundinaceum*. Fungal Genet Biol 53:22–33

Martzy R, Mach-Aigner AR (2021) The potential of synthetic biology for *Trichoderma reesei*. In: Mach-Aigner AR, Martzy R (eds) Trichoderma reesei: methods and protocols. Springer US, New York

Mendoza JLH, Pérez MIS, Prieto JMG, Vel Squez JDQ, Olivares JGG, Langarica HRG (2015) Antibiosis of *Trichoderma* spp. strains native to northeastern Mexico against the pathogenic fungus *Macrophomina phaseolina*. Braz J Microbiol 46:1093–1101

Mendoza-Mendoza A, Pozo MJ, Grzegorski D, Martínez P, García JM, Olmedo-Monfil V, Cortés C, Kenerley C, Herrera-Estrella A (2003) Enhanced biocontrol activity of *Trichoderma* through inactivation of a mitogen-activated protein kinase. Proc Natl Acad Sci 100:15965–15970

Montero-Barrientos M, Hermosa R, Cardoza RE, Gutiérrez S, Monte E (2011) Functional analysis of the *Trichoderma harzianum nox1* gene, encoding an NADPH oxidase, relates production of reactive oxygen species to specific biocontrol activity against *Pythium ultimum*. Appl Environ Microbiol 77:3009–3016

Moreno-Ruiz D, Salzmann L, Fricker MD, Zeilinger S, Lichius A (2021) Stress-activated protein kinase signalling regulates mycoparasitic hyphal-hyphal interactions in *Trichoderma atroviride*. J Fungi 7:365

Mukherjee PK, Kenerley CM (2010) Regulation of morphogenesis and biocontrol properties in *Trichoderma virens* by a VELVET protein, Vel1. Appl Environ Microbiol 76:2345–2352

Mukherjee PK, Latha J, Hadar R, Horwitz BA (2003) TmkA, a mitogen-activated protein kinase of *Trichoderma virens*, is involved in biocontrol properties and repression of conidiation in the dark. Eukaryot Cell 2:446–455

Mukherjee M, Mukherjee PK, Kale SP (2007) cAMP signalling is involved in growth, germination, mycoparasitism and secondary metabolism in *Trichoderma virens*. Microbiology 153:1734–1742

Mukherjee AK, Sampath Kumar A, Kranthi S, Mukherjee PK (2014) Biocontrol potential of three novel *Trichoderma* strains: isolation, evaluation and formulation. 3 Biotech 4:275–281

Nguyen T, Inkong N, Faysee N (2020) Role of local institutions in the transition towards sustainable agriculture: the case study of Thailand. WIT Trans Ecol Environ 245:135–143

Nygren K, Dubey M, Zapparata A, Iqbal M, Tzelepis GD, Durling MB, Jensen DF, Karlsson M (2018) The mycoparasitic fungus *Clonostachys rosea* responds with both common and specific gene expression during interspecific interactions with fungal prey. Evol Appl 11:931–949

Pacheco JC, Poltronieri AS, Porsani MV, Zawadneak MAC, Pimentel IC (2017) Entomopathogenic potential of fungi isolated from intertidal environments against the cabbage aphid Brevicoryne brassicae (Hemiptera: aphididae). Biocontrol Sci Tech 27:496–509

Pang G, Sun T, Yu Z, Yuan T, Liu W, Zhu H, Gao Q, Yang D, Kubicek CP, Zhang J, Shen Q (2020) Azaphilones biosynthesis complements the defence mechanism of *Trichoderma guizhouense* against oxidative stress. Environ Microbiol 22:4808–4824

Pozo MAJ, Baek J-M, García JM, Kenerley CM (2004) Functional analysis of tvsp1, a serine protease-encoding gene in the biocontrol agent *Trichoderma virens*. Fungal Genet Biol 41:336–348

Priyadarshini P, Abhilash PC (2020) Policy recommendations for enabling transition towards sustainable agriculture in India. Land Use Policy 96:104718

Proctor RH, Mccormick SP, Kim H-S, Cardoza RE, Stanley AM, Lindo L, Kelly A, Brown DW, Lee T, Vaughan MM, Alexander NJ, Busman M, Gutiérrez S (2018) Evolution of structural diversity of trichothecenes, a family of toxins produced by plant pathogenic and entomopathogenic fungi. PLoS Pathog 14:e1006946

Reithner B, Brunner K, Schuhmacher R, Peissl I, Seidl V, Krska R, Zeilinger S (2005) The G protein α subunit Tga1 of *Trichoderma atroviride* is involved in chitinase formation and differential production of antifungal metabolites. Fungal Genet Biol 42:749–760

Reithner B, Schuhmacher R, Stoppacher N, Pucher M, Brunner K, Zeilinger S (2007) Signaling via the Trichoderma atroviride mitogen-activated protein kinase Tmk1 differentially affects mycoparasitism and plant protection. Fungal Genet Biol 44:1123–1133

Reithner B, Mach-Aigner AR, Herrera-Estrella A, Mach RL (2014) The transcriptional regulator Xyr1 of *Trichoderma atroviride* supports the induction of systemic resistance in plants. Appl Environ Microbiol:AEM.00930-14

Rocha-Ramirez V, Omero C, Chet I, Horwitz BA, Herrera-Estrella A (2002) *Trichoderma atroviride* G-protein α-subunit gene *tga1* is involved in mycoparasitic coiling and conidiation. Eukaryot Cell 1:594–605

Rodríguez MA, Cabrera G, Gozzo FC, Eberlin MN, Godeas A (2011) *Clonostachys rosea* BAFC3874 as a *Sclerotinia sclerotiorum* antagonist: mechanisms involved and potential as a biocontrol agent. J Appl Microbiol 110:1177–1186

Rosmana A, Samuels GJ, Ismaiel A, Ibrahim ES, Chaverri P, Herawati Y, Asman A (2015) *Trichoderma asperellum*: a dominant endophyte species in cacao grown in Sulawesi with potential for controlling vascular streak dieback disease. Trop Plant Pathol 40:19–25

Rosmana A, Nasaruddin N, Hendarto H, Hakkar AA, Agriansyah N (2016) Endophytic Association of *Trichoderma asperellum* within *Theobroma cacao* suppresses vascular streak dieback incidence and promotes side graft growth. Mycobiology 44:180–186

Rubio MB, Pardal AJ, Cardoza RE, Gutiérrez S, Monte E, Hermosa R (2017) Involvement of the transcriptional coactivator ThMBF1 in the biocontrol activity of *Trichoderma harzianum*. Front Microbiol 8

Ruiu L (2018) Microbial biopesticides in agroecosystems. Agronomy 8:235

Sarrocco S, Matarese F, Baroncelli R, Vannacci G, Seidl-Seiboth V, Kubicek CP, Vergara M (2017) The constitutive endopolygalacturonase TvPG2 regulates the induction of plant systemic resistance by *Trichoderma virens*. Phytopathology 107:537–544

Schaeffer HJ, Weber MJ (1999) Mitogen-activated protein kinases: specific messages from ubiquitous messengers. Mol Cell Biol 19:2435–2444

Schmidt AR, Dörfelt H, Perrichot V (2007) Carnivorous fungi from cretaceous Amber. Science 318:1743–1743

Seiboth B, Karimi RA, Phatale PA, Linke R, Hartl L, Sauer DG, Smith KM, Baker SE, Freitag M, Kubicek CP (2012) The putative protein methyltransferase LAE1 controls cellulase gene expression in *Trichoderma reesei*. Mol Microbiol 84:1150–1164

Seidl-Seiboth V, Ihrmark K, Druzhinina I, Karlsson M (2014) Chapter 5 - Molecular evolution of *Trichoderma* Chitinases. In: Gupta VK, Schmoll M, Herrera-Estrella A, Upadhyay RS, Druzhinina I, Tuohy MG (eds) Biotechnology and biology of Trichoderma. Elsevier, Amsterdam

Sharon E, Bar-Eyal M, Chet I, Herrera-Estrella A, Kleifeld O, Spiegel Y (2001) Biological control of the root-knot nematode *Meloidogyne javanica* by *Trichoderma harzianum*. Phytopathology 91:687–693

Shentu X, Yao J, Yuan X, He L, Sun F, Ochi K, Yu X (2018) *Tri11*, *tri3*, and *tri4* genes are required for trichodermin biosynthesis of *Trichoderma brevicompactum*. AMB Express 8:58

Suarez B, Rey M, Castillo P, Monte E, Llobell A (2004) Isolation and characterization of PRA1, a trypsin-like protease from the biocontrol agent *Trichoderma harzianum* CECT 2413 displaying nematicidal activity. Appl Microbiol Biotechnol 65:46–55

Sun Z-B, Sun M-H, Zhou M, Li S-D (2017) Transformation of the endochitinase gene Chi67-1 in *Clonostachys rosea* 67-1 increases its biocontrol activity against *Sclerotinia sclerotiorum*. AMB Express 7:1

Sun Z-B, Wang Q, Zhang J, Jiang W-Z, Wang Q, Li S-D, Ma G-Z, Sun M-H (2018) The transcription factor-encoding gene *crtf* is involved in *Clonostachys chloroleuca* mycoparasitism on *Sclerotinia sclerotiorum*. Microbiol Res 210:6–11

Sung G-H, Poinar GO, Spatafora JW (2008) The oldest fossil evidence of animal parasitism by fungi supports a cretaceous diversification of fungal–arthropod symbioses. Mol Phylogenet Evol 49:495–502

Szekeres A, Kredics L, Antal Z, Kevei F, Manczinger L (2004) Isolation and characterization of protease overproducing mutants of *Trichoderma harzianum*. FEMS Microbiol Lett 233:215–222

Tijerino A, Elena Cardoza R, Moraga J, Malmierca MG, Vicente F, Aleu J, Collado IG, Guti Rrez S, Monte E, Hermosa R (2011) Overexpression of the trichodiene synthase gene *tri5* increases trichodermin production and antimicrobial activity in *Trichoderma brevicompactum*. Fungal Genet Biol 48:285–296

Toyama H, Yamagishi N, Toyama N (2002) Construction of cellulase hyper-producers of *Trichoderma reesei* Rut C-30 for utilization of waste paper using colchicine and benomyl. J Mol Catal B-Enz 17:175–178

Tsror L (2010) Biology, epidemiology and management of *Rhizoctonia solani* on potato. J Phytopathol 158:649–658

Ujor VC, Adukwu EC, Okonkwo CC (2018) Fungal wars: the underlying molecular repertoires of combating mycelia. Fungal Biol 122:191–202

Vargas WA, Mukherjee PK, Laughlin D, Wiest A, Moran-Diez ME, Kenerley CM (2014) Role of gliotoxin in the symbiotic and pathogenic interactions of *Trichoderma virens*. Microbiology 160:2319–2330

Vinale F, Marra R, Scala F, Ghisalberti EL, Lorito M, Sivasithamparam K (2006) Major secondary metabolites produced by two commercial *Trichoderma* strains active against different phytopathogens. Lett Appl Microbiol 43:143–148

Vinale F, Ghisalberti EL, Sivasithamparam K, Marra R, Ritieni A, Ferracane R, Woo S, Lorito M (2009) Factors affecting the production of *Trichoderma harzianum* secondary metabolites during the interaction with different plant pathogens. Lett Appl Microbiol 48:705–711

Vogelgsang S, Sulyok M, Bänziger I, Krska R, Schuhmacher R, Forrer H-R (2008) Effect of fungal strain and cereal substrate on in vitro mycotoxin production by *Fusarium poae* and *Fusarium avenaceum*. Food Addit Contam 25:745–757

Woo SL, Ruocco M, Vinale F, Nigro M, Marra R, Lombardi N, Pascale A, Lanzuise S, Manganiello G, Lorito M (2014) Trichoderma-based products and their widespread use in agriculture. Open Mycol J 8

Yadav AN, Singh J, Singh C, Yadav N (2020) Current trends in microbial biotechnology for sustainable agriculture. Springer

Yang P (2017) The gene *task1* is involved in morphological development, mycoparasitism and antibiosis of *Trichoderma asperellum*. Biocontrol Sci Tech 27:620–635

Yang E, Xu L, Yang Y, Zhang X, Xiang M, Wang C, An Z, Liu X (2012) Origin and evolution of carnivorism in the Ascomycota (fungi). Proc Natl Acad Sci 109:10960–10965

Yu F-M, Jayawardena RS, Liu J, Hyde KD, Zhao Q (2020) *Hypomyces pseudolactifluorum* sp. nov. (Hypocreales: Hypocreaceae) on *Russula* sp. from Yunnan, PR China. Biodiver Data J 8

Zahran Z, Mohamed Nor NMI, Dieng H, Satho T, Ab Majid AH (2017) Laboratory efficacy of mycoparasitic fungi (*Aspergillus tubingensis* and *Trichoderma harzianum*) against tropical bed bugs (*Cimex hemipterus*) (Hemiptera: Cimicidae). Asian Pac J Trop Biomed 7:288–293

Zeilinger S, Reithner B, Scala V, Peissl I, Lorito M, Mach RL (2005) Signal transduction by Tga3, a novel G protein α subunit of *Trichoderma atroviride*. Appl Environ Microbiol 71:1591–1597

Zeng ZQ, Zhuang WY (2019) Two new species and a new Chinese record of Hypocreaceae as evidenced by morphological and molecular data. Mycobiology 47:280–291

Zhang J, Bayram Akcapinar G, Atanasova L, Rahimi MJ, Przylucka A, Yang D, Kubicek CP, Zhang R, Shen Q, Druzhinina IS (2016) The neutral metallopeptidase NMP1 of *Trichoderma guizhouense* is required for mycotrophy and self-defence. Environ Microbiol 18:580–597

Zhang S, Gan Y, Ji W, Xu B, Hou B, Liu J (2017) Mechanisms and characterization of *Trichoderma longibrachiatum* T6 in suppressing nematodes (*Heterodera avenae*) in wheat. Front Plant Sci 8

Zhang J, Miao Y, Rahimi MJ, Zhu H, Steindorff A, Schiessler S, Cai F, Pang G, Chenthamara K, Xu Y, Kubicek CP, Shen Q, Druzhinina IS (2019) Guttation capsules containing hydrogen peroxide: an evolutionarily conserved NADPH oxidase gains a role in wars between related fungi. Environ Microbiol 21:2644–2658

Zhu H, Zhang J, Gao Q, Pang G, Sun T, Li R, Yu Z, Shen Q (2021) A new atypical short-chain dehydrogenase is required for interfungal combat and conidiation in *Trichoderma guizhouense*. Environ Microbiol n/a

Trichoderma Secondary Metabolites Involved in Microbial Inhibition

Yael González, Sergio de los Santos-Villalobos, and Ernestina Castro-Longoria

Contents

1 Introduction

Primary metabolic pathways (e.g., glycolysis, amino acid or nucleotide synthesis) are fundamental for all species of life, where coordination between anabolism and catabolism is the principle that sustains cellular functions. Besides these fundamental pathways, there are secondary specialized pathways – particular to

Y. González · E. Castro-Longoria (✉)
Departamento de Microbiología, Centro de Investigación Científica y de Educación Superior de Ensenada (CICESE), Ensenada, Baja California, Mexico
e-mail: ecastro@cicese.mx

S. de los Santos-Villalobos
Laboratorio de Biotecnología del Recurso Microbiano, Instituto Tecnológico de Sonora (ITSON), Ciudad Obregón, Sonora, Mexico

N. Amaresan et al. (eds.), *Advances in* Trichoderma *Biology for Agricultural Applications*, Fungal Biology, https://doi.org/10.1007/978-3-030-91650-3_3

taxonomically restricted groups of organisms – not essential for growth, development, or reproduction; but they likely confer an adaptive advantage to the producer organism (O'Connor 2015). The outcome molecules or factors of these secondary pathways are the compounds referred to as secondary metabolites (SMs) (Jenke-Kodama et al. 2008). As shown previously (Cary 2004), fungi are recognized as excellent producers of SMs that help them to adapt and survive in their natural environment. The enormous structural diversity and variation displayed by fungal SMs reflects a wide range of biological activities (Demain and Fang 2000) highly valuable for medicine (Weber 2010; Zhong and Xiao 2009), biotechnology (Arora et al. 2004), and agriculture (Contreras-Cornejo et al. 2018; Guzmán-Guzmán et al. 2018).

Trichoderma (phylum, *Ascomycota*; class, *Sordariomycetes*; order, *Hypocreales*; family, *Hypocreaceae*) is a genus of necrotrophic, mycoparasitic unspecialized fungi. *Trichoderma* members are proficient to establish beneficial associations with plants through the rhizosphere (Contreras-Cornejo et al. 2018; Guzmán-Guzmán et al. 2018). In addition, *Trichoderma* species are capable of antagonizing and parasitizing other fungi for its nutrition (Karlsson et al. 2017). Once the contact with a potential prey is made, *Trichoderma* coils its hyphae around the mycelium of the host, forming hook-like structures at the points of contact. The host range of *Trichoderma* is usually wide; they kill its fungal prey by invasion of the host hyphae, the secretion of damaging SMs, and the incorporation of the released nutrients (Karlsson et al. 2017; Lorito et al. 1996b). These traits render this group of fungi as biocontrol agents with great value for crop protection from infectious diseases (Benítez et al. 2004; Howell 2003; Vinale 2014). The protective effects of *Trichoderma* against phytopathogens have been observed and well documented; some examples are next shown. The membrane-channeling antibiotics, the peptaibols (trichorzianines), and cell wall-degrading enzymes from *T. harzianum* act synergistically for inhibiting spore germination and hyphal extension in the phytopathogen fungus *Botrytis cinerea* (Schirmböck et al. 1994). *T. asperellum* T8a produces enzymes with lytic activity against the mycelium of *Colletotrichum gloeosporioides*, the causal agent of the anthracnose in mango (*Mangifera indica*) (de los Santos-Villalobos et al. 2012, 2013). Volatile and non-volatile SMs secreted by *T. harzianum*, *T. virens*, and *T. viride* reduce significantly the hyphal growth of *Fusarium moniliforme* var. *subglutinans*; this phytopathogen is the causal agent of the malformation disease also in mango (Dennis and Webster 1971a, b; Kumar et al. 2012). The molecules T22azaphilone and harzianopyridone isolated from *T. afroharzianum* T22 (formerly *T. harzianum*) (Chaverri et al. 2015) and T39, respectively, completely inhibited the growth of fungal phytopathogens *Leptosphaeria maculans*, *Phytophthora cinnamomi*, and *B. cinerea* (Vinale et al. 2009). This work is a critical review focused on SMs produced by *Trichoderma*, due to their complex action mode. Thus, this understanding can lead not only to discover new molecules but it opens the possibility to design sustainable strategies for crop and post-harvested protection or even biomedical applications (Keswani et al. 2014).

2 Different Classes of SMs Produced by *Trichoderma*

2.1 Non-ribosomal Peptides

The non-ribosomal peptides (NRPs) are short amino acid chains from 3 to 21 residues long, with a molecular mass of 500 to 2100 Da., they may contain proteinogenic and non-proteinogenic amino acids, and their molecular structure can be linear or cyclic (Zeilinger et al. 2016). Peptaibiotics and peptaibols exhibit antifungal, antibacterial, and anticancer properties, arising from their membrane insertion and pore-forming abilities due to their amphipathic nature and linear structure (Daniel and Rodrigues Filho 2007; Hermosa et al. 2014). These molecules are assembled by the consecutive condensation of amino acid residues, which is achieved by the multidomain, multimodular enzymes named non-ribosomal peptide synthetases (NRPSs) (Daniel and Rodrigues Filho 2007; Finking and Marahiel 2004; Keller et al. 2005; Walsh 2008). The NRPSs are formed by several modules, and each module is composed by specific domains; during the process of synthesis, the amino acid residues are translocated from one domain to the other. The first step in the synthesis reaction corresponds to the adenylation (A) domain, which recognizes and activates the amino acid residue in the form of an aminoacyl-AMP. Once in the active state, the amino acid is transferred from the A domain to the pantothenylation/peptidyl carrier (P) domain. In this step, the amino acid is covalently bonded, by a thioester, to a 4′-phosphopantetheine cofactor, which is attached to a conserved serine in the P domain. Afterward, the condensation/peptide bond formation (C) domain catalyzes the formation of a peptide bond between amino acid residues linked onto adjacent modules. Finally, the complete peptide is released from the synthetase by the action of a thioesterase (TE) domain (Fig. 1) (Finking and Marahiel 2004; Keller et al. 2005, p. 20; Strieker et al. 2010). The analysis of three genome-sequenced *Trichoderma* species revealed high numbers of genes involved in NRPSs, 28 for *T. virens*, 16 for *T. atroviride*, and 10 for *T. reesei* (Kubicek et al. 2011). This kind of genome analysis is supported by databases like AntiSMASH (https://fungismash.secondarymetabolites.org/#!/start), which analyzes the fungal genome, in search for biosynthesis of genes coding for SM synthases like NRPSs, polyketide synthetases (PKSs), and terpenes (Blin et al. 2019). The major groups of NRPs from *Trichoderma* are peptaibols and peptaibiotics, followed by epidithiodioxopiperazines (ETPs) and siderophores (Zeilinger et al. 2016).

2.1.1 Peptaibols and Peptaibiotics

This is the most abundant group of NRPs produced by many fungi, including *Trichoderma* (Krause et al. 2006). These molecules present an alkyl-N-terminal residue and a C-terminal hydroxyl-amino acid residue (1,2-amino alcohol like phenylalaninol, valinol, leucinol, isoleucinol, or tryptophanol) and contain non-proteinogenic amino acids like α-aminoisobutyric acid (Aib), isovaline (Iva), or the

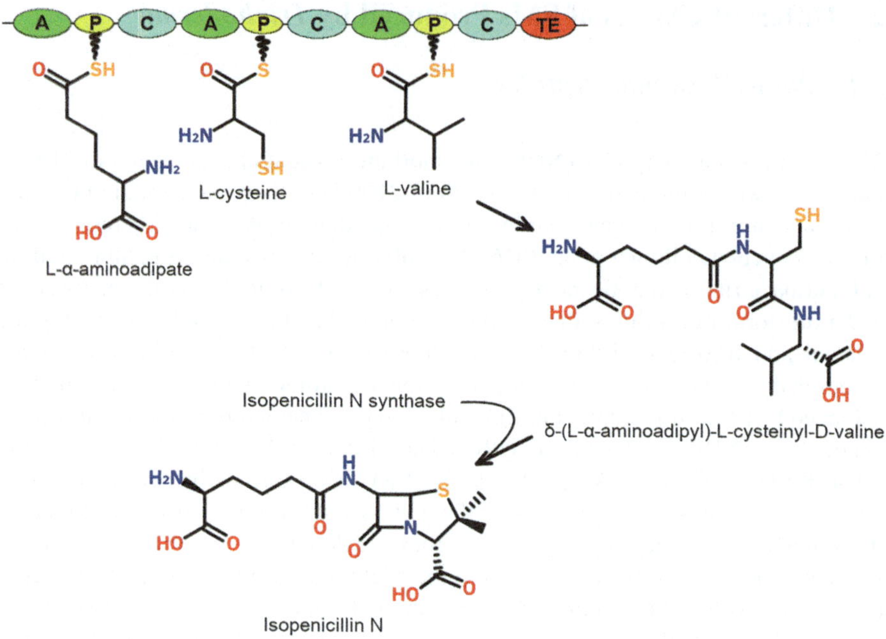

Fig. 1 Isopenicillin N synthesis from the non-ribosomal tripeptide δ-(l-α-aminoadipyl)-l-cysteinyl-d-valine (ACV). The ACV synthetase, from *Penicillium notatum*, is a trimodular non-ribosomal peptide synthetase (NRPS) responsible for the first step in penicillin and cephalosporin biosynthesis. In the NRPS structure are shown the adenylation (A) domain, the pantothenylation/peptidyl carrier (P) domain, and the condensation/peptide bond formation (C) domain. The tripeptide is cyclized by the isopenicillin N synthase to form the β-lactam ring. (Modified from Keller et al. (2005) and complemented with information from Rabe et al. (2018))

lipoamino acid 2-amino-6-hydroxy-4-methyl-8-oxodecanoic acid (AHMO). In the case of peptaibols, the N-terminal is acetylated, and in the case of peptaibiotics, the N-terminal is acylated by a short fatty acid chain; the acyl chain ranges from 8 to 15 carbon atoms (Daniel and Rodrigues Filho 2007; Neuhof et al. 2007; Toniolo et al. 2001; Vinale et al. 2014). Peptaibiotics and peptaibols can be classified upon their length and chemical differences; they are grouped into subfamilies 1, 4, 5, and 9 (Hermosa et al. 2014). Subfamily 1 includes peptaibols with 18, 19, and 20 residues long, the N-terminal is acetylated, and the C-terminal has an amino alcohol. Subfamily 4 has 11- or 14-residue-long peptaibols; they are also acetylated at the N-terminal and present the amino alcohol at the C-terminal. Subfamily 5 is the group of peptaibiotics or lipopeptaibols; these molecules present 7, 10, or 11 residues; the N-terminal is acylated by octanoic, decanoic, or (Z)-dec-4-enoic acid; and the C-terminal keeps the amino alcohol residue (Degenkolb et al. 2006). Finally, subfamily 9 encloses the lipoaminopeptides; they are 10 residues long, have an acyl chain at the N-terminal (MDA, 2-methyldecanoic acid), and present a lipoamino acid in the second position of the peptide chain; the C-terminal is occupied by a reduced form of the dipeptide Ala-Sar-OH (where Sar is sarcosine or N-methylglycine), AMAE (2-[(2′-aminopropyl)-methylamino]-ethanol), or trichodiaminol (Tdol) (Neuhof et al. 2007; Toniolo et al. 2001). This classification and several examples of peptaibols and peptaibiotics are summarized in Table 1, and the general structure of these molecules is shown in Fig. 2.

It has been observed that β-glucan synthase activity, on isolated plasma membranes of *B. cinerea*, was inhibited in vitro by the peptaibols trichorzianin TA IIA and TB IIa (subfamily 1, 19 residues) produced by *T. harzianum*; these molecules can prevent the association of β-glucan synthase with the plasma membrane; this causes disruption in the cell wall formation (Lorito et al. 1996a). These results suggest that peptaibols act synergistically with the cell wall-degrading enzymes from *Trichoderma* to interfere with the growth of fungal pathogens (Schirmböck et al. 1994).

The peptaibol alamethicin (subfamily 1), produced by *T. viride*, is a 20-amino-acid-long peptide that forms voltage-dependent channels across membranes of artificial liposomes, which became permeable to Ca^2, Mn^{2+}, and Ni^{2+}. The leak of these ionic species, by the membrane-permeabilizing activity of alamethicin, leads to the loss of osmotic balance and cell death. Alamethicin is effective against fungi and Gram-positive bacteria but seems ineffective against the Gram-negative ones (Bortolus et al. 2013; Duclohier and Wróblewski 2001; Fonteriz et al. 1991).

The harzianin HCI (subfamily 4), extracted from *T. harzianum*, has antagonistic activity against the bacterium *Staphylococcus aureus* and the fungus *Sclerotium rolfsii*. The alteration in the growth of both pathogens is related to the membrane depolarization due to the pore-forming properties of this peptaibol (Rebuffat et al. 1995). The lipopeptaibols trichogin GA IV and trikoningin (subfamily 5), both purified from *T. longibrachiatum*, present antibacterial activity against *S. aureus* at concentration from 1.5 μg/pit to 3 μg/pit and from 6.2 μg/pit to 11 μg/pit, respectively; however, both molecules are inactive against *Escherichia coli* (Toniolo et al. 2001). Trichopolyn VI (subfamily 9), isolated from *T. brevicompactum*, specifically disrupt the function of the ADP/ATP carrier (AAC) protein from the insects *Tribolium castaneum* and *Acyrthosiphon pisum*. The AAC protein transports the ATP from the mitochondrial matrix to the intermembrane spaces through the mitochondrial inner membrane; thus, ATP can be utilized in the cytosol (Ruprecht et al. 2019). AAC from both insect models were heterologously expressed in a Δ*aac Saccharomyces cerevisiae* (Suga et al. 2015). Trichopolyns have shown immunosuppressive properties with a mode of action different from that of cyclosporin A. On the other hand, the IC_{50} (nM) for proliferation of lymphocytes under allogeneic mixed lymphocyte reaction in mice were trichopolyn I = 5.2, trichopolyn II = 10.7, and cyclosporin A = 7.5 (Toniolo et al. 2001).

Table 1 Representative examples from each peptaibol subfamly

Subfamily	Sequence	Organism	References
Subfamily 1 Peptaibols with 18 to 20 residues in length.	**Trichorzin HA** (18 residues). Ac-Aib-Gly-Ala-Aib--Aib-Gln-Aib-Val-Aib-Gly-Leu-Aib-Pro-Leu-Aib-Aib-Gln-Leu-OH	*T. harzianum*, *T. virens*	Hermosa et al. (2014), Whitmore (2004), and Wiest et al., (2002)
	Tricholongin BI (19 residues). Ac-Aib-Gly-Phe-Aib--Aib-Gln-Aib-Aib-Aib-Ser-Leu-Aib-Pro-Val-Aib-Aib-Gln-Gln-Leu-OH	*T. longibrachiatum*, *T. strigosum*	Hermosa et al. (2014) and Whitmore (2004)
	Trichorzianin TA IIA (19 residues) Ac-Aib-Ala-Ala-Aib--Aib-Gln-Aib-Aib-Aib-Ser-Leu-Aib-Pro-Leu-Aib-Ile-Gln-Gln-Trp-OH	*T. harzianum*	Lorito et al. (1996a)
	Trichorzianin TB IIa (19 residues) Ac-Aib-Ala-Ala-Aib--Aib-Gln-Aib-Aib-Aib-Ser-Leu-Aib-Pro-Leu-Aib-Ile-Gln-Glu-Trp-OH	*T. harzianum*	Lorito et al. (1996a)
	Alamethicin (20 residues) Ac-Aib-Pro-Aib-Ala--Aib-Aib-Gln-Aib-Val-Aib-Gly-Leu-Aib-Pro-Val-Aib-Aib-Glu-Gln-Phe-OH	*T. viride*, *T. hamatum*, *T. brevicompactum*	Bortolus et al. (2013) and Hermosa et al. (2014)

(continued)

Table 1 (continued)

Subfamily	Sequence	Organism	References
Subfamily 4 Peptaibols with 11 or 14 residues in length.	**Harzianin HBI** (11 residues) Ac-Aib-Asn-Ile-Ile--Aib-Pro-Iva-Leu-Aib-Pro-Leu-OH	*T. harzianum*	Hermosa et al. (2014) and Neuhof et al. (2007)
	Harzianin HCI (14 residues) Ac-Aib-Asn-Leu-Aib--Pro-Ser-Val-Aib-Pro-Aib-Leu-Aib-Pro-Leu-OH	*T. harzianum*	Rebuffat et al. (1995)
	Trichovirin I (14 residues) Ac-Aib-Asn-Leu-Aib--Pro-Ser-Val-Aib-Pro-Aib-Leu-Aib-Pro-Leu-OH	*T. viride*	Hermosa et al. (2014)
Subfamily 5 Peptaibiotics or lipopeptaibols, have 7, 10 or 11 residues in length and a lipophilic acyl chain at the N-terminal.	**Trichodecenin I** (7 residues) Fat-Gly-Gly-Leu-Aib--Gly-Ile-Leu-OH	*T. viride*	Toniolo et al. (2001)
	Lipopubescins (10 residues) Fat-Aib-Gly-Val-Aib--Val-Leu-Aib-Gly-Leu-Leu-OH	*T. pubescens*	Hermosa et al. (2014)
	Lipostrigocin (11 residues) Fat-Aib-Gly-Leu-Aib--Gly-Gly-Leu-Aib-Gly-Ile-Leu-OH	*T. strigosum* *T. pubescens*	Hermosa et al. (2014)
	Trichogin GA IV (11 residues) Fat-Aib-Gly-Leu-Aib--Gly-Gly-Leu-Aib-Gly-Ile-Leu-OH	*T. longibrachiatum*	Toniolo et al. (2001)
	Trikoningin KBI (11 residues) Fat-Aib-Gly-Val-Aib--Gly-Gly-Val-Aib-Gly-Ile-Leu-OH	*T. konongii*	Toniolo et al. (2001)

(continued)

Table 1 (continued)

Subfamily	Sequence	Organism	References
Subfamily 9 Lipoaminopeptides have an acyl chain at the N-terminal, a lipoamino acid (AHMOD) near the N-terminal and a reduced form of the dipeptide Ala-Sar-OH (AMAE).	**Trichopolyn I** (10 residues) MDA-Pro-AHMOD-Ala-Aib-Aib-Ile-Ala-Aib-Aib-AMAE	*T. polysporum*	Hermosa et al. (2014), Suga et al. (2015), and Toniolo et al. (2001)
	Trichopolyn VI (10 residues) MDA-Pro-AMOD-Ala-Aib-Aib-Ile-Ala-Aib-Aib-AMAE	*T. brevicompactum*	Suga et al. (2015)

Sequences are given in standard three-letter code; Ac = Acetyl-, Aib = α-Aminobutyric acid, Iva = Isovaline, Fat = octanoic, decanoic or *cis*-dec-4-enoic acid, OH = C-terminal amino alcohol, MDA = 2-methyldecanoic acid, AHMOD = 2-amino-6-hydroxy-4-methyl-8-oxo-decanoic acid, AMOD = 2-amino-4-methyl-8-oxodeca-6-enoic acid, AMAE = 2-[(2′-aminopropyl)-methylamino]-ethanol. Information was taken from Hermosa et al. (2014), complemented with information from (Krause et al. 2006; Neuhof et al. 2007; Suga et al. 2015; Toniolo et al. 2001) and from the Peptaibol Database (http://peptaibol.cryst.bbk.ac.uk/home.shtml) (Whitmore 2004)

2.1.2 Epipolythiodioxopiperazines

Epipolythiodioxopiperazine (ETP) molecules are characterized by a diketopiperazine ring (DKP) derived from a cyclic dipeptide (Fig. 3, panel B); although the first isolated and best characterized ETP is the gliotoxin (Fig. 3, panel C) from *T. virens* (formerly *Gliocladium fimbriatum*), at least other 13 known compounds of this group have been isolated from several fungi of the genera *Leptosphaeria*, *Chaetomium*, or *Hyalodendron*, among others (Gardiner et al. 2005; Zeilinger et al. 2016). These are highly reactive fungal SMs whose toxicity is due to the presence of an internal disulfide bridge, which can bind to susceptible thiol groups in protein residues, inactivating the whole protein. A second toxicity mechanism is the generation of reactive oxygen species by the redox cycling between the reduced (dithiol) and oxidized (disulfide) forms of the DKP (Gardiner et al. 2005). *Trichoderma virens* gliotoxin is synthesized by 8 genes of the *gli* cluster, where 6 genes of the *gli* cluster were found in *T. reesei* genome and 12 genes of the same cluster were found in *Aspergillus fumigatus* (Fig. 3, panel A) (Mukherjee et al. 2012b; Zeilinger et al. 2016).

The mechanism of action of gliotoxin over fungal pathogens has not been studied in depth. Gliotoxin was described as an antibiotic causing cytoplasmic leakage in *Rhizoctonia solani* hyphae (Lewis et al. 1991) and inhibiting the germination of sporangia and mycelial growth of *Pythium ultimum* (Roberts 1990). Synergistic interaction between gliotoxin and an endochitinase from *T. virens* inhibited spore germination of *B. cinerea* (Lorito et al. 1994). The biosynthesis of gliotoxin (Fig. 4) starts with the formation of the DKP scaffold by the NRPS GliP. This multimodular enzyme catalyzes the condensation of phenylalanine and serine to the Phe-Ser dipeptide, followed by an intramolecular cyclization; deletion of the *gliP* gene results in abrogation of gliotoxin biosynthesis. The next step is the di-hydroxylation

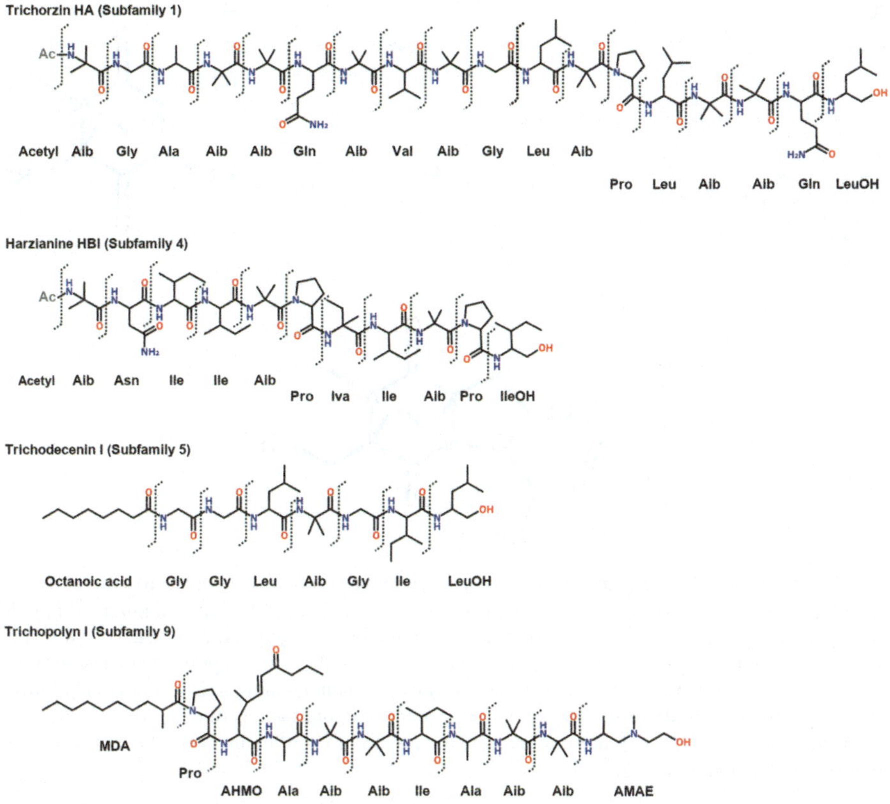

Fig. 2 Peptaibols' and peptaibiotics' general structure. Sequences are given in standard three-letter code; Ac = acetyl-, Aib = α-aminobutyric acid, Iva = isovaline, OH=C-terminal amino alcohol, MDA = 2-methyldecanoic acid, AHMO = 2-amino-6-hydroxy-4-methyl-8-oxo-decanoic acid, AMAE = 2-[(2′-aminopropyl)-methylamino]-ethanol. (Information taken from Hermosa et al. (2014))

of the carbon atoms of the DKP scaffold by the cytochrome P450 monooxygenases GliC and GliF. Next, the introduction of sulfur in the gliotoxin precursor is mediated by the glutathione-S-transferase GliG. This reaction couples glutathione to the DKP backbone. The glutamic acid is removed from the glutathione by the γ-glutamate cyclotransferase GliK, leaving a Gly-Cys dipeptide; afterward, the glycine residue is eliminated by the specialized dipeptidase GliJ. Once the glycine residue has been removed, the C-S bond of the remaining cysteine and the DKP scaffold is cleaved by the pyridoxal 5′-phosphate-dependent C-S bond lyse GliI. As mentioned previously, the toxic effect of gliotoxin is due to the internal disulfide bridge; the formation of this motif is catalyzed by the thioredoxin reductase, GliT. Finally, after a series of not described oxidation reactions, the N-methyltransferase GliN or the O-methyltransferase GliM renders the active form of the gliotoxin. The secretion of gliotoxin to the extracellular media depends on the transporter GliA (Wang

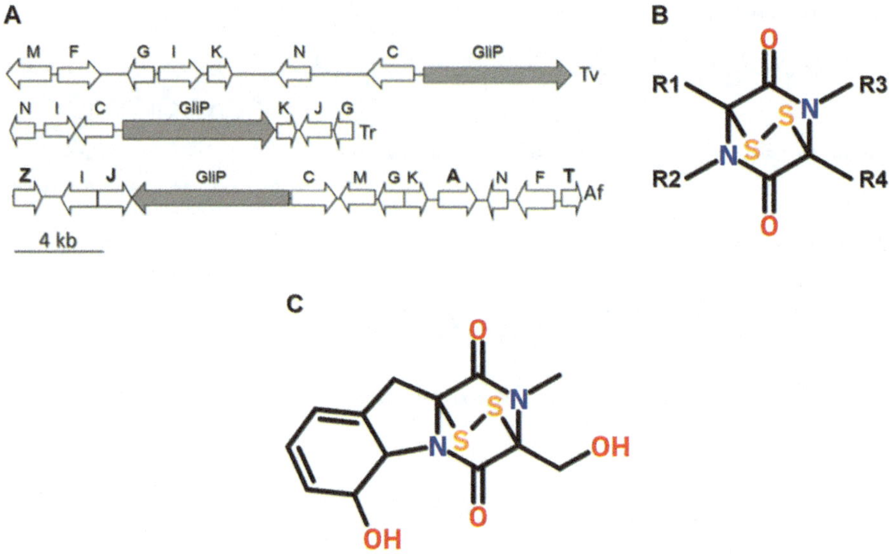

Fig. 3 Biosynthetic cluster and fundamental structures in gliotoxin molecule. (**a**) Gliotoxin biosynthesis cluster in *T. virens* (Tv), *T. reesei* (Tr), and *A. fumigatus* (Af); genes highlighted in bold in *A. fumigatus gli* cluster are missing or not yet identified in *T. virens* and *T. reesei*. The GliP coding gene is pointed in gray in every cluster; Glip is the non-ribosomal peptide synthetase responsible for the formation of the diketopiperazine ring by phenylalanine and serine condensation. (**b**) Core structure of the diketopiperazine ring with the internal disulfide bridge. R1/R2/R3/R4 = any atom or group. (**c**) Gliotoxin molecular structure. (Modified from Gardiner et al. (2005) and Mukherjee et al. (2012b))

et al. 2014). The genes coding for the GliA, GliZ, GliJ, and GliT proteins are not present in the *gli* cluster of *T. virens* or *T. reesei* (Fig. 3, panel A) (Fox and Howlett 2008; Gardiner et al. 2005; Scharf et al. 2016).

2.1.3 Siderophores

Siderophores are low-molecular-weight SMs with Fe^{3+} chelating activity. These molecules are involved in extracellular iron acquisition and intracellular protection from oxidative stress (Mukherjee et al. 2012b; Zeilinger et al. 2016). Fungal siderophores belong to the hydroxamate group that includes the fusarinines, coprogens, ferrichromes, and rhodotorulic acid. Siderophores of the hydroxamate group are built by Nδ-acyl-Nδ-hydroxyornithine structural units, produced by the acylation of the non-proteinogenic amino acid hydroxy-L-ornithine (derived by hydroxylation of L-ornithine). The acylation can be done with acetyl or more complex groups such as anhydromevalonyl (Renshaw et al. 2002). To increase affinity for Fe^{3+}, most fungal siderophores include three of these building units linked by ester or peptide bonds to form hexadentate structures coordinating the metal ion. Cyclization of the siderophore, as observed in ferrichromes and some fusarinines, improves the

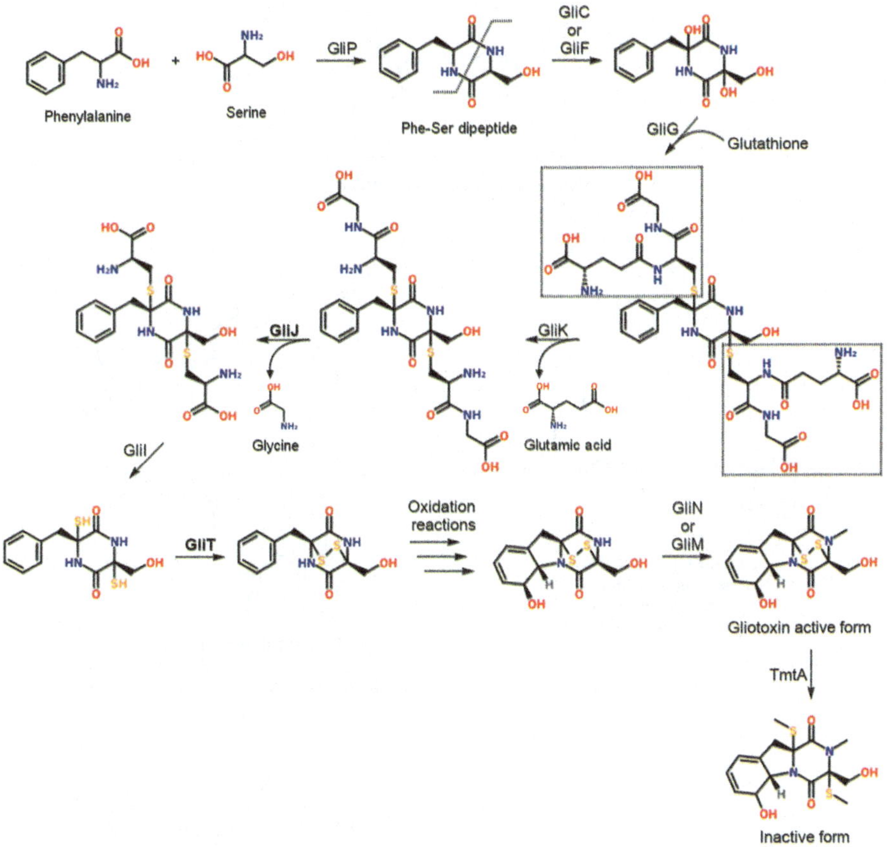

Fig. 4 Gliotoxin biosynthesis pathway. In *A. fumigatus* and *T. virens*, gliotoxin biosynthesis starts with phenylalanine and serine. The coding genes for GliT and GliJ (in bold) are missing from *T. virens gli* cluster. Pointed squares address for glutathione addition. Methylation of the disulfide bonds in gliotoxin by the specialized S-methyltransferase TmtA eliminates the toxic effects of the molecule. (Modified from Scharf et al. (2016) and complemented with information from Fox and Howlett (2008) and Gardiner et al. (2005))

chemical stability (Fig. 5) (Haas 2014; Zeilinger et al. 2016). Nine strains of the genus *Trichoderma*, including *T. harzianum*, *T. hamatum*, *T. viride*, *T. koningii*, *T. longibrachiatum*, and *T. pseudokoningii*, produce fusarinine B, fusarinine C (fusigen), coprogen, coprogen B, and ferricrocin (ferrichrome type) (Anke et al. 1991; Lehner et al. 2013; Renshaw et al. 2002; Vinale et al. 2014).

It has been documented that harzianic acid, extracted from *T. harzianum*, presents iron chelating activity (Mo et al. 2014; Vinale et al. 2013). This molecule is derived from the tetramic acid (pyrrolidine-2,4-dione) (Fig. 6), which arises from the assembly of an amino acid and an activated acyl entity. This biosynthesis is performed by mixed polyketide synthase and non-ribosomal peptide synthetase pathways (Mo et al. 2014; Schobert 2007). In the case of harzianic acid, the amino

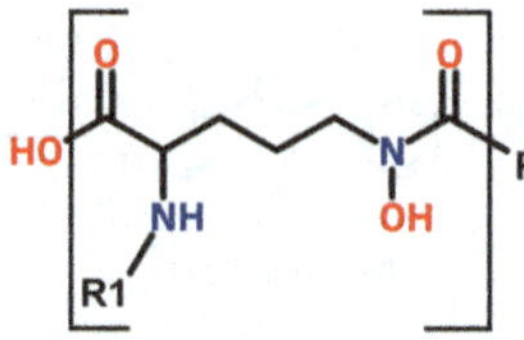

Nδ-acyl-Nδ-hydroxyornithine

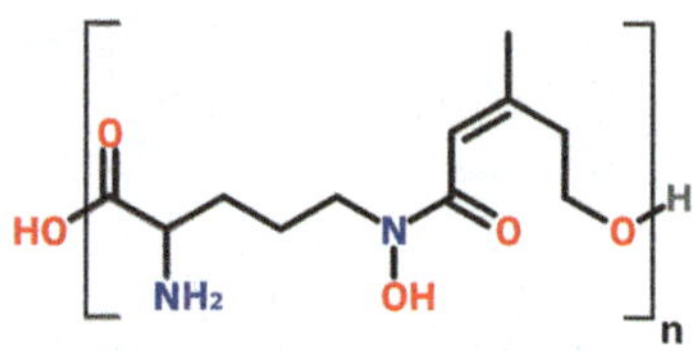

Linear Fusarinine
n=1 cis-fusarinine
n=2 fusarinine A
n=3 fusarinine B

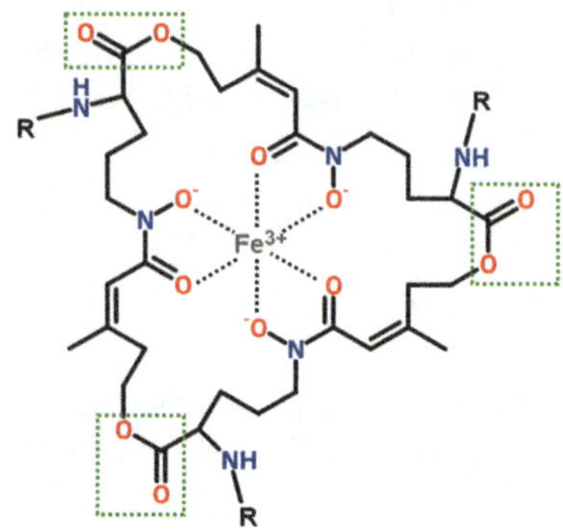

Cyclic Fusarinine (n=3)
R=H fusarinine C (fusigen)
R=Acetyl N,N',N"-triacetylfusarinine C (triacetylfusigen)

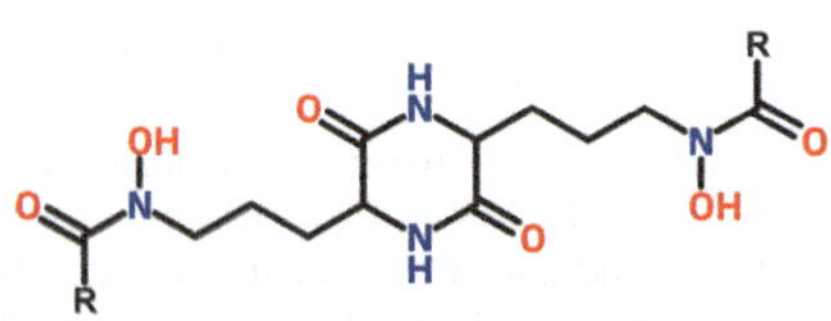

R=CH_3 Rhodoturulic acid
R=acyl Dimerum acid

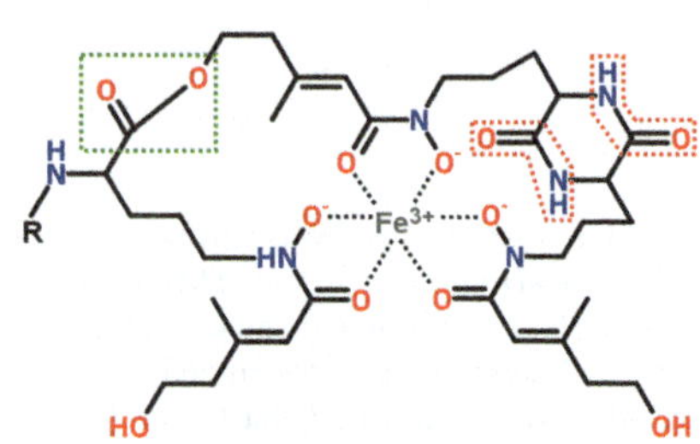

R=Acetyl Coprogen
R=H Coprogen B

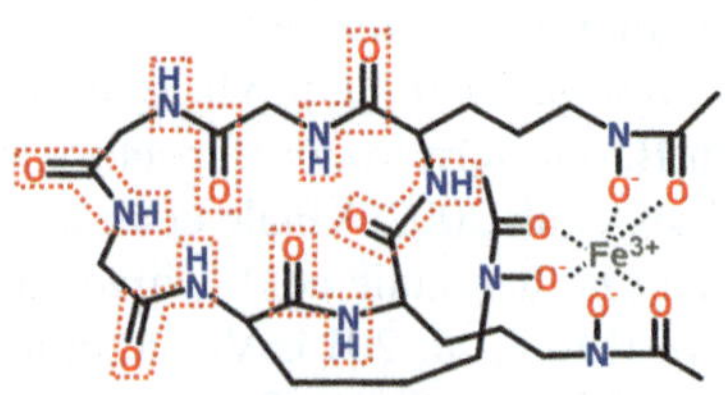

Ferrichome

Fig. 5 Molecular structures of the siderophores present in *Trichoderma* species. Siderophores are made from Nδ-acyl-Nδ-hydroxyornithine structural units. Fusarinines can be monomers, linear dimers or trimers, or cyclic trimers. In the coprogen molecules, dimers of the Nδ-acyl-Nδ

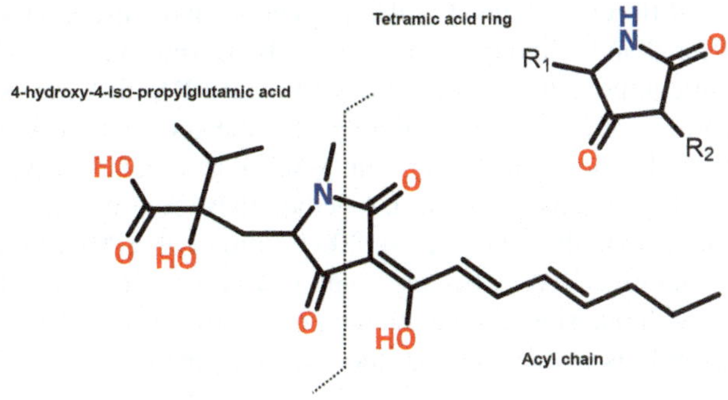

Fig. 6 Harzianic acid molecular structure. Harzianic acid is formed by the condensation of 4-hydroxy-4-iso-propylglutamic acid and an acyl chain; this union assembles the tetramic acid ring (pyrrolidine-2,4-dione), shown in the subfigure panel. (Modified from Healy et al. (2015) and Mo et al. (2014))

acid unit in the tetramic acid core is a modified 4,4-disubstituted form of glutamic acid, the 4-hydroxy-4-*iso*-propylglutamic acid (Healy et al. 2015).

According to studies with a *T. virens* mutant strain, the elimination of the non-ribosomal peptide synthetase gene *tex10* (involved in intracellular siderophore ferricrocin biosynthesis) causes enhanced growth rate, reduced conidiation, and hypersensitivity to oxidative stress as compared to the wild-type strain. In addition, this mutant showed reduced levels of gliotoxin and dimethyl gliotoxin but enhanced ability to colonize maize seedling roots. The mutant strain was also impaired in the activation of the induced systemic resistance (ISR) in maize against the foliar pathogen *Cochliobolus heterostrophus* (Mukherjee et al. 2018).

2.2 Polyketides

Polyketides are the most abundant fungal SMs described so far. The best-characterized polyketides include the yellow *Aspergillus nidulans* spore pigment intermediate naphthopyrone (WA), the carcinogen aflatoxin, and the commercially important cholesterol-lowering compound lovastatin; fungal polyketides are synthesized by type I polyketide synthases (PKSs), a group of multidomain proteins related to eukaryotic fatty acid synthases (Keller et al. 2005). In the PKS synthesis reaction, short-chain carboxylic acids, usually acetyl-coenzyme A (acetyl-CoA), propionyl-CoA, or methylmalonyl-CoA, are the building blocks condensed to form

Fig. 5 (continued) hydroxyornithine units are joined head to head forming a diketopiperazine ring (highlighted in red pointed lines). Finally, the ferrichrome molecules are cyclic hexapeptides consisting of three Nδ-acyl-Nδ-hydroxyornithines and three additional amino acids; two of these residues can be alanine, serine, or glycine, and the third residue is always a glycine. Peptide bonds, ester bonds, and Fe^{3+} are shown in red pointed lines, green pointed lines, and gray letters, respectively. (Modified from Haas (2014) and complemented with information from Renshaw et al. (2002))

carbon chains of different lengths. For the polyketide chain, the full reduction of the β-carbon is optional and mandatory for fatty acids (Hermosa et al. 2014).

The essential components for a functional fungal PKS, that can be observed in the naphthopyrone synthase from *A. nidulans*, are the ketoacyl-CoA synthase (KS), acyltransferase (AT), acyl carrier protein (ACP), and Claisen-type cyclization (CYC) domains (Fig. 7, superior panel); together, these four minimal domains conform an elongation module (Cheng et al. 2009; Fujii et al. 2001). Fungal PKSs are considered as "iterative PKSs" because they can carry out repeated biosynthetic reaction rounds with one single elongation module – this is how a molecule as complex as a hexaketide is constructed. In the first step, the carboxylic acid from the

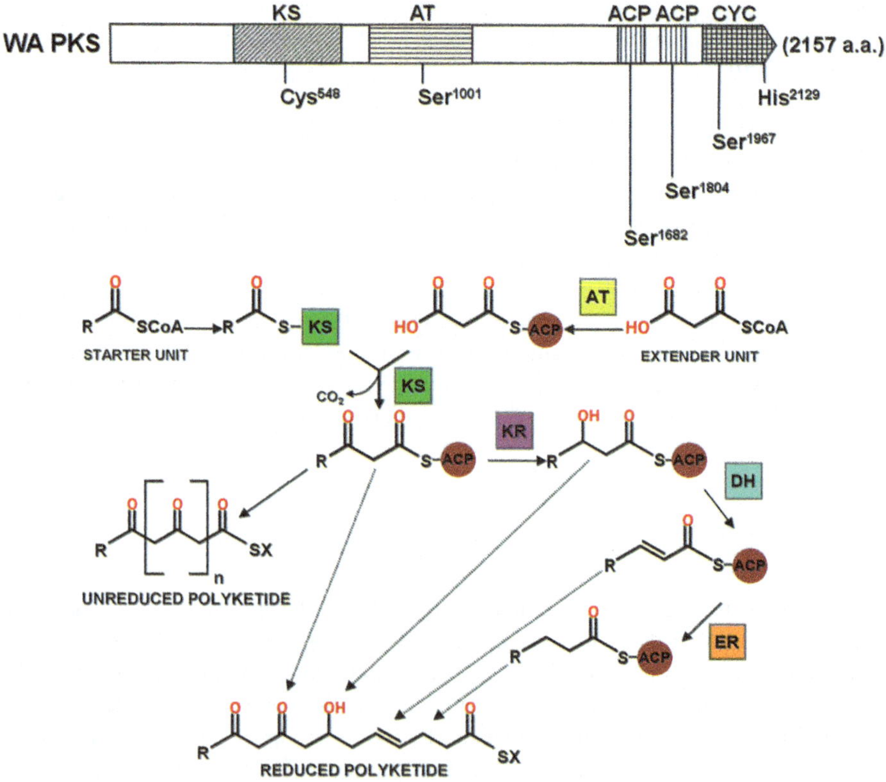

Fig. 7 Domain organization of prototypical polyketide synthase and polyketide biosynthesis reactions. Superior panel: Protein architecture of naphthopyrone synthase from *A. nidulans* (WA), with the reactive amino acids at the catalytic sites: Claisen-type cyclization domain (CYC). Inferior panel: Polyketide biosynthesis path. Polyketide chain is synthesized by the basic PKS domains: the ketoacyl-CoA synthase (KS), the acyltransferase (AT), and the acyl carrier protein (ACP); this reaction produces an unreduced polymer. After chain elongation, the nascent polyketide chain can be the target of modifying enzymes: the ketoreductase (KR), the dehydratase (DH), and the enoyl reductase (ER). This image represents the function of each basic domain in the PKS and the functions of the modifying enzymes. (Modified from Fujii et al. (2001) and Gokhale et al. (2007))

acyl-CoA is loaded into the KS domain; this first acyl chain is the starter unit. In the next step, a second carboxylic acid from an acyl-CoA is loaded into the ACP domain by the AT domain; this second acyl chain is the extender unit. In the third step, the KS domain catalyzes the chain elongation involving a decarboxylative condensation reaction. The new elongated chain can be loaded again in the KS domain for a next round of elongation by the addition of extender units. Notice that the elongated chain can be subjected to modifications by accessory domains like ketoreductase (KR), dehydratase (DH), and enoyl reductase (ER). Finally, the acyl chain is cyclized and released from the PKS by the C-terminal region of the enzyme; this domain is responsible for a Claisen-type cyclization (CYC) reaction that renders the formation of aromatic polyketides (Fig. 7, inferior panel). How the number of cycles of condensation is regulated to stop the elongation is not understood for this or for any other fungal type I PKS enzyme. The wide diversity of fungal polyketide results from several factors: the number of iteration cycles, the number of reduction reactions, the type of extender unit, and, in the case of aromatic polyketides, the cyclizations of the nascent polyketide chain. Further variety is achieved by the many different post-polyketide synthesis modifications (Fujii et al. 2001; Gokhale et al. 2007; Keller et al. 2005; Khosla 2009; Khosla et al. 2014). Genome sequence analysis in *T. virens*, *T. atroviride*, and *T. reesei* reveals the presence of 18, 18, and 11 PKS-coding genes, respectively (Kubicek et al. 2011).

The anthraquinones pachybasin, chrysophanol, and emodin, isolated from *T. viride*, are polyketides mainly related to pigmentation, but it was also observed that these compounds decrease the linear growth rate of the fungal model *Fomes annosus*, which possesses both monoamine oxidase and tyrosine kinase inhibiting activities. They act as antimicrobial, antineoplastic, and cathartic agents and exhibit a remarkable bacteriostatic effect on Gram-positive bacteria, especially toward *S. aureus*. Trichodermaol, another anthraquinone from *T. viride*, exhibits antibacterial activity against *B. subtilis* and *S. aureus* at 50 μg/ml (Fig. 8) (Reino et al. 2008).

The compounds koninginins and koningiopisins isolated from *T. koningii* and *T. koningiopsis* are polyketides with antibiotic activity, e.g., koninginin D affected the growth of the plant pathogens *Gaeumannomyces graminis*, *Rhizoctonia solani*, *Phytophthora cinnamomi*, *Pythium middletonii*, *Fusarium oxysporum*, and *Bipolaris sorokiniana* (Reino et al. 2008). Koningiopisin C showed antimicrobial activities

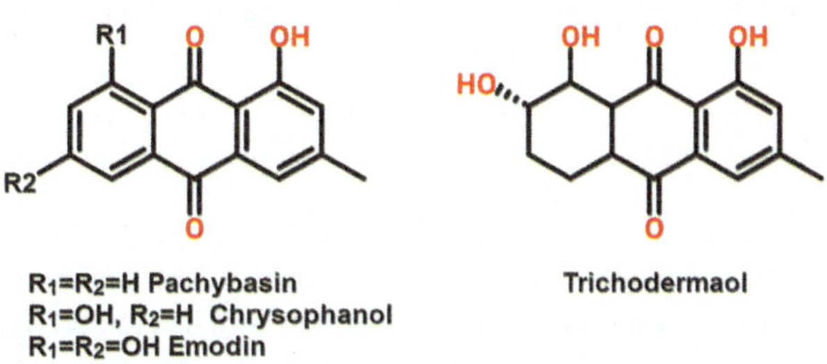

Fig. 8 Aromatic polyketide structure. Anthraquinones isolated from *T. virens*. (Modified from Reino et al. (2008))

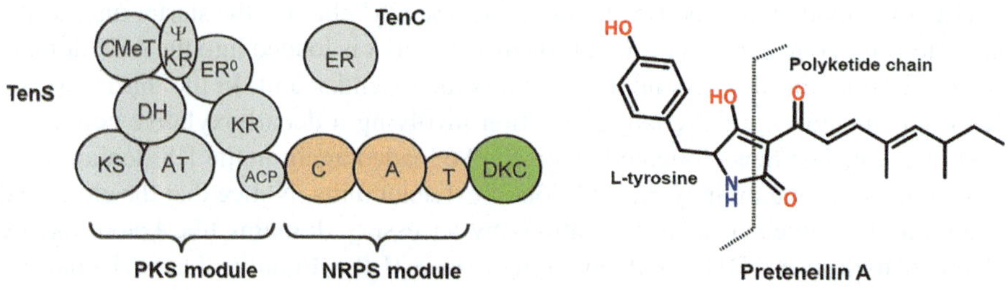

Fig. 9 Domain architecture and biosynthesis product of TenS. The protein TenS from *B. bassiana* is an example of a fungal iterative hybrid PKS-NRPS. The tetramic ring in pretenellin A is formed by the condensation of L-tyrosine (shown in orange) and a polyketide chain. Domains of the PKS module are shown in gray as KS (ketosynthase), AT (acyltransferase), DH (dehydratase), CmeT (C-methyltransferase), ΨKR (a structural domain of the ketoreductase), ER (enoylreductase), KR (ketoreductase), and ACP (acyl carrier protein); the domains of the NRPS module are shown in orange as C (condensation), A (adenylation), and T (thiolation, also known as PCP (peptidyl carrier protein)). The additional DKC (Dieckmann cyclase) domain, required to release the natural product from the enzyme, is shown in green. (Modified from Fisch (2013))

against *S. aureus*, *F. oxysporum*, *A. panax*, *F. solani*, and *Plectosphaerella cucumerina* with MICs at 64, 32, 64, 32, and 16 μg/ml, respectively (Liu et al. 2016). A study using a *pks4* mutant strain from *T. reesei*, shown that this PKS is not only responsible for the green conidial pigmentation, due to the loss of Pks4 resulted in pigmentation alteration of teleomorph structures, a decrease of the conidial cell wall stability and the antagonistic abilities of *T. reesei* against other fungi, possibly by the reduction in the formation of inhibitory SMs (Atanasova et al. 2013a).

Further domain analysis in *T. virens*, *T. atroviride*, and *T. reesei* showed the presence of PKS-NRPS hybrid enzymes in the genomes (Mukherjee et al. 2012b). The PKS-NRPS hybrid enzymes are proteins that consist of a PKS module fused alongside with a NRPS module. The function of these enzymes is the synthesis of a different kind of polyketides incorporating an amino acid residue. The *Beauveria bassiana* protein TenS condenses the pigment precursor pretenellin A from L-tyrosine and a polyketide chain by forming a tetramic acid ring (Fisch 2013) (Fig. 9). The PKS-NRPS Tex13 from *T. virens* participates in induction of the defense-related gene *pal1* of the maize plant; the *T. virens* mutant strain for *tex13* fails to upregulate the *pal1* gene expression in the plant (Mukherjee et al. 2012a).

2.3 Terpenes

Fungal terpenes are molecules composed of several C5 isoprene units as building blocks. These units are the isopentenyl diphosphate and its isomer, the dimethylallyl diphosphate. These compounds can be linear or cyclic and saturated or unsaturated and also are the target of several modifications; some examples of this group of molecules are the carotenoids, gibberellins, indole-diterpenes, and trichothecenes

(Keller et al. 2005). Isopentenyl diphosphate (IPP) and dimethylallyl diphosphate (DMAPP) are precursors derived from the acetyl-CoA through the mevalonate pathway; the first enzyme in the mevalonate biosynthesis pathway is the hydroxymethylglutaryl-coenzyme A reductase (HMGR) (Cardoza et al. 2007). In the first step of terpene biosynthesis, a molecule of IPP is transformed to DMAPP by an isomerase; this DMAPP molecule is the starter unit. In the second step, DMAPP is fused to a second molecule of IPP by an isoprenyl diphosphate synthase (IDS); this second IPP molecule is the extender unit. The successive head-to-tail 1′-4 addition of IPP units forms a set of linear polyprenyl diphosphates, which are intermediates of many primary and secondary metabolism molecules. These linear polyprenyl diphosphates, the geranyl diphosphate (GPP, C10), farnesyl diphosphate (FPP, C15), and geranylgeranyl diphosphate (GGPP, C20) (Fig. 10), function as a branching point from which the monoterpenes (from GPP), sesquiterpenes (from FPP), and diterpenes (from GGPP) diverge in structure through the action of the terpenoid cyclases and other modification enzymes; longer chains (C30 triterpenes and C40 tetraterpenes) are formed by a 1′-1 head-to-head condensation of two FPP units or two GGPP units (Keller et al. 2005; Leeper and Vederas 2000; Schmidt-Dannert 2015).

The diterpenes koninginols A and B, obtained from *T. koningiopsis* A729, exhibited a significant antibacterial activity against *B. subtilis* with MIC values of 10 and 2 μg/mL, respectively (Chen et al. 2019). In *T. harzianum* and *T. erinaceum*, the diterpene harziandione has been described; although its function is not clear, it has been addressed that harziandione has no cytotoxic activity against various cancer cell lines (Ghisalberti et al. 1992; Xie et al. 2013).

Trichothecenes constitute a group of sesquiterpenes with mycotoxin activity, which is present in several members of the order *Hypocreales*. Trichothecenes have the ability to inhibit protein synthesis (Bennett and Klich 2003) and/or to induce apoptosis in eukaryotic cells and also can act as immunosuppressors, neurotoxins, and phytotoxic agents. In *T. arundinaceum* and *T. brevicompactum*, a trichothecene biosynthetic gene (TRI) cluster has been identified. This cluster encodes transport, regulatory, and synthesis enzymes required for the formation of the mycotoxins trichodermin and harzianum A; the former displays antifungal and phytotoxic activity, while the last presents growth inhibition of phytopathogenic fungi (Cardoza et al. 2011; Malmierca et al. 2012; Tijerino et al. 2011). Lignoren, a sesquiterpene from *T. lignorum*, shows moderate antimicrobial activities; during a diffusion assay, 100 μg per agar well (9 mm diameter) caused 25 mm diameter of inhibition zone of *B. subtilis* ATCC 6633, 22 mm of *Mycobacterium smegmatis* SG 987, and 16 mm of *P. aeruginosa* K 599/WT, and the same amount of lignoren resulted in 18 mm and 13 mm diameter of inhibition zone of the yeasts *Sporobolomyces salmonicolor* SBUG 549 and *Rhodotorula rubra* IMET 25030, respectively. No activity was found against *C. albicans*, *Penicillium notatum* JP36, and *Fusarium culmorum* JP15 (Berg et al. 2004). Another example of a sesquiterpene with antifungal activity is the daucane sesquiterpene, 3,4-dihydroxycarotane, isolated from *T. virens* and *T. viride*. This metabolite belongs to a rare group of compounds, which, to a large extent, are

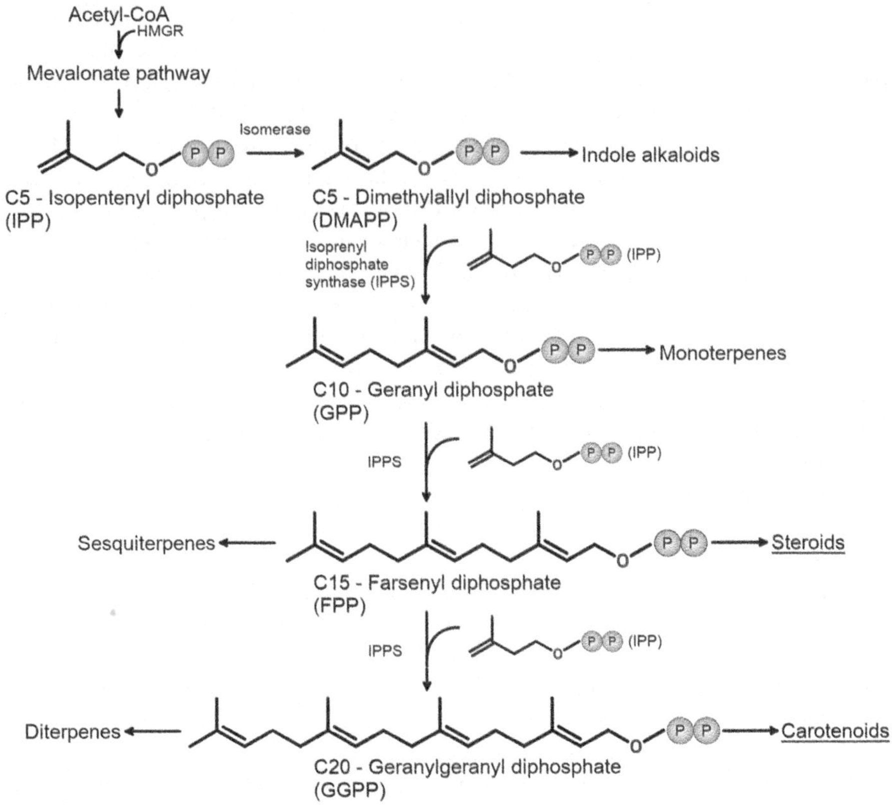

Fig. 10 Common terpene biosynthesis path. In this metabolic route, the isopentenyl diphosphate (IPP) and the dimethylallyl diphosphate (DMAPP), precursors derived from the mevalonate pathway led by the enzyme hydroxymethylglutaryl-coenzyme A reductase (HMGR), are the building blocks for the synthesis of several linear polyprenyl diphosphates. Geranyl diphosphate (GPP, C10), farnesyl diphosphate (FPP, C15), and geranylgeranyl diphosphate (GGPP, C20) are the key intermediates for monoterpene, sesquiterpene, and diterpene biosynthesis, respectively. Underlined molecules belong to primary metabolism. (Modified from Keller et al. (2005) and complemented with information from Cardoza et al. (2007), Schmidt-Dannert (2015), and Tudzynski et al. (1999))

characteristics of the Umbelliferae family of plants (Cardoza et al. 2005; Reino et al. 2008). The triterpenes (C30) ergokinins A and B, from *T. longibrachiatum*, are steroid antibiotics with evident antifungal activity against the genus *Candida* and *Aspergillus*. Especially the ergokonin A, a sulfate carboxysteroid antibiotic, induced alterations in the hyphal morphology of *A. fumigatus* and has been found to inhibit the (1,3)-β-D-glucan synthase (Cardoza et al. 2005; Vicente et al. 2001). Chemical structures of the terpenes described here are shown in Fig. 11.

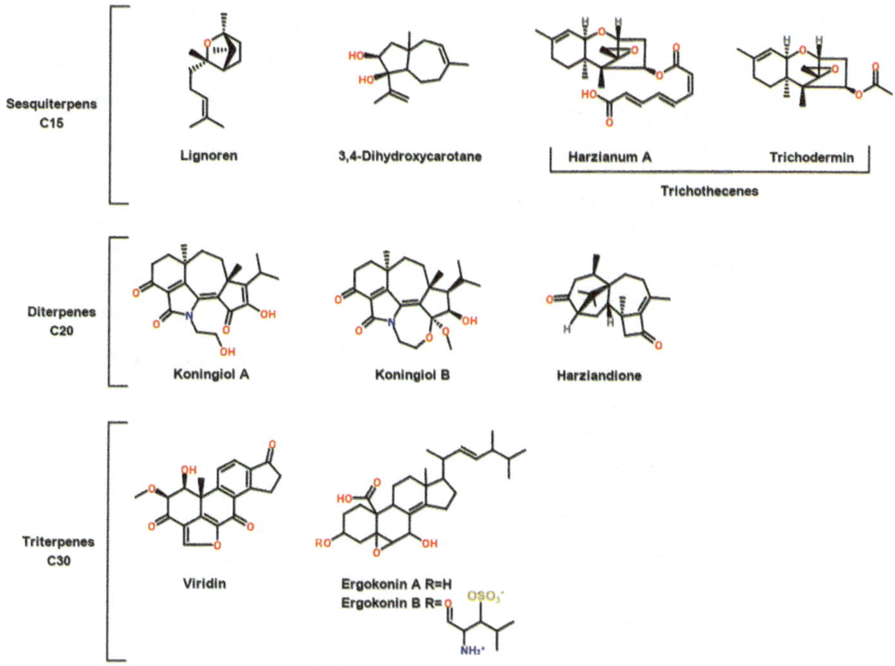

Fig. 11 Molecular structure of representative terpenes from genus *Trichoderma*; these compounds show antimicrobial activity. (Modified from Hermosa et al. (2014) and complemented with information from Chen et al. (2019))

2.4 Pyrones

Pyrones belong to a chemically diverse group of low-molecular-weight metabolites classified as volatile organic compounds (VOC); this is due to their high vapor pressure at room temperature and low water solubility; this compound is responsible for the coconut aroma associated with *Trichoderma* strains (Reino et al. 2008; Sivasithamparam and Ghisalberti 1998; Zeilinger et al. 2016). The 6-pentyl-2H-pyran-2-one also known as 6-pentyl-α-pyrone (6-PP) (Fig. 12) is one of the first VOC isolated from the *Trichoderma* genus, specifically *T. viride*, but it can be also found in other species. In *T. harzianum*, the 6-PP and its closely related analogue 6-pentenyl-2H-pyran-2-one (Fig. 12, right side) can inhibit, in a significant way, the growth of *R. solani* (Claydon et al. 1987; Dennis and Webster 1971b). In other work with *T. koningii*, the 6-PP inhibited the growth of *Gaeumannomyces graminis*, *R. solani*, *Phytophthora cinnamomi*, *Pythium middletonii*, *F. oxysporum*, and *Bipolaris sorokiniana*. The possible mechanism of action includes the disruption of the plasma membrane and the cell wall; also 6-PP is involved in mitochondrial deterioration explained by respiratory inhibition (Ismaiel and Ali 2017; Simon et al. 1988). Cytosporone S (Fig. 12, right side) was isolated from a fermentation broth of *Trichoderma* sp. FKI-6626. Its chemical structure was determined primarily by

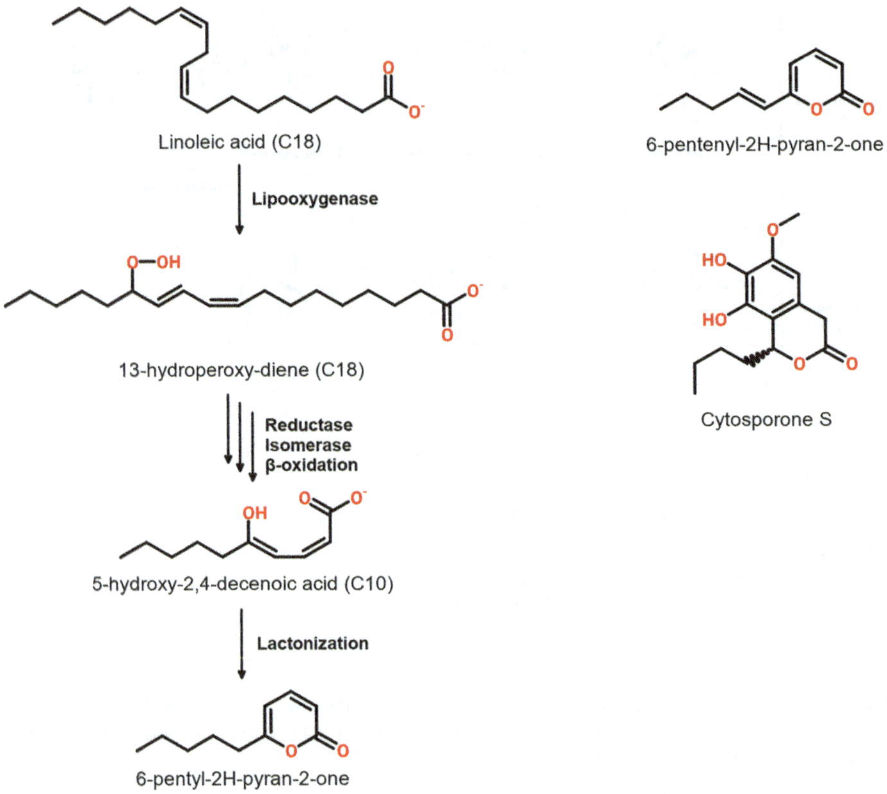

Fig. 12 Hypothetical pathway for 6-pentyl-2H-pyran-2-one (6-PP) biosynthesis. (**a**) This path is based on the expression profile of a lipoxygenase gene (ID 33350) from *T. atroviride*, observed when this strain is interacting directly with the phytopathogen *R. solani*. (**b**) Molecular structure of the analogue 6-pentenyl-2H-pyran-2-one isolated from *T. harzianum* and the cytosporone S isolated from *Trichoderma* sp. FKI-6626. (Modified from Serrano-Carreon et al. (1993) and Ishii et al. (2013); complemented with information from Claydon et al. (1987))

NMR spectroscopy and mass spectrometry; this compound showed antimicrobial activity against several Gram-positive and Gram-negative bacteria and fungi (Ishii et al. 2013).

The biosynthesis pathway for 6-PP is not clear; some authors consider pyrones are derived from fatty acid metabolism, directly from linoleic acid as precursor. In this hypothetical pathway, the first step would be the oxidation of linoleic acid to a 13-hydroperoxy-diene form by a lipoxygenase reaction. Next, this compound is subjected to several rounds of β-oxidation and isomerization to form the intermediate 5-hydroxy-2,4-decenoic acid; the final esterification results in 6-PP (Serrano-Carreon et al. 1993) (Fig. 12, left side). Although the exact enzymes and the respective coding genes for 6-PP biosynthesis in *Trichoderma* remain unclear, the hypothesis of the fatty acid pathway is supported by a comparative transcriptomic approach, where the expression of a lipoxygenase gene (ID 33350) from *T.*

atroviride is upregulated when this strain is interacting directly with *R. solani* (Atanasova et al. 2013b; Kubicek et al. 2011).

On the other side, the hypothesis of a 6-PP derived from mevalonate and the terpene pathway as a monoterpene (C10) is supported by the function of the G protein α subunit Tga1 in *T. atroviride*. In the Δ*tga1* mutant strain, the production of 6-PP and of metabolites with sesquiterpene structure was significantly reduced. Also *tga1* gene deletion resulted in a complete loss of mycoparasitic overgrowth and lysis of *R. solani*, *B. cinerea*, and *Sclerotinia sclerotiorum* during direct confrontation, although infection structure formation was unaffected. At the same time, other low-molecular-weight antifungal metabolites were also overproduced in this mutant strain (Reithner et al. 2005). The transcription factor THCTF1 of *T. harzianum* is related to 6-PP biosynthesis; the *Thctf1* null mutant did not produce two secondary metabolites derived from 6-PP and shows loss of pigmentation and a reduced antimicrobial capacity (Hermosa et al. 2014; Rubio et al. 2009).

3 Conclusions and Perspectives

For more than 30 years, hundreds of SMs produced by beneficial fungi have been isolated and characterized. Here we focused on those compounds produced by the *Trichoderma* genus, which have been involved in the interactions with phytopathogenic agents, showing a positive outcome for agriculture and even medicine (Lee et al. 2005). The isolation and characterization of these molecules is the first step in the way to achieve the full potential of these biocontrol agents. It has been documented the application of harzianic acid and 6-pentyl-α-pyrone, isolated from *T. harzianum* M10 or *T. atroviride* P1, directly over *Vitis vinifera* leaves by foliar spray or drenching in field experiments, with remarkable outcomes. The results of the assay showed that the purified molecules were able to reduce significantly the impact of the powdery mildew disease caused by *Uncinula necator*; the effects of the isolated SMs were comparable with those obtained by using the living *Trichoderma* strains (Pascale et al. 2017). In a similar way, 6-pentyl-α-pyrone used as treatment was successful in significantly reducing the incidence of kiwifruit storage rots by *B. cinerea*; this in both inoculated and naturally infected fruit (Poole et al. 1998). These practical examples show the usefulness SMs have. The complete understanding of these molecules requires detailed information of the biosynthesis pathways with precursors and their synthetic enzymes, the corresponding coding genes, and the regulation mechanisms. However, a lot of work is still needed to reach a complete understanding of the SMs and their functions, and it is clear that this understanding can lead not only to discover new molecules but it opens the possibility to the design of these very same molecules in a different way to chemical synthesis. The PKSs and NRPSs are modular proteins, and the principles of their operations are not fully understood. The manipulation of these principles can lead to the biological synthesis of new polyketides and peptaibols not produced in natural conditions.

Synthesis of biologically active molecules by fungi is the result of natural selection, growth conditions, precursor's availability, or the nature of the interaction with plant or other fungi; all of these variables lead to a process with almost infinite possibilities. The development of these findings represents an opportunity to overcome the world's antibiotic and pesticide crisis and the beginning of new therapeutic procedures.

Acknowledgments *This work was supported by a CONACYT grant from the National Postdoctoral Program (application 2019-000019-01NACV-00037) to Yael González.*

References

Anke H, Kinn J, Bergquist K-E, Sterner O (1991) Production of siderophores by strains of the genus *Trichoderma*: isolation and characterization of the new lipophilic coprogen derivative, palmitoylcoprogen. Biol Met 4:176–180. https://doi.org/10.1007/BF01141311

Arora DK, Bridge PD, Bhatnagar D (eds) (2004) Handbook of fungal biotechnology, 2nd ed., rev. expanded. ed, Mycology series. Marcel Dekker, New York

Atanasova L, Knox BP, Kubicek CP, Druzhinina IS, Baker SE (2013a) The polyketide synthase gene *pks4* of *Trichoderma reesei* provides pigmentation and stress resistance. Eukaryot Cell 12:1499–1508. https://doi.org/10.1128/EC.00103-13

Atanasova L, Le Crom S, Gruber S, Coulpier F, Seidl-Seiboth V, Kubicek CP, Druzhinina IS (2013b) Comparative transcriptomics reveals different strategies of *Trichoderma* mycoparasitism. BMC Genomics 14:121. https://doi.org/10.1186/1471-2164-14-121

Benítez T, Rincón AM, Limón MC, Codón AC (2004) Biocontrol mechanisms of *Trichoderma* strains. Int Microbiol 7:249–260

Bennett JW, Klich M (2003) Mycotoxins. Clin Microbiol Rev 16:497–516. https://doi.org/10.1128/cmr.16.3.497-516.2003

Berg A, Wangun HVK, Nkengfack AE, Schlegel B (2004) Lignoren, a new sesquiterpenoid metabolite from *Trichoderma lignorum* HKI 0257. J Basic Microbiol 44:317–319. https://doi.org/10.1002/jobm.200410383

Blin K, Shaw S, Steinke K, Villebro R, Ziemert N, Lee SY, Medema MH, Weber T (2019) antiSMASH 5.0: updates to the secondary metabolite genome mining pipeline. Nucleic Acids Res 47:W81–W87. https://doi.org/10.1093/nar/gkz310

Bortolus M, De Zotti M, Formaggio F, Maniero AL (2013) Alamethicin in bicelles: orientation, aggregation, and bilayer modification as a function of peptide concentration. Biochim Biophys Acta Biomembr 1828:2620–2627. https://doi.org/10.1016/j.bbamem.2013.07.007

Cardoza R, Hermosa R, Vizcaino J, Sanz L, Monte E, Gutiérrez S (2005) Secondary metabolites produced by *Trichoderma* and their importance in the biocontrol process. In: Microorganisms for industrial enzymes and biocontrol, pp 1–22

Cardoza RE, Hermosa MR, Vizcaíno JA, González F, Llobell A, Monte E, Gutiérrez S (2007) Partial silencing of a hydroxy-methylglutaryl-CoA reductase-encoding gene in *Trichoderma harzianum* CECT 2413 results in a lower level of resistance to lovastatin and lower antifungal activity. Fungal Genet Biol 44:269–283. https://doi.org/10.1016/j.fgb.2006.11.013

Cardoza RE, Malmierca MG, Hermosa MR, Alexander NJ, McCormick SP, Proctor RH, Tijerino AM, Rumbero A, Monte E, Gutiérrez S (2011) Identification of loci and functional characterization of trichothecene biosynthesis genes in filamentous fungi of the genus *Trichoderma*. Appl Environ Microbiol 77:4867–4877. https://doi.org/10.1128/AEM.00595-11

Cary J (2004) Secondary metabolic gene clusters in filamentous fungi. In: Handbook of fungal biotechnology. Marcel Dekker, New York, pp 81–94

Chaverri P, Branco-Rocha F, Jaklitsch W, Gazis R, Degenkolb T, Samuels GJ (2015) Systematics of the *Trichoderma harzianum* species complex and the re-identification of commercial biocontrol strains. Mycologia 107:558–590. https://doi.org/10.3852/14-147

Chen S, Li H, Chen Y, Li S, Xu J, Guo H, Liu Z, Zhu S, Liu H, Zhang W (2019) Three new diterpenes and two new sesquiterpenoids from the endophytic fungus Trichoderma koningiopsis A729. Bioorg Chem 86:368–374. https://doi.org/10.1016/j.bioorg.2019.02.005

Cheng Y-Q, Coughlin JM, Lim S-K, Shen B (2009) Type I polyketide synthases that require discrete acyltransferases. Meth Enzymol 459:165–186. https://doi.org/10.1016/S0076-6879(09)04608-4

Claydon N, Allan M, Hanson JR, Avent AG (1987) Antifungal alkyl pyrones of *Trichoderma harzianum*. Trans Br Mycol Soc 88:503–513. https://doi.org/10.1016/S0007-1536(87)80034-7

Contreras-Cornejo HA, Macías-Rodríguez L, del-Val E, Larsen J (2018) Interactions of *Trichoderma* with plants, insects, and plant pathogen microorganisms: chemical and molecular bases. In: Merillon J-M, Ramawat KG (eds) Co-evolution of secondary metabolites, reference series in phytochemistry. Springer International Publishing, Cham, pp 1–28. https://doi.org/10.1007/978-3-319-76887-8_23-1

Daniel JF d S, Rodrigues Filho E (2007) Peptaibols of *Trichoderma*. Nat Prod Rep 24:1128. https://doi.org/10.1039/b618086h

de los Santos-Villalobos S, Hernandez-Rodriguez LE, Villasenor-Ortega F, Pena-Cabriales JJ (2012) Production of spores *Trichoderma asperellum* T8a by a "home-made" solid-state fermentation of mango industrial wastes. Bioresources 7:4938–4951. https://doi.org/10.15376/biores.7.4.4938-4951

de los Santos-Villalobos S, Guzmán-Ortiz DA, Gómez-Lim MA, Délano-Frier JP, de-Folter S, Sánchez-García P, Peña-Cabriales JJ (2013) Potential use of *Trichoderma asperellum* (Samuels, Liechfeldt et Nirenberg) T8a as a biological control agent against anthracnose in mango (*Mangifera indica* L.). Biol Control 64:37–44. https://doi.org/10.1016/j.biocontrol.2012.10.006

Degenkolb T, Gräfenhan T, Berg A, Nirenberg HI, Gams W, Brückner H (2006) Peptaibiomics: screening for polypeptide antibiotics (peptaibiotics) from plant-protective *Trichoderma* species. C&B 3:593–610. https://doi.org/10.1002/cbdv.200690063

Demain AL, Fang A (2000) The natural functions of secondary metabolites. In: Fiechter A (ed) History of modern biotechnology I. Springer, Berlin/Heidelberg, pp 1–39. https://doi.org/10.1007/3-540-44964-7_1

Dennis C, Webster J (1971a) Antagonistic properties of species-groups of *Trichoderma*: I. production of non-volatile antibiotics. Trans Br Mycol Soc 57:25-IN3. https://doi.org/10.1016/S0007-1536(71)80077-3

Dennis C, Webster J (1971b) Antagonistic properties of species-groups of *Trichoderma*: II. Production of volatile antibiotics. Trans Br Mycol Soc 57:41-IN4. https://doi.org/10.1016/S0007-1536(71)80078-5

Duclohier H, Wróblewski H (2001) Voltage-dependent pore formation and antimicrobial activity by alamethicin and analogues. J Membr Biol 184:1–12. https://doi.org/10.1007/s00232-001-0077-2

Finking R, Marahiel MA (2004) Biosynthesis of nonribosomal peptides. Annu Rev Microbiol 58:453–488. https://doi.org/10.1146/annurev.micro.58.030603.123615

Fisch KM (2013) Biosynthesis of natural products by microbial iterative hybrid PKS–NRPS. RSC Adv 3:18228–18247. https://doi.org/10.1039/C3RA42661K

Fonteriz RI, López MG, Garcia-Sancho J, Garcia AG (1991) Alamethicin channel permeation by Ca2+, Mn2+ and Ni2+ in bovine chromaffin cells. FEBS Lett 283:89–92. https://doi.org/10.1016/0014-5793(91)80560-P

Fox EM, Howlett BJ (2008) Biosynthetic gene clusters for epipolythiodioxopiperazines in filamentous fungi. Mycol Res 112:162–169. https://doi.org/10.1016/j.mycres.2007.08.017

Fujii I, Watanabe A, Sankawa U, Ebizuka Y (2001) Identification of Claisen cyclase domain in fungal polyketide synthase WA, a naphthopyrone synthase of *Aspergillus nidulans*. Chem Biol 8:189–197

Gardiner DM, Waring P, Howlett BJ (2005) The epipolythiodioxopiperazine (ETP) class of fungal toxins: distribution, mode of action, functions and biosynthesis. Microbiology (Reading, Engl) 151:1021–1032. https://doi.org/10.1099/mic.0.27847-0

Ghisalberti EL, Hockless DCR, Rowland C, White AH (1992) Harziandione, a new class of diterpene from Trichoderma harzianum. J Nat Prod 55:1690–1694. https://doi.org/10.1021/np50089a023

Gokhale RS, Sankaranarayanan R, Mohanty D (2007) Versatility of polyketide synthases in generating metabolic diversity. Curr Opin Struct Biol 17:736–743. https://doi.org/10.1016/j.sbi.2007.08.021

Guzmán-Guzmán P, Porras-Troncoso MD, Olmedo-Monfil V, Herrera-Estrella A (2018) *Trichoderma* species: versatile plant symbionts. Phytopathology 109:6–16. https://doi.org/10.1094/PHYTO-07-18-0218-RVW

Haas H (2014) Fungal siderophore metabolism with a focus on *Aspergillus fumigatus*. Nat Prod Rep 31:1266–1276. https://doi.org/10.1039/C4NP00071D

Healy AR, Izumikawa M, Slawin AMZ, Shin-ya K, Westwood NJ (2015) Stereochemical assignment of the protein-protein interaction inhibitor JBIR-22 by total synthesis. Angew Chem Int Ed 54:4046–4050. https://doi.org/10.1002/anie.201411141

Hermosa R, Cardoza RE, Rubio MB, Gutiérrez S, Monte E (2014) Secondary metabolism and antimicrobial metabolites of *Trichoderma*. In: Biotechnology and biology of *Trichoderma*. Elsevier, pp 125–137. https://doi.org/10.1016/B978-0-444-59576-8.00010-2

Howell CR (2003) Mechanisms employed by *Trichoderma* species in the biological control of plant diseases: the history and evolution of current concepts. Plant Dis 87:4–10. https://doi.org/10.1094/PDIS.2003.87.1.4

Ishii T, Nonaka K, Suga T, Masuma R, Ōmura S, Shiomi K (2013) Cytosporone S with antimicrobial activity, isolated from the fungus *Trichoderma* sp. FKI-6626. Bioorg Med Chem Lett 23:679–681. https://doi.org/10.1016/j.bmcl.2012.11.113

Ismaiel AA, Ali DMI (2017) Antimicrobial properties of 6-pentyl-α-pyrone produced by endophytic strains of *Trichoderma koningii* and its effect on aflatoxin B1 production. Biologia 72. https://doi.org/10.1515/biolog-2017-0173

Jenke-Kodama H, Müller R, Dittmann E (2008) Evolutionary mechanisms underlying secondary metabolite diversity. In: Petersen F, Amstutz R (eds) Natural compounds as drugs, vol I. Birkhäuser Basel, Basel, pp 119–140. https://doi.org/10.1007/978-3-7643-8117-2_3

Karlsson M, Atanasova L, Jensen DF, Zeilinger S (2017) Necrotrophic mycoparasites and their genomes. Microbiol Spectr 5. https://doi.org/10.1128/microbiolspec.FUNK-0016-2016

Keller NP, Turner G, Bennett JW (2005) Fungal secondary metabolism — from biochemistry to genomics. Nat Rev Microbiol 3:937–947. https://doi.org/10.1038/nrmicro1286

Keswani C, Mishra S, Sarma BK, Singh SP, Singh HB (2014) Unraveling the efficient applications of secondary metabolites of various *Trichoderma* spp. Appl Microbiol Biotechnol 98:533–544. https://doi.org/10.1007/s00253-013-5344-5

Khosla C (2009) Structures and mechanisms of polyketide synthases. J Org Chem 74:6416–6420. https://doi.org/10.1021/jo9012089

Khosla C, Herschlag D, Cane DE, Walsh CT (2014) Assembly line polyketide synthases: mechanistic insights and unsolved problems. Biochemistry 53:2875–2883. https://doi.org/10.1021/bi500290t

Krause C, Kirschbaum J, Brückner H (2006) Peptaibiomics: an advanced, rapid and selective analysis of peptaibiotics/peptaibols by SPE/LC-ES-MS. Amino Acids 30:435–443. https://doi.org/10.1007/s00726-005-0275-9

Kubicek CP, Herrera-Estrella A, Seidl-Seiboth V, Martinez DA, Druzhinina IS, Thon M, Zeilinger S, Casas-Flores S, Horwitz BA, Mukherjee PK, Mukherjee M, Kredics L, Alcaraz LD, Aerts A, Antal Z, Atanasova L, Cervantes-Badillo MG, Challacombe J, Chertkov O, McCluskey K, Coulpier F, Deshpande N, von Döhren H, Ebbole DJ, Esquivel-Naranjo EU, Fekete E, Flipphi M, Glaser F, Gómez-Rodríguez EY, Gruber S, Han C, Henrissat B, Hermosa R, Hernández-Oñate M, Karaffa L, Kosti I, Le Crom S, Lindquist E, Lucas S, Lübeck M, Lübeck PS, Margeot

A, Metz B, Misra M, Nevalainen H, Omann M, Packer N, Perrone G, Uresti-Rivera EE, Salamov A, Schmoll M, Seiboth B, Shapiro H, Sukno S, Tamayo-Ramos JA, Tisch D, Wiest A, Wilkinson HH, Zhang M, Coutinho PM, Kenerley CM, Monte E, Baker SE, Grigoriev IV (2011) Comparative genome sequence analysis underscores mycoparasitism as the ancestral life style of *Trichoderma*. Genome Biol 12:R40. https://doi.org/10.1186/gb-2011-12-4-r40

Kumar P, Misra AK, Modi DR, Gupta VK (2012) Biocontrol potential of *Trichoderma* species against mango malformation pathogens. Arch Phytopathol Plant Protect 45:1237–1245. https://doi.org/10.1080/03235408.2012.664711

Lee H b, Kim Y, Jin H z, Lee J j, Kim C-J, Park J y, Jung H s (2005) A new *Hypocrea* strain producing harzianum A cytotoxic to tumour cell lines. Lett Appl Microbiol 40:497–503. https://doi.org/10.1111/j.1472-765x.2005.01719.x

Leeper FJ, Vederas JC (eds) (2000) Biosynthesis: aromatic polyketides, isoprenoids, alkaloids, topics in current chemistry. Springer-Verlag, Berlin/Heidelberg

Lehner SM, Atanasova L, Neumann NKN, Krska R, Lemmens M, Druzhinina IS, Schuhmacher R (2013) Isotope-assisted screening for iron-containing metabolites reveals a high degree of diversity among known and unknown siderophores produced by *Trichoderma* spp. Appl Environ Microbiol 79:18–31. https://doi.org/10.1128/AEM.02339-12

Lewis JA, Roberts DP, Hollenbeck MD (1991) Induction of cytoplasmic leakage from *Rhizoctonia solani* hyphae by *Gliocladium virens* and partial characterization of a leakage factor. Biocontrol Sci Tech 1:21–29. https://doi.org/10.1080/09583159109355182

Liu K, Yang Y, Miao C-P, Zheng Y-K, Chen J-L, Chen Y-W, Xu L-H, Guang H-L, Ding Z-T, Zhao L-X (2016) Koningiopisins A-H, polyketides with synergistic antifungal activities from the endophytic fungus *Trichoderma koningiopsis*. Planta Med 82:371–376. https://doi.org/10.1055/s-0035-1558228

Lorito M, Peterbauer C, Hayes CK, Harman GE (1994) Synergistic interaction between fungal cell wall degrading enzymes and different antifungal compounds enhances inhibition of spore germination. Microbiology 140:623–629. https://doi.org/10.1099/00221287-140-3-623

Lorito M, Farkas V, Rebuffat S, Bodo B, Kubicek CP (1996a) Cell wall synthesis is a major target of mycoparasitic antagonism by *Trichoderma harzianum*. J Bacteriol 178:6382–6385. https://doi.org/10.1128/jb.178.21.6382-6385.1996

Lorito M, Mach RL, Sposato P, Strauss J, Peterbauer CK, Kubicek CP (1996b) Mycoparasitic interaction relieves binding of the Cre1 carbon catabolite repressor protein to promoter sequences of the *ech42* (endochitinase-encoding) gene in *Trichoderma harzianum*. Proc Natl Acad Sci U S A 93:14868–14872

Malmierca MG, Cardoza RE, Alexander NJ, McCormick SP, Hermosa R, Monte E, Gutiérrez S (2012) Involvement of *Trichoderma* trichothecenes in the biocontrol activity and induction of plant defense-related genes. Appl Environ Microbiol 78:4856–4868. https://doi.org/10.1128/AEM.00385-12

Mo X, Li Q, Ju J (2014) Naturally occurring tetramic acid products: isolation, structure elucidation and biological activity. RSC Adv 4:50566–50593. https://doi.org/10.1039/C4RA09047K

Mukherjee PK, Buensanteai N, Moran-Diez ME, Druzhinina IS, Kenerley CM (2012a) Functional analysis of non-ribosomal peptide synthetases (NRPSs) in *Trichoderma virens* reveals a polyketide synthase (PKS)/NRPS hybrid enzyme involved in the induced systemic resistance response in maize. Microbiology 158:155–165. https://doi.org/10.1099/mic.0.052159-0

Mukherjee PK, Horwitz BA, Kenerley CM (2012b) Secondary metabolism in *Trichoderma*--a genomic perspective. Microbiology (Reading, Engl.) 158:35–45. https://doi.org/10.1099/mic.0.053629-0

Mukherjee PK, Hurley JF, Taylor JT, Puckhaber L, Lehner S, Druzhinina I, Schumacher R, Kenerley CM (2018) Ferricrocin, the intracellular siderophore of *Trichoderma virens*, is involved in growth, conidiation, gliotoxin biosynthesis and induction of systemic resistance in maize. Biochem Biophys Res Commun 505:606–611. https://doi.org/10.1016/j.bbrc.2018.09.170

Neuhof T, Dieckmann R, Druzhinina IS, Kubicek CP, von Dohren H (2007) Intact-cell MALDI-TOF mass spectrometry analysis of peptaibol formation by the genus *Trichoderma/Hypocrea*:

can molecular phylogeny of species predict peptaibol structures? Microbiology 153:3417–3437. https://doi.org/10.1099/mic.0.2007/006692-0

O'Connor SE (2015) Engineering of secondary metabolism. Annu Rev Genet 49:71–94. https://doi.org/10.1146/annurev-genet-120213-092053

Pascale A, Vinale F, Manganiello G, Nigro M, Lanzuise S, Ruocco M, Marra R, Lombardi N, Woo SL, Lorito M (2017) *Trichoderma* and its secondary metabolites improve yield and quality of grapes. Crop Prot 92:176–181. https://doi.org/10.1016/j.cropro.2016.11.010

Poole PR, Ward BG, Whitaker G (1998) The effects of topical treatments with 6-pentyl-2-pyrone and structural analogues on stem end postharvest rots in kiwifruit due to *Botrytis cinerea*. J Sci Food Agric 77:81–86. https://doi.org/10.1002/(SICI)1097-0010(199805)77:1<81::AID-JSFA6>3.0.CO;2-5

Rabe P, Kamps JJAG, Schofield CJ, Lohans CT (2018) Roles of 2-oxoglutarate oxygenases and isopenicillin N synthase in β-lactam biosynthesis. Nat Prod Rep 35:735–756. https://doi.org/10.1039/c8np00002f

Rebuffat S, Goulard C, Bodo B (1995) Antibiotic peptides from *Trichoderma harzianum*: harzianins HC, proline-rich 14-residue peptaibols. J Chem Soc Perkin Trans 1:1849–1855. https://doi.org/10.1039/P19950001849

Reino JL, Guerrero RF, Hernández-Galán R, Collado IG (2008) Secondary metabolites from species of the biocontrol agent *Trichoderma*. Phytochem Rev 7:89–123. https://doi.org/10.1007/s11101-006-9032-2

Reithner B, Brunner K, Schuhmacher R, Peissl I, Seidl V, Krska R, Zeilinger S (2005) The G protein α subunit Tga1 of Trichoderma atroviride is involved in chitinase formation and differential production of antifungal metabolites. Fungal Genet Biol 42:749–760. https://doi.org/10.1016/j.fgb.2005.04.009

Renshaw JC, Robson GD, Trinci APJ, Wiebe MG, Livens FR, Collison D, Taylor RJ (2002) Fungal siderophores: structures, functions and applications. Mycol Res 106:1123–1142. https://doi.org/10.1017/S0953756202006548

Roberts DP (1990) Effect of extracellular metabolites from *Gliocladium virens* on germination of sporangia and mycelial growth of *Pythium ultimum*. Phytopathology 80:461. https://doi.org/10.1094/Phyto-80-461

Rubio MB, Hermosa R, Reino JL, Collado IG, Monte E (2009) Thctf1 transcription factor of *Trichoderma harzianum* is involved in 6-pentyl-2H-pyran-2-one production and antifungal activity. Fungal Genet Biol 46:17–27. https://doi.org/10.1016/j.fgb.2008.10.008

Ruprecht JJ, King MS, Zögg T, Aleksandrova AA, Pardon E, Crichton PG, Steyaert J, Kunji ERS (2019) The molecular mechanism of transport by the mitochondrial ADP/ATP carrier. Cell 176:435–447.e15. https://doi.org/10.1016/j.cell.2018.11.025

Scharf DH, Brakhage AA, Mukherjee PK (2016) Gliotoxin – bane or boon? Environ Microbiol 18:1096–1109. https://doi.org/10.1111/1462-2920.13080

Schirmböck M, Lorito M, Wang Y-L, Hayes CK, Arisan-Atac I, Scala F, Harman GE, Kubicek CP (1994) Parallel formation and synergism of hydrolytic enzymes and peptaibol antibiotics, molecular mechanisms involved in the antagonistic action of *Trichoderma harzianum* against phytopathogenic fungi. Appl Environ Microbiol 60:4364–4370

Schmidt-Dannert C (2015) Biosynthesis of terpenoid natural products in fungi. In: Schrader J, Bohlmann J (eds) Biotechnology of isoprenoids, advances in biochemical engineering/biotechnology. Springer International Publishing, Cham, pp 19–61. https://doi.org/10.1007/10_2014_283

Schobert R (2007) Domino syntheses of bioactive tetronic and tetramic acids. Naturwissenschaften 94:1–11. https://doi.org/10.1007/s00114-006-0152-8

Serrano-Carreon L, Hathout Y, Bensoussan M, Belin J-M (1993) Metabolism of linoleic acid or mevalonate and 6-pentyl-α-pyrone biosynthesis by *Trichoderma* species. Appl Environ Microbiol 59:2945–2950

Simon A, Dunlop RW, Ghisalberti EL, Sivasithamparam K (1988) Trichoderma koningii produces a pyrone compound with antibiotic properties. Soil Biol Biochem

Sivasithamparam K, Ghisalberti E (1998) Secondary metabolism in *Trichoderma* and *Gliocladium*. In: Kubicek CP, Harman GE (eds) *Trichoderma* and *Gliocladium*. Taylor & Francis, London, pp 139–191

Strieker M, Tanović A, Marahiel MA (2010) Nonribosomal peptide synthetases: structures and dynamics. Curr Opin Struct Biol 20:234–240. https://doi.org/10.1016/j.sbi.2010.01.009

Suga T, Asami Y, Hashimoto S, Nonaka K, Iwatsuki M, Nakashima T, Watanabe Y, Sugahara R, Shiotsuki T, Yamamoto T, Shinohara Y, Ichimaru N, Murai M, Miyoshi H, Ōmura S, Shiomi K (2015) Trichopolyn VI: a new peptaibol insecticidal compound discovered using a recombinant *Saccharomyces cerevisiae* screening system. J Gen Appl Microbiol 61:82–87. https://doi.org/10.2323/jgam.61.82

Tijerino A, Cardoza RE, Moraga J, Malmierca MG, Vicente F, Aleu J, Collado IG, Gutiérrez S, Monte E, Hermosa R (2011) Overexpression of the trichodiene synthase gene tri5 increases trichodermin production and antimicrobial activity in *Trichoderma brevicompactum*. Fungal Genet Biol 48:285–296. https://doi.org/10.1016/j.fgb.2010.11.012

Toniolo C, Crisma M, Formaggio F, Peggion C, Epand RF, Epand RM (2001) Lipopeptaibols, a novel family of membrane active, antimicrobial peptides. CMLS, Cell Mol Life Sci 58:1179–1188. https://doi.org/10.1007/PL00000932

Tudzynski P, Hölter K, Correia T, Arntz C, Grammel N, Keller U (1999) Evidence for an ergot alkaloid gene cluster in *Claviceps purpurea*. Mol Gen Genet 261:133–141. https://doi.org/10.1007/s004380050950

Vicente MF, Cabello A, Platas G, Basilio A, Díez MT, Dreikorn S, Giacobbe RA, Onishi JC, Meinz M, Kurtz MB, Rosenbach M, Thompson J, Abruzzo G, Flattery A, Kong L, Tsipouras A, Wilson KE, Peláez F (2001) Antimicrobial activity of ergokonin A from *Trichoderma longibrachiatum*. J Appl Microbiol 91:806–813

Vinale F (2014) Biopesticides and biofertilizers based on fungal secondary metabolites. J Biofert Biopest 05. https://doi.org/10.4172/2155-6202.1000e119

Vinale F, Ghisalberti EL, Sivasithamparam K, Marra R, Ritieni A, Ferracane R, Woo S, Lorito M (2009) Factors affecting the production of *Trichoderma harzianum* secondary metabolites during the interaction with different plant pathogens. Lett Appl Microbiol 48:705–711. https://doi.org/10.1111/j.1472-765X.2009.02599.x

Vinale F, Nigro M, Sivasithamparam K, Flematti G, Ghisalberti EL, Ruocco M, Varlese R, Marra R, Lanzuise S, Eid A, Woo SL, Lorito M (2013) Harzianic acid: a novel siderophore from Trichoderma harzianum. FEMS Microbiol Lett 347:123–129. https://doi.org/10.1111/1574-6968.12231

Vinale F, Sivasithamparam K, Ghisalberti EL, Woo SL, Nigro M, Marra R, Lombardi N, Pascale A, Ruocco M, Lanzuise S, Manganiello G, Lorito M (2014) *Trichoderma* secondary metabolites active on plants and fungal pathogens. Open Mycol J 8:127–139. https://doi.org/10.2174/1874437001408010127

Walsh CT (2008) The chemical versatility of natural-product assembly lines. Acc Chem Res 41:4–10. https://doi.org/10.1021/ar7000414

Wang D-N, Toyotome T, Muraosa Y, Watanabe A, Wuren T, Bunsupa S, Aoyagi K, Yamazaki M, Takino M, Kamei K (2014) GliA in *Aspergillus fumigatus* is required for its tolerance to gliotoxin and affects the amount of extracellular and intracellular gliotoxin. Med Mycol 52:506–518. https://doi.org/10.1093/mmy/myu007

Weber T (2010) Antibiotics: biosynthesis, generation of novel compounds. In: Encyclopedia of industrial biotechnology. Wiley, Hoboken, p eib040. https://doi.org/10.1002/9780470054581.eib040

Whitmore L (2004) The Peptaibol Database: a database for sequences and structures of naturally occurring peptaibols. Nucleic Acids Res 32:593D–594D. https://doi.org/10.1093/nar/gkh077

Wiest A, Grzegorski D, Xu B-W, Goulard C, Rebuffat S, Ebbole DJ, Bodo B, Kenerley C (2002) Identification of peptaibols from *Trichoderma virens* and cloning of a peptaibol synthetase. J Biol Chem 277:20862–20868. https://doi.org/10.1074/jbc.M201654200

Xie Z-L, Li H-J, Wang L-Y, Liang W-L, Liu W, Lan W-J (2013) Trichodermaerin, a new diterpenoid lactone from the marine fungus *Trichoderma erinaceum* associated with the sea star *Acanthaster planci*. Nat Prod Commun 8:67–68

Zeilinger S, Gruber S, Bansal R, Mukherjee PK (2016) Secondary metabolism in *Trichoderma* – chemistry meets genomics. Fungal Biol Rev 30:74–90. https://doi.org/10.1016/j.fbr.2016.05.001

Zhong J-J, Xiao J-H (2009) Secondary metabolites from higher fungi: discovery, bioactivity, and bioproduction. In: Zhong J-J, Bai F-W, Zhang W (eds) Biotechnology in China I: from bioreaction to bioseparation and bioremediation, advances in biochemical engineering / biotechnology. Springer, Berlin/Heidelberg, pp 79–150. https://doi.org/10.1007/10_2008_26

Genes Involved in the Secondary Metabolism of *Trichoderma* and the Biochemistry of These Compounds

Sabrina Sarrocco, Isabel Vicente, Alessia Staropoli, and Francesco Vinale

Contents

S. Sarrocco (✉) · I. Vicente
Department of Agriculture, Food and Environment, University of Pisa, Pisa, Italy
e-mail: sabrina.sarrocco@unipi.it

A. Staropoli
Department of Agricultural Sciences, University of Naples Federico II, Portici, Italy

Institute for Sustainable Plant Protection, National Research Council, Portici, Italy

F. Vinale (✉)
Institute for Sustainable Plant Protection, National Research Council, Portici, Italy

Department of Veterinary Medicine and Animal Production, University of Naples Federico II, Naples, Italy
e-mail: frvinale@unina.it

N. Amaresan et al. (eds.), *Advances in* Trichoderma *Biology for Agricultural Applications*, Fungal Biology, https://doi.org/10.1007/978-3-030-91650-3_4

1 Role of Secondary Metabolites in *Trichoderma* spp. Ecophysiology: An Introduction

The wide adaptability of fungi to survive in different environmental conditions is generally recognized as a consequence of the huge and versatile repertoire of secondary metabolites (SMs) they can produce. Even if not indispensable for their survival, SMs give an important contribution to fungal lifestyles being actively involved in the interactions established with the environment as well as with other organisms (Singh and Chandra 2019; Sarrocco 2016). These compounds, whose production is highly susceptible to the environment, can help to guarantee the adaptation to local changes by triggering reproduction, conidia differentiation, or production of virulence factors (Mahmood et al. 2010) as well as the defense of the territory. Fungi evolved the ability to release SMs with antimicrobial activities that allow the impairment or inhibition of competitors'/predators' growth and development, finalized to substrate colonization or niche maintenance (Künzler 2018). In addition to antimicrobial compounds, fungi can produce mycotoxins that can also, but not only, play a crucial role in fungus-microbe and fungus-plant/animal interactions, by acting as pathogenicity or virulence factors (Susca et al. 2017). However, some fungal SMs are able to enhance plant growth – by mimicking phytohormones – or to elicit plant defense responses (Pusztahelyi et al. 2015; Alfiky and Weisskopf 2021).

Among all genera included in the Fifth Kingdom, those belonging to *Trichoderma* genus (*Hypocreaceae* family) are characterized by a high versatility to be adapted to different ecological conditions. Frequently isolated in soil or on decaying wood as well as in many other substrates, isolates belonging to this genus demonstrated a high opportunistic potential and a huge adaptability to changing environmental conditions in addition to a wide range of lifestyles (Druzhinina et al. 2011), with mycoparasitism considered as the ancestral one (Kubicek et al. 2011). Mycoparasitic attitude is supported by the ability to produce and release hydrolytic enzymes (such as chitinases, glucanases, and proteases) that digest fungal (and oomycete) cell walls (Howell 2003). From an enzymatic point of view, *Trichoderma* spp. are also able to release an arsenal of cellulose- and xylan-degrading enzymes that enabled species such as *T. reesei* to be considered of high industrial value (Druzhinina et al. 2011). This plethora of enzymes, together with antimicrobial compounds *Trichoderma* can produce, improves the competitive ability for nutrients and space these fungi evolved (Sarrocco et al. 2009; Saravanakumar et al. 2016; Sarrocco et al., 2021; Jona Lasinio et al. 2021). However, in addition to the interaction with other fungi (Zeilinger et al. 2016; Zapparata et al. 2021), *Trichoderma* spp. can establish a positive intimate relationship with plants resulting in growth promotion and/or induction of resistance to abiotic and biotic stresses (Harman et al. 2004; Sarrocco et al., 2017; Rai et al. 2019). Thanks to all these positive features, *Trichoderma* spp. are one of the most popular genera of fungi commercially marketed as biopesticides, biofertilizers, and soil amendments (Keswani et al. 2014; Sarrocco and Vannacci 2018; Sarrocco et al. 2019a, b).

This array of lifestyles is likely supported by the diversity of their SM inventory that consists in more than 800 molecules, including volatile and non-volatile compounds (Shenouda and Cox 2021). In the last years, the central role played by *Trichoderma* SMs in the interaction with plants, insects, nematodes, and microorganisms has been extensively studied (Contreras-Cornejo et al. 2018; Patil et al. 2016; Salwan et al. 2019; Rai et al. 2019; Li et al. 2019; Khan et al. 2020). From an antimicrobial point of view, *Trichoderma* SMs confer the ability to reduce or suppress the growth of a huge amount of plant pathogens, such as those belonging to *Botrytis*, *Fusarium*, *Phytium*, *Rhizoctonia*, *Phytophthora*, *Colletotrichum*, and *Sclerotinia* genera. Metabolomic studies also demonstrated the enormous contribution of SMs on the establishment of beneficial *Trichoderma*-plant interactions (Vinale et al. 2012; Contreras-Cornejo et al. 2016; Ramírez-Valdespino et al. 2019). Compounds such as indoleacetic acid (IAA) can help plant growth and improve adaptation to saline stress (Contreras-Cornejo et al. 2009; Waqas et al. 2012); harzianolide, 6-pentyl-pyrone (6-PP), harzianic acid, and aspinolides can elicit plant defense responses (Vinale et al. 2008; Malmierca et al. 2015; Manganiello et al. 2018); siderophores facilitate iron sequestration by plants and by *Trichoderma* itself (Kubicek et al. 2011). Of the great interest is the fact that some *Trichoderma* SMs can show more than one activity as in the case of harzianic acid that is classified both as a siderophore and also as an antifungal compound (Vinale et al. 2013).

In the following sections of this chapter, the reader will be driven into the exploration of the role of SMs in the ecophysiology of *Trichoderma*, focusing on both genes and gene regulation as well as on the biochemistry of the most important compounds produced by these fungi.

2 Genes Involved in *Trichoderma* Secondary Metabolite Biosynthesis and Major Regulators

One hypothesis about the biological sense of SM origin could be they originated as the result of the recycling of primary metabolism residues, a process where they underwent enzymatic modifications conferring them new bioactive properties. This resulted in advantageous skills for fungal communication and defense that evolution has been shaping for over 500 million years (Brakhage 2013). Indeed, the high diversity of *Trichoderma* SMs is originated from molecules derived from few primary metabolic pathways (Zeilinger et al. 2016).

Enzymes responsible for building SMs can be divided into core enzymes and tailoring enzymes. Core enzymes transform simple precursors in structurally different backbone molecules that are further remodeled by tailoring enzymes (i.e., p450 monooxygenases, hydrolases, etc.), conferring different bioactive properties and generating a wide variety of SMs (Keller 2019). *Trichoderma* produces SMs in a strain-dependent manner (Yu and Keller 2005), and there are evidences that the intra- and interspecific variability observed on the core gene inventory size is the result of the phylogenetic distribution of the species, the niches occupied by each strain, and their lifestyle divergences (Vicente 2020).

Typically, genes involved in the biosynthesis of a given SM are arranged in cluster (Keller et al. 2005), where specific efflux transporters can also be present (Rokas et al. 2018). Advantages of clustering genes include co-inheritance, co-transcriptional regulation, or coordinated management of post-transcriptional processes (Chavali and Rhee 2017). Depending on the species, from 42 to 59% of the SM core genes is included in clusters in *Trichoderma* (Vicente 2020). Thus, approximately half of the core enzymes is involved in specific SM pathways, as they require co-expression with tailoring enzymes. Those not clustered can potentially cooperate as donors of SM precursors in different pathways. This brings out the complexity of the SM arsenal and reflects an enormous plasticity of the SM genetic machinery of *Trichoderma* spp.

Trichoderma produces a rich variety of SMs. Genes involved in the biosynthesis of non-ribosomal peptides, polyketides, and terpenes are the most represented SM synthase genes in the genomes of this genus (Kubicek et al. 2011; Vicente 2020). The following sections describe the structure of the main SM synthases and their respective encoding genes and associated gene clusters that have been identified in *Trichoderma*, as well as the regulation of secondary metabolism in such fungi.

2.1 Non-ribosomal Peptide Biosynthetic Genes

Non-ribosomal peptides (NRP), mostly derived from condensation of both proteinogenic and non-proteinogenic amino acids, are synthesized by the large multimodular enzymes NRP synthases (NRPSs). In NRPS, each module catalyzes the addition of a single amino acid (Marahiel 2009) and contains the adenylation (A), the pantothenylation/peptidyl carrier (P), and the condensation/peptide bond formation (C) as core domains, along with several specialized thioesterase C-terminal domains responsible for chain termination (Te). Additional modifying domains such as epimerization (E) and methyltransferase (M), among others, can be present as well (Bushley and Turgeon 2010) (Fig. 1). NRPS products can be cyclic or linear showing different lengths, depending on the NRPS structure and on their eventual tailoring modifications, being peptaibols, siderophores, and epidithiodioxopiperazines (ETP) the most relevant NRP produced by *Trichoderma.*

Peptaibols are membrane-active compounds characterized by the presence of an α-aminoisobutyric acid residue. They are able to form voltage-dependent ion channels in lipidic membranes, modifying the membrane permeability and inducing cell death (Molle et al. 1987). Although their length can vary from 4 to 21 amino acids, 11-residue peptaibols are the most common and broadly distributed in *Trichoderma* spp. (Degenkolb et al. 2012). However, only 7-, 14-, and 18–20-module peptaibol synthases have been identified in the genomes of *T. virens*, *T. atroviride*, and *T. reesei* (Wiest et al. 2002; Komon-Zelazowska et al. 2007; Mukherjee et al. 2011; Kubicek et al. 2011; Degenkolb et al. 2012). Disruption of a 14-module peptaibol synthase

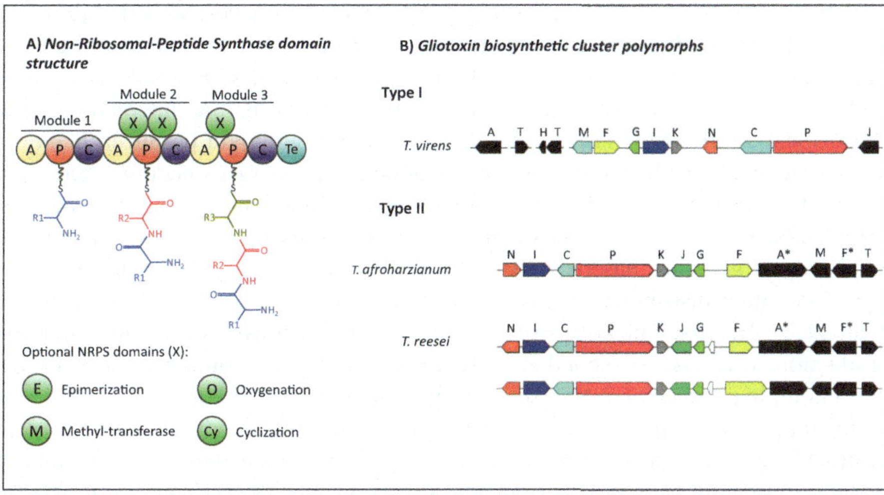

Fig. 1 Non-ribosomal peptide synthase domain structure and gliotoxin biosynthetic cluster polymorphs in *Trichoderma*. **(a)** Example of non-ribosomal peptide synthase domain structure: the minimal structure contains the adenylation (A), pantothenylation/peptidyl carrier (P), condensation (C), and thioesterase (Te) domains. Additional but not essential domains, such as epimerization (E), methyltransferase (M), oxygenation (O), and cyclization (Cy) domains, among others, can also be present. **(b)** Gliotoxin biosynthetic cluster polymorphs found in *Trichoderma* genomes: type I organization contains genes encoding two-module NRPS (P), O-methyltransferase (M), two cytochrome P450 monooxygenases (C and F), glutathione-S-transferase (G), C-S bond lyase (I), γ-glutamate cyclotransferase (K), and N-methyltransferase (N), which are distributed in four loci; type II organization consists of a single-locus cluster lacking on M but containing a dipeptidase-encoding gene (J). Genes in black color represent orthologs of the *Aspergillus fumigatus* GliP genes and adjacent conserved hypothetical proteins that have been predicted in *Trichoderma* genomes: major facilitator superfamily transporter (A), ATP-binding cassette (A*), gliotoxin oxidase (T), conserved hypothetical protein (H), and F putative ortholog (F*) (Adapted from Bulgari et al. 2020)

(*tex2*) in these three species suppressed the biosynthesis of both 14- and 11-residue peptaibols (Mukherjee et al. 2011; Degenkolb et al. 2012). In silico analysis of the active-site residues and conserved domains revealed the structural diversity of peptaibol synthases, and module skipping was then proposed as the mechanism by which a single NRPS enzyme can lead to different peptaibolic products (Mukherjee et al. 2011; Degenkolb et al. 2012). Mining of the clusters associated with 14- and 20-module peptaibol synthases in the genomes of several *Trichoderma* spp. belonging to the Longibrachiatum clade revealed the presence of genes encoding a proline-specific permease, suggesting a possible role of these genes in peptaibol secretion (Marik et al. 2019).

ETP are characterized by the presence of a diketopiperazine ring, and their toxicity relies in the presence of a disulfide bridge that can inactivate proteins by binding thiol groups and by generating reactive oxygen species (Gardiner 2005). The gliotoxin biosynthetic cluster and its variants constitute the most studied gene clusters associated with ETP biosynthesis in *Trichoderma* (Fig. 1). In *T. virens*, the cluster contains seven genes associated with a core NRPS (*gliP*), whose encoded proteins were identified as glutathione-S-transferase, dipeptidase, N-methyltransferase, two cytochrome P450 monooxygenases, O-methyltransferase, C-S bond lyase, and γ-glutamate cyclotransferase, all of them induced during the interaction with *Rhizoctonia solani* (Mukherjee et al. 2012). The non-gliotoxin producer *T. reesei* has a truncated GliP cluster missing on the O-methyltransferase-encoding gene, and their genes are not expressed, suggesting a lack of functionality due to gene loss (Mukherjee et al. 2012). Assessment of the GliP cluster polymorphism on ten *Trichoderma* genomes has recently revealed two types of organization, in which genes are either distributed across four different loci or in a single locus (Bulgari et al. 2020). The first type is only present in *T. virens* and is associated with gliotoxin production, whereas the second is more conserved and is commonly found in the non-gliotoxin producer *Trichoderma* species (Bulgari et al. 2020).

In fungal siderophore biosynthesis, NRPS binds N5-acyl-N5-hydroxy-L--ornithine through covalent bonds to linear or cyclic oligomers that are further modified generating a variety of siderophores (Renshaw et al. 2002; Lehner et al. 2012). Only two NRPSs have been linked to siderophore biosynthesis in the mycoparasites *T. virens* and *T. atroviride* and in the saprotroph *T. reesei* (Mukherjee et al. 2012, 2013). The three species share the ferricrocin biosynthetic cluster (Mukherjee et al. 2012), and a NPS6-type NRPS is involved in the biosynthesis of most of the siderophores secreted by *T. virens* (Mukherjee et al. 2013). An additional, but not yet functionally characterized, putative gene cluster for siderophore biosynthesis is also present in the genomes of *T. virens* and *T. reesei* (Mukherjee et al. 2012).

2.2 Polyketide Biosynthetic Genes

Similar to NRP, polyketides are synthesized by large multi-modular proteins called polyketide synthases (PKS), which condense acyl-coA thioesters in carbon skeletons varying in both chain length and reduction level. Domains of ketoacyl synthase (KS), acyltransferase (AT), and acyl carrier protein (ACP) result indispensable for polyketide biosynthesis, but additional domains such as dehydratase (DH), ketoreductase (KR), enoylreductase (ER), and thioesterase (Te) can be either present or absent in fungal PKS (Keller et al. 2005; Schümann and Hertweck 2006) (Fig. 2).

Different from what was observed in other *Trichoderma* SM core genes, the number of clustered PKS genes is significantly higher (48–92%, depending on the species) (Vicente 2020). A phylogenomic analysis of PKS genes revealed that most of them occur as orthologs in the genomes of *T. virens*, *T. atroviride*, and *T. reesei*; thus, the heterogenicity of polyketides likely relies on the diversity of tailoring enzymes clustered with PKS (Baker et al. 2012). Most *Trichoderma* PKS genes are frequently clustered with genes encoding cytochrome P450 monooxygenases, epimerases, and short-chain dehydrogenases/reductases (Schmoll et al. 2016), but multicopper oxidases can be also found accompanying PKS genes (Baker et al. 2012). The need to compensate the scarce variability of PKS could explain a greater

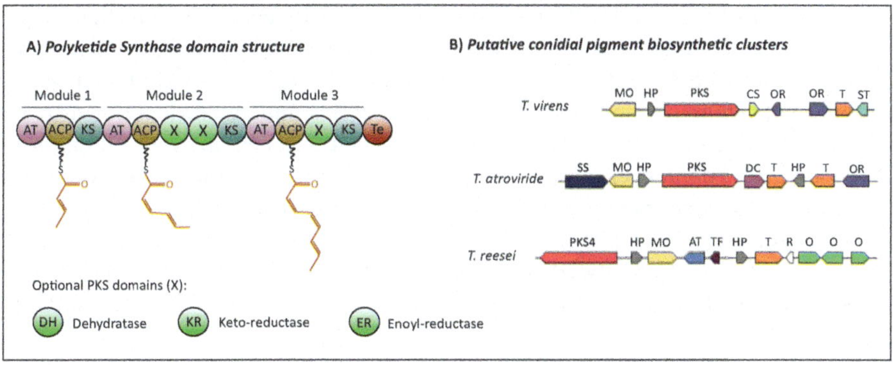

Fig. 2 Polyketide synthase domain structure and putative biosynthetic clusters associated with conidia pigmentation in *Trichoderma*. **(a)** Example of polyketide synthase domain structure: the minimal structure contains the acyltransferase (AT), acyl carrier protein (ACP), and ketoacyl synthase (KS) domains. Additional but not essential domains, such as dehydratase (DH), ketoreductase (KR), enoylreductase (ER), and thioesterase (Te) domains, can also be present. **(b)** Putative conidial pigment biosynthetic clusters found in *Trichoderma* genomes, containing genes encoding polyketide synthase (PKS), multicopper oxidase (MO), hypothetical protein (HP), cystathionine b synthase (CS), oxidoreductase (OR), major facilitator superfamily transporter (T), S-adenosylmethyltransferase (ST), salicylic acid synthase (SS), decarboxylase (DC), acetyltransferase (AT), transcription factor (TF), RTA1-like protein (R), and cytochrome P450 monooxygenase (O) (Adapted from Zeilinger et al. 2016)

tendency of clustering them with a variety of accessory enzymes, leading to many genomic combinations and expanding the catalogue of *Trichoderma* polyketides (Vicente 2020).

Some *Trichoderma* PKS genes have been linked to pigment biosynthesis, since they are included in clusters containing the tailoring genes required for aurofusarin, bikaverin, and DHN melanin biosynthesis of *Fusarium* spp. and *Aspergillus* spp., respectively (Baker et al. 2012) (Fig. 2). Deletion of *pks4* in *T. reesei* confirmed this gene is responsible for the green pigmentation of conidia and also its involvement in supporting conidial cell wall stability and the antagonistic activity of *T. reesei* exerted against other fungi (Baker et al. 2012). Although not functionally characterized, other 20 PKS clusters have been predicted in the genomes of *T. virens*, *T. atroviride*, and *T. reesei*, some of them including their own regulatory proteins (Bansal and Mukherjee 2016). No PKS genes have been yet linked to the biosynthesis of 6-pentyl-α-pyrone (6-PP), one of the most ecologically relevant polyketide derivatives of *Trichoderma*. However, it has been proposed that 6-PP derives from linoleic acid and that its biosynthesis involves the action of a lipoxygenase enzyme (Serrano-Carreon et al. 1992). There are evidences that the lipoxygenase gene ID 33350 of *T. atroviride* could be responsible for 6-PP biosynthesis, since this gene is induced during the interaction of this species with *Rhizoctonia solani* and it is absent in the genomes of the non-6-PP producers *T. virens* and *T. reesei* (Kubicek et al. 2011; Atanasova et al. 2013).

Modules of PKS and NRPS constitute functionally independent units that can be exchangeable, leading to the emergence of hybrid PKS-NRPS enzymes during evolution. PKS-NRPS hybrids consist of a PKS module fused to a C-terminal NRPS module and synthetize amidated polyketide chains that undergo further downstream modifications (Zhu et al. 2021). Hybrid PKS-NRPS genes are very common in *Trichoderma* genomes, especially in species of Harzianum clade (Vicente 2020), and there are evidences that some species have an expansion of these hybrids as a result of recent gene duplications (Kubicek et al. 2011). Recently, heterologous expression of a silent PKS-NRPS gene cluster (*thn*) from *T. harzianum* in *Aspergillus nidulans* enabled the identification of six new tretonate products (Zhu et al. 2021).

2.3 Terpenoid Biosynthetic Genes

Terpenoids are synthesized from isopentenyl pyrophosphate and its isomer dimethylallyl pyrophosphate by terpene synthase (TS) enzymes. TSs bind polyprenyl pyrophosphates to their active center via Mg^{2+}, triggering either substrate cyclization (by terpene cyclases), substrate condensation with another polyprenyl pyrophosphate (by prenyltransferases (PT)), or substrate transference to a non-isoprenoid-derived molecule (by aromatic PT) (Fig. 3). TSs are usually classified based on the mechanism triggering the formation of new carbon-carbon bonds (Class I, Class II, or ABBA), which derives from differences in protein sequence and structure, but they are also classified based on their substrate specificity, which rely on the length of the prenyl pyrophosphate they accept (mono-, sesqui-, di-, tri-, tetra-TS, ecc) (Pérez-Gil et al. 2019). Homology of TS is structural rather than at sequence level, and mutations on the active-site residues confer enormous plasticity to these enzymes (Shaw et al. 2015). Indeed, most TS can be considered as "promiscuous" enzymes able to generate terpenoid blends containing up to 52 different products (Christianson 2008).

Trichoderma spp. are reported to produce all types of terpenoids (Pachauri et al. 2019), including volatile compounds. The TS gene inventory of *Trichoderma* outnumbers that found in other fungi considered as rich SM producers such as *Aspergillus* spp. (De Vries et al. 2017; Kubicek et al. 2019), demonstrating that terpenoid biosynthesis significantly contributes to the SM complexity in *Trichoderma*. The TS family size across *Trichoderma* spp. is very homogeneous, ranging from 15 to 23 TS per each genome analyzed (Vicente et al. 2020). A survey on 387 proteins from 21 *Trichoderma* genomes revealed 15 functional groups of TS and enabled the identification of clade-specific TS (mostly sesquiTS and diTS) (Vicente et al. 2020). Thus, despite their similar terpenoid biosynthetic potential, *Trichoderma* spp. have evolved different terpenoid chemotypes according to their evolutive history and adaptation to different environmental challenges (Vicente et al. 2020).

The first SM-related biosynthetic cluster was identified in *T. virens* (Mukherjee et al. 2006), and disruption of its TS gene (*vir4*) showed it is required for the

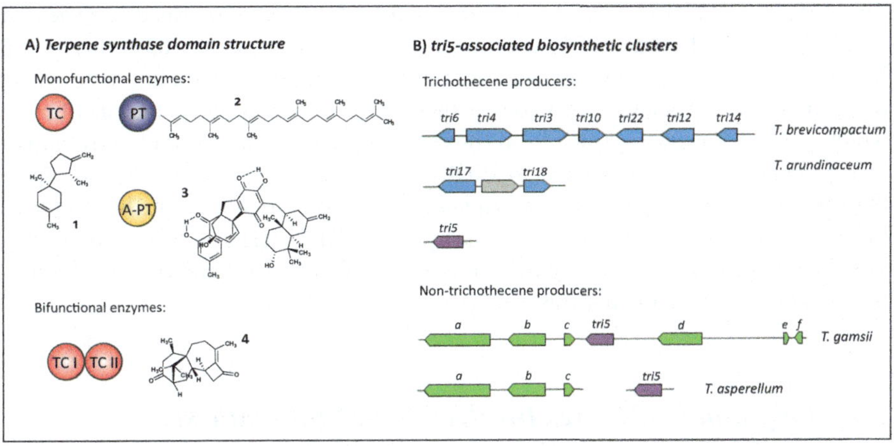

Fig. 3 Terpene synthase domain structure and *tri5*-associated loci in *Trichoderma*. (**a**) Terpene synthase domain structure: monofunctional enzymes containing a cyclase (TC), prenyltransferase (PT), or aromatic prenyltransferase (A-PT) domains and an example of their biosynthetic products trichodiene (**1**), squalene (**2**), and neomacrophorin (**3**); bifunctional enzyme containing Class I (TCI) and Class II (TCII) cyclase domains biosynthesizing products such as harziandione (**4**). (**b**) Trichodiene synthase (TRI5) orthologs and associated clusters found in *Trichoderma* genomes: loci found on trichothecene producer species contain genes encoding trichodiene synthase (TRI5), Zn_2His_2 transcription factor (TRI6), C2-C11-C13 monooxygenase (TRI4), C15 acetyltransferase (TRI3), transcription factor (TRI10), C4 hydroxylase (TRI22), major facilitator superfamily transporter (TRI12), hypothetical protein (TRI14), polyketide synthase (TRI17), and acyl−/acetyltransferase (TRI18); loci found associated with TRI5 in non-trichothecene producer species contain genes putatively encoding Zn_2C_6 transcription factor (A), oxygenase (B and D), a-b hydrolase (B), carbonic anhydrase (F), and major facilitator superfamily transporter (E) (Adapted from Proctor et al. 2018 and Vicente et al. 2020)

biosynthesis of mono- and sesquiterpenes (Crutcher et al. 2013). In *T. reesei*, overexpression of a farnesyl pyrophosphate synthase (*erg-20*) affected the activities of enzymes of the dolichol and sterol biosynthetic pathways, modifying the ergosterol levels (Pilsyk et al. 2013). More recently, in *T. viride*, a combination of genome mining, heterologous expression, and metabolomic approaches enabled the detection of a novel sesquiTS synthase and its associated products (Sun et al. 2019).

Nevertheless, functional characterization of TS genes in *Trichoderma* has been mainly focused on the trichodiene synthase (TRI5)-encoding gene, which catalyzes the first committed step in the biosynthesis of trichothecenes trichodermin and harzianum A (HA), and its associated genes belonging to the TRI cluster (Cardoza et al. 2011; Proctor et al. 2018). The example of the TRI cluster illustrates how metabolic pathways governed by the same genes can diversify, resulting in the biosynthesis of structurally similar products with different bioactive properties. Trichodermin produced by *T. brevicompactum* has an acetyl group at carbon 4 (C-4) and exerts toxic effects on plants, while HA produced by *T. arundinaceum* has an octa-2,4,6-trienedioic acid at C-4 and lacks phytotoxic effects, providing plant protection against *Botrytis cinerea* (Tijerino et al. 2011; Malmierca et al. 2012, 2013).

The *tri5* gene is also present in some non-trichothecene producer species of *Trichoderma* (Gallo et al. 2005; Tijerino et al. 2011; Vicente et al. 2020) (Fig. 3). In *T. gamsii*, *tri5* expression is strongly upregulated during the interaction with wheat roots, and it is associated with some other genes encoding tailoring enzymes in which putative functions differ from those of the TRI enzymes found in trichothecene producer species (Vicente et al. 2020). This suggests that *tri5* could participate in different metabolic pathways in *Trichoderma* beyond trichothecene biosynthesis (Vicente et al. 2020). Another particularity of the TRI cluster is that its transcriptional regulation is governed by two transcription factors (*tri6* and *tri10*) that interact to induce *tri* gene expression (Lindo et al. 2018, 2019).

2.4 Regulation of Genes Involved in SM Biosynthesis

Biosynthesis of fungal SMs is a fine-tuned regulated process, mostly influenced by nutrient availability, temperature, pH, light, redox balance, developmental transitions, and interaction with other organisms (Calvo et al. 2002; Macheleidt et al. 2016; Keller 2019).

Approximately half of fungal SM gene clusters is governed by global transcription factors (Macheleidt et al. 2016), but several *Trichoderma* SM clusters encode their own specific regulatory proteins as well (Zeilinger et al. 2016).

The global regulator PacC responds to environmental pH changes, and its regulating role in SM biosynthesis and iron transport has been demonstrated in deletion mutants of *T. virens* (Mukherjee et al. 2012; Trushina et al. 2013). Ras GTPases constitute another example of proteins regulating primary cellular processes with important roles in secondary metabolism, as shown for TBRG-1, a negative regulator of SM biosynthesis in *T. virens* (Dautt-Castro et al. 2019).

The velvet complex is a heterotrimeric global regulator, including the methyltransferase LaeA and the two velvet proteins VeA and VelB that synchronize sexual development and SM biosynthesis in response to light (Zeilinger et al. 2016). The LaeA ortholog of *T. reesei* (*lae1*) is involved in the regulation of SM gene clusters, and it is essential for the expression of lignocellulose-degrading enzymes (Seiboth et al. 2012; Karimi-Aghcheh et al. 2013). Deletion of *lae1* in *T. atroviride* reduced the expression of PKS genes and the production of water-soluble metabolites and VOCs, suppressing the mycoparasitic activity of the fungus (Aghcheh et al. 2013). Similar effects were observed in *T. longibrachiatum* Δ*lae1* strains that were affected in growth, conidiation, and peptaibol production, since expression of two peptaibol synthase genes *tlx1* and *tlx2* was significantly reduced in the mutant (Shi et al. 2020). In *T. afroharzianum*, *lae1* overexpression led to the isolation of two structurally new polyketides with antifungal effects (Ding et al. 2020). Lae1 protein seems to regulate also 6-PP production (Aghcheh et al. 2013), which is in turn associated with the Thctf1 transcription factor: *T. afroharzianum* deletion mutants of this gene are unable to produce 6-PP derivatives and present alterations in their antimicrobial activity (Rubio et al. 2009). Recently, involvement of lipoxygenase LOX1 in 6-PP biosynthesis was confirmed by gene deletion, showing this gene is also required for the biosynthesis of several SMs including oxylipins and volatile compounds (Speckbacher et al. 2020). In the same way, the *veA* gene ortholog *vel1* has been studied through gene deletion, being involved in regulating several SM clusters and mating partner sensing in *T. virens* and *T. reesei*, respectively (Mukherjee and Kenerley 2010; Bazafkan et al. 2015).

Global regulation of *Trichoderma* SM biosynthesis also relies in G-protein-/cAMP-mediated signaling (Reithner et al. 2005; Zeilinger et al. 2005; Omann and Zeilinger 2010). Deletion of genes encoding the adenylyl cyclase-inhibiting Gα subunit (*tga1*) and the adenylyl cyclase-stimulating Gα subunit (*tga3*) altered 6-PP and peptaibol production in *T. atroviride* (Reithner et al. 2005; Zeilinger et al. 2005; Komon-Zelazowska et al. 2007). Similarly, the *T. atroviride* adenylate cyclase-encoding gene *tac1* and cAMP-dependent signaling have proven a key role in regulating SM biosynthesis, growth, germination, and mycoparasitism in this species (Mukherjee et al. 2007). cAMP is known to activate protein kinase A (PKA), in which catalytic subunit (PKA1) activity negatively regulates the expression of *lae1* in *T. reesei* (Mukherjee et al. 2007). Another kinase (*usk1*) has positive effects on *vel1* transcript levels and is involved in the regulation of several SM clusters in *T. reesei* (Beier et al. 2020). Mitogen-activated protein kinase (MAPK)-encoding gene *tmk1* deletion mutants of *T. atroviride* have enhanced peptaibol and 6-PP biosynthesis (Reithner et al. 2007), but deletion of its homolog in *T. virens* (*tmkA*/*tvk1*) showed no effects in SM production (Mendoza-Mendoza et al. 2003; Mukherjee et al. 2003). The complexity of the SM biosynthesis regulating network mediated by G-protein, cAMP, and MAPK has been recently described in *T. reesei* (Hinterdobler et al. 2020). The major carbon catabolite repressor CRE1 regulates a G-protein-coupled receptor (*gpr8*) in a light-dependent manner, triggering a signaling cascade involving MAPK and cAMP that activates YPR2 and SOR7 transcription factors, affecting the expression of several genes and gene clusters involved in SM biosynthesis (Hinterdobler et al. 2020).

3 Biochemistry of *Trichoderma* Secondary Metabolites

SMs are a heterogeneous group of natural compounds, belonging to diverse chemical classes and characterized by low molecular weight. SMs are mostly synthesized by microorganisms and plants and are typically genera, species, or strain specific. Secondary metabolism is associated with the shift from biomass production to metabolite biosynthesis in order to provide a benefit for the producer. SM biosynthesis starts from few primary metabolites (i.e., amino acids, acetyl-CoA) that follow specialized pathways and conduct to different natural compounds (Vinale and Sivasithamparam 2020).

Several strains of the genus *Trichoderma* are well-known producers of a plethora of secondary metabolites (SMs) with biological activity; these natural products are considered one of the elements that contributes to the positive effects exerted in agriculture. Such molecules may be involved in antibiosis and act synergistically with other compounds to promote plant growth and to induce systemic resistance (Vinale et al. 2008).

Trichoderma metabolites have been widely described over the years and classified according to their structure or the producing species. Here, SMs will be grouped through biological activity and applications in agriculture.

3.1 Antibiotics (Against Phytopathogens)

Trichoderma is used as a biocontrol agent worldwide due its antimicrobial activity against a broad spectrum of both bacterial and fungal phytopathogens. One of the reasons for this great efficiency is the chemical diversity of the secondary metabolites produced by *Trichoderma* strains (Keswani et al. 2014).

Ghisalberti and Sivasithamparam (1991) classified metabolites with antimicrobial activity into three groups: volatile compounds, non-volatile compounds, and peptaibols.

The chemical structures could be the key to understand the mechanism of action of those metabolites. For example, low molecular weight and volatile compounds can have a relatively long-distance range of influence on the microbial community. On the other hand, peptaibols and non-volatile molecules can achieve a short-distance effect acting close to the producing hyphae (Vinale et al. 2008).

Lorito et al. (1996) demonstrated that peptaibols produced by *T. harzianum* inhibit β-glucan synthase activity in the host fungus in order to prevent the assembling of the pathogen cell wall, while the endogenous β-glucanases perform their disruptive action.

The action of trichokonins VI, VII, and VIII (from *Trichoderma koningii*) against a broad range of pathogens, such as *Staphylococcus aureus*, *Bacillus subtilis*, *Streptococcus faecalis*, *Clavibacter* spp., *Fusarium oxysporum* f. sp. *phaseoli*, *F. oxysporum* f. sp. *niveum*, *F. oxysporum* f. sp. *vasinfectum*, *Botrytis cinerea*, *Rhizoctonia solani*, *Curvularia lunata*, *Bipolaris sorokiniana*, and *Colletotrichum lagenarium*, has been reported by Xiao-Yan et al. (2006) with agar disk diffusion assays.

Gliotoxin (Fig. 4 – **1**) and gliovirin (Fig. 4 – **2**) belong to the class of diketopiperazine and were isolated from *Trichoderma virens*; according to the molecule produced, strains of *T. virens* are divided into a "Q" group of strains able to produce gliotoxin and a "P" group of strains that produce gliovirin instead (Howell et al. 1996). Gliotoxin has a broad spectrum of antibiotic activity also against the human pathogenic fungus *Aspergillus fumigatus* (Scharf et al. 2016). Gliovirin is a specific oomycete inhibitor and has a positive effect in biocontrolling *Pythium ultimum* damping-off of cotton (Howell 1998).

6-Pentyl-α-pyrone (6-PP) (Fig. 4 – **3**) is a flavoring agent responsible for coconut aroma and belongs to the chemical group of pyrones, low molecular weight and volatile molecules. This compound is active against *R. solani* and *F. oxysporum* f.

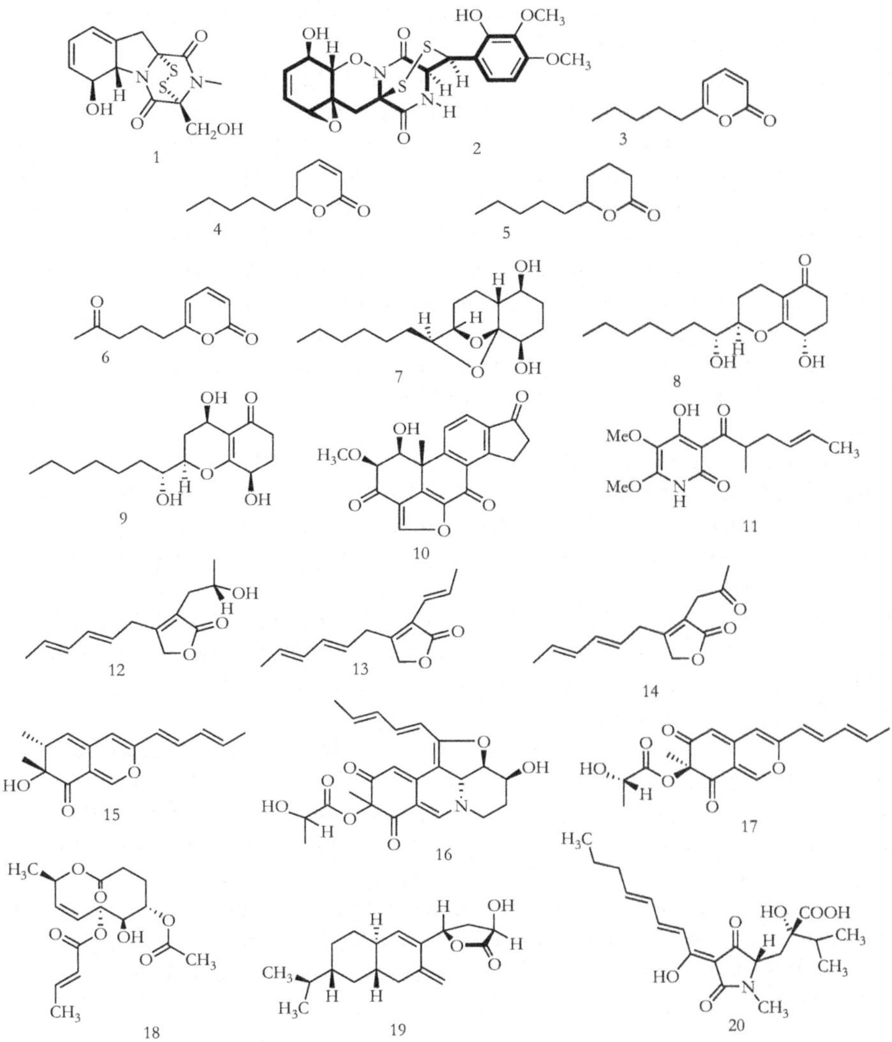

Fig. 4 Chemical structures of secondary metabolites produced by *Trichoderma* spp. with biological activity against plant pathogens. Gliotoxin (**1**); gliovirin (**2**); 6-pentyl-α-pyrone (6-PP) (**3**); massoilactone (**4**); δ-decanolactone (**5**); viridepyronone (**6**); koninginins A and B (**7–8**); koninginin D (**9**); viridin (**10**); harzianopyridone (**11**); harzianolide (**12**); dehydro-harzianolide (**13**); T39butenolide (**14**); harziphilone (**15**); fleephilone (**16**); T22azaphilone (**17**); cremenolide (**18**); cerinolactone (**19**); harzianic acid (**20**)

sp. *lycopersici*, reducing the growth of 31.7% and 69.6%, respectively (Scarselletti and Faull 1994).

Two hydrogenated derivatives of 6-PP, massoilactone (Fig. 4 – **4**) and δ-decanolactone (Fig. 4 – **5**), are active against *B. cinerea*, *Phytophthora* spp., *Aspergillus niger*, and *Candida albicans* (Kishimoto et al. 2005). Another analogue of 6-PP, viridepyronone (Fig. 4 – **6**), showed antagonistic activity against *Sclerotium rolfsii* (Evidente et al. 2003).

Complex pyrones such as koninginins A and B (Fig. 4 – **7–8**) show antifungal activity against *Gaeumannomyces graminis* var. *tritici* (Ghisalberti and Rowland 1993), while koninginin D (Fig. 4 – **9**) inhibited the growth of *Bipolaris sorokiniana, Pythium middleonii, F. oxysporum, Phytophthora cinnamomi*, and *R. solani* (Dunlop et al. 1989). Similarly, viridin (Fig. 4 – **10**), a broad-spectrum antifungal compound, prevents spore germination of *Stachybotrys atra, A. niger, Penicillium expansum, Colletotrichum lini, Fusarium caeruleum*, and *Botrytis allii* (Brian and McGowan 1945).Harzianopyridone (Fig. 4 – **11**) belongs to the class of pyridines and contains a pyridine ring system with a 2,3-dimethoxy-4-pyridinol pattern. Application of the racemic form shows a strong antifungal activity against *R. solani, B. cinerea* (Dickinson et al. 1989), *P. ultimum*, and *G. graminis* var. *tritici* (Vinale et al. 2006).

Antifungal butenolides such as harzianolide (Fig. 4 – **12**), its dehydro-derivative (Fig. 4 – **13**), and T39butenolide (Fig. 4 – **14**) are produced by specific strains of *T. harzianum* and are all active against *G. graminis* var. *tritici* (Almassi et al. 1991). Additionally, harzianolide and T39butenolide showed in vitro growth inhibition of *R. solani* and *P. ultimum*.

Harziphilone (Fig. 4 – **15**), fleephilone (Fig. 4 – **16**), and T22azaphilone (Fig. 4 – **17**) belong to the class of azaphilones and are all produced by *Trichoderma harzianum*. They are significantly active against *G. graminis* var. *tritici, R. solani*, and *P. ultimum* (Vinale et al. 2006).

The ten-member lactone cremenolide (Fig. 4 – **18**) showed antifungal activities against *R. solani, B. cinerea*, and *F. oxysporum* (Vinale et al. 2016). Furthermore, Vinale et al. (2011) showed 100%, 41%, and 28% inhibition of *P. ultimum, R. solani*, and *B. cinerea*, respectively, at 100 μg/plug concentration of cerinolactone (Fig. 4 – **19**), another molecule belonging to lactone class.

Harzianic acid (Fig. 4 – **20**), a tetramic acid produced by *T. harzianum* strain M10, showed complete inhibition of *Pythium irregulare* and *Sclerotinia sclerotiorum* at 10 μg, while at 100 μg, it completely arrested the growth of *R. solani* (Vinale et al. 2009).

3.2 Plant Growth Regulators

In addition to direct activity against phytopathogens, some *Trichoderma* spp. produce compounds that can cause substantial changes in the metabolism of the host plant. One of the consequences is related to an enhanced production of plant biomass or growth promotion of roots. Several strains can stimulate plant development by activating an auxin-dependent mechanism and/or by producing indole-3-acetic acid (IAA) or auxin analogues (Vinale et al. 2012).

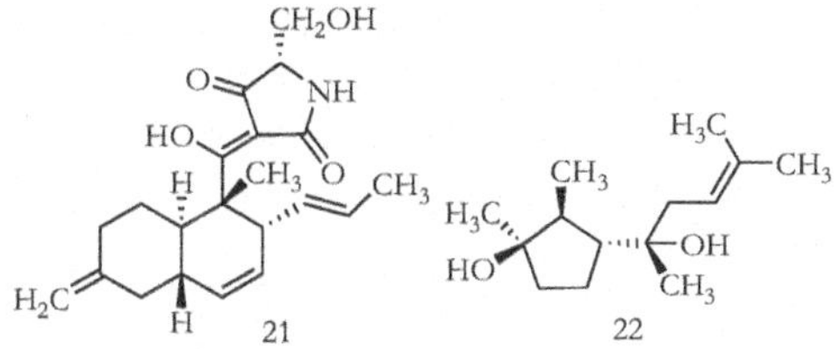

Fig. 5 Chemical structures of secondary metabolites produced by *Trichoderma* spp. with plant growth-promoting activity. Trichosetin (**21**), cyclonerodiol (**22**)

Koninginins A–C, E, and G from *Trichoderma koningii* and 6-pentyl-α-pyrone (**3**) significantly inhibit the growth of etiolated wheat coleoptiles when used at high concentration, 10^{-3} M (Parker et al. 1997).

Harzianic acid (Fig. 4 – **20**) from *T. harzianum* is a nitrogen heterocyclic compound with growth promotion activity in a concentration-dependent manner; specifically, in canola seedlings, inhibition up to 45 and 33% in stem length is caused after application of harzianic acid at a concentration of 100 and 10 μg/seed respectively, while at a concentration of 100, 10, and 1 ng/seed, stem length increases by 42, 44, and 52%, respectively, compared to control (Vinale et al. 2009).

Treatments with harzianolide (Fig. 4 – **12**) from *T. harzianum* at a concentration of 1 mg/l have positive effect on the growth in *Brassica napus* and *Solanum lycopersicum* seedlings (Vinale et al. 2008).

The dual culture of *T. harzianum* and calli of *Catharathus roseus* produces a compound derived of tetramic acid named trichosetin (Fig. 5 – **21**) that affects the root and shoot growth of several plant species such as *Oryza sativa*, *Vigna radiata*, *Medicago sativa*, *Capsicum frutescens*, and *Lycopersicum esculentum* (Marfori et al. 2003).

Cyclonerodiol (Fig. 5 – **22**), a simple sesquiterpene isolated from *T. koningii* and *T. harzianum*, exhibits growth regulatory effects at low concentration (Cutler et al. 1991).

Cerinolactone (Fig. 4 – **19**) has been isolated from culture filtrates of *Trichoderma cerinum*, together with other known butenolides, and altered the growth of tomato seedlings 3 days after treatment (Vinale et al. 2012).

3.3 Siderophores

Living organisms require nutrients for their development and reproduction, and the competition for organic/inorganic compounds is extremely important during the interaction with microorganisms and plants.

Iron is a transition metal that can exist in two oxidation states, ferrous (Fe^{2+}) and ferric (Fe^{3+}), with essential nutrient properties. Although iron is one of the most abundant elements on earth, its bioavailability is low in aerobic environments at neutral pH and in presence of oxygen, mostly because Fe^{3+} reacts with oxygen to form insoluble ferric hydroxides. Thus, several organisms have developed regulated strategies for the control of iron uptake, utilization, and storage. One of these strategies involves the production of metabolites, named siderophores, able to bind Fe^{3+}. These compounds are released into the extra-hyphal space to solubilize, bind, and take up iron. This behavior is beneficial for plants because the complex Fe-siderophore can be easily taken up by plants to provide iron, and the production of microbial siderophores can also suppress the growth of pathogens by depriving iron.

Coprogen, coprogen B, and fusarinine C are siderophores produced by *Trichoderma atroviride*, *Trichoderma gamsii*, *Trichoderma asperellum*, *Trichoderma hamatum*, *Trichoderma virens*, and *Trichoderma harzianum* belonging to the group of hydroxamate siderophores that share the structural unit N5-acyl-N5-hydroxyornithine.

Harzianic acid (Fig. 4 – **20**), named in previous paragraphs for its outstanding characteristic, is also able to chelate Fe^{3+} due to its structure derived from tetramic acid (Zeilinger et al. 2016; Vinale et al. 2012).

3.4 Plant Defense Response Inducers

Fungi of the genus *Trichoderma*, similar to other plants' beneficial microorganisms, are able to release elicitor-like substances that induce a systemic or localized resistance response during the interaction with plants (Harman et al. 2004). These substances can be grouped into three classes: proteins with enzymatic activity; avirulence-like gene products able to induce defense reactions in plants; and low molecular weight compounds released from either fungal or plant cell walls by the activity of *Trichoderma* enzymes.

For example, 6-PP not only has an antifungal and plant growth-promoting activity but is also able to induce systemic resistance. It has been shown that tomato plants treated with 6-PP produce more γ-aminobutyric acid and acetylcholine that help plants to fight against pathogens (Mazzei et al. 2016). Garnica-Vergara et al. (1991) also demonstrated that 6-PP interferes with the pathway of auxins and ethylene in plants and induces the formation of lateral shoots, by modulating the expression of genes encoding for auxin transporters.

Peptaibols are another important class of plant defense elicitor. In particular, the 20-residue alamethicin F30 induces jasmonic acid- and salicylic acid-mediated resistance in lima bean (Engelberth et al. 2000). The 11-, 14-, and 18-mer peptaibols produced by *Trichoderma virens* are able to induce the production of salicylic acid and camalexin in *Arabidopsis thaliana* (Velázquez-Robledo et al. 2011).

Finally, the *Trichoderma arundinaceum* polyketide, aspinolide C (Fig. 6 – **23**), induced in plant the expression of genes involved in the signaling pathway mediated by salicylic acid. On the contrary, the aspinolide B (Fig. 6 – **24**) (an aspinolide C derivative) moderately repressed salicylic acid-related genes, while the influence on the expression of jasmonic acid-related genes was not homogeneous for both the two polyketides (Malmierca et al. 2015).

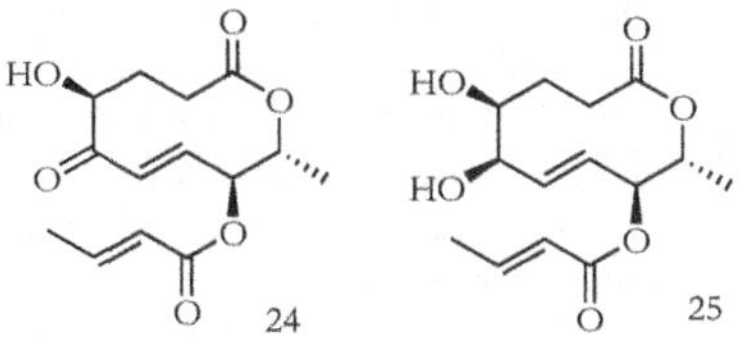

Fig. 6 Chemical structures of elicitor-like metabolites produced by *Trichoderma* spp. Aspinolide C (**23**); aspinolide B (**24**)

4 Conclusions and Future Perspectives

Despite many studies have been herewith performed on the secondary metabolism of *Trichoderma*, its products, and their role in the ecophysiology of this important group of fungi, many are the information still lacking to complete this framework. Actually, the availability of the so-called NGS as well as the new "omics" techniques represents important tools to support investigations on complex network of plant-microbe (including *Trichoderma*) interactions. Particularly, metabolomics can be of help to improve knowledge about *Trichoderma* SMs and their role in modulating the beneficial activities against plant pathogens in favor of the plant host.

Improving the knowledge of *Trichoderma* spp. and their SMs is fundamental to improve formulations based on living fungi and/or their bioactive compounds representing the main ingredients of new biopesticides.

References

Aghcheh RK, Druzhinina IS, Kubicek CP (2013) The putative protein methyltransferase LAE1 of *Trichoderma atroviride* is a key regulator of asexual development and mycoparasitism. PLoS One 8(6):e67144

Alfiky A, Weisskopf L (2021) Deciphering *Trichoderma*–plant–pathogen interactions for better development of biocontrol applications. J Fungi 7(1):61

Almassi F, Ghisalberti EL, Narbey MJ, Sivasithamparam K (1991) New antibiotics from strains of *Trichoderma harzianum*. J Nat Prod 54:396–402

Atanasova L, Crom SL, Gruber S, Coulpier F, Seidl-Seiboth V, Kubicek CP, Druzhinina IS (2013) Comparative transcriptomics reveals different strategies of *Trichoderma* mycoparasitism. BMC Genomics 14(1):121

Baker SE, Perrone G, Richardson NM, Gallo A, Kubicek CP (2012) Phylogenomic analysis of polyketide synthase-encoding genes in *Trichoderma*. Microbiology 158:147–154

Bansal R, Mukherjee PK (2016) The terpenoid biosynthesis toolkit of *Trichoderma*. Nat Prod Commun 11:1934578X1601100

Bazafkan H, Dattenböck C, Böhmdorfer S, Tisch D, Stappler E, Schmoll M (2015) Mating type-dependent partner sensing as mediated by VEL1 in *Trichoderma reesei*. Mol Microbiol 96(6):1103–1118

Beier S, Hinterdobler W, Monroy AA, Bazafkan H, Schmoll M (2020) The Kinase USK1 regulates cellulase gene expression and secondary metabolite biosynthesis in *Trichoderma reesei*. Front Microbiol 11

Brakhage AA (2013) Regulation of fungal secondary metabolism. Nat Rev Microbiol 11(1):21–32

Brian PW, McGowan JC (1945) Viridin a highly fungistatic substance produced by *Trichoderma viride*. Nature 156:144–145

Bushley KE, Turgeon BG (2010) Phylogenomics reveals subfamilies of fungal nonribosomal peptide synthetases and their evolutionary relationships. BMC Evol Biol 10:26

Calvo AM, Wilson RA, Bok JW, Keller NP (2002) Relationship between secondary metabolism and fungal development. Microbiol Mol Biol Rev 66:447–459

Cardoza RE, Malmierca MG, Hermosa MR, Alexander NJ, McCormick SP, Proctor RH et al (2011) Identification of loci and functional characterization of trichothecene biosynthesis genes in filamentous fungi of the genus *Trichoderma*. Appl Environ Microbiol 77(14):4867–4877

Chavali AK, Rhee SY (2017) Bioinformatics tools for the identification of gene clusters that biosynthesize specialized metabolites. Brief Bioinform 19:1022–1034

Christianson DW (2008) Unearthing the roots of the terpenome. Curr Opin Chem Biol 12:141–150

Contreras-Cornejo HA, Macias-Rodriguez L, Cortes-Penagos C, Lopez-Bucio J (2009) *Trichoderma virens*, a plant beneficial fungus, enhances biomass production and promotes lateral root growth through an auxin-dependent mechanism in *Arabidopsis*. Plant Physiol 149:1579–1592

Contreras-Cornejo HA, Macías-Rodríguez L, del-Val E, Larsen J (2016) Ecological functions of *Trichoderma* spp. and their secondary metabolites in the rhizosphere: interactions with plants. FEMS Microbiol Ecol 92:fiw036

Contreras-Cornejo HA, Del-Val E, Macías-Rodríguez L, Alarcón A, González-Esquivel CE, Larsen J (2018) *Trichoderma atroviride*, a maize root associated fungus, increases the parasitism rate of the fall armyworm *Spodoptera frugiperda* by its natural enemy *Campoletis sonorensis*. Soil Biol Biochem 122:196–202

Crutcher FK, Parich A, Schuhmacher R, Mukherjee PK, Zeilinger S, Kenerley CM (2013) A putative terpene cyclase, *vir4*, is responsible for the biosynthesis of volatile terpene compounds in the biocontrol fungus *Trichoderma virens*. Fungal Genet Biol 56:67–77

Cutler HG, Jacyno JM, Phillips RS, von Tersch RL, Cole PD, Montemurro N (1991) Cyclonerodiol from a novel source, *Trichoderma koningii*: plant growth regulatory activity. Agric Biol Chem 55:243–244

Dautt-Castro M, Estrada-Rivera M, Olguin-Martínez I, del Carmen Rocha-Medina M, Auxiliadora Islas-Osuna M, Casas-Flores S (2019) TBRG-1 a ras-like protein in *Trichoderma virens* involved in conidiation, development, secondary metabolism, mycoparasitism, and biocontrol unveils a new family of Ras-GTPases. Fungal Genet Biol 103292

De Vries RP, Riley R, Wiebenga A, Aguilar-Osorio G, Amillis S, Uchima CA et al (2017) Comparative genomics reveals high biological diversity and specific adaptations in the industrially and medically important fungal genus *Aspergillus*. Genome Biol 18:28

Degenkolb R, Aghcheh K, Dieckmann R, Neuhof T, Baker SE, Druzhinina IS et al (2012) The production of multiple small peptaibol families by single 14-module peptide synthetases in *Trichoderma/Hypocrea*. Chem Biodivers 2012(9):499–535

Dickinson JM, Hanson JR, Hitchcock PB, Claydon N (1989) Structure and biosynthesis of harzianopyridone, an antifungal metabolite of *Trichoderma harzianum*. J Chem Soc Perkin Trans 1:1885–1887

Ding Z, Wang X, Kong FD, Huang HM, Zhao YN, Liu M et al (2020) Overexpression of global regulator Talae1 leads to the discovery of new antifungal polyketides from endophytic fungus *Trichoderma afroharzianum*. Front Microbiol 11

Druzhinina IS, Seidl-Seiboth V, Herrera-Estrella A, Horwitz BA, Kenerley CM, Monte E, Mukherjee PK, Zeilinger S, Grigoriev IG, Kubicek CP (2011) *Trichoderma*: the genomics of opportunistic success. Nat Rev Microbiol 9:749–759

Dunlop RW, Simon A, Sivasithamparam K, Ghisalberti EL (1989) An antibiotic from *Trichoderma koningii* active against soilborne plant pathogens. J Nat Prod 52:67–74

Engelberth J, Koch T, Schuler G, Bachmann N, Rechtenbach J, Boland W (2000) Ion channel-forming alamethicin is a potent elicitor of volatile biosynthesis and tendril coiling. Cross talk between jasmonate and salicylate signaling in lima bean. Plant Physiol 125:369–377

Evidente A, Cabras A, Maddau L, Serra DA, Andolfi A, Motta A (2003) Viride Pyronone, a new antifungal 6 substituted 2H-pyran-2-one produced by *Trichoderma viride*. J Agric Food Chem 51:6957–6960

Gallo A, Mulè G, Favilla M, Altomare C (2005) Isolation and characterisation of a trichodiene synthase homologous gene in *Trichoderma harzianum*. Physiol Mol Plant Pathol 65:11–20

Gardiner DM (2005) The epipolythiodioxopiperazine (ETP) class of fungal toxins: distribution, mode of action, functions and biosynthesis. Microbiology 151:1021–1032

Garnica-Vergara A, Barrera-Ortiz S, Muñoz-Parra E, Raya-González J, Ghisalberti EL, Sivasithamparam K (1991) Antifungal antibiotics produced by *Trichoderma* spp. Soil Biol Biochem 23:1011–1020

Ghisalberti EL, Rowland CY (1993) Antifungal metabolites from *Trichoderma harzianum*. J Nat Prod 56:1799–1804

Harman GE, Howell CR, Viterbo A, Chet I, Lorito M (2004) *Trichoderma* species - opportunistic, avirulent plant symbionts. Nat Rev Microbiol 2:43–56

Hinterdobler W, Beier S, Monroy AA, Berger H, Dattenböck C, Schmoll M (2020) The G-protein coupled receptor GPR8 regulates secondary metabolism in *T. reesei*. Front Bioeng Biotechnol 8:558996

Howell CR (1998) The role of antibiosis in biocontrol. In: Harman GE, Kubicek CP (eds) *Trichoderma* and *Gliocladium*. Enzymes, biological control and commercial application, vol 2. Taylor & Francis Ltd., London, pp 173–183

Howell CR (2003) Mechanisms employed by *Trichoderma* species in the biological control of plant diseases: the history and evolution of current concepts. Plant Dis 87:4–10

Howell CR, Stipanovic RD, Lumsden RD (1996) Antibiotic production by strains of *Gliocladium virens* and its relation to the biocontrol of cotton seedling diseases. Biocontrol Sci Tech 3(4):435–441

Jona Lasinio G, Pollice A, Pappalettere L, Vannacci G, Sarrocco S (2021) A new statistical protocol to describe differences among nutrient utilization patterns of *Fusarium* spp. and *Trichoderma gamsii*. Plant Pathol. https://doi.org/10.1111/ppa.13362

Karimi-Aghcheh R, Bok JW, Phatale PA, Smith KM, Baker SE, Lichius A et al (2013) Functional analyses of *Trichoderma reesei* LAE1 reveal conserved and contrasting roles of this regulator. G3 Genes|Genomes|Genetics 3(2):369–378

Keller NP (2019) Fungal secondary metabolism: regulation, function and drug discovery. Nat Rev Microbiol 17:1671–1680

Keller NP, Turner G, Bennett JW (2005) Fungal secondary metabolism — from biochemistry to genomics. Nat Rev Microbiol 3:937–947

Keswani C, Mishra S, Sarma BK, Singh SP, Sinh HB (2014) Unraveling the efficient applications of secondary metabolites of various *Trichoderma* spp. Appl Microbiol Biotechnol 98:533–544

Khan RAA, Najeeb S, Hussain S, Xie B, Li Y (2020) Bioactive secondary metabolites from *Trichoderma* spp. against phytopathogenic fungi. Microorganisms 8:817

Kishimoto N, Sugihara S, Mochida K, Fujita T (2005) In vitro antifungal and antiviral activities of c- and d-lactone analogs utilized as food flavoring. Biocontrol Sci 10:31–36

Komon-Zelazowska M, Neuhof T, Dieckmann R, von Döhren H, Herrera-Estrella A, Kubicek CP, Druzhinina IS (2007) Formation of atroviridin by *Hypocrea atroviridis* is conidiation associated and positively regulated by blue light and the G protein GNA3. Eukaryot Cell 6(12):2332–2342

Kubicek CP, Herrera-Estrella A, Seidl-Seiboth V, Martinez DA, Druzhinina IS, Thon M et al (2011) Comparative genome sequence analysis underscores mycoparasitism as the ancestral life style of *Trichoderma*. Genome Biol 2011(12):R40

Kubicek CP, Steindorff AS, Chenthamara K, Manganiello G, Henrissat B, Zhang J et al (2019) Evolution and comparative genomics of the most common *Trichoderma* species. BMC Genomics 20:485

Künzler M (2018) How fungi defend themselves against microbial competitors and animal predators. PLoS Pathog 14:e1007184

Lehner SM, Atanasova L, Neumann NKN, Krska R, Lemmens M, Druzhinina IS, Schuhmacher R (2012) Isotope-assisted screening for iron-containing metabolites reveals a high degree of diversity among known and unknown siderophores produced by *Trichoderma* spp. Appl Environ Microbiol 79(1):18–31

Li MF, Li GH, Zhang KQ (2019) Non-volatile metabolites from *Trichoderma* spp. Meta 9:58
Lindo L, McCormick SP, Cardoza RE, Brown DW, Kim HS, Alexander NJ et al (2018) Effect of deletion of a trichothecene toxin regulatory gene on the secondary metabolism transcriptome of the saprotrophic fungus *Trichoderma arundinaceum*. Fungal Genet Biol 119:29–46
Lindo L, McCormick SP, Cardoza RE, Kim HS, Brown DW, Alexander NJ et al (2019) Role of *Trichoderma arundinaceum tri10* in regulation of terpene biosynthetic genes and in control of metabolic flux. Fungal Genet Biol 122:31–46
Lorito M, Farkas V, Rebuffat S, Bodo B, Kubicek CP (1996) Cell wall synthesis is a major target of mycoparasitic antagonism by *Trichoderma harzianum*. J Bacteriol 178(21):6382–6385
Macheleidt J, Mattern DJ, Fischer J, Netzker T, Weber J, Schroeckh V et al (2016) Regulation and role of fungal secondary metabolites. Annu Rev Genet 50:371–392
Mahmood ZA, Ahmed SW, Azhar I, Sualeh M, Baig MT, Zoha S (2010) Bioactive alkaloids produced by fungi I. Updates on alkaloids from the species of the genera *Boletus*, *Fusarium* and *Psilocybe*. J Pharm Sci 23:349–357
Malmierca MG, Cardoza RE, Alexander NJ, McCormick SP, Hermosa R, Monte E, Gutiérrez S (2012) Involvement of *Trichoderma* trichothecenes in the biocontrol activity and induction of plant defense-related genes. Appl Environ Microbiol 78(14):4856–4868
Malmierca MG, Cardoza RE, Alexander NJ, McCormick SP, Collado IG, Hermosa R et al (2013) Relevance of trichothecenes in fungal physiology: disruption of *tri5* in *Trichoderma arundinaceum*. Fungal Genet Biol 53:22–33
Malmierca MG, Barua J, McCormick SP, Izquierdo-Bueno I, Cardoza RE, Alexander NJ et al (2015) Novel aspinolide production by *Trichoderma arundinaceum* with a potential role *in Botrytis cinerea* antagonistic activity and plant defence priming. Environ Microbiol 17:1103–1118
Manganiello G, Sacco A, Ercolano MR, Vinale F, Lanzuise S, Pascale A, Napolitano M, Lombardi N, Lorito M, Woo SL (2018) Modulation of tomato response to *Rhizoctonia solani* by *Trichoderma harzianum* and its secondary metabolite Harzianic acid. Front Microbiol 9:1966
Marahiel MA (2009) Working outside the protein-synthesis rules: insights into non-ribosomal peptide synthesis. J Pept Sci 15(12):799–807
Marfori EC, Kajiyama S, Fukusaki E, Kobayashi A (2003) Phytotoxicity of the tetramic acid metabolite trichosetin. Phytochemistry 62:715–721
Marik T, Tyagi C, Balázs D, Urbán P, Szepesi Á, Bakacsy L et al (2019) Structural diversity and bioactivities of peptaibol compounds from the Longibrachiatum clade of the filamentous fungal genus *Trichoderma*. Front Microbiol 10:1434
Mazzei P, Vinale F, Woo SL, Pascale A, Lorito M, Piccolo A (2016) Metabolomics by proton high-resolution magic-angle-spinning nuclear magnetic resonance of tomato plants treated with two secondary metabolites isolated from *Trichoderma*. J Agric Food Chem 64(18):3538–3545
Mendoza-Mendoza A, Pozo MJ, Grzegorski D, Martinez P, Garcia JM, Olmedo-Monfil V et al (2003) Enhanced biocontrol activity of *Trichoderma* through inactivation of a mitogen-activated protein kinase. Proc Natl Acad Sci 100(26):15965–15970
Molle G, Doclohier H, Spach G (1987) Voltage dependent and multistate ionic channels induced by trichorzianines, antifungal peptides related to alamethicin. FEBS Lett 224:208–212
Mukherjee PK, Kenerley CM (2010) Regulation of morphogenesis and biocontrol properties in *Trichoderma virens* by a VELVET protein, Vel1. Appl Environ Microbiol 76(7):2345–2352
Mukherjee PK, Latha J, Hadar R, Horwitz BA (2003) TmkA, a mitogen-activated protein kinase of *Trichoderma virens*, is involved in biocontrol properties and repression of conidiation in the dark. Eukaryot Cell 2(3):446–455
Mukherjee M, Horwitz BA, Sherkhane PD, Hadar R, Mukherjee PK (2006) A secondary metabolite biosynthesis cluster in *Trichoderma virens*: evidence from analysis of genes underexpressed in a mutant defective in morphogenesis and antibiotic production. Curr Genet 50:193–202
Mukherjee M, Mukherjee PK, Kale SP (2007) cAMP signalling is involved in growth, germination, mycoparasitism and secondary metabolism in *Trichoderma virens*. Microbiology 153(6):1734–1742
Mukherjee PK, Horwitz BA, Kenerley CM (2011) Secondary metabolism in *Trichoderma* – a genomic perspective. Microbiology 158:35–45

Mukherjee PK, Buensanteai N, Morán-Diez ME, Druzhinina I, Kenerley CM (2012) Functional analysis of non-ribosomal pep-tide synthetases (NRPSs) in *Trichoderma virens* reveals a polyketide synthase (PKS)/NRPS hybrid enzyme involved in the induced systemic resistance response in maize. Microbiology 158:155–165

Mukherjee PK, Horwitz BA, Herrera-Estrella A, Schmoll M, Kenerley CM (2013) *Trichoderma* research in the genome era. Annu Rev Phytopathol 51(1):105–129

Omann M, Zeilinger S (2010) How a mycoparasite employs G-protein signaling: using the example of *Trichoderma*. J Signal Transduct 1–8

Pachauri S, Sherkhane PD, Mukherjee P (2019) Secondary metabolism in *Trichoderma*: chemo- and geno-diversity. In: Satyanarayana T, Das SK, Johri BN (eds) Microbial diversity in ecosystem sustainability and biotechnological applications. Springer, Singapore, pp 441–456

Parker SR, Cutler HG, Jacyno JM, Hill RA (1997) Biological activity of 6-pentyl-2*H*-pyran-2-one and its analogs. J Agric Food Chem 45:2774–2776

Patil AS, Patil SR, Paikrao HM (2016) *Trichoderma* secondary metabolites: their biochemistry and possible role in disease management. In: Choudhary DK, Varma A (eds) Microbial-mediated induced systemic resistance in plants. Springer, Singapore, pp 69–102

Pérez-Gil J, Rodríguez-Concepción M, Vickers CE (2019) Formation of isoprenoids. In: Geiger O (ed) Biogenesis of fatty acids, lipids and membranes. Handbook of hydrocarbon and lipid microbiology, pp 57–85

Proctor RH, McCormick SP, Kim HS, Cardoza RE, Stanley AM, Lindo L et al (2018) Evolution of structural diversity of trichothecenes, a family of toxins produced by plant pathogenic and entomopathogenic fungi. PLoS Pathog 14:e1006946

Pusztahelyi T, Holb IJ, Pócsi I (2015) Secondary metabolites in fungus-plant interactions. Front Plant Sci 6:573

Rai S, Solanki MK, Solanki AC, Surapathrudu K (2019) Biocontrol potential of *Trichoderma* spp.: current understandings and future outlooks on molecular techniques. In: Ansari RA, Mahmood I (eds) Plant health under biotic stress. Springer, Singapore, pp 129–160

Ramírez-Valdespino CA, Casas-Flores S, Olmedo-Monfil V (2019) *Trichoderma* as a model to study effector-like molecules. Front Plant Sci 10:1030

Reithner B, Brunner K, Schuhmacher R, Peissl I, Seidl V, Krska R, Zeilinger S (2005) The G protein α subunit Tga1 of *Trichoderma atroviride* is involved in chitinase formation and differential production of antifungal metabolites. Fungal Genet Biol 42(9):749–760

Reithner B, Schuhmacher R, Stoppacher N, Pucher M, Brunner K, Zeilinger S (2007) Signaling via the *Trichoderma atroviride* mitogen-activated protein kinase Tmk1 differentially affects mycoparasitism and plant protection. Fungal Genet Biol 44(11):1123–1133

Renshaw JC, Robson GD, Trinci APJ, Wiebe MG, Livens FR, Collison D, Taylor RJ (2002) Fungal siderophores: structures, functions and applications. Mycol Res 106(10):1123–1142

Rokas A, Wisecaver JH, Lind AL (2018) The birth, evolution and death of metabolic gene clusters in fungi. Nat Rev Microbiol 16:731–744

Rubio MB, Hermosa R, Reino J, Collado I, Monte E (2009) Thctf1 transcription factor of *Trichoderma harzianum* is involved in 6-pentyl-2H-pyran-2-one production and antifungal activity. Fungal Genet Biol 46(1):17–27

Salwan R, Rialch N, Sharma V (2019) Bioactive volatile metabolites of *Trichoderma:* an overview. In: Singh HB, Keswani C, Reddy MS, Sansinenea E, García-Estrada C (eds) Secondary metabolites of plant growth promoting Rhizomicroorganisms. Springer, Singapore, pp 87–111

Saravanakumar K, Yu C, Dou K, Wang M, Li Y, Chen J (2016) Synergistic effect of *Trichoderma*-derived antifungal metabolites and cell wall degrading enzymes on enhanced biocontrol of *Fusarium oxysporum* f. sp. *cucumerinum*. Biol Control 94:37–46

Sarrocco S (2016) Dung-inhabiting fungi: a potential reservoir of novel secondary metabolites for the control of plant pathogens. Pest Manag Sci 72:643–652

Sarrocco S, Vannacci G (2018) Preharvest application of beneficial fungi as a strategy to prevent postharvest mycotoxin contamination: a review. Crop Prot 110:160–170

Sarrocco S, Guidi L, Fambrini S, Degl'Innocenti E, Vannacci G (2009) Competition for cellulose exploitation between *Rhizoctonia solani* and two *Trichoderma* isolates in the decomposition of wheat straw. J Plant Pathol 91:331–338

Sarrocco S, Mauro A, Battilani P (2019a) Use of competitive filamentous fungi as an alternative approach for mycotoxin risk reduction in staple cereals: state of art and future perspectives. Toxins 11:701

Sarrocco S, Valenti F, Manfredini S, Esteban P, Bernardi R, Puntoni G, Baroncelli R, Haidukowski M, Moretti A, Vannacci G (2019b) Is exploitation competition *involved in a multitrophic strategy for the biocontrol of Fusarium Head Blight?* Phytopathology 109:560–570

Sarrocco S, Esteban P, Vicente I, Bernardi R, Plainchamp T, Domenichini S, Vicente Munoz I, Puntoni G, Baroncelli R, Vannacci G (2021) Dufresne M Straw competition and wheat root endophytism of *Trichoderma gamsii* T6085 as useful traits in the biocontrol of Fusarium Head Blight. Phytopathology. https://doi.org/10.1094/PHYTO-09-20-0441-R

Scarselletti R, Faull JL (1994) In vitro activity of 6-pentyl-α-pyrone, a metabolite of *Trichoderma harzianum*, in the inhibition of *Rhizoctonia solani* and *Fusarium oxysporum* f. sp. *Lycopersici*. Mycol Res 98(10):1207–1209

Scharf DH, Brakhage AA, Mukherjee PK (2016) Gliotoxin--bane or boon? Environ Microbiol 18(4):1096–1109

Schmoll M, Dattenböck C, Carreras-Villaseñor N, Mendoza-Mendoza A, Tisch D, Alemán MI et al (2016) The genomes of three uneven siblings: footprints of the lifestyles of three *Trichoderma* species. Microbiol Mol Biol Rev 80(1):205–327

Schümann J, Hertweck C (2006) Advances in cloning, functional analysis and heterologous expression of fungal polyketide synthase genes. J Biotechnol 124:690–703

Seiboth B, Karimi RA, Phatale PA, Linke R, Hartl L, Sauer DG et al (2012) The putative protein methyltransferase LAE1 controls cellulase gene expression in *Trichoderma reesei*. Mol Microbiol 84(6):1150–1164

Serrano-Carreon L, Hathout Y, Bensoussan M, Belin JM (1992) Production of 6-pentyl-α-pyrone by *Trichoderma harzianum* from 18:n fatty acid methyl esters. Biotechnol Lett 14(11):1019–1024

Shaw JJ, Berbasova T, Sasaki T, Jefferson-George K, Spakowicz DJ, Dunican BF et al (2015) Identification of a fungal 1,8-cineole synthase from *Hypoxylon* sp. with specificity determinants in common with the plant synthases. J Biol Chem 290:8511–8526

Shenouda ML, Cox RJ (2021) Molecular methods unravel the biosynthetic potential of *Trichoderma* species. RSC Adv 11:3622

Shi JC, Shi WL, Zhou YR, Chen XL, Zhang YZ, Zhang X et al (2020) The putative methyltransferase TlLAE1 is involved in the regulation of peptaibols production in the biocontrol fungus *Trichoderma longibrachiatum* SMF2. Front Microbiol 11

Singh E, Chandra P (2019) Fungal community for novel secondary metabolites. In: Yadav A, Singh S, Mishra S, Gupta A (eds) Recent advancement in white biotechnology through fungi. Fungal biology. Springer, Cham, pp 249–283

Speckbacher V, Ruzsanyi V, Martinez-Medina A, Hinterdobler W, Doppler M, Schreiner U et al (2020) The lipoxygenase Lox1 is involved in light- and injury-response, conidiation, and volatile organic compound biosynthesis in the mycoparasitic fungus *Trichoderma atroviride*. Front Microbiol 11

Sun X, Cai YS, Yuan Y, Bian G, Ye Z, Deng Z, Liu T (2019) Genome mining in *Trichoderma viride* J1-030: discovery and identification of novel sesquiterpene synthase and its products. Beilstein J Org Chem 15:2052–2058

Susca A, Moretti A, Logrieco AF (2017) Mycotoxin biosynthetic pathways: a window on the evolutionary relationships among toxigenic fungi. In: Varma A, Sharma A (eds) Modern tools and techniques to understand microbes. Springer, Cham, pp 135–148

Tijerino A, Hermosa R, Cardoza RE, Moraga J, Malmierca MG, Aleu J et al (2011) Overexpression of the *Trichoderma brevicompactum tri5* gene: effect on the expression of the trichodermin biosynthetic genes and on tomato seedlings. Toxins 3:1220–1232

Trushina N, Levin M, Mukherjee PK, Horwitz B (2013) PacC and pH–dependent transcriptome of the mycotrophic fungus *Trichoderma virens*. BMC Genomics 14:138

Velázquez-Robledo R, Contreras-Cornejo HA, Macías-Rodrígez L, Hernández-Morales A, Aguirre J, Casas-Flores S, Herrera-Estrella A (2011) Role of the 4-phosphopantetheinyl transferase of *Trichoderma virens* in secondary metabolism and induction of plant defense response. Mol Plant-Microbe Interact 24:1459–1471

Vicente I (2020) Computational and molecular study of terpene synthase genes in *Trichoderma*. Ph.D. Thesis, University of Pisa, Pisa (Italy)

Vicente I, Baroncelli R, Morán-Diez ME, Bernardi R, Puntoni G, Hermosa R et al (2020) Combined comparative genomics and gene expression analyses provide insights into the terpene synthases inventory in *Trichoderma*. Microorganisms 8:1603

Vinale F, Sivasithamparam K (2020) Beneficial effects of *Trichoderma* secondary metabolites on crops. Phytother Res 34:2835–2842

Vinale F, Marra R, Scala F, Ghisalberti EL, Lorito M, Sivasithamparam K (2006) Major secondary metabolites produced by two commercial *Trichoderma* strains active against different phytopathogens. Lett Appl Microbiol 43:143–148

Vinale F, Sivasithamparam K, Ghisalberti EL, Marra R, Woo SL, Lorito M (2008) *Trichoderma*–plant–pathogen interactions. Soil Biol Biochem 40(1):1–10

Vinale F, Flematti G, Sivasithamparam K, Lorito M, Marra R, Skelton BW, Ghisalberti EL (2009) Harzianic acid, an antifungal and plant growth promoting metabolite from *Trichoderma harzianum*. J Nat Prod 72:2032–2035

Vinale F, Girona IA, Nigro M, Mazzei P, Piccolo A, Ruocco M, Woo S, Rosa RD, Herrera CL, Lorito M (2011) Cerinolactone, a hydroxylactone derivative from *Trichoderma Cerinum*. J Nat Prod 75:103–106

Vinale F, Sivasithamparam K, Ghisalberti EL, Ruocco M, Wood S, Lorito M (2012) *Trichoderma* secondary metabolites that affect plant metabolism. Nat Prod Commun 7(11):1545–1550

Vinale F, Nigro M, Sivasithamparam K, Flematti G, Ghisalberti EL, Ruocco M, Varlese R, Marra R, Lanzuise S, Eid A, Woo SL, Lorito M (2013) Harzianic acid: a novel siderophore from *Trichoderma harzianum*. FEMS Microbiol Lett 347:123–129

Waqas M, Khan AL, Kamran M, Hamayun M, Kang SM, Kim YH, Lee YJ (2012) Endophytic fungi produce gibberellins and indoleacetic acid and promotes host-plant growth during stress. Molecules 17:10754–10773

Wiest A, Grzegorski D, Xu BW, Goulard C, Rebuffat S, Ebbole DJ et al (2002) Identification of peptaibols from *Trichoderma virens* and cloning of a peptaibol synthetase. J Biol Chem 277(23):20862–20868

Xiao-Yan S, Qing-Tao S, Shu-Tao X, Xiu-Lan C, Cai-Yun S, Yu-Zhong Z (2006) Broad-spectrum antimicrobial activity and high stability of Trichokonins from *Trichoderma koningii* SMF2 against plant pathogens. FEMS Microbiol Lett 260(1):119–125

Yu JH, Keller N (2005) Regulation of secondary metabolism in filamentous fungi. Annu Rev Phytopathol 2005(43):437–458

Zapparata A, Baroncelli R, Brandstrom Durling M, Kubicek CP, Karlsson M, Vannacci G, Sarrocco S (2021) Fungal cross-talk: an integrated approach to study distance communication. Fungal Genet Biol 148:103518. https://doi.org/10.1016/j.fgb.2021.103518

Zeilinger S, Reithner B, Scala V, Peissl I, Lorito M, Mach RL (2005) Signal transduction by Tga3, a novel G protein subunit of *Trichoderma atroviride*. Appl Environ Microbiol 71(3):1591–1597

Zeilinger S, Gruber S, Bansal R, Mukherjee PK (2016) Secondary metabolism in *Trichoderma* – chemistry meets genomics. Fungal Biol Rev 30(2):74–90

Zhu Y, Wang J, Mou P, Yan Y, Chen M, Tang Y (2021) Genome mining of cryptic tetronate natural products from a PKS-NRPS encoding gene cluster in *Trichoderma harzianum*. Org Biomol Chem 19:1985–1990

New Insights on the Duality of *Trichoderma* as a Phytopathogen Killer and a Plant Protector Based on an Integrated Multi-omics Perspective

Mitzuko Dautt-Castro, Saúl Jijón-Moreno, Nicolás Gómez-Hernández, María del Carmen González-López, Eyra Judith Hernández-Hernández, Maria Montserrat Rosendo-Vargas, Oscar G. Rebolledo-Prudencio, and Sergio Casas-Flores

Contents

Mitzuko Dautt-Castro, Saúl Jijón-Moreno and Sergio Casas-Flores contributed equally with all other contributors.

M. Dautt-Castro (✉) · S. Jijón-Moreno (✉) · N. Gómez-Hernández ·
M. del Carmen González-López · E. J. Hernández-Hernández · M. M. Rosendo-Vargas ·
O. G. Rebolledo-Prudencio · S. Casas-Flores (✉)
División de Biología Molecular, IPICyT, San Luis Potosí, Mexico
e-mail: mitzuko.dautt@ipicyt.edu.mx; saul.jijon@ipicyt.edu.mx; scasas@ipicyt.edu.mx

N. Amaresan et al. (eds.), *Advances in* Trichoderma *Biology for Agricultural Applications*, Fungal Biology, https://doi.org/10.1007/978-3-030-91650-3_5

1 Introduction

The rapidly growing world population is increasing demands on farm producers, which requires efficient green strategies for pest control and diseases in crops and to reduce the impact of abiotic stress to which plants are constantly exposed (Gouda et al. 2018; Khan et al. 2019). The damages provoked by the causal agents of plant diseases are attributed mainly to poor agricultural practices, particularly monocultures and the excessive use of pesticides and chemical fertilizers. Together, these promote the occurrence of resistant and more virulent plant pathogenic microorganisms, which severely affect crop yields and indigenous wild species. Furthermore, the use of chemical pesticides and fertilizers is harmful for human health, wildlife, water resources, and the environment (Khan et al. 2019; Bhattacharyya et al. 2020). These negative effects have led to search for alternative eco-friendly and cost-effective solutions that not only improve crop yields but also ensure agricultural sustainability (Adnan et al. 2019; Bramlett et al. 2019; Bhattacharyya et al. 2020).

Trichoderma-based biostimulants have acquired significant interest among farmers and in the research community because these fungal species provide several positive effects to plants directly or indirectly. Among the beneficial effects, *Trichoderma* species impact positively on plant development, promote plant growth, improve seed germination, increase nutrient uptake, and trigger resistance against biotic and abiotic stresses, which together will increase crop yields. All of these mechanisms broaden the scope of potential applications of *Trichoderma* spp.-based formulations for agriculture (Woo et al. 2014).

During the last two decades, the development of "omics," including genomics, transcriptomics, proteomics, and metabolomics, has meant great progress for biological sciences, since they allow identifying, characterizing, and quantifying molecules such as DNA, RNA, proteins, and metabolites, respectively, which are involved in the structure, function, and dynamics of the cell, tissue, or organism (Vailati-Riboni et al. 2017). For the *Trichoderma* spp. research community, these tools have had a significant bearing, helping to understand the mechanisms of action used by these fungi and expanding the knowledge, either as biocontrol agents or as beneficial microorganisms to plants.

Among such mechanisms, mycophagy plays a pivotal role. This mechanism includes various events, such as the recognition of the prey, the attachment and coiling around the host hyphae, and killing them to finally feed on its host. These events occur by a combination of a wide variety of molecules secreted by the mycoparasite (Olmedo-Monfil and Casas-Flores 2014; Rebolledo-Prudencio et al. 2020). Transcriptomic and proteomic studies, using DNA microarrays and RNA sequencing (RNA-seq) as well as 2DE-LC-MS/MS and 2DE-MALDI-TOF-MS, respectively, in some species of *Trichoderma* including *T. atroviride*, *T. virens*, and *T. harzianum* have revealed that cell wall-degrading enzymes (CWDEs) recognized as carbohydrate-active enzymes (CAZymes), especially chitinases, glucanases, and proteases, are highly expressed in the presence of cell walls of plant pathogenic fungi, such as *Verticillium dahliae*, *Rhizoctonia solani*, and *Botrytis cinerea*, highlighting their importance during mycophagy (Grinyer et al. 2005; Monteiro et al. 2010; Morán-Diez et al. 2019; Halifu et al. 2020).

Trichoderma spp. are highly efficient competitors in the rhizosphere for uptaking photosynthates secreted by the plant roots (Ahmad and Baker 1987). *Trichoderma* spp. utilize root exudates as their main nutrient sources and have the capability of mobilizing and feeding on nutrients not readily available in the soil like copper, phosphorus, iron, manganese, and sodium (Harman et al. 2004), turning them more efficient and competitive than many other soil microbes. Genomic analyses have shown that *Trichoderma* spp. rely on their genomes' pH-responsive genes, such as the transcription factor PACC, that allow them to adapt to changes in pH, giving these fungi advantages over their competitors (Moreno-Mateos et al. 2007). In this regard, a microarray analysis of *T. virens* wt (wild-type) and a Δ*pacc* strain growing in medium at pH 8 or pH 4 shows that 650 genes are differentially regulated in response to this cue. Accordingly, Δ*pacc* mutant was impaired in their capability to grow on *R. solani* and *Sclerotium rolfsii*, whereas the constitutively active $PACC^c$ strain overgrows *R. solani* at the same extent as the wt (Trushina et al. 2013).

Several *Trichoderma* species can establish a long-lasting and robust relationship with plants, colonizing their roots intercellularly forming appressorium-like structures, although it has been reported that they can also colonize intracellularly in some plant species (Chacón et al. 2007; Salas-Marina et al. 2011; Nogueira-Lopez et al. 2018). The success of these fungal species to colonize the plant roots depends on the suppression of the immune response and the evasion of antimicrobial compounds or proteins that are secreted by the plant in response to the colonizer (Salas-Marina et al. 2011; Estrada-Rivera et al. 2019; Hermosa et al. 2012). CWDEs also play pivotal roles during this process. For instance, the characterization of *T. virens* secretome during its interaction with *Zea mays* seedlings shows that glycosyl hydrolases are secreted into the plant apoplast putatively for the hydrolysis of the plant cell wall (Lamdan et al. 2015; Nogueira-Lopez et al. 2018). In agreement with this, the transcriptomic response of *T. virens* to the presence of *Arabidopsis thaliana* seedlings shows a fine-tuning regulation of CWDEs to avoid excessive damage to the plant tissue, especially at the early stages of the interaction (Estrada-Rivera et al. 2020).

Furthermore, *Trichoderma koningiopsis*, *T. atroviride*, and *T. harzianum* T6776 (recently renamed as *T. afroharzianum* by Cai and Druzhinina 2021) can enhance plant growth by different mechanisms, including phosphate solubilization and nutrient uptake (Tandon et al. 2020), production of phytohormones (Salas-Marina et al. 2011), and increasing the photosynthetic rate (Fiorini et al. 2016; Rebolledo-Prudencio et al. 2020). In this sense, the transcriptomic analysis of *Solanum lycopersicum* roots in interaction with *T. afroharzianum* (formerly *T. harzianum*; Cai and Druzhinina 2021) indicates that the fungus induces the expression of genes related to nutrient transport and provokes the downregulation of the transcriptional regulator SIMYB93, which negatively regulates lateral root development (De Palma et al. 2019).

Trichoderma spp. are also capable of bestowing protection to plants against biotic and abiotic stresses. Experimental data suggest that the main signal transduction pathways triggered in plants by *T. atroviride* are related to jasmonate/ethylene (JA/ET) and salicylic acid (SA) (Salas-Marina et al. 2011; Villalobos-Escobedo et al. 2020). RNA-seq data of *A. thaliana* during its interaction with *T. atroviride* IMI 206040 reveals the induction of genes encoding proteins of the JA pathway such as *PAD3*, *JAZ1*, *JAZ6*, and *LOX1* (Villalobos-Escobedo et al. 2020). Furthermore, a proteomic analysis shows that a pathogenesis-related (PR)-like protein and an acyl-CoA-binding protein (ACBP) are upregulated in roots of *Phaseolus vulgaris* challenged with *T. harzianum* (Pereira et al. 2014). Particularly, PR proteins are key components of the plant systemic acquired resistance (SAR) and are induced by plant pathogenic organisms as well as by some species of *Trichoderma* (Salas-Marina et al. 2011; Ali et al. 2018; Estrada-Rivera et al. 2020).

Moreover, *Trichoderma* spp. also confer protection to plants by indirect mechanisms through the production of a plethora of secondary metabolites (SMs) of high and low molecular weight, which are crucial for the antagonism of plant pathogenic microorganisms highlighting nonribosomal peptides (NRPs), polyketides, terpenoids, and siderophores, among others. According to a *Trichoderma* genome analysis, SM-related genes are distributed in clusters (Mukherjee et al. 2012, 2013), and, thus, the number of clusters and associated genes correlates with the production of SMs and the mycoparasitic capability of each *Trichoderma* strain (Kubicek et al. 2011). Furthermore, volatile organic compounds (VOCs) are SMs that have been used to control plant pathogenic microorganisms and for plant growth stimulation. Metabolomic analyses support the production of different kinds and amounts of VOCs by *Trichoderma* species alone or in contact with other organisms as *Arabidopsis* and *Oryza sativa* plants, as well as plant pathogenic microorganisms including *Sclerotium rolfsii* Sacc., *Macrophomina phaseolina*, and *Fusarium* and the ectomycorrhizal fungus *Laccaria bicolor*. Nowadays, more than 700 VOCs have been described for *Trichoderma* spp. Exposure of plants to *Trichoderma* VOCs induced increased expression of defense-related genes, such as the plant defensin 1.2 (*PDF1.2*) and the pathogenesis-related 1 (*PR-1a*) genes, as well as the accumulation of defense-related compounds like H_2O_2, anthocyanins, and camalexin and an increase in trichomes (Kottb et al. 2015).

To improve their association with plants, *Trichoderma* spp. synthesize an arsenal of protein that can act as elicitors and effector-like proteins. A bioinformatics analysis of the predicted proteomes for *T. virens*, *T. atroviride*, and *Trichoderma reesei* identified 233 effector-like proteins. This study includes members of the LysM repeats, serine proteases, hydrophobins, thioredoxins, CFEM domain, and cerato-platanin families. Besides, 16 genes encoding effector-like proteins from *T. virens* and *T. atroviride* were upregulated during their interaction with *Arabidopsis* (Guzmán-Guzmán et al. 2017).

In this review, we will describe, from an omics perspective, the molecular mechanisms that confer *Trichoderma* spp. their capability to act as mycophagous as well as to establish a mutualistic relationship with plants.

2 The Mycoparasitic Lifestyle of *Trichoderma* spp.

Members of the *Trichoderma* genus (*Ascomycota*, *Hypocreales*) have been widely recognized for their ability of feeding on a wide variety of substrates, including other fungi, animals, and decaying wood. In this regard, evolutionary analyses have shown that this genus derived from an ancestor with low cellulolytic capacity; however, apparently a lateral transfer of genes has provided *Trichoderma* spp. with enzymes and proteins to degrade the cell wall of both plants and other microorganisms, conferring them symbiotic capabilities such as mutualism and mycoparasitism (Druzhinina et al. 2018). Thus, the ability to antagonize plant pathogenic fungi and oomycetes is one of the main mechanisms used by *Trichoderma* spp. when used as biocontrol agents. The different phases during the mycoparasitic process of *Trichoderma* have been well described. The first event, which implies the recognition of the host or prey, takes place even without contact between the mycophagous and its host, through the recognition of secreted molecules by both, the mycophagous and the prey (Fig. 1). In this sense, *Trichoderma* spp. secrete proteases that degrade the host proteins, producing low-molecular-weight byproducts that are sensed by the mycophagous, triggering a signaling cascade, where G-proteins and MAP kinases (MAPK) play important roles. These molecular responses induce the expression of genes encoding CWDEs and secondary metabolism-related proteins (Vinale et al. 2008a, b; Druzhinina et al. 2011). For instance, during the phase of host sensing (physically non-interacting), *T. atroviride* secretes a huge amount of peptaibols of 11 and 18 amino acids in length that migrate toward the fungal prey *R. solani* (Holzlechner et al. 2016).

Once the mycophagous reaches its host, some morphological changes are observed, being the attachment and coiling around the hyphal prey the most evident (Fig. 1). During this process, lectins from the fungal prey and proteins containing cellulose-binding modules play important roles. In counterpart, the prey produces reactive oxygen species (ROS) and SMs as a defense mechanism (Inbar and Chet 1996; Druzhinina et al. 2011). Granulation is a morphological change induced by *T. harzianum* and *T. viride* T12 in the entomopathogenic fungi *Beauveria bassiana* and *Metarhizium anisopliae*. Furthermore, *T. viride* T12 causes vacuolation and distortion of these entomopathogens (Banerjee et al. 2016). Dual confrontation of *Trichoderma cerinum* Gur1 with *Fusarium oxysporum* leads to changes in hyphal morphology and the presence of large vacuoles in the cytoplasm (Khare et al. 2018).

In agreement with this, abnormalities of fungal mycelia of the plant pathogens *Corynespora cassiicola* and *Curvularia aeria* are also observed when treated with crude extracts of *Lactuca sativa* previously inoculated with a spore suspension of *Trichoderma asperellum* T1 (Baiyee et al. 2019).

The last event occurring during *Trichoderma* spp. mycoparasitism on their hosts is killing their prey, resulting from the action of lytic enzymes and a plethora of SMs secreted by the mycophagous. Interestingly, not only the direct contact but also the VOCs produced by *Trichoderma* spp. strains are able to degrade the prey's cell walls inhibiting their growth (Al-Naemi et al. 2016) (Fig. 1).

2.1 *Lytic Enzymes of* Trichoderma *spp. Are Essential for Mycoparasitism*

As key players of the carbon cycle, fungi arise as skilled degraders of plant biomass and have evolved into an efficient machinery for the degradation of the plant cell wall (Glass et al. 2013). In fungi, the cell wall is a physical barrier that protects the cell against environmental fluctuation or host infection. The fungal cell wall is involved in adhesion to surfaces and is composed of a branched β-1,3-glucan cross-linked to chitin (Latgé 2010). For the mycoparasitic activity, *Trichoderma* spp. need to activate the synthesis of lytic enzymes, which degrade the fungal cell wall during mycophagy. For this purpose, the mycophagous must release enzymes with chitinase, protease, glucanase, and N-acetylglucosaminidase activities to the medium.

Fig 1 (continiued) monic acid (JA) and ethylene (ET). Activation of phytohormone signaling pathways can be extended systematically at distal sites of the plant enabling an enhanced resistance to plant pathogenic microorganisms (upper panel). *Trichoderma* spp. colonize and penetrate the plant root, growing into the epidermis between the intercellular spaces. To penetrate the root epidermis, *Trichoderma* secrete CWDEs that hydrolyze components of the plant cell wall. In response, the plant cell secretes protease inhibitors into the apoplast, which probably inactivate some CWDEs secreted by the fungus to avoid an excessive cell wall damage. Once the fungus has been successfully established in the plant root, several events take place for the establishment of a beneficial relationship with the host plant. In this regard, some *Trichoderma* spp. can avoid the inhibition of cell elongation and root growth (a deleterious effect that occurs in plants because of high ethylene levels) through the production of the enzyme 1-aminocyclopropane-1-carboxylate deaminase (ACCD), which hydrolyzes the 1-aminocyclopropane-1-carboxylic acid (ACC), the precursor of ethylene, to produce α-ketobutyrate and ammonia. Some *Trichoderma* strains produce and secrete phytohormones, including indole-3-acetic acid (IAA), which play key roles in plant growth and development. In addition, during interaction with the plant, *Trichoderma* secrete phytases and acid phosphatases that solubilize organic phosphate sources (e.g., phytic acid) releasing inorganic phosphorus (P_i), which can be taken up by the plant and the fungus for their metabolism. Additionally, VOCs (e.g., 6-pentyl-2H-pyran-2-one) emitted by *Trichoderma* contribute to plant growth promotion. All these mechanisms result in an increased number of plant roots and leaves, root branching, and total chlorophyll content. SAM, S-adenosyl-L-methionine. Question marks (?) indicate aspects that remain to be clarified

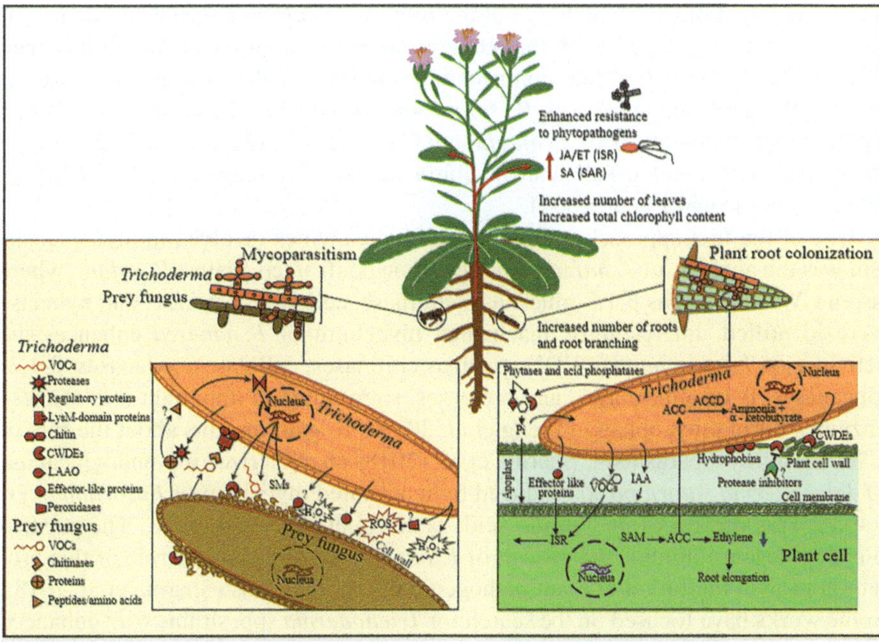

Fig. 1 Main mechanisms involved in *Trichoderma* spp. as biocontrol agents and plant mutualistic fungi. *Trichoderma* spp. confer a variety of benefits to their host plants, including stimulation of growth (right side panel) and control of soilborne plant pathogenic microorganisms by acting as mycophagous (left side panel). The first event of the mycoparasitic process consists in the recognition of the prey fungus, which takes place even without contact between the mycophagous and its host, through the recognition of secreted molecules by both, the mycophagous and the prey. *Trichoderma* spp. secrete proteases that degrade extracellular host-derived proteins, producing low-molecular-weight byproducts that may be sensed by the mycophagous, triggering an intracellular signaling cascade that leads to the expression of genes encoding cell wall-degrading enzymes (CWDEs), as well as genes related to the biosynthesis of volatile organic compounds (VOCs) and secondary metabolites (SMs). CWDEs are secreted to the extracellular milieu and hydrolyze components of the host cell wall, whereas VOCs migrate toward the prey fungus and contribute to the degradation of the host cell wall. Once *Trichoderma* reach their host, attachment and coiling around the hyphae occurs. Effector-like proteins secreted by *Trichoderma* during interaction with the prey fungus may play a role in the hyphal coiling through a yet unknown mechanism. As an additional strategy, *Trichoderma* spp. produce H_2O_2 to inhibit the growth of their fungal prey. In counterpart, the host fungus produces reactive oxygen species (ROS), SMs, and proteases as a defense mechanism. Probably, *Trichoderma* spp. overcome the effects of ROS by secreting ROS-scavenging enzymes (e.g., peroxidases). In addition, *Trichoderma* spp. secrete LysM domain proteins that protect their hyphae from degradation by host chitinases through binding to chitin of the cell wall. Additionally, *Trichoderma* secrete the enzyme L-amino acid oxidase (LAAO), which probably contributes to the lysis of the prey fungus by binding to cell wall proteins of the host fungus, which leads to the oxidation of the target proteins and produces an increase in the concentration of H_2O_2 and the subsequent apoptosis of the prey fungus. Thus, the last event occurring during *Trichoderma* spp. mycoparasitism on their hosts is killing of the prey. On the other hand, elicitors derived from *Trichoderma* spp., including effector-like proteins and VOCs, trigger the activation of the systemic acquired resistance (SAR) that results in an increased level of salicylic acid (SA) and the induced systemic resistance (ISR) which is related to the accumulation of jas

Some investigations have pointed out the role of hydrolytic enzymes in the mycoparasitic process, as well as in the recognition of potential preys, which has been demonstrated employing dual confrontation cultures of the mycoparasite against plant pathogenic fungi such as *R. solani* and *B. cinerea* (Reithner et al. 2011). Furthermore, the analysis of *T. harzianum* CECT 2413 grown in medium supplemented with *B. cinerea* cell wall or chitin revealed an upregulation of CWDE-related genes (Suárez et al. 2007).

One of the first approaches looking for lytic enzymes involved in mycoparasitism was the analysis of *T. harzianum* secretome confronted against *R. solani*, where seven CWDEs such as β-1,3-glucanase, chitinase, cellulase, protease, and xylanase were identified. Interestingly, inactivated mycelium of *B. cinerea* enhances the activities of *T. harzianum* CWDEs, such as chitinases, cellulases, xylanases, β-1,3-glucanases, β-1,6-glucanases, and proteases, suggesting an important role of these enzymes during mycophagy (Tseng et al. 2008). To gain insights about the role of CWDEs in mycoparasitism, Sharma et al. (2018) characterized a β-endoglucanase of *Trichoderma saturniporum* induced by inactivated mycelium of *Fusarium oxysporum*. This protein of 347 amino acids is active at pH 5 and 60 °C. The purified β-endoglucanase inhibits the growth of *F. oxysporum*, indicating a role of this protein in growth inhibition of plant pathogenic microorganisms (Sharma et al. 2018). Some works have focused on the search for *Trichoderma* spp. strains with enhanced mycoparasitic activity for their potential use in agriculture. In this sense, Geraldine et al. (2013) showed that two isolates of *T. asperellum* effectively reduce the density of apothecia and the severity of the disease caused by the fungal phytopathogen *Sclerotinia sclerotiorum* under field conditions. They analyzed the CWDEs in different isolates, finding that the N-β-acetylglucosaminidase and β-1,3-glucanase activities are central components of the *Trichoderma* isolates that show activity against *S. sclerotiorum*. This analysis suggests that CWDEs can be used as markers to select new biocontrol strains for agricultural use (Geraldine et al. 2013). A different study of *T. harzianum* confirmed that the cell wall of *Fusarium solani* induces the expression of *bgn*, *chit*, and *endo* genes, which code for a β-endoglucanase, a chitinase 33, and an endochitinase 42, respectively. These genes are highly expressed mainly after contact, which is in agreement with an enhanced cell wall degradation (Vieira et al. 2013). Interestingly, the proteins β-endoglucanase, chitinase 33, and endochitinase 42 have a central role in the mycoparasitism on the plant pathogens *R. solani*, *F. oxysporum*, *S. sclerotiorum*, and *S. rolfsii* (Sharma et al. 2011; Troian et al., 2014). Moreover, *T. viride* (NBAII Tv 23) shows higher chitinase and protease activities when grown in synthetic medium supplemented with *S. rolfsii* cell walls (Parmar et al. 2015).

On the other hand, the mutant strains *tvmh-9* of *T. virens*, *nas-k1m25* of *Trichoderma koningii* (NAS-K1), and *rp698* of *T. reesei*, which were generated using the mutagen ethyl methanesulfonate, cobalt-60, and UV light, respectively, overproduce chitinase, N-acetylglucosaminidase, and cellulose, respectively. These mutants are more effective in inhibiting the plant pathogenic fungi *Macrophomina phaseolina* and *Scytalidium thermophilum* (Goharzad et al. 2020; Silva et al. 2020; Vyawahare et al. 2019), highlighting the relevance of CWDEs during mycophagy.

2.2 *Mycophagy-Related Genes in* Trichoderma *spp.*

Besides genes coding for CWDEs in *Trichoderma* spp., there are several genes and their products associated with mycophagy. For instance, in *T. atroviride*, GPR1, a seven-transmembrane protein of the cAMP receptor-like family, participates in mycoparasitism-related processes, since *gpr1*-silenced transformants are unable to attach to the host hyphae. Furthermore, *grp1*-silenced strains are affected in their antagonistic activity because they are unable to parasitize *R. solani*, *S. sclerotiorum*, and *B. cinerea*. These phenotypes correlate with the fact that *grp1*-silenced transformants could not induce the mycoparasitism-related genes *nag1* and *ech42* (chitinases) and *prb1* (protease) in the presence of *R. solani*, contrary to wt, where a significant induction of these genes is observed upon contact with the host. Interestingly, the addition of cAMP to the confrontation plates restored the Δ*grp1*-silenced strain attachment and coiling around *R. solani* but not the growth over the host (Omann et al. 2012). Moreover, mutants of *sfp2*, a member of the Sur7 superfamily, whose upregulation under mycoparasitic conditions is dependent on GPR1, show significantly reduced mycoparasitic activity, whereas their overexpression causes enhanced overgrowth and killing of the prey (Atanasova et al. 2018). Also, in *T. atroviride* TGF-1, the orthologous gene of the histone acetyltransferase Gcn5p from *Saccharomyces cerevisiae* is involved in the capacity of the fungus to grow over *R. solani*, but not in coiling. Apparently, this protein regulates negatively mycoparasitism-related genes such as *ech-42* and *prb-1* in the absence of the prey, probably through its histone acetyltransferase activity or by acetylation of the promoters of negative regulators, which, in consequence, repress indirectly these genes and whose transcript is absent in Δ*tgf-1*. Moreover, in confrontation assays against *R. solani*, *ech-42* and *prb-1* are downregulated in Δ*tgf-1*, indicating that TGF-1 is required for transcription of these genes to respond to the presence of the host (Gómez-Rodríguez et al. 2018). In *T. virens*, mutants in *pgy1* and *ecm33*, which code a proline-glycine-tyrosine-rich protein (PGYRP) and a GPI-anchored cell wall protein, respectively, failed to grow over their hosts *S. rolfsii* and *R. solani*, indicating that their products are involved in the antagonistic process (Bansal et al. 2019).

Furthermore, proteins classified as *Trichoderma* spp. effector-like proteins are also involved in mycoparasitism (Fig. 1). For instance, overexpression of *hydii1*, a gene encoding a class II hydrophobin, improved the growth of *T. virens* over *R. solani* AG2, demonstrating its role during its antagonistic activity against this plant pathogenic fungus (Guzmán-Guzmán et al. 2017). Additionally, a member of the cerato-platanin family, EPL1 of *T. harzianum*, which is an elicitor of plant disease resistance, is involved in hyphal coiling during mycoparasitism since mutants in *epl1* are unable to coil around *S. sclerotiorum*. Consistently, an opposite expression pattern of mycoparasitism-related genes between the wt and Δ*epl1* is observed, mainly after hyphal contact between *T. atroviride* and its host (Gomes et al. 2015). Also, *tal6*, a LysM effector of *T. atroviride*, increases its expression levels during and after contact with *R. solani* AG2 and AG5 strains. Accordingly, *tal6*-OE1.1 strain overgrows better its host compared to the wt and Δ*tal6–4.2* strains, which indicates that this putative effector plays a role in the mycoparasitic capability of *T. atroviride* (Romero-Contreras et al. 2019). Recently, TBRG-1, the founding member of a new subfamily of big Ras GTPase of *T. virens*, was reported as a negative regulator of mycoparasitism. Δ*tbrg-1* mutants show enhanced antagonistic effects against *R. solani*, *F. oxysporum*, and *S. rolfsii*. Furthermore, the expression of the mycoparasitism-related genes *sp1* (protease) and *cht1* (chitinase) was upregulated in the Δ*tbrg-1* strain compared to the wt in the presence or in the absence of its hosts, supporting the role of TBRG-1 during the mycoparasitic process (Dautt-Castro et al. 2020, 2021).

3 Competition for Nutrients Between *Trichoderma* and Microorganisms in the Rhizosphere

As previously described, *Trichoderma* spp. have the capability of antagonizing other microorganisms under different conditions, in several niches, and by different mechanisms. Another important mechanism of *Trichoderma* spp. in the rhizosphere is the competition for space and nutrients as well as for plant root exudates (Ahmad and Baker 1987). Plant root exudates, which are composed mainly of carbohydrates, amino acids, lipids, organic acids, vitamins, and minerals, can be influenced by the surrounding microbiota, including other *Trichoderma* species, and vice versa (Bais et al. 2006). For example, *T. atroviride* grown in medium amended with root exudates of *S. lycopersicum* improves the colony growth and the formation of aerial mycelium, suggesting that root exudates are a nutrient source for the fungus. Intriguingly, *T. atroviride* modified the proportions of carbohydrates in *S. lycopersicum* root exudates, and sucrose was secreted only by roots colonized by the fungus (Macías-Rodríguez et al. 2018). These behaviors are relevant for biocontrol because the plant can improve the beneficial fungi growth allowing them to compete more efficiently for space and nutrients with the inhabitant microbiota in the rhizosphere. In this regard, zoospores of the phytopathogenic oomycete *Phytophthora cinnamomi* are not detected when co-cultured with *T. atroviride*. Contrastingly, conidia of *T. atroviride* increase when this co-culture is carried out in a medium with *S. lycopersicum* root exudates compared to the treatment without *P. cinnamomi* (Macías-Rodríguez et al. 2018). Moreover, under biotic and abiotic stresses, the root exudates act as a chemoattractant for *T. afroharzianum*, and promote its growth, but not for their host *F. oxysporum*, strengthening the mutualistic relationship (Lombardi et al. 2018).

The high success rate of competition of *Trichoderma* spp. over other microbes is largely due to its capability to mobilize and uptake soil nutrients like copper (Cu), phosphorus (P), iron (Fe), manganese (Mn), and sodium (Na) (Harman et al. 2004). In this regard, Fe uptake is essential for the viability of most filamentous fungi. Because iron is normally present in the soil as an insoluble form, most fungi, including *Trichoderma* spp., can produce Fe^{3+}-chelating complexes named siderophores, which are SMs of low molecular weight. This mechanism facilitates the conversion of Fe to a soluble form, helping plants to uptake it and consequently depriving plant pathogenic microorganisms from Fe suppressing their growth (Sood et al. 2020). *T. virens* genome contains three genes that putatively code for enzymes involved in the synthesis of siderophores, *tex10*, *tex20* (*sidd*), and *tex21* (*nps6*), which are induced under iron depletion conditions (Mukherjee et al. 2012, 2018). Regarding the regulation of siderophore synthesis in *T. reesei*, YPR2, a transcription factor implicated in the regulation of the SOR cluster, is involved in the biosynthesis of the SMs responsible for the yellow pigment with antimicrobial activities, called sorbicillinoids, and regulates siderophore production in a light-specific way (Derntl et al. 2016; Hitzenhammer et al. 2019). In this sense, a regulation of the whole cluster of siderophore synthesis-related genes and a coregulation of NRPS (nonribosomal peptide synthetase), one of the two major pathways of the biosynthesis of such SMs, were observed in a transcriptomic analysis of Δ*ypr2* (Hitzenhammer et al. 2019). For instance, *ftr1b*, a gene that codes for an iron permease, is downregulated in Δ*ypr2* in darkness and upregulated under light conditions, showing a similar regulation to that of the siderophore cluster. Also, the genes *sidd* (*nrps* homologue), *sidf* (transacylase), *sidj* (siderophore biosynthesis lipase/esterase), *sitt* (ABC multidrug transporter), *sidf* (hydroxyornithine transacylase), *sidh* (enoyl-CoA hydratase/isomerase family protein), and *mirb* (siderophore iron transporter) were coregulated in this mutant genetic background. Since *T. reesei* is a non-root colonizer fungus but a sibling of *T. vires* and *T. atroviride*, two root colonizer species, suggests that this kind of regulation could be occurring in other species of *Trichoderma* that are beneficial to plants.

A key factor for nutrient availability in the rhizosphere is the soil pH; hence, alkalinization or acidification represents the strongest known predictor of microbial community composition and abundance in soils. Soil acidity affects the availability of some elements like P, calcium (Ca), and molybdenum (Mo) and determines the toxicity of Fe, Al, and Mn (Muthukumar et al. 2014). In this sense, the *Trichoderma* genomes contain consensus sequences that code for the pH-responsive transcription factor PACC, which allowed them to adapt to a wide range of pH and to compete efficiently with pathogens (Benítez et al. 2004). In *T. harzianum*, PAC1 regulates antagonism-related genes like *chit42* (chitinase), *papa* (protease), *gtt1* (glucose permease), and *qid74* (cell wall protein). Furthermore, PAC1 appears to be a positive regulator of parasitism, because *pac1* null mutants are unable to grow over their hosts *B. cinerea*, *R. solani*, *Rhizoctonia meloni*, *Phytophthora citrophthora*, and *Fusarium fujikuroi*, compared to wt. However, Δ*pac1* strains produce metabolites that inhibit more efficiently the growth of *R. solani* and *B. cinerea*, suggesting that the lack of *pac1* improves the production of some SMs responsible for the growth inhibition of these plant pathogenic microorganisms, but, on the other hand, its absence impairs its ability to parasitize them (Moreno-Mateos et al. 2007). Similarly, in *T. virens*, Δ*pacc1* and Δ*pacc2* strains show a lesser growth at alkaline pH and slower growth over *R. solani* compared to wt. Moreover, in confrontation with *S. rolfsii*, Δ*pacc1* and Δ*pacc2* are not able to cover the sclerotia produced by the fungal host, compared to the wt strain.

All these data along with the fact that *Trichoderma* spp. have faster growth cycles than most of the plant pathogenic microorganisms and a more proficient capability to both mobilize and utilize nutrients make them more efficient and competitive than many other soil microbes.

4 Analysis of *Trichoderma* spp. Genomes Unveils Their Potential as Biocontrol Agents

Genomics has emerged as a tool for getting massive information to ensemble the organism genomes, which has revolutionized many research areas, including that related with *Trichoderma* spp. Until July 2020, according to Cai and Druzhinina (2021), a total of 42 whole genomes of different *Trichoderma* strains had been deposited in public databases.

The first *Trichoderma* genome sequenced was that of *T. reesei* (Martinez et al. 2008), thus establishing the beginning of the genomic era for *Trichoderma*. As mentioned before, *T. reesei* is not a root colonizer but is important in the industry because of its excellent capability to degrade polysaccharides. Sequencing of *T. reesei* genome has allowed to unveil many important characteristics of the *Trichoderma* genus. For instance, comparative genomic analysis with other *Trichoderma* species revealed that the secondary metabolism-related genes are distributed in clusters, many of which were acquired by horizontal transfer (Martinez et al. 2008; Mukherjee et al. 2012, 2013). The *Trichoderma* genus is characterized by being a great producer of SMs, like NRPs, polyketides, terpenoids, and siderophores, among other compounds, that play important roles in its antagonistic activity against many plant pathogenic microorganisms with different lifestyles, as well as in the beneficial effects this genus confers to plants (Zeilinger et al. 2016). Consistently, these characteristics are present in the genomes of the different *Trichoderma* species, where a correlation exists between the number of clusters and the genes associated with the production of SMs and their mycoparasitic capability. In this regard, *T. virens* and *T. atroviride* genomes, two mycoparasitic species, contain more polyketide synthases (PKSs) and NRPS than *T. reesei*. Interestingly, 1273 genes are shared between *T. virens* and *T. atroviride* but not with *T. reesei*, some of which could be associated with SM production, thus putatively giving them a better mycoparasitic capability (Kubicek et al. 2011). In agreement with this, those genes that encode for CAZymes are arranged in clusters between regions of synteny and other *Sordariomycetes* (a class of fungi of the *Pezizomycotina* subdivision of the ascomycete division) (Martinez et al. 2008). Particularly, the fungal CWDEs, chitinases and glucanases, which are essential for the mycoparasitic activity, are increased in *T. virens* and *T. atroviride* genomes. In this sense, the family of glycoside hydrolases (GHs) in the genome of *T. reesei* contains 193 GH-encoding genes, whereas *T. virens* and *T. atroviride* contain 259 and 258, respectively (Schmoll et al. 2016; Kubicek et al. 2011). Similarly, the analysis of CAZymes in *T. harzianum* derived from genome information revealed that its genome contains a total of 430 CAZymes, including 259 GHs, being the GH18 the most represented (Baroncelli et al. 2015; Ferreira Filho et al. 2017).

After sequencing the genomes of *T. reesei*, *T. virens*, and *T. atroviride*, which have been the most studied, several other species of *Trichoderma* were also sequenced. For instance, the genome of *Trichoderma* cf. *atrobrunneum* was sequenced by Fanelli et al. (2018), where one of the main goals was to study the genes implicated in biocontrol. In this regard, the Pfam domain analysis of the families of genes associated with antagonistic activities showed that the GHs were the biggest with 247 members, only behind *T. virens*, *T. harzianum* B97, and *T. afroharzianum* T6776 strains, according to comparison with other 20 species of *Trichoderma*. Regarding the secondary metabolism, the genome of *T.* cf. *atrobrunneum* contains 18 putative PKS, 8 NRPS, 5 PKS-NRPS, and 5 terpenoid synthase (TS) genes (Fanelli et al. 2018). Proteases are another group of proteins secreted by *Trichoderma* that play important roles in mycoparasitism, causing the release of small molecules from the prey, such as peptides, which are then recognized by the mycoparasite. Also, these proteins can hydrolyze other types of proteins whose byproducts serve as nutritional sources for *Trichoderma* (Druzhinina et al. 2011). In *T.* cf. *atrobrunneum*, a predicted proteinase showed high percentages of identity (91–93%) with proteases from *T. atroviride*, *T. harzianum*, *T. virens*, *T. viride*, and *T. reesei* (Fanelli et al. 2018). Through a multi-omics approach, Wu et al. (2017) described that *T. asperellum* GDFS1009 genome possesses 16 chitinase-, 8 protease-, and 11 glucanase-encoding genes. This fungus also contains 16 clusters of PKSs and 16 NRPSs. Besides biocontrol-related proteins, *Trichoderma* also secrete elicitors to induce immune resistance in plants against plant pathogenic microorganisms (Nawrocka and Małolepsza 2013). In this regard, *T. asperellum* GDFS1009 has 12 potential elicitor-encoding genes, including 2 endopolygalacturonases (endo-PGs), 2 EPl, 2 hydrophobins, 1 PG, 1 swollenin, and 4 xylanases (Wu et al. 2017).

Recently, Kubicek et al. (2019) performed a comparative genomic analysis of 12 species of *Trichoderma*. Their phylogenetic analysis showed that the most common species of *Trichoderma* are distributed in different groups named clade *Harzianum/Virens* (HV: *T. harzianum*, *T. guizhouense (T. sp. NJAU4742)*, *T. afroharzianum*, *T. virens*), the section of *Longibrachiatum* (SL: *T. reesei*, *T. longibrachiatum*, *T.* cf. *citrinoviride*, *T. parareesei*), and the section of *Trichoderma* (ST: *T. atroviride*, *T. gamsii*, *T. asperellum*, *T. hamatum*). They also show that *Trichoderma* evolved 66.5 (±15) million years ago (mya) next to the Cretaceous-Paleogene (K-Pg) extinction event, where a massive extinction of plants and animals occurred. Furthermore, SL and ST sections and HV clade apparently arose 21–25 mya, where the section ST is the oldest, whereas SL and HV evolved later and have a common ancestor. Furthermore, the origin of *Trichoderma* is characterized by a significant gene expansion (Kubicek et al. 2019). The genome sizes and the number of predicted genes of the 12 species ranged from 31 to 41 Mb and from 9292 to 14,095 predicted genes, where the section of SL is the smallest in both categories. In agreement with other genomes of *Trichoderma* spp., the core genome of these 12 species showed enrichment in genes encoding glycoside hydrolases (GHs). GHs represent more than 50%

of the CAZymes, where those acting on chitin and β-glucan comprised the most abundant (Kubicek et al. 2019). Regarding secondary metabolism, the authors found 10 to 25 PKSs, 12 to 34 NRPSs, and 6 to 14 TSs in the core genome, being the clade of *Harzianum/Virens* the one with the greatest amount of these proteins (Kubicek et al. 2019). With respect to proteases, the screening of the 12 *Trichoderma* genomes showed the presence of A1 aspartyl proteases, G1 eqolisins, C13 legumain-type cysteine proteases, 8 metalloprotease families, and 6 families of serine proteases (Kubicek et al. 2019).

All the information obtained by genome sequencing has increased the general knowledge about the *Trichoderma* genus but has also allowed to study specific genes, thus predicting the functions of many proteins that were previously unknown. Furthermore, the generation of *Trichoderma* mutant strains and their subsequent genome sequencing have helped to characterize strains with enhanced capabilities for biocontrol and to understand the mechanisms in those strains with negative phenotypes. In this regard, the importance of *T. reesei* in the industry has led it to be the target of mutagenesis programs in the search for strains that produce more cellulase activity. For instance, the original isolate QM6a^T has been mutated by UV light, irradiation by linear particle accelerators, and/or *N*-methyl-*N*′-nitro-*N*-nitrosoguanidine (NTG), which has resulted in strains with high production of cellulase activity such as NG14, Rut-C30, QM9123, and QM9414, as well as strains with cellulase-negative phenotypes like QM9136, QM9978, and QM9979 (Le Crom et al. 2009; Vitikainen et al. 2010; Lichius et al. 2015; Ivanova et al. 2017). The strains NG-14 and Rut-C30 have a high number of mutations like single nucleotide variants (SNVs), deletions, and insertions, which, consequently, affect 18 genes in NG14 and 25 more in Rut-C30 putatively involved in RNA metabolism, protein secretion, and vacuolar targeting and transcription factors (Le Crom et al. 2009). Contrastingly, strain QM9136 has a low number of mutagenesis events, which allowed the authors to identify a truncation of 140 amino acids in the C-terminal of the transcription factor XYR1. This mutation caused no cellulase- or xylanase-related gene expression in QM9136 and Δ*xyr1* strain, whereas complementation of QM9136 with the wild-type *xyr1* allele recovered completely the production of cellulases, confirming the role of XYR1 in this process (Lichius et al. 2015). Similarly, Ivanova et al. (2017) identified the transcription factor VIB1 as a key regulator of cellulases by analyzing the QM9978 strain. The authors found a translocation between chromosomes V and VII upstream of *vib1*, which suppressed its expression. Interestingly, deletion of *vib1* in QM9414 and Rut-C30 reduced the expression of cellulase, and the complementation and overexpression of *vib1* in QM9978 restored cellulase expression. On the other hand, in the *T. virens*, the mutant M7 strain generated by gamma ray resulted impaired in morphogenesis, secondary metabolism, and mycoparasitism. For instance, M7 does not overgrow or coil on plant pathogen *R. solani* and *Pythium aphanidermatum* hyphae. M7 does not produce important VOCs related to *Trichoderma* antagonism such as viridin, viridol, and heptelidic acid; it only produces 13 of the 73 VOCs present in the wt strain. The authors related these phenotypes with the mutations in M7, which include the deletion of PKS6 and Tex9 clusters related to SM, 8 transcription factors including 2 SM-related, 5 genes for carbohydrate metabolism, 2 genes associated with cell signaling, and 11 oxidoreductases (Pachauri et al. 2020).

All this above discussed genomic-derived knowledge can potentially help to improve the use of *Trichoderma* as a biological control in a more effective way.

5 Transcriptomic Analysis Unveils Genes Associated with Biocontrol in *Trichoderma* spp.

Nowadays, gene expression analysis is one of the most used tools in biological sciences. Getting these data in a massive way has undoubtedly represented a great advancement in many areas of knowledge. Some of the approaches that established the beginnings in this field were the expressed sequence tag (EST), which consists of small cDNA sequences to identify gene transcripts; suppression subtractive hybridization (SSH), which allows the amplification of differentially expressed cDNAs between control and activated transcriptome; and DNA arrays (macro and micro), which are used to quantify the gene expression using membranes or chips that contain a collection of DNA sequences arrayed in a matrix. This technology also uses fluorescence and image analysis to obtain the data. Macroarrays contain a collection of ESTs, whereas microarrays contain large datasets or complete genomes, being latter the most used during the last two decades among all techniques to analyze massively gene expression (Lorito et al. 2010).

Lorito and coworkers published in 2010 a review where they discuss the most important advances in this field described up to that moment. All these works allowed identifying which genes are expressed in *Trichoderma* under different conditions, such as biocontrol (interaction with plant pathogens), nutritional stress, plant root colonization, and light response, among others. Some of the most represented genes under biocontrol conditions were the cell wall protein QID3, oxidoreductases, HEX1, hydrophobin, cyclophilins, and a subtilisin-like serine protease, among others, whereas for plant interaction conditions, genes involved in lipid metabolism, degradation, and synthesis of the cell wall, redox metabolism, and energy-related processes, among others, were upregulated (Lorito et al. 2010). After the works revised by Lorito and coworkers, many other investigations using these approaches have been published with very interesting results, which will be discussed here.

The differences at the genome level of *Trichoderma* species with different lifestyles, previously discussed, are also reflected in their transcriptome. In this regard, an analysis of microarray data under mycoparasitic conditions of *T. reesei*, *T. atroviride*, and *T. virens* with the plant pathogen *R. solani* showed that all *Trichoderma* species present different responses. *T. atroviride* exhibits differential expression patterns of several genes coding for proteases, oligopeptide transporters, C-type lectins, small-secreted cysteine-rich proteins (SSCPs), PTH11 receptors, and β-glucanases of the GH16 family. *T. virens* shows increased levels mainly of gliotoxin biosynthesis-related genes and heat shock protein-encoding genes, whereas *T. reesei* showed an upregulation of genes that code for cellulolytic and hemicellulolytic CAZymes, ribosomal proteins, several transporter proteins, and one PKS. These transcriptional responses in the three species were mainly observed before hyphal contact with the host. Additionally, among differentially expressed genes (DEGs), only 9 orthologous genes increased their expression in the 3 species, whereas 29 upregulated genes were shared by the 2 mycoparasitic species, most of them with unknown functions (Atanasova et al. 2013). The transcriptional profiles of *Trichoderma* under mycoparasitic conditions have been extensively studied. Besides the differences at transcriptional levels among *Trichoderma* species, specific responses are also displayed depending on the host's lifestyles. For instance, Morán-Diez et al. (2019) designed a microarray of 385,000 probes to identify mycoparasitism-related genes of *T. atroviride* T11 during its interaction with *V. dahliae*, finding that the highest transcriptional activation occurs during the growth of the mycoparasite over its host. Among the 143 DEGs, 128 were upregulated, being the CAZymes the most enriched, especially those encoding for hydrolytic enzymes such as glucanases and peptidases. Also, SM-associated genes encoding for oxidoreductases and monooxygenases were upregulated. Moreover, the upregulation of *cpa1*, which codes for a protein that belongs to the M14 family of metallocarboxypeptidases, has an important role during mycoparasitism against *V. dahliae*, since strains overexpressing such gene show significantly higher inhibitory effect against its host (Table 1) (Morán-Diez et al. 2019). As an additional example, *npm1* from *T.* sp. NJAU4742, which encodes a metallopeptidase, is induced in the presence of phytopathogenic fungi, including *Alternaria alternata* and *F. oxysporum*. Although the purified NMP1 protein does not inhibit the growth of other fungi, an insertional mutagenesis approach indicates that this protein has a major role in mycotrophic interactions and defense against other fungi (Zhang et al. 2019). Using the SSH technique, Rabinal and Bhat (2020) found that, in the presence of *S. rolfsii*, *T. koningii* induces the expression of genes related to cell wall hydrolysis such as *β*-1,4-D-glucan cellobiohydrolases, chitinases, glycosyltransferases, serine endopeptidases, and xylanase-2, among others (Rabinal and Bhat 2020). Also, an EST analysis carried out with mRNA from *T. harzianum* (ALL42) grown in medium containing cell walls of *F. solani* shows that from a total of 1450 unigenes identified, a putative QID74 cell wall protein-encoding gene is the most represented, followed by CFEM domain-containing protein, a Woronin body major protein HEXA, and an exochitinase-encoding gene, all of them are potentially related to mycoparasitism (Table 1) (Trushina et al. 2013).

As described above, *Trichoderma* fungi are good rhizosphere-competent microorganisms, and pH plays a key role in this skill. In agreement with this, *Trichoderma* spp. contain in their genomes pH-responsive genes that allow them to adapt to pH changes, such as the transcription factor PACC (Benítez et al. 2004; Moreno-Mateos et al. 2007). To identify genes dependent on PACC and pH, a DNA microarray using cDNA from *T. virens* wt and Δ*pacc* strains, grown under pH 8 or pH 4, showed that 650 genes are differentially regulated in response to pH. In the mutant strain Δ*pacc*, a set of genes related to carbohydrates and inorganic ion transport and metabolism

Table 1 Overview of recent studies on *Trichoderma* spp. that used a transcriptomic approach to unravel genes related to biocontrol

Biocontrol agent	Experimental design	Method and technology	Main findings	References
T. harzianum ALL42	ESTs were constructed with mRNA from mycelium of *T. harzianum* ALL42 grown in a medium containing cell wall from *F. solani*	Expressed sequence tag (EST)	1450 unigenes were identified. A putative QID74 cell wall protein was the most represented gene, followed by CFEM domain-containing protein, a Woronin body major protein HEXA, and an exochitinase	Trushina et al. (2013)
T. virens	SSH libraries were constructed using RNA from *T. virens* wt and M7 mutant strain, in which the latter is impaired in conidiation	Suppression subtractive hybridization (SSH)	12 unigenes putatively related to conidiation were identified. *pgy1* (proline-glycine-tyrosine-rich protein) and *ecm33* (GPI-anchored cell wall protein) were downregulated in the mutant M7. Δ*pgy1* and Δ*ecm33* showed a slow growth and reduced conidiation and were not able to overgrow *S. rolfsii*; their filtrates did not cause inhibition of *P. aphanidermatum* and produced low antimicrobial viridin	Bansal et al. (2019)
T. atroviride T11	Microarrays were designed with 385,000 probes to identify mycoparasitism-related genes of *T. atroviride* T11 in interaction with *V. dahliae*	Microarray	The highest transcriptional activation occurred in overgrowing condition; 128 DEGs were upregulated, where the CAZyme-encoding genes were the most enriched, mainly for glucanases and peptidases. SM-associated genes oxidoreductases and monooxygenases were upregulated. *cpa1* gene (M14 family of metallocarboxypeptidases) was upregulated and found to have an important role during mycoparasitism against *V. dahliae*	Morán-Diez et al. (2019)
T. harzianum Tr-92	Transcriptome of *T. harzianum* Tr-92 under chlamydospore-producing condition was obtained	RNA-seq	Chlamydospore-based formulations showed higher biocontrol capability against *B. cinerea* compared to the conidia-based formulation. Genes as glutathione-S-transferases, an oxidoreductase, a glycosyltransferase, and a peroxidase were downregulated, and a protein kinase, a chitinase, and an intracellular serine protease were upregulated in *T. harzianum* Tr-92 under the chlamydospore-producing condition	Yuan et al. (2019b)

(continued)

Table 1 (continued)

Biocontrol agent	Experimental design	Method and technology	Main findings	References
T. viride	The biofilm produced by *Trichoderma viride* and *Azotobacter chroococcum* (Az) in co-culture was used to isolate the RNA and obtain their transcriptome	RNA-seq	Genes related to biofilm biosynthesis, such as *alg8*, *sipw*, *pssa*, *fadd*, *purb*, *phob*, and *glgp*, were identified. The most induced gene was an RNA-dependent RNA polymerase (RdRP), suggesting that the gene regulation mediated by sRNAs has a very important role during biofilm formation between *T. virens* and *A. chroococcum*	Velmourougane et al. (2019)
T. virens ZT05	Transcriptome of *T. virens* ZT05 was sequenced under antagonistic conditions with *R. solani*	RNA-seq	Genes associated with prey recognition and signal transduction (two extracellular proteases, one protease, one belonging to oligopeptide transporters, and four G-protein coupled receptors), as well as those related to hyperparasitism (six chitinases, six glucanases, and one proteasome gene), were induced. Thirty antibiotic and stress resistance genes were identified, including 9 reductases, 2 tetracycline resistance genes, 8 heat shock response genes, 2 multidrug resistance transporters, 8 ABC efflux transporters, and 1 oxidative stress response gene, many of them were highly induced. *T. virens* was able to inhibit the growth, coiling, and penetration of the mycelium of *R. solani.* The volatile and nonvolatile metabolites of *T. virens* inhibited the growth of the prey as well	Halifu et al. (2020)

are enriched compared to wt. Genes related with SM biosynthesis, transport, and catabolism and with nucleotide transport and metabolism were downregulated. Accordingly, Δ*pacc* mutants were affected in their ability to compete against *R. solani* and *S. rolfsii*, whereas constitutively active pacc^c strain overgrows *R. solani* to the same extent as *T. virens* wt (Trushina et al. 2013). The authors suggest that this information could be used to genetically manipulate *T. virens* to enhance its capabilities for biocontrol. Indeed, one of the most important implications of this type of study is that they allow the selection of specific genes that are induced or repressed under the conditions of interest in a more specific way. The subsequent characterization of these genes and their products could have a significant impact on science and finally in the field, as just exemplified with PACC. Another clear example was recently published, aimed at studying genes related to conidiation; SSH libraries were constructed using RNA from *T. virens* wt and M7 strain, which the latter is not able to conidiate. The authors identified 12 unigenes apparently related to conidiation, of which 2 were notably downregulated in the mutant growing on potato dextrose agar. One of these genes codes for a novel PGYRP (proline-glycine-tyrosine-rich protein) named PGY1 and the other for a GPI-anchored cell wall protein named ECM33. Mutant strains Δ*pgy1* and Δ*ecm33* showed slow growth and reduced conidiation. Furthermore, these mutants are not able to grow over *S. rolfsii*, and their mycelium-free culture filtrates do not cause inhibition of *Pythium aphanidermatum*, contrary to the wt. The latter is related with low production of the antimicrobial viridin by the mutant strains. The fact that PGY1 and ECM33 are related with conidiation and biocontrol represents an excellent opportunity to select them as possible targets to improve genetically *Trichoderma* strains and then use them commercially (Table 1) (Bansal et al. 2019).

Although the methods discussed here have represented very good approaches for understanding the transcriptome of any living organism, they also have limitations because they require a previous knowledge of the genome. SSH technique, for example, could present high background in cross-hybridization; some are expensive and not quantitative as ESTs, among others. However, in the last years, the new technology of RNA sequencing (RNA-seq) has emerged, which has become the method of choice and will be discussed below.

6 Transcriptomics of *Trichoderma* by RNA-seq Approaches

Toward 2008, the first works using the RNA-seq method were published for *S. cerevisiae*, *S. pombe*, *A. thaliana*, *Mus musculus*, and *Homo sapiens* cells. This technology, known as next-generation sequencing (NGS), allows for the identification of the complete set of transcripts in a cell at the moment of a specific condition, as well as to quantify them. Also, small RNAs, splicing isoforms, and gene fusion transcripts can be sequenced as well. Furthermore, de novo annotation is used to find novel transcripts from unannotated genes (Wang et al. 2009; Martin and Wang 2011).

In the *Trichoderma* research field, this new approach has allowed to broaden the knowledge of and to understand better the mechanisms used by some species of this genus that are useful for biological control, as well as to know the mechanisms they use to establish a beneficial relationship with plants, which will be addressed later.

Wu and coworkers reported the characterization of *T. asperellum* GDFS1009 strain, using different tools to draw a more complete picture of its capabilities for biocontrol. In in vitro dual culture assays of *T. asperellum* GDFS1009 against the plant pathogen *F. oxysporum* f. sp. *cucumerinum* Owen, growth was inhibited up to 80.82%, whereas in greenhouse conditions, this inhibition rate reached up to 86.34%. The inhibition caused by mycelium-free culture filtrates of *T. asperellum* GDFS1009 against *F. oxysporum* f. sp. *cucumerinum* Owen was 67.59% and 100% for *F. graminearum*. The transcriptome of *T. asperellum* GDFS1009 grown on potato dextrose medium and collected at 24 and 48 h shows that one glucanase-, one protease-, and one chitinase-encoding genes are highly induced, as well as three elicitor-encoding genes including two hydrophobins and EPL1. These results as well as the fungus-host interaction assays indicate that during the interaction of *T. asperellum* GDFS1009 with its host, the transcriptome is more dynamic and induces numerous genes related with its capabilities as biocontrol agent (Wu et al. 2017). On the other hand, the transcriptome of the well-characterized fungus *T. virens* ZT05 was determined under antagonistic conditions with *R. solani*. Several genes related to its mycoparasitic and antagonistic capabilities are induced, including eight genes associated with host recognition and signal transduction (two extracellular proteases, one protease, one belonging to oligopeptide transporters, and four G-protein-coupled receptors). Genes related to hyperparasitic genes, including six chitinases, six glucanases, and one proteasome, were also induced. With respect to antibiotic and stress resistance genes, 30 were identified, including 9 reductases, 2 tetracycline resistance genes, 8 heat shock response genes, 2 multidrug resistance transporters, 8 ABC efflux transporters, and 1 oxidative stress response gene, most of them were highly induced. RNA-seq data are consistent with the growth inhibition of *R. solani* by *T. virens*, which can coil and penetrate the phytopathogen mycelium. Also, the volatile and nonvolatile metabolites of *T. virens* inhibit the growth of the prey (Table 1) (Halifu et al. 2020). All these data point out that under antagonistic conditions, *Trichoderma* is capable to turn on most of its vast repertoire of genes necessary to be successful against its prey.

As mentioned above, SMs have a primordial role during the interaction of *Trichoderma* spp. with their preys. 6-PP is one of the main VOCs produced by many fungi of this genus, which possesses antifungal activity and is the compound responsible for the coconut odor in the producer strains (Vinale et al. 2008a). To learn more about the molecular mechanisms involved in 6-PP activity, Jin et al. (2020) analyzed the transcriptome of *Cylindrocarpon destructans* (one of the most devastating diseases of *Panax notoginseng*) exposed to 6-PP, the main metabolite secreted by *T. atroviride* T2, similar to that reported by other authors (Estrada-Rivera et al. 2019). Among the DEGs, 83 were enriched in 16 KEGG pathways, highlighting those involved in amino acid metabolism, such as valine, leucine, lysine, and isoleucine degradation, arginine and proline metabolism, and beta-alanine metabolism, indicating that 6-PP significantly affected these processes. Interestingly, a co-expression network analysis revealed that *echs1* (enoyl coenzyme A hydratase, short-chain 1, mitochondria) is the hub gene correlated with 6-PP stress. RNA-seq and RT-qPCR analyses also show that *echs1* is downregulated in the presence of 6-PP. Intriguingly, studies in hepatocellular carcinoma have shown that autophagy is present by silencing *echs1* (Xu et al. 2015), a phenotype that was observed in hyphae of *C. destructans* treated with 6-PP. Therefore, the mechanism of action of 6-PP produced by *T. atroviride* T2 could include the downregulation of *echs1* to induce autophagy in *C. destructans*.

Many SMs produced by *Trichoderma* have been reported; however, there is a group named ribosomally synthesized and post-translationally modified peptides (RiPPs) scarcely studied in fungi. RiPPs have been well studied in bacteria, including 20 different classes of compounds, whereas in fungi, only 6 have been described. Like many other SMs, RiPPs are organized in biosynthetic gene clusters and have potential bioactive properties (Arnison et al. 2013; Luo and Dong 2019). Using computational tools, Vignolle et al. (2020) predicted 6, 110, 222, and 92 putative RiPPs for *T. reesei*, *Trichoderma citrinoviride*, *T. harzianum*, and *T. brevicompactum*, respectively. Characterization of the cluster 55 of RiPPs in *T. reesei* shows homology with *T. citrinoviride* cluster 75. Also, cluster 55 of *T. reesei* possesses 22 predicted genes and 2 possible pseudogenes, which potentially code a putative major facilitator superfamily (MFS, gene D), a sulfatase (gene F), a putative hydrolase (gene L), an acid phosphatase (gene N), a cytochrome P450 (gene P), and a peptidase (gene S). The RNA-seq data and their alignment with the *T. reesei* genome were used to eliminate the RiPP false-positive precursors. Consequently, only those putative precursor peptide genes that aligned to RNA-seq data are considered as true positives. Transcriptome expression analyses show that the putative RiPP precursor peptide from cluster 55 is transcribed at low levels, revealing that it is present in the genome of *T. reesei* (Vignolle et al. 2020). Although further analyses of RiPPs in biocontrol are necessary, this work opened a new opportunity in the field.

RNA-seq approach has been used also to study indirectly some aspects related to biocontrol that could help when applying *Trichoderma* spp. in the field. For instance, the transcriptome of *T. harzianum* Tr-92 under chlamydospore-producing condition has been analyzed to know the molecular mechanisms implicated in this process that, in the future, could help to propose new and better commercial formulations (Yuan et al. 2019b). Until now, most *Trichoderma*-based biopesticides are made using conidia; however, they usually have short shelf life (Swaminathan et al. 2016; Li et al. 2016). However, besides conidia, *Trichoderma* spp. grow as mycelia and produce chlamydospores, a thick-walled spore produced vegetatively by mycelia that is more resistant to adverse conditions. Yuan et al. (2019b) showed that chlamydospore-based formulations present higher biocontrol capability against *B. cinerea* compared to the conidia-based formulation. Transcriptome analysis shows that genes that code for a glutathione-S-transferase, an oxidoreductase, a glycosyltransferase, and a peroxidase are downregulated in *T. harzianum* Tr-92 under the chlamydospore-producing condition and that a protein kinase, a chitinase, and an intracellular serine protease-encoding gene were upregulated (Table 1).

On the other hand, the association between different organisms in the microbiomes could enhance their capabilities during biocontrol. In this regard, co-cultures of *T. viride* with *Azotobacter chroococcum* (Az), a gram-negative beneficial bacterium that fixes atmospheric nitrogen, enhance biofilm formation and aggregation of the microbial partners, which represent interesting characteristics of multi species bioinoculants for its use in agriculture (Triveni et al. 2013; Velmourougane et al. 2017). The transcriptomic analysis of *T. viride-A. chroococcum* biofilm allows identifying biofilm biosynthesis-related genes such as *alg8*, *sipw*, *pssa*, *fadd*, *purb*, *phob*, and *glgp*, which code for an alpha-1,3-glucosyltransferase, a signal peptidase, a CDP-diacylglycerol-serine O-phosphatidyltransferase, a long-chain acyl-CoA synthetase, an adenylosuccinate lyase, an alkaline phosphatase, and a glycogen phosphorylase, respectively. Interestingly, the highest induced gene codes for an RNA-dependent RNA polymerase (RdRP) (Table 1). RdRPs use a single-strand RNA to generate double-strand RNAs, which then are processed by Dicer to generate sRNAs, thus having a very important role in RNA silencing (Willmann et al. 2011). This result indicates that gene regulation mediated by sRNAs has a pivotal role during biofilm formation between *T. virens* and *A. chroococcum* (Velmourougane et al. 2019), as observed for other microorganisms (Mika and Hengge 2013; Chambers and Sauer 2013).

It is noticeable that after the arrival of NGS technologies, science has advanced faster. Moreover, NGS not only has allowed having a bigger picture about what is occurring in an organism under specific conditions but, also, they allow selecting those genes that could have a relevant participation under such conditions, leading to more accurate hypotheses.

7 Proteomics for the Discovery of Proteins Potentially Involved in Mycoparasitism

Proteomic technologies enable the identification of proteins' associated peptides in a sample through the use of mass spectrometry (MS) instrumentation and the analysis of a whole set of proteins that are differentially expressed under different biological conditions between comparative samples (Aslam et al. 2017). The determination of differential expression of a protein can be done by counting its peptides that are detected by mass spectrometers and comparing them between two conditions. Additionally, data generated from expression proteomics experiments can be used as the basis of hypothesis-driven research followed by functional studies of selected proteins (Lippolis et al. 2019). In the past few decades, protein separation and identification have been achieved mainly through two-dimensional gel electrophoresis (2DE) coupled to matrix-assisted laser desorption/ionization time-of-flight mass spectrometry (MALDI-TOF MS) or liquid chromatography-tandem mass spectrometry (LC-MS/MS). In the 2DE technique, proteins are resolved into individual protein spots based on their isoelectric point and molecular weight, and then they can be excised from the gels, proteolytically digested, and identified by

MS (Lee et al. 2020). During the last decades, these methods have been used by several authors to gain insight into the knowledge of mechanisms involved in the biocontrol of plant pathogenic fungi by *Trichoderma* strains (Table 2). For example, Suárez et al. (2005) used 2DE-MALDI-TOF-MS to analyze the extracellular proteome of *T. harzianum* CECT 2413 in the presence of fungal cell walls. In particular, an aspartic protease was found to be significantly accumulated in a medium supplemented with *B. cinerea* or *R. solani* cell walls relative to medium supplemented with chitin or glucose as carbon source (Suárez et al. 2005). Aspartic proteases of *Trichoderma* have been found to be involved in mycoparasitism on plant pathogenic fungi by hydrolyzing host cell wall proteins (Deng et al. 2018; Yang et al. 2013). 2DE-MALDI-TOF-MS method was also used by Grinyer et al. (2005) to identify *T. harzianum* P1 proteins secreted in response to *R. solani* cell walls and compared with proteins secreted in a medium containing glucose as carbon source. Among the proteins that are upregulated in response to *R. solani* cell walls, known fungal CWDEs including an endochitinase and a *N*-acetyl-β-D-glucosaminidase, as well as three proteases and a superoxide dismutase, were identified (Grinyer et al. 2005). Additionally, 2DE-MALDI-TOF-MS method was used by Monteiro et al. (2010) to analyze the secretome of *T. harzianum* ALL42 after it was grown in liquid medium supplemented with purified cell walls of the plant pathogenic fungi *R. solani* or *Fusarium* sp. Analysis of the gel spots by MS identified a set of *T. harzianum* CWDEs secreted in response to each type of cell wall of the host, including an α-1,3-glucanase, a carboxypeptidase 2, and a glucosidase I induced by the presence of *R. solani* cell walls, as well as an endochitinase and a carboxypeptidase induced when *T. harzianum* was grown in the presence of *Fusarium* sp. cell walls (Monteiro et al. 2010). Extracellular α-1,3-glucanases of *Trichoderma* strains hydrolyze intrachain glycosidic linkages of α-1,3-glucan, a major cell wall polysaccharide in filamentous fungi, releasing β-glucose residues in a progressive manner (Grün et al. 2006; Soler et al. 2001), whereas endochitinases are known to hydrolyze glycosidic bonds in chitin, the major component of fungal cell walls (Du et al. 2020). In fact, the biological control of diverse fungal plant pathogens by some *Trichoderma* species has been directly related to their extracellular chitinase activity (Aoki et al. 2020; Loc et al. 2020). Carboxypeptidases are enzymes that remove C-terminal amino acid residues from peptides or proteins (Drzymała and Bielawski 2009). The role of *Trichoderma* carboxypeptidases in the biological control of plant pathogenic organisms was demonstrated by Morán-Diez et al. (2019). Using an overexpression approach, authors found that the carboxypeptidase CPA1 of *T. atroviride* T11 plays a role in the antifungal activity of this pathogen against *V. dahliae* (Morán-Diez et al. 2019).

Yang et al. (2009) used 2DE-LC-MS/MS to identify *T. harzianum* ETS 323 proteins that were secreted in response to deactivated *B. cinerea* mycelium. Among the identified proteins, an L-amino acid oxidase (LAAO) is accumulated in the medium containing deactivated *B. cinerea* mycelium as the sole carbon source (Yang et al. 2009). In a previous study by the same group, Tseng et al. (2008) also used 2DE-LC-MS/MS to identify *T. harzianum* ETS 323 proteins that were secreted in media containing either glucose or glucose plus deactivated *R. solani* hyphae as

Table 2 Overview of studies on *Trichoderma* spp. that used a proteomic approach to unravel proteins related to biocontrol

Biocontrol agent	Experimental design	Method and technology	Main findings	References
T. harzianum CECT 2413	Grown in medium supplemented with either glucose, chitin, or *B. cinerea* or *R. solani* cell walls as carbon sources	2DE-MALDI-TOF-MS	Identification of an extracellular aspartic protease of *T. britannicum* that was induced in medium supplemented with *B. cinerea* or *R. solani* cell walls relative to medium supplemented with chitin or glucose	Suárez et al. (2005)
T. harzianum P1	Grown in medium containing either glucose or cell walls of *R. solani* as carbon sources	2DE-MALDI-TOF-MS	Identification of an extracellular endochitinase and an *N*-acetyl-β-D-glucosaminidase as well as three proteases and a superoxide dismutase of *T. harzianum* P1 that were induced in medium supplemented with cell walls of *R. solani* relative to the glucose-supplemented medium	Grinyer et al. (2005)
T. harzianum ETS 323	Grown in medium containing either glucose or a mixture of glucose and deactivated *R. solani* hyphae as carbon sources	2DE-LC-MS/MS	Identification of an L-amino acid oxidase of *T. harzianum* ETS 323 that was found to be enriched only in medium containing glucose plus deactivated *R. solani* hyphae	Tseng et al. (2008)
T. harzianum ETS 323	Grown in medium containing either glucose or a mixture of glucose and deactivated *B. cinerea* mycelia as carbon sources	2DE-LC-MS/MS	Identification of one extracellular L-amino acid oxidase (LAAO) and two endochitinases of *T. harzianum* ETS 323 that were induced only in medium containing glucose plus deactivated *B. cinerea* mycelia	Yang et al. (2009)
T. harzianum ALL42	Grown in medium containing either glucose or cell walls of *R. solani* or cell wall of *Fusarium* sp. as carbon sources	2DE-MALDI-TOF-MS	Identification of a set of *T. harzianum* ALL42 secreted proteins, including an α-1,3-glucanase, a carboxypeptidase 2, and a glucosidase I that were induced by the presence of *R. solani* cell walls and an endochitinase and a carboxypeptidase that were induced when *T. harzianum* was grown in the presence of *Fusarium* sp. cell walls relative to the glucose-supplemented medium	Monteiro et al. (2010)

(continued)

Table 2 (continued)

Biocontrol agent	Experimental design	Method and technology	Main findings	References
T. atroviride T17	Grown in medium containing either glucose or *Guignardia citricarpa* GC3 inactivated mycelium	2DE-LC-MS/MS	Identification of 59 *T. atroviride* T17 extracellular proteins that were differentially expressed in medium containing *G. citricarpa* GC3 inactivated mycelium compared with medium containing glucose	de Lima et al. (2016)

carbon sources. Among the identified proteins, a LAAO enzyme was found to be enriched only in media containing glucose plus deactivated *R. solani* hyphae compared to glucose only (Tseng et al. 2008). LAAO are enzymes that catalyze the oxidative deamination of L-amino acid substrates and, thereby, produce hydrogen peroxide, ammonia, and the corresponding α-keto acid (Butzke et al. 2005). A hypothetical mechanism of the antagonistic effect of *T. harzianum* ETS 323 against *R. solani* mediated by LAAO was proposed in a study from Yang et al. (2011). The authors suggest that monomeric *T. harzianum* LAAO (Th-LAAO) may bind to hyphal lysis (apoptosis)-related cell wall proteins of *R. solani*, causing dysfunction of these proteins by altering their structures. Furthermore, Th-LAAO may induce oxidation of target proteins and produce an increase in the concentration of H_2O_2, which cause the apoptosis of *R. solani* (Yang et al. 2011) (Fig. 1). In fact, recent evidence indicates that H_2O_2 production may be a strategy of *Trichoderma* to inhibit the growth of their fungal prey (Fig. 1). In this regard, Zhang et al. (2019) found that, during interaction with its host the fungus *Fusarium oxysporum* f. sp. *cubense* 4 (renamed as *F. odoratissimum* by Maryani et al. (2019)), *T. guizhouense* produces an excessive amount of H_2O_2 that is stored in microscopic guttation droplets hanging on the contacting hyphae. Additionally, authors found that a strain of *T. guizhouense* with a deletion in the NADPH oxidase gene (Δ*nox1*) and a strain with a deletion in the NADPH oxidase regulator gene (Δ*nor1*), which are affected in the production of H_2O_2, lost their ability to efficiently grow over *F. odoratissimum*, indicating that H_2O_2 produced by *T. guizhouense* is required for its combative interaction with *Fusarium* (Zhang et al. 2019). On the other hand, a study from de Lima et al. (2016), integrating 2DE and LC-MS/MS method, led to the identification of a set of extracellular proteins that may have a role in the mycoparasitic activity of *T. atroviride* T17 against the plant pathogenic fungus *Guignardia citricarpa* GC3. Authors identified a total of 59 *T. atroviride* T17 proteins that were differentially expressed in a medium containing *G. citricarpa* GC3 inactivated mycelium compared with a medium containing glucose as carbon source. Among these, a set of glycoside hydrolases, two carboxylic ester hydrolases, an acid phosphatase, and a putative N,O-diacetyl muramidase were induced in response to the inactivated mycelium of the host fungus (de Lima et al. 2016).

8 Metabolomics as a Tool for the Discovery of SMs in *Trichoderma* Species

Fungi are a rich source of biological active compounds because they produce large amounts of SMs with biological activity. SMs comprise a very valuable group of chemical compounds that have a wide spectrum of application including drugs (immunosuppressants, antitumor agents, and antibiotics), biofuels (squalene and oleoresin), food additives (essential oils, carotenoids, and flavonoids), and agrochemicals (insecticides, pesticides, and antifeedants). Fungi of the *Trichoderma* genus are not the exception because they are also considered a rich source of novel SMs of agricultural, industrial, and medical interest.

Metabolomics has been defined as the in-depth quantitative and qualitative analysis of all small molecules (molecular weight < 3 kDa) in biological systems, being either a sample of cell, body fluids, tissues, or an entire organism (Fiehn 2001). The metabolome is the final step in the omics, at the biochemical and molecular levels; hence, it is most closely related to the phenotype of the organism, providing a better comprehension of its biological function (Van der Werf et al. 2005).

The development of new instrumentation, analytical technologies, and specialized software has allowed for the identification and description of metabolomes (Scalbert et al. 2009). MS has been one of the most used platforms in metabolomic studies, because of its flexibility in experimental design, its high sensitivity, and its capability to quantify low-abundance metabolites, and its high accuracy (Dettmer et al. 2007). MS is an approach used to determine the mass-to-charge ratio of ions, whose results are shown as mass spectrum, that is, a plot intensity as a function of the mass-to-charge ratio. Such spectra are used to figure out the isotopic form of a sample, the masses of particles and molecules, and to determine the structure or chemical identity of chemical compounds. The MS approach is applied in several research areas to heterogeneous or defined samples. MS is commonly and successfully applied in microbiological samples together with liquid chromatography (LC-MS), gas chromatography (GC-MS), and capillarity electrophoresis (CE-MS). LC-MS is a highly sensitive approach that allows examining at the same time large amounts of metabolites demanding a small sample volume for analysis. GC-MS is flawless for the identification and quantification of small molecular metabolites (< 650 Da), using chemical derivatization to volatilize these compounds for GC. EC-MS is the result of hyphenating a separation approach based on the movement of ions under electrophoretic and/or electro-osmotic forces produced by the application of an electric field with a mass spectrometer. This approach is suited for polar and ionic compounds in complex polar matrices, augmenting the metabolite hedge of LC-MS and GC-MS.

Metabolomics is used often to identify and quantify SM of orphan SM gene cluster's products unveiled by genome mining (Challis 2008; Scherlach and Hertweck 2009; Fischbach and Voigt 2010; Medema et al. 2011), as well as of desired products of a genetically modified strain. Comparative metabolomics is really promising in the discovering of new SM by comparing the metabolomic profile of a wild-type strain and its derivative mutants. This also works when comparing the same strain subjected to different growth conditions; by applying these approaches, unknown SM can be identified simply by estimating differentially abundant masses from the different samples.

For instance, the analysis of the histone deacetylase *hda-2* mutant of *T. atroviride* metabolome revealed that it regulates multiple responses in the model *A. thaliana* plant, including the stimulation of growth by some VOCs of the fungus (Estrada-Rivera et al. 2019). This work reveals that *hda-2* has a dual role in the regulation of secondary metabolism-related genes. In a different work, LC/ESI-MS methods were developed and validated using different standards of peptaibols (11-, 14-, and 20-amino acid residues) to quantify them in extracts of 13 different marine *Trichoderma* strains according to their chain length. Based on the optimal culture time for higher concentration of such peptaibols, three strains seemed to be good candidates as potential new biological control agents (Van Bohemen et al. 2016).

9 Exploring the Mutualistic Relationship Between *Trichoderma* spp. and Plants in the "Omics" Era

9.1 Plant Root Colonization

The initial processes in root colonization include the recognition between *Trichoderma* and the plant and the attachment of the fungus, which is essential to succeed in the mutualistic relationship. Most *Trichoderma* spp. colonize and penetrate the plant roots, growing into the epidermis, the first few cortical cell layers, intercellularly and then limited mostly to the apoplast, whereas vessels remain intact or minimally altered (Salas-Marina et al. 2011; Vargas et al. 2009). However, *T. virens* is able to surpass the intercellular spaces and is able to grow inside cells of *Z. mays* roots (Nogueira-Lopez et al. 2018).

Attachment of *Trichoderma* hyphae to the plant roots requires the action of extracellular hydrophobins (Mendoza-mendoza et al. 2017) (Fig. 1). Fungal hydrophobins are small proteins localized on the surfaces of aerial hyphae and spores and function by decreasing water surface tension and helping in the attachment of fungal hyphae to surfaces (Cai et al. 2020; Wu et al. 2017). The role of *Trichoderma* hydrophobins in plant root colonization has been reported in some studies. For instance, overexpression of the gene that encodes the hydrophobin *hydii1* in *T. virens* increases the capability of the fungus to colonize *A. thaliana* roots, whereas its deletion reduces it (Guzmán-Guzmán et al. 2017). Viterbo and Chet (2006) found that the deletion of the *hyd1* gene that encodes for the hydrophobin from *T. asperellum* reduces the capability of the fungus to colonize *Cucumis sativus* roots (Viterbo and Chet 2006). Interestingly, treatment of *Lotus japonicus* cells with the purified hydrophobin HYTLO1 from *T. longibrachiatum* revealed that the protein localizes at the plant cell surface, where it forms a protein film covering the plant cell wall

(Moscatiello et al. 2018). In this regard, it has been proposed that *Trichoderma* hydrophobins might protect the growing hyphae of the fungus from locally synthesized plant defense compounds during early stages of plant interaction allowing these beneficial fungi to colonize the plant root (Viterbo and Chet 2006).

In addition to hydrophobins, the colonization of plant roots by *Trichoderma* requires the action of extracellular expansin-like proteins, swollenins (Cosgrove 2017). The role of *Trichoderma* swollenins in plant root colonization was demonstrated by Brotman et al. (2008). Authors found that silencing of the gene that encodes a swollenin in *T. asperellum* by RNA interference reduces *C. sativus* root colonization, whereas its overexpression increases it remarkably. They speculate that swollenins could facilitate the access of extracellular cellulolytic enzymes of *T. asperellum* to less accessible areas of the plant cell wall, leading to its disruption and the subsequent root colonization (Brotman et al. 2008). Additionally, Meng et al. (2019) found that the purified expansin-like protein SWO from *Trichoderma* sp. NJAU4742 modifies the morphology and root architecture of *C. sativus* seedlings. Furthermore, the authors observed a high number of NJAU4742 spores attached to the plant root surface pretreated with SWO, suggesting that this protein aids the fungus to colonize the plant root (Meng et al. 2019).

During the initial processes of interaction with the plant root, *Trichoderma* secretes an array of hydrolytic enzymes, with a potential role in the degradation of plant cell wall polysaccharides, including cellulases, xyloglucan-specific endo-β-1,4-glucanases, and endo-1,4-β-xylanases, among other CWDEs (Gonzalez-et al., under review) (Fig. 1). All these enzymes are glycoside hydrolases (EC 3.2.1) that break down glycosidic bonds in polysaccharides and have been classified as carbohydrate-active enzymes (CAZy) in the Carbohydrate Enzyme Database (CAZy; http://www.cazy.org) (Lombard et al. 2014). Some of these CWDEs have been studied in plant pathogenic fungi. For instance, disruption of the endo-beta-1,4-xylanase gene *xyn11A* in *B. cinerea* causes pronounced negative effect on virulence, reducing the average damage in *S. lycopersicum* leaves by more than 70% (Brito et al. 2006), whereas the endo-1,4-glucanases XEG12A and XEG5A from *A. oryzae* contribute to the degradation of xyloglucan polysaccharides (Matsuzawa et al. 2020).

In addition to glycoside hydrolases, during their interaction with plants, *Trichoderma* secrete acetylxylan esterases and cutinases, which may play a role during colonization of plant root (Gonzales López et al., under review). Acetylxylan esterases and cutinases are carboxylic ester hydrolases (EC 3.1.1) involved in the breakdown of the plant cell wall components, xylan and cutin, respectively. Acetylxylan esterase from *T. reesei* is able to deacetylate both mono- and double-acetylated xylan residues (Hakulinen et al. 2000). A cutinase gene from *T. harzianum* T34 was characterized by Rubio et al. (2008). The authors overexpressed the *cut1* gene from *T. harzianum* T34 in *Pichia pastoris*, observing a high level of esterase activity when the recombinant strain of *P. pastoris* was cultured in a medium containing *p*-nitrophenyl acetate as substrate, revealing the cutinase nature of CUT1. Furthermore, *cut1* mRNA was highly induced when *T. harzianum* T34 was grown in the presence of the cutin monomer, 16-hydroxyhexadecanoic acid. The authors suggested that *Trichoderma* cutinases could facilitate the access of other

CWDEs to cell wall polymers during the plant cell wall degradation process (Rubio et al. 2008).

9.2 *Insights from Proteomic Studies on the Plant Root Colonization by* Trichoderma *spp.*

With the advances in instrumentation, in the last few years, new methods have been used to characterize the extracellular proteome of *Trichoderma* strains during their interaction with plants, including the gel-free LC-MS/MS-based profiling method, which allows for a global comparison of a whole set of proteins between two conditions. For example, Nogueira-Lopez et al. (2018) used a gel-free shotgun proteomic approach to characterize the secretome of *T. virens* during its interaction with *Z. mays* roots. A set of glycosyl hydrolases was particularly secreted into the plant apoplast, which are probably involved in the plant cell wall hydrolysis (Nogueira-Lopez et al. 2018). In a different study, using a gel-free LC-MS/MS approach, Lamdan et al. (2015) identified 32 *T. virens* secreted proteins that were enriched during co-culture with *Z. mays* seedlings in a hydroponic culture system as compared with the fungus grown alone. Among these, ten glycoside hydrolases with putative roles in the plant cell wall degradation were found in higher abundance when the fungus was grown in the plant's presence (Lamdan et al. 2015). Additionally, in a recent study from our laboratory, using a gel-free LC-MS/MS approach led to the identification of extracellular proteins of *T. atroviride* IMI 206040 and *A. thaliana* during their interaction, most of which were predicted to have a putative enzymatic function. A set of putative CWDEs was identified in the secretome of *T. atroviride*, which increased in abundance when the fungus was grown in the presence of *A. thaliana* seedlings in a semi-hydroponic system (González-López et al. 2021). Based on these findings, it has been proposed that, to interact with the plant, *T. virens* and *T. atroviride* secrete CWDEs that disrupt the plant cell wall facilitating the penetration of these fungi to the internal tissues (Lamdan et al. 2015; Nogueira-Lopez et al. 2018; González-López et al. 2021).

9.3 *Using "Omics" to Understand How Plants Respond to the Colonization by* Trichoderma *spp.*

During the initial steps of interaction with *Trichoderma*, plants secrete proteins related to defense probably to avoid an excessive damage to the roots caused by extracellular CWDEs. This hypothesis is reinforced by transcriptomic and proteomic studies that have analyzed the whole set of genes or proteins, respectively, that were differentially upregulated or downregulated during the interaction of *Trichoderma* strains with plants. In this regard, Estrada-Rivera et al. (2020) assessed the transcriptomic response of *T. virens* in the presence of *Arabidopsis* plants by

RNA-seq. The main response of the fungus to the presence of the plant was the repression of genes that encode CWDEs mainly at early time points of interaction (48 and 72 h). Authors suggest that *T. virens* downregulates the expression of CWDE-encoding genes at the early step of plant root colonization to avoid the excessive damage to the plant tissue (Estrada-Rivera et al. 2020). Additionally, proteomic analysis of *T. virens* secretome revealed that a set of CWDEs, which have been predicted to be specific to fungal cell wall degradation, decrease in abundance when the fungus was grown in the presence of *Z. mays* plants (Lamdan et al. 2015). It is tempting to speculate that the downregulation of *Trichoderma* CWDEs is provoked, at least in part, due to the action of defense proteins that are secreted by the host plant during the early stages of interaction, as a response to the presence of *Trichoderma* (Fig. 1). This idea is reinforced by the fact that, during interaction with *T. atroviride*, *A. thaliana* secretes protease inhibitors with known role in plant defense response, such as the unusual serine protease inhibitor UPI and the Kunitz trypsin inhibitor KTI1 (González-López et al. 2021). Additionally, a serine-type endopeptidase inhibitor and the proteinase inhibitor Pis7 were found to be secreted by *Z. mays* plants in response to the presence of *T. virens*. These proteins are probably secreted by the plant to target proteins secreted by *T. virens* in the apoplast (Nogueira-Lopez et al. 2018).

10 Induction of Plant Growth by *Trichoderma* spp.

Many *Trichoderma* strains have been identified as being able to stimulate plant growth. For instance, *Phaseolus vulgaris* plants inoculated with *Trichoderma* strains isolated from *P. vulgaris* soil in the field show higher hypocotyl diameter, dry weight of aerial parts, and a more developed root system (Mayo-Prieto et al. 2020). Barley plants inoculated with *Trichoderma* strains isolated from barley crop areas showed increased aerial and radicular dry weight and chlorophyll content compared with non-inoculated plants (Moya et al. 2020). Additionally, a greater stem diameter and soluble sugar and protein content were observed in *S. lycopersicum* seedlings treated with *T. asperellum* TaspHu1, a strain isolated from *Juglans mandshurica* rhizosphere (Yu et al. 2021).

Several mechanisms by which *Trichoderma* may promote plant growth have been proposed, including phosphate solubilization, production of VOCs and phytohormones, as well as production of 1-aminocyclopropane-1-carboxylate (ACC) deaminase (Viterbo et al. 2010; Estrada-Rivera et al. 2019) (Fig. 1). Phosphate solubilization has been related to the extracellular phytase and acid phosphatase activities in several *Trichoderma* strains (Saravanakumar et al. 2013; Tandon et al. 2020). Phytases are phosphatases that hydrolyze phytic acid (the primary form of organic phosphate in many soils) preferentially into inositol and inorganic phosphorus (P_i) that plants and microorganisms can take up for their metabolism (Corrêa and de Araújo 2020). Acid phosphatases have been purified from *Trichoderma* strains, including *T. asperellum* Q1 (Zhao et al. 2017) and *T. harzianum* ALL42 (Souza et al. 2016). Acid phosphatases hydrolyze phosphomonoester and amide substrates, thereby transforming organic phosphate into a soluble inorganic form (Schenk et al. 2013). *Trichoderma* sp. TSK8, a phosphate-solubilizing strain isolated from the rhizosphere of mangrove (*Avicennia marina*), enhances total mangrove seedlings' biomass when inoculated in the plant roots in the presence of soluble super phosphate (KH_2PO_4) as a phosphate source (Saravanakumar et al. 2013).

Furthermore, the ability of *Trichoderma* to promote plant growth has been attributed to the capability of this genus to produce VOCs (Fig. 1). The exposure to VOCs emitted by *T. asperellum* T1 increases numbers of *L. sativa* roots and leaves, plant biomass, and total chlorophyll content (Wonglom et al. 2020), and the VOC 6-PP enhances root branching of *A. thaliana* (Estrada-Rivera et al. 2019). Additionally, VOCs emitted by *T. azevedoi* CEN1241 promote plant growth and increase the content of carotenoids and chlorophyll in *L. sativa* plants (Silva et al. 2021).

Gravel et al. (2007) proposed that the synthesis of IAA through tryptophan-dependent pathways by *T. atroviride* affects the growth of *S. lycopersicum* seedlings, and the regulation in the concentration of IAA in the rhizosphere promotes the fruit yield and stem growth of *S. lycopersicum* plants (Gravel et al. 2007). Moreover, *T. atroviride* and *T. virens* produce indole-3-acetic acid-related compounds, which are involved in plant growth promotion (Gravel et al. 2007; Salas-Marina et al. 2011) (Fig. 1). *Trichoderma* sp. DEMTkZ3A0, a strain isolated from a healthy rye rhizosphere, produces indoleacetic acid and gibberellic acid in liquid medium and colonizes the rhizoplane of *Triticum aestivum* seedlings and causes an increase of the stem weight of *T. aestivum* seedlings (Jaroszuk-ściseł et al. 2019).

On the other hand, it has been observed that the inoculation of plants with different strains of *Trichoderma* could result in reduced ethylene (ET) production, due to the activity of the ACC deaminase, which degrades its precursor 1-aminocycloprop ane-1-carboxylic acid (ACC) to produce α-ketobutyrate and ammonia (Todorovic and Glick 2008) (Fig. 1). High levels of ethylene in plants stimulate auxin biosynthesis, which is then distributed toward the elongation zone of the root tip, where it causes inhibition of cell elongation and overall root growth (Růžička et al. 2007). The role of *Trichoderma* ACC deaminase in plant growth promotion was demonstrated in a study by Viterbo et al. (2010). Authors found that silencing the ACC gene *acds* in *T. asperellum* T203 by RNAi results in a decreased ability of the mutant strain to promote root elongation of canola (*Brassica napus*) seedlings (Viterbo et al. 2010).

10.1 Genes and Proteins Related to Plant Growth Stimulated by Trichoderma *spp.*

Recent transcriptomic- and proteomic-based studies revealed the induction of genes and proteins of the host plant as a response to the colonization by *Trichoderma* spp. For example, De Palma et al. (2019) characterized the transcriptomes expressed in

S. lycopersicum roots during a course of interaction with *T. afroharzianum* and compared them with non-inoculated control plants using the RNA-seq method. A total of 1243 *S. lycopersicum* transcripts were found to be differentially expressed across experimental time points of the interaction with *T. afroharzianum*. Singularly, the fungus triggered the expression of a set of genes encoding proteins involved in transport of nutrients and the downregulation of a gene encoding the transcriptional regulator SlMYB93 (De Palma et al. 2019). In *Arabidopsis*, MYB93 is a negative regulator of lateral root development, and its expression is induced by auxin in the basal meristem of the primary root (Gibbs et al. 2014). Membrane transport proteins play crucial roles in the uptake of nutrients and water from the soil, supporting diverse biological processes in plants including photosynthesis and plant growth (Zelazny and Vert 2014). Induction of transporter genes in the host plants by *T. afroharzianum* indicates a strong effect of these fungi on plant nutrition process that is part of its growth-promoting ability (De Palma et al. 2019). Coppola et al. (2019) studied the transcriptomic changes in *S. lycopersicum* plants induced by *T. afroharzianum* colonization. They observed a wide transcriptomic reprograming in the plant induced by *T. afroharzianum* associated with several biological processes. Among them, genes associated with the "photosynthesis-related mechanism" process were found to be induced, particularly genes related to photosynthesis and chlorophyll biosynthesis (Coppola et al. 2019). Also, the transcript that encodes for the small subunit of the key CO_2 fixation enzyme ribulose-1,5-bisphosphate carboxylase (rubisco) and the oxygen-evolving enhancer protein 3–1 was increased in leaves of *Z. mays* colonized by *T. virens*, consequently improving the photosynthetic rate. This process is dependent on *T. virens* invertase, a glycoside hydrolase involved in sucrose degradation (Vargas et al. 2009). Interestingly, proteomic analysis of the *A. thaliana* secretome root inoculated with *T. atroviride* revealed that, in response to the fungus, the plant secretes enzymes related to photorespiration, including rubisco and the glutamate:glyoxylate aminotransferase 1 (GGAT1), which is involved in the detoxification of glyoxylate (González-López et al. 2021). However, why these proteins would be secreted by *Arabidopsis* during interaction with *T. atroviride* remains unclear.

11 Priming for Plant Defense Induced by *Trichoderma* spp.

Many strains of *Trichoderma* are well-known for their ability to protect host plants from plant pathogenic microorganisms by triggering the expression of genes related to the plant defense response both locally and systemically (Fig. 1). Induction of defense-related genes by *Trichoderma* species is related with a physiological phenomenon known as priming. Priming consists in the preactivation of molecular mechanisms of plant defense, which can be extended systematically at distal sites of the plant, enabling a faster and robust plant response to subsequent attacks by plant pathogenic microorganisms (Gupta and Bar 2020). In this regard, Aamir et al.

(2019) found that *S. lycopersicum* plants primed with *Trichoderma erinaceous* show an enhanced accumulation of defense-related *WRKY* transcripts in roots and leaves and a higher number of lignified cell layers related to the reinforcement of plant cell wall (Aamir et al. 2019). In *S. lycopersicum* plants, *T. atroviride* induces both SAR- and induced systemic resistance (ISR)-related genes, as well as systemic protection against some plant pathogenic microorganisms including *A. solani*, *B. cinerea*, and *P. syringae* (Salas-Marina et al. 2015). SAR and ISR are two forms of induced resistance in plants that are activated in response to plant pathogenic microorganisms and beneficial microbes, respectively (Pieterse et al. 2014). Activation of SAR results in local and systemic increased levels of the hormone SA and the subsequent activation of genes encoding pathogenesis-related (PR) proteins (Pieterse et al. 2014). ISR is activated without the accumulation of SA but requires JA and ET signaling pathways and is related to the expression of defense-related genes, including *PDF1.2* (Van Oosten et al. 2008). Activation of genes related to phytohormone signaling by *Trichoderma* strains has been reported in several studies. For instance, inoculation of *T. longibrachiatum* H9 to *C. sativus* roots triggers the expression of JA-, ET-, and SA-related genes (Yuan et al. 2019a), whereas inoculation of *T. virens* to *A. thaliana* plants induces the expression of the SA- and JA/ET-responsive genes *PR-1a* and *PDF1.2*, respectively (Estrada-Rivera et al. 2020), and *T. afroharzianum* triggers the expression of the JA–/ET-responsive marker *PR3* in roots of sugar beet plants (Schmidt et al. 2020). Additionally, *T. harzianum* T-78 primes SA- and JA-dependent defenses in roots of *S. lycopersicum* plants, which limits root invasion by the nematode *Meloidogyne incognita* (Martínez-Medina et al. 2017a).

In the last years, microarray and RNA-seq technologies have been used in diverse studies to investigate the whole set of genes of the host plants that are differentially expressed during their interaction with a number of *Trichoderma* species, including genes related to the defense against an increasing number of plant pathogenic microorganisms (Sharma et al. 2017). Rubio et al. (2019) used a microarray approach to investigate the early global transcriptomic changes in *Triticum aestivum* seedlings' roots induced by *T. harzianum* T34. Authors found that, among the genes that were differentially modulated, several genes related to plant defense were upregulated in plant roots at early stages of the interaction with *T. harzianum*, including two genes encoding ethylene response factors (ERFs) and a gene encoding a xylanase inhibitor (Rubio et al. 2019). ERFs represent one of the largest families of transcription factors in plants that regulate molecular response to plant pathogens' attack and to abiotic stresses (Müller and Munné-Bosch 2015). In *A. thaliana*, ERF8 has a positive role in the resistance against *P. syringae* and induces cell death through its transcriptional repression activity (Cao et al. 2018). Plant xylanase inhibitors inhibit the activity of microbial xylanases, which are glycoside hydrolases that break down the plant cell wall polysaccharide xylan and have been related to plant defense against plant pathogenic microorganisms, including *B. cinerea* (Tundo et al. 2020) and *Magnaporthe oryzae* (Sun et al. 2018). In another study, RNA-seq was used to assess global gene expression profile of *A. thaliana* during the establishment of the interaction with the beneficial fungus *T. atroviride* IMI 206040. This approach revealed an increased expression level of genes encoding proteins of the JA pathway such as *PAD3*, *JAZ1*, *JAZ6*, and *LOX1* compared with non-inoculated plants. Interestingly, the authors found that all these genes are upregulated at a higher level when the plant is grown in the presence of a mutant strain of *T. atroviride* affected in its regulatory subunit of NADPH oxidase (Δ*noxr*) compared with plants inoculated with the wt strain, indicating that NOXR mediates the host defense response in *A. thaliana* (Villalobos-Escobedo et al. 2020). In phytopathogenic fungi, NADPH oxidases are required for pathogenicity through the production of ROS (Egan et al. 2007). In *T. atroviride*, NOXR plays a role in the response of the fungus to stress conditions, including osmotic, oxidative, membrane, and cell wall stresses, and in the antagonistic activity of *T. atroviride* against *R. solani* and *S. sclerotiorum* (van Zijll de Jong et al. 2019).

In addition to transcriptomic analysis, some studies have used proteomic approaches to analyze the expression of plant proteins induced by *Trichoderma* with potential roles in the plant response to plant pathogenic microorganisms. Pereira et al. (2014) used 2DE-MALDI-TOF-MS to evaluate the ability of *T. harzianum* ALL42 to promote a defense response of *Phaseolus vulgaris* plants in the presence or in the absence of *R. solani* and *Fusarium solani*. These authors constructed proteomic maps using roots and leaves of plants challenged with *T. harzianum* in the presence or absence of a plant pathogen. MS analysis of 2DE-derived spots led to the identification of a PR-like protein and an acyl-CoA-binding protein (ACBP), which were found to be upregulated in the root maps of *P. vulgaris* challenged with *T. harzianum* in comparison with unchallenged plant roots. Furthermore, *T. harzianum* induced the expression of a cinnamoyl-CoA reductase (CCR) in leaves of *P. vulgaris* plants. When plants were double-challenged with *T. harzianum* and *R. solani*, a higher expression level of a histone acetyltransferase complex component and a NAC1 domain protein was observed in *P. vulgaris* roots in comparison with that obtained for plants in the presence of *R. solani* alone. These findings may suggest that *T. harzianum* potentiates *P. vulgaris* response to plant pathogenic microorganisms through the induction of defense-related proteins (Pereira et al. 2014). The role of ACBPs in the defense response of plants has been reported in some studies. In *A. thaliana*, the acyl-CoA-binding proteins ACBP3, ACBP4, and ABCP6 are required for cuticle development, as well as for the defense against bacterial and fungal plant pathogens (Xia et al. 2012). Transgenic *A. thaliana* lines overexpressing the acyl-CoA-binding protein 5 (ACBP5) from *O. sativa* show enhanced resistance against *P. syringae* and fungal necrotrophs (De Oliveira et al. 2011; Panthapulakkal Narayanan et al. 2019). NAC1 domain proteins belong to the NAC family of transcription factors and play critical roles in plant immune response against plant pathogenic microorganisms by acting as positive or negative regulators of immunity-related genes (Yuan et al. 2019c).

12 *Trichoderma* Proteins that Act as Elicitors and Effector-like Proteins

To colonize and establish an association with plants, microorganisms such as bacteria, fungi, and oomycetes must deliver a suite of molecules known as effectors. These molecules include proteins, SMs, and nucleic acids (i.e., small RNAs) (Collemare et al. 2019; Liu et al. 2019; Rebolledo-Prudencio et al. 2020, 2021). Effectors are molecules that manipulate the host cell physiology to suppress its immunity (Hadar et al. 2020; Snelders et al. 2018). Effectors can be attached to the fungal cell wall forming a barrier that protects the fungal hyphae from degradation by plant hydrolytic enzymes, while others can act in the plant apoplast acting at the extracellular host-microbe interface. Some effectors can even translocate to distinct subcellular locations within the host cell where they target proteins that participate in diverse processes such as transcription, signaling, cellular trafficking, metabolism, and protein regulation, among other processes, which generally entail the suppression of host immunity (He et al. 2020; Liu et al. 2019; Rocafort et al. 2020). Compared with fungal plant pathogens, relatively few of these proteins have been functionally characterized in beneficial fungi. Because the structural characteristics of beneficial fungal effectors resemble those of pathogenic fungi, like the presence of known effector motifs, the small size, and the presence of several cysteine residues, these are termed here as effector-like proteins.

With the advent of omics and the creation of new bioinformatics tools, important advances have been made to identify potential effector-like proteins in beneficial fungi (Guzmán-Guzmán et al. 2017; Sperschneider et al. 2018), including the prediction of their subcellular localization and action into the host cell (Sperschneider et al. 2017). Genome sequence analyses of 3 *Trichoderma* species revealed a total of 233 effector-like proteins grouped into 18 families. In particular, the *lysm1* gene (encoding a LysM repeat family protein) from *T. virens* and the *trx2* gene from *T. atroviride* (encoding a thioredoxin family protein), among others, were found to be upregulated in the presence of *A. thaliana* (Guzmán-Guzmán et al. 2017).

Lysin motifs (LysM) are carbohydrate-binding modules that are present in diverse organisms including bacteria, plants, and mammals. They bind to N-acetylglucosamine (GlcNAc)-containing carbohydrates, such as chitin and peptidoglycan (Akcapinar et al. 2015; Liu et al. 2019). Experimental evidence of the participation of LysM proteins from *T. atroviride* during its association with plants has been reported recently. Tal6 is an effector protein produced by *T. atroviride* that binds to chitin oligomers localized on the fungal cell wall, conferring a protective barrier against plant chitinases and enhancing its mycoparasitic activity. Plants inoculated with *T. atroviride* Tal6-overexpressing strain showed better root and stem growth compared to the plants grown in the presence of *tal6* mutant or the wt strain (Romero-Contreras et al. 2019). Furthermore, the upregulation of *Trichoderma* spp. genes encoding LysM proteins, as well as their identification in secretomes during interaction with their host plant, supports the hypothesis that *Trichoderma* spp. also employ LysM proteins as effectors to enhance their interaction with plants

(Fig. 1) (Guzmán-Guzmán et al. 2017; González-López et al. 2021). All these findings suggest that the strategies used by beneficial fungi to interact with host plants are very similar to those used by fungal plant pathogens (Romero-Contreras et al. 2019).

The cerato-platanin family (CPF) proteins are fungal small-secreted cysteine-rich proteins (SSCPs) that act as effectors or elicitors of defense responses (Gao et al. 2020; Yang et al. 2017). A particular feature of all CPF proteins is the presence of four cysteine residues involved in the formation of two disulfide bonds (Luti et al. 2017). Experimental data suggest that CPF proteins play important roles during *Trichoderma*-plant interaction working as pathogen-associated molecular patterns (PAMPs). Depending on the species, *Trichoderma* spp. encode several CPF proteins (i.e., three in *T. atroviride* IMI 206040 and three in *T. virens* Gv29–8) that have been identified by broad-scale genome and secretome analysis (Cai et al. 2020; Gao et al. 2020; Kubicek et al. 2019). In this regard, a genome analysis for CPF protein-encoding genes in *Trichoderma* spp. shows that *epl1*, *epl2*, and *epl3* are conserved in 37 *Trichoderma* genomes. Particularly, *epl1* from *Trichoderma* sp. NJAU4742 is expressed during its interaction with *S. lycopersicum* seedlings. Its corresponding mutant strain increases its capability to colonize the plant roots compared to the wild-type strain (Gao et al. 2020). CPF proteins have been studied with considerable details in *Trichoderma* spp. For instance, SM1 from *T. virens*, EPL1 from *T. atroviride*, and EPL1 from *T. harzianum* induce the upregulation of defense-related genes in the plant, like *PAL* (Phe ammonia-lyase), *LOX1* (lipoxygenase), *GLU1* (β-1,3-glucanase), and α-*DOX1* (α-dioxygenase) genes (Djonović et al. 2007; Gomes et al. 2015; Salas-Marina et al. 2015).

12.1 Identification of Effector-Like Proteins in Trichoderma spp. by Using "Omics" Approaches

Considerable progress has been made in the study of *Trichoderma*-plant interaction with the use of transcriptomic and proteomic technologies. For example, using a gel-free shotgun proteomic approach, Lamdan et al. (2015) identified 280 proteins that are secreted by *T. virens* during its interaction with *Z. mays* seedlings. Of these proteins, 32 were upregulated by the presence of plant roots. In addition, 34 proteins were downregulated, including 13 SSCPs, which probably act as negative effector-like proteins (Lamdan et al. 2015). In a different study of *T. virens* Gv29–8 apoplastic secretome, during its interaction with *Z. mays* plants under hydroponic conditions, 43 proteins secreted into the apoplast were identified, which included effector-like proteins, hydrolytic enzymes, and proteins that participate in secondary metabolism, among others (Nogueira-Lopez et al. 2018). More recently, using a gel-free shotgun proteomic approach, 77 proteins were identified to be upregulated in *T. virens* in response to banana (*Musa* spp.) roots, including glycoside hydrolases and SSCPs (Muthukathan et al. 2020). In addition, global transcriptomic analyses of *T. virens* revealed that genes encoding effector-like proteins were upregulated during interaction with *Z. mays* and *A. thaliana* roots (Estrada-Rivera et al. 2020; Malinich et al. 2019). In our group, a considerable number of predicted effector-like proteins have been identified in the secretome of *T. atroviride* during its interaction with *A. thaliana* seedlings grown in a semi-hydroponic system. These proteins belong to N-acetyltransferase (NAT), cerato-platanin, chaperonin, fungal hydrophobin, LysM domain, redoxin, thioredoxin, endoribonuclease L-PSP, exonuclease, and single-strand binding protein families, among others (González-López et al. 2021).

13 Secondary Metabolites of *Trichoderma* spp. that Act as Elicitors

When interacting with plants, *Trichoderma* spp. produce a wide variety of SMs including NRPs, pyrones, and terpenoids (Pascale et al. 2017; Wu et al. 2017). Application of *T. atroviride* or its SMs induces significant promotion of plant growth (Estrada-Rivera et al. 2019; Contreras-Cornejo et al. 2020). In addition to plant growth promotion, SMs could also act as elicitors/effectors stimulating or modulating the immune response in plants that result in resistance against the attack by plant pathogenic microorganisms.

In this regard, the 6-PP is a flavoring agent that has been reported to have plant growth-promoting properties (Estrada-Rivera et al. 2019; Pascale et al. 2017). For example, treatment of *Vitis vinifera* plants with both *T. afroharzianum* strain and 6-PP improved crop yield. It increased the total amount of polyphenols and antioxidant activity in grapes, which indicated that 6-PP produces also positive effects similar to the application of live fungus (Pascale et al. 2017). In addition, it has been reported that 6-PP has the potential to elicit plant resistance against plant pathogenic microorganisms (Farh and Jeon 2020; Hamrouni et al. 2019). The ability of 6-PP produced by *T. atroviride* (strain P1), *T. afroharzianum*, and *T. harzianum* (T39 and A6 strains) to act as an elicitor *in planta* was investigated by Vinale et al. (2008a). In their study, an overexpression of genes encoding pathogenesis-related (PR) proteins was detected in *S. lycopersicum* seedlings treated with 6-PP. Moreover, disease symptoms caused by *B. cinerea* are significantly reduced in 6-PP-treated plants (Vinale et al. 2008a, b). Interestingly, the exposure of *A. thaliana* roots to 6-PP modulates the expression of genes encoding PIN family of auxin transporters, including *PIN1*, *PIN2*, *PIN3*, and *PIN7*, indicating that root responses to 6-PP involve components of auxin transport and signaling (Garnica-Vergara et al. 2016). Upregulation of phytohormone-related genes, including *PIN3* and *PIN7*, was also observed in *Arabidopsis* plants inoculated with a 6-PP-overproducing strain of *T. atroviride* with a deletion of the histone deacetylase-2 (*hda-2*) gene (Estrada-Rivera et al. 2019). In another study, Dini et al. (2020) reported that olive trees (*Olea europaea*) exposed to 6-PP increased the concentration of polyphenols in both leaves and oil (Dini et al. 2020).

Besides 6-PP, *Trichoderma* also secrete other SMs including peptaibols. Peptaibols are short peptides of 5–21 residues that are biosynthesized by NRPSs. These membrane-active compounds form linear helical structures, some of which form ion channels, creating holes in the lipid bilayer membranes making them leaky and inducing cell death (Marik et al. 2019). It has been found that the peptaibol alamethicin (ALA) produced by *T. viride* induces systemic resistance in plants. Engelberth et al. (2001) found that ALA from this fungus triggers the production of VOCs in lima bean (*Phaseolus lunatus*). Particularly, two homoterpenes were induced by ALA, 4,11-dimethylnona-1,3,7-triene and 4,8,12-trimethyltrideca-1,3,7,11-tetraene, as well as methyl salicylate and JA (Engelberth et al. 2001).

Because peptaibols induce cell lysis when aggregated in the cell membranes, it is necessary that plant cells protect themselves from fungal peptaibols to establish a successful relationship with *Trichoderma* strains. In this regard, Dotson et al. (2018) found that *Arabidopsis* root apical meristem and epidermis were permeabilized by alamethicin from *T. arundinaceum*, but not if pretreated with cellulase. Authors hypothesized that plant roots contain cells sensitive to peptaibols that detect cellulases from *Trichoderma* strains, which in turn induce protection against alamethicin (Dotson et al. 2018).

Another SM produced by *Trichoderma* spp. is harzianolide, which has been investigated for its effect on plant growth promotion and induction of ISR in plants (Vinale and Sivasithamparam 2020). Cai et al. (2020) found that purified harzianolide increases the expression of genes involved in the SA and JA/ET signaling pathways and promotes *S. lycopersicum* seedlings' growth. These results indicate that harzianolide also plays a role in both plant growth regulation and plant defense responses (Fu et al. 2013). All these findings suggest that some SMs produced by *Trichoderma* may play important roles as signals to modulate plant morphogenesis and immunity. In the future, molecular, physiological, and biochemical studies of SMs will allow for a better understanding of the mechanisms by which plants sense these fungal metabolites to establish a symbiotic association with *Trichoderma*.

13.1 The Role of VOCs in Plant Immunity During Trichoderma*-Plant Interaction: A Metabolomic Approach*

A plethora of VOCs has been identified in *Trichoderma*, showing a great variability in structure and function. It has been shown that some VOCs produced by *Trichoderma* spp. have a dual role promoting plant growth and playing critical roles in mycoparasitism as well. GC-MS analysis of 20 *Trichoderma* strains identified as much as 141 unique VOCs per strain (Lee et al. 2016). Moreover, during the *T. longibrachiatum-O. sativa* interaction, the fungus produces up to 138 different VOCs. These compounds inhibit the growth of the plant pathogens *S. rolfsii* and *M.*

phaseolina (Sridharan et al. 2020). Even when *Trichoderma* are not in contact with another organism, they can produce VOCs. Using proton transfer reaction time-of-flight MS and GC-MS, VOC production of *T. harzianum*, *T. hamatum*, *T. reesei*, and *T. velutinum* was quantified in vitro. These *Trichoderma* species produced monoterpenes, sesquiterpenes, alcohols, acyclic alkenes, alkanes, aldehydes, ketones, acids, benzenoids, and esters (Guo et al. 2020). To date, a complete picture for the VOC production and synthesis regulation is not available; however, progress has been made in this field in the last years. In this regard, *T. atroviride* can produce a great variety of VOCs both in free life and when associated with plants. Using GC-MS, the volatile compounds produced by *T. atroviride* wt strain and a mutant in the histone deacetylase (Δ*hda*-2) were identified when grown on PDA media. A total of 28 VOCs were identified, and the compounds were assigned to alcohols; ketones; mono-, di-, and sesquiterpenes; alkanes; and pyrones. The wt strain produced the ketones 2-heptanone and 3-octanone as the most abundant compounds, whereas for the mutant strain, it was 6-PP. Four VOCs were exclusive for the wt strain, including 3-octanone, 2-heptanol, 3-octanol, and 1-octen-3-ol, whereas 17 volatiles were unique for the Δ*hda*-2, including 6 new terpenes, 2-octanone, 3 unknown ketones, γ-terpinene, α-zingiberene, β-sesquiphellandrene, 2 unknown alcohols, 1 unknown alkane, and 1 unknown phenol. Compounds such as 2-heptanone, 2-pentylfuran, 2-nonanone, unknown terpene, 2-undecanone, 6-PP, and β-curcumene were common to both the wild-type and Δ*hda-2* strains (Estrada-Rivera et al. 2019). Furthermore, VOCs can regulate diverse processes, such as iron uptake and triggering ISR in plants. The transcription factor MYB72 from *A. thaliana* is one of the main regulators of ISR in plants. In addition, MYB72 regulates iron uptake when plants are in limiting conditions of this metal. *Arabidopsis* roots were exposed to the volatile compounds of *T. asperellum* or *T. harzianum* in a split-plate system, showing that the VOCs produced by these strains can induce the expression of *MYB72*, thereby regulating the induction of the ISR, as well as the regulation of the iron uptake in the plant (Martínez-Medina et al. 2017b).

It is expected that under interaction with its prey or another related fungus, *Trichoderma* might synthesize different sets and quantities of VOCs. Under this condition, an important number of SMs might be exchanged among fungi. In soil, *Trichoderma* species can share their niche with fungi of the same genus; however, the molecular mechanisms of the interaction remain unclear. During the co-cultures of *T. harzianum*, *T. hamatum*, and *T. velutinum* with the ectomycorrhizal fungus *L. bicolor*, each *Trichoderma* species shows different VOC profiles, without shared compounds among the three strains. The most abundant compounds detected were the oxygenated monoterpene "tetrahydrocarvone" and the monoterpene "γ-terpinene" for *T. harzianum* and *T. hamatum*, respectively. *T. velutinum* emitted the lowest number of VOCs without detecting a majority compound. On the other hand, *T. harzianum* is the strongest and most diverse VOC producer followed by *T. hamatum*. These results suggest that organisms from the same genera can interact differently with other fungal mutualistic species of plants (Guo et al. 2019).

14 Concluding Remarks

Over the past few decades, omics technologies, including genomics, transcriptomics, proteomics, and metabolomics, have been particularly useful for the identification of genes, proteins, and SMs of *Trichoderma* strains that would potentially be involved in mycoparasitism, as well as in the interaction with their host plant. Genome sequencing of *Trichoderma* species has led to the enrichment of transcriptomic and proteomic databases, which have been used to identify many genes and proteins of *Trichoderma* spp. involved in the recognition, attachment, and degradation of structural components of the prey fungus, as well as proteins related to plant root colonization and biological processes associated with plant growth promotion. RNA-seq is becoming the most utilized method to analyze the whole set of *Trichoderma* spp. genes that are differentially expressed under different biological conditions, particularly in response to a determined host fungus, or during interaction with the host plant. Using this method, many studies have identified a large amount of genes related to the biosynthesis of SMs and VOCs and genes that encode regulatory proteins, including hydrophobins, glycoside hydrolases, and effector-like proteins. Many of these genes may play important roles in the mycoparasitic activity of *Trichoderma* strains against plant pathogenic microorganisms and in plant growth promotion or activation of plant defense responses against diverse plant pathogenic microorganisms by acting as elicitors. During the last few decades, 2DE and gel-free shotgun methods coupled to MALDI-TOF-MS or LC-MS/MS technologies have been used mainly to characterize the extracellular proteome of *Trichoderma* strains during growth in presence of components of their host fungus (i.e., cell walls) or during interaction with plants, revealing a number of extracellular proteins related mainly with the hydrolysis of components of both plant and fungal cell walls, as well as proteins involved in plant growth promotion (i.e., phytases and acid phosphatases that solubilize organic phosphate sources from soil). In the future, these sophisticated technologies can be used to unveil the whole set of intracellular proteins that are differentially expressed during *Trichoderma*-plant-pathogen interaction and, then, select relevant proteins to elucidate their biological role through functional studies. The latter studies will allow us to better understand how these beneficial fungi confer benefits to their host plant and how they control diverse soilborne plant pathogenic microorganisms.

References

Aamir M, Kashyap SP, Zehra A, Dubey MK, Singh VK, Ansari WA, Upadhyay RS, Singh S (2019) Trichoderma erinaceum bio-priming modulates the WRKYs defense programming in tomato against the Fusarium oxysporum f. sp. lycopersici (Fol) challenged condition. Front Plant Sci 10. https://doi.org/10.3389/fpls.2019.00911

Adnan M, Islam W, Shabbir A, Khan KA, Ghramh HA, Huang Z, Chen HYH, Lu G (2019) Plant defense against fungal pathogens by antagonistic fungi with Trichoderma in focus. Microb Pathog 129:7–18

Ahmad JS, Baker R (1987) Rhizosphere competence of Trichoderma harzianum. Phytopathology 77:182–189

Akcapinar GB, Kappel L, Sezerman OU, Seidl-Seiboth V (2015) Molecular diversity of LysM carbohydrate - binding motifs in fungi. Curr Genet 61:103–113. https://doi.org/10.1007/s00294-014-0471-9

Ali S, Ganai BA, Kamili AN, Bhat AA, Mir ZA, Bhat JA, Tyagi A, Islam ST, Mushtaq M, Yadav P (2018) Pathogenesis-related proteins and peptides as promising tools for engineering plants with multiple stress tolerance. Microbiol Res 212:29–37

Al-Naemi FA, Ahmed TA, Nishad R, Radwan O (2016) Antagonistic effects of Trichoderma harzianum isolates against Ceratocystis radicicola: pioneering a biocontrol strategy against black scorch disease in date palm trees. J Phytopathol 164:464–475

Aoki Y, Haga S, Suzuki S, Llorens E (2020) Direct antagonistic activity of chitinase produced by Trichoderma sp. SANA20 as biological control agent for grey mould caused by Botrytis cinerea. Cogent Biol 6:1747903. https://doi.org/10.1080/23312025.2020.1747903

Arnison PG, Bibb MJ, Bierbaum G, Bowers AA, Bugni TS, Bulaj G, Camarero JA, Campopiano DJ, Challis GL, Clardy J (2013) Ribosomally synthesized and post-translationally modified peptide natural products: overview and recommendations for a universal nomenclature. Nat Prod Rep 30:108–160

Aslam B, Basit M, Nisar MA, Khurshid M, Rasool MH (2017) Proteomics: technologies and their applications. J Chromatogr Sci 55:182–196. https://doi.org/10.1093/chromsci/bmw167

Atanasova L, Le Crom S, Gruber S, Coulpier F, Seidl-Seiboth V, Kubicek CP, Druzhinina IS (2013) Comparative transcriptomics reveals different strategies of Trichoderma mycoparasitism. BMC Genomics 14:121

Atanasova L, Gruber S, Lichius A, Radebner T, Abendstein L, Münsterkötter M, Stralis-Pavese N, Łabaj PP, Kreil DP, Zeilinger S (2018) The Gpr1-regulated Sur7 family protein Sfp2 is required for hyphal growth and cell wall stability in the mycoparasite Trichoderma atroviride. Sci Rep 8:12064

Bais HP, Weir TL, Perry LG, Gilroy S, Vivanco JM (2006) The role of root exudates in rhizosphere interactions with plants and other organisms. Annu Rev Plant Biol 57:233–266

Baiyee B, Ito S, Sunpapao A (2019) Trichoderma asperellum T1 mediated antifungal activity and induced defense response against leaf spot fungi in lettuce (Lactuca sativa L.). Physiol Mol Plant Pathol 106:96–101

Banerjee S, Pal S, Mukherjee S, Podder D, Mukherjee A, Nandi A, Debnath P, Sur PK, Ghosh SK (2016) Cellular abnormalities induced by Trichoderma spp. during in vitro interaction and control of white muscardine (Beauveria bassiana) and green muscardine (Metarhizium anisopliae) disease of silkworm Bombyx mori. J Biopest 9:104

Bansal R, Mukherjee M, Horwitz BA, Mukherjee PK (2019) Regulation of conidiation and antagonistic properties of the soil-borne plant beneficial fungus Trichoderma virens by a novel proline-, glycine-, tyrosine-rich protein and a GPI-anchored cell wall protein. Curr Genet:1–12

Baroncelli R, Piaggeschi G, Fiorini L, Bertolini E, Zapparata A, Pè ME, Sarrocco S, Vannacci G (2015) Draft whole-genome sequence of the biocontrol agent Trichoderma harzianum T6776. Genome Announc 3

Rubio, M.B., Martinez de Alba, A.E., Nicolás, C., Monte, E., Hermosa, R., 2019. Early root transcriptomic changes in wheat seedlings colonized by Trichoderma harzianum under different inorganic nitrogen supplies. Front. Microbiol. 2444. https://doi.org/10.3389/fmicb.2019.02444

Benítez T, Rincón AM, Limón MC, Codon AC (2004) Biocontrol mechanisms of Trichoderma strains. Int Microbiol 7:249–260

Bhattacharyya C, Roy R, Tribedi P, Ghosh A, Ghosh A (2020) Biofertilizers as substitute to commercial agrochemicals. In: Agrochemicals detection, treatment and remediation. Elsevier, pp 263–290

Bramlett M, Plaetinck G, Maienfisch P (2019) RNA-based biocontrols—a new paradigm in crop protection. Engineering

Brito N, Espino JJ, González C (2006) The endo-β-1,4-xylanase Xyn11A is required for virulence in Botrytis cinerea. Mol Plant-Microbe Interact 19:25–32. https://doi.org/10.1094/MPMI-19-0025

Brotman Y, Briff E, Viterbo A, Chet I (2008) Role of swollenin, an expansin-like protein from Trichoderma, in plant root colonization. Plant Physiol 147:779–789. https://doi.org/10.1104/pp.108.116293

Butzke D, Hurwitz R, Thiede B, Goedert S, Rudel T (2005) Cloning and biochemical characterization of APIT, a new L-amino acid oxidase from Aplysia punctata. Toxicon 46:479–489. https://doi.org/10.1016/j.toxicon.2005.06.005

Cai F, Druzhinina IS (2021) In honor of John Bissett: authoritative guidelines on molecular identification of Trichoderma. Fungal Divers:1–69

Cai F, Gao R, Zhao Z, Ding M, Jiang S, Yagtu C, Zhu H, Zhang J, Ebner T, Philipp MM, Komal K, Akcapinar GB, Shen Q, Druzhinina IS (2020) Evolutionary compromises in fungal fitness: hydrophobins can hinder the adverse dispersal of conidiospores and challenge their survival. ISME J. https://doi.org/10.1038/s41396-020-0709-0

Cao FY, DeFalco TA, Moeder W, Li B, Gong Y, Liu XM, Taniguchi M, Lumba S, Toh S, Shan L, Ellis B, Desveaux D, Yoshioka K (2018) Arabidopsis ETHYLENE RESPONSE FACTOR 8 (ERF8) has dual functions in ABA signaling and immunity. BMC Plant Biol 18:1–16. https://doi.org/10.1186/s12870-018-1402-6

Chacón MR, Rodríguez-galán O, Benítez T, Sousa S, Rey M, Llobell A, Delgado-jarana J (2007) Microscopic and transcriptome analyses of early colonization of tomato roots by Trichoderma harzianum. Int Microbiol 10:19–27. https://doi.org/10.2436/20.1501.01.4

Challis GL (2008) Mining microbial genomes for new natural products and biosynthetic pathways. Microbiology 154:1555–1569

Chambers JR, Sauer K (2013) Small RNAs and their role in biofilm formation. Trends Microbiol 21:39–49

Collemare J, O'Connell R, Lebrun MH (2019) Nonproteinaceous effectors: the terra incognita of plant–fungal interactions. New Phytol 223:590–596. https://doi.org/10.1111/nph.15785

Contreras-Cornejo HA, Macías-Rodríguez L, Del-Val E, Larsen J (2020) Interactions of Trichoderma with plants, insects, and plant pathogen microorganisms: chemical and molecular bases. In: Co-evolution of secondary metabolites, pp 263–290

Coppola M, Diretto G, Digilio MC, Woo SL, Giuliano G, Molisso D, Pennacchio F, Lorito M, Rao R (2019) Transcriptome and metabolome reprogramming in tomato plants by trichoderma harzianum strain t22 primes and enhances defense responses against aphids. Front Physiol 10. https://doi.org/10.3389/fphys.2019.00745

Corrêa TLR, de Araújo EF (2020) Fungal phytases: from genes to applications. Braz J Microbiol 51:1009–1020. https://doi.org/10.1007/s42770-020-00289-y

Cosgrove DJ (2017) Microbial expansins. Annu Rev Microbiol 71:479–497. https://doi.org/10.1146/annurev-micro-090816-093315

Dautt-Castro M, Estrada-Rivera M, Olguin-Martínez I, Rocha-Medina M d C, Islas-Osuna MA, Casas-Flores S (2020) TBRG-1 a Ras-like protein in Trichoderma virens involved in conidiation, development, secondary metabolism, mycoparasitism, and biocontrol unveils a new family of Ras-GTPases. Fungal Genet Biol 136:103292. https://doi.org/10.1016/J.FGB.2019.103292

Dautt-Castro M, Rosendo-Vargas M, Casas-Flores S (2021) The Small GTPases in Fungal Signaling Conservation and Function. Cells 10(5):1039. https://doi.org/10.3390/cells10051039

De Oliveira AL, Gallo M, Benedetti CE, Cappugi G, Scala A, Pantera B, Pertinhez TA, Daniel O, De Oliveira AL, Gallo M, Pazzagli L, Benedetti CE, Cappugi G, Scala A (2011) The structure of the elicitor Cerato-platanin (CP), the first member of the CP fungal protein family, reveals a double-barrel fold and carbohydrate binding. J Biol Chem 286:17560–17568. https://doi.org/10.1074/jbc.M111.223644

De Palma M, Salzano M, Villano C, Aversano R, Lorito M, Ruocco M, Docimo T, Piccinelli AL, D'Agostino N, Tucci M (2019) Transcriptome reprogramming, epigenetic modifications and alternative splicing orchestrate the tomato root response to the beneficial fungus Trichoderma harzianum. Hortic Res 6:5

Deng JJ, Huang WQ, Li ZW, Lu DL, Zhang Y, Luo X, Chun (2018) Biocontrol activity of recombinant aspartic protease from Trichoderma harzianum against pathogenic fungi. Enzym Microb Technol 112:35–42. https://doi.org/10.1016/j.enzmictec.2018.02.002

Derntl C, Rassinger A, Srebotnik E, Mach RL, Mach-Aigner AR (2016) Identification of the main regulator responsible for synthesis of the typical yellow pigment produced by Trichoderma reesei. Appl Environ Microbiol 82:6247–6257

Dettmer K, Aronov PA, Hammock BD (2007) Mass spectrometry-based metabolomics. Mass Spectrom Rev 26:51–78

Dini I, Graziani G, Fedele FL, Sicari A, Vinale F, Castaldo L, Ritieni A (2020) Effects of Trichoderma biostimulation on the phenolic profile of extra-virgin olive oil and olive oil by-products. Antioxidants 9:284

Djonović S, Howell CR, Pozo MJ, Dangott LJ, Kenerley CM (2007) Sm1, a Proteinaceous elicitor secreted by the biocontrol fungus Trichoderma virens induces plant defense responses and systemic resistance. Mol Plant-Microbe Interact 19:838–853. https://doi.org/10.1094/mpmi-19-0838

Dotson BR, Soltan D, Schmidt J, Areskoug M, Rabe K, Swart C, Widell S, Rasmusson AG (2018) The antibiotic peptaibol alamethicin from Trichoderma permeabilises Arabidopsis root apical meristem and epidermis but is antagonised by cellulase-induced resistance to alamethicin. BMC Plant Biol:1–11

Druzhinina IS, Seidl-Seiboth V, Herrera-Estrella A, Horwitz BA, Kenerley CM, Monte E, Mukherjee PK, Zeilinger S, Grigoriev IV, Kubicek CP (2011) Trichoderma: the genomics of opportunistic success. Nat Rev Microbiol 9:749. https://doi.org/10.1038/nrmicro2637

Druzhinina IS, Chenthamara K, Zhang J, Atanasova L, Yang D, Miao Y, Rahimi MJ, Grujic M, Cai F, Pourmehdi S, Salim KA, Pretzer C, Kopchinskiy AG, Henrissat B, Kuo A, Hundley H, Wang M, Aerts A, Salamov A, Lipzen A, Labutti K, Barry K, Grigoriev IV, Shen Q, Kubicek CP (2018) Massive lateral transfer of genes encoding plant cell wall-degrading enzymes to the mycoparasitic fungus Trichoderma from its plant-associated hosts

Drzymała A, Bielawski W (2009) Isolation and characterization of carboxypeptidase III from germinating triticale grains. Acta Biochim Biophys Sin Shanghai 41:69–78. https://doi.org/10.1093/abbs/gmn008

Du J, Duan S, Miao J, Zhai M, Cao Y (2020) Purification and characterization of chitinase from Paenibacillus sp. Biotechnol Appl Biochem 1–12. https://doi.org/10.1002/bab.1889

Egan MJ, Wang ZY, Jones MA, Smirnoff N, Talbot NJ (2007) Generation of reactive oxygen species by fungal NADPH oxidases is required for rice blast disease. Proc Natl Acad Sci U S A 104:11772–11777. https://doi.org/10.1073/pnas.0700574104

Engelberth J, Koch T, Schüler G, Bachmann N, Rechtenbach J, Boland W (2001) Ion channel-forming alamethicin is a potent elicitor of volatile biosynthesis and tendril coiling. Cross talk between jasmonate and salicylate signaling in lima bean. Plant Physiol 125:369–377. https://doi.org/10.1104/pp.125.1.369

Estrada-Rivera M, Rebolledo-Prudencio OG, Pérez-Robles DA, del Carmen Rocha-Medina M, del Carmen González-López M, Casas-Flores S (2019) Trichoderma histone deacetylase HDA-2 modulates multiple responses in Arabidopsis. Plant Physiol:01092

Estrada-Rivera M, Hernández-Oñate MÁ, Dautt-Castro M, Gallardo-Negrete J d J, Rebolledo-Prudencio OG, Uresti-Rivera EE, Arenas-Huertero C, Herrera-Estrella A, Casas-Flores S (2020) IPA-1 a putative chromatin Remodeler/helicase-related protein of Trichoderma virens plays important roles in antibiosis against Rhizoctonia solani and induction of Arabidopsis systemic disease resistance. Mol Plant-Microbe Interact 33:808–824. https://doi.org/10.1094/MPMI-04-19-0092-R

Fanelli F, Liuzzi VC, Logrieco AF, Altomare C (2018) Genomic characterization of Trichoderma atrobrunneum (T. harzianum species complex) ITEM 908: insight into the genetic endowment of a multi-target biocontrol strain. BMC Genomics 19:662

Farh ME, Jeon J (2020) Roles of fungal volatiles from perspective of distinct lifestyles in filamentous fungi. Plant Pathol 36:193–203

Ferreira Filho JA, Horta MAC, Beloti LL, dos Santos CA, de Souza AP (2017) Carbohydrate-active enzymes in Trichoderma harzianum: a bioinformatic analysis bioprospecting for key enzymes for the biofuels industry. BMC Genomics 18:779

Fiehn O (2001) Combining genomics, metabolome analysis, and biochemical modelling to understand metabolic networks. Comp Funct Genomics 2:155–168

Fiorini L, Guglielminetti L, Mariotti L, Curadi M, Picciarelli P, Scartazza A, Sarrocco S, Vannacci G (2016) Trichoderma harzianum T6776 modulates a complex metabolic network to stimulate tomato cv. Micro-Tom growth. Plant Soil 400:351–366

Fischbach M, Voigt CA (2010) Prokaryotic gene clusters: a rich toolbox for synthetic biology. Biotechnol J 5:1277–1296

Fu L, Cai F, Shen Q, Chen W, Wei Z, Wang P, Yu G (2013) Harzianolide, a novel plant growth regulator and systemic resistance elicitor from Trichoderma harzianum. Plant Physiol Biochem 73:106–113. https://doi.org/10.1016/j.plaphy.2013.08.011

Gao R, Ding M, Jiang S, Zhao Z, Chenthamara K, Shen Q, Feng C, Druzhinina I (2020) The evolutionary and functional paradox of cerato-platanins in the mycoparasitic fungus Trichoderma: high diversity, stabilizing selection, and a minor role in biotic interactions. Appl Environ Microbiol 83. https://doi.org/10.1128/AEM.00696-20

Garnica-Vergara A, Barrera-Ortiz S, Muñoz-Parra E, Raya-González J, Méndez-Bravo A, Macías-Rodríguez L, Ruiz-Herrera LF, López-Bucio J (2016) The volatile 6-pentyl-2H-pyran-2-one from Trichoderma atroviride regulates Arabidopsis thaliana root morphogenesis via auxin signaling and ETHYLENE INSENSITIVE 2 functioning. New Phytol 209:1496–1512

Geraldine AM, Brandão RS, Barbosa ET, Rodrigues AR, Carvalho DDC, Lopes FAC, Lobo Junior M, Ulhoa CJ (2013) Cell wall-degrading enzymes and parasitism of sclerotia are key factors on field biocontrol of white mold by Trichoderma spp. Biol Control 67:308–316. https://doi.org/10.1016/j.biocontrol.2013.09.013

Gibbs DJ, Voß U, Harding SA, Fannon J, Moody LA, Yamada E, Swarup K, Nibau C, Bassel GW, Choudhary A, Lavenus J, Bradshaw SJ, Stekel DJ, Bennett MJ, Coates JC (2014) AtMYB93 is a novel negative regulator of lateral root development in Arabidopsis. New Phytol 203:1194–1207. https://doi.org/10.1111/nph.12879

Glass NL, Schmoll M, Cate JHD, Coradetti S (2013) Plant Cell Wall deconstruction by Ascomycete fungi. Annu Rev Microbiol 67:477–498. https://doi.org/10.1146/annurev-micro-092611-150044

Goharzad F, Tajick Ghanbary MA, Shahbazi S, Askari H (2020) Two–dimensional and chitinase activity analysis of a novel mutant of Trichoderma koningii for biodegradation of Macrophomina phaseolina cell walls. Mycol Iran 0:73–87. https://doi.org/10.22043/MI.2020.121450

Gomes EV, do Nascimento Costa M, De Paula RG, De Azevedo RR, Da Silva FL, Noronha EF, Ulhoa CJ, Monteiro VN, Cardoza RE, Gutiérrez S (2015) The Cerato-Platanin protein Epl-1 from Trichoderma harzianum is involved in mycoparasitism, plant resistance induction and self cell wall protection. Sci Rep 5:17998

Gómez-Rodríguez EY, Uresti-Rivera EE, Patrón-Soberano OA, Islas-Osuna MA, Flores-Martínez A, Riego-Ruiz L, Rosales-Saavedra MT, Casas-Flores S (2018) Histone acetyltransferase TGF-1 regulates Trichoderma atroviride secondary metabolism and mycoparasitism. PLoS One 13:e0193872. https://doi.org/10.1371/journal.pone.0193872

González-López, M. del C., Jijón-Moreno, S., Dautt-Castro, M., Ovando-Vázquez, C., Ziv, T., Horwitz, B.A., Casas-Flores, S., (2021). Secretome analysis of Arabidopsis–Trichoderma atroviride interaction unveils new roles for the plant glutamate: glyoxylate aminotransferase GGAT1 in plant growth induced by the fungus and resistance against Botrytis cinerea. Int. J. Mol. Sci. 22, 6804. https://doi.org/10.3390/ijms22136804

Gouda S, Kerry RG, Das G, Paramithiotis S, Shin H-S, Patra JK (2018) Revitalization of plant growth promoting rhizobacteria for sustainable development in agriculture. Microbiol Res 206:131–140

Gravel V, Antoun V, Tweddell RJ (2007) Growth stimulation and fruit yield improvement of greenhouse tomato plants by inoculation with Pseudomonas putida or Trichoderma atroviride: possible role of indole acetic acid (IAA). Soil Biol Biochem 39:1968–1977. https://doi.org/10.1016/j.soilbio.2007.02.015

Grinyer J, Hunt S, McKay M, Herbert BR, Nevalainen H (2005) Proteomic response of the biological control fungus Trichoderma atroviride to growth on the cell walls of Rhizoctonia solani. Curr Genet 47:381–388. https://doi.org/10.1007/s00294-005-0575-3

Grün CH, Dekker N, Nieuwland AA, Klis FM, Kamerling JP, Vliegenthart JFG, Hochstenbach F (2006) Mechanism of action of the endo-(1 → 3)-α-glucanase MutAp from the mycoparasitic fungus Trichoderma harzianum. FEBS Lett 580:3780–3786. https://doi.org/10.1016/j.febslet.2006.05.062

Guo Y, Ghirardo A, Weber B, Schnitzler J, Benz JP, Silva R (2019) Trichoderma species differ in their volatile profiles and in antagonism toward Ectomycorrhiza Laccaria bicolor. 10:1–15. https://doi.org/10.3389/fmicb.2019.00891

Guo Y, Jud W, Ghirardo A, Rosenkranz M, Antritter F, Benz JP (2020) Methods sniffing fungi – phenotyping of volatile chemical diversity in Trichoderma species, pp 244–259. https://doi.org/10.1111/nph.16530

Gupta R, Bar M (2020) Plant immunity, priming, and systemic resistance as mechanisms for Trichoderma spp. biocontrol. Springer, Singapore, pp 81–110. https://doi.org/10.1007/978-981-15-3321-1_5

Guzmán-Guzmán P, Alemán-Duarte MI, Delaye L, Herrera-Estrella A, Olmedo-Monfil V (2017) Identification of effector-like proteins in Trichoderma spp. and role of a hydrophobin in the plant-fungus interaction and mycoparasitism. BMC Genet 18:16

Hadar Y, Feldman D, Yarden O (2020) Seeking the roles for fungal small-secreted proteins in affecting saprophytic lifestyles. Front Microbiol. https://doi.org/10.3389/fmicb.2020.00455

Hakulinen N, Tenkanen M, Rouvinen J (2000) Three-dimensional structure of the catalytic core of acetylxylan esterase from Trichoderma reesei: insights into the deacetylation mechanism. J Struct Biol 132:180–190. https://doi.org/10.1006/jsbi.2000.4318

Halifu S, Deng X, Song X, Song R, Liang X (2020) Inhibitory mechanism of Trichoderma virens ZT05 on Rhizoctonia solani. Plan Theory 9:912

Hamrouni R, Molinet J, Mich L, Carbou Q, Dupuy N, Masmoudi A, Roussos S (2019) Production of coconut aroma in solid-state cultivation: screening and identification of Trichoderma strains for 6-Pentyl-alpha-pyrone and conidia production. J Chem. https://doi.org/10.1155/2019/8562384

Harman GE, Howell CR, Viterbo A, Chet I, Lorito M (2004) Trichoderma species—opportunistic, avirulent plant symbionts. Nat Rev Microbiol 2:43. https://doi.org/10.1038/nrmicro797

He Q, Mclellan H, Boevink PC, Birch PRJ (2020) All roads lead to susceptibility: the many modes of action of fungal and oomycete intracellular effectors. Plant Commun 1:100050. https://doi.org/10.1016/j.xplc.2020.100050

Hermosa, R., Viterbo, A., Chet, I., & Monte, E. (2012). Plant-beneficial effects of Trichoderma and of its genes. Microbiology, 158(1), 17-25. https://doi.org/10.1099/mic.0.052274

Hitzenhammer E, Büschl C, Sulyok M, Schuhmacher R, Kluger B, Wischnitzki E, Schmoll M (2019) YPR2 is a regulator of light modulated carbon and secondary metabolism in Trichoderma reesei. BMC Genomics 20:211

Holzlechner M, Reitschmidt S, Gruber S, Zeilinger S, Marchetti-Deschmann M (2016) Visualizing fungal metabolites during mycoparasitic interaction by MALDI mass spectrometry imaging. Proteomics 16:1742–1746

Inbar J, Chet I (1996) The role of lectins in recognition and adhesion of the mycoparasitic fungus Trichoderma spp. to its host. In: Toward anti-adhesion therapy for microbial diseases. Springer, pp 229–231

Ivanova C, Ramoni J, Aouam T, Frischmann A, Seiboth B, Baker SE, Le Crom S, Lemoine S, Margeot A, Bidard F (2017) Genome sequencing and transcriptome analysis of Trichoderma reesei QM9978 strain reveals a distal chromosome translocation to be responsible for loss of vib1 expression and loss of cellulase induction. Biotechnol Biofuels 10:1–15

Jaroszuk-ściseł J, Tyśkiewicz R, Nowak A, Ozimek E, Majewska M, Hanaka A, Tyśkiewicz K, Pawlik A, Janusz G (2019) Phytohormones (Auxin, gibberellin) and ACC deaminase in vitro synthesized by the mycoparasitic trichoderma DEMTKZ3A0 strain and changes in the level of auxin and plant resistance markers in wheat seedlings inoculated with this strain conidia. Int J Mol Sci 20:1–35. https://doi.org/10.3390/ijms20194923

Jin X, Guo L, Jin B, Zhu S, Mei X, Wu J, Liu T, He X (2020) Inhibitory mechanism of 6-Pentyl-2H-pyran-2-one secreted by Trichoderma atroviride T2 against Cylindrocarpon destructans. Pestic Biochem Physiol 170:104683

Jian, Zhang Youzhi, Miao Mohammad Javad, Rahimi Hong, Zhu Andrei, Steindorff Sabine, Schiessler Feng, Cai Guan, Pang Komal, Chenthamara Yu, Xu Christian P., Kubicek Qirong, Shen Irina S., Druzhinina (2019) (2019) Guttation capsules containing hydrogen peroxide: an evolutionarily conserved NADPH oxidase gains a role in wars between related fungi. Environmental Microbiology 21(8) 2644–2658 https://doi.org/10.1111/1462-2920.14575

Khan MR, Haque Z, Rasool F, Salati K, Khan U, Mohiddin FA, Zuhaib M (2019) Management of root-rot disease complex of mungbean caused by Macrophomina phaseolina and Rhizoctonia solani through soil application of Trichoderma spp. Crop Prot 119:24–29

Khare E, Kumar S, Kim K (2018) Role of peptaibols and lytic enzymes of Trichoderma cerinum Gur1 in biocontrol of Fusarium oxysporum and chickpea wilt. Environ Sustain 1:39–47

Kottb M, Gigolashvili T, Großkinsky DK, Piechulla B (2015) Trichoderma volatiles effecting Arabidopsis: from inhibition to protection against phytopathogenic fungi. Front Microbiol 6:1–14. https://doi.org/10.3389/fmicb.2015.00995

Kubicek CP, Herrera-Estrella A, Seidl-Seiboth V, Martinez DA, Druzhinina IS, Thon M, Zeilinger S, Casas-Flores S, Horwitz BA, Mukherjee PK (2011) Comparative genome sequence analysis underscores mycoparasitism as the ancestral life style of Trichoderma. Genome Biol 12:1. https://doi.org/10.1186/gb-2011-12-4-r40

Kubicek C, Steindorff AS, Chenthamara K, Manganiello G, Henrissat B, Zhang J, Cai F, Kopchinskiy A, Kubicek E, Kuo A, Baroncelli R, Sarrocco S, Noronha E, Vannacci G, Shen Q, Grigoriev I, Druzhinina I (2019) Evolution and comparative genomics of the most common Trichoderma species. BMC Genomics 20:485

Lamdan N, Shalaby S, Ziv T, Kenerley CM, Horwitz BA (2015) Secretome of Trichoderma interacting with maize roots : role in induced systemic resistance * □. Mol Cell Proteomics 14:1054–1063. https://doi.org/10.1074/mcp.M114.046607

Latgé JP (2010) Tasting the fungal cell wall. Cell Microbiol 12:863–872. https://doi.org/10.1111/j.1462-5822.2010.01474.x

Le Crom S, Schackwitz W, Pennacchio L, Magnuson JK, Culley DE, Collett JR, Martin J, Druzhinina IS, Mathis H, Monot F (2009) Tracking the roots of cellulase hyperproduction by the fungus Trichoderma reesei using massively parallel DNA sequencing. Proc Natl Acad Sci 106:16151–16156

Lee S, Yap M, Behringer G, Hung R, Bennett JW (2016) Volatile organic compounds emitted by Trichoderma species mediate plant growth. Fungal Biol Biotechnol 3:7

Lee PY, Saraygord-Afshari N, Low TY (2020) The evolution of two-dimensional gel electrophoresis - from proteomics to emerging alternative applications. J Chromatogr A 1615:460763. https://doi.org/10.1016/j.chroma.2019.460763

Li Y-q, Song K, Li Y-c, Chen J (2016) Statistical culture-based strategies to enhance chlamydospore production by Trichoderma harzianum SH2303 in liquid fermentation. J Zhejiang Univ B 17:619–627

Lichius A, Bidard F, Buchholz F, Le Crom S, Martin J, Schackwitz W, Austerlitz T, Grigoriev IV, Baker SE, Margeot A (2015) Genome sequencing of the Trichoderma reesei QM9136 mutant identifies a truncation of the transcriptional regulator XYR1 as the cause for its cellulase-negative phenotype. BMC Genomics 16:326

Lima FB d, Félix C, Osório N, Alves A, Vitorino R, Domingues P, Correia A, da Silva Ribeiro RT, Esteves AC (2016) Secretome analysis of Trichoderma atroviride T17 biocontrol of Guignardia citricarpa. Biol Control 99:38–46. https://doi.org/10.1016/j.biocontrol.2016.04.009

Lippolis JD, Powell EJ, Reinhardt TA, Thacker TC, Casas E (2019) Symposium review: Omics in dairy and animal science—promise, potential, and pitfalls. J Dairy Sci 102:4741–4754. https://doi.org/10.3168/jds.2018-15267

Liu L, Xu L, Jia Q, Pan R, Oelmüller R, Zhang W, Wu C (2019) Arms race: diverse effector proteins with conserved motifs. Plant Signal Behav 14. https://doi.org/10.1080/15592324.2018.1557008

Loc NH, Huy ND, Quang HT, Lan TT, Thu Ha TT (2020) Characterisation and antifungal activity of extracellular chitinase from a biocontrol fungus, Trichoderma asperellum PQ34. Mycology 11:38–48. https://doi.org/10.1080/21501203.2019.1703839

Lombard V, Golaconda Ramulu H, Drula E, Coutinho PM, Henrissat B (2014) The carbohydrate-active enzymes database (CAZy) in 2013. Nucleic Acids Res 42:490–495. https://doi.org/10.1093/nar/gkt1178

Lombardi N, Vitale S, Turrà D, Reverberi M, Fanelli C, Vinale F, Marra R, Ruocco M, Pascale A, d'Errico G (2018) Root exudates of stressed plants stimulate and attract Trichoderma soil fungi. Mol Plant-Microbe Interact 31:982–994

Lorito M, Woo SL, Harman GE, Monte E (2010) Translational research on Trichoderma: from 'omics to the field. Annu Rev Phytopathol 48:395–417

Luo S, Dong S-H (2019) Recent advances in the discovery and biosynthetic study of eukaryotic RiPP natural products. Molecules 24:1541

Luti S, Martellini F, Bemporad F, Mazzoli L, Paoli P, Pazzagli L (2017) A single amino acid mutation affects elicitor and expansins-like activities of cerato-platanin, a non-catalytic fungal protein. PLoS One 12:1–18

Macías-Rodríguez L, Guzmán-Gómez A, García-Juárez P, Contreras-Cornejo HA (2018) Trichoderma atroviride promotes tomato development and alters the root exudation of carbohydrates, which stimulates fungal growth and the biocontrol of the phytopathogen Phytophthora cinnamomi in a tripartite interaction system. FEMS Microbiol Ecol 94:fiy137

Malinich EA, Wang K, Mukherjee PK, Kolomiets M, Kenerley CM (2019) Differential expression analysis of Trichoderma virens RNA reveals a dynamic transcriptome during colonization of Zea mays roots. BMC Genomics 20:280

Marik T, Tyagi C, Balázs D, Urbán P, Szepesi Á, Bakacsy L, Endre G, Rakk D, Szekeres A, Andersson M, Salonen H, Druzhinina I, Vágvölgyi C, Kredics L (2019) Structural diversity and bioactivities of peptaibol compounds from the Longibrachiatum clade of the filamentous fungal genus Trichoderma. Front Microbiol 10. https://doi.org/10.3389/fmicb.2019.01434

Martin JA, Wang Z (2011) Next-generation transcriptome assembly. Nat Rev Genet 12:671–682

Martinez D, Berka RM, Henrissat B, Saloheimo M, Arvas M, Baker SE, Chapman J, Chertkov O, Coutinho PM, Cullen D (2008) Genome sequencing and analysis of the biomass-degrading fungus Trichoderma reesei (syn. Hypocrea jecorina). Nat Biotechnol 26:553–560

Martínez-Medina A, Fernandez I, Lok GB, Pozo MJ, Pieterse CMJ, Van Wees SCM (2017a) Shifting from priming of salicylic acid- to jasmonic acid-regulated defences by Trichoderma protects tomato against the root knot nematode Meloidogyne incognita. New Phytol 213:1363–1377. https://doi.org/10.1111/nph.14251

Martínez-Medina A, Van Wees SCM, Pieterse CMJ (2017b) Airborne signals from Trichoderma fungi stimulate iron uptake responses in roots resulting in priming of jasmonic acid-dependent defences in shoots of Arabidopsis thaliana and Solanum lycopersicum. Plant Cell Environ 40:2691–2705

Maryani N, Lombard L, Poerba YS, Subandiyah S, Crous PW, Kema GHJ (2019) Phylogeny and genetic diversity of the banana Fusarium wilt pathogen Fusarium oxysporum f. sp. cubense in the Indonesian centre of origin. Stud Mycol 92:155–194

Matsuzawa T, Kameyama A, Nakamichi Y, Yaoi K (2020) Identification and characterization of two xyloglucan-specific endo-1,4-glucanases in Aspergillus oryzae. Appl Microbiol Biotechnol 104:8761–8773. https://doi.org/10.1007/s00253-020-10883-7

Mayo-Prieto S, Campelo MP, Lorenzana A, Rodríguez-González A, Reinoso B, Gutiérrez S, Casquero PA (2020) Antifungal activity and bean growth promotion of Trichoderma strains isolated from seed vs soil. Eur J Plant Pathol 158:817–828. https://doi.org/10.1007/s10658-020-02069-8

Medema MH, Blin K, Cimermancic P, de Jager V, Zakrzewski P, Fischbach MA, Weber T, Takano E, Breitling R (2011) antiSMASH: rapid identification, annotation and analysis of secondary metabolite biosynthesis gene clusters in bacterial and fungal genome sequences. Nucleic Acids Res 39:W339–W346

Mendoza-mendoza A, Zaid R, Lawry R, Monte E, Horwitz BA, Prasun K (2017) Molecular dialogues between Trichoderma and roots: role of the fungal secretome. Fungal Biol Rev 1–24. https://doi.org/10.1016/j.fbr.2017.12.001

Meng X, Miao Y, Liu Q, Ma L, Guo K, Liu D, Ran W, Shen Q (2019) TgSWO from Trichoderma guizhouense NJAU4742 promotes growth in cucumber plants by modifying the root morphology and the cell wall architecture. Microb Cell Factories 18:1–15. https://doi.org/10.1186/s12934-019-1196-8

Mika F, Hengge R (2013) Small regulatory RNAs in the control of motility and biofilm formation in E. coli and Salmonella. Int J Mol Sci 14:4560–4579

Monfil VO, Casas-Flores S (2014) Molecular mechanisms of biocontrol in Trichoderma spp. and their applications in agriculture. In: Biotechnology and biology of Trichoderma. Elsevier, pp 429–453

Monteiro VN, do Nascimento Silva R, Steindorff AS, Costa FT, Noronha EF, Ricart CAO, de Sousa MV, Vainstein MH, Ulhoa CJ (2010) New insights in Trichoderma harzianum antagonism of fungal plant pathogens by secreted protein analysis. Curr Microbiol 61:298–305. https://doi.org/10.1007/s00284-010-9611-8

Morán-Diez ME, Carrero-Carrón I, Rubio MB, Jiménez-Díaz RM, Monte E, Hermosa R (2019) Transcriptomic analysis of Trichoderma atroviride overgrowing plant-wilting Verticillium dahliae reveals the role of a new M14 metallocarboxypeptidase CPA1 in biocontrol. Front Microbiol 10:1120

Moreno-Mateos MA, Delgado-Jarana J, Codon AC, Benítez T (2007) pH and Pac1 control development and antifungal activity in Trichoderma harzianum. Fungal Genet Biol 44:1355–1367

Moscatiello R, Sello S, Ruocco M, Barbulova A, Cortese E, Nigris S, Baldan B, Chiurazzi M, Mariani P, Lorito M, Navazio L (2018) The hydrophobin HYTLO1 secreted by the biocontrol fungus Trichoderma longibrachiatum triggers a NAADP-mediated calcium signalling pathway in Lotus japonicus. Int J Mol Sci 19. https://doi.org/10.3390/ijms19092596

Moya P, Barrera V, Cipollone J, Bedoya C, Kohan L, Toledo A, Sisterna M (2020) New isolates of Trichoderma spp. as biocontrol and plant growth–promoting agents in the pathosystem Pyrenophora teres-barley in Argentina. Biol Control 141:104152. https://doi.org/10.1016/j.biocontrol.2019.104152

Mukherjee PK, Horwitz BA, Kenerley CM (2012) Secondary metabolism in Trichoderma–a genomic perspective. Microbiology 158:35–45. https://doi.org/10.1099/mic.0.053629-0

Mukherjee PK, Horwitz BA, Herrera-Estrella A, Schmoll M, Kenerley CM (2013) Trichoderma research in the genome era. Annu Rev Phytopathol 51:105–129

Mukherjee PK, Hurley JF, Taylor JT, Puckhaber L, Lehner S, Druzhinina I, Schumacher R, Kenerley CM (2018) Ferricrocin, the intracellular siderophore of Trichoderma virens, is involved in growth, conidiation, gliotoxin biosynthesis and induction of systemic resistance in maize. Biochem Biophys Res Commun 505:606–611

Müller M, Munné-Bosch S (2015) Ethylene response factors: a key regulatory hub in hormone and stress signaling. Plant Physiol 169:32–41. https://doi.org/10.1104/pp.15.00677

Muthukathan G, Mukherjee P, Salaskar D, Pachauri S, Tak H, Ganapathi TR, Mukherjee PK (2020) Secretome of Trichoderma virens induced by banana roots - identification of novel fungal proteins for enhancing plant defence. Physiol Mol Plant Pathol 110:101476. https://doi.org/10.1016/j.pmpp.2020.101476

Muthukumar T, Priyadharsini P, Uma E, Jaison S, Pandey RR (2014) Role of arbuscular mycorrhizal fungi in alleviation of acidity stress on plant growth. In: Use of microbes for the alleviation of soil stresses

Nawrocka J, Małolepsza U (2013) Diversity in plant systemic resistance induced by Trichoderma. Biol Control 67:149–156. https://doi.org/10.1016/j.biocontrol.2013.07.005

Nogueira-Lopez G, Greenwood DR, Middleditch M, Winefield C, Eaton C, Steyaert JM, Mendoza-Mendoza A (2018) The apoplastic secretome of Trichoderma virens during interaction with maize roots shows an inhibition of plant defence and scavenging oxidative stress secreted proteins. Front Plant Sci 9:409

Omann MR, Lehner S, Rodríguez CE, Brunner K, Zeilinger S (2012) The seven-transmembrane receptor Gpr1 governs processes relevant for the antagonistic interaction of Trichoderma atroviride with its host. Microbiology 158:107

Pachauri S, Sherkhane PD, Kumar V, Mukherjee PK (2020) Whole genome sequencing reveals major deletions in the genome of M7, a gamma ray-induced mutant of Trichoderma virens that is repressed in conidiation, secondary metabolism, and Mycoparasitism. Front Microbiol 11:1030

Panthapulakkal Narayanan S, Liao P, Taylor PWJ, Lo C, Chye M (2019) Overexpression of a monocot acyl-CoA-binding protein confers broad-spectrum pathogen protection in a dicot. Proteomics 19:1800368. https://doi.org/10.1002/pmic.201800368

Parmar HJ, Bodar NP, Lakhani HN, Patel SV, Umrania VV, Hassan MM (2015) Production of lytic enzymes by Trichoderma strains during in vitro antagonism with Sclerotium rolfsii, the causal agent of stem rot of groundnut. Afr J Microbiol 9:365–372. https://doi.org/10.5897/AJMR2014.7330

Pascale A, Vinale F, Manganiello G, Nigro M, Lanzuise S, Ruocco M, Marra R, Lombardi N, Woo SL, Lorito M (2017) Trichoderma and its secondary metabolites improve yield and quality of grapes. Crop Prot 92:176–181. https://doi.org/10.1016/j.cropro.2016.11.010

Pereira JL, Queiroz RML, Charneau SO, Felix CR, Ricart CAO, Lopes Da Silva F, Steindorff AS, Ulhoa CJ, Noronha EF (2014) Analysis of Phaseolus vulgaris response to its association with Trichoderma harzianum (ALL-42) in the presence or absence of the phytopathogenic fungi Rhizoctonia solani and Fusarium solani. PLoS One 9. https://doi.org/10.1371/journal.pone.0098234

Pieterse CMJ, Zamioudis C, Berendsen RL, Weller DM, Van Wees SCM, Bakker PAHM (2014) Induced systemic resistance by beneficial microbes. Annu Rev Phytopathol 52:347–375. https://doi.org/10.1146/annurev-phyto-082712-102340

Rabinal C, Bhat S (2020) Identification of differentially expressed genes in Trichoderma koningii IABT1252 during its interaction with Sclerotium rolfsii. Curr Microbiol 77:396–404

Rebolledo-Prudencio OG, Dautt-Castro M, Estrada-Rivera M, González-López M d C, Jijón-Moreno S, Casas-Flores S (2020) Trichoderma in the rhizosphere: an approach toward a long and successful symbiosis with plants. In: Gupta VK, Zeilinger S, Singh HB, Druzhinina I (eds) New and future developments in microbial biotechnology and bioengineering: recent developments in Trichoderma research, pp 3–38

Rebolledo-Prudencio, O. G., Estrada-Rivera, M., Dautt-Castro, M., Arteaga-Vazquez, M. A., Arenas-Huertero, C., Rosendo-Vargas, M. M., Hailing, J., & Casas-Flores, S. The small RNA-mediated gene silencing machinery is required in Arabidopsis for stimulation of growth, systemic disease resistance, and suppression of the nitrile-specifier gene NSP4 by Trichoderma atroviride. The Plant Journal. https://doi.org/10.1111/tpj.15599

Rebolledo-Prudencio, O. G., Estrada-Rivera, M., Dautt-Castro, M., Arteaga-Vazquez, M. A., Arenas-Huertero, C., Rosendo-Vargas, M. M., Hailing, J., & Casas-Flores, S. (2021). The small RNA-mediated gene silencing machinery is required in Arabidopsis for stimulation of growth, systemic disease resistance, and suppression of the nitrile-specifier gene NSP4 by Trichoderma atroviride. The Plant Journal. https://doi.org/10.1111/tpj.15599

Reithner B, Ibarra-Laclette E, Mach RL, Herrera-Estrella A (2011) Identification of mycoparasitism-related genes in Trichoderma atroviride. Appl Environ Microbiol 77:4361–4370. https://doi.org/10.1128/AEM.00129-11

Rocafort M, Fudal I, Mesarich CH (2020) Apoplastic effector proteins of plant-associated fungi and oomycetes. Curr Opin Plant Biol 56:9–19. https://doi.org/10.1016/j.pbi.2020.02.004

Romero-Contreras YJ, Ramírez-Valdespino CA, Guzmán-Guzmán P, Macías-Segoviano JI, Villagómez-Castro JC, Olmedo-Monfil V (2019) Tal6 from Trichoderma atroviride is a LysM effector involved in mycoparasitism and plant association. Front Microbiol 10:2231

Rubio MB, Cardoza RE, Hermosa R, Gutiérrez S, Monte E (2008) Cloning and characterization of the Thcut1 gene encoding a cutinase of Trichoderma harzianum T34. Curr Genet 54:301–312. https://doi.org/10.1007/s00294-008-0218-6

Růžička K, Ljung K, Vanneste S, Podhorská R, Beeckman T, Friml J, Benková E (2007) Ethylene regulates root growth through effects on auxin biosynthesis and transport-dependent auxin distribution. Plant Cell 19:2197–2212. https://doi.org/10.1105/tpc.107.052126

Salas-Marina MA, Silva-Flores MA, Uresti-Rivera EE, Castro-Longoria E, Herrera-Estrella A, Casas-Flores S (2011) Colonization of Arabidopsis roots by Trichoderma atroviride promotes growth and enhances systemic disease resistance through jasmonic acid/ethylene and salicylic acid pathways. Eur J Plant Pathol 131:15–26

Salas-Marina MA, Isordia-jasso MI, Islas-osuna MA, Delgado-sánchez P (2015) The Epl1 and Sm1 proteins from Trichoderma atroviride and Trichoderma virens differentially modulate systemic disease resistance against different life style pathogens in Solanum lycopersicum. Front Plant Sci 6:1–13. https://doi.org/10.3389/fpls.2015.00077

Saravanakumar K, Arasu VS, Kathiresan K (2013) Effect of Trichoderma on soil phosphate solubilization and growth improvement of Avicennia marina. Aquat Bot 104:101–105. https://doi.org/10.1016/j.aquabot.2012.09.001

Scalbert A, Brennan L, Fiehn O, Hankemeier T, Kristal BS, van Ommen B, Pujos-Guillot E, Verheij E, Wishart D, Wopereis S (2009) Mass-spectrometry-based metabolomics: limitations and recommendations for future progress with particular focus on nutrition research. Metabolomics 5:435–458

Schenk G, Mitić NŠ, Hanson GR, Comba P (2013) Purple acid phosphatase: a journey into the function and mechanism of a colorful enzyme. Coord Chem Rev 257:473–482. https://doi.org/10.1016/j.ccr.2012.03.020

Scherlach K, Hertweck C (2009) Triggering cryptic natural product biosynthesis in microorganisms. Org Biomol Chem 7:1753–1760

Schmidt J, Dotson BR, Schmiderer L, van Tour A, Kumar B, Marttila S, Fredlund KM, Widell S, Rasmusson AG (2020) Substrate and plant genotype strongly influence the growth and gene expression response to trichoderma afroharzianum T22 in sugar beet. Plan Theory 9:1–14. https://doi.org/10.3390/plants9081005

Schmoll M, Dattenböck C, Carreras-Villaseñor N, Mendoza-Mendoza A, Tisch D, Alemán MI, Baker SE, Brown C, Cervantes-Badillo MG, Cetz-Chel J (2016) The genomes of three uneven siblings: footprints of the lifestyles of three Trichoderma species. Microbiol Mol Biol Rev 80:205–327. https://doi.org/10.1128/MMBR.00040-15

Sharma P, Vignesh Kumar P, Ramesh R, Saravanan K, Deep S, Sharma M (2011) Biocontrol genes from Trichoderma species : a review. 10:19898–19907. https://doi.org/10.5897/AJBX11.041

Sharma V, Salwan R, Sharma PN, Gulati A (2017) Integrated translatome and proteome: approach for accurate portraying of widespread multifunctional aspects of Trichoderma. Front Microbiol 8:1–13. https://doi.org/10.3389/fmicb.2017.01602

Sharma V, Salwana R, Shanmugamb V (2018) Molecular characterization of β -endoglucanase from antagonistic Trichoderma saturnisporum isolate GITX-Panog (C) induced under mycoparasitic conditions. Pestic Biochem Physiol 0–1. https://doi.org/10.1016/j.pestbp.2018.06.001

Silva JCR, Salgado JCS, Vici AC, Ward RJ, Polizeli MLTM, Guimarães LHS, Furriel RPM, Jorge JA (2020) A novel Trichoderma reesei mutant RP698 with enhanced cellulase production. Braz J Microbiol 51:537–545. https://doi.org/10.1007/s42770-019-00167-2

Silva LR d, Valadares-Inglis MC, Peixoto GHS, Luccas BEG d, Muniz PHPC, Magalhães DM, Moraes MCB, Mello SCM d (2021) Volatile organic compounds emitted by Trichoderma azevedoi promote the growth of lettuce plants and delay the symptoms of white mold. Biol Control 152:104447. https://doi.org/10.1016/j.biocontrol.2020.104447

Snelders N, Kettles G, Rudd J, Thomma B (2018) Plant pathogen effector proteins as manipulators of host microbiomes? Mol Plant Pathol 19:257–259. https://doi.org/10.1111/mpp.12628

Soler S, Rey M, Ait-lahsen H, Monte E, Llobell A (2001) Biocontrol fungus Trichoderma harzianum. Society 67:5833–5839. https://doi.org/10.1128/AEM.67.12.5833

Sood M, Kapoor D, Kumar V, Sheteiwy MS, Ramakrishnan M, Landi M, Araniti F, Sharma A (2020) Trichoderma: the "secrets" of a multitalented biocontrol agent. Plan Theory 9:762

Souza AA, Leitão VO, Ramada MH, Mehdad A, Georg RDC, Ulhôa CJ, De Freitas SM (2016) Trichoderma harzianum produces a new thermally stable acid phosphatase, with potential for biotechnological application. PLoS One 11:1–18. https://doi.org/10.1371/journal.pone.0150455

Sperschneider J, Catanzariti A, Deboer K, Petre B, Gardiner DM, Singh KB, Dodds PN, Taylor JM (2017) OPEN LOCALIZER : subcellular localization prediction of both plant and effector proteins in the plant cell. Nat Publ Gr 1–14. https://doi.org/10.1038/srep44598

Sperschneider J, Dodds P, Gardiner D, Singh K, Taylor J (2018) Improved prediction of fungal effector proteins from secretomes with EffectorP 2.0. Mol Plant Pathol 19:2094–2110. https://doi.org/10.1111/mpp.12682

Sridharan AP, Thankappan S, Karthikeyan G, Uthandi S (2020) Comprehensive profiling of the VOCs of Trichoderma longibrachiatum EF5 while interacting with Sclerotium rolfsii and Macrophomina phaseolina. Microbiol Res 236:126436. https://doi.org/10.1016/j.micres.2020.126436

Suárez MB, Sanz L, Chamorro MI, Rey M, González FJ, Llobell A, Monte E (2005) Proteomic analysis of secreted proteins from Trichoderma harzianum: identification of a fungal cell wall-induced aspartic protease. Fungal Genet Biol 42:924–934. https://doi.org/10.1016/j.fgb.2005.08.002

Suárez MB, Vizcaíno JA, Llobell A, Monte E (2007) Characterization of genes encoding novel peptidases in the biocontrol fungus Trichoderma harzianum CECT 2413 using the TrichoEST functional genomics approach. Curr Genet 51:331–342. https://doi.org/10.1007/s00294-007-0130-5

Sun RJ, Xu Y, Hou CX, Zhan YH, Liu MQ, Weng XY (2018) Expression and characteristics of rice xylanase inhibitor OsXIP, a member of a new class of antifungal proteins. Biol Plant 62:569–578. https://doi.org/10.1007/s10535-018-0787-2

Swaminathan J, Van Koten C, Henderson HV, Jackson TA, Wilson MJ (2016) Formulations for delivering Trichoderma atroviridae spores as seed coatings, effects of temperature and relative humidity on storage stability. J Appl Microbiol 120:425–431

Tandon A, Fatima T, Anshu, Shukla D, Tripathi P, Srivastava S, Singh PC (2020) Phosphate solubilization by Trichoderma koningiopsis (NBRI-PR5) under abiotic stress conditions. J King Saud Univ Sci 32:791–798. https://doi.org/10.1016/j.jksus.2019.02.001

Todorovic B, Glick BR (2008) The interconversion of ACC deaminase and D-cysteine desulfhydrase by directed mutagenesis. Planta 229:193–205

Triveni S, Prasanna R, Shukla L, Saxena AK (2013) Evaluating the biochemical traits of novel Trichoderma-based biofilms for use as plant growth-promoting inoculants. Ann Microbiol 63:1147–1156

Troian RF, Steindorff AS, Ramada MH, Arruda W, Ulhoa CJ. Mycoparasitism studies of Trichoderma harzianum against Sclerotinia sclerotiorum: evaluation of antagonism and expression of cell wall-degrading enzymes genes. Biotechnol Lett. 2014 Oct;36(10):2095-101. doi: 10.1007/s10529-014-1583-5. Epub 2014 Jun 26. PMID: 24966041.

Trushina N, Levin M, Mukherjee PK, Horwitz BA (2013) PacC and pH–dependent transcriptome of the mycotrophic fungus Trichoderma virens. BMC Genomics 14:138

Tseng SC, Liu SY, Yang HH, Lo CT, Peng KC (2008) Proteomic study of biocontrol mechanisms of Trichoderma harzianum ETS 323 in response to Rhizoctonia solani. J Agric Food Chem 56:6914–6922. https://doi.org/10.1021/jf703626j

Tundo, S., Paccanaro, M.C., Elmaghraby, I., Moscetti, I., Ovidio, R.D., Favaron, F., Sella, L., 2020. The xylanase inhibitor TAXI-I increases plant

Vailati-Riboni M, Palombo V, Loor JJ (2017) What are omics sciences? In: Ametaj BN (ed) Periparturient diseases of dairy cows: a systems biology approach. Springer International Publishing, Cham, pp 1–7. https://doi.org/10.1007/978-3-319-43033-1_1

Van Bohemen A-I, Zalouk-Vergnoux A, Poirier L, Phuong NN, Inguimbert N, Salah KBH, Ruiz N, Pouchus YF (2016) Development and validation of LC–MS methods for peptaibol quantification in fungal extracts according to their lengths. J Chromatogr B 1009:25–33

Van Oosten VR, Bodenhausen N, Reymond P, Van Pelt JA, Van Loon LC, Dicke M, Pieterse CMJ (2008) Differential effectiveness of microbially induced resistance against herbivorous insects in Arabidopsis. Mol Plant-Microbe Interact 21:919–930. https://doi.org/10.1094/MPMI-21-7-0919

van Zijll de Jong E, Stewart A, Loguercio LL, Kandula J, Cruz-Magalhães V, Herrera-Estrella A, Steyaert JM, Kandula D, Rostás M, Nieto-Jacobo MF, Mendoza-Mendoza A, Hampton J, Padilla-Arizmendi F (2019) The NADPH oxidases Nox1 and Nox2 differentially regulate volatile organic compounds, fungistatic activity, plant growth promotion and nutrient assimilation in Trichoderma atroviride. Front Microbiol 9:1–15. https://doi.org/10.3389/fmicb.2018.03271

Vargas WA, Mandawe JC, Kenerley CM (2009) Plant-derived sucrose is a key element in the symbiotic association between Trichoderma virens and maize plants. Plant Physiol 151:792–808

Velmourougane K, Prasanna R, Saxena AK, Singh SB, Chawla G, Kaushik R, Ramakrishnan B, Nain L (2017) Modulation of growth media influences aggregation and biofilm formation between Azotobacter chroococcum and Trichoderma viride. Appl Biochem Microbiol 53:546–556

Velmourougane K, Prasanna R, Supriya P, Ramakrishnan B, Thapa S, Saxena AK (2019) Transcriptome profiling provides insights into regulatory factors involved in Trichoderma viride-Azotobacter chroococcum biofilm formation. Microbiol Res 227:126292

Vieira PM, Siqueira A, Coelho G, Steindorff AS, José S, De Siqueira L, Silva N, Ulhoa CJ (2013) Identification of differentially expressed genes from Trichoderma harzianum during growth on cell wall of Fusarium solani as a tool for biotechnological application. BMC Genomics 1–11. https://doi.org/10.1186/1471-2164-14-177

Vignolle GA, Mach RL, Mach-Aigner AR, Derntl C (2020) Novel approach in whole genome mining and transcriptome analysis reveal conserved RiPPs in Trichoderma spp. BMC Genomics 21:1–12

Villalobos-Escobedo JM, Esparza-Reynoso S, Pelagio-Flores R, López-Ramírez F, Ruiz-Herrera LF, López-Bucio J, Herrera-Estrella A (2020) The fungal NADPH oxidase is an essential element for the molecular dialog between Trichoderma and Arabidopsis. Plant J 0–2. https://doi.org/10.1111/tpj.14891

Vinale F, Sivasithamparam K (2020) Beneficial effects of Trichoderma secondary metabolites on crops. Phyther Res. https://doi.org/10.1002/ptr.6728

Vinale F, Sivasithamparam K, Ghisalberti EL, Marra R, Barbetti MJ, Li H, Woo SL, Lorito M (2008a) A novel role for Trichoderma secondary metabolites in the interactions with plants. Physiol Mol Plant Pathol 72:80–86

Vinale F, Sivasithamparam K, Ghisalberti EL, Marra R, Woo SL, Lorito M (2008b) Trichoderma–plant–pathogen interactions. Soil Biol Biochem 40:1–10. https://doi.org/10.1016/j.soilbio.2007.07.002

Viterbo A, Chet I (2006) TasHyd1, a new hydrophobin gene from the biocontrol agent Trichoderma asperellum, is involved in plant root colonization. Mol Plant Pathol 7:249–258. https://doi.org/10.1111/j.1364-3703.2006.00335.x

Viterbo A, Landau U, Kim S, Chernin L, Chet I (2010) Characterization of ACC deaminase from the biocontrol and plant growth-promoting agent Trichoderma asperellum T203. FEMS Microbiol Lett 305:42–48

Vitikainen M, Arvas M, Pakula T, Oja M, Penttilä M, Saloheimo M (2010) Array comparative genomic hybridization analysis of Trichoderma reesei strains with enhanced cellulase production properties. BMC Genomics 11:441

Vyawahare MP, Ingle ST, Gathe AG (2019) Morphological variation and chitinase production ability of Trichoderma viride mutants. Int J Curr Microbiol Appl Sci 8:447–455. https://doi.org/10.20546/ijcmas.2019.801.047

Wang Z, Gerstein M, Snyder M (2009) RNA-Seq: a revolutionary tool for transcriptomics. Nat Rev Genet 10:57–63

Werf MJ v d, Jellema RH, Hankemeier T (2005) Microbial metabolomics: replacing trial-and-error by the unbiased selection and ranking of targets. J Ind Microbiol Biotechnol 32:234–252

Willmann MR, Endres MW, Cook RT, Gregory BD (2011) The functions of RNA-dependent RNA polymerases in Arabidopsis. Arab Book/Am Soc Plant Biol 9

Wonglom P, Ito S i, Sunpapao A (2020) Volatile organic compounds emitted from endophytic fungus Trichoderma asperellum T1 mediate antifungal activity, defense response and promote plant growth in lettuce (Lactuca sativa). Fungal Ecol 43:100867. https://doi.org/10.1016/j.funeco.2019.100867

Woo SL, Ruocco M, Vinale F, Nigro M, Marra R, Lombardi N, Pascale A, Lanzuise S, Manganiello G (2014) Trichoderma -based products and their widespread use in agriculture. Open Mycol J 8:71–126

Wu Q, Sun R, Ni M, Yu J, Li Y, Yu C (2017) Identification of a novel fungus, Trichoderma asperellum GDFS1009, and comprehensive evaluation of its biocontrol efficacy, pp 1–20

Xia Y, Yu K, Gao QM, Wilson EV, Navarre D, Kachroo P, Kachroo A (2012) Acyl CoA binding proteins are required for cuticle formation and plant responses to microbes. Front Plant Sci 3:1–18. https://doi.org/10.3389/fpls.2012.00224

Xu WJ, Chen LG, Chen X, Liu YS, Zheng TH, Song JJ, Xu W, Li P, Zhang MQ, Xiao CX (2015) Silencing ECHS1 attenuates the proliferation and induces the autophagy of hepatocellular carcinoma via impairing cell metabolism and activating AMPK. Neoplasma 62:872–880

Yang HH, Yang SL, Peng KC, Lo CT, Liu SY (2009) Induced proteome of Trichoderma harzianum by Botrytis cinerea. Mycol Res 113:924–932. https://doi.org/10.1016/j.mycres.2009.04.004

Yang CA, Cheng CH, Lo CT, Liu SY, Lee JW, Peng KC (2011) A novel L-amino acid oxidase from Trichoderma harzianum ETS 323 associated with antagonism of Rhizoctonia solani. J Agric Food Chem 59:4519–4526. https://doi.org/10.1021/jf104603w

Yang X, Cong H, Song J, Zhang J (2013) Heterologous expression of an aspartic protease gene from biocontrol fungus Trichoderma asperellum in Pichia pastoris. World J Microbiol Biotechnol 29:2087–2094. https://doi.org/10.1007/s11274-013-1373-6

Yang G, Tang L, Gong Y, Xie J, Fu Y, Jiang D, Li G, Collinge DB, Chen W, Cheng J (2017) A cerato-platanin protein SsCP1 targets plant PR1 and contributes to virulence of Sclerotinia sclerotiorum. New Phytol 217:739–755. https://doi.org/10.1111/nph.14842

Yu Z, Wang Z, Zhang Y, Wang Y, Liu Z (2021) Biocontrol and growth-promoting effect of Trichoderma asperellum TaspHu1 isolate from Juglans mandshurica rhizosphere soil. Microbiol Res 242:126596. https://doi.org/10.1016/j.micres.2020.126596

Yuan M, Huang Y, Ge W, Jia Z, Song S, Zhang L, Huang Y (2019a) Involvement of jasmonic acid, ethylene and salicylic acid signaling pathways behind the systemic resistance induced by Trichoderma longibrachiatum H9 in cucumber. BMC Genomics 20:144

Yuan M, Huang Y, Jia Z, Ge W, Zhang L, Zhao Q, Song S, Huang Y (2019b) Whole RNA-sequencing and gene expression analysis of Trichoderma harzianum Tr-92 under chlamydospore-producing condition. Genes Genomics 41:689–699

Yuan X, Wang H, Cai J, Li D, Song F (2019c) NAC transcription factors in plant immunity. Phytopathol Res 1:1–13. https://doi.org/10.1186/s42483-018-0008-0

Zeilinger S, Gruber S, Bansal R, Mukherjee PK (2016) Secondary metabolism in Trichoderma - chemistry meets genomics. Fungal Biol Rev 30:74–90. https://doi.org/10.1016/j.fbr.2016.05.001

Zhang J, Miao Y, Rahimi MJ, Zhu H, Steindorff A, Schiessler S, Cai F, Pang G, Chenthamara K, Xu Y, Kubicek CP, Shen Q, Druzhinina IS (2019) Guttation capsules containing hydrogen peroxide: an evolutionarily conserved NADPH oxidase gains a role in wars between related fungi. Environ Microbiol 21:2644–2658. https://doi.org/10.1111/1462-2920.14575

Part III
Biology and Genetics of *Trichoderma* Interactions with Plants

The Role of Secondary Metabolites in Rhizosphere Competence of *Trichoderma*

Hexon Angel Contreras-Cornejo, Lourdes Macías-Rodríguez, and John Larsen

Contents

H. A. Contreras-Cornejo (✉)
Instituto de Investigaciones en Ecosistemas y Sustentabilidad, Universidad Nacional Autónoma de México, Morelia, Michoacán, Mexico

Instituto de Investigaciones Químico-Biológicas, Universidad Michoacana de San Nicolás de Hidalgo, Morelia, Michoacán, Mexico
e-mail: hcontreras@cieco.unam.mx

L. Macías-Rodríguez
Instituto de Investigaciones Químico-Biológicas, Universidad Michoacana de San Nicolás de Hidalgo, Morelia, Michoacán, Mexico

J. Larsen
Instituto de Investigaciones en Ecosistemas y Sustentabilidad, Universidad Nacional Autónoma de México, Morelia, Michoacán, Mexico

N. Amaresan et al. (eds.), *Advances in* Trichoderma *Biology for Agricultural Applications*, Fungal Biology, https://doi.org/10.1007/978-3-030-91650-3_6

1 Introduction

Trichoderma species[1] are soilborne fungi that have a prolific growth in soils rich in organic matter although some strains have been isolated from aquatic ecosystems (Shi et al. 2020a; Haouhach et al. 2020). Around 1920, these fungi were recognized as microorganisms for biological control of plant pathogens (Harman 2006; Zeilinger et al. 2016). To date, a broad number of species have been molecularly typified. Most frequently isolated fungal species are *Trichoderma atroviride*, *Trichoderma virens*, and *Trichoderma harzianum* (Macías-Rodríguez et al. 2020). In addition to the mentioned fungal strains, also *T. viride*, *T. parareesei*, *T. longibrachiatum*, *T. gamsii*, *T. citrinoviride*, and *T. asperellum* are among the species frequently investigated due to their importance in medicine, biotechnology, and agriculture (Schuster and Schmoll 2010; Mukherjee et al. 2013; Crutcher et al. 2013). *Trichoderma* spp. while growing in their natural niche produce several secondary metabolites that include nitrogen and sulfur-containing metabolites that naturally possess different chemical structures. Important efforts have been made to understand the effects of *Trichoderma* in the rhizosphere when interacting with other plants, microorganisms, and arthropods (Martínez-Medina et al. 2017a, b; Contreras-Cornejo et al. 2019).

Under natural or in vitro conditions *Trichoderma* interact with plants via the emission of volatile organic compounds (VOCs) or establish physical contact with the root that can result in the formation of close associations either endophytic or in the rhizosphere. Depending of the mode of interaction, the fungus may induce biochemical changes in its host and induce the accumulation of key metabolites (i.e., phytoalexins and phytohormones), which are a potent weapon for plant immunity which result effective against multiple pathogens and arthropod pests (Manganiello et al. 2018; Villalobos-Escobedo et al. 2020). Furthermore, *Trichoderma* cause profound plant phenotypic changes and can alter endogenous signaling events that lead to the modulation of gene expression and biochemical events (De Palma et al. 2019). Chemical studies with plants as *Arabidopsis thaliana*, tomato, and maize revealed that *Trichoderma* spp. can modulate the levels of phytohormones, carbohydrates, amino acids, and green leaf and other plant volatiles (Battaglia et al. 2013; Macías-Rodríguez et al. 2020).

One of the most outstanding activities of *Trichoderma* in the rhizosphere is the biocontrol of plant pathogens. This is a very complex phenomenon that involves the physical contact of *Trichoderma* with its fungal prey (mycoparasitism) and the activity of hydrolytic enzymes. The biocontrol of *Trichoderma* involves also the production of antibiotic compounds (antibiosis). Due to the multiple beneficial effects of *Trichoderma* on plants of economic importance, different fungal strains have been used in formulations as bioinoculants to be applied in soils of culturable lands and improve the plant yield. This book chapter describes the main findings about the performance of *Trichoderma* spp. on the rhizosphere and their effects on other organisms of ecological relevance.

2 Natural Habitats of *Trichoderma*

These fungi can be found in the environment as sexual telemorphic stage and as asexual or anamorphic stage (Druzhinina et al. 2011). These ascomycete fungi are present in both terrestrial and aquatic ecosystems (Gal-Hemed et al. 2011; Carreras-Villaseñor et al. 2012; Guo et al. 2020). For example, *T. asperellum* cf44-2 is a marine strain isolated as endophyte from the algae *Sargassum* sp. and other 29 strains isolated from Mediterranean *Psammocinia* sp. sponges (Song et al. 2018). Some strains have been isolated from tree bark as the case of *T. atroviride* that was associated with *Acacia* sp. and *T. harzianum* CICR-E isolated from *Eucalyptus* sp. (Mukherjee et al. 2014). To date, at least 300 species of *Trichoderma* are recognized (Bissett et al. 2015; Marik et al. 2019). Although *Trichoderma* spp. are cosmopolitan fungi, some species such as *T. parareesei*, *T. reesei*, *T. pseudokoningii*, or *T. novae-zelandiae* are geographically restricted (Druzhinina et al. 2012).

Trichoderma spp. are fungi highly active in the rhizosphere, where they fulfill multiple functions. Figure 1 shows the kind of interactions that these fungi establish in the environment. A predominant feature of *Trichoderma* is fast growth on the substrate that is colonized and the high production of conidia (Dautt-Castro et al. 2020; Mota et al. 2019). A number of strains have been isolated from soils rich in organic matter where they thrive as saprotrophs, but they can have a biotrophic lifestyle (Zachow et al. 2016). In fact, the presence of *Trichoderma* in the rhizosphere is considered as an indicator of soil health (Vandenkoornhuyse et al. 2002; Van der Heijden et al. 2008; Meincke et al. 2009; Harman et al. 2012; Lange 2014). The estimated population density of *Trichoderma* in tropical soils is approximately 10^1 to 10^3 viable propagules per gram (Harman et al. 2004).

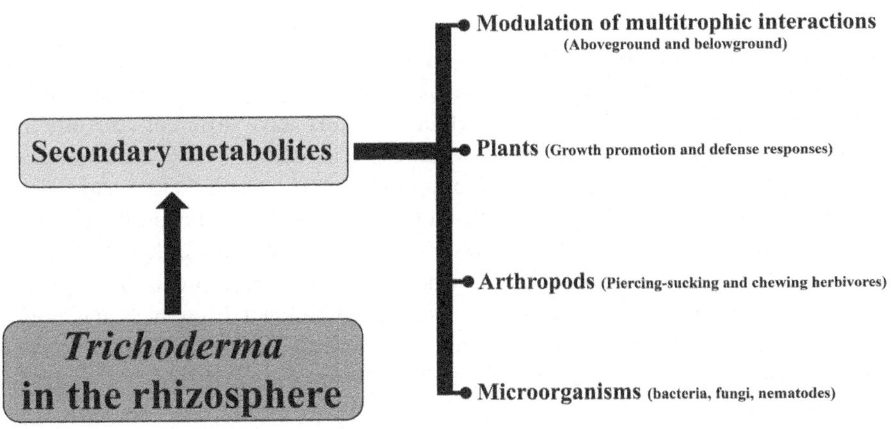

Fig. 1 Schematic representation of *Trichoderma* as a rhizosphere microorganism and the kind of organisms with which it cointeracts

Belowground, these fungi are highly interactive with plant roots as reported by Zachow et al. (2016) who studied the *Trichoderma* existing communities among maize (*Zea mays*) a cosmopolite plant and *Aeonium* sp., *Diospyros* sp., *Hebe* sp., and *Rhododendron* sp., endemic plants in the Northwest Africa to New Zealand via the European Alps and Madagascar. Among the tested plants, no clear quantitative differences in *Trichoderma* communities were found. In contrast, qualitative differences were identified between *Trichoderma* communities harbored in endemic plants, where fungal strains were very specific and different, especially in the samples tested in Madagascar (i.e., *H. andinensis,*[1] *H. lixii/T. harzianum, H. virens/T. virens, T. hamatum, T. konilangbra, T. longibrachiatum*, and *T. spirale*) and New Zealand (i.e., *H. hunua, H. pachybasioides/T. polysporum, H. stellate, T. fertile, T. paucisporum, T. petersenii, T. stromaticum*). However, maize plants formed associations with ~66% of the identified fungal strains. Interestingly, it was also identified a fungal core where *T. koningii* and *T. koningiopsis* were predominant in the analyzed plants.

Trichoderma spp. can also grow under adverse environmental conditions caused by different pollutants including toxic agrochemicals (Blaya et al. 2013; Tripathi et al. 2013). Certain fungal strains can resist the negative effects caused by heavy metals as Cd and Cu or establish associations with plants grown in either acidic or alkaline soils, while other species remain in soils with high concentrations of Co and Ni (Lo et al. 1996; Rosatto et al. 2019). *Trichoderma* species have been also been reported as opportunistic human pathogens. For example, *T. longibrachiatum* caused pericarditis in a farmer that received a lung transplant (Recio et al. 2019).

3 The Metabolome of *Trichoderma*

Trichoderma spp. are prolific producers of secondary metabolites of high and low molecular weight (Reino et al. 2008; Leylaie and Zafari 2018). Fungal metabolites comprise volatiles and non-volatiles, which naturally have different structure and physicochemical properties (Rajani et al. 2021). The identified *Trichoderma* compounds are alkaloids, terpenes, peptides, polyketides, organic acids, and siderophores (Reino et al. 2008; Mukherjee et al. 2012a, b; Crutcher et al. 2013; Macías-Rodríguez et al. 2020). Naturally, among fungal strains, there are differences in the production of metabolites whose biosynthesis is directly related with the proteins encoded in the genome. Interestingly, significant differences at genomic and proteomic levels among fungal species as indicated in parenthesis have been reported for *T. reesei* QM6a (32.68 Mb, 9, 114), *T. atroviride* IMI 206040 (36.40 Mb, 11, 815), and *T. virens* Gv29-8 (40.52 Mb, 12, 389), respectively (Kubicek et al. 2011; Mendoza-Mendoza et al. 2018).

[1]* Species identities are cited as initially published, and the current taxonomic status of each species requires verification. Note that the generic name "*Hypocrea*" is not in use.

It has been estimated that *Trichoderma* spp. secrete ~100 organic compounds that may be involved in key functions in the rhizosphere during their interactions with other organisms (Sharma et al. 2017). Through analytical techniques, at least 479 fungal volatiles have been identified (Siddiquee 2014). The profiles of volatiles released by *Trichoderma* differ between species and strains (Guo et al. 2019). Some fungal species produce rich blends in C_{10} and C_{15} terpene derived from the isoprenoid pathway, which produce a number of compounds of a structurally diverse family with different stereochemistry, whereas other fungal strains release C_8 compounds as 1-octen-3-ol, 3-octanone, and 3-octanol, which are considered as the end compounds of the metabolism of fatty acids, sharing acetyl-CoA as precursor (Contreras-Cornejo et al. 2019).

Among the compounds produced by these rhizosphere fungi trichoviridin, a cyclopentyl isocyanide metabolite was found to be produced by *T. koningii* and *T. viride* (Yamano et al. 1970; Tamura et al. 1975; Nobuhara et al. 1976). In *T. virens* Gv29.8 4-phosphopantetheinyl transferase that participates in the biosynthesis of 11, 14, and 18 mer peptaibols has been reported (Velázquez-Robledo et al. 2011). Some peptaibols produced by *Trichoderma* are constituted by 7–20 amino acids and characteristically possess an acylated N-terminal group and C-terminal amino alcohol and contain 2-amino-isobutyric acid (Aib) (Reino et al. 2008; Mukherjee et al. 2011). *T. longibrachiatum* SMF2 also produces several peptaibols with a range of biological activities, which increases the value of such strain and its compounds; interestingly an otholog of the putative methyltransferase LAE1 is a modulator for the production of secondary metabolites that includes fungal peptaibols (Shi et al. 2020b).

Trichodermamides A and B, which are two modified dipeptides produced by *T. virens*, should be further characterized (Reino et al. 2008). In *T. virens* Gv29.8, the gene *TvCyt2* that encodes a homologous protein of the p450 monooxygenase has been identified. Interestingly, *TvCyt2* was downregulated at the early stages of the plant-fungus interaction. Chemical analyses revealed that TvCyt2 participates in the production of the metabolites pyrazine [1,2-a] indole-1, 4-diene, 2,3-dihydro-2-methyl-3-methylene, α-cadinol, tau-muurolol, and viridiflorol, which are elicitors of plant defense (Ramírez-Valdespino et al. 2018). In *T. viride* a steroidal metabolite named viridiol that displayed phytotoxic activity has been identified (Reino et al. 2008).

6-Pentyl-2*H*-pyran-2-one (6-PP) is a volatile metabolite produced by *T. atroviride* IMI 206040, *T. viride*, and *T. harzianum* (Contreras-Cornejo et al. 2018a, b). 6-PP is a compound derived from linoleic acid though a mechanism that requires the oxidation of such fatty acid to 13-hydroperoxide-diene and the subsequent production of the intermediary 5-hydroxy-2,4-decenoic acid by β-oxidation and isomerization, and a final step of esterification, results in the formation of 6-PP (Serrano-Carreon et al. 1993; Zeilinger et al. 2016). In the environment, such compound has multiple biological functions.

In the extracts of *T. asperellum* cf44-2, seven new fungal molecules were elucidated including three bisabolane-derived compounds, three cyclonerane sesquiterpene-related compounds, and an unreported harzine diterpene (Song et al.

2018). More recently, eight highly oxygenated polyketides such as koninginin were reported from *T. koningiopsis* QA-3 (Shi et al. 2020c). In the case of *T. atroviride* D16, the production of a heteropolysaccharide whose molecular weight is 36.13 kDa and is constituted by galactose, glucose, and mannose was detected (Wu et al. 2019).

Both fumaric and gluconic acids are also organic compounds produced by *Trichoderma*, and when released in the environment, they can decrease soil pH and might increase the solubilization of phosphate and mineral cations like Fe, Mg, and Mn (Contreras-Cornejo et al. 2019; Vinale et al. 2008; Reino et al. 2008; Vinale et al. 2014). In addition, among the substances produced by *Trichoderma* spp., hydrolytic enzymes are present; they have several natural functions (i.e., biocontrol of plant pathogenic microorganisms and plant root colonization). Wortmannolone another compound produced by *T. virens* is a metabolite with potential importance in medicine due to its inhibition of phosphatidylinositol 3-kinase, which could be employed against human neoplasms (Dodge et al. 1995; Reino et al. 2008). Figure 2 shows the molecular structures of *Trichoderma*-derived metabolites.

4 Properties of *Trichoderma* Used for Biocontrol

As a soil habitant, *Trichoderma* interacts with other microorganisms exerting positive, neutral (without apparent effect), or negative effects (microbial growth inhibition). Mechanisms of biocontrol based on *Trichoderma* involve antibiosis, mycoparasitism (ability to parasitize other fungi), and competition for space and nutrients (Contreras-Cornejo et al. 2016; Macías-Rodríguez et al. 2020). Since starvation causes possible soil microbial death, competition for nutrients is considered a biocontrol mechanism established by *Trichoderma* against other microorganisms (Benítez et al. 2004). Indeed, it is known that *Trichoderma* compete for space and plant-derived nutrients (Harman et al. 2004). Plant roots exude a number or organic compounds that include amino acids, organic acids, lipids, and carbohydrates (Lombardi et al. 2018; Macías-Rodríguez et al. 2018). *T. britannicum* CECT 2413 (previously known as *T. harzianum)* possess a gene named *Gtt1* that encodes a high-affinity glucose transporter, which is expressed under low concentrations of glucose, which may occur during competition with other soil microorganisms (Delgado-Jarana et al. 2003; Benítez et al. 2004). Another important finding is the characterization of an intracellular invertase in *T. virens* (TvInv), which drives the hydrolysis of sucrose suggesting that plant-released sucrose is a nutritional source to the fungus (Vargas et al. 2009).

Siderophores are another group of compounds produced by soil microorganisms including *Trichoderma* spp. that chelate iron and, in that way, deprive Fe for other microorganisms such as plant pathogens and hence inhibit their growth (Contreras-Cornejo et al. 2019). Importantly, rhizosphere-competent microorganisms can take up sequestered iron charged in the siderophore via ferric-chelate transporters (Benítez et al. 2004; Vinale et al. 2014). Coprogen, ferricrocin, fusigen, and fusarine A are siderophores produced by *T. harzianum*, *T. reesei*, *T. asperellum*, *T. gamsii*,

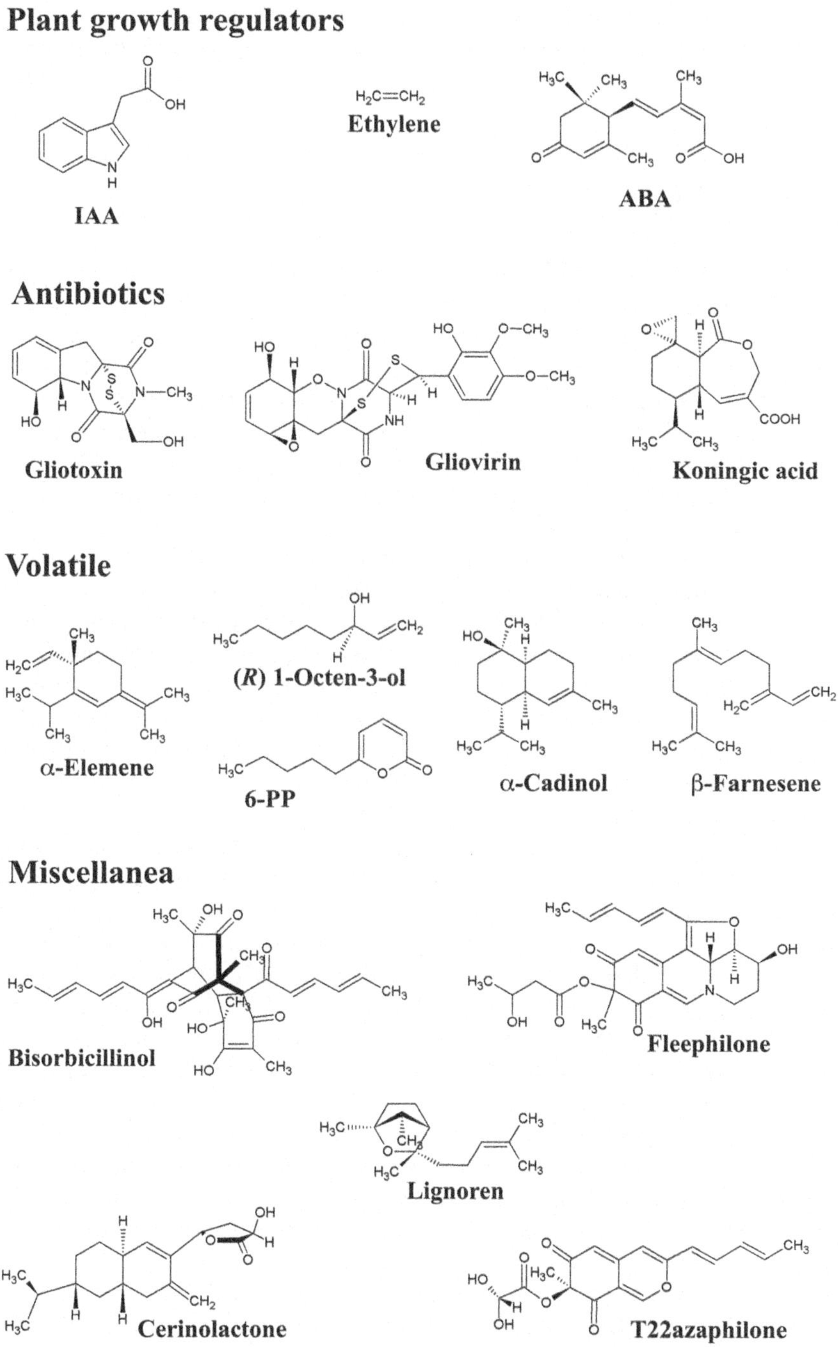

Fig. 2 Metabolites produced by *Trichoderma* spp.

T. atroviride, *T. hamatum*, *T. virens*, and *T. polysporum* (Lehner et al. 2013). Antagonistic activity of *Trichoderma* has been observed on *Botrytis*, *Fusarium*, *Rhizoctonia*, *Pytium*, *Phytophthora*, and *Sclerotinia* (Ji et al. 2019; Mota et al. 2019).

4.1 Antibiosis

It is well-known that *Trichoderma* spp. produce several secondary metabolites that inhibit the growth of plant pathogens (Brain and McGowan 1945; Vinale et al. 2008; Graczyk et al. 2020). Fugal metabolites with antibiotic activity include peptaibols, non-ribosomal peptides, steroids, diketopiperazines, isonitriles, polyketides, sesquiterpenes, and alkyl pyrones (Sivasithamparam and Ghisalberti 1998; Wiest et al. 2002; Contreras-Cornejo et al. 2019). T22azaphilone is an antibiotic produced by *T. afroharzianum*; such metabolite has an oxygenated bicyclic core and can inhibit the growth of *Rhizoctonia solani*, *Pytium ultimum*, and *Gaeumannomyces graminis* (Vinale et al. 2006).

6-PP is a volatile pyrone produced by *T. atroviride* IMI 206040, *T. harzianum* 38, which has antimicrobial activity against the fungus *Cylindrocarpon destructans*. The mechanism of action of 6-PP targets the metabolic processes of the pathogen because the pyrone alters the homeostasis of amino acids leading to autophagy to the cells under 6-PP treatment (Jin et al. 2020). Koningic acid is a sesquiterpenoid produced by three different fungal strains isolated from soil: *Chaetomium globosum*, *Gliocladium virens* (renamed *Trichoderma virens*), and *T. viride*. Trichosetin is other fungal metabolite that displayed antibacterial activity against *Staphylococcus aureus* and *Bacillus subtilis* (Marfori et al. 2003).

T. virens produces peptaibols (11-, 14-, and 18- mer) that have inhibitory activity against the plant pathogens *Alternaria solani*, *Phytophthora capsici*, *Fusarium oxysporum*, *Sclerotium rolfsii*, *S. cepivorum* and *R. solani*, and *Fusarium* spp. (Velázquez-Robledo et al. 2011). Gliotoxin is a small non-ribosomal peptide that is also considered as a diketopiperazine produced by *T. virens*, which showed immune-suppressive properties and has antiviral, antibacterial, and fungistatic activities (Mukherjee et al. 2013; Contreras-Cornejo et al. 2016). The cyclonerodiol-derived compound, lignoren, is a metabolite from a strain identified as *T. lignorum* (the taxonomic name is not in use) that has moderate antibacterial activity against *Pseudomonas aeruginosa* and *Bacillus subtilis*, but did not show any effect against fungi including *Penicillium notatum*, *Fusarium culmorum*, or *Candida albicans* (Berg et al. 2004). Antifungal activity has also been reported for the hydroxyl-lactone produced by *T. cerinum* against *Botrytis cinerea*, *P. ultimum*, and *R. solani* (Vinale et al. 2011).

A more recently report showed that VOCs released by *Trichoderma* spp. can inhibit the growth of the plant pathogenic fungi *F. oxysporum*-CFO, *S. rolfsii*-CSR, and *Sclerotinia sclerotiorum*-TSS (Rajani et al. 2021). Chemical analyses by gas-chromatography mass spectrometry revealed that *T. longibrachiatum* strain 2 produced a volatile blend constituted by alcohols, aldehydes, ethers, esters, hydrocarbon, ketones, and terpenes. Importantly, during the antagonism exerted by *Trichoderma* against *F. oxysporum*-CFO or *Macrophina phaseolina*-CMP via volatile emission, several compounds with antibiotic and cytotoxic activity were detected (Rajani et al. 2021).

4.2 Mycoparasitism

At least 75 mycoparasitic species of *Trichoderma* has been identified with the ability to restrict the proliferation of plant pathogenic fungi such as *R. solani*, *S. sclerotiorum*, *B. cinerea*, and *Alternaria alternata* and the oomycetes *Phytophthora* spp. and *Pythium* spp. (Harman et al. 2004; Druzhinina et al. 2011). Table 1 shows some specific features used for biocontrol based on these fungi. *Trichoderma* activates this mechanism when perceiving the fungal pathogen and grows tropically toward

Table 1 Mechanisms of biocontrol activated by *Trichoderma* against pathogens[a]

Strain	Plant pathogen	Description of the interaction	Reference
T. asperellum TR356	*S. Sclerotiorum*, *F. oxysporum*	Pathogen growth inhibition and in TR356 depending on the stage of interaction between microorganisms alter the expression of genes that encode for heat shock proteins (*TaHsp26b*, *TaHsp26c*, *TaHsp70a*, *TaHsp70b*, *TaHsp70c*, *TaHsp90*, *TaHsp104a*, *TaHsp104b*)	Mota et al. (2019)
T. gamsii B21	*F. verticillioides*	The biocontrol agent inhibited 74 ± 1% the colony growth of the plant pathogen and 66 ± 4% the root colonization. In addition, B21 reduced the production of the mycotoxin fumonisinin a liquid medium culture	Galletti et al. (2020)
Trichoderma sp.	*R. solani*	6-PP produced by the mycoparasitic fungus inhibits the growth of the plant pathogen	Dennis and Webster (1971)
T. harzianum isolates CICR-G and CICR-E	*S. delphinii*	Outcompetetion of the pathogen	Mukherjee et al. (2014)
T. viride	*S. rolfsii*	The secondary metabolite compound viridepyronone showed antagonic activity against the pathogen	Evidente et al. (2003)
T. Guizhouense	*F. odoratissimum* (FOC4)	The mycoparasitic fungus attacks the aerial hypha of Foc4 through production of H_2O_2 produced via NADPH oxidases	Pang et al. (2020)
T. atroviride	*B. cinerea*	The biocontrol activity involves the participation of Tal6, a LysM protein, which increases the mycoparasitism	Romero-Contreras et al. (2019)
	Colletotrichum lindemuthianum		

[a]Species identities are cited as initially published, and the current taxonomic status of each species requires verification

its prey. Then, hyphae of the biocontrol agent are coiled in a process mediated by lectin that strongly stick to the pathogen and via the combined action of the hydrolytic enzymes chitinases and glucanases holes in the hypha of pathogen occur. Furthermore, *Trichoderma* can degrade pectinases and other enzymes that are used by plant pathogenic microorganisms to colonize the tissues (Contreras-Cornejo et al. 2016).

Mycoparasitism is a complex process that also involves the production of antibiotic compounds, which in a fine-tuned mechanism results in killing the pathogen (Harman et al. 2004). Atpenins are compounds that participate specifically inhibiting the mitochondrial metabolism of the prey (Miyadera et al. 2003). Gliotoxin is an antibiotic compound biosynthesized by Q strains of *T. virens* though a gene cluster with the enzymatic core NRP GliP regulated by LAEA/VEA (Mukherjee et al. 2013). Importantly, the members of this cluster are modulated during mycoparasitism. The saprotrophic fungus *T. reesei* has part of that cluster, but it was not regulated in confrontation with other fungi and did not induce gliotoxin production (Mukherjee et al. 2012a, 2013; Atanasova et al. 2013). During mycoparasitism fungal compounds that are involved in the biocontrol of the pathogen are secreted. Bivertinolone has antibiotic properties and inhibits the biosynthesis of β-(1, 6)-glucan. Pachybasin is a metabolite that augment the number of coils of *Trichoderma* on its prey (Macías-Rodríguez et al. 2020).

5 *Trichoderma*-Arthropod Interactions

Due to the presence of saprotrophic fungi and arthropods in the soil, both kinds of organisms can interact in different terrestrial ecosystems. Depending of the lifestyle of rhizospheric fungi, some of them can benefit or cause damage to their hosts (Foo et al. 2017). Natural interactions among *Penicillium citrinum* and *T. harzianum* with the bioluminescent firefly *Pteroptyx bearni* have been detected (Foo et al. 2017). It has also been reported that soil fungi augment the decomposition of organic matter in the rhizosphere by stimulating the activity of wood-feeding termites. It was observed that wood consumption of the termite *Coptotermes formosanus* was increased when *T. viride* was inoculated to the soil (Xiong et al. 2018). A more recent finding shows that *Trichoderma* species protect termites against the entomopathogenic fungus *Metarhizium anisopliae* (Metschn) Sorokin by inducing pathogen-avoiding behaviors of the termites (Wen et al. 2020). Also biocontrol of the bird cherry-oat aphid *Rhopalosiphum padi* with *T. citrinoviride* ITEM 4484 has been reported revealing a key role for long-chain and linear and unbranched alcohols (C_{15}-C_{17}) produced by ITEM 4484 altering the feeding preference in *R. padi* (Ganassi et al. 2016). To date, interactions of ecological relevance of *Trichoderma* with arthropods are not well documented nor the metabolites that govern such interactions. *Trichoderma*-associations of economic importance that occur in agroecosystems have a great interest currently.

6 *Trichoderma*-Plant Interactions

Trichoderma spp. are considered as plant beneficial fungi due to their plant growth-promoting effects on plants (Villalobos-Escobedo et al. 2020). Figure 3 shows the typical beneficial effects of *Trichoderma* induced in vitro. Interactions of these fungi with plants occur even before the physical contact between the mycelia and the roots through the exchange of low molecular weight compounds from both organisms (Ramírez-Valdespino et al. 2019). Some of the fungal compounds serve as signals for plants to coordinate their growth or trigger defense responses (Estrada-Rivera et al. 2019). In counterpart, plants release some compounds toward the rhizosphere that serve to attract the fungus in proximity with roots and promote the root colonization. For example, during the interaction of *T. atroviride* IMI 206040 with tomato plants (*Lycopersicon esculentum* L. cv. Río Grande), the fungus slightly modulated the root exudation of carbohydrates. Chemical analyses revealed that a fraction of the tomato root exudates is composed by arabinose, xylose, glucose,

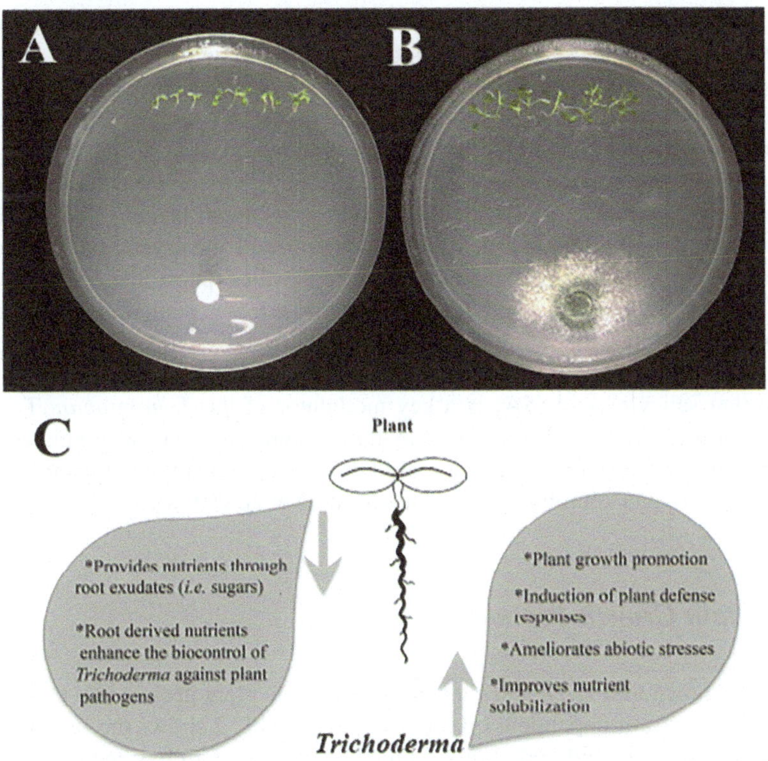

Fig. 3 In vitro interaction system to investigate the effects of *Trichoderma* on plants. Images show *Arabidopsis thaliana* ecotype Columbia-0 interacting with *T. atroviride* IMI 206040 during 5 days. Notice that the fungus increases the plant growth as reflected by the increased growth of the shoot and the formation of lateral roots

myo-inositol, fructose, and sucrose (a disaccharide constituted by glucose and fructose). Particularly, sucrose was not detected in the pre-colonization stage of the interaction. In contrast, such carbohydrate was exuded at a concentration of 0.29 ± 0.04 μg per sample when *T. atroviride* IMI 206040 colonized the tomato root system (Macías-Rodríguez et al. 2018). The uptake of sucrose by *Trichoderma* when interacting with plants seems possible since an intracellular invertase that can drive the hydrolysis of sucrose in *T. virens* Gv29.8 was identified (Vargas et al. 2009). On the other hand, many of the observed plant beneficial effects of *Trichoderma* involve modulations of plant hormones (Macías-Rodríguez et al. 2020).

Trichoderma colonizes the root system of a broad number of host plants (mono- and dicotyledonous), and such event induces profound changes in plant metabolism modifying the content of amino acids, soluble carbohydrates, phytohormones, and photosynthetic rate and altering the leaf transpiration and water content (Yedidia et al. 2003; Bae et al. 2009; Brotman et al. 2012; Contreras-Cornejo et al. 2020). Thus, during root colonization important events for plant fitness, as activation of plant immunity, growth promotion, and induction of abiotic stress, take place (Shoresh and Harman 2008; Contreras-Cornejo et al. 2014a; Fiorini et al. 2016). Plant root colonization involves the participation of fungal hydrophobins that are required in the host recognition and adhesion (Viterbo and Chet 2006). *T. harzianum* contain the gene *qid74*, which encodes a cysteine-rich cell wall, and it is required in adherence to the tomato root system (Samolski et al. 2012). *Trichoderma* requires the participation of plant cell wall degrading enzymes during the root colonization. The endopolygalacturonase ThPG1 from *T. harzianum* is included in this process (Morán-Diez et al. 2009). Swollenin is a fungal protein that contains a cellulose-binding module and disrupts the structure of the crystalline cellulose of the cell walls and likely as expansins facilitate cell wall expansion in roots (Brotman et al. 2008). It has been reported that *Trichoderma* releases plant defense elicitors in the root cell apoplast including proteins related with scavenging of reactive oxygen species and cell wall hydrolysis (Nogueira-Lopez et al. 2018). Interestingly, it was shown that salicylic acid (SA) is a key modulator of the *T. harzianum* T-38 root colonization in *A. thaliana* because in the mutant *sid2* that accumulates lower amounts of SA, the fungal strain increased the root colonization compared with the wild-type ecotype Columbia-0 (Martínez-Medina et al. 2017a).

6.1 Plant Growth Promotion

It is well-known that *Trichoderma* spp. can promote plant growth (Yu et al. 2021). This effect is strain-dependent on the host plant. Table 2 shows the fungal beneficial effects in the inoculated plants. Enhanced plant growth has been attributed to the activity of fungal compounds of low molecular weight that impact on key endogenous signaling programs (Garnica-Vergara et al. 2016). Since *Trichoderma* spp. naturally associate with plants roots here the main part of the plant perceives the fungal signals. Bioactive *Trichoderma* metabolites can induce profound

Table 2 Beneficial effects of *Trichoderma* spp. on plants[a]

Strain	Plant	Effect	Reference
T. virens Gv 29.8	*A. thaliana*	Shoot growth promotion and root branching	Contreras-Cornejo et al. (2009, 2014a, b, c)
T. atroviride IMI 206040			
T. Harzianum	Melon (*Cucumis melo* cv. Giotto)	Shoot biomass accumulation	Martínez-Medina et al. (2011)
T. asperellum TaspHu1	Tomato (*Solanum lycopersicum*)	Greater height, stem diameter, soluble protein content and soluble sugar content.	Yu et al. (2021)
T. harzianum	Cucumber (*Cucumis sativus*)	Promotion of leaf size, steam diameter base, plant height, root length, root number, and increases in the chlorophyll content when plants were cultured under salt stress	Zhang et al. (2019)
T-soybean			
T. asperellum T42	Rice (*Oryza sativa*) var. CO-51	Increased shoot and root dry weights	Singh et al. (2020)
T. atroviride IMI 206040	Tomato (*Lycopersicon esculentum* cv. Río Grande)	Promotion of both hypocotyl and primary root lengths	Macías-Rodríguez et al. (2018)
T. harzianum 38	*Maize (Zea mays)*	Increased the shoot nitrogen content of plants cultured in field conditions	Contreras-Cornejo et al. (2020)

[a]Species identities are cited as initially published, and the current taxonomic status of each species requires verification

changes in the plant phenotype as shoot development, root branching, and cell differentiation for the formation of root hairs (Contreras-Cornejo et al. 2009; Ming et al. 2013; Wu et al. 2019). The association of *T. afroharzianum* Rifai strain 22 (T22) with the roots of maize plants induced changes at proteomic level in the shoots (Shoresh and Harman 2008). In cucumber, *T. asperellum* T34 modulated different classes of proteins related with protein synthesis and folding; secondary metabolism, defense, and stress responses; metabolism; and energy (Segarra et al. 2007). Modifications in root architecture have been related with improved acquisition of soil nutrients. *T. harzianum* T-203, first characterized as *T. asperellum*, and after renamed as *T. asperelloides* increased 30% and 90% the concentrations of Fe and P, respectively. Plant growth promotion induced by T-203 was also related with elevations of the concentrations of Cu, Fe, Mn, Na, P, and Zn in roots. In addition, the fungal inoculation resulted in increased shoot content of Mn, P, and Zn with 70, 30, and 25%, respectively (Yedidia et al. 2001). Similarly, *T. afroharzianum* T22 can solubilize several soil nutrients like rock phosphate, Zn^0, Mn^{4+}, Fe^{3+}, and Cu^{2+}, which could be limited for plants (Altomare et al. 1999). Chemical analyses revealed that T22 biosynthesizes metabolites that can reduce the elements Cu (II) and Fe (III) as determined by the formation of the complexes

Cu(I)-Na_2–2,9-dimethyl-4,7-diphenyl-1,10-phenanthrolinedisulfonic acid and Fe(II)-Na_2_2,9-,bathophenanthrolinedisulfonic acid (Altomare et al. 1999). Recently, it was reported that *T. afroharzianum* T22 produces tricholignan, a polyketide that is a redox-active ortho-hydroquinone that reduce Fe (III), and its production improves plant growth under iron deficiency conditions (Chen et al. 2019).

The inoculation of the model plant *A. thaliana* with *T. virens* Gv29.8 induced the accumulation of ~50% more foliar biomass and increased ~3.5-fold the formation of lateral roots compared to the control treatment without fungal inoculation. Plant growth promotion activated by the fungus was a consequence of the auxin-signaling activation as evidenced by the increased expression of the auxin-responsive marker *DR5::GUS* and by the reduced or null response of the mutants *aux1-7*, *eir1-1* (affected in the auxin transport at the influx and efflux carriers, respectively), and *axr1-3* (affected in auxin response) to the promoting effects induced by *T. virens* Gv29.8. Also chemical analyses with gas chromatography mass spectrometry of the exuded fungal metabolites in the acidic fraction revealed the presence of an indolic metabolite at 10.81 min to a concentration of ~13.48 ± 0.97 $\mu g \cdot l^{-1}$ that corresponded to indole-3-acetic acid (IAA). In the neutral fraction of the *T. virens* Gv29.8 exudates, both indole-3-acetaldehyde (IAAld) at 8.83 min to a concentration of ~59.4 ± 4.47 $\mu g \cdot l^{-1}$ and indole-3-ethanol (IEt) at 9.97 min with a proportion of ~72.33 ± 1.41 $\mu g \cdot l^{-1}$ were detected. More interestingly, all the three indolic compounds increased their concentrations when L-tryptophan was added to the culture medium of the fungal strain (Contreras-Cornejo et al. 2009). Furthermore, pharmacological bioassays with the pure compounds IAA, IAAld, and IEt at concentrations in the range of nano- to micromolar showed that such indolic molecules can activate plant growth promotion as revealed by the induction of lateral roots and shoot biomass accumulation in a dose-dependent manner (Contreras-Cornejo et al. 2009).

T. atroviride IMI 206040 modulated the growth of tomato dependent of the stage of the interaction. Fungal inoculation enhanced hypocotyl length and primary root growth without physical contact. On the contrary, when *T. atroviride* IMI 206040 colonized the root system, the root branching was induced as evidenced by the number of lateral roots formed (Macías-Rodríguez et al. 2018). In *A. thaliana*, this same fungal strain promoted lateral root formation and root hair induction through a fine-tuned hormonal cross talk signaling mechanism that requires both auxin and ethylene and the activity of the mitogen activated protein kinase 6 (MPK6). The bioactive metabolites released by the fungus were perceived quickly in the plant; for example, plants exposed to 1 μM of indole-3-acetic acid during 15 min increased the MPK6 activity. In the interaction *T. atroviride* IMI 206040 and *A. thaliana* MPK6 seems to be a negative modulator of the primary root growth, lateral root formation, and root hair induction. Furthermore, genetic evidence revealed that root branching in *A. thaliana* modulated by the fungus involved the ethylene elements ETR1 and EIN2 (Contreras-Cornejo et al. 2015).

Plant growth promotion enhanced by rhizosphere microorganisms via ET signaling has been well documented. Several plant beneficial microorganisms produce ACC deaminase (ACCD) an enzyme involved in the production of ET by cleaving

ACC (Todorovic and Glick 2008). In *T. asperellum* T203 a gene that encodes an ACCD was identified, which resulted to be involved in the elongation of roots of rape plants (Viterbo et al. 2010). ET production in the range of nanograms was detected in *T. atroviride* IMI 206040. Importantly, ET levels increased dramatically when the fungus was grown in the presence of the amino acid L-methionine and *A. thaliana* plants inoculated with such fungal strains developed a phenotype related with the effects induced by ET (Contreras-Cornejo et al. 2015).

6.2 Induction of Plant Abiotic Stress Resistance

Since *Trichoderma* spp. can survive under unfavorable growth conditions as saline environments or in the presence of heavy metals, these fungi have been tested for bioremediation purposes (Contreras-Cornejo et al. 2014b; Li et al. 2019). In the model plant, *A. thaliana* that *Trichoderma* spp. induced salt stress tolerance through the elimination of Na^+ via root exudates and increases the content of key metabolites that included L-proline, which is an osmolite that acts into cells suffering water deficit; ascorbic acid a molecule that has multiple functions and particularly in this condition is a potent antioxidant compound that detoxifies cells of reactive species generated after the oxidative burst caused by salinity. Clearly, *Trichoderma* spp. alleviated the plant stress caused by salinity as evidenced by the normalized levels of ABA accumulated in *A. thaliana* (Contreras-Cornejo et al. 2014b). Similarly, the T-soybean isolate *T. harzianum* confers salt stress resistance by modulating the activity of GR (glutathione reductase), CAT (catalase), SOD (superoxide dismutase), APX (ascorbate peroxidase), POD, phenylalanine ammonia-lyase, and polyphenol oxidase and by increasing the production and accumulation of sugars, proline, ascorbic acid, and chlorophyll. In soybean, this same strain regulated the ratios of glutathione (GSH) to oxidized glutathione (GSSG) and AsA to dehydroascorbate (DHA) and the expression of *CsAPX* that form part of the AsA-GSH cycle. In addition, *T. harzianum* increased the levels of K^+ but reduced the levels of Na^+ (Zhang et al. 2019).

During the *Trichoderma*-plant interactions, it has been observed that fungal inoculation modulated the stomata closure and hence water loss from these structures. It is known that stomatal aperture is controlled by undulating concentrations of ABA. In *A. thaliana*, two mutants *abi1* and *abi2* impaired in their response to ABA were isolated. The *abi1* and *abi2* loci encode semi-dominant mutations in two different 2C protein phosphatases like enzymes (Allen et al. 1999). Interestingly, *abi1-1* and *abi2-1* do not close their stomata in response to fungal inoculation, which strongly suggested that *Trichoderma* spp. produce ABA. Chemical analyses through gas chromatography-mass spectrometry revealed that the presence of ABA in the fungal exuded metabolites. Interestingly, the microbial-derived extracts induced the expression of *abi4::GUS*, a transgenic marker that possesses the promoter of *ABI4*, a transcription factor of the *AP2* (*APETALA 2*) family (Finkelstein et al. 1998). ABI4 participates in seed development, sugar signaling, and salt responses (Arroyo et al. 2003).

Under salt stress, *Trichoderma* spp. induced root branching by providing auxin sensitivity as evidenced by the increased expression of the IAA-responsive marker *DR5::GUS*. Undoubtedly, that effect caused by fungal inoculation favors plants to improve water and nutrient uptake (Contreras-Cornejo et al. 2014b). Salt stress tolerance induced by *Trichoderma* has also been associated with the ACCD activity. This important finding was obtained from experiments where seeds of cucumber and *A. thaliana* were germinated under both normal and saline conditions induced by watering the seeds with solutions of NaCl at concentrations from 75 to 125 mM. In that work, in normal conditions, the strain wild type of *T. asperelloides* T203 and their ACCD silenced mutants (ΔACC#2/3) impacted in a similar manner of the germination of both model plants. Importantly, under salt conditions, higher germination rate was observed in the treatments with the wild type than those seeds treated with the mutants. This indicates that *Trichoderma* triggers specific molecular mechanisms via ET that provides salt stress tolerance (Brotman et al. 2013).

The effect of *Trichoderma* on plants grown under osmotic stress has been tested. In vitro experiments showed that −0.2 and −0.3 MPa affected severely the germination rate of tomato seeds. Whereas, under lower water potential, *T. afroharzianum* T22 broke seed dormancy and induced a homogeneous germination rate compared with the control treatment (Mastouri et al. 2010). In cacao plants, water deficit altered both the plant metabolism and expression of various genes like *TcSOT* (sorbitol transporter) and *TcTPP* (trehalose-6-phosophate phosphatase). However, negative effects caused by drought on cacao plants that included changes in net photosynthesis and stomatal conductance were delayed by *Trichoderma hamatum* isolate DIS 219b, which suggested that plants treated with this fungus suffered less than those uninoculated (Bae et al. 2009).

6.2.1 *Trichoderma*-Mediated Bioremediation

Waste waters spills, fertilizers, oils, and chemical pesticides contaminate soils (Cristaldi et al. 2020). Pesticides have as target of action damaging bacteria, fungi, arthropods, and weeds (Chen et al. 2020). It was calculated that pesticides prevent ~80% of crop yield loss (Oerke 2005). Indiscriminate applications of pesticides result in their accumulation in the environment and adverse effects on non-target organisms (Pimentel 1995; Maltby et al. 2009). Since *Trichoderma* are resistant to various pesticides and in some cases participate in the degradation of such compounds, they can be applied as bioinoculants to agroecosystems (Contreras-Cornejo et al. 2016). More recently, it was showed that *T. asperellum* TM detoxifies tomato roots of the organophosphate pesticide phoxim through three stages of the xenobiotic metabolism that comprise i. conversion, ii. conjugation (i.e., GSH-phoxim), and iii. sequestration (Chen et al. 2020). In this same work, TM increased the expression of *ABC2* (a member of the secondary transport system), *CYP724B2* (a member of the P450 superfamily), *GPX* (*glutathione peroxidase*), and *GR* (Chen et al. 2020).

Similarly, it was reported that *Trichoderma* spp. have the ability to biodegrade the organic compounds alachlor and metolachlor both chloroacetanilide herbicides. That report showed that after 7 days, the assayed fungal strains transformed from 80% to 99% of the doses tested of alachlor and from 40% to 79% in the case of metolachlor. The biotransformation of the chlorinated compounds was a consequence of the reactions of dechlorination and hydroxylation. The conversion of metolachlor resulted at least in four by-products. In addition, it was found that after 7 days of the inoculation of rapeseed seedlings with *T. koningii* IM 0956, *T. citrinoviride* IM 6325, *T. harzianum* KKP 534, *T. viride* KKP 792, and *T. virens* DSM 1963 and also treated with both chloroacetanilide herbicides, the fungal inoculation promoted the seedling growth (Nykiel-Szymańska et al. 2020).

Furthermore, the exposition of *Trichoderma* spp. to alachlor and metolachlor induced fungal oxidative stress as evidenced by the accumulation of the superoxide anion $[O_2 \cdot ^-]$. The resistance of the fungal strains to chloroacetanilides seems to be related with changes in the content of phosphatidic acid, phosphatidylcholines, and phosphatidylethanolamine (Nykiel-Szymańska et al. 2019). This feature of *Trichoderma* spp. to degradate toxic compounds underlines their importance as bioinoculants in agroecosystems.

The presence of heavy metals in water and soils is a frequent threat to ecosystem and human health. However, some fungal strains are resistant to pollutants and hence show strong potential in bioremediation. A clear example is the strain *Trichoderma* sp. MG that is tolerant to heavy metals, which can also tolerate high concentrations of As (500 mg/l) and Pb (650 mg/l). The strain MG induced soil bioremediation to both metals by precipitating those elements with calcite and ureoloytic activity (Govarthanan et al. 2019). Precipitation of metallic compounds with $CaCO_3$ or calcites is a well-known activity in algae, bacteria, fungi, and Protista (Gadd 2010; Qian et al. 2015). Another strain is *Trichoderma lixii* CR700, an electroplating wastewater isolate, which can tolerate 2000 mg/l of As, 1200 mg/l of Cu, 1000 mg/l of Cr, 1500 mg/l of Ni, 1200 mg/l of Zn, and 100 mg/l of Pb and Cd (Kumar and Dwivedi 2019). *T. lixii* CR700 has an extraordinary ability to detoxify $Cr_{(VI)}$ via reduction and a mechanism of remotion in the fungus itself. Such processes resulted beneficial for *Cicer arietium* and *Vigna radiata* when seedlings were watered with a solution of 100 mg/l $Cr_{(VI)}$ since the fungus reduced the negative effects on the plant growth caused by $Cr_{(VI)}$ (Kumar and Dwivedi 2019).

6.3 Activation of Plant Immunity

Fungal-induced plant defense confers protection against a broad spectrum of aggressors that include nematodes, bacteria, and fungi, which belowground attack roots, but is also effective against foliar herbivores and virus (Martínez-Medina et al. 2017b; Contreras-Cornejo et al. 2016, 2018a, b, 2020). *Trichoderma* spp. can trigger plant defense responses after the perception of fungal molecules released in the rhizosphere and atmosphere and during the root colonization (Velázquez-Robledo et al. 2011; Contreras-Cornejo et al. 2014c). Plants sense microorganisms by perceiving microorganism-associated molecular patterns (MAMPs), which for pathogens are named pathogen-associated molecular patters (PAMPs) that include chitin, ergosterol, flagellin, glycoproteins, and lipopolysaccharides, which are recognized by plant resistance proteins (Jones and Dangl 2006; Chisholm et al. 2006; Göhre and Robatzek 2008; Gimenez-Ibanez et al. 2009; Nürnberger and Kemmerling 2009).

The first identified MAMP derived from *Trichoderma* was an ET-inducing xylanase (Xyn2/Eix), which triggered plant defense responses in tomato and tobacco (Sharon et al. 1993). Later, in other *Trichoderma* strains, several effectors of plant defense such as alamethicin, cellulases, endopolygalacturonase ThPG1, 18 mer peptaibols, swollenin TasSwo and Sm1, and a small peptide of the cerato-platanin family were identified (Sharon et al. 1993; Engelberth et al. 2001; Martinez et al. 2001; Djonovic et al. 2006; Viterbo et al. 2007; Vargas et al. 2008; Brotman et al. 2008; Morán-Diez et al. 2009; Hermosa et al. 2012). Sm1 exogenously applied on cotton (*Gossypium hirsutum*) roots increased the expression of several genes related with plant defense such as *GhLOX1* (*Lipoxygenase 1*, an element for the biosynthesis of JA), *CHT* (chitinase), *GLU* (β-1,3-glucanase), and *POD6* (peroxidase). In addition, Sm1 applied at a concentration of 0.5 nmol on the cotyledons of cotton-induced plant resistance against the attack of *Colletotrichum* sp. (Djonovic et al. 2006). *T. atroviride* also produces a Sm1 homologous protein called Epl1, which in the presence of maize roots is released as dimer, but, in its monomeric form, Epl1 triggers defense responses against *Colletotrichum graminicola* (Vargas et al. 2008). In *T. formosa*, a proteinaceous elicitor that is a 12 kDa peptide and is homologous of Epl11 has also been identified (Cheng et al. 2018).

Mitogen-activated protein kinases (MAPKs) transfer cell information into the plant perceived by sensors and depending of the perceived cue generates cellular responses (Koornneef and Pieterse 2008). Root inoculation of cucumber plants with *T. asperellum* activated a protein kinase homologous with the MPK3 of *A. thaliana* that is necessary to confer resistance against the plant pathogen bacterium *Pseudomonas syringae* pv. *lachrymans* (Shoresh et al. 2006). Similarly, *T. atroviride* IMI 206040 increased the activity of MPK6 in *A. thaliana* during the pre- and root colonization, which was related with the modulation of some phenotypic responses mediated by ET (Contreras-Cornejo et al. 2015). Plant defense elicited by rhizosphere microorganisms is the result of fine-tuned endogenous signaling events where early messengers participate and involve changes of intracellular Ca^{+2}, production of H_2O_2 and NO, and accumulation of phytohormones SA and JA/ET and the subsequent modulation of their responsive genes (Kong et al. 2020).

Inoculation of melon (*Cucumis melo* cv. Giotto) plants with *T. harzianum* resulted in accumulation of SA, JA, and ACC (1-aminocyclopropane-1-carboxylic acid, ET donor) in the shoots (Martínez-Medina et al. 2011). The colonization of *A. thaliana* roots by *T. asperellum* T203 increased the expression of the transcription factors *WRKY18*, *WRKY33*, *WRKY40*, and *WRKY60* and *WRKY33* that activate JA-dependent responses. Particularly, *WRKY18*, *WRKY33*, and *WRKY60* are induced by pathogens and encode proteins that play a key role in the JA-signaling mechanism (Brotman et al. 2013). *T. asperellum* TaspHu1 increased the expression of the hormonal-responsive genes *JAR1* (*jasmonic acid resistance*), *MYC2* (transcription factor *MYC2*), *NPR1* (*Non-expressor of pathogenesis-related*), and *PR1* (*Pathogenesis related 1* that encodes a plant defensin with antimicrobial activity) and provided resistance against *A. alternata* (Yu et al. 2021). Similarly, in cucumber plants, *T. asperellum* T203 regulated the expression of a set of defense-related genes such as *PAL1* (*Phenylalanine ammonia lyase 1*, a component for SA biosynthesis); *LOX1*; and *CTR1* (*Constitutive triple response 1*) and *ETR1* (*Ethylene receptor 1*) both elements of the ethylene (ET) signaling pathway (Shoresh et al. 2005).

More recently, it was shown that plant defense against the mycotoxigenic fungal pathogen *Fusarium verticillioides* is specific strain-dependent because *T. gamsii* (IMO5) modulated the expression of the genes *ZmLOX10* (a lipoxygenase involved in green leaf volatile production), *ZmAOS* (*allene oxide synthase* required for jasmonate synthesis), and *ZmHPL* (*hydroperoxide lyase* involved in green leaf volatile production) typical genes of the induced systemic resistance (ISR) a process modulated by JA and ET. In contrast, *T. gamsii* B21 enhanced the transcript levels of *ZmPR1* in the maize leaves (that encodes a plant defensin with antimicrobial activity) and *ZmPR5* (*pathogenesis-related 5*, with antifungal activity); both genes modulated during the systemic acquired resistance (SAR) an immune response modulated by SA (Galletti et al. 2020).

Inoculation of *A. thaliana* with an inoculum of $1x10^6$ spores of *T. virens* Gv29.8 and *T. atroviride* IMI 206040 increased the content of foliar JA and induced changes in the expression of the reporter gene *LOX2::GUS*, whose promoter belongs to the LIPOXYGENASE 2. Similarly, 8 days after *A. thaliana* root inoculation, the levels of free SA in leaves increased and consequently also induced the expression of the reporter gene *PR-1a::GUS*. SA also fulfills an important role in modulating the *T. harzianum* root colonization in *A. thaliana* (Alonso-Ramírez et al. 2014). More interestingly, *T. harzianum* T39 failed to induce systemic resistance in *A. thaliana* mutants affected in the signaling pathways mediated by JA and ET, which suggested that such canonical plant hormones are modulated by the beneficial fungus to confer protection; particularly, the mutants *ein2*, *eto2*, *eto3*, and *npr1-5* of *A. thaliana* infected with *B. cinerea* and co-inoculated with *T. harzianum* were severely affected by the necrotrophic fungus, which confirmed that JA and ET signaling pathways are activated by *Trichoderma* (Korolev et al. 2008).

Phytoalexins are low molecular weight compounds induced in plants as part of defensive mechanisms serving as antimicrobial compounds to restrict the proliferation of plant pathogens (Blechert et al. 1995). Simultaneously, during the interaction of *A. thaliana* roots with *Trichoderma* spp. and compared with uninoculated plants, camalexin an indole-derived phytoalexin increased at least ~9-fold. This compound is involved in the resistance against the infection caused by the necrotrophic fungus *B. cinerea* (Contreras-Cornejo et al. 2011). It has also been found that *T. atroviride* induced changes in the expression of the gene *PAD3 (Phytoalexin deficient 3)*, which encodes the last enzyme for the biosynthesis of camalexin in *A. thaliana* (Salas-Marina et al. 2011).

The conferred plant protection induced by *Trichoderma* against pathogens also occurs via fungal volatiles. This effect was observed when *A. thaliana* plants were exposed to the VOCs emitted by *T. virens* Gv29.8 and infected with spores of *B. cinerea*. Airborne VOCs released by this fungus considerably reduced the percentage of both symptomatic and dead plants. In contrast, *A. thaliana* leaves without exposition of VOCs from *T. virens* Gv29.8 were severely damaged as evidenced by the presence of the fungal infection in tissues. The perception of fungal VOCs caused changes in the expression of *LOX2::GUS*, a gene activated by JA and a oxidative burst evidenced by the accumulation of H_2O_2. Undoubtedly, the modulation of *A. thaliana* immunity after the perception of *T. virens* Gv29.8 also reinforced the resistance to the *B. cinerea* infection. Chemical analyses revealed that *T. virens* Gv29.8 released a rich blend of VOCs constituted principally by terpenes that included the monoterpenes β-myrcene and linalool and the sesquiterpenes β-caryophyllene, δ-cadinene, and copaene. Furthermore, *B. cinerea* exposed to Gv29.8 VOCs showed reduced growth when compared with its respective control (Contreras-Cornejo et al. 2014c). It has also been detected that cellulysin produced by *T. viride* promoted the emission of several VOCs including 4,8-dimethylnona-1,3,7-triene, linalool, β-ocimene, (3Z)-hexenyl acetate, indole, and *cis*-jasmone of the octadecanoid pathway (Piel et al. 1997).

7 *Trichoderma* and Its Role in Multitrophic Interactions

Due to the broad distribution of *Trichoderma* in the environment and their interactions with plants and arthropods, recent findings have demonstrated that these fungi participate in multiple interactions at higher trophic levels (Macías-Rodríguez et al. 2020). Several strains of *Trichoderma* confer protection against the attack of arthropods belonging to the Thysanoptera and Hemiptera (Battaglia et al. 2013; Muvea et al. 2014; Coppola et al. 2019a, b). It was observed that root colonization of onion (*Allium cepa* cv. 'Red Creole') by *Trichoderma* conferred protection against the attack of *Thrips tabaci* as evidenced by the lower number of feeding punctures caused by the herbivore (Muvea et al. 2014). *Trichoderma harzianum* activated indirect defense responses that resulted in the attraction of *Aphidius ervi* the parasitoid of the aphid *Macrosiphum euphorbiae* (Coppola et al. 2017). Importantly, *T. longibrachiatum* modulated the emission of VOCs released by tomato plants affecting the performance of the aphid *Macrolophus pygmaeus* and its natural enemy *A. ervi* (Battaglia et al. 2013). Tomato plants inoculated with *Trichoderma* and infested with aphids displayed important transcriptomic changes that included 484 transcripts that were upregulated and 850 downregulated. During the aphid-plant interaction, the expression of the genes *OSM* (*Osmotin*), *SST* (*sesquiterpene synthase*), *SAM* (*S-Adenosyl-L-methionine salicylic acid carboxyl methyltransferase*), and *GDS* (*Germacrene-d-synthase*) were upregulated (Coppola et al. 2017). Interestingly, *T. afroharzianum* T22 decreased the survival of aphids feeding on tomato plants and upregulated the expression of the genes *Photosystem I reaction* canter subunit VI, *photosystem II reaction center W*, and *Photosystem II subunit S* involved in photosynthesis and the 30S, 40S, and 50S ribosomal proteins involved in biosynthesis (Coppola et al. 2019b). On the contrary, aphids also induced down-regulation of the genes *LOXA* and *LOXC* in the lipoxygenase pathway; ethylene response factor A.1 and ethylene-responsive transcription factor; and the JA- responsive genes *Threonine deaminase*, *Serine carboxypeptidase*, *Protease inhibitor I*, and *Pin-II type proteinase inhibitor 69* (Coppola et al. 2019b).

Plants are also attacked by a broad number of chewing insects. *Spodoptera frugiperda* commonly identified as the fall armyworm is a severe pest that damage maize shoots. Root association of maize plants with *T. atroviride* IMI 206040 conferred protection against the attack of *S. frugiperda* resulting in reduced the foliar herbivory (Contreras-Cornejo et al. 2018a). *T. atroviride* IMI 206040 in association with maize roots triggered the emission of (1*S*)-α-pinene, β-myrcene, *p*-cymene, γ-terpineol, β-caryophyllene, and α-humulene and the accumulation of JA in shoots. The fungal blend of VOCs released from the soil and the volatile oxylipins 1-octen-3-ol and 6-PP played a key role as antifeedant metabolites because such compounds reduced the attack of *S. frugiperda* on maize leaves when assayed at a concentration of 60 μg (Contreras-Cornejo et al. 2018b).

Under natural conditions, *S. frugiperda* is endoparasitized by female wasps of *Campoletis sonorensis*. It was detected that maize roots colonized by *T. atroviride* IMI 206040 and infested with *S. frugiperda* increased the parasitism rate of the herbivore when compared with plants that were not colonized by *T. atroviride* IMI 206040 (54.16 ± 9.00% vs. 29.16 ± 7.33%, respectively). Importantly, 6-PP released by *T. atroviride* IMI 206040 served as specific signaling molecule to attract to *C. sonorensis* toward leaves that were feeding on maize leaves, resulting in biocontrol of the chewing herbivore by modulating the behavior of *C. sonorensis*. Such information revealed that *Trichoderma* could be a key modulator of the maize interactions with economic importance (Contreras-Cornejo et al. 2018b).

A recent work performed by Contreras-Cornejo et al. (2020) under field conditions revealed that inoculation of *T. harzianum* 38 on the maize root system caused modulations in the abundance of native arthropods associated with shoots. Maize plants harbored at least 13 orders, 23 families of arthropods. Naturally, some of these arthropods are pests and other beneficial such as natural pest enemies. Pest arthropods included members of piercing-sucking insects belonging to the families Aphididae and Cicadidae, but also included individuals of the families Acrididae, Curculionidae, Chrysomelidae, and Noctuidae with feeding guild habits as chewers. The agricultural importance of *T. harzianum* 38 inoculation on maize roots resulted in the conferred protection of maize leaves against the attack of piercing-sucking insects. Such protection conferred by the fungus was related with increases in the content of (*Z*)-3-hexen-1-ol, a volatile metabolite that attracts natural enemies of the herbivores. More importantly, *T. harzianum* increased the presence of predators belonging to the families Forficulidae and Nitidulidae, which can modulate the proliferation of pest arthropods. In addition, it was reported that under adequate conditions *Trichoderma* sp. produces chitinases that attack the insect cuticle and negatively affect the peritrophic matrix of silkworms (Berini et al. 2016).

Belowground, *Trichoderma* can also confer protection against parasitic nematodes. For example, the inoculation of *T. harzianum* T78 on roots of tomato (*Solanum lycopersicon* cv. 'Moneymaker') was effective against the attack of *Meloidogyne incognita*. Depending on the nematode progress infection, the fungus primed plant defense responses through the activity of the canonical plant hormones. *T. harzianum* primed SA-modulated defenses regulating the expression of the genes *PR-1a* and *PR-P6* that encode for components that contribute to restricting the root pathogens. *T. harzianum* T78 also stimulates the JA-dependent defense responses modulating the expression of the genes *PI-II* (*proteinase inhibitor II*) and *MC* (*multicystatin*), which counteracts the repression of such signaling pathway induced by the nematode, and this undulation among phytohormones negatively impacts on the galling fecundity of *M. incognita* (Martínez-Medina et al. 2017b). Likewise, *T. citrinoviride* Snef1910 conferred protection against *M. incognita* in tomato roots, displaying high virulence against second-stage juveniles (J2s) of the nematode. Derived metabolites from Snef1910 diminished the number of egg masses, J2s, and root galls (Fan et al. 2020).

8 Conclusions

Trichoderma spp., with plant beneficial traits promoting crop health and nutrition, are natural inhabitants in most agroecosystems, where they are involved in multiple phytobiome interactions at all trophic levels with other microorganisms, arthropods, and plants. Secondary matabolites from *Trichoderma* spp. are key in these interactions as "signaling molecules." Antimicrobial substances from *Trichoderma* spp. such as T22azaphilone inhibit a broad spectrum of plant pathogens, and also volatile (i.e., terpenes, ET, and 6-PP) and no volatile (auxins, ABA, harzianic acid) metabolites can alter plant growth and development, fitness, and productivity. The association of *Trichoderma* with roots induces the accumulation of the phytohormones SA and JA, which in turn modulate the expression of their responsive genes. *Trichoderma* also induces changes in the primary and secondary plant metabolism that might alter the interactions with shoot and root associated arthropods. Recent research has been focused in discovering the biological activity of the fungal metabolites in the rhizosphere, yet a broad number of secondary metabolites released by *Trichoderma* remain to be studied. Also future research should focus on the integration of applied and basic mechanistic research, for the successful application of *Trichoderma* in agroecosystems.

Acknowledgments We thank N. Amaresan, A. Sankaranarayanan, Mitesh Dwivedi, and Irina S. Druzhinina for the kind invitation to write this chapter.

References

Allen GJ, Kuchitsu K, Chu SP, Murata Y, Schroeder JI (1999) Arabidopsis *abi1-1* and *abi2-1* phosphatase mutations reduce abscisic acid-induced cytoplasmic calcium rises in guard cells. Plant Cell 11:1785–1798. https://doi.org/10.1105/tpc.11.9.1785

Alonso-Ramírez A, Poveda J, Martín I, Hermonsa R, Monte E, Nicolás C (2014) Salicylic acid prevents *Trichoderma harzianum* from entering the vascular system of roots. Mol Plant Pathol 15:823–831. https://doi.org/10.1111/mpp.12141

Altomare C, Norvell WA, Björkman T, Harman GE (1999) Solubilization of phosphates and micronutrients by the plant-growth-promoting and biocontrol fungus *Trichoderma harzianum* Rifai 1295-22. Appl Environ Microbiol 65:2926–2933. https://doi.org/10.1128/aem.65.7.2926-2933.1999

Arroyo A, Bossi F, Finkelstein RR, León P (2003) Three genes that affects sugar sensing (Abscisic acid insensitive 4, abscisic acid insensitive 5, and constitutive triple response 1) are differentially regulated by glucose in *Arabidopsis*. Plant Physiol 133:1–12. https://doi.org/10.1104/pp.103.021089

Atanasova L, Le Crom S, Gruber S, Coulpier F, Seidl-Seiboth V, Kubicek CP, Druzhinina IS (2013) Comparative transcriptomics reveals different strategies of *Trichoderma* mycoparasitism. BMC Genomics 14:121. https://doi.org/10.1186/1471-2164-14-121

Bae H, Sicher RC, Kim MS, Kim SH, Strem MD, Melnick RL, Bailey BA (2009) The beneficial endophyte *Trichoderma hamatum* isolate DIS 219b promotes growth and delays the onset of the drought response in *Theobroma cacao*. J Exp Bot 60:3279–3295. https://doi.org/10.1093/jxb/erp165

Battaglia D, Bossi S, Cascone P, Digilio MC, Duran Prieto J, Fanti P, Guerrieri E, Iodice L, Lingua G, Lorito M, Maffei ME, Massa N, Ruocco M, Sasso R, Trotta V (2013) Tomato below ground-above ground interactions: *Trichoderma longibrachiatum* affects the performance of *Macrosiphum euphorbiae* and its natural antagonists. Mol Plant-Microbe Interact 26:1249–1256. https://doi.org/10.1094/mpmi-02-13-0059-r

Benítez T, Rincón AM, Limón MC, Codón AC (2004) Biocontrol mechanisms of *Trichoderma* strains. Int Microbiol 7:249–260

Berg A, Wangun HVK, Nkengfack AE, Schlegel B (2004) Lignoren, a new sesquiterpenoid metabolite from *Trichoderma lignorum* HKI 0257. J Basic Microbiol 44:317–319. https://doi.org/10.1002/jobm.200410383

Berini F, Caccia S, Franzetti E, Congiu T, Marinelli F, Casartelli M, Tettamanti G (2016) Effects of *Trichoderma viride* chitinases on the peritrophic matrix of Lepidoptera. Pest Manag Sci 72:980–989. https://doi.org/10.1002/ps.4078

Bissett J, Gams W, Jaklitsch W, Samuels GJ (2015) Accepted *Trichoderma* names in the year 2015. IMA Fungus 6:263–295. https://doi.org/10.5598/imafungus.2015.06.02.02

Blaya J, López-Mondéjar R, Lloret E, Pascual JA, Ros M (2013) Changes induced by *Trichoderma harzianum* in suppressive compost controlling *Fusarium* wilt. Pestic Biochem Physiol 107:112–119. https://doi.org/10.1016/j.pestbp.2013.06.001

Blechert S, Brodschelm W, Hölder S, Kammerer L, Kutchan TM, Mueller MJ, Xia ZQ, Zenk (1995) The octadecanoic pathway: signal molecules for the regulation of secondary pathways. Proc Natl Acad Sci U S A 92:4099–4105. https://doi.org/10.1073/pnas.92.10.4099

Brain PW, McGowan JC (1945) Viridin: a highly fungistatic substance produced by *Trichoderma viride*. Nature 156:144–145

Brotman Y, Briff E, Viterbo A, Chet I (2008) Role of swollenin, an expansin-like protein from *Trichoderma*, in plant root colonization. Plant Physiol 147:779–789. https://doi.org/10.1104/pp.108.116293

Brotman Y, Lisec J, Méret M, Chet I, Willmitzer L, Viterbo A (2012) Transcript and metabolite analysis of the *Trichoderma*-induced systemic resistance response to *Pseudomonas syringae* in *Arabidopsis thaliana*. Microbiology 158:139–146

Brotman Y, Landau U, Cuadros-Inostroza A, Tohge T, Fernie AR, Chet I, Viterbo A, Willmitzer L (2013) *Trichoderma*-plant root colonization: escaping early plant defense responses and activation of the antioxidant machinery for saline stress tolerance. PLoS Pathog 9:e1003221. https://doi.org/10.1371/journal.ppat.1003221

Carreras-Villaseñor N, Sánchez-Arreguín JA, Herrera-Estrella AH (2012) *Trichoderma:* sensing the environment for survival and dispersal. Microbiology 158:3–16. https://doi.org/10.1099/mic.0.052688-0

Chen M, Liu Q, Gao SS, Young AE, Jacobsen SE, Tang Y (2019) Genome mining and biosynthesis of a polyketide from a biofertilizer fungus that can facilitate reductive iron assimilation in plant. Proc Natl Acad Sci U S A 116:5499–5504. https://doi.org/10.1073/pnas.1819998116

Chen S, Yan WY, Wu M, Mao Q, Chen Y, Ren J, Liu A, Lin X, Ahammed GJ (2020) *Trichoderma asperellum* reduces phoxim residue in roots by promoting plant detoxification potential in *Solanum lycopersicum* L. Environ Pollut 259:113893. https://doi.org/10.1016/j.envpol.2019.113893

Cheng CH, Shen BN, Shang QW, Liu LYD, Peng KC, Chen YH, Chen FF, Hu SF, Wang YT, Wang HC, Wu HY, Lo CT, Lin SS (2018) Gene-to-gene network analysis of the mediation of plant innate immunity by the eliciting plant response-like 1 (Epl1) elicitor of *Trichoderma formosa*. Mol Plant-Microbe Interact 31:683. https://doi.org/10.1094/MPMI-01-18-0002-TA

Chisholm ST, Coaker G, Day B, Staskawicz BJ (2006) Host-microbe interactions: shaping the evolution of the plant immune response. Cell 124:803–814. https://doi.org/10.1016/j.cell.2006.02.008

Contreras-Cornejo HA, Macías-Rodríguez L, Cortés-Penagos C, López-Bucio J (2009) *Trichoderma virens*, a plant beneficial fungus enhances biomass production and promotes lateral root growth through an auxin dependent mechanism in Arabidopsis. Plant Physiol 149:1579–1592. https://doi.org/10.1104/pp.108.130369

Contreras-Cornejo HA, Macías-Rodríguez L, Beltrán-Peña E, Herrera-Estrella A, López-Bucio J (2011) *Trichoderma*-induced plant immunity likely involves both hormonal and camalexin dependent mechanisms in *Arabidopsis thaliana* and confers resistance against necrotrophic fungi *Botrytis cinerea*. Plant Signal Behav 6:1554–1563. https://doi.org/10.4161/psb.6.10.17443

Contreras-Cornejo HA, Macías-Rodríguez L, López-Bucio J (2014a) Enhanced plant immunity using *Trichoderma*. In: Gupta VK (ed) Biotechnology and biology of *Trichoderma*. Elsevier, Oxford, pp 495–504

Contreras-Cornejo HA, Macías-rodríguez LI, Alfaro-Cuevas R, López-Bucio J (2014b) *Trichoderma* improves growth of Arabidopsis seedlings under salt stress through enhanced root development, osmolite production and Na^+ elimination through root exudates. Mol Plant-Microbe Interact 27:503–514. https://doi.org/10.1094/MPMI-09-13-0265-R

Contreras-Cornejo HA, Macías-Rodríguez LI, Herrera-Estrella A, López-Bucio J (2014c) The 4-phosphopantetheinyl transferase of *Trichoderma virens* plays a role in plant protection against *Botrytis cinerea* through volatile organic compound emission. Plant Soil 379:261–274. https://doi.org/10.1007/s11104-014-2069-x

Contreras-Cornejo HA, López-Bucio JS, Méndez-Bravo A, Macías-Rodríguez L, Ramos-Vega M, Guevara-García AA, López-Bucio J (2015) Mitogen-activated protein kinase 6 and ethylene and auxin signaling pathways are involved in *Arabidopsis* root-system architecture alterations by *Trichoderma atroviride*. Mol Plant-Microbe Interact 28:701–710. https://doi.org/10.1094/MPMI-01-15-0005-R

Contreras-Cornejo HA, Macías-Rodríguez L, del-Val E, Larsen J (2016) Ecological functions of *Trichoderma* spp. and their secondary metabolites in the rhizosphere: interactions with plants. FEMS Microbiol Ecol:92. https://doi.org/10.1093/femsec/fiw036

Contreras-Cornejo HA, del-Val E, Macías-Rodríguez L, Alarcón A, González-Esquivel CE, Larsen J (2018a) *Trichoderma atroviride*, a maize root associated fungus, increases the parasitism rate of the fall armyworm *Spodoptera frugiperda* by its natural enemy *Campoletis sonorensis*. Soil Biol Biochem 122:196–202. https://doi.org/10.1016/j.soilbio.2018.04.013

Contreras-Cornejo HA, Macías-Rodríguez L, del-Val E, Larsen J (2018b) The root endophytic fungus *Trichoderma atroviride* induces foliar herbivory resistance in maize plants. Appl Soil Ecol 124:45–53. https://doi.org/10.1016/j.apsoil.2017.10.004

Contreras-Cornejo HA, Macías-Rodríguez L, del-Val E, Larsen E (2019) Interactions of *Trichoderma* with plants, insects, and plant pathogen microorganisms: chemical and molecular bases. In: Mérillon J-M, Ramawat KG (eds) Co-evolution of secondary metabolites, reference series in phytochemistry. Springer Nature Switzerland AG. https://doi.org/10.1007/978-3-319-76887-8_23-1

Contreras-Cornejo HA, Viveros-Bremauntz F, del-Val E, Macías-Rodríguez L, López-Carmona DA, Alarcón A, González-Esquivel CE, Larsen J (2020) Alterations of foliar arthropod communities in a maize agroecosystem induced by the root-associated fungus *Trichoderma harzianum*. J Pest Sci. https://doi.org/10.1007/s10340-020-01261-3

Coppola M, Cascone P, Chiusano ML, Colantuono C, Lorito M, Pennacchio F, Rao R, Woo SL, Guerrieri E, Digilio MC (2017) *Trichoderma harzianum* enhances tomato indirect defense against aphids. Insect Sci 24:1025–1033

Coppola M, Cascone P, Di Lelio I, Woo SL, Lorito M, Rao R, Pennacchio F, Guerrieri E, Digilio MC (2019a) *Trichoderma atroviride* P1 colonization of tomato plants enhances both direct and indirect defense barriers against insects. Front Physiol 10:813. https://doi.org/10.3389/fphys.2019.00813

Coppola M, Diretto G, Digilio MC, Woo SL, Giuliano G, Molisso D, Pennacchio F, Lorito M, Rao R (2019b) Transcriptome and metabolome reprogramming in tomato plants by *Trichoderma harzianum* primes and enhances defense responses against aphids. Front Physiol 10:745

Cristaldi A, Oliveri Conti G, Luciano Cosentino S, Mauromicale G, Copat C, Grasso A, Zuccarello P, Fiore M, Restuccia C, Ferrante M (2020) Phytoremediation potential of *Arundo donax* (Giant Reed) in contaminated soil by heavy metals. Environ Res 185:109427. https://doi.org/10.1016/j.envres.2020.109427

Crutcher FK, Parich A, Schuhmacher R, Mukherjee PS, Zeilinger S, Kenerley CM (2013) A putative terpene cyclase, vir4, is responsible for the biosynthesis of volatile terpene compounds in the biocontrol fungus *Trichoderma virens*. Fungal Genet Biol 56:67–77. https://doi.org/10.1016/j.fgb.2013.05.003

Dautt-Castro M, Estrada-Rivera M, Olguin-Martínez I, del Carmen Rocha-Medina M, Auxiliadora Islas-Osuna M, Casas-Flores S (2020) TBRG-1 a ras-like protein in *Trichoderma virens* involved in conidiation, development, secondary metabolism, mycoparasitism, and biocontrol unveils a new family of Ras- GTPases. Fungal Genet Biol 136:103292. https://doi.org/10.1016/j.fgb.2019.103292

De Palma M, Salzano M, Villano C, Aversano R, Lorito M, Ruocco M, Docimo T, Piccinelli AL, D'Agostino N, Tucci M (2019) Transcriptome reprogramming, epigenetic modifications and alternative splicing orchestrate the tomato root response to the beneficial fungus *Trichoderma harzianum*. Hort Res 6:5. https://doi.org/10.1038/s41438-018-0079-1

Delgado-Jarana J, Moreno-Mateos MA, Benítez T (2003) Glucose uptake in *Trichoderma harzianum*: role of *gtt1*. Euk Cell 2:708–717. https://doi.org/10.1128/ec.2.4.708-717.2003

Dennis C, Webster J (1971) Antagonistic properties of species-groups of *Trichoderma* II. Production of volatile antibiotics. Trans Br Mycol Soc 57:41–48. https://doi.org/10.1016/S0007-1536(71)80078-5

Djonovic S, Pozo MJ, Dangott LJ, Howell CR, Kenerley CM (2006) Sm1, a proteinaceous elicitor secreted by the biocontrol fungus *Trichoderma virens* induces plant defense responses and systemic resistance. Mol Plant-Microbe Interact 19:838–853. https://doi.org/10.1094/MPMI-19-0838

Dodge JA, Sato M, Vlahos CJ (1995) Inhibition of phosphatidylinositol 3-kinase with viridin and analogs thereof. Eur Patent Appl 648492:19

Druzhinina IS, Seidl-Seiboth V, Herrera-Estrella A, Horwitz BA, Kenerley CM, Monte E, Mukherjee PK, Zeilinger S, Grigoriev IV, Kubicek CP (2011) *Trichoderma*: the genomics of opportunistic success. Nat Rev Microbiol 9:749–759. https://doi.org/10.1038/nrmicro2637

Druzhinina IS, Komon-Zelazowska M, Ismaiel A, Jaklitsch W, Mullaw T, Samuels GJ, Kubicek CP (2012) Molecular phylogeny and species delimitation in the section Longibrachiatum of *Trichoderma*. Fungal Genet Biol 49:358–368. https://doi.org/10.1016/j.fgb.2012.02.004

Engelberth J, Koch T, Schüler G, Bachmann N, Rechtenbach J, Boland W (2001) Ion channel-forming alamethicin is a potent elicitor of volatile biosynthesis and tendril coiling. Cross talk between jasmonate and salicylate signaling in lima bean. Plant Physiol 125:369–377. https://doi.org/10.1104/pp.125.1.369

Estrada-Rivera M, Rebolledo-Prudencio OG, Pérez-Robles DA, Rocha-Medina MC, González-López MC, Casas-Flores S (2019) *Trichoderma* histone deacetylase HDA-2 modulates multiple responses in Arabidopsis. Plant Physiol 179:1343–1361. https://doi.org/10.1104/pp.18.01092

Evidente A, Cabras A, Maddau L, Serra S, Andolfi A, Motta A (2003) Viridepyronone, a new antifungal 6-substituted 2H-pyran-2-one produced by *Trichoderma viride*. J Agric Food Chem 51:6957–6960. https://doi.org/10.1021/jf034708j

Fan H, Yao M, Wang H, Zhao D, Zhu X, Wang Y, Liu X, Duan Y, Chen L (2020) Isolation and effect of *Trichoderma citrinoviride* Snef1910 for the biological control of root-knot nematode, *Meloidogyne incognita*. BMC Microbiol 20:299. https://doi.org/10.1186/s12866-020-01984-4

Finkelstein RR, Wang ML, Lynch TJ, Rao S, Goodman HM (1998) The Arabidopsis abscisic acid response locus *ABI4* encodes an APETALA 2 domain protein. Plant Cell 10:1043–1054. https://doi.org/10.1105/tpc.10.6.1043

Fiorini L, Guglielminetti L, Mariotti L, Curadi M, Picciarelli P, Scartazza A, Sarrocco S, Vannacci G (2016) *Trichoderma harzianum* T6776 modulates a complex metabolic network to stimulate tomato cv. Micro-Tom growth. Plant Soil 400:351–366. https://doi.org/10.1007/s11104-015-2736-6

Foo K, Seelan JSS, Dawood MM (2017) Microfungi associated with *Pteroptyx bearni* (Coleoptera: Lampyridae) eggs and larvae from kawang river, sabah (Northern Borneo). Insects:Pii:E66. https://doi.org/10.3390/insects8030066

Gadd GM (2010) Metals, minerals and microbes: geomicrobiology and bioremediation. Microbiology 156:609–643. https://doi.org/10.1099/mic.0.037143-0

Gal-Hemed I, Atanasova L, Komon-Zelazowska M, Druzhinina IS, Viterbo A, Yarden O (2011) Marine isolates of *Trichoderma* spp. as potential halotolerant agents of biological control for arid-zone agriculture. Appl Environ Microbiol 77:5100–5109. https://doi.org/10.1128/AEM.00541-11

Galletti S, Paris R, Cianchetaa S (2020) Selected isolates of *Trichoderma gamsii* induce different pathways of systemic resistance in maize upon *Fusarium verticillioides* challenge. Microbiol Res 233:126406. https://doi.org/10.1016/j.micres.2019.126406

Ganassi S, Grazioso P, De Cristofaro A, Fiorentini F, Sabatini MA, Evidente A, Altomare C (2016) Long chain alcohols produced by *Trichoderma citrinoviride* have phagodeterrent activity against the bird cherry-oat aphid *Rhopalosiphum padi*. Front Microbiol 7:297. https://doi.org/10.3389/fmicb.2016.00297

Garnica-Vergara A, Barrera-Ortiz S, Muñoz-Parra E, Raya-Gonzalez J, Méndez-Bravo A, Macías-Rodríguez L, Ruiz-Herrera LF, López-Bucio J (2016) The volatile 6-pentyl-2H-pyran-2-one from *Trichoderma atroviride* regulates *Arabidopsis thaliana* root morphogenesis via auxin signaling and ETHYLENE INSENSITIVE 2 functioning. New Phytol 209:1496–1512. https://doi.org/10.1111/nph.13725

Gimenez-Ibanez S, Ntoukakis V, Rathjen JP (2009) The LysM receptor kinase CERK1 mediates bacterial perception in *Arabidopsis*. Plant Signal Behav 4:539–541. https://doi.org/10.4161/psb.4.6.8697

Göhre V, Robatzek S (2008) Breaking the barriers: microbial effector molecules subvert plant immunity. Annu Rev Phytopathol 46:189–215. https://doi.org/10.1146/annurev.phyto.46.120407.110050

Govarthanan M, Mythili R, Kamala-Kannan S, Selvankumar T, Srinivasan P, Kim H (2019) *In-vitro* bio-mineralization of arsenic and lead from aqueous solution and soil by wood rot fungus, *Trichoderma* sp. Ecotoxicol Environ Saf 174:699–705. https://doi.org/10.1016/j.ecoenv.2019.03.034

Graczyk S, Perlińska-Lenart U, Górka-Nieć W, Lichota R, Piłsyk S, Zembek P, Lenart J, Bernat P, Gryz E, Augustyniak J, Palamarczyk G, Kruszewska JS (2020) Increased activity of the sterol branch of the mevalonate pathway elevates glycosylation of secretory proteins and improves antifungal properties of *Trichoderma atroviride*. Fungal Genet Biol 137:10334. https://doi.org/10.1016/j.fgb.2020.103334

Guo Y, Ghirardo A, Weber B, Schnitzler J-P, Benz JP, Rosenkranz M (2019) *Trichoderma* species differ in their volatile profiles and in antagonism toward ectomycorrhiza *Laccaria bicolor*. Front Microbiol 10:891. https://doi.org/10.3389/fmicb.2019.00891

Guo YW, Gong BQ, Yuan J, Li HJ, Mahmud T, Huang Y, Li JF, Yang DP, Lan WJ (2020) L-Phenylalanine alters the privileged secondary metabolite production in the marine-derived fungus *Trichoderma erinaceum* F1-1. J Nat Prod 83:79–87. https://doi.org/10.1021/acs.jnatprod.9b00710

Haouhach S, Karkachi N, Oguiba B, Sidaoui A, Chamorro I, Kihal M, Monte E (2020) Three new reports of *Trichoderma* in Algeria: *T. atrobrunneum*, (South) *T. longibrachiatum* (South), and *T. afroharzianum* (Northwest). Microorganisms 8:1455. https://doi.org/10.3390/microorganisms8101455

Harman GE (2006) Overview of mechanisms and uses of *Trichoderma* spp. Phytopathology 96:190–194. https://doi.org/10.1094/PHYTO-96-0190

Harman GE, Howell CR, Viterbo A, Chet I, Lorito M (2004) *Trichoderma* species-opportunistic, avirulent plant symbionts. Nat Rev Microbiol 2:43–56. https://doi.org/10.1038/nrmicro797

Harman GE, Herrera-Estrella AH, Horwitz BA, Lorito M (2012) Special issue: *Trichoderma*-from basic biology to biotechnology. Microbiology 158:1–2. https://doi.org/10.1099/mic.0.056424-0

Hermosa R, Viterbo A, Chet I, Monte E (2012) Plant-beneficial effects of *Trichoderma* and of its genes. Microbiology 158:17–25. https://doi.org/10.1099/mic.0.052274-0

Ji S, Liu Z, Liu B, Wang Y (2019) Comparative analysis of biocontrol agent *Trichoderma asperellum* ACCC30536 transcriptome during its interaction with *Populus davidiana* × *P. alba* var. *pyramidalis*. Microbiol Res 227:126294. https://doi.org/10.1016/j.micres.2019.126294

Jin X, Guo L, Jin B, Zhu S, Mei X, Wu J, Liu T, He X (2020) Inhibitory mechanism of 6-pentyl-2H-pyran-2-one secreted by *Trichoderma atroviride* T2 against *Cylindrocarpon destructans*. Pestic Biochem Physiol 170:104683. https://doi.org/10.1016/j.pestbp.2020.104683

Jones JDG, Dangl JL (2006) The plant immune system. Nature 444:323–329. https://doi.org/10.1038/nature05286

Kong X, Zhang C, Zheng H, Min S, Zhang F, Zhang M, Cui F, Lv D, Liu L, Guo S, Zhang Y, Yuan X, Zhao S, Tian H, Ding Z (2020) Antagonistic interaction between auxin and SA signaling pathways regulates bacterial infection through lateral root in Arabidopsis. Cell Rep 32:108060. https://doi.org/10.1016/j.celrep.2020.108060

Koornneef A, Pieterse CM (2008) Cross talk in defense signaling. Plant Physiol 146:839–844. https://doi.org/10.1104/pp.107.112029

Korolev N, Rav D, Elad DY (2008) The role of phytohormones in basal resistance and *Trichoderma*-induced systemic resistance to *Botrytis cinerea* in *Arabidopsis thaliana*. BioControl 53:667–683. https://doi.org/10.1007/s10526-007-9103-3

Kubicek CP, Herrera-Estrella A, Seidl-Seiboth V, Martinez DA, Druzhinina IS, Thon M, Zeilinger S, Casas-Flores S, Horwitz BA, Mukherjee PK, Mukherjee M, Kredics L, Alcaraz LD, Aerts A, Antal Z, Atanasova L, Cervantes-Badillo MG, Challacombe J, Chertkov O, McCluskey K, Coulpier F, Deshpande N, von Döhren H, Ebbole DJ, Esquivel-Naranjo EU, Fekete E, Flipphi M, Glaser F, Gómez-Rodríguez EY, Gruber S, Han C, Henrissat B, Hermosa R, Hernández-Oñate M, Karaffa L, Kosti I, Le Crom S, Lindquist E, Lucas S, Lübeck M, Lübeck PS, Margeot A, Metz B, Misra M, Nevalainen H, Omann M, Packer N, Perrone G, Uresti-Rivera EE, Salamov A, Schmoll M, Seiboth B, Shapiro H, Sukno S, Tamayo-Ramos JA, Tisch D, Wiest A, Wilkinson HH, Zhang M, Coutinho PM, Kenerley CM, Monte E, Baker SE, Grigoriev IV

(2011) Comparative genome sequence analysis underscores mycoparasitism as the ancestral life style of *Trichoderma*. Genome Biol 12:R40. https://doi.org/10.1186/gb-2011-12-4-r40
Kumar V, Dwivedi SK (2019) Hexavalent chromium stress response, reduction capability and bioremediation potential of *Trichoderma* sp. isolated from electroplating wastewater. Ecotoxicol Environ Saf 185:109734. https://doi.org/10.1016/j.ecoenv.2019.109734
Lange L (2014) The importance of fungi and mycology for addressing major global challenges. IMA Fungus 5:463. https://doi.org/10.5598/imafungus.2014.05.02.10
Lehner SM, Atanasova L, Neumann NK, Krska R, Lemmens M, Druzhinina IS, Schuhmacher R (2013) Isotope-assisted screening for iron-containing metabolites reveals a high degree of diversity among known and unknown siderophores produced by *Trichoderma* spp. Appl Environ Microbiol 79:18–31. https://doi.org/10.1128/AEM.02339-12
Leylaie S, Zafari D (2018) Antiproliferative and antimicrobial activities of secondary metabolites and phylogenetic study of endophytic *Trichoderma* species from Vinca plants. Front Microbiol 9:1484. https://doi.org/10.3389/fmicb.2018.01484
Li X, Zhang X, Wang X, Yang X, Cui Z (2019) Bioaugmentation-assisted phytoremediation of lead and salinity co-contaminated soil by *Suaeda salsa* and *Trichoderma asperellum*. Chemosphere 224:716–725. https://doi.org/10.1016/j.chemosphere.2019.02.184
Lo CT, Nelson EB, Harman GE (1996) Biological control of turfgrass diseases with a rhizosphere competent strain of *Trichoderma harzianum*. Plant Dis 80:736–741. https://doi.org/10.1094/PD-80-0736
Lombardi N, Vitale S, Turrà D, Reverberi M, Fanelli C, Vinale F, Marra R, Ruocco M, Pascale A, d'Errico G, Woo SL, Lorito M (2018) Root exudates of stressed plants stimulate and attract *Trichoderma* soil fungi. Mol Plant-Microbe Interact 31:982. https://doi.org/10.1094/MPMI-12-17-0310-R
Macías-Rodríguez L, Guzmán-Gómez A, García-Juárez P, Contreras-Cornejo HA (2018) *Trichoderma atroviride* promotes tomato development and alters the root exudation of carbohydrates, which stimulates fungal growth and the biocontrol of the phytopathogen Phytophthora cinnamomi in a tripartite interaction system. FEMS Microbiol Ecol. https://doi.org/10.1093/femsec/fiy137
Macías-Rodríguez L, Contreras-Cornejo HA, Adame-Garnica SG, del-Val E, Larsen J (2020) The interactions of *Trichoderma* at multiple trophic levels: inter-kingdom communication. Microbiol Res 240:126552. https://doi.org/10.1016/j.micres.2020.126552
Maltby L, Brock TCM, van den Brink PJ (2009) Fungicide risk assessment for aquatic ecosystems: importance of interspecific variation, toxic mode of action, and exposure regime. Environ Sci Technol 43:7556e7563. https://doi.org/10.1021/es901461c
Manganiello G, Sacco A, Ercolano MR, Vinale F, Lanzuise S, Pascale A, Napolitano M, Lombardi N, Lorito M, Woo SL (2018) Modulation of tomato response to *Rhizoctonia solani* by *Trichoderma harzianum* and its secondary metabolite harzianic acid. Front Microbiol 9:1966. https://doi.org/10.3389/fmicb.2018.01966
Marfori EC, Kajiyama S, Fukusaki E, Kobayashi A (2003) Phytotoxicity of the tetramic acid metabolite trichosetin. Phytochemistry 62:715–721. https://doi.org/10.1016/S0031-9422(02)00629-5
Marik T, Tyagi C, Balázs D, Urbán P, Szepesi Á, Bakacsy L, Endre G, Rakk D, Szekeres A, Andersson MA, Salonen H, Druzhinina IS, Vágvölgyi C, Kredics L (2019) Structural diversity and bioactivities of peptaibol compounds from the Longibrachiatum Clade of the filamentous fungal genus *Trichoderma*. Front Microbiol 10:1434. https://doi.org/10.3389/fmicb.2019.01434
Martinez C, Blanc F, Le Claire E, Besnard O, Nicole M, Baccou JC (2001) Salicylic acid and ethylene pathways are differentially activated in melon cotyledons by active or heat-denatured cellulase from *Trichoderma longibrachiatum*. Plant Physiol 127:334–344. https://doi.org/10.1104/pp.127.1.334
Martínez-Medina A, Roldán A, Albacete A, Pascual JA (2011) The interaction with arbuscular mycorrhizalfungi or *Trichoderma harzianum* alters the shoot hormonal profile in melon plants. Phytochemistry 72:223–229. https://doi.org/10.1016/j.phytochem.2010.11.008

Martínez-Medina A, Appels FVW, Van Wees SCM (2017a) Impact of salicylic acid and jasmonic acid-regulated defences on root colonization by *Trichoderma harzianum* T-78. Plant Signal Behav 12:e1345404. https://doi.org/10.1080/15592324.2017.1345404

Martínez-Medina A, Fernandez I, Lok GB, Pozo MJ, Pieterse CMJ, Van Wees SCM (2017b) Shifting from priming of salicylic acid-to jasmonic acid-regulated defences by *Trichoderma* protects tomato against the root knot nematode *Meloidogyne incognita*. New Phytol 213:1363–1377. https://doi.org/10.1111/nph.14251

Mastouri F, Björkman T, Harman GE (2010) Seed treatment with *Trichoderma harzianum* alleviates biotic, abiotic, and physiological stresses in germinating seeds and seedlings. Phytopathology 100:1213–1221. https://doi.org/10.1094/phyto-03-10-0091

Meincke R, Weinert N, Radl V, Schloter M, Smalla K, Berg G (2009) Development of a molecular approach to describe the composition of *Trichoderma* communities. J Microbiol Method 80:63–69. https://doi.org/10.1016/j.mimet.2009.11.001

Mendoza-Mendoza A, Zaid R, Lawry R, Hermosa R, Monte E, Horwitz BA, Mukherjee PK (2018) Molecular dialogues between *Trichoderma* and roots: role of the fungal secretome. Fungal Biol Rev 32:62–85. https://doi.org/10.1016/j.fbr.2017.12.001

Ming Q, Su C, Zheng C, Jia M, Zhang Q, Zhang H, Rahman K, Han T, Qin L (2013) Elicitors from the endophytic fungus *Trichoderma atroviride* promote *Salvia miltiorrhiza* hairy root growth and tanshinone biosynthesis. J Exp Bot 64:5687–5694. https://doi.org/10.1093/jxb/ert342

Miyadera H, Shiomi K, Ui H, Yamaguchi Y, Masuma R, Tomoda H, Miyoshi H, Osanai A, Kita K, Omura S (2003) Atpenins, potent and specic inhibitors of mitochondrial complex II (succinate-ubiquinone oxidoreductase). Proc Natl Acad Sci U S A 100:473–477. https://doi.org/10.1073/pnas.0237315100

Morán-Diez E, Hermosa R, Ambrosino P, Cardoza RE, Gutiérrez S, Lorito M, Monte E (2009) The ThPG1 endopolygalacturonase is required for the *Trichoderma harzianum*–plant beneficial interaction. Mol Plant-Microbe Interact 22:1021–1031. https://doi.org/10.1094/MPMI-22-8-1021

Mota TM, Oshiquiri LH, Viana Lopes EC, Barbosa Filho JR, Ulhoa CJ, Castro Georg R (2019) Hsp genes are differentially expressed during *Trichoderma asperellum* self-recognition, mycoparasitism and thermal stress. Microbiol Res 227:126296. https://doi.org/10.1016/j.micres.2019.126296

Mukherjee PK, Wiest A, Ruiz N, Keightley A, Moran-Diez ME, McCluskey K, François Pouchus Y, Kenerley CM (2011) Two classes of new peptaibols are synthesized by a single non-ribosomal peptide synthetase of *Trichoderma virens*. J Biol Chem 286:4544–4554. https://doi.org/10.1074/jbc.M110.159723

Mukherjee PK, Buensanteai N, Moran-Diez ME, Druzhinina IS, Kenerley CM (2012a) Functional analysis of non-ribosomal peptide synthetases (NRPSs) in *Trichoderma virens* reveals a polyketide synthase (PKS)/NRPS hybrid enzyme involved in the induced systemic resistance response in maize. Microbiology 158:155–165. https://doi.org/10.1099/mic.0.052159-0

Mukherjee PK, Horwitz BA, Kenerley CM (2012b) Secondary metabolism in *Trichoderma* – a genomic perspective. Microbiology 158:35–45

Mukherjee PK, Horwitz BA, Herrera-Estrella A, Schmoll M, Kenerley CM (2013) *Trichoderma* research in the genome era. Annu Rev Phytopathol 51:105–129. https://doi.org/10.1146/annurev-phyto-082712-102353

Mukherjee AK, umar AS, Kranthi S, Mukherjee PK (2014) Biocontrol potential of three novel *Trichoderma* strains: isolation, evaluation and formulation. 3 Biotech 4:275–281. https://doi.org/10.1007/s13205-013-0150-4

Muvea AM, Meyhöfer R, Subramanian S, Poehling HM, Ekesi S, Maniania NK (2014) Colonization of onion roots by endophytic fungi and their impacts on the biology of *Thrips tabaci*. PLoS One 9:e108242. https://doi.org/10.1371/journal.pone.0108242

Nobuhara M, Tazima H, Shudo K, Itai A, Okamoto T, Iitaka Y (1976) A fungal metabolite, novel isocyano epoxide. Chem Pharm Bull 24:832–834. https://doi.org/10.1248/cpb.24.832

Nogueira-Lopez G, Greenwood DR, Middleditch M, Winefield C, Eaton C, Steyaert JM, Mendoza-Mendoza A (2018) The apoplastic secretome of *Trichoderma virens* during interaction with maize roots shows an inhibition of plant defence and scavenging oxidative stress secreted proteins. Front Plant Sci 9:409. https://doi.org/10.3389/fpls.2018.00409

Nürnberger T, Kemmerling B (2009) Pathogen-associated molecular patterns (PAMP) and PAMP-triggered immunity. Annu Rev Plant Biol 34:16–47. https://doi.org/10.1002/9781119312994.apr0362

Nykiel-Szymańska J, Różalska S, Bernat P, Słaba M (2019) Assessment of oxidative stress and phospholipids alterations in chloroacetanilides-degrading *Trichoderma* spp. Ecotoxicol Environ Saf 184:109629. https://doi.org/10.1016/j.ecoenv.2019.109629

Nykiel-Szymańska J, Bernat P, Słaba M (2020) Biotransformation and detoxication of chloroacetanilide herbicides by *Trichoderma* spp. with plant growth-promoting activities. Pestic Biochem Physiol 163:216–226. https://doi.org/10.1016/j.pestbp.2019.11.018

Oerke EC (2005) Crop losses to pests. J Agric Sci 144:31e43. https://doi.org/10.1017/S0021859605005708

Pang G, Sun T, Yu Z, Yuan T, Liu W, Zhu H, Gao Q, Yang D, Kubicek CP, Zhang J, Shen Q (2020) Azaphilones biosynthesis complements the defence mechanism of *Trichoderma guizhouense* against oxidative stress. Environ Microbiol 22:4808–4824. https://doi.org/10.1111/1462-2920.15246

Piel J, Atzorn R, Gäbler R, Kühnemann F, Boland W (1997) Cellulysin from the plant parasitic fungus *Trichoderma viride* elicits volatile biosynthesis in higher plants via the octadecanoid signalling cascade. FEBS Lett 416:143–148. https://doi.org/10.1016/s0014-5793(97)01169-1

Pimentel D (1995) Amounts of pesticides reaching target pests: environmental impacts and ethics. J Agr Environ Ethics 8:17–29. https://doi.org/10.1007/BF02286399

Qian XY, Zhang Q, Wilkinson S, Achal V (2015) Cleaning of historic monuments: looking beyond the conventional approach? J Clean Prod 101:180–181. https://doi.org/10.1016/j.jclepro.2015.03.089

Rajani P, Rajasekaran C, Vasanthakumari MM, Olsson SB, Ravikanth G, Shaanker RU (2021) Inhibition of plant pathogenic fungi by endophytic *Trichoderma* spp. through mycoparasitism and volatile organic compounds. Microbiol Res 242:126595. https://doi.org/10.1016/j.micres.2020.126595

Ramírez-Valdespino CA, Porras-Troncoso MD, Corrales-Escobosa AR, Wrobel K, Martínez-Hernández P, Olmedo-Monfil V (2018) Functional characterization of TvCyt2, a member of the p450 monooxygenases from *Trichoderma virens* relevant during the association with plants and mycoparasitism. Mol Plant-Microbe Interact 31:289–298. https://doi.org/10.1094/MPMI-01-17-0015-R

Ramírez-Valdespino CA, Casas-Flores S, Olmedo-Monfil V (2019) *Trichoderma* as a model to study effector-like molecules. Front Microbiol 10:1030. https://doi.org/10.3389/fmicb.2019.01030

Recio R, Meléndez-Carmona MÁ, Martín-Higuera MC, Pérez V, López E, López-Medrano F, Pérez-Ayala A (2019) *Trichoderma longibrachiatum*-an unusual pathogen of fungal pericarditis. Clin Microbiol Infect. https://doi.org/10.1016/j.cmi.2019.02.006

Reino JL, Guerrero RF, Hernández-Galán R, Collado IG (2008) Secondary metabolites from species of the biocontrol agent *Trichoderma*. Phytochem Rev 7:89–123

Romero-Contreras YJ, Ramírez-Valdespino CA, Guzmán-Guzmán P, Macías-Segoviano JI, Villagómez-Castro JC, Olmedo-Monfil V (2019) Tal6 from *Trichoderma atroviride* is a LysM effector involved in mycoparasitism and plant association. Front Microbiol 10:2231. https://doi.org/10.3389/fmicb.2019.02231

Rosatto S, Roccotiello E, Di Piazza S, Cecchi G, Greco G, Zotti M, Vezzulli L, Mariotti M (2019) Rhizosphere response to nickel in a facultative hyperaccumulator. Chemosphere 232:243–253. https://doi.org/10.1016/j.chemosphere.2019.05.193

Salas-Marina MA, Silva-Flores MA, Uresti-Rivera EE, Castro-Longoria E, Herrera-Estrella A, Casas-Flores S (2011) Colonization of *Arabidopsis* roots by *Trichoderma atroviride* promotes growth and enhances systemic disease resistance through jasmonic acid/ethylene and salicylic acid pathways. Eur J Plant Pathol 131:15–26. https://doi.org/10.1007/s10658-011-9782-6

Samolski I, Rincón A, Pinzón LM, Viterbo A, Monte E (2012) The *qid74* gene from *Trichoderma harzianum* has a role in root architecture and plant biofertilization. Microbiology 158:129–138. https://doi.org/10.1099/mic.0.053140-0

Schuster A, Schmoll M (2010) Biology and biotechnology of *Trichoderma*. Appl Microbiol Biotechnol 87:787–799. https://doi.org/10.1007/s00253-010-2632-1

Segarra G, Casanova E, Bellido D, Odena MA, Oliveira E, Trillas I (2007) Proteome, salicylic acid and jasmonic acid changes in cucumber plants inoculated with *Trichoderma asperellum* strain T34. Proteomics 7:3943–3952. https://doi.org/10.1002/pmic.200700173

Serrano-Carreon L, Hathout Y, Bensoussan M, Belin JM (1993) Production of 6-pentyl-α-pyrone by *Trichoderma harzianum* from 18:n fatty acid methyl esters. Biotechnol Lett 14:1019–1024. https://doi.org/10.1007/BF01021051

Sharma V, Salwan R, Sharma PN (2017) The comparative mechanistic aspects of *Trichoderma* and probiotics: scope for future research. Physiol Mol Plant Pathol 100:84e96. https://doi.org/10.1016/j.pmpp.2017.07.005

Sharon A, Fuchs Y, Anderson JD (1993) The elicitation of ethylene biosynthesis by a *Trichoderma* xylanase is not related to the cell wall degradation activity of the enzyme. Plant Physiol 102:1325–1329. https://doi.org/10.1104/pp.102.4.1325

Shi T, Shao C-L, Liu Y, Zhao D-L, Cao F, Fu X-M, Yu J-Y, Wu J-S, Zhang Z-K, Wang C-Y (2020a) Terpenoids from the coral-derived fungus *Trichoderma harzianum* (XS-20090075) induced by chemical epigenetic manipulation. Front Microbiol 11:572. https://doi.org/10.3389/fmicb.2020.00572

Shi JC, Shi WL, Zhou YR, Chen XL, Zhang YZ, Zhang X, Zhang WX, Song XY (2020b) The putative methyltransferase *Tl*LAE1 is involved in the regulation of peptaibols production in the biocontrol fungus *Trichoderma longibrachiatum* SMF2. Front Microbiol 11:1267. https://doi.org/10.3389/fmicb.2020.01267

Shi XS, Li HL, Li XM, Wang DJ, Li X, Meng LH, Zhou XW, Wang BG (2020c) Highly oxygenated polyketides produced by *Trichoderma koningiopsis* QA-3, an endophytic fungus obtained from the fresh roots of the medicinal plant *Artemisia argyi*. Bioorg Chem 94:103448. https://doi.org/10.1016/j.bioorg.2019.103448

Shoresh M, Harman GE (2008) The molecular basis of shoot responses of maize seedlings to *Trichoderma harzianum* T22 inoculation of the root: a proteomic approach. Plant Physiol 147:2147–2163. https://doi.org/10.1104/pp.108.123810

Shoresh M, Yedidia I, Chet I (2005) Involvement of jasmonic acid/ethylene signaling pathway in the systemic resistance induced in cucumber by *Trichoderma asperellum* T203. Phytopathology 95:76–84. https://doi.org/10.1094/phyto-95-0076

Shoresh M, Gal-On A, Leibman D, Chet I (2006) Characterization of a mitogen-activated protein kinase gene from cucumber required for *Trichoderma*-conferred plant resistance. Plant Physiol 142:1169–1179. https://doi.org/10.1104/pp.106.082107

Siddiquee S (2014) Recent advancements on the role and analysis of volatile compounds (VOCs) from *Trichoderma*. In: Gupta VK, Schmoll M, Herrera-Estrella A (eds) Biotechnology and biology of *Trichoderma*. Elsevier B.V, Amsterdam

Singh DP, Singh V, Shukla R, Sahu P, Prabha GA, Sarma BK, Gupta VK (2020) Stage-dependent concomitant microbial fortification improves soil nutrient status, plant growth, antioxidative defense system and gene expression in rice. Microbiol Res 239:126538. https://doi.org/10.1016/j.micres.2020.126538

Sivasithamparam K, Ghisalberti EL (1998) Secondary metabolism in *Trichoderma* and *Gliocladium*. In: Kubicek CP, Harman GE (eds) Trichoderma and Gliocladium. Taylor & Francis, London, pp 139–191

Song YP, Liu XH, Shi ZZ, Miao FP, Fang ST, Ji NY (2018) Bisabolane, cyclonerane, and harziane derivatives from the marine-alga-endophytic fungus *Trichoderma asperellum* cf44-2. Phytochemistry 152:45–52. https://doi.org/10.1016/j.phytochem.2018.04.017

Tamura A, Kotani H, Naruto S (1975) Trichoviridin and dermadin from *Trichoderma* sp. TK- 1. J Antibiot 28:161–162. https://doi.org/10.7164/antibiotics.28.161

Todorovic B, Glick BR (2008) The interconversion of ACC deaminase and D-cysteine desulfhydrase by directed mutagenesis. Planta 229:193–205. https://doi.org/10.1007/s00425-008-0820-3

Tripathi P, Singh PC, Mishra A, Chauhan PS, Dwivedi S, Bais RT, Tripathi RD (2013) *Trichoderma*: a potential bioremediator for environmental clean up. Clean Techn Environ Policy 15:541–550. https://doi.org/10.1007/s10098-012-0553-7

Van der Heijden MGA, Bardgett RD, van Straalen NM (2008) The unseen majority: soil microbes as drivers of plant diversity and productivity in terrestrial ecosystems. Ecol Lett 11:296–310. https://doi.org/10.1111/j.1461-0248.2007.01139.x

Vandenkoornhuyse P, Baldauf SL, Leyval C, Straczek J, Young JPW (2002) Extensive fungal diversity in plant roots. Science 295:2051. https://doi.org/10.1126/science.295.5562.2051

Vargas WA, Djonović S, Sukno SA, Kenerley CM (2008) Dimerization controls the activity of fungal elicitors that trigger systemic resistance in plants. J Biol Chem 283:19804–19815. https://doi.org/10.1074/jbc.M802724200

Vargas WA, Mandawe JC, Kenerley CM (2009) Plant-derived sucrose is a key element in the symbiotic association between *Trichoderma virens* and maize plants. Plant Physiol 151:792–808. https://doi.org/10.1104/pp.109.141291

Velázquez-Robledo R, Contreras-Cornejo HA, Macías-Rodríguez L, Hernández-Morales A, Aguirre J, Casas-Flores S, López-Bucio J, Herrera-Estrella A (2011) Role of the 4-phosphopantetheinyl transferase of *Trichoderma virens* in secondary metabolism, and induction of plant defense responses. Mol Plant-Microbe Interact 24:1459–1471. https://doi.org/10.1094/MPMI-02-11-0045

Villalobos-Escobedo JM, Esparza-Reynoso S, Pelagio-Flores R, López-Ramírez F, Ruíz-Herrera FL, López-Bucio J, Herrera-Estrella A (2020) The fungal NADPH oxidase is an essential element for the molecular dialog between Trichoderma and Arabidopsis. Plant J 103:2178–2192. https://doi.org/10.1111/tpj.14891

Vinale F, Marra R, Scala F, Ghisalberti EL, Lorito M, Sivasithamparam K (2006) Major secondary metabolites produced by two commercial *Trichoderma* strains active against different phytopathogens. Lett Appl Microbiol 43:143–148. https://doi.org/10.1111/j.1472-765X.2006.01939.x

Vinale F, Sivasithamparam K, Ghisalberti EL, Marra R, Woo SL, Lorito M (2008) *Trichoderma*-plant-pathogen interactions. Soil Biol Biochem 40:1–10. https://doi.org/10.1016/j.soilbio.2007.07.002

Vinale F, Arjona Girona I, Nigro M, Mazzei P, Piccolo A, Ruocco M, Woo S, Ruano Rosa R, López Herrera C, Lorito M (2011) Cerinolactone, a hydroxyl-lactone derivative from *Trichoderma cerinum*. J Nat Prod 75:103–106. https://doi.org/10.1021/np200577t

Vinale F, Sivasithamparam K, Ghisalberti EL, Woo SL, Nigro M, Marra R, Lombardi N, Pascale A, Ruocco M, Lanzuise S, Manganiello G, Lorito M (2014) *Trichoderma* secondary metabolites active on plants and fungal pathogens. Open Mycol J 8:127–139. https://doi.org/10.2174/1874437001408010127

Viterbo A, Chet I (2006) *TasHyd1*, a new hydrophobin gene from the biocontrol agent *Trichoderma asperellum*, is involved in plant root colonization. Mol Plant Pathol 7:249–258. https://doi.org/10.1111/j.1364-3703.2006.00335.x

Viterbo A, Wiest A, Brotman Y, Chet I, Kenerley C (2007) The 18mer peptaibols from *Trichoderma virens* elicit plant defence responses. Mol Plant Pathol 8:737–746. https://doi.org/10.1111/j.1364-3703.2007.00430.x

Viterbo A, Landau U, Kim S, Chernin L, Chet I (2010) Characterization of ACC deaminase from the biocontrol and plant growth-promoting agent *Trichoderma asperellum* T203. FEMS Microbiol Lett 305:42–48. https://doi.org/10.1111/j.1574-6968.2010.01910.x

Wen C, Xiong H, Wen J, Wen X, Wang C (2020) *Trichoderma* species attract *Coptotermes formosanus* and antagonize termite pathogen *Metarhizium anisopliae*. Front Microbiol 11:653. https://doi.org/10.3389/fmicb.2020.00653

Wiest A, Crzegorski D, Xu BW, Goulard C, Rebuffat S, Ebbole DJ, Bodo B, Kenerley C (2002) Identification of peptaibols from *Trichoderma virens* and cloning of a peptaibol synthetase. J Biol Chem 277:20862–20868. https://doi.org/10.1074/jbc.m201654200

Wu J, Ming Q, Zhai WS, Zhu B, Zhang Q, Xu Y, Shi S, Wang S, Zhang Q, Han T, Qin L (2019) Structure of a polysaccharide from *Trichoderma atroviride* and its promotion on tanshinones production in *Salvia miltiorrhiza* hairy roots. Carbohydr Polym 223:115125. https://doi.org/10.1016/j.carbpol.2019.115125

Xiong H, Xue K, Qin W, Chen X, Wang H, Shi X, Ma T, Sun Z, Chen W, Tian X, Lin W, Wen X, Wang C (2018) Does soil treated with conidial formulations of *Trichoderma* spp. attract or repel subterranean termites? J Econ Entomol 111:808–816. https://doi.org/10.1093/jee/toy021

Yamano T, Hemmi S, Yamamoto I, Tsubaki K (1970) Trichoviridin, a new antibiotic. Jpn. Tokkyo Koho, JP Patent 45015435, 29 May 1970

Yedidia I, Srivastva AK, Kapulnik Y, Chet I (2001) Effect of *Trichoderma harzianum* on microelement concentrations and increased growth of cucumber plants. Plant Soil 235:235–242

Yedidia I, Shoresh M, Kerem Z, Benhamou N, Kapulnik Y, Chet I (2003) Concomitant induction of systemic resistance to *Pseudomonas syringae* pv. lachrymans in cucumber by *Trichoderma asperellum* (T-203) and accumulation of phytoalexins. Appl Environ Microbiol 69:7343–7353. https://doi.org/10.1128/aem.69.12.7343-7353.2003

Yu Z, Wang Z, Zhang Y, Wang Y, Liu Z (2021) Biocontrol and growth-promoting effect of *Trichoderma asperellum* TaspHu1 isolate from *Juglans mandshurica* rhizosphere soil. Microbiol Res 242:126596. https://doi.org/10.1016/j.micres.2020.126596

Zachow C, Berg C, Müller H, Monk J, Berg G (2016) Endemic plants harbour specific *Trichoderma* communities with an exceptional potential for biocontrol of phytopathogens. J Biotechnol 235:162–170. https://doi.org/10.1016/j.jbiotec.2016.03.049

Zeilinger S, Gruber S, Bansal R, Mukherjee PK (2016) Secondary metabolism in *Trichoderma*-chemistry meets genomics. Fungal Biol Rev 30:74–90. https://doi.org/10.1016/j.fbr.2016.05.001

Zhang F, Wang Y, Liu C, Chen F, Ge H, Tian F, Yang T, Ma K, Zhang Y (2019) *Trichoderma harzianum* mitigates salt stress in cucumber via multiple responses. Ecotoxicol Environ Saf 170:436–445. https://doi.org/10.1016/j.ecoenv.2018.11.084

Elicitor Proteins from *Trichoderma* for Biocontrol Products

Kandasamy Saravanakumar, Anbazhagan Sathiyaseelan, Arokia Vijaya Anand Mariadoss, and Myeong-Hyeon Wang

Contents

1 Introduction

Plants are the primary producers of food for human consumption, but their productivity is mainly affected by plant disease caused by an array of pathogens (bacteria, fungi, insects, vertebrate herbivores, and viruses). The available chemical fungicides in the market are dangerous to the human food chain and environments. Therefore, fungi from the genus *Trichoderma* (Ascomycota, Hypocreales) are considered as an alternative to chemical products for protection of plants from the various abiotic and biotic stress and improve the annual yield. *Trichoderma* spp. are a multitalented, which inhabited in the various ecological niches, and utilized in the various applications including agriculture, food, and medicine (Lorito et al. 2010).

K. Saravanakumar · A. Sathiyaseelan · A. V. A. Mariadoss · M.-H. Wang (✉)
Department of Bio-Health Convergences, Kangwon National University, Chuncheon, Gangwon Do, South Korea
e-mail: mhwang@kangwon.ac.kr

N. Amaresan et al. (eds.), *Advances in* Trichoderma *Biology for Agricultural Applications*, Fungal Biology, https://doi.org/10.1007/978-3-030-91650-3_7

Trichoderma-based biocontrol agents received a remarkable impact in agriculture plant diseases management for increasing the crop productivity. *Trichoderma* spp.-mediated activities occurred through mycoparasitism, antibiosis, competition, and host-microbe interactions. *Trichoderma* spp. directly provide the beneficial effects to the plants by enhancing the growth and inducing the defense system against various abiotic and biotic stress (Hermosa et al. 2012; Shoresh et al. 2010). Moreover, the *Trichoderma* strains also directly protect the plants through mycoparasitism of plant pathogens (Benítez et al. 2004). Among the various mechanisms, plants immunity is triggered by *Trichoderma*-host plants and pathogen interactions via microbial elicitors and/or plant receptors (Harman et al. 2012; Contreras-Cornejo et al. 2011). These interactions activate the plant immunity through altering the biochemical and physiological properties of plants toward various biotic and abiotic stressors (Contreras-Cornejo et al. 2011; Baker et al. 2012; Mathys et al. 2012; López-Mondéjar et al. 2011). Although the *Trichoderma* spp. colonization of the plant could increase the signalizing pathway related to the plant immunity, the colonization is restricted by salicylic acid (SA) signaling transduction (Alonso-Ramírez et al. 2014). Therefore, *Trichoderma* colonization in plant root cortex induces plant immunity. For the instance, the researchers have developed an SA knockout mutant (*sid2*) *Arabidopsis thaliana* to scientifically evidence how SA involved in the *T. harzianum* colonization in plant root. The result reveals that the colonization of *Trichoderma* in the *sid2* mutant was higher than the wild-type *Arabidopsis thaliana*. Although there are known or unknown plants molecules are restricts the colonization of the *Trichoderma* spp. in the plant vascular tissue, the support of SA is important for the colonization of the fungi (*Trichoderma* spp.) (Alonso-Ramírez et al. 2014).

The association between the beneficial microbes and plants led the more effective improvement in plant growth, nutrient uptake, and tolerance toward abiotic and biotic stressors (Schirawski and Perlin 2018). These interactions are established through manipulation of the SA/jasmonic acid (JA) pathways (Berens et al. 2017). Also, these plant/microbe interactions were modulated through the plant hormones such as ethylene (Et), auxins, cytokinins (CKs), gibberellins (GAs), and abscisic acid (ABA)(Zamioudis and Pieterse 2012). It is noteworthy that the plant hormones are not only involved in establishing the plant beneficial interactions but also trigger the plant growth, developments, and reproduction and deal with abiotic and biotic stress (Guzmán-Guzmán et al. 2019). Among the various beneficial microbial groups, the *Trichoderma* spp. play an important role in the health of ecosystem. *Trichoderma* is a native microflora inhabited in agriculture soils, colonize in the plant roots and aerial parts, and even grow as an endophyte. *Trichoderma* spp. are one of the major bio-pesticides because ~60% of registered commercial bio-pesticides contain any one of *Trichoderma* spp. (Mukherjee et al. 2013). Overall, it is evidenced that the *Trichoderma*/plant interactions mutually beneficial for plant growth and immunity developments. Perhaps the *Trichoderma*-plant-pathogens interactions occurred through elicitor from *Trichoderma* and receptor plants. According to current research, result confirms that *Trichoderma* spp. are known to produce the volatile compounds, secondary metabolites, small RNA, and proteins which acted as effector/elicitor to moderate the interaction and promote the plant growth and immunity (Ramírez-Valdespino et al. 2019). However, there is no work summarizing the elicitor like proteins from the *Trichoderma* response to plant-microbe interactions. Therefore, this book chapter aimed to summarize the earlier presented research on *Trichoderma* elicitors proteins and molecular mechanism for plants beneficial effects.

2 *Trichoderma*-Plant Interactions

Trichoderma spp. are spore forming fast-growing fungi, a cell factory for production of enzymes (cellulases, glucanases, chitinases, and cell wall degrading enzymes (CWDEs)), secondary metabolites including antibiotics. Moreover, it impacts in agriculture and environments as biodegrader for hydrocarbons, polysaccharides, xenobiotic pesticides, and chlorophenolic compounds used in agriculture (Harman and Kubicek 1998; Harman et al. 2004). *Trichoderma* spp. are opportunistic avirulent symbionts which colonize in plants' root providing the beneficial effects. *Trichoderma* inoculation moderates the plant growth, through modifying the soil properties including microbiome, pH, and micro and macro nutrient (Tandon et al. 2020). For example, *T. harzianum* (CCTCC-RW0024) inoculation prevents the *Fusarium* stalk rot disease in maize through establishing root colonization and augmenting the plant growth-promoting bacteria (Saravanakumar et al. 2017). Similarly, another study evidenced that the soil or seed application of the *Trichoderma* spp. has trigger the mangroves' growth through enhancing the soil phosphate solubilization (Saravanakumar et al. 2013, 2018a). Also, *Trichoderma* has reportedly promote the biomass of the wheat (*Triticum aestivum* L.) under the salt stress through increasing the water uptake and photosynthesis efficacy (Oljira et al. 2020). Similarly, several studies have revealed that the inoculation of the *Trichoderma* strains acted bio-stimulants and bio-fertilizer to promote the plants under growth and immunity under the stress (Sun et al. 2021; Mukherjee et al. 2012; Elkelish et al. 2020; Sousa et al. 2020).

Trichoderma improves the root development and branching which accelerate the nutrient uptake by the plants (Mukherjee et al. 2013). The communication between the plants and *Trichoderma* is achieved via chemical signaling transfer activated by their various factors/elicitors, which alter the proteomic, transcriptomic, and metabolomics of plants (Contreras-Cornejo et al. 2009; Bonfante and Genre 2010). *Trichoderma* spp. owned as potential agent through production of effector/elicitor molecules such as volatile compounds, secondary metabolites, small RNA, and proteins promote the plant growth and immunity (Ramírez-Valdespino et al. 2019). *Trichoderma* promotes the plants growth and immunity through accelerating the signaling transfer through SA/JA pathways. However, the elicitor protein from the *Trichoderma* spp. is a key factor for the plant/fungi or organism's interactions medicated plant beneficial effect. According to the recently research, a total of 100 potential effector/elicitor proteins were documented from *Trichoderma* genome

(Mendoza-Mendoza et al. 2018; Nogueira-Lopez et al. 2018; Guzmán-Guzmán et al. 2017). In the early interaction, the JA/SA-related genes are downregulated in *A. thaliana* for favoring root colonization of *Trichoderma*. After the complete colonization, the *Trichoderma* activates plant defense-related gene expressions both systematically and locally. This indicated that *Trichoderma* root colonization is not threat to the host plants (Morán-Diez et al. 2012). The interaction of *Trichoderma* spp. with host plants provides several beneficial effects such as stress tolerance, root modification, soil quality improvement, solubilizing nutrients and increasing the nutrient uptake, inhibit the pathogenic organism colonization by mycoparasitism or antibiosis (Fig. 1). In the initial step, *Trichoderma* colonize in the root and penetrate through anchoring proteins such as cysteine-rich proteins (hydrophobin) (Viterbo and Chet 2006). *Trichoderma* enables the hormonal balancing in plant, which results in plant resistance and defense against pathogens, increasing the nutrient absorption (blue circles) (Guzmán-Guzmán et al. 2019).

3 *Trichoderma* Elicitor Proteins

The term of "*Trichoderma* elicitor" was proposed that the molecules might involve in triggering plant immunity by altering the physiology and causing structural changes. The plant-microbe's interactions can cause the negative or positive effect to the plants. The molecule-derived pathogenic organism can establish the plant-pathogenic organism interactions which cause the plant infection or trigger the plant defense in a nonspecific fashion (Alba et al. 2011). Overall, these molecule-associated plant-pathogenic organism mechanisms are defined as pathogen-associated molecular patterns (PAMPs), and in case of the nonpathogenic microbes, it is called as microorganism-associated molecule patterns (MAMPs) (Jones and Dangl 2006). Recently, these molecular patterns are referred as PAMPs/MAMPs due to the wide distribution of the elicitors and their properties. Moreover, in some cases, the difference between the PAMPs and MAMPs is not well described because it is completely related to the microbial genera or whether it is pathogenic organism or nonpathogenic beneficial microbes. However, the classification of the PAMPs and MAMPs involved in the plant-microbe interactions remains open. However, the *Trichoderma* spp. are known to trigger the plant immunity through the MAMPs as it nonpathogenic microbes. The microbes triggered plant immunity classified as induced systemic resistance (ISR), which mediated through ethylene (ET) and jasmonic acid (JA)(Loon et al. 2006). Also, the systemic acquired resistance (SAR) mediated by salicylic acid, results in the expression of pathogenesis-related genes (Bari and Jones 2009). It is reported that the small molecular weight proteins from the *Trichoderma* spp. is known to induce the plant immunity as elicitor (Yu et al. 2018). The *Trichoderma* spp., secreting small molecular weight proteins, are containing the four or more cysteine residues named as *Trichoderma* generated small secreted cysteine-rich proteins (SSCPs) and divided into four groups: (i) elicitor-like proteins, (ii) hydrophobins and hydrophobin-like proteins, (iii) SSCP with no

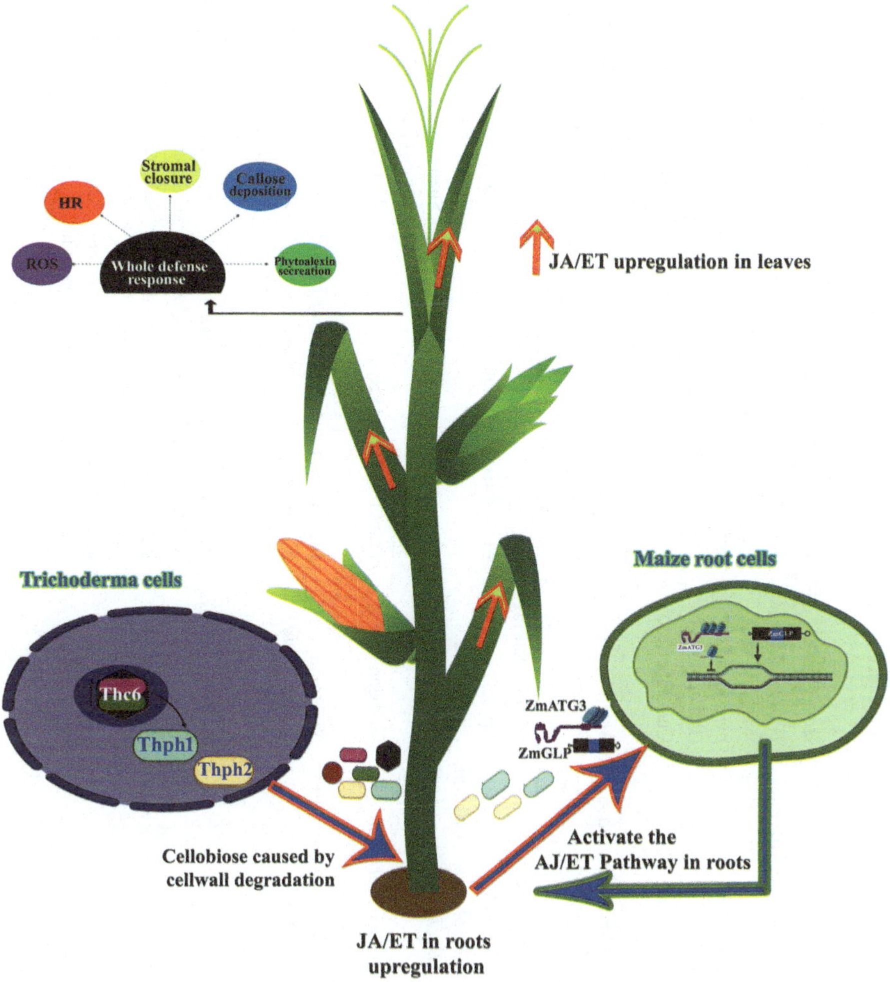

Fig. 1 The cellulase (THPH1 and THPH2) from the *T. harzianum* interacted with the autophagocytosis associated protein (ZmATG3) and germin-like protein (ZmGLP) respectively which trigger the plant immunity against pathogenic organism through upregulation of genes related to JA and ET signaling pathway. (Adapted from [39] Attribution 4.0 International (CC BY 4.0))

attribution to functional category, and (iv) MAP kinase repressed secreted protein 1 (MRSP1) (Kubicek et al. 2011; Druzhinina et al. 2012). The eliciting plant response protein (Epl1) also known as Sm1 from the *T. virens* Gv 29-8 induced the plant immunity locally and systematically in the cotton seedlings (Djonović et al. 2006). Similarly, the EPL1-Tas (Eliciting plant response protein) identified from the *T. asperellum* elicit the plant defense response against *Alternaria alternata (*Yu et al. 2018*)*. The list of elicitor proteins identified from the *Trichoderma* spp., which trigger the plant immune response, is summarized in Table 1. Some of research reports

Table 1 Elicitor proteins from *Trichoderma* sp., responsible for activation of plant immunity and their receptor/signaling pathway

Species	Elicitor	Targeted pathogenic organism	Host plant	Receptors/signaling pathway	Reference
T. asperellum	EPL1-Tas (Eliciting plant response protein)	*Alternaria alternata*	*Populus davidiana* × *P. alba* var. *pyramidalis* (PdPap)	Induction of genes related to SA, JA, and auxin signal transduction pathway	Yu et al. (2018)
T. asperellum	Protease (aspartyl protease)	*R. solani*	Cucumber roots	–	Viterbo et al. (2004)
T. virens	Transferase (4-phosphopantetheinyl transferase)	*Botrytis cinerea*	*Arabidopsis thaliana*	Salicylic acid (SA), jasmonic acid pathway	Velázquez-Robledo et al. (2011)
T. virens	Serine protease (TVSP1)	*R. solani*	Cotton seedlings	–	Pozo et al. (2004)
T. atroviride, *T. virens*, *T. harzianum*	Chitinase (endochitinase)	*Venturia inaequalis*, *R. solani*, *A. radicina*, *B. cinerea*	Apple, carrot, cotton, rice	Plant resistance-related pathway	Faize et al. (2003), Kumar et al. (2009), Shah et al. (2008), Baranski et al. (2008)
T. asperellum T4	EPLt4 proteinaceous elicitor	*Cercosporidium sofinum*	Soybean	Plant resistance (PR)-protein genes	Wang et al. (2013)
T. formosa	EPL1 (Eliciting plant response protein)	Tomato mosaic virus (ToMV)	*Nicotiana benthamiana*	Induction of genes related to JA and ET signaling, SA signaling, leucine-rich repeats, transcription factors, and histone variants	Cheng et al. (2018)
T. longibrachiatum	HYTLO1	–	*Lotus japonicus*	Activation of defense-related genes (MPK3, WRK33, and CP450)	Moscatiello et al. (2018)
T. harzianum	THPH1 and THPH2 (Cellulase)	*Cochliobolus lunatus*	*Zea mays*	Upregulation of genes related to JA and ET signaling pathway, autophagocytosis-associated protein (ZmATG3), and germin-like protein (ZmGLP)	Saravanakumar et al. (2016, 2018b)

Species	Elicitor	Targeted pathogenic organism	Host plant	Receptors/signaling pathway	Reference
T. virins and *T. atroviride*	SM1 and EPL1	*Alternaria solani*, *Botrytis cinerea*, and *Pseudomonas syringae*	*Solanum lycopersicum*	Induce the expression of peroxidase and an a α-dioxygenase encoding genes which increase in disease resistance	Salas-Marina et al. (2015)
T. harzianum	EPL1	*Sclerotinia sclerotiorum*	*Phaseolus vulgaris*	Increase plant resistance and mycoparasitism	Gomes and Silva (2015)
T. asperellum	HFB2–6 (hydrophobin gene)	*Alternaria alternata*	Poplar	Root colonization, upregulated JA, and SA signal transduction pathway	Huang et al. (2015)
T. longibrachiatum	HYTLO1	*Botrytis cinerea*	*Solanum lycopersicum*	Increase plant resistance against pathogens and activated defense related response	Ruocco et al. (2015)
T. virens	SM2 (Small protein 2)	*Colletotrichum graminicola*	*Zea mays*	Root colonization in maize.	Crutcher et al. (2015)
T. harzianum	Endopolygalacturonase (ThPG1)	*Botrytis cinerea*	*Solanum lycopersicum*	Activation of root colonization and plant defense system	Morán-Diez et al. (2009)
T. asperellum	Swollenin (Expansin-like protein)	*Botrytis cinerea* and *Pseudomonas syringae* pv. *lachrymans*	*Cucumis sativus*	Stimulating local defense response against plant pathogenic organism	Brotman et al. (2008)
T. asperellum	TasHYD1 (hydrophobin-like clone)	*Rhizoctonia solani*	*Cucumis sativus*	Retained root colonization and against plant pathogen *R. solani*	Viterbo and Chet (2006)
T. virens	SM1 (Small protein 1)	*Colletotrichum graminicola*	*Gossypium hirsutum* (dicot); *Zea mays* (monocot)	Triggers the production of ROS thereby induction of JA and green leaf volatile-biosynthetic genes	Djonović et al. (2007)

(continued)

Table 1 (continued)

Species	Elicitor	Targeted pathogenic organism	Host plant	Receptors/signaling pathway	Reference
T. viride	EIX (ethylene-inducing xylanase)	–	*Nicotiana* and *Solanum lycopersicum*	Elicits hypersensitive responses and other plant defense responses	Roblat et al. (2002)
T. longibrachiatum	Cellulase (active and heat-denatured)	*Sphaerotheca fuliginea*	*Cucumis melo*	Activation of plant defense mechanisms to the reduction of powdery mildew disease	Martinez et al. (2001)

evidenced the elicitor proteins and their receptor, but most of the work deals with only the elicitor. For example, a work reported the elicitor and receptor from the maize (Fig. 1). This work reports that the cellulases like Thph1 and Thph2 from *T. harzianum* regulate the plant immune system through upregulating JA- and ET-related gene expression (Wang et al. 2013), followed by the same group identifying the receptor responsible for the interaction with elicitor Thph1 and Thph2 as autophagocytosis associated protein (ZmATG3) and germin-like protein (ZmGLP), respectively, by Y2H assay (Saravanakumar et al. 2018b).

4 *Trichoderma* Proteins Altering the Physiology and Causing Structural Changes in Plants

The root colonization of the *Trichoderma* in plants cause some structural changes such as callose deposition, tylose formation, and cell wall thickening in cork layers or xylem vessels (Bolton 2009). The plants known to show the ROS production increased regulation of resistance-related proteins, plant molecules, and secondary metabolites (phytoalexins) due to the plant-microbe interactions. Moreover, these expressing are considered as marker or indicator for the activation of the plant defense reaction toward plant disease management, which is evidenced through the various experimental models (Harman et al. 2012; Salas-Marina et al. 2011), in which one of the work demonstrated that *Trichoderma* colonization in the *Arabidopsis* roots induced the systemic disease resistance through activation of the molecules corresponding to the jasmonic acid/ethylene and salicylic acid pathways (Salas-Marina et al. 2011). Moreover, the appropriate recognition between the microbe-derived elicitors and plant receptors can activate the ion channels and plasma membrane potential, which enable the transient influx of $Ca2^+$ and H^+ and exflux of K^+ and Cl^- (Luo et al. 2010). *Trichoderma* colonization also alters the phytohormone secretion, which has the important role in the plant immunity and metabolism. Among the ROS, the hydrogen peroxide received the more attention because that triggers the cell wall expansion and lignification (Bolton 2009; Ahmad et al. 2008). For example, the *T. harzianum* colonization triggers the H_2O_2 in the cucumber and tomato plants (Nawrocka et al. 2012). Moreover, it is noteworthy that the ROS is accompanied due to the process of cell wall damage its occurred through suppression of OH' (hydroxyl radicals) by enzymes such as peroxidase or catalases (Nawrocka et al. 2012). The earlier reports revealed that the *Trichoderma* interactions with plants activate or regulate the several secondary metabolites involved in the various plant secondary metabolic signaling pathways such as ethylene (ET), jasmonic acid (JA), and salicylic acid (SA)(Harman et al. 2012; Contreras-Cornejo et al. 2011). These studies indicated that *Trichoderma* induce the plant defense response through activation of various metabolic signals and gene/protein expression in the plants against various abiotic and biotic stress.

5 *Trichoderma* Elicitor Proteins Trigger the Plant Immunity

Although the molecular studies are report the *Trichoderma* colonization is beneficially activates the plant defense against various stresses. The specific nature of the resistance activation in the plants is unclear. *Trichoderma* inoculation induces the plant hormone-dependent metabolic molecules characteristic in systemic acquired resistance (SAR) and induced systemic resistance (ISR) (Hermosa et al. 2012). *Trichoderma* induced plant immunity classified into ISR and SAR, which is not different with the phenotypic of plants (Contreras-Cornejo et al. 2011). The SAR accompanied through exposure of the host plants with hemibiotrophic and biotrophic as well as nonpathogenic microbes (Salas-Marina et al. 2011). This reaction would have occurred through endogenous accumulation of SA (salicylic acid) by hypersensitive reaction, phytotoxicity, or necrotic in hormonal and camalexin-dependent mechanisms in the *Arabidopsis thaliana* (Contreras-Cornejo et al. 2011). The ISR is triggered by JA and ET pathway though interacting of the nonpathogenic microbes, root associated rhizosphere microbes such as plant growth-promoting rhizobacteria (PGPR) and insects (Salas-Marina et al. 2011). The reports claimed that the *Trichoderma* also elicit the plant immunity through ISR mechanism by activation of ET o JA associated signal molecules similar to PGPR (Hermosa et al. 2012). The possible mechanism of the *Trichoderma* elicitor regulated plant immunity is presented in Fig. 2. *Trichoderma* activate the plant receptor protein kinase (TIPK) cascade through *Trichoderma* exposure. Followed by trigger the defeance genes such as MPKKK/MEKK, MPKKs which trigger the phosphorylate MPK. The induction of MKP (mitogenic kinase pathway) trigger the plant defense-related genes (Schuster and Schmoll 2010). *Trichoderma* interaction activates plant defense mechanism comprising the secretion of the plant molecules against the pathogenic organisms, which includes the cationic peptides widely reported from the all type of plants (Cui et al. 2018). These kinds of antimicrobial molecules synthesized from the plants are proved to be an effective inhibitor for plant pathogens (Seo et al. 2014; Nawrocka and Małolepsza 2013). For instance, the *Trichoderma*-induced plant resistance contains the compounds of SAR or ISR expected at molecular level. Further the expression of the transcription factor from the same group of the family may affect through the response of the ET, JA, and SA signaling pathways (Contreras-Cornejo et al. 2011). The accumulation of the compounds such as enzymes, terpenoids, phenol, and PR proteins changed in relation to the encoding gene expression of JA/ET when exposed to the *Trichoderma* (Mathys et al. 2012). Also *Trichoderma* induce the plant defense system through modulation of the other molecules such as abscisic acid, sucrose, gibberellins, cytokinins, auxins, and peptide hormones by JA, ET, and SA signaling pathway (Vargas et al. 2009). These reports indicated that the plant defense system is not activated only through a specific molecule or pathway; it occurred through synergic changes in the plant cellular, structural, and biochemical parameters (Contreras-Cornejo et al. 2011).

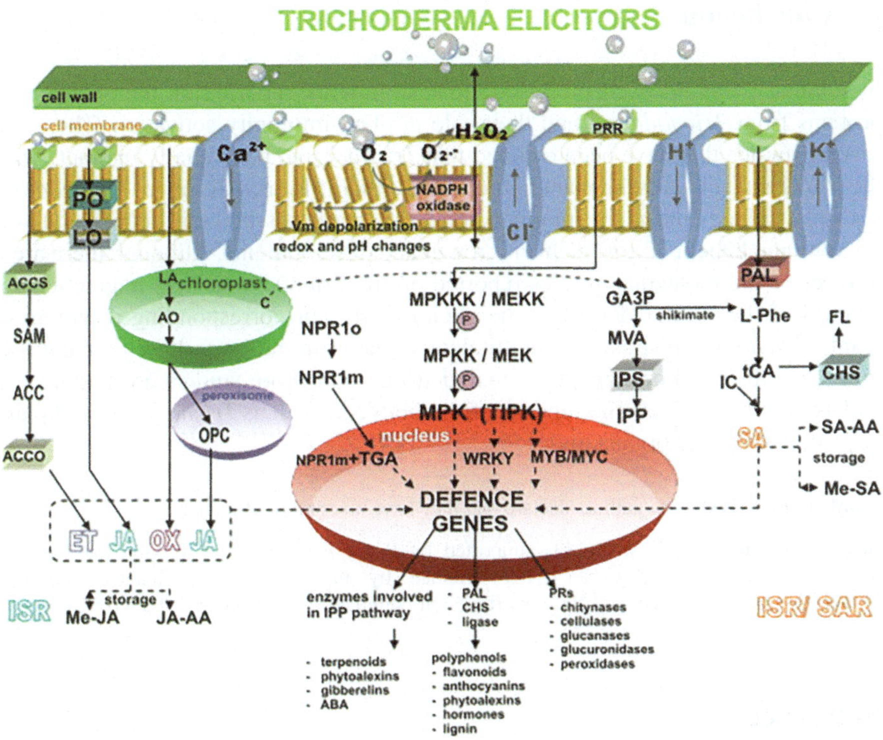

Fig. 2 Possible reactions at the biochemical and molecular level in plant cells related to defense response and development of resistance induced by *Trichoderma*. Membrane depolarization, pH changes, ROS accumulation and signaling pathways are activated in response to *Trichoderma* elicitors-plant receptors interaction and lead to defense gene expression. In plants, enzyme activity increase, as well as accumulation of phenol compounds and phytoalexins are more often observed. MAMPs-microbe associated molecular patterns, MPKKK, MEKK, MPKK,MEK, MPK-mitogen-activated protein kinases of MPK cascade, TIPK-*Trichoderma*-induced protein kinase PRR-pattern recognition receptor, WRKY-transcription co-activators, SA-salicylic acid, Me-SA-salicilic acid metyl, SA-AA-link form of salicilic acid, ET-ethylene, JA-jasmonate acid, Me-JA-jasmonate acid methyl, JA-AA-link form of jasmonate acid, ABA-abscisic acid, PRs-pathogenic organism related proteins, PAL-phenylalanine ammonia lyase, CHS-chalcone synthase, l-Phe-l-phenylalanine, t-CA-trans-cinnamic acid, IC-isochorismate, NPRo/m-NPR protein oligomer/monomer, MYB/MYC-co-activators of *Trichodrema* induced resistance, OX oxylipins, TGA-transcription factor, LA-linolenic acid, AO-allene oxide, OPDA-12-oksofitodien acid, OPC- 3-oxo-2-(2′-pentenyl)-cyclopentane-1-octanoic, PO- phospholipase, LO – lipoxygenase, C – carbohydrates, CHS – chalconesynthase,FL–flavonoidsandderivatives,GA3P–glyceraldehyde-3-phosphate,MVA–mevalonic acid, IPP – isoprenoids, IPS – isoprenoid synthase, ACC – 1-aminocyclopropane- 1-carboxylic amid, SAM – S-adenosyl methionine, ACCS – synthase, ACCO – oxidase. (Reprinted from [73] Biological control, 67 (2), 2013, 149–156. Diversity in plant systemic resistance induced by *Trichoderma*, Biological Control with permission from Elsevier. Copyright (2013), Elsevier. License number: 5043941510594)

6 Conclusion

In summary, this book chapter consolidates the work reported on the elicitor-like proteins from *Trichoderma* and their role in plant immunity activation. There are several studies that have been claimed that *Trichoderma* beneficially interacts with the plants and activates the plant immunity in either ISR or SAR mode. *Trichoderma* not only induces the plant immunity that also increases the soil quality and soil microbiome beneficial for the plant growth and developments. Although there are a number of the molecular studies reported on the *Trichoderma*-host interactions, a handful of the studies described the elicitor and their corresponding receptors in plants. *Trichoderma* spp. acted multitalented player in the agriculture crop cultivation. Overall, this book chapter concluded that still more studies are required to understand the elicitors and reporters from both end such as *Trichoderma* and plant or plant pathogenic organisms.

Conflict of Interest The authors declare that they have no conflict of interest.

Acknowledgments This work was supported by Brain Pool Program through the National Research Foundation of Korea (NRF) funded by the Ministry of Science and ICT (2017H1D3A1A01052610) and National Research Foundation of Korea (2019R1A1055452).

References

Ahmad P, Sarwat M, Sharma S (2008) Reactive oxygen species, antioxidants and signaling in plants. J Plant Biol 51(3):167–173

Alba JM, Glas JJ, Schimmel BCJ, Kant MR (2011) Avoidance and suppression of plant defenses by herbivores and pathogens. J Plant Interact 6(4):221–227

Alonso-Ramírez A, Poveda J, Martín I, Hermosa R, Monte E, Nicolás C (2014) Salicylic acid prevents Trichoderma harzianum from entering the vascular system of roots. Mol Plant Pathol 15(8):823–831

Baker SE, Perrone G, Richardson NM, Gallo A, Kubicek CP (2012) Phylogenomic analysis of polyketide synthase-encoding genes in *Trichoderma*. Microbiology 158(1):147–154

Baranski R, Klocke E, Nothnagel T (2008) Chitinase CHIT36 from Trichoderma harzianum enhances resistance of transgenic carrot to fungal pathogens. J Phytopathol 156(9):513–521

Bari R, Jones JDG (2009) Role of plant hormones in plant defence responses. Plant Mol Biol 69(4):473–488

Benítez T, Rincón AM, Limón MC, Codon AC (2004) Biocontrol mechanisms of *Trichoderma* strains. Int Microbiol 7(4):249–260

Berens ML, Berry HM, Mine A, Argueso CT, Tsuda K (2017) Evolution of hormone signaling networks in plant defense. Annu Rev Phytopathol 55(1):401–425

Bolton MD (2009) Current review: primary metabolism and plant defense-fuel for the fire. Mol Plant-Microbe Interact 22(5):487–497

Bonfante P, Genre A (2010) Mechanisms underlying beneficial plant–fungus interactions in mycorrhizal symbiosis. Nat Commun 1(1):48

Brotman Y, Briff E, Viterbo A, Chet I (2008) Role of swollenin, an expansin-like protein from Trichoderma, in plant root colonization. Plant Physiol 147(2):779–789

Cheng C-H, Shen B-N, Shang Q-W, Liu L-YD, Peng K-C, Chen Y-H, Chen F-F, Hu S-F, Wang Y-T, Wang H-C, Wu H-Y, Lo C-T, Lin S-S (2018) Gene-to-gene network analysis of the mediation of plant innate immunity by the eliciting plant response-like 1 (Epl1) elicitor of Trichoderma formosa. Mol Plant-Microbe Interact 31(7):683–691

Contreras-Cornejo HA, Macías-Rodríguez L, Cortés-Penagos C, López-Bucio J (2009) Trichoderma virens, a plant beneficial fungus, enhances biomass production and promotes lateral root growth through an auxin-dependent mechanism in Arabidopsis. Plant Physiol 149(3):1579–1592

Contreras-Cornejo HA, Macías-Rodríguez L, Beltrán-Peña E, Herrera-Estrella A, López-Bucio J (2011) *Trichoderma*-induced plant immunity likely involves both hormonal- and camalexin-dependent mechanisms in *Arabidopsis thaliana* and confers resistance against necrotrophic fungi Botrytis cinerea. Plant Signal Behav 6(10):1554–1563

Crutcher FK, Moran-Diez ME, Ding S, Liu J, Horwitz BA, Mukherjee PK, Kenerley CM (2015) A paralog of the proteinaceous elicitor SM1 is involved in colonization of maize roots by Trichoderma virens. Fungal Biol 119(6):476–486

Cui J, Jiang N, Meng J, Hou X, Yang G, Luan Y (2018) Identification and characterization of defensin genes conferring Phytophthora infestans resistance in tomato. Physiol Mol Plant Pathol 103:28–35

Djonović S, Pozo MJ, Dangott LJ, Howell CR, Kenerley CM (2006) Sm1, a proteinaceous elicitor secreted by the biocontrol fungus Trichoderma virens induces plant defense responses and systemic resistance. Mol Plant-Microbe Interact 19(8):838–853

Djonović S, Vargas WA, Kolomiets MV, Horndeski M, Wiest A, Kenerley CM (2007) A proteinaceous elicitor Sm1 from the beneficial fungus *Trichoderma virens* is required for induced systemic resistance in maize. Plant Physiol 145(3):875–889

Druzhinina IS, Shelest E, Kubicek CP (2012) Novel traits of Trichoderma predicted through the analysis of its secretome. FEMS Microbiol Lett 337(1):1–9

Elkelish AA, Alhaithloul HAS, Qari SH, Soliman MH, Hasanuzzaman M (2020) Pretreatment with Trichoderma harzianum alleviates waterlogging-induced growth alterations in tomato seedlings by modulating physiological, biochemical, and molecular mechanisms. Environ Exp Bot 171:103946

Faize M, Malnoy M, Dupuis F, Chevalier M, Parisi L, Chevreau E (2003) Chitinases of Trichoderma atroviride induce scab resistance and some metabolic changes in two cultivars of apple. Phytopathology 93(12):1496–1504

Gomes EV, Costa MdN, de Paula RG, de Azevedo RR, da Silva FL, Noronha EF, Ulhoa CJ, Monteiro VN, Cardoza RE, Gutiérrez S, Silva RN (2015) The Cerato-Platanin protein Epl-1 from Trichoderma harzianum is involved in mycoparasitism, plant resistance induction and self cell wall protection. Sci Rep 5(1):17998

Guzmán-Guzmán P, Alemán-Duarte MI, Delaye L, Herrera-Estrella A, Olmedo-Monfil V (2017) Identification of effector-like proteins in Trichoderma spp. and role of a hydrophobin in the plant-fungus interaction and mycoparasitism. BMC Genet 18(1):16

Guzmán-Guzmán P, Porras-Troncoso MD, Olmedo-Monfil V, Herrera-Estrella A (2019) Trichoderma species: versatile plant Symbionts. Phytopathology 109(1):6–16

Harman GE, Kubicek CP (1998) Trichoderma and gliocladium, volume 2: Enzymes, biological control and commercial applications. CRC Press

Harman GE, Howell CR, Viterbo A, Chet I, Lorito M (2004) Trichoderma species - opportunistic, avirulent plant symbionts. Nat Rev Microbiol 2(1):43–56

Harman GE, Herrera-Estrella AH, Horwitz BA, Lorito M (2012) Special issue: *Trichoderma* – from basic biology to biotechnology. Microbiology 158(1):1–2

Hermosa R, Viterbo A, Chet I, Monte E (2012) Plant-beneficial effects of *Trichoderma* and of its genes. Microbiology 158(1):17–25

Huang Y, Mijiti G, Wang Z, Yu W, Fan H, Zhang R, Liu Z (2015) Functional analysis of the class II hydrophobin gene HFB2-6 from the biocontrol agent Trichoderma asperellum ACCC30536. Microbiol Res 171:8–20

J.D. Jones, J.L. Dangl, The plant immune system, nature 444(7117) (2006) 323–329

Kubicek CP, Herrera-Estrella A, Seidl-Seiboth V, Martinez DA, Druzhinina IS, Thon M, Zeilinger S, Casas-Flores S, Horwitz BA, Mukherjee PK, Mukherjee M, Kredics L, Alcaraz LD, Aerts A, Antal Z, Atanasova L, Cervantes-Badillo MG, Challacombe J, Chertkov O, McCluskey K, Coulpier F, Deshpande N, von Döhren H, Ebbole DJ, Esquivel-Naranjo EU, Fekete E, Flipphi M, Glaser F, Gómez-Rodríguez EY, Gruber S, Han C, Henrissat B, Hermosa R, Hernández-Oñate M, Karaffa L, Kosti I, Le Crom S, Lindquist E, Lucas S, Lübeck M, Lübeck PS, Margeot A, Metz B, Misra M, Nevalainen H, Omann M, Packer N, Perrone G, Uresti-Rivera EE, Salamov A, Schmoll M, Seiboth B, Shapiro H, Sukno S, Tamayo-Ramos JA, Tisch D, Wiest A, Wilkinson HH, Zhang M, Coutinho PM, Kenerley CM, Monte E, Baker SE, Grigoriev IV (2011) Comparative genome sequence analysis underscores mycoparasitism as the ancestral life style of Trichoderma. Genome Biol 12(4):R40–R40

Kumar V, Parkhi V, Kenerley CM, Rathore KS (2009) Defense-related gene expression and enzyme activities in transgenic cotton plants expressing an endochitinase gene from Trichoderma virens in response to interaction with Rhizoctonia solani. Planta 230(2):277–291

Loon LCv, Rep M, Pieterse CMJ (2006) Significance of inducible defense-related proteins in infected plants. Annu Rev Phytopathol 44(1):135–162

López-Mondéjar R, Ros M, Pascual JA (2011) Mycoparasitism-related genes expression of Trichoderma harzianum isolates to evaluate their efficacy as biological control agent. Biol Control 56(1):59–66

Lorito M, Woo SL, Harman GE, Monte E (2010) Translational research on *Trichoderma*: from 'omics to the field. Annu Rev Phytopathol 48(1):395–417

Luo Y, Zhang D-D, Dong X-W, Zhao P-B, Chen L-L, Song X-Y, Wang X-J, Chen X-L, Shi M, Zhang Y-Z (2010) Antimicrobial peptaibols induce defense responses and systemic resistance in tobacco against tobacco mosaic virus. FEMS Microbiol Lett 313(2):120–126

Martinez C, Blanc F, Le Claire E, Besnard O, Nicole M, Baccou JC (2001) Salicylic acid and ethylene pathways are differentially activated in melon cotyledons by active or heat-denatured cellulase from Trichoderma longibrachiatum. Plant Physiol 127(1):334–344

Mathys J, De Cremer K, Timmermans P, Van Kerkhove S, Lievens B, Vanhaecke M, Cammue B, De Coninck B (2012) Genome-wide characterization of ISR induced in Arabidopsis thaliana by *Trichoderma hamatum* T382 against *Botrytis cinerea* infection. Front Plant Sci 3:108

Mendoza-Mendoza A, Zaid R, Lawry R, Hermosa R, Monte E, Horwitz BA, Mukherjee PK (2018) Molecular dialogues between Trichoderma and roots: role of the fungal secretome. Fungal Biol Rev 32(2):62–85

Morán-Diez E, Hermosa R, Ambrosino P, Cardoza RE, Gutiérrez S, Lorito M, Monte E (2009) The ThPG1 endopolygalacturonase is required for the Trichoderma harzianum–plant beneficial interaction. Mol Plant-Microbe Interact 22(8):1021–1031

Morán-Diez E, Rubio B, Domínguez S, Hermosa R, Monte E, Nicolás C (2012) Transcriptomic response of Arabidopsis thaliana after 24h incubation with the biocontrol fungus Trichoderma harzianum. J Plant Physiol 169(6):614–620

Moscatiello R, Sello S, Ruocco M, Barbulova A, Cortese E, Nigris S, Baldan B, Chiurazzi M, Mariani P, Lorito M, Navazio L (2018) The Hydrophobin HYTLO1 secreted by the biocontrol fungus Trichoderma longibrachiatum triggers a NAADP-mediated calcium signalling pathway in Lotus japonicus. Int J Mol Sci 19(9):2596

Mukherjee M, Mukherjee PK, Horwitz BA, Zachow C, Berg G, Zeilinger S (2012) Trichoderma-plant-pathogen interactions: advances in genetics of biological control. Indian J Microbiol 52(4):522–529

Mukherjee PK, Horwitz BA, Herrera-Estrella A, Schmoll M, Kenerley CM (2013) Trichoderma research in the genome era. Annu Rev Phytopathol 51(1):105–129

Nawrocka J, Małolepsza U (2013) Diversity in plant systemic resistance induced by Trichoderma. Biol Control 67(2):149–156

Nawrocka J, Snochowska M, Gajewska E, Pietrowska E, Szczech M, Małolepsza U (2012) Activation of defense responses in cucumber and tomato plants by selected polish Trichoderma strains. J Fruit Ornament Plant Res 75(1):105–116

Nogueira-Lopez G, Greenwood DR, Middleditch M, Winefield C, Eaton C, Steyaert JM, Mendoza-Mendoza A (2018) The apoplastic secretome of Trichoderma virens during interaction with maize roots shows an inhibition of plant defence and scavenging oxidative stress secreted proteins. Front Plant Sci 9(409)

Oljira AM, Hussain T, Waghmode TR, Zhao H, Sun H, Liu X, Wang X, Liu B (2020) Trichoderma enhances net photosynthesis, water use efficiency, and growth of wheat (Triticum aestivum L.) under salt stress. Microorganisms 8(10):1565

Pozo MaJ, Baek J-M, García JM, Kenerley CM (2004) Functional analysis of tvsp1, a serine protease-encoding gene in the biocontrol agent Trichoderma virens. Fungal Genet Biol 41(3):336–348

Ramírez-Valdespino CA, Casas-Flores S, Olmedo-Monfil V (2019) Trichoderma as a model to study effector-like molecules. Front Microbiol 10(1030)

Rotblat B, Enshell-Seijffers D, Gershoni JM, Schuster S, Avni A (2002) Identification of an essential component of the elicitation active site of the EIX protein elicitor. Plant J 32(6):1049–1055

Ruocco M, Lanzuise S, Lombardi N, Woo SL, Vinale F, Marra R, Varlese R, Manganiello G, Pascale A, Scala V, Turrà D, Scala F, Lorito M (2015) Multiple roles and effects of a novel Trichoderma hydrophobin. Mol Plant-Microbe Interact 28(2):167–179

Salas-Marina MA, Silva-Flores MA, Uresti-Rivera EE, Castro-Longoria E, Herrera-Estrella A, Casas-Flores S (2011) Colonization of Arabidopsis roots by Trichoderma atroviride promotes growth and enhances systemic disease resistance through jasmonic acid/ethylene and salicylic acid pathways. Eur J Plant Pathol 131(1):15–26

Salas-Marina MA, Isordia-Jasso MI, Islas-Osuna MA, Delgado-Sánchez P, Jiménez-Bremont JF, Rodríguez-Kessler M, Rosales-Saavedra MT, Herrera-Estrella A, Casas-Flores S (2015) The Epl1 and Sm1 proteins from Trichoderma atroviride and Trichoderma virens differentially modulate systemic disease resistance against different life style pathogens in Solanum lycopersicum. Front Plant Sci 6(77)

Saravanakumar K, Shanmuga Arasu V, Kathiresan K (2013) Effect of Trichoderma on soil phosphate solubilization and growth improvement of Avicennia marina. Aquat Bot 104:101–105

Saravanakumar K, Fan L, Fu K, Yu C, Wang M, Xia H, Sun J, Li Y, Chen J (2016) Cellulase from Trichoderma harzianum interacts with roots and triggers induced systemic resistance to foliar disease in maize. Sci Rep 6(1):35543

Saravanakumar K, Li Y, Yu C, Wang Q-q, Wang M, Sun J, Gao J-x, Chen J (2017) Effect of Trichoderma harzianum on maize rhizosphere microbiome and biocontrol of Fusarium Stalk rot. Sci Rep 7(1):1771

Saravanakumar K, MubarakAli D, Kathiresan K, Wang M-H (2018a) An evidence of fungal derived 1-aminocyclopropane-1-carboxylate deaminase promoting the growth of mangroves. Beni-Suef Univ J Basic Appl Sci 7(4):446–451

Saravanakumar K, Wang S, Dou K, Lu Z, Chen J (2018b) Yeast two-hybrid and label-free proteomics based screening of maize root receptor to cellulase of Trichoderma harzianum. Physiol Mol Plant Pathol 104:86–94

Schirawski J, Perlin MH (2018) Plant–microbe interaction 2017—the good, the bad and the diverse. Int J Mol Sci 19(5):1374

Schuster A, Schmoll M (2010) Biology and biotechnology of Trichoderma. Appl Microbiol Biotechnol 87(3):787–799

Seo H-H, Park S, Park S, Oh B-J, Back K, Han O, Kim J-I, Kim YS (2014) Overexpression of a defensin enhances resistance to a fruit-specific anthracnose fungus in pepper. PLoS One 9(5):e97936

Shah JM, Raghupathy V, Veluthambi K (2008) Enhanced sheath blight resistance in transgenic rice expressing an endochitinase gene from Trichoderma virens. Biotechnol Lett 31(2):239

Shoresh M, Harman GE, Mastouri F (2010) Induced systemic resistance and plant responses to fungal biocontrol agents. Annu Rev Phytopathol 48(1):21–43

Sousa TPd, Chaibub AA, Silva GBd, Filippi MCCd (2020) Trichoderma asperellum modulates defense genes and potentiates gas exchanges in upland rice plants. Physiol Mol Plant Pathol 112:101561

Sun X, Sun M, Chao Y, Wang H, Pan H, Yang Q, Cui X, Lou Y, Zhuge Y (2021) Alleviation of lead toxicity and phytostimulation in perennial ryegrass by the Pb-resistant fungus *Trichoderma asperellum* SD-5. Funct Plant Biol 48(3):333–341

Tandon A, Fatima T, Anshu D, Shukla P, Tripathi S, Srivastava PCS (2020) Phosphate solubilization by Trichoderma koningiopsis (NBRI-PR5) under abiotic stress conditions. J King Saud Univ - Sci 32(1):791–798

Vargas WA, Mandawe JC, Kenerley CM (2009) Plant-derived sucrose is a key element in the symbiotic association between Trichoderma virens and maize plants. Plant Physiol 151(2):792–808

Velázquez-Robledo R, Contreras-Cornejo HA, Macías-Rodríguez L, Hernández-Morales A, Aguirre J, Casas-Flores S, López-Bucio J, Herrera-Estrella A (2011) Role of the 4-phosphopantetheinyl transferase of Trichoderma virens in secondary metabolism and induction of plant defense responses. Mol Plant-Microbe Interact 24(12):1459–1471

Viterbo A, Chet I (2006) TasHyd1, a new hydrophobin gene from the biocontrol agent Trichoderma asperellum, is involved in plant root colonization. Mol Plant Pathol 7(4):249–258

Viterbo A, Harel M, Chet I (2004) Isolation of two aspartyl proteases from Trichoderma asperellum expressed during colonization of cucumber roots. FEMS Microbiol Lett 238(1):151–158

Wang Y, Song J, Wu Y, Odeph M, Liu Z, Howlett BJ, Wang S, Yang P, Yao L, Zhao L, Yang Q (2013) Eplt4 proteinaceous elicitor produced in Pichia pastoris has a protective effect against Cercosporidium sofinum infections of soybean leaves. Appl Biochem Biotechnol 169(3):722–737

Yu W, Mijiti G, Huang Y, Fan H, Wang Y, Liu Z (2018) Functional analysis of eliciting plant response protein Epl1-Tas from Trichoderma asperellum ACCC30536. Sci Rep 8(1):7974

Zamioudis C, Pieterse CM (2012) Modulation of host immunity by beneficial microbes. Mol Plant-Microbe Interact 25(2):139–150

Genus *Trichoderma*: Its Role in Induced Systemic Resistance of Plants Against Phytopathogens

Kishor Chand Kumhar, Azariah Babu, John Peter Arulmarianathan, Abhay Kumar Pandey, and Bhabesh Deka

Contents

1 Introduction

Food is one of the basic requirements of human existence by providing proper nourishment and energy to carry out various physiological and biological activities which are necessary for survival and adequate development (Shewry and Hey 2015). The major part of human food is being derived from the agricultural sector (Pretty et al. 2010). Cereals, pulses, vegetables, spices and condiments, oilseed, fruits, etc.

K. C. Kumhar (✉)
Department of Plant Pathology, College of Agriculture, CCS Haryana Agricultural University, Hisar, Haryana, India

A. Babu · A. K. Pandey · B. Deka
Tea Research Association, North Bengal Regional Research and Development Centre, Nagrakata, Jalpaiguri, West Bengal, India

J. P. Arulmarianathan
Varsha Bioscience and Technology India Pvt Ltd, Hyderabad, Telangana, India

N. Amaresan et al. (eds.), *Advances in* Trichoderma *Biology for Agricultural Applications*, Fungal Biology, https://doi.org/10.1007/978-3-030-91650-3_8

are primary components of human food. The success of crop production depends on several factors such as weather conditions, agricultural practices, attack of weeds, insect pests and diseases, plant protection measures, and so on. Plant diseases are considered an important constraint in deciding the quantity as well as the quality of crop production to a greater extent. According to an estimate, plant diseases cause crop loss worth rupees 60,000 crores per annum (Tripathi et al. 2020).

The phytopathogens are broadly categorized, based on their nature of survival, as soil-dwelling, found in association with seed and airborne. They can effectively be controlled through adopting different management strategies at an appropriate time. Seed-borne disease can effectively and economically be controlled through seed treatment; similarly, soil- and airborne species can be controlled through soil treatment and foliar application, respectively, with suitable plant protection measures. The phytopathogens such as fungi, bacteria, viruses, etc. are the important biotic challenges for the health of plants. Since long back, these could be managed through the gamut of practices including chemical fungicides at top priority owing to immediate and long-lasting effect. The excess and frequent usage of chemical fungicides are risky in terms of soil health deterioration, unhealthy environment, imbalanced ecosystem, etc. (Wightwick et al. 2010).

However, nature has already provided an inbuilt mechanism for plants to resist or tolerate unfavorable biotic challenges or stresses. In the current scenario, agriculturists may be growers; researchers, as well as consumers, are diverting their choices toward non-chemical means of disease management to maintain environmental health in a very good status and having chemical-free food items. The "plant disease resistance" phenomenon is of immense importance owing to its natural existence and safety aspects.

Certain microorganisms are capable of initiating resistance in plants to take care of phytopathogenic microorganisms. The importance of biological control agents (BCAs) in managing a wide range of phytopathogens through induced resistance is well recognized, which may be local or systemic (Carreras-Villaseñor et al. 2012; Hermosa et al. 2012). The induced resistance could be one of the long-lasting and effective plant protection strategies. When the plants come in contact with microorganisms, the plant receptors recognize the microbial elicitors and activate the numerous signaling pathways. Interaction of both plants and microbes results in different changes in plant physiology and its biochemical reactions which protect the plants from the attack of the pathogen (Contreras-Cornejo et al. 2011). Rhizosphere microorganisms induce resistance in plants to suppress the disease (Keswani et al. 2013).

Genus *Trichoderma* has been established as the superstar among the beneficial microorganisms of agricultural importance owing to its multiple modes of actions. It is estimated that bioagents contribute 2.5% of the total pesticide market in India worth rupees 690 crores (Tripathi et al. 2020). This chapter is focused on this antagonist, considering its various aspects of defense mechanisms.

2 Consequences of Disease Management Through ISR

***Trichoderma* species:** They are opportunistic, avirulent plant symbionts, as well as an antagonist for other fungi. Certain strains establish vigorous and long-lasting colonization of root surfaces and penetrate the epidermis. They release a variety of compounds to induce localized or systemic resistance responses and cause considerable changes to the plant proteome and metabolism. A root colonization phenomenon of *Trichoderma* spp. also enhances root growth and development, crop productivity, resistance to abiotic stresses, and the uptake and utilization of nutrients. Soilborne disease-causing microorganisms mostly attack the seeds or seedlings of various crops. Species of *Trichoderma* can be applied for seed treatment, biopriming, and furrow treatment to reduce crop losses.

Owing to the manifold increase in its demand, several *Trichoderma*-based products have been commercialized in India and other countries. *Trichoderma* products enhance biocontrol potency through the manipulation of the rhizosphere and phylloplane environment. Its integrated application provides synergistic effects to other alternative methods that alone do not provide adequate protection (Sanjeev 2013).

Genus *Trichoderma* has come into sight as a multiple featured beneficial fungal candidate. Its peculiar characters include antagonism, competition for nutrients, and induction of systemic resistance, plant growth promotion, and improvement of abiotic stresses. It can perform better than fungicides for the control of soilborne fungal diseases and can sustain longer for a longer period. Besides, it reduces health risks, costs, and environmental damage due to the overburden of fungicide usages. Its formulated products can be used in many ways, viz., seed treatment and direct application to the soil before planting and along with organic fertilizers (Ha 2010). Among the promising biological control agents (BCAs), which provide a protective umbrella to crops against several phytopathogens, the genus *Trichoderma* is considered the most popular one. It is efficient in controlling a gamut of phytopathogens (Kumhar et al. 2015, 2020; Kumhar and Babu 2019, 2020; Harman 2006; Kumar et al. 2017) in various ways. Out of about hundreds of species, *T. viride*, *T. harzianum*, *T. asperellum*, *T. atroviride*, *T. hamatum*, etc. are on the list of potential candidates.

Among multifarious actions of this genus as an ideal disease manager, it also helps in the development of resistance (Meller Harel et al. 2014; Alizadeh et al. 2013). Induced systemic resistance (ISR) develops after infection of phytopathogens; certain chemicals or non-pathogenic microorganisms can also initiate it. Induction of defense response through *Trichoderma* spp. in different plants against various fungal and bacterial phytopathogens was enlisted by researchers (Bisen et al. 2016). When *Trichoderma* spp. are inefficient in controlling the pathogens directly, under such situations, the induced resistance is the most viable tactic that strongly protects the plants from airborne phytopathogens. A setup of various defense pathways in host plants activated owing to the participation of *Trichoderma* spp. Such defense has been extensively established in the lab as well as field conditions against a wide variety of soilborne phytopathogens (Singh 2014; Harman et al. 2012).

3 Sequential Events in ISR

At the time of disease development, plants undergo a series of activities or processes to make up for the incurred losses. This process includes regulation of specific genes; change in the level of reactive oxygen species (ROS) requires in plant defense pathway activation of the specific transcription factor, defense regulation genes, increased transport of macromolecules, enzymes, and phytohormones (Bari and Jones 2009; Vitti et al. 2013). Phytohormones such as jasmonic acid, salicylic acid, abscisic acid, ethylene, auxins, cytokinins, and gibberellic acid are dominant signal in regulating the local as well as systemic defense in the plant system. The pathogen-induced systemic acquired resistance is governed by the salicylic acid signaling pathway (Métraux 2013; Conrath 2006). The induced systemic resistance (ISR) due to beneficial microbes is regulated by jasmonic acid signaling (Hermosa et al. 2013; Manganiello et al. 2018).

Comprehensive attention toward the research related to the significance of fungal BCAs in ISR was paid during the past decade, and it was proved that plant growth promotional fungi (PGPF) are efficient to elicit ISR in plants. Evaluation of antagonistic fungi for this aspect showed positive responses in the plant (Keswani et al. 2013; Singh 2014). The role of *Trichoderma* on this aspect was first reported in *Vitis vinifera* (Calder6n et al. 1993); subsequently, this phenomenon was noted in the case of *T. longibrachiatum* which induced resistance against *P. parasitica* in *Nicotiana tabacum* owing to the induced resistance of the tobacco plants exhibiting increased pathogenesis-related genes (Chang et al. 1997).

Being a successful colonizer of plant roots, *Trichoderma* spp. can induce resistance in an array of monocot and dicot plants. Its root colonization is a must to stimulate the defense response. Soil inoculation of *T. harzianum* T-39 triggered a defense response in *Phaseolus vulgaris* to protect the crop from some fungal pathogens (Bigirimana et al. 1997). Tomato plants exhibited local and systemic resistance against early blight pathogen when *T. harzianum* was applied (Howell 2003). Soil and foliar application of *T. harzianum*T-39 could efficiently reduce disease incidence owing to induced ISR against gray mold fungus (De Meyer et al. 1998). *T. harzianum* strain T-203 could enhance the defense system in cucumber (Yedidia et al. 2000). Similarly, cottonseed treated with such antagonist controlled soilborne fungus (Howell et al. 2000).

4 Communication Between *Trichoderma* and Host Plant While ISR

The interconnected hormone signaling pathways govern the growth and immunity aspects of plants (Pieterse et al. 2009) to maintain their proper health. In this process, JA/ET, SA, abscisic acid pathways, and other signaling cascades are of great importance. In plants and *Trichoderma* interaction, the ACC deaminase (ACCD) action suppresses the level of 1-aminocyclopropane-1-carboxylic acid (ACC) which is essential for the biosynthesis of ET. ET is helpful in the signaling of gibberellins which enhance plant growth and degrading the DELLA proteins. Straightway, the gibberellin is responsible for the initiation of jasmonic and salicylic acid-oriented resistance. Biosynthesis of indole acetic acid (IAA) and ethylene (ET) can take place by each other in the plant system (Stepanova et al. 2007).

5 Root Colonization by *Trichoderma* spp.

The secretion of the plant roots acts as a bridge between *Trichoderma* and plant roots (Bais et al. 2006). Polysaccharides secreted by roots assist the *Trichoderma* to grow well; however, sucrose secreted in such exudates is a major performer (Contreras-Cornejo et al. 2009; Vargas et al. 2009). Numerous proteins like *TasHydl* protein, cell wall protein, *SwolleninTasSwo* protein, and endopolygalacturonase *thpg1* are useful for this antagonist to colonize the roots successfully (Viterbo and Chet 2006; Samolski et al. 2012; Brotman et al. 2008; Eugenia et al. 2009). The successful colonization of the roots leads to certain alterations in plant system (Zhang et al. 2013; Mukherjee et al. 2012).

6 Plant Defense Elicitors Secreted by *Trichoderma* spp.

In the process of systemic resistance, the plant receptors recognize the specific components from the cell surface of microorganisms. The microbial cell surface may be pathogen-associated molecular patterns (PAMPs) in the case of phytopathogens or other microorganisms; it may be microbe-associated molecular patterns (MAMPs) (Schwessinger and Zipfel 2008). The secondary metabolites, i.e., antibiotics and some other compounds produced by *Trichoderma* spp., activate a systemic resistance reaction in plants (Lorito et al. 2010). A nice description of *Trichoderma*-originated elicitors involved in the plant defense activity toward the phytopathogens was made by some workers (Bisen et al. 2016).

Among numerous recognized fungal BCAs, merely *Trichoderma* spp. are actively involved in systemic resistance activity through MAMPs (Vinale et al. 2008). Ethylene-inducing xylanase (Xyn2/Eix) is the first identified MAMP responsible for bringing out defense reactions in plants (Rotblat et al. 2002). Cellulases produced by *Trichoderma* spp. activate the ethylene and salicylic acid pathways for the initiation of defense response in plants (Martinez et al. 2001). Certain *Trichoderma*-originated proteins, viz., *swollenin* TasSWO, endopolygalacturonase *th*PG1, and EPL1/SM1, also play a vital role in plant defense responses (Brotman et al. 2008; Eugenia et al. 2009; Djonović et al. 2006; Seidl et al. 2006). *Trichoderma*-based secondary metabolites at low doses act as MAMPs, helping in resistance initiation. Harzianolide, 6-pentyl-o-pyrone, harzianopyridone alamethicin, enzyme hybrid, and peptaibol reported initiating systemic resistance in tomato, pea, and canola, bean, and maize (Engelberth et al. 2001; Viterbo et al. 2007; Luo et al. 2010; Druzhinina et al. 2011; Mukherjee et al. 2012).

Trichoderma, a ubiquitous antagonist, has occupied an apparent position among the beneficial microbes because of its manifold features in taking care of plant health in terms of vegetative growth promotion, management of phytopathogens, enhancing resistance development in plants against phytopathogens, etc. It produces pathogenesis-related proteins (PRPs), phytoalexins, terpenoids, as well as enzymes such as phenylalanine ammonia-lyase, peroxidase, polyphenol oxidase, and lipoxygenase which determine the resistance in the plants against pathogens (Chakraborty et al. 2020).

7 Defense Signaling Pathways Activated During ISR

T. harzianum has been reported to encourage systemic resistance in tomato, lettuce, pepper, bean, tobacco, and cucumber seedlings (De Meyer et al. 1998; Yedidia et al. 1999). The induced response is a time as well as a concentration-dependent phenomenon. Root colonization causes different sequential changes in a very systematic way. Plants react with *Trichoderma* while invading following the signaling to activate the synthesis of salicylic acid, jasmonic acid, and ethylene to ultimately induce the resistance responses in the entire plant against the phytopathogens. At higher concentrations, *T. asperellum* triggers an SA-mediated SAR response in the plants (Segarra et al. 2007; Contreras-Cornejo et al. 2011; Salas-Marina et al. 2011; Yoshioka et al. 2012).

Application of *T. asperellum* strain T-34 and *T. harzianum* showed a SAR response to a quick increase in peroxidase, activity, and level of jasmonic as well as salicylic acid in cucumber and potato plant to resist against the plant disease-causing microbes (Segarra et al. 2007; Gallou et al. 2009). Biomass suspension and culture filtrate of *T. asperellum* SKT-1 could play an active role in such resistance against the bacterial disease of tomato following several alterations in the plant system (Yoshioka et al. 2012). The role of *T. atroviride* in resistance development was proven against the root pathogen of *Arabidopsis* by earlier researchers (Salas-Marina et al. 2011). *T. asperellum* SKT-1 induced systemic resistance against cucumber mosaic virus in *Arabidopsis thaliana* (Elsharkawy et al. 2013). This antagonist is capable of resistance initiation for the control of *B. cinerea* through different modifications in the plant system (Tucci et al. 2011; Mathys et al. 2012). *T. asperellum* application laid enhancement of phytoalexin buildup, which helped in the induction of systemic resistance toward bacterial pathogen in cucumber (Yedidia et al. 2003). Inoculation of *T. asperellum* and *T. harzianum* in cucumber, potato, and grapevine caused genetic changes in support of resistance development (Shoresh et al. 2005; Gallou et al. 2009; Perazzolli et al. 2008, 2011).

8 Application of Genus *Trichoderma*

As a plant protection strategy, it can be used for seed and soil treatment, foliar spray, enrichment of vermicompost, and/or farmyard manure.

8.1 Soil Application and ISR

T. harzianum-treated tomato induced salicylic acid signaling pathway and ethylene biosynthesis against root-knot nematode (Leonetti et al. 2017). Tomato seed treatment with *T. virens* plus its soil application expressed defense-related enzymes such as peroxidase (PO), polyphenol oxidase (PPO), and phenylalanine ammonia-lyase (PAL) playing a role in the induction of systemic resistance against wilt-causing fungus, *Fusarium oxysporum* f. sp. *lycopersicae*. The enzymatic activity increased after a week and reached the maximum level after 2 weeks (Christopher et al. 2010). Treatment of oil palm seedlings with *T. virens* for the control of basal stem rot (BSR) encouraged the activities of plant defense-related enzymes (peroxidase, polyphenol oxidase, superoxide dismutase, and phenylalanine lyase) in the leaves (Paudzai et al. 2019).

Wheat seed treatment with *T. harzianum* for spot blotch pathogen, *Bipolaris sorokiniana* management in combination with the application of methyl jasmonate enhanced indole acetic acid production in the rhizosphere the activities of defense-related enzymes, viz., catalase, ascorbate peroxidase, ascorbate oxidized, phenylalanine ammonia-lyase, cinnamic acid, and peroxidase (Singh et al. 2019). Maize seed treatment with *T. harzianum* strain-T22 for the management of damping-off caused by *Pythium ultimum* resulted in the accumulation of defense-associated protein such as endochitinase, pathogenesis-related protein, GTP-binding protein, isoflavone reductase, and other proteins (Chen et al. 2005).

Chickpea, *Cicer arietinum*, seed treatment with *T. viride* followed by its foliar sprays induced systemic resistance against *C. campestris* infestation through the increased production of defense enzymes (Kannan et al. 2014). Incorporation of *T. harzianum* OTPB3 cell suspension into pots containing tomato seed had significantly increased the level of indole-3-acetic acid (IAA) and gibberellic acid (GA3) in roots of treated seedlings. The antagonist enhanced the levels of defense-related enzymes, i.e., peroxidase, polyphenol oxidase, and superoxide dismutase (Chowdappa et al. 2013).

8.2 Foliar Spray and ISR

Application of live and dead biomass of *T. harzianum* T-39 on the roots and leaves of cucumber plants, respectively, induced the local and systemically induced resistance for the control of powdery mildew caused by *Pseudoperonospora cubensis* and *Sphaerotheca fusca* under greenhouse conditions (Elad 2000). *T. harzianum* T-39 induced resistance in grapevine against downy mildew, and it was not affected by exposure to heat or drought (Roatti et al. 2013).

Five species of *Trichoderma*, viz., *T. harzianum*, *T. asperellum*, *T. viride*, *T. virens*, and *T. aureoviride*, are the main candidates playing a crucial role in managing plant diseases via resistance development (Chakraborty et al. 2020). Soil treatment as well as the foliar application of *T. harzianum* (10^7 cells/ml) induced systemic resistance in cucumber plants (cv. Beit-Alpha) against the cucumber mosaic virus (Helmy and Maklad 2002). Soil application of *T. harzianum* produced pathogenesis-related proteins, i.e., chitinases, and hence induced systemic resistance in tomato plants (Ene et al. 2013).

T. harzianum, an isolate of the safflower rhizosphere, was tested for effectiveness in controlling the root-rot of safflower caused by *M. phaseolina*. The seed was treated with talc formulation at different concentrations. *T. harzianum* (10 g/kg) was found effective in controlling disease under laboratory, greenhouse, and field conditions as good as carbendazim. This antagonist could lead the higher activity of peroxidase, phenylalanine ammonia-lyase, chitinase, polyphenol oxidase, and β-1,3-glucanase which ultimately induced systemic resistance and physiological changes for plant defense mechanisms (Govindappa et al. 2010). Tomato plants grown in *T. harzianum*-treated soil had expressed genes for priming of salicylic acid and ethylene in plant leaves and such plants expressed induced resistance toward *B. cinerea* (Elada 2018).

Cucumber seed treatment with liquid *T. harzianum* formulation developed the systemic acquired resistance in cucumber plant cucumber mosaic virus Cucumovirus (CMV)(El-Dougdoug et al. 2013). Seed treatment at different concentrations of talc formulation of *T. harzianum*, an isolate of cotton rhizosphere soil, is found effective in controlling bacterial blight of cotton caused by *Xanthomonas campestris* pv. *malvacearum* (Xcm) (Raghavendra et al. 2013).

Grapevine exhibited a resistance response against *B. cinerea*, a gray mold pathogen (Calderón et al. 1993) while the application of *T. viride*. The resistance was due to the exclusive role of peroxidase in the elicitor-mediated formation of ROPs. In this context, elicitor treatment produced an increase in the level of extracellular peroxidases and the appearance of a new basic peroxidase isoenzyme, B3, which was correlated with the formation of resveratrol oxidation products (ROPs) (Calderón et al. 1994).

T. harzianum T-39 controls the foliar pathogens, *B. cinerea*, *Pseudoperonospora cubensis*, *Sclerotinia sclerotiorum*, and *Sphaerotheca fusca* in cucumber under commercial greenhouse conditions through local and systemic induced resistance. Cells of the beneficial microorganisms applied to the roots and dead cells applied to the leaves of cucumber plants induced the control of powdery mildew (Elad 2000). *T. harzianum* isolates T-30 and T-78 expressed genes encoding for NAGases (*exc1* and *exc2*), chitinases (*chit42* and *chit33*), proteases (*prb1*), and β-glucanases (*bgn13*.1) which showed the greatest mycoparasitic potential against *F. oxysporum* (López-Mondéjar et al. 2011).

T. harzianum, *T. asperellum*, *T. koningiopsis*, *T. longibrachiatum*, and *T. aureoviride* promoted plant growth; reduced disease incidence of *Sclerotium rolfsii*, *Sclerotinia sclerotiorum*, and *Colletotrichum capsici*; and increased tolerance against biotic in chickpea plants (Saxena et al. 2015). *T. harzianum*-P1 mutant exhibited differential control potency against *Pythium ultimum* and *R. solani* indicating different mechanisms of interaction with various fungal phytopathogen (Woo et al. 1999). Application of *T. harzianum* with *Glomus mosseae* could increase the production of catalase and peroxidase enzymes which induced the resistance in *Cucumis sativus* plants, and ultimately these bioagents protected the crop from the attack of *A. alternata*, an opportunistic wilt pathogen (Matrood et al. 2020).

References

Alizadeh H, Behboudi K, Ahmadzadeh M, Javan-Nikkhah M, Zamioudis C, Pieterse CMJ, Bakker PAHM (2013) Induced systemic resistance in cucumber and *Arabidopsis thaliana* by the combination of *Trichoderma harzianum* Tr6 and *Pseudomonas* sp. Ps14. Biol Control 65:14–23. https://doi.org/10.1016/j.biocontrol.2013.01.009

Bais HP, Weir TL, Perry LG, Gilroy S, Vivanco JM (2006) The role of root exudates in rhizosphere interactions with plants and other organisms. Annu Rev Plant Biol. https://doi.org/10.1146/annurev.arplant.57.032905.105159

Bari R, Jones JDG (2009) Role of plant hormones in plant defence responses. Plant Mol Biol. https://doi.org/10.1007/s11103-008-9435-0

Bigirimana J, De Meyer G, Poppe J, Elad Y, Höfte M (1997) Induction of systemic resistance on bean (Phaseolus vulgaris) by Trichoderma harziamum. Meded VAN Fac Landbouwkd EN TOEGEPASTE Biol Wet Univ GENT 62:1001–1007

Bisen K, Keswani C, Patel JS, Sarma BK, Singh HB (2016) *Trichoderma* spp.: efficient inducers of systemic resistance in plants. In: Microbial-mediated induced systemic resistance in plants, pp 185–195. https://doi.org/10.1007/978-981-10-0388-2_12

Brotman Y, Briff E, Viterbo A, Chet I (2008) Role of swollenin, an expansin-like protein from *Trichoderma*, in plant root colonization. Plant Physiol 147:779–789. https://doi.org/10.1104/pp.108.116293

Calderón AA, Zapata JM, Muñoz R, Pedreño MA, Barceló AR (1993) Resveratrol production as a part of the hypersensitive like response of grapevine cells to an elicitor from *Trichoderma viride*. New Phytol 124:455–463. https://doi.org/10.1111/j.1469-8137.1993.tb03836.x

Calderón AA, Zapata JM, Ros Barceló A (1994) Peroxidase-mediated formation of resveratrol oxidation products during the hypersensitive-like reaction of grapevine cells to an elicitor from *Trichoderma viride*. Physiol Mol Plant Pathol 44:289–299. https://doi.org/10.1016/S0885-5765(05)80031-1

Carreras-Villaseñor N, Sánchez-Arreguín JA, Herrera-Estrella AH (2012) *Trichoderma*: sensing the environment for survivaland dispersal. Microbiology. https://doi.org/10.1099/mic.0.052688-0

Chakraborty BN, Chakraborty U, Sunar K (2020) Induced immunity developed by *Trichoderma* species in plants, pp 125–147. https://doi.org/10.1007/978-981-15-3321-1_7

Chang PFL, Xu Y, Narasimhan ML, Cheah KT, D'Urzo MP, Damsz B, Kononowicz AK, Abad L, Hasegawa PM, Bressan RA (1997) Induction of pathogen resistance and pathogenesis-related genes in tobacco by a heat-stable *Trichoderma* mycelial extract and plant signal messengers. Physiol Plant 100:341–352. https://doi.org/10.1034/j.1399-3054.1997.1000217.x

Chen J, Harman GE, Comis A, Cheng GW (2005) Proteins related to the biocontrol of Pythium damping-off in maize with *Trichoderma harzianum* rifai. J Integr Plant Biol 47:988–997. https://doi.org/10.1111/j.1744-7909.2005.00053.x

Chowdappa P, Mohan Kumar SP, Jyothi Lakshmi M, Upreti KK (2013) Growth stimulation and induction of systemic resistance in tomato against early and late blight by *Bacillus subtilis* OTPB1 or *Trichoderma harzianum* OTPB3. Biol Control. https://doi.org/10.1016/j.biocontrol.2012.11.009

Christopher DJ, Raj TS, Rani SU, Udhayakumar R (2010) Role of defense enzymes activity in tomato as induced by *Trichoderma virens* against Fusarium wilt caused by *Fusarium oxysporum* f sp. *lycopersici*. J Biopest 3:158–162

Conrath U (2006) Systemic acquired resistance. Plant Signal Behav. https://doi.org/10.4161/psb.1.4.3221

Contreras-Cornejo HA, Macías-Rodríguez L, Cortés-Penagos C, López-Bucio J (2009) Trichoderma virens, a plant beneficial fungus, enhances biomass production and promotes lateral root growth through an auxin-dependent mechanism in arabidopsis. Plant Physiol 149:1579–1592. https://doi.org/10.1104/pp.108.130369

Contreras-Cornejo HA, Macías-Rodríguez L, Beltrán-Peña E, Herrera-Estrella A, López-Bucio J (2011) *Trichoderma*-induced plant immunity likely involves both hormonal- and camalexindependent mechanisms in *Arabidopsis thaliana* and confers resistance against necrotrophic fungus *Botrytis cinerea*. Plant Signal Behav 6:1554–1563. https://doi.org/10.4161/psb.6.10.17443

De Meyer G, Bigirimana J, Elad Y, Höfte M (1998) Induced systemic resistance in *Trichoderma harzianum* T39 biocontrol of *Botrytis cinerea*. Eur J Plant Pathol 104:279–286. https://doi.org/10.1023/A:1008628806616

Djonović S, Pozo MJ, Dangott LJ, Howell CR, Kenerley CM (2006) Sm1, a proteinaceous elicitor secreted by the biocontrol fungus *Trichoderma virens* induces plant defense responses and systemic resistance. Mol Plant-Microbe Interact 19:838–853. https://doi.org/10.1094/MPMI-19-0838

Druzhinina IS, Seidl-Seiboth V, Herrera-Estrella A, Horwitz BA, Kenerley CM, Monte E, Mukherjee PK, Zeilinger S, Grigoriev IV, Kubicek CP (2011) *Trichoderma*: the genomics of opportunistic success. Nat Rev Microbiol. https://doi.org/10.1038/nrmicro2637

Elad Y (2000) Biological control of foliar pathogens by means of *Trichoderma harzianum* and potential modes of action. Crop Prot 19:709–714. https://doi.org/10.1016/S0261-2194(00)00094-6

Elada Y (2018) Disease management: disease suppression by cultural means and through biocontrol. Acta Hortic:105–113. https://doi.org/10.17660/ActaHortic.2018.1207.14

El-Dougdoug KA, Megahed AA, Othman AA, Lashin SM, Ibrahim MA, Attitalla IH (2013) Induction of salicylic acid in cucumber plants against cucumber mosaic cucumovirus using biotic inducers. Am J Biochem Mol Biol 3:248–255. https://doi.org/10.3923/ajbmb.2013.248.255

Elsharkawy MM, Shimizu M, Takahashi H, Ozaki K, Hyakumachi M (2013) Induction of systemic resistance against cucumber mosaic virus in *Arabidopsis thaliana* by *Trichoderma asperellum* SKT-1. Plant Pathol J 29:193–200. https://doi.org/10.5423/PPJ.SI.07.2012.0117

Ene M, Alexandru M, Neguţ DC, Şesan TE (2013) Class II acidic chitinase expression in tomatoes treated with *Trichoderma* and AM fungi. Rom Biotechnol Lett 18:8645–8656

Engelberth J, Koch T, Schüler G, Bachmann N, Rechtenbach J, Boland W (2001) Ion channel-forming alamethicin is a potent elicitor of volatile biosynthesis and tendril coiling. Cross talk between jasmonate and salicylate signaling in lima bean. Plant Physiol 125:369–377. https://doi.org/10.1104/pp.125.1.369

Eugenia MD, Hermosa R, Ambrosino P, Cardoza RE, Gutiérrez S, Lorito M, Monte E (2009) The ThPG1 endopolygalacturonase is required for the *Trichoderma harzianum* plant beneficial interaction. Mol Plant-Microbe Interact 22:1021–1031. https://doi.org/10.1094/MPMI-22-8-1021

Gallou A, Cranenbrouck S, Declerck S (2009) *Trichoderma harzianum* elicits defence response genes in roots of potato plantlets challenged by *Rhizoctonia solani*. Eur J Plant Pathol 124:219–230. https://doi.org/10.1007/s10658-008-9407-x

Govindappa M, Lokesh S, Rai VR, Naik VR, Raju SG (2010) Induction of systemic resistance and management of safflower *Macrophomina phaseolina* root rot disease by biocontrol agents. Arch Phytopathol Plant Prot 43:26–40. https://doi.org/10.1080/03235400701652227

Ha TN (2010) Using *Trichoderma* species for biological control of. J Issaas 16:17–21

Harman GE (2006) Overview of mechanisms and uses of *Trichoderma* spp. In: Phytopathology, pp 190–194. https://doi.org/10.1094/PHYTO-96-0190
Harman GE, Herrera-Estrella AH, Horwitz BA, Lorito M (2012) Special issue: *Trichoderma* from basic biology to biotechnology. Microbiology 158:1–2. https://doi.org/10.1099/mic.0.056424-0
Helmy YI, Maklad FM (2002) Induced systemic resistance in cucumber plants under plastic houses against cucumber mosaic virus using biotic inducers. Egypt J Hortic 29:1–16
Hermosa R, Viterbo A, Chet I, Monte E (2012) Plant-beneficial effects of *Trichoderma* and of its genes. Microbiology. https://doi.org/10.1099/mic.0.052274-0
Hermosa R, Belén Rubio M, Cardoza RE, Nicolás C, Monte E, Gutiérrez S (2013) The contribution of *Trichoderma* to balancing the costs of plant growth and defense. Int Microbiol 16:69–80. https://doi.org/10.2436/20.1501.01.181
Howell CR (2003) Mechanisms employed by *Trichoderma* species in the biological control of plant diseases: the history and evolution of current concepts. Plant Dis. https://doi.org/10.1094/PDIS.2003.87.1.4
Howell CR, Hanson LE, Stipanovic RD, Puckhaber LS (2000) Induction of terpenoid synthesis in cotton roots and control of *Rhizoctonia solani* by seed treatment with *Trichoderma virens*. Phytopathology 90:248–252. https://doi.org/10.1094/PHYTO.2000.90.3.248
Kannan C, Kumar B, Aditi P, Gharde Y (2014) Effect of native *Trichoderma viride* and *Pseudomonas fluorescens* on the development of *Cuscuta campestris* on chickpea, *Cicer arietinum*. J Appl Nat Sci. https://doi.org/10.31018/jans.v6i2.544
Keswani C, Singh SP, Singh HB (2013) A superstar in biocontrol enterprise: *Trichoderma* spp. Biotech Today 3:27–30. https://doi.org/10.5958/2322-0996.2014.00005.2
Kumar G, Maharshi A, Patel J, Mukherjee A, Singh HP, Sarma BK (2017) *Trichoderma* : a potential fungal antagonist to control plant diseases. Satsa (Mb)
Kumhar KC, Babu A (2019) Biocontrol potency of *Trichoderma* isolates against tea (*Camellia* sp.) pathogens and their susceptibility towards fungicides. Int J Chem Stud 7:4192–4195
Kumhar KC, Babu A (2020) Economically important diseases of tea (*Camellia* sp.) and their management. In: Diseases of fruits and vegetable crops, pp 435–459. https://doi.org/10.1201/9780429322181-24
Kumhar KC, Babu A, Bordoloi M, Banerjee P, Dey T (2015) Biological and chemical control of *Fusarium solani*, causing dieback disease of tea *Camellia sinensis* (L): an i*n vitro* study. Int J Curr Microbiol Appl Sci 4:955–963
Kumhar KC, Babu A, Arulmarianathan JP, Deka B, Bordoloi M, Rajbongshi H, Dey P (2020) Role of beneficial fungi in managing diseases and insect pests of tea plantation. Egypt J Biol Pest Control 30. https://doi.org/10.1186/s41938-020-00270-9
Leonetti P, Zonno MC, Molinari S, Altomare C (2017) Induction of SA-signaling pathway and ethylene biosynthesis in *Trichoderma harzianum* treated tomato plants after infection of the root-knot nematode *Meloidogyne incognita*. Plant Cell Rep 36:621–631. https://doi.org/10.1007/s00299-017-2109-0
López-Mondéjar R, Ros M, Pascual JA (2011) Mycoparasitism-related genes expression of *Trichoderma harzianum* isolates to evaluate their efficacy as biological control agent. Biol Control 56:59–66. https://doi.org/10.1016/j.biocontrol.2010.10.003
Lorito M, Woo SL, Harman GE, Monte E (2010) Translational research on *Trichoderma*: from 'omics to the field. Annu Rev Phytopathol 48:395–417. https://doi.org/10.1146/annurev-phyto-073009-114314
Luo Y, Zhang DD, Dong XW, Zhao PB, Chen LL, Song XY, Wang XJ, Chen XL, Shi M, Zhang YZ (2010) Antimicrobial peptaibols induce defense responses and systemic resistance in tobacco against tobacco mosaic virus. FEMS Microbiol Lett. https://doi.org/10.1111/j.1574-6968.2010.02135.x
Manganiello G, Sacco A, Ercolano MR, Vinale F, Lanzuise S, Pascale A, Napolitano M, Lombardi N, Lorito M, Woo SL (2018) Modulation of tomato response to *Rhizoctonia solani* by *Trichoderma harzianum* and its secondary metabolite harzianic acid. Front Microbiol 9. https://doi.org/10.3389/fmicb.2018.01966

Martinez C, Blanc F, Le Claire E, Besnard O, Nicole M, Baccou JC (2001) Salicylic acid and ethylene pathways are differentially activated in melon cotyledons by active or heat-denatured cellulase from *Trichoderma longibrachiatum*. Plant Physiol 127:334–344. https://doi.org/10.1104/pp.127.1.334

Mathys J, De Cremer K, Timmermans P, Van Kerckhove S, Lievens B, Vanhaecke M, Cammue BPA, De Coninck B (2012) Genome-wide characterization of ISR induced in *Arabidopsis thaliana* by *Trichoderma hamatum* T382 against *Botrytis cinerea* infection. Front Plant Sci 3. https://doi.org/10.3389/fpls.2012.00108

Matrood AAA, Khrieba MI, Okon OG (2020) Synergistic interaction of *Glomus mosseae* T. and *Trichoderma harzianum* R. and the induction of systemic resistance of *Cucumis sativus* L. and *Alternaria alternata* (Fr.) K. Plant Sci Today 7:101–108. https://doi.org/10.14719/PST.2020.7.1.629

Meller Harel Y, Haile Mehari Z, Rav-David D, Elad Y (2014) Systemic resistance to gray mold induced in tomato by benzothiadiazole and *Trichoderma harzianum* T39. Phytopathology 104:150–157. https://doi.org/10.1094/PHYTO-02-13-0043-R

Métraux JP (2013) Systemic acquired resistance. In: Brenner's encyclopedia of genetics, 2nd edn, pp 627–629. https://doi.org/10.1016/B978-0-12-374984-0.01509-6

Mukherjee PK, Buensanteai N, Moran-Diez ME, Druzhinina IS, Kenerley CM (2012) Functional analysis of non-ribosomal peptide synthetases (NRPSs) in *Trichoderma virens* reveals a polyketide synthase (PKS)/NRPS hybrid enzyme involved in the induced systemic resistance response in maize. Microbiology 158:155–165. https://doi.org/10.1099/mic.0.052159-0

Paudzai FAM, Sundram S, Yusof MT, Angel LPL, Hashim AM, Abdullah SNA (2019) Induced systemic resistance and promotion of plant growth in oil palm seedlings by endophytic *Trichoderma virens*. J Oil Palm Res 31:572–581. https://doi.org/10.21894/jopr.2019.0031

Perazzolli M, Dagostin S, Ferrari A, Elad Y, Pertot I (2008) Induction of systemic resistance against *Plasmopara viticola* in grapevine by *Trichoderma harzianum* T39 and benzothiadiazole. Biol Control 47:228–234. https://doi.org/10.1016/j.biocontrol.2008.08.008

Perazzolli M, Roatti B, Bozza E, Pertot I (2011) Trichoderma harzianum T39 induces resistance against downy mildew by priming for defense without costs for grapevine. Biol Control 58:74–82. https://doi.org/10.1016/j.biocontrol.2011.04.006

Pieterse CMJ, Leon-Reyes A, Van Der Ent S, Van Wees SCM (2009) Networking by small-molecule hormones in plant immunity. Nat Chem Biol. https://doi.org/10.1038/nchembio.164

Pretty J, Sutherland WJ, Ashby J, Auburn J, Baulcombe D, Bell M, Bentley J, Bickersteth S, Brown K, Burke J, Campbell H, Chen K, Crowley E, Crute I, Dobbelaere D, Edwards-Jones G, Funes-Monzote F, Godfray HCJ, Griffon M, Gypmantisiri P, Haddad L, Halavatau S, Herren H, Holderness M, Izac AM, Jones M, Koohafkan P, Lal R, Lang T, McNeely J, Mueller A, Nisbett N, Noble A, Pingali P, Pinto Y, Rabbinge R, Ravindranath NH, Rola A, Roling N, Sage C, Settle W, Sha JM, Shiming L, Simons T, Smith P, Strzepeck K, Swaine H, Terry E, Tomich TP, Toulmin C, Trigo E, Twomlow S, Vis JK, Wilson J, Pilgrim S (2010) The top 100 questions of importance to the future of global agriculture. Int J Agric Sustain 8:219–236. https://doi.org/10.3763/ijas.2010.0534

Raghavendra VB, Siddalingaiah L, Sugunachar NK, Nayak C, Ramachandrappa NS (2013) Induction of systemic resistance by biocontrol agents against bacterial blight of cotton caused by *Xanthomonas campestris* pv. *Malvacearum*. Int J Phytopathol 2:59–69. https://doi.org/10.33687/phytopath.002.01.0071

Roatti B, Perazzolli M, Gessler C, Pertot I (2013) Abiotic stresses affect *Trichoderma harzianum* T39- induced resistance to downy mildew in grapevine. Phytopathology 103:1227–1234. https://doi.org/10.1094/PHYTO-02-13-0040-R

Rotblat B, Enshell-Seijffers D, Gershoni JM, Schuster S, Avni A (2002) Identification of an essential component of the elicitation active site of the EIX protein elicitor. Plant J 32:1049–1055. https://doi.org/10.1046/j.1365-313X.2002.01490.x

Salas-Marina MA, Silva-Flores MA, Uresti-Rivera EE, Castro-Longoria E, Herrera-Estrella A, Casas-Flores S (2011) Colonization of *Arabidopsis* roots by *Trichoderma atroviride* promotes

growth and enhances systemic disease resistance through jasmonic acid/ethylene and salicylic acid pathways. Eur J Plant Pathol 131:15–26. https://doi.org/10.1007/s10658-011-9782-6
Samolski I, Rincón AM, Pinzón LM, Viterbo A, Monte E (2012) The qid74 gene from *Trichoderma harzianum* has a role in root architecture and plant biofertilization. Microbiology 158:129–138. https://doi.org/10.1099/mic.0.053140-0
Sanjeev K (2013) *Trichoderma*: a biological weapon for managing plant diseases and promoting sustainability. Int J Agric Sci Vetinary Med 1:106–121
Saxena A, Raghuwanshi R, Singh HB (2015) *Trichoderma* species mediated differential tolerance against biotic stress of phytopathogens in *Cicer arietinum* L. J Basic Microbiol 55:195–206. https://doi.org/10.1002/jobm.201400317
Schwessinger B, Zipfel C (2008) News from the frontline: recent insights into PAMP-triggered immunity in plants. Curr Opin Plant Biol. https://doi.org/10.1016/j.pbi.2008.06.001
Segarra G, Casanova E, Bellido D, Odena MA, Oliveira E, Trillas I (2007) Proteome, salicylic acid, and jasmonic acid changes in cucumber plants inoculated with *Trichoderma asperellum* strain T34. Proteomics 7:3943–3952. https://doi.org/10.1002/pmic.200700173
Seidl V, Marchetti M, Schandl R, Allmaier G, Kubicek CP (2006) Epl1, the major secreted protein of *Hypocrea atroviridis* on glucose, is a member of a strongly conserved protein family comprising plant defense response elicitors. FEBS J 273:4346–4359. https://doi.org/10.1111/j.1742-4658.2006.05435.x
Shewry PR, Hey SJ (2015) The contribution of wheat to human diet and health. Food Energy Secur. https://doi.org/10.1002/FES3.64
Shoresh M, Yedidia I, Chet I (2005) Involvement of jasmonic acid/ethylene signaling pathway in the systemic resistance induced in cucumber by *Trichoderma asperellum* T203. Phytopathology 95:76–84. https://doi.org/10.1094/PHYTO-95-0076
Singh HB (2014) Management of plant pathogens with microorganisms. Proc Indian Natl Sci Acad. https://doi.org/10.16943/ptinsa/2014/v80i2/55120
Singh UB, Malviya D, Singh S, Kumar M, Sahu PK, Singh HV, Kumar S, Roy M, Imran M, Rai JP, Sharma AK, Saxena AK (2019) *Trichoderma harzianum*-and methyl jasmonate-induced resistance to *Bipolaris sorokiniana* through enhanced phenylpropanoid activities in bread wheat (*Triticum aestivum* L.). Front Microbiol 10. https://doi.org/10.3389/fmicb.2019.01697
Stepanova AN, Yun J, Likhacheva AV, Alonso JM (2007) Multilevel interactions between ethylene and auxin in *Arabidopsis* roots. Plant Cell 19:2169–2185. https://doi.org/10.1105/tpc.107.052068
Tripathi AN, Meena BR, Pandey KK, Singh J (2020) Microbial bioagents in agriculture: current status and prospects. In: New frontiers in stress management for durable agriculture, pp 331–368. https://doi.org/10.1007/978-981-15-1322-0_20
Tucci M, Ruocco M, De Masi L, De Palma M, Lorito M (2011) The beneficial effect of *Trichoderma* spp. on tomato is modulated by the plant genotype. Mol Plant Pathol 12:341–354. https://doi.org/10.1111/j.1364-3703.2010.00674.x
Vargas WA, Mandawe JC, Kenerley CM (2009) Plant-derived sucrose is a key element in the symbiotic association between *Trichoderma virens* and maize plants. Plant Physiol 151:792–808. https://doi.org/10.1104/pp.109.141291
Vinale F, Sivasithamparam K, Ghisalberti EL, Marra R, Barbetti MJ, Li H, Woo SL, Lorito M (2008) A novel role for *Trichoderma* secondary metabolites in the interactions with plants. Physiol Mol Plant Pathol 72:80–86. https://doi.org/10.1016/j.pmpp.2008.05.005
Viterbo A, Chet I (2006) TasHyd1, a new hydrophobin gene from the biocontrol agent *Trichoderma asperellum*, is involved in plant root colonization. Mol Plant Pathol 7:249–258. https://doi.org/10.1111/j.1364-3703.2006.00335.x
Viterbo A, Wiest A, Brotman Y, Chet I, Kenerley C (2007) The 18mer peptaibols from *Trichoderma virens* elicit plant defence responses. Mol Plant Pathol 8:737–746. https://doi.org/10.1111/j.1364-3703.2007.00430.x
Vitti A, Nuzzaci M, Scopa A, Tataranni G, Remans T, Vangronsveld J, Sofo A (2013) Auxin and cytokinin metabolism and root morphological modifications in *Arabidopsis thaliana* seed-

lings infected with cucumber mosaic virus (CMV) or exposed to cadmium. Int J Mol Sci 14:6889–6902. https://doi.org/10.3390/ijms14046889

Wightwick A, Walters R, Allinson G, Reichman S, Menzies N (2010) Environmental risks of fungicides used in horticultural production systems. In: Fungicides. https://doi.org/10.5772/13032

Woo SL, Donzelli B, Scala F, Mach R, Harman GE, Kubicek CP, Del Sorbo G, Lorito M (1999) Disruption of the ech42 (Endochitinase-encoding) gene affects biocontrol activity in *Trichoderma harzianum* P1. Mol Plant-Microbe Interact 12:419–429. https://doi.org/10.1094/MPMI.1999.12.5.419

Yedidia I, Benhamou N, Chet I (1999) Induction of defense responses in cucumber plants (Cucumis sativus L.) by the biocontrol agent *Trichoderma harzianum*. Appl Environ Microbiol 65:1061–1070. https://doi.org/10.1128/aem.65.3.1061-1070.1999

Yedidia I, Benhamou N, Kapulnik Y, Chet I (2000) Induction and accumulation of PR proteins activity during early stages of root colonization by the mycoparasite *Trichoderma harzianum* strain T-203. Plant Physiol Biochem 38:863–873. https://doi.org/10.1016/S0981-9428(00)01198-0

Yedidia I, Shoresh M, Kerem Z, Benhamou N, Kapulnik Y, Chet I (2003) Concomitant induction of systemic resistance to *Pseudomonas syringae* pv. *lachrymans* in cucumber by *Trichoderma asperellum* (T-203) and accumulation of phytoalexins. Appl Environ Microbiol 69:7343–7353. https://doi.org/10.1128/AEM.69.12.7343-7353.2003

Yoshioka Y, Ichikawa H, Naznin HA, Kogure A, Hyakumachi M (2012) Systemic resistance induced in *Arabidopsis thaliana* by *Trichoderma asperellum* SKT-1, a microbial pesticide of seedborne diseases of rice. Pest Manag Sci 68:60–66. https://doi.org/10.1002/ps.2220

Zhang F, Yuan J, Yang X, Cui Y, Chen L, Ran W, Shen Q (2013) Putative *Trichoderma harzianum* mutant promotes cucumber growth by enhanced production of indole acetic acid and plant colonization. Plant Soil 368:433–444. https://doi.org/10.1007/s11104-012-1519-6

Role of *Trichoderma* in Plant Growth Promotion

Sivagami Subramaniam, Nur Ain Izzati Mohd Zainudin, Asma Aris, and Zainap Ab Easa Hasan

Contents

1 Introduction

Trichoderma sp. is a widespread soilborne ascomycete that bears green spores commonly classified as avirulent symbionts for plants, but is well-known for its antagonistic and mycoparasitism mechanisms toward fungal diseases (Nakkeeran et al.

S. Subramaniam · Nur Ain Izzati Mohd Zainudin (✉) · A. Aris
Department of Biology, Faculty of Science, Universiti Putra Malaysia, Serdang, Selangor, Malaysia
e-mail: ainizzati@upm.edu.my

Z. Ab Easa Hasan
Department of Biology, Faculty of Science, Universiti Putra Malaysia, Serdang, Selangor, Malaysia

Department of Botany, Faculty of Science, Omar Al Mukhtar University, Al Bayda, Libya

N. Amaresan et al. (eds.), *Advances in* Trichoderma *Biology for Agricultural Applications*, Fungal Biology, https://doi.org/10.1007/978-3-030-91650-3_9

2018). Ninety percent of the *Trichoderma* species application has been experimentally carried out from various strains (Kredics et al. 2018). Lately, the search to develop *Trichoderma* as a promising chemical-free fertilizer is overwhelming due to its capability to boost plant growth. Additionally, *Trichoderma* sp. is an omnipresent colonizer of cellulose substances and can usually be found where decomposing plants are available and in the vicinity of the plant roots, causing the induction of systemic resistance in plants against plant pathogenic organisms (Nawrocka and Małolepsza 2013). During the infection and in application of *Trichoderma* species for disease management, the responses that occur include the systemic acquired resistance (SAR) induced by plant pathogenic organisms and the induced systemic resistance (ISR) mediated by *Trichoderma*. Generally, when there is a plant pathogen invasion, the plant gives precedence to its defense mechanisms despite its growth that causes stunt (Jiang et al. 2017). Therefore, the *Trichoderma* assists the plants in their defense mechanism toward diseases through systemic resistance induction primarily to enhance their growth.

Trichoderma is claimed to function antagonistically toward plant pathogenic organisms via various mechanisms including competition, enzyme secretion, hyphal interactions, and mycoparasitism (Bastakoti et al. 2017). Besides, *Trichoderma* is also potent in diminishing pathogenic activity of these pathogens by priming for an improved defense mechanism of the host plants (Hermosa et al. 2012; Kottb et al. 2015). Literally, an induced resistance indicates a state of a plant where it becomes infrequently infected compared to normal healthy plants that are non-induced (Loon 2007). Unlike systemic acquired resistance (SAR) that is salicylic acid-dependent, ISR relies on the pathways coordinated by jasmonic acid and ethylene ratio (Elsharkawy et al. 2013). Furthermore, *T. harzianum* has depicted systematic inhibition in tomato plants infected with *Cucumber mosaic virus* (CMV). After 3 months of observation since the virus has been inoculated in the host plant, the jasmonic acid and ethylene ratio was witnessed as the highest in the leaves of the plant inoculated with CMV after a week of *T. harzianum* inoculation, comparatively with the infected plant with no treatment, the non-infected plant with *Trichoderma* alone, and the healthy plant (Vitti et al. 2016). Similarly, Yuan et al. (2019) also proclaimed that *T. longibrachiatum* exhibited increased activation of both systemic resistance pathways via phytohormones secretion such as jasmonic acid, ethylene ratio, and salicylic acid (SAR) to suppress *Botrytis cinerea* in cucumber. Due to the activated plant mechanisms in the presence of arbuscular mycorrhizal, the defense mechanisms were plant-mediated. Usually, the mechanism of ISR results in an improved defensing capacity of a plant (Loon 2007). Therefore, upon the presence of the plant pathogen, the host manifests this boosted protective inclination, thus resulting in a reduced disease development rate, lessened disease incidence, and severity.

In contradiction with the resistance gene in highly specific plants, ISR is a nonspecific mechanism that efficiently exhibits its antagonism toward a vast number of plant pathogenic organisms comprising insects and nematodes (Miyashita and Takahashi 2015; Guo and Ge 2017). Although a wide range of pathogens is inhibited, the expression might defer depending on the nature of the plant pathogen with the plant and inducing inoculum. After induction, this ISR resistance of plant protects the plant and persists for a more prolonged period or at least a few months stably (Nawrocka and Małolepsza 2013).

On the other hand, SAR is another defense regulatory pathway like ISR but dependent on a different growth-regulating hormone of plants named salicylic acid (SA). Generally, this SA-mediated mechanism grants enduring protection against a vast array of plant pathogenic microbes (Durrant and Dong 2004). In recent research conducted by Yang et al. (2013), hydrogen peroxide (H_2O_2) conferred to be the second messenger of SAR. In agreement with that, Tewari and Paek (2011) reported that the accumulation of H_2O_2 in the adventitious root of *Panax ginseng* after 40 days of treatment was due to the SA-triggered pathway. Similarly, when there is an increment of toxicity in the concentration of lead ion, the leaves of the SA pretreated rice seedlings had a greater level of H_2O_2. The production of H_2O_2 has simultaneously depicted an elevation with the increased toxicity produced by the SA mechanism (Jing et al. 2007). However, there are some contradictory observations inscribed from other researchers for the chronology of H_2O_2 presence in the SA-influenced pathway. León et al. (1995) and Chamnongpol et al. (1998) advocated that H_2O_2 initiates the biosynthesis of SA in *Nicotiana tabacum* leading to an enhanced plant tolerance against abiotic stresses. At all events, the production of SA and H_2O_2 are interrelated in the systemically acquired defensive pathway in a plant to intensify its immune system. When an observation of H_2O_2 is proclaimed to be due to the presence of *Trichoderma*, that could be inferred from the SAR of the plants against the plant pathogens (Yang et al. 2013).

2 *Trichoderma* spp. Reduce Disease Severity and Promotes Plant Growth

Trichoderma is well-known for its ability to alter various plant physiological processes such as stomatal conductance, transpiration rate, net photosynthesis rate, carbon dioxide regulation, water, and nutrient uptake (Doni et al. 2014a), which can promote plant growth (Table 1). This genus enhances the nutrient uptake that is subsequently increasing the physiological activities within treated plants. *Trichoderma* facilitates nutrient availability in the plant through solubilization and chelation of minerals engaged with plant metabolism. This leads to the enhancement of physiological activities (Harman et al. 2004). It also enhances plant growth such as shoot and root length, water content, number of leaves and flowers, chlorophyll content, and plants' photosynthetic efficiencies (Table 1).

Trichoderma could reduce the disease severity of several plant diseases such as *Fusarium* wilt disease in banana (Sharifah, 2018) and cucumber (Asma, 2019) (Fig. 1). An investigation has been conducted using *T. harzianum* against damping-off caused by *Phytophthora melonis* in cucumber, which depicted an accelerated exudation of polyphenol oxidase and peroxidase enzymes (Table 1).

The presence of peroxidase enzyme that is responsible for the disintegration of H_2O_2 justified the secretion of the second messenger of the SA-mediated SAR

Table 1 Role of *Trichoderma* in plant growth promotion in infected plants

No.	*Trichoderma* species	Plant	Disease	Plant pathogenic organism or a virus	Effects	Reference
1	*T. asperelloides*	*Arabidopsis*	*Fusarium* wilt	*Fusarium oxysporum*	Reduce disease severity	Gupta et al. (2014)
2	*T. asperellum*	*Arabidopsis thaliana*	Black spot and gray mold	*Alternaria brassicicola* and *Botrytis cinerea*	Reduce disease severity; enhance plant growth (leaf area, root length, and fresh weight); increase production of defense molecules (ROS, camalexin, anthocyanins, and the SA-dependent plant hormone pathways)	Kottb et al. (2015)
3	*T. asperellum*	*Arabidopsis thaliana*	Cucumber mosaic	*Cucumber mosaic virus*	Reduce disease severity; enhance defense mechanism by producing SA inducible genes	Elsharkawy et al. (2013)
4	*T. asperellum*	Barley	Net blotch	*Drechslera teres*	Reduce disease severity; enhance plant growth (kernel weight, grain yield per plot); increase catalase, peroxidase, and polyphenol oxidase activities	Hafez et al. (2019)
5	*T. asperellum*	Olive	*Verticillium* wilt	*Verticillium dahliae*	Reduce disease severity	Carrero-carron et al. (2016)
6	*T. asperellum*	Tomato	Gray mold	*Botrytis cinerea*	Reduce disease severity; decrease ROS levels	Herrera-Tellez et al. (2019)
7	*T. asperellum*	Tomato	*Fusarium* wilt	*F. oxysporum* f. sp. *lycopersici*	Reduce the severity of *Fusarium* wilt; enhance all plant growth (shoot and root length, water content, number of leaves and flowers, and chlorophyll content); enhance the quality and increase nutrient contents of tomato fruit (lycopene, sugar, K, N, Ca, P, and Mg); enhance nutrient uptake by plants via mineral solubilization	Zainap et al. (2020) Zainap et al. (2020) Li et al. (2018)
8	*T. asperellum*	Tomato	Damping-off and root-rot	*Pythium aphanidermatum*	Reduce disease severity; stimulate systemic defense response by activating defense enzymes (peroxidase, polyphenol oxidase, and chitinase); enhance plant growth (chlorophyll content)	Elshahawy and El-Mohamedy (2019)

9	*T. atroviride*	Beans	Gray mold rots	*Botrytis cinerea*	Reduce disease severity; increase plant resistance; provide glucose oxidase molecules	Brunner et al. (2005)
10	*T. atroviride*	Beans	Seedling death, root and hypocotyl rot, stem cankers, and pod rot	*Rhizoctonia solani*	Reduce disease severity; increase plant resistance; provide glucose oxidase molecules	Brunner et al. (2005)
11	*T. atroviride*	Rice	Brown spot	*Bipolaris oryzae*	Reduce disease severity; enhance plant growth (root and stem dry weight, root and stem wet weight, root length and shoot height)	Khalili et al. (2012)
12	*T. atroviride*	Beans	Root rot	*Pythium ultimum*	Reduce disease severity; increase plant resistance; provide glucose oxidase molecules	Brunner et al. (2005)
13	*T. harzianum*	Cocoa plant	Witches' broom disease	*M. perniciosa*	Reduce disease severity	Marco and Felix (2002)
14	*T. harzianum*	*Elaeis guineensis*	Basal stem rot disease	*Ganoderma boninense*	Reduce disease severity; increase systemic resistance of plant; enhance production of phytohormones and mineral solubilization; enhance plant growth (secondary roots, biomass, and root extension)	Nusaibah and Musa (2019)
15	*T. harzianum*	Grapevine	Grey mold	*B. cinerea*	Reduce disease severity	O'Neill et al. (1995)
16	*T. harzianum*	Maize	Late wilt disease	*Cephalosporium maydis*	Reduce disease severity; enhance plant growth (plant height, dry weight, and yield) (number and weight of kernel, ear length and diameter, and ratio of weight of grains to plot and plant)	Elshahawy and El-Sayed (2018)

(continued)

Table 1 (continued)

No.	*Trichoderma* species	Plant	Disease	Plant pathogenic organism or a virus	Effects	Reference
17	*T. harzianum*	Oil palm	Leaf spot disease	*Curvularia oryzae*	Reduce disease severity; enhance the activities of phenylalanine ammonia-lyase, peroxidase, and polyphenol oxidase enzymes	Sunpapao et al. (2018)
18	*T. harzianum*	Onion	Damping-off	*F. solani*	Reduce disease severity; increase plant growth (seedling and root length, water content, and number of leaves)	Dabire et al. (2016)
19	*T. harzianum*	Onion	Damping-off	*F. oxysporum*	Reduce disease severity; increase growth parameters (seedling and root length, water content, and number of leaves)	Dabire et al. (2016)
20	*T. harzianum*	Onion	Basal rot	*F. oxysporum* f. sp. *Cepae*	Reduce the severity of basal rot	Akhtar and Javaid (2016)
21	*T. harzianum*	Potato	Stem canker and black scurf	*Rhizoctonia solani*	Improves sprouting; increase sprout weight and vigor and root weight; decrease disease severity of stem canker and black scurf; increase polyphenol oxidase, peroxidase total phenol, and chlorophyll content	Al-askar et al. (2016)
22	*T. harzianum*	Rice	Brown spot	*Bipolaris oryzae*	Reduce disease severity; enhance plant growth (root and stem dry weight, root and stem wet weight, root length, and shoot height)	Khalili et al. (2012)
23	*T. harzianum*	Sugarcane	Red rot	*Colletotrichum falcatum*	Reduce disease severity; enhance nutrient uptake; enhance plant growth (ratoon initiation, number of tillers, cane height and weight, girth, number and length of internode, and number of millable cane and average yield)	Singh et al. (2010)
24	*T. harzianum*	Tomato	Cucumber mosaic	*Cucumber mosaic virus*	Reduce disease severity; enhance plant growth (plant height, photosynthesis, total chlorophyll content, and plant gas exchange); induce systemic resistance	Vitti et al. (2016)

25	*T. harzianum*	Tomato	Fusarium wilt	*F. oxysporum* f. sp. *lycopersici*	Reduce the severity of *Fusarium* wilt; enhance all plant growth (shoot and root length, water content, number of leaves, chlorophyll content, and IAA); solubilize phosphate	Bader et al. (2020)
26	*T. harzianum*	Tomato	Damping-off	*Rhizoctonia solani*	Reduce disease severity; increase the vigor index of the plants; enhance nutrient uptake of plant; enhance plant growth (shoot length, fresh and dry weight, root length, fresh and dry weight, chlorophyll content, and yield)	Singh et al. (2014)
27	*T. harzianum*	Broccoli	Root rot	*Pythium ultimum*	Reduce disease severity	El-Mohamedy (2012)
28	*T. harzianum*	Cucumber	Damping-off	*Phytophthora melonis*	Reduce disease severity; increase total phenolic content, polyphenol oxidase activity, and peroxidase enzyme activity	Sabbagh et al. (2017)
29	*T. harzianum*	Cucumber and pepper	Damping-off disease	*Pythium* spp. and *Rhizoctonia solani*	Reduce disease severity; enhance plant growth (seedling height, leaf area, plant dry weight, and chlorophyll content)	Inbar et al. (1994)
30	*T. harzianum*	Pea plant	Damping-off disease	*Pythium ultimum*	Reduce disease severity; enhance plant growth parameters (number of lateral roots, wet root weights, shoot and root length; decrease the activity of C, N, and P cycle enzymes)	Naseby et al. (2000)
31	*T. harzianum*	Potato and tomato	Blight disease	*Phytophthora infestans*	Reduce disease severity	Kerroum et al. (2015)
32	*T. harzianum*	Tomato	Damping-off and root-rot	*Pythium aphanidermatum*	Reduce disease severity; stimulate systemic defense response by activating defense enzymes (peroxidase, polyphenol oxidase, and chitinase); enhance growth parameters (chlorophyll content)	Elshahawy and El-Mohamedy (2019)

(continued)

Table 1 (continued)

No.	*Trichoderma* species	Plant	Disease	Plant pathogenic organism or a virus	Effects	Reference
33	*T. koningii*	*Elaeis guineensis*	Basal stem rot disease	*Ganoderma boninense*	Reduce disease severity; increase systemic resistance of plant; enhance production of phytohormones, mineral solubilization; enhance plant growth (secondary roots, biomass, and root extension)	Nusaibah and Musa (2019)
34	*T. koningii*	Maize	Late wilt disease	*Cephalosporium maydis*	Reduce disease severity; enhance plant growth (plant height, dry weight, and yield, number and weight of kernel, ear length and diameter, and ratio of weight of grains to plot and plant)	Elshahawy and El-Sayed (2018)
35	*T. longibrachiatum*	Cucumber	Gray mold	*Botrytis cinerea*	Reduce disease severity; enhance plant growth (root length, plant height, and fresh weight); induce systemic resistance by initiating phytohormones and secondary metabolite secretion	Yuan et al. (2019)
36	*T. polysporum*	Cereals	Snow rot	*Pythium iwayamai*	Reduce disease severity	Kamo et al. (2016)
37	*T. virens*	Maize	Late wilt disease	*Cephalosporium maydis*	Reduce disease severity; enhance plant growth (plant height, dry weight and yield, number and weight of kernel, ear length and diameter, and ratio of weight of grains to plot and plant)	Elshahawy and El-Sayed (2018)
38	*T. virens*	Rice	Brown spot	*Bipolaris oryzae*	Reduce disease severity; enhance plant growth (root length)	Khalili et al. (2012)
39	*T. virens*	Tomato	Damping-off and root-rot	*Pythium aphanidermatum*	Reduce disease severity; stimulate systemic defense response by activating defense enzymes (peroxidase, polyphenol oxidase, and chitinase); enhance chlorophyll content	Elshahawy and El-Mohamedy (2019)

40	*T. viride*	Lemon	Citrus canker	*Xanthomonas citri*	Reduce disease severity	Jatav et al. (2018)
41	*T. viride*	Maize	Late wilt disease	*Cephalosporium maydis*	Reduce disease severity; enhance plant growth (plant height, dry weight and yield, number and weight of kernel, ear length and diameter, and ratio of weight of grains to plot and plant)	Elshahawy and El-Sayed (2018)
42	*T. viride*	Broccoli	Root rot	*Pythium ultimum*	Reduce disease severity	El-Mohamedy (2012)

Fig. 1 *Trichoderma* reduces disease severity of several plant diseases such as Fusarium wilt disease in banana and cucumber. Cucumber plants after 30 days post inoculation, inoculated plants treated with *T. asperellum* (**a**), plant infected with *F. solani* (**b**), Infected banana plant (**c**) and treated banana plant with *Trichoderma*. (**a–b** are reproduced from Asma, 2019; **c–d** are reproduced from Sharifah, 2018 with permission)

pathway. Decisively, the plants treated with *Trichoderma* exhibited a lesser percentage (40%) contrasted with the control of 100% disease severity and an increased length and weight of the root with lower necrosis (Sabbagh et al. 2017). Similar results were regarded when several *Trichoderma* isolates (*T. asperellum, T. harzianum,* and *T. virens*) were evaluated for *Pythium aphanidermatum,* the damping-off causal agent of tomato (Elshahawy and El-Mohamedy 2019), by suppressing the reactive oxygen species production. Indeed, under the biotic and abiotic stresses, plants inoculated with *Trichoderma* sp. are to display a remarkable rise in antioxidants like polyphenols along with the SAR (Sabbagh et al. 2017; Herrera-Tellez et al. 2019; Sood et al. 2020).

3 *Trichoderma* spp. Enhance Photosynthetic Performance of Plants

Photosynthesis is an essential process of green plants for their survival in which the light (photo-) is synthesized into chemical energy. This process resides in the chloroplast of the plant where the light or photon is captured by the thylakoid membrane of grana and yields a reducing agent named nicotinamide adenine dinucleotide phosphate (NADPH). This process involves the photoreduction of water molecules into H^+ ions and oxygen gas (O_2) as well as photophosphorylation, which converts adenosine diphosphate (ADP) to adenosine triphosphate (ATP), an energy-storing molecule. The continuation of this process occurs independently of light named Calvin cycle that is catalyzed by ribulose-1,5 biphosphate carboxylase (Rubisco) for carbon fixation to generate carbon dioxide and glucose (Harman et al. 2019). Magnesium (Mg) is one of the essential chlorophyll constituents that involves in the gene regulation, and enzymatic activity associated with photosynthesis and *Trichoderma* spp. can enhance the Mg uptake of the plants (Sood et al. 2020). When a plant absorbs a complete package of nutrients required, photosynthesis is reported to escalate the growth of the plant by one-third of its performance percentage, thus boosting the plant growth (Kirschbaum 2011). Additionally, similar outcomes have been recorded by Doni et al. (2014a) when *Trichoderma* significantly increased the photosynthetic rate by threefold of the rice seedlings compared to the seedling applied with common nitrogen-phosphorus-potassium (N:P:K) fertilizer.

Stewart and Hill (2014) described that in a research conducted by Inbar et al. (1994), plant dry weight, leaf area, plant health, and the chlorophyll content of pepper and cucumber seedlings inoculated with *T. harzianum* were increased than those of uninoculated seedlings in a greenhouse. These potentials of *Trichoderma* were also noticed when the plants acquired sufficient nutrients for their growth and no variation in the content of nitrogen and phosphorus levels of the plants (Harman and Bjorkman 1998). The role of *Trichoderma* is significant although a plant is in its optimal conditions. Similarly, in a Romaine lettuce tested with *Trichoderma* under optimal condition without N:P:K with the absence of any plant pathogenic organisms, an increased greenness, size, and several leaves that ensure the ability of arbuscular mycorrhiza to enhance the plant's photosynthetic capacity was observed (Rouphael et al. 2020).

As mentioned earlier, the volatile organic compounds secreted through secondary metabolism can also improve the plant's defensive mechanism in abiotic and biotic strains as well as promote plant growth via the escalation of photosynthesis. According to that, an investigation with *the Arabidopsis thaliana* model system was performed where the seeds were grown in a shared atmosphere with a non-direct physical contact with *T. viride*. Starting the third week of exposure, a significant increase in the total chlorophyll concentration comparatively with the controls was noted. In the fourth week of observation, there was about 58% increment in the chlorophyll content and 45% increase in the fresh weight of the plant (aboveground). The volatile organic compounds of *T. viride* have been evaluated through thermal desorption-gas chromatography-mass spectrometry (TD-GC-MS) and discerned that a vast number of them were present including isobutyl alcohol, 3-methylbutanal, and isopentyl alcohol (Hung et al. 2013).

A similar kind of experiment with non-direct physical contact with *Trichoderma* toward *A. thaliana* has also been conducted by Lee et al. (2015). In the experiment, the age of the seedlings was taken into account to evaluate the chlorophyll content. When 7-day-old seedlings were exposed with 5-day-old *T. atroviride*, there was a drastic increase observed in the chlorophyll content compared to control after 2 weeks of treatment. The volatile organic compounds that resulted from this analysis included alcohols, aldehydes, alkenes, aromatics, and ketones. Generally, the volatile organic compounds are secreted by a variety of *Trichoderma* species comprising *T. viride*, *T. pseudokoningii*, *T. longibrachiatum*, *T. harzianum*, *T. asperellum*, and *T. aggressivum* that have been declared to elevate the total chlorophyll content of *A. thaliana* (Hesham et al. 2020). Even with a non-direct physical contact of *T. atroviride*, *T. citrinoviride*, *T. harzianum*, *T. koningii*, *T. longibrachiatum*, *T. viride*, and *T. viridescence*, they have depicted an increase in chlorophyll content followed by fresh shoot and root weights (Jalali et al. 2017). The photosynthetic rate and stomatal conductance of infectious plants can be enhanced as much as 45%. Besides, the chlorophyll content as well as relative greenness of infected plants can be increased by *Trichoderma* treatment (Shukla et al. 2012).

4 *Trichoderma* spp. Improve Stomatal Conductance

Stomatal conductance is the potential of stomata to permit the gaseous exchange of carbon dioxide (CO_2) and transpiration (Clavijo-herrera et al. 2018). Generally, the conductance of stomata is positively correlated with the photosynthetic rate. Since most of the *Trichoderma* spp. are potent to increase the photosynthetic rate, they regulate the conduction of stomata. If a plant is detected to face stress in some scenarios like drying soil or drought, the capacity of leaves to hold water would decrease leading to the closing of stomata. Some journals have reported that even when the water level in the leaves is not interrupted, the stomatal conductance and leaf growth rate are still able to decline due to the drying soil (Passioura 2002).

It has been confirmed that *Trichoderma* undeniably regulates photosynthesis as well as gaseous exchange and stomatal conductance (Sood et al. 2020). Additionally, Sood et al. (2020) reported on the capability of *Trichoderma* in enhancing the photosynthetic rate and stomatal conductance by threefold compared to the plants fertilized with general chemical fertilizer. Sood et al. (2020) also highlighted that the regulation and conduction of stomata during the investigation with *T. atroviride* and *T. virens* are induced by abscisic acid (ABA), a phytohormone that regulates the growth of a plant by performing various physiological processes through the stomatal aperture (Chen et al. 2019). When the maize and rice seedlings are subjected to *T. harzianum* treatment under various salinity conditions by sodium chloride (NaCl), the stomatal conductance and water content computed are better than the seedlings without the fungal inoculation (Yasmeen and Siddiqui 2017). Besides, the increased concentration of carbon dioxide can alter the conduction of stomata, especially during water stress conditions. During a high gaseous exchange, the concentration of CO_2 absorbed is recorded to be higher and can concurrently boost photosynthesis and stomata aperture (Kirschbaum 2011). This assertion is also supported by Aguera and Haba (2018) in their study where the escalated CO_2 promoted photosynthetic activity by its involvement in the carbon fixation that enhanced the activity of Rubisco and reduced photorespiration. Citric acid that is yielded via mineral solubilization of *T. harzianum* is also capable of enhancing the conduction of stomata by promoting gaseous exchange and photosynthesis in lead-exposed bean plants

(Mallhi et al. 2019). It is also discerned that when there is declination in the photosynthesis rate due to phytopathogen presence, the stomatal conductance decreased. Nonetheless, the reverse mechanism where the stomatal induction affects the photosynthetic rate has been also noticed. During the disease development of CMV, the photosynthetic activity is diminished due to the stomatal limitations. Also, a non-stomatal inhibition by the CMV has been reported in 3-month-old cucumber without the inoculation of *Trichoderma* (Vitti et al. 2016).

Resulting from the increased photosynthesis rate and stomatal conductance, the root mass and canopy mass of the rice plant grown through the system of rice intensification (SRI) method with inoculation of *T. asperellum* were significantly higher. This enhancement was probably contributed by the chlorophyll concentration and stomatal density (Harman 2019). As the *Trichoderma* sp. has been noted to be a potent mineral solubilizer resulting in nitrogen oxide responsible for functionality and regulation of stomata, the *Trichoderma* indirectly plays a vital role in the stomatal conduction (Gupta et al. 2014). Additionally, Nusaibah and Musa (2019) also denoted the potential of *Trichoderma* in delaying the effects of droughts including emission of green fluorescence, stomatal conductance, and photosynthesis rate in an ameliorated water status. Furthermore, in a study aimed to utilize the sugarcane and corn bagasse as a *Trichoderma* carrier, a greater value of stomatal conductance was reported from the plant with *Trichoderma* inoculant than the plant fertilized with NPK fertilizer. The correlation of the stomatal conductance with photosynthesis is explained further with the content of water and CO_2 where these elements are required to pass through the stomata openings to enter the chloroplast in the stroma (Doni et al. 2014b).

5 *Trichoderma* spp. Enhance Nutrient Uptake

Trichoderma, as a plant growth regulator, solubilizes minerals via acidification (organic acids), chelation (siderophores), redox (ferric reductase), and hydrolysis (phytase) (Hesham et al. 2020). These properties are supported by Li et al. (2015) where the *T. harzianum* solubilized the iron (III) oxide, metallic zinc, phytate, and copper (II) oxide and led to the production of organic acids like tartaric acid, succinic acid, lactic acid, and citric acid that enhances plant growth (Mallhi et al. 2019).

In addition, ten isolates of *Trichoderma* have been subjected to solubilize the phosphorus from calcium phosphate found in the rhizosphere of *Avicennia marina*, and all the isolates have depicted positive solubilization corresponding to the phytase activity. The more the extracellular phytase activity, the higher the phosphorus solubilizing potential of the *Trichoderma* isolates (Saravanakumar et al. 2013). Since the *Trichoderma* can boost plant resistance through ISR and SAR as well as secreting secondary metabolites, the proteome and transcriptome of the plants are modified for enhancing nutrient uptake (Harman 2005). For example, in a study demonstrating sugarcane that was infected with red rot, the *T. harzianum* boosted the nutrient in the soil via solubilization with an average of 42.96% for all the available soil nutrients comprising nitrogen, carbon, potassium, phosphorus, copper, manganese, zinc, and iron (Singh et al. 2010). Li et al. (2018) also supported this assertion as the *T. asperellum* enhanced the nutrient uptake of the *Fusarium* wilt diseased tomato plants by improving the systemic resistance of the plant, making the nutrients available by mineral solubilization and increasing the surface area for a promoted root growth. Besides, *T. harzianum* is potent in increasing the performance of mineral solubilization of micronutrients, iron, manganese, and magnesium by decreasing the soil pH level, thus sustaining the soil fertility with adequate nutrients (Fiorentino et al. 2018). The reduced pH level causes an enhanced the solubility of insoluble compounds and availability of micronutrients for the accessibility of the plants (Azarmi et al. 2011).

It has been also emphasized that the efficiency of nutrient uptake can be boosted by increasing the root surface area, suggesting the ability of the *Trichoderma* in promoting the growth of the lateral roots of a plant (Lamont 1982). In a research conducted by Contreras-Cornejo et al. (2014), the *T. virens* and *T. atroviride* remarkably enriched the root hair and lateral root growth via indole-3-acetic acid biosynthesis in ideal-conditioned and stressed plants. In agreement to that, Hesham et al. (2020) also mentioned the significant growth of the lateral root of the tomato seedlings by *T. viride*. In general, the intensified mineral solubilization, carbohydrate metabolism, and photosynthetic activity, as well as the rooting depth, lead to an improved root growth for a plant to withstand drought conditions (Nusaibah and Musa 2019).

6 *Trichoderma* spp. Delay Senescence

Plant senescence or ageing is a process that begins with chlorophyll degradation and revelation of carotenoids like xanthophylls and anthocyanin causing the leaves to transform their color from green to yellow and red (Keskitalo et al. 2005). Generally, senescence in plants can be induced by stress and age. One of the causal agents of senescence is the low content of nitrogen that alters the gene expression, aspects of photosynthesis, protein content and production, as well as nitrogen and sugar metabolism (Agüera and Haba 2018). El-Katatny (2010) described that *T. harzianum* assists the nitrogen-fixation by promoting the growth of *A. brasilense*, a soil-borne bacterium commonly found in the rhizosphere. Decreased photosynthetic activity and chlorophyll content also urge senescence in plants. As discussed earlier, since the *Trichoderma* sp. can boost the chlorophyll content, it also increases the lifespan of the leaves and plants. Besides, *Trichoderma* naturally possesses high melatonin levels that are claimed to delay the senescence of a plant (Arnao and Hernandez-Ruiz 2017). Recent reports suggested that *Trichoderma* spp. have extensive potential in degrading the cellulose, thus releasing a notable amount of nitrogen in the rhizosphere of the rice plant. Again, this escalated content of nitrogen possesses a positive correlation for prolonged photosynthetic activity, enhanced root growth, as well as delayed senescence (Doni et al. 2014a).

Although the nitrogen fixation is evident by *Trichoderma* sp., a short-term nitrogen oxide has also been detected from the interaction between *T. asperelloids* and *A. thaliana* (Gupta et al. 2014). Anyhow, more articles are supporting the nitrogen fixation of *Trichoderma* benefitting plants in various aspects. For example, Singh et al. (2019) recorded the role of *T. asperellum* as one of the synthesizers of nitrogen oxide resulted from the interrelation with the root of *Nicotiana tobaccum.* Lately, the nitrogen oxide known as a gaseous reactive oxygen species has been recognized for its crucial role in plant physiological processes including photosynthesis, root organogenesis, plant pathogen defense, seed germination, hypocotyl growth, floral regulation, and finally senescence.

In a study conducted to identify the role of *Trichoderma* in drought tolerance of rice plants, a low number of scorched leaves were observed in response to the *T. harzianum* colonized plant. The delayed senescence in rice plants is positively correlated with its drought tolerance. When the transformation of *psag12-ipt* gene holds the leaf senescence of rice, the leaf treatment of *T. harzianum* focuses on promoting the total leaf photosynthetic pigments. Meanwhile, the lesser number of scorching leaves is assumed to be due to the prominent drought tolerance induction by *T. harzianum* (Pandey et al. 2016).

7 *Trichoderma* spp. Mediate Drought Tolerance

Drought in plants can be defined as water deficit in plants, which is one of the important stresses faced by plants. In general, there are more than 80% of non-woody plant biomass occupied by water content, and if there is a reduction in the content, the plants will be pushed to stress. Usually, water stress in plants is caused by low rainfall, variance in temperature, salinity, and high light intensity. On the other hand, the event where there is sufficient water supply and the plants possess a reduced potential to uptake water is called pseudo-drought. Injuries and phytopathogens are accountable for this scenario (Salehi-Lisar and Bakhshayeshan-Agdam 2016). Usually, water scarcity leads to leaf size reduction, stem elongation suspension, decreased root colonization, increased water functionality disruption, and interference of abundance biochemical and physiological responses, thus seizing the plant growth (Farooq et al. 2009).

This phenomenon in plants can be fixed by escalating their tolerance against drought through an increased water-retaining capacity. For instance, a gene named aquaglyceroporin from *T. harzianum* isolated and overexpressed in *N. tabacum* plants subjected to the high salinity of NaCl has been observed to improve the efficiency of water use and drought tolerance. It has also been deduced that the upregulation of *aqgp* gene enhances the stomatal conductance, transpiration rate, efficient photosynthesis, turgor recovery, cellular water status, and transportation of CO_2 supported by water under water stress (Vieira et al. 2017). At the same time, the colonization of *Trichoderma* with the root of rice seedlings under drought conditions delayed the wilting of the plants due to the boosted leaf greenness, stomatal conductance, and photosynthesis rate. Also, it has been emphasized that *Trichoderma* is potent to maximize the drought tolerance of plants at 40% even after 9 days of drought stress. The drought condition in rice plants led to a boosted production of stress metabolites and declination in the membrane stability index (MSI). However, the colonization of the root of the plants by *T. harzianum* causes a reduction in the malondialdehyde, H_2O_2, and proline contents as well as a promotion in the phenolic compounds and MSI (Shukla et al. 2012). Similar observations have been noted with the seed biopriming of wheat, *Triticum aestivum*, where the *T. harzianum* regulated the osmotic pressure and ameliorates the root vigor as well as enhanced the physiological defense mechanism of plants against oxidative stress through increased scavenging and L-PAL level (Shukla et al. 2014). In addition to that, *T. harzianum* has been also discerned to cause similar effects on the drought-stressed rice plants with further modulation of *dhn/aqu* transcript level, lipid peroxidation yield, superoxide dismutase, and the growth properties of stressed plants (Pandey et al. 2016). Not only with the perennial grasses (rice and wheat) but a homogenous effect by *T. harzianum*, *T. virens*, *T. atroviride*, *T. parareesei*, and *T asperellum* has also been observed in the shrubs (*Brassica napus, A. thaliana*) and herbaceous plants (*Solanum lycopersicum, Zea mays*) (Contreras-cornejo et al. 2014; Guler et al. 2016; Alwhibi et al. 2017; Hidangmayum and Dwivedi 2018; Poveda 2020; Estévez-Geffriaud et al. 2020).

8 *Trichoderma* spp. Induce Secretion of Phytohormones

Phytohormones are also known as plant hormones, which are a group of little quantities of growth-regulators, encompassing auxins, abscisic acid (ABA), ethylene, gibberellic acid (GA), and cytokinin (Wani et al. 2016). A substantial number of studies exhibited that *Trichoderma* is proficient in enhancing plant growth by inducing growth-promoting hormone secretion (Chepsergon et al. 2014).

Indole-3-acetic acid (IAA) is the most abundant and fundamental auxin natively found with multifunctions in the plants. It usually produces the majority of auxin effects in the entire plant and is declared as the most effective natural auxin. In general, auxin functions in the coordination, growth, and development of various tissues, cytoplasmic streaming, growth and proliferation of cells, adventitious and lateral root initiation, as well as the emergence of shoot and phototropism (Chepsergon et al. 2014). On this basis, the *Trichoderma* strains comprising *T. brevicompactum*, *T. gamsii*, and *T. harzianum* were evaluated for their capacity to secrete the IAA by solubilizing the phosphate. A significant amount of orthophosphate ions ranging from 215.80 to 288.18 mg/ml and IAA production of about 13.38 to 21.24 mg/ml were detected evidencing these strains as good mineral solubilizers and growth hormone producers. Those *Fusarium* wilt diseased tomato plants inoculated with *T. harzianum* did not only inhibit the growth of *Fusarium oxysporum* by 10 to 30% but also enhanced the shoot length (40.99–139.62%), leaf area (22.55–42.16%), chlorophyll content, and dry and fresh weight of roots and shoots.

Therefore, it can be inferred that the plant growth is positively correlated with the IAA content and the IAA production is to soluble phosphorus content (Bader et al. 2020). In another experiment on cucumber with *T. harzianum*, the IAA was boosted together with the chlorophyll content and biomass of the plant in greenhouse and hydroponics. Thus, *T. harzianum* is not solely dependent on the soil but also acts as endophytic fungi colonizing the root directly with hydroponic plants. Even after a month of inoculation in the soil, the population of fungus was noted to be stable together with the IAA production. Thus, it is concluded that the total producing content of IAA is dependent on the colonized population along with time (Zhang et al. 2013). Similarly, the *T. viride* isolated from a mangrove exhibited an enormous production of IAA (115 μg mL^{-1}) with L-tryptophan (0.5%) and few secondary metabolites in laboratory conditions. Meanwhile, the seeds of *Vigna radiata*, *Vigna mungo,* and *Sesamum indicum* immersed in the IAA supernatant recorded induced germination and increased growth percentage (Kumar et al. 2017).

In Colombia, from the total sampled 101 isolates of *Trichoderma*, only 60 of them were able to produce IAA. However, these compounds were not positively correlated with the growth enhancement on the bean seedlings. Additionally, only seven strains out of nine significantly increased the growth of the seedlings, and not all of them produced the growth-promoting metabolites. Therefore, the results depicted that the growth enhancement of *Trichoderma* does not singly rely on the growth-promoting compounds but the rhizosphere competence. The strains were also taken into consideration as these characteristics are strain-specific and not the fixed characteristic for each species (Hoyos-carvajal et al. 2009).

On the other hand, ethylene is also one of the phytohormones that accelerates the senescence, abscission, and ripening and improves the resistance by regulating SA and jasmonic acid pathways (Sood et al. 2020). One of the cyclic α-amino acids named 1-aminocyclopropane-1-carboxylic acid (ACC) responsible for the biosynthesis of ethylene also involves in the growth mechanism of the plant. During the experiment of wheat with *T. longibrachiatum* at variant salinity levels, the fungal strain has shown an elevated tolerance at higher concentrations (10 mg ml^{-1}) of NaCl. At boosted stress conditions, the IAA and ACC-deaminase (26% at 10 mg ml^{-1}, 31% at 20 mg ml^{-1}) concentration were enhanced substantially. The ACC-deaminase is an enzyme that catalyzes sequestering and cleaving of ACC produced by plants, thus reducing the ethylene level. Because the high concentration of ethylene would escalate the senescence and abscission, it would cause the aging of the plant (Glick 2005). Therefore, the promoted concentration of ACC declined the ACC oxidase (12%), ACC synthase (13%) activity, and the content of ACC (22%) and ethylene (12%). The boosted concentrations of ACC-deaminase and IAA probably serve as a vital cue to ease the unfavorable effect of salinity stress on growth. These hormonal changes caused an increased wheat growth with high salinity tolerance (Zhang et al. 2019). Nevertheless, a contradictory assertion has been shown and explained by Martínez-Medina et al. (2011). With the inoculation of *T. harzianum* in melon plants, they possess an increment of IAA, ABA, ACC, and zeatin contents with elevated shoot growth. These researchers also inferred that the increased ethylene modifies the root and shoot growth via the auxin biosynthesis stimulation. They also indicated a strong correlation between ACC and IAA levels to boost the shoot growth as well as the parallelism of these two compound concentrations.

Ultimately, gibberellic acid is a plant growth regulator that majorly functions to stimulate leaf and stem elongation and cell division. GA promotes growth in plants by degrading the growth-inhibiting DELLA proteins (aspartic acid-glutamic acid-leucine-leucine-alanine) and reducing the ethylene level (Vera-sirera et al. 2016). Sofo et al. (2011) experimented on the phytohormone level by *T. harzianum* in the cherry rootstock. After 7 days of mycorrhizal fungi in the inducing medium in in vitro conditions, the shoot (61%) and root (76%) have extensively grown, therefore increasing the plant growth hormones in the shoot (IAA: 49%; GA: 71%) and root (IAA: 40%; GA: 143%). The higher shoot growth cues the acceleration of leaf production and plant hardening by stem lignification. However, no difference in ABA levels was noticed. The significantly increased secretion of GA by *T. harzianum* suppresses the *R. solani*, encourages plant growth, and heightens the yield production in potatoes. Hence, GA can also be used in plant pathogenic organism controls (Al-askar et al. 2016).

9 Conclusion

There are various *Trichoderma* spp. that can promote plant growth comprising a variety of herbaceous and woody plants, shrubs, and herbs even grasses. Sometimes, their interaction is strain-specific as their mechanism and secondary metabolites secretion could vary along with strains. In general, *Trichoderma* can extensively increase the growth of plants by promoting lateral root production, chlorophyll content, disease resistance, drought tolerance, stomatal conductance, and increased biomass. The validating study is abundant to broaden the multifunctional *Trichoderma* researches. It is also highlighted that promoting plant growth physical association is non-mandatory. Even in the presence of non-physical contact or without soil, *Trichoderma* can enhance plant growth. Not only in agriculture but the role of *Trichoderma* in commercial purposes is also overwhelming.

Declaration of Interest The authors declare no declarations of interest.

References

Agüera E, Haba PDL (2018) Leaf senescence in response to elevated atmospheric CO_2 concentration and low nitrogen supply. Biol Plant 62(3):401–408. https://doi.org/10.1007/s10535-018-0798-z

Akhtar R, Javaid A (2016) Biological management of basal rot of onion by *Trichoderma harzianum* and *Withania somnifera*. Planta Daninha. https://doi.org/10.1590/s0100-83582018360100009

Al-askar AA, Ezzat AA, Ghoneem K, Saber WIA (2016) *Trichoderma harzianum* WKY5 and its gibberellic acid control of *Rhizoctonia solani*, improve sprouting, growth and productivity of potato. Egypt J Biol Pest Co 26(4):787–796

Alwhibi MS, Hashem A, Abd_Allah EF, Alqarawi A, Soliman DWK, Wirth S, Egamberdieva D (2017) Increased resistance of drought by *Trichoderma harzianum* fungal treatment correlates with increased secondary metabolites and proline content. J Integr Agric 16(8):1751–1757. https://doi.org/10.1016/S2095-3119(17)61695-2

Arnao MB, Hernandez-Ruiz J (2017) Melatonin and its relationship to plant hormones. Ann Bot 00:1–13. https://doi.org/10.1093/aob/mcx114

Asma A (2019) Characterization of *Fusarium* species associated with wilt disease of cucurbits and effectiveness of *Trichoderma asperellum* as biological enhancer. PhD Thesis, Universiti Putra Malaysia

Azarmi R, Hajieghrari B, Giglou A (2011) Effect of *Trichoderma* isolates on tomato seedling growth response and nutrient uptake. Afr J Biotechnol 10(31):5850–5855. https://doi.org/10.5897/AJB10.1600

Bader AN, Salerno GL, Covacevich F, Consolo VF (2020) Native *Trichoderma harzianum* strains from Argentina produce indole-3 acetic acid and phosphorus solubilization, promote growth and control wilt disease on tomato (*Solanum lycopersicum* L.). J King Saud Univ Sci 32(1):867–873. https://doi.org/10.1016/j.jksus.2019.04.002

Bastakoti S, Belbase S, Manandhar S, Arjyal C (2017) *Trichoderma* species as biocontrol agent against soil borne fungal pathogens. Nepal J Biotechnol 5(1):39–45

Brunner K, Zeilinger S, Ciliento R, Woo SL, Lorito M, Kubicek CP, Mach RL (2005) Improvement of the fungal biocontrol agent *Trichoderma atroviride* to enhance both antagonism and induction of plant systemic disease resistance. Appl Environ Microbiol 71(7):3959–3965. https://doi.org/10.1128/AEM.71.7.3959

Carrero-carron I, Trapero-Casas JL, Olivares-Garcia C, Monte E, Hermosa R, Jimenez-Diaz RM (2016) *Trichoderma asperellum* is effective for biocontrol of Verticillium wilt in olive caused by the defoliating pathotype of *Verticillium dahliae*. Crop Prot 88:45–52. https://doi.org/10.1016/j.cropro.2016.05.009

Chamnongpol S, Willekens H, Moeder W, Langebartels C, Sandermann H, Montagu MV, Inze D, Camp WV (1998) Defense activation and enhanced pathogen tolerance induced by H_2O_2 in transgenic tobacco. Proc Natl Acad Sci U S A 95:5818–5823

Chen K, Li G-J, Bressan RA, Song C-P, Zhu J-K, Zhao Y (2019) Abscisic acid dynamics, signalling, and functions in plants. J Integr Plant Biol 62(1). https://doi.org/10.1111/jipb.12899

Chepsergon J, Mwamburi L, Kassim MK (2014) Mechanism of drought tolerance in plants using *Trichoderma* spp. Int J Sci Res 3(11):1592–1595

Clavijo-herrera J, Santen EV, Gomez C (2018) Growth, water-use efficiency, stomatal conductance, and nitrogen uptake of two lettuce cultivars grown under different percentages of blue and red light. Horticulturae 4(16). https://doi.org/10.3390/horticulturae4030016

Contreras-cornejo HA, Macias-Rodriguez L, Vergara AG, Lopez-Bucio J (2014) *Trichoderma* modulates stomatal aperture and leaf transpiration through an abscisic acid-dependent mechanism in *Arabidopsis*. J Plant Growth Regul 34(2):425–432. https://doi.org/10.1007/s00344-014-9471-8

Dabire TG, Bonzi S, Somda I, Legreve A (2016) Evaluation of the potential of *Trichoderma harzianum* as a plant growth promoter and biocontrol agent against *Fusarium* damping-off in onion in Burkina Faso. Asian J Plant Pathol 10:49–60. https://doi.org/10.3923/ajppaj.2016.49.60

Doni F, Isahak A, Zain CRCM, Ariffin SM, Mohamad WNW, Yusoff WMW (2014a) Formulation of *Trichoderma* sp. SL2 inoculants using different carriers for soil treatment in rice seedling growth. Springerplus 3(532):3–7

Doni F, Isahak A, Zain CRCM, Yusoff WMW (2014b) Physiological and growth response of rice plants (*Oryza sativa* L.) to *Trichoderma* spp. inoculants. AMB Express 4(45):1–7

Durrant WE, Dong X (2004) Systemic acquired resistance. Annu Rev Phytopathol 42:185–209. https://doi.org/10.1146/annurev.phyto.42.040803.140421

El-katatny MH (2010) Enzyme production and nitrogen fixation by free, immobilized and coimmobilized inoculants of *Trichoderma harzianum* and *Azospirillum brasilense* and their possible role in growth promotion of tomato. Food Technol Biotechnol 48(2):161–174

El-Mohamedy RSR (2012) Biological control of *Pythium* root rot of broccoli plants under greenhouse condition. Int J Agric Technol 8(3):1017–1028

Elshahawy IE, El-Mohamedy RS (2019) Biological control of *Pythium* damping-off and root-rot diseases of tomato using *Trichoderma* isolates employed alone or in combination. J Plant Pathol 101:597–608

Elshahawy IE, El-Sayed AEB (2018) Maximizing the efficacy of *Trichoderma* to control *Cephalosporium maydis,* causing maize late wilt disease, using freshwater microalgae extracts. Egypt J Biol Pest Co 28(48):1–11

Elsharkawy MM, Shimizu M, Takahashi H, Ozaki K, Hyakumachi M (2013) Induction of systemic resistance against *Cucumber mosaic virus* in *Arabidopsis thaliana* by *Trichoderma asperellum* SKT-1. Plant Pathol J 29(2):193–200

Estévez-Geffriaud V, Vicente R, Vergara-Díaz O, Reinaldo JJN, Trillas MI (2020) Application of *Trichoderma asperellum* T34 on maize *(Zea mays)* seeds protects against drought stress. Planta 252(1):1–12. https://doi.org/10.1007/s00425-020-03404-3

Farooq M, Wahid A, Kobayashi N, Fujita D, Basra SMA (2009) Plant drought stress: effects, mechanisms and management. Agron Sustain Dev 29:185–212

Fiorentino N, Ventorino V, Woo SL, Pepe O, Rosa AD, Gioia L, Romano I, Lombardi N, Napolitano M, Colla G, Rouphael Y (2018) *Trichoderma* -based biostimulants modulate rhizosphere microbial populations and improve n uptake efficiency, yield, and nutritional quality of leafy vegetables. Front Plant Sci 9(743):1–15. https://doi.org/10.3389/fpls.2018.00743

Glick BR (2005) Modulation of plant ethylene levels by the bacterial enzyme ACC deaminase. FEMS Microbiol Lett 251:1–7. https://doi.org/10.1016/j.femsle.2005.07.030

Guler NS, Pehlivan N, Karaoglu SA, Sule G, Bozdeveci A (2016) *Trichoderma atroviride* ID20G inoculation ameliorates drought stress-induced damages by improving antioxidant defence in maize seedlings. Acta Physiol Plant 38(132). https://doi.org/10.1007/s11738-016-2153-3

Guo H, Ge F (2017) Root nematode infection enhances leaf defence against whitefly in tomato. Arthropod Plant Interact 11(1):23–33. https://doi.org/10.1007/s11829-016-9462-8

Gupta KJ, Mur LAJ, Brotman Y (2014) *Trichoderma asperelloides* suppresses nitric oxide generation elicited by *Fusarium oxysporum* in *Arabidopsis* Roots. Mol Plant-Microbe Interact 27(4):307–314

Hafez Y, Abdelfatah A, El-Nashar F, Badr M, Elkady S (2019) Management of barley net blotch using *Trichoderma asperellum* (T34), eugenol, non-traditional compounds and fungicides. Egypt J Biol Pest Co 29(88):848–856

Harman GE (2005) Overview of mechanisms and uses of *Trichoderma* spp. Phytopathology 96(2):190–194. https://doi.org/10.1094/phyto-96-0190

Harman GE, Doni F, Khadka RB, Uphoff N (2019) Endophytic strains of *Trichoderma* increase plants' photosynthetic capability. J App Microbiol. https://doi.org/10.1111/jam.14368

Harman GE, Bjorkman T (1998) Potential and existing uses of *Trichoderma* and *Gliocladium* for plant disease control and plant growth enhancement. In: *Trichoderma* and *Gliocladium.* Volume 2: Enzymes, biological control and commercial uses, 1st edn. CRC Press, London, pp 229–265

Harman GE, Howell CR, Viterbo A, Chet I, Lorito M (2004) *Trichoderma* species opportunistic, avirulent plant symbionts. Nat Rev Microbiol 2:43–56

Hermosa R, Viterbo A, Chet I, Monte E (2012) Mini-review plant-beneficial effects of *Trichoderma* and of its genes. Microbiol 158:17–25. https://doi.org/10.1099/mic.0.052274-0

Herrera-Tellez VI, Cruz-olmedo AK, Plasencia J, Gavilanes-Ruiz M, Arce-Cervantes O, Hernnandez-Leon S, Saucedo-Garcia M (2019) The protective effect of *Trichoderma asperellum* on tomato plants against *Fusarium oxysporum* and *Botrytis cinerea* diseases involves inhibition of reactive oxygen species production. Int J Mol Sci 20:1–13

Hesham AE-L, Upadhyay RS, Sharma GD, Manoharachary C, Gupta VK (2020) Fungal biotechnology and bioengineering. Springer. https://doi.org/10.1007/978-3-030-41870-0

Hidangmayum A, Dwivedi P (2018) Plant responses to *Trichoderma* spp. and their tolerance to abiotic stresses: a review. J Pharmacogn Phytochem 7(1):758–766

Hoyos-carvajal L, Orduz S, Bissett J (2009) Growth stimulation in bean (*Phaseolus vulgaris* L.) by *Trichoderma*. Biol Control 51(3):409–416. https://doi.org/10.1016/j.biocontrol.2009.07.018

Hung R, Lee S, Bennett JW (2013) *Arabidopsis thaliana* as a model system for testing the effect of *Trichoderma* volatile organic compounds. Fungal Ecol 6(1):19–26. https://doi.org/10.1016/j.funeco.2012.09.005

Inbar J, Abramsky M, Cohen D, Chet I (1994) Plant growth enhancement and disease control by *Trichoderma harzianum* in vegetable seedlings grown under commercial conditions. Eur J Plant Pathol 100:337–346

Jalali F, Zafari D, Salari H (2017) Volatile organic compounds of some *Trichoderma* spp. increase growth and induce salt tolerance in *Arabidopsis thaliana*. Fungal Ecol 29:67–75. https://doi.org/10.1016/j.funeco.2017.06.007

Jatav P, Ahirwar SS, Gupta A, Kushwaha K, Jatav S (2018) Antagonistic activity of cellulase enzyme produced by *Trichoderma viride* against *Xanthomonas citri*. Indian J Agric Res 52(5):497–504. https://doi.org/10.18805/IJARe.A-4999

Jiang C, Liu X, Liu X, Zhang H, Yu Y, Liang Z-W (2017) Stunted growth caused by blast disease in rice seedlings is associated with changes in phytohormone signalling pathways. Front Plant Sci 8(1558):1–12. https://doi.org/10.3389/fpls.2017.01558

Jing C, Cheng Z, Li-Ping L, Zhong-Yang S, Xue-bo P (2007) Effects of exogenous salicylic acid on growth and H_2O_2-metabolizing enzymes in rice seedlings under lead stress. J Environ Sci 19:44–49

Kamo M, Tojo M, Yamazaki Y, Itabashi T, Takeda H, Wakana D, Hosoe T (2016) Isolation of growth inhibitors of the snow rot pathogen *Pythium iwayamai* from an arctic strain of *Trichoderma polysporum*. J Antibiot 69:451–455. https://doi.org/10.1038/ja.2015.130

Kerroum F, Noureddine K, Henni JE, Mebrouk K (2015) Antagonistic effect of *Trichoderma harzianum* against *Phytophthora infestans* in the North-west of Algeria. Int J Agron Agri Res 6(4):44–53

Keskitalo J, Bergquist G, Gardestrom P, Jansson S (2005) A cellular timetable of autumn senescence. Plant Physiol 139:1635–1648

Khalili E, Sadravi M, Naeimi S, Khosravi V (2012) Biological control of rice brown spot with native isolates of three *Trichoderma* species. Braz J Microbiol:297–305

Kirschbaum MUF (2011) Does enhanced photosynthesis enhance growth? Lessons learned from CO_2 enrichment studies [W]. Plant Physiol 155:117–124. https://doi.org/10.1104/pp.110.166819

Kottb M, Gigolashvili T, Großkinsky DK, Piechulla B (2015) *Trichoderma* volatiles effecting *Arabidopsis:* from inhibition to protection against phytopathogenic *fungi*. Front Microbiol 6(995):1–14. https://doi.org/10.3389/fmicb.2015.00995

Kredics L, Chen L, Edves O, Buchner R, Hatvani L, Allaga H, Nagy VD, Haled JM, Alharbi NS, Vagvolgyi C (2018) Molecular tools for monitoring *Trichoderma* in agricultural environments. Front Microbiol 9:1599. https://doi.org/10.3389/fmicb.2018.01599

Kumar NV, Rajam KS, Rani ME (2017) Plant growth promotion efficacy of indole acetic acid (IAA) produced by a mangrove associated *fungi- Trichoderma viride* VKF3. Int J Curr Microbiol Appl Sci 6(11):2692–2701. https://doi.org/10.20546/ijcmas.2017.611.xxx

Lamont B (1982) Mechanisms for enhancing nutrient uptake in plants, with particular reference to mediterranean South Africa and Western Australia. Bot Rev 48(3):597–689

Lee S, Hung R, Yap M, Bennett JW (2015) Age matters: the effects of volatile organic compounds emitted by *Trichoderma atroviride* on plant growth. Arch Microbiol 197(5):723–727. https://doi.org/10.1007/s00203-015-1104-5

León J, Lawton MA, Raskin I (1995) Hydrogen peroxide stimulates salicylic acid biosynthesis in tobacco. Plant Physiol 108:1673–1678

Li R, Cai F, Pang G, Shen Q, Li R, Chen W (2015) Solubilisation of phosphate and micronutrients by *Trichoderma harzianum* and its relationship with the promotion of tomato plant growth. PLoS One 10(6):1–16. https://doi.org/10.5061/dryad.p460c

Li Y, Hwang S, Huang Y, Huang C (2018) Effects of *Trichoderma asperellum* on nutrient uptake and Fusarium wilt of tomato. Crop Prot 110:275–282. https://doi.org/10.1016/j.cropro.2017.03.021

Loon LCV (2007) Plant responses to plant growth-promoting rhizobacteria. Eur J Plant Pathol 119:243–254. https://doi.org/10.1007/s10658-007-9165-1

Mallhi ZI, Rizwan M, Mansha A, Ali Q, Asim S, Ali S, Hussain A, Alrokayan SH, Khan HA, Alam P, Ahmad P (2019) Citric acid enhances plant growth, photosynthesis, and phytoextraction of lead by alleviating the oxidative stress in castor beans. Plan Theory 8(525). https://doi.org/10.3390/plants8110525

Marco JLD, Felix CR (2002) Characterization of a protease produced by a *Trichoderma harzianum* isolate which controls cocoa plant witches' broom disease. BMC Biochem 3:1–7

Martínez-medina A, Roldán A, Albacete A, Pascual JA (2011) Phytochemistry the interaction with arbuscular mycorrhizal *fungi* or *Trichoderma harzianum* alters the shoot hormonal profile in melon plants. Phytochemistry 72:223–229. https://doi.org/10.1016/j.phytochem.2010.11.008

Miyashita S, Takahashi H (2015) 植物のR遺伝子によるウイルス抵抗性 (R-gene-mediated resistance to plant viruses). Uirusu 65(2):199–208. https://doi.org/10.2222/jsv.65.199

Nakkeeran S, Vinodkumar S, Priyanka R, Renukadevi P (2018) Mode of action of *Trichoderma* spp. in biological control of plant diseases. In: Biocontrol of soil borne pathogens and nematodes. Jaya Publishing House, Delhi, pp 81–95

Naseby DC, Pascual JA, Lynch JM (2000) Effect of biocontrol strains of *Trichoderma* on plant growth, *Pythium ultimum* populations, soil microbial communities and soil enzyme activities. J App Microbiol 88:161–169

Nawrocka J, Małolepsza U (2013) Diversity in plant systemic resistance induced by *Trichoderma*. Biol Control 67(2):149–156. https://doi.org/10.1016/j.biocontrol.2013.07.005

Nusaibah SA, Musa H (2019) A review report on the mechanism of *Trichoderma spp.* as biological control agent of the basal stem rot (BSR) disease of *Elaeis guineensis*. IntechOpen. https://doi.org/10.5772/intechopen.84469

O'Neill TM, Elad Y, Shtienberg D, Cohen A (1995) Control of grapevine grey mould with *Trichoderma harzianum* T39. Biocontrol Sci Tech 6:139–146. https://doi.org/10.1080/09583159650039340

Pandey V, Ansari MW, Tula S, Yadav S, Sahoo RK, Shukla N, Bains G, Badal S, Chandra S, Gaur AK, Kumar A, Shukla A, Kumar J, Tuteja N (2016) Dose-dependent response of *Trichoderma harzianum* in improving drought tolerance in rice genotypes. Planta 243(5):1251–1264. https://doi.org/10.1007/s00425-016-2482-x

Passioura JB (2002) Soil conditions and plant growth. Plant Cell Environ 25:311–318. https://doi.org/10.1046/j.0016-8025.2001.00802.x

Poveda J (2020) *Trichoderma parareesei* favors the tolerance of Rapeseed (*Brassica napus* L.) to salinity and drought due to a chorismate mutase. Agronomy 10(118). https://doi.org/10.3390/agronomy10010118

Rouphael Y, Carillo P, Colla G, Fiorentino N, Sabatino L, El-Nakhel C, Giordano M, Pannico A, Corollo V, Shabani E, Cozzolino E, Lombardi N, Napolitano M, Woo SL (2020) Appraisal of combined applications of *Trichoderma virens* and a biopolymer-based biostimulant on lettuce agronomical, physiological, and qualitative properties under variable n regimes. Agronomy 10(196). https://doi.org/10.3390/agronomy10020196

Sabbagh SK, Roudini M, Panjehkeh N (2017) Systemic resistance induced by *Trichoderma harzianum* and *Glomus mossea* on cucumber damping-off disease caused by *Phytophthora melonis*. Arch Phytopathol Pflanzenschutz. https://doi.org/10.1080/03235408.2017.1317953

Salehi-Lisar SY, Bakhshayeshan-Agdam H (2016) Causes, consequences, and tolerance. In: Hossain MA et al (eds) Drought stress in plants, vol 1. Springer International Publishing Switzerland. https://doi.org/10.1007/978-3-319-28899-4

Saravanakumar K, Arasu VS, Kathiresan K (2013) Effect of *Trichoderma* on soil phosphate solubilization and growth improvement of *Avicennia marina*. Aquat Bot 104:101–105. https://doi.org/10.1016/j.aquabot.2012.09.001

Sharifah SMSAR (2018) Characterization of *Trichoderma* species isolated from soil and efficacy of *Trichoderma asperellum* as biocontrol agent of Fusarium wilt disease of banana. MSc. Thesis, Universiti Putra Malaysia

Shukla N, Awasthi RP, Rawat L, Kumar J (2012) Biochemical and physiological responses of rice (*Oryza sativa* L.) as influenced by *Trichoderma harzianum* under drought stress. Plant Physiol Biochem 54:78–88. https://doi.org/10.1016/j.plaphy.2012.02.001

Shukla N, Awasthi RP, Rawat L, Kumar J (2014) Seed biopriming with drought tolerant isolates of *Trichoderma harzianum* promote growth and drought tolerance in *Triticum aestivum*. Ann Appl Biol. https://doi.org/10.1111/aab.12160

Singh V, Singh PN, Yadav RL, Awasthi SK, Joshi BB, Singh RK, Lal RJ, Duttamajumder SK (2010) Increasing the efficacy of *Trichoderma harzianum* for nutrient uptake and control of red rot in sugarcane. J Hortic For 2(4):66–71

Singh SP, Singh HB, Singh D, Rakshit A (2014) *Trichoderma* -mediated enhancement of nutrient uptake and reduction in incidence of *Rhizoctonia solani* in tomato. Egypt J Biol 16:29–38. https://doi.org/10.4314/ejb.v16i1.4

Singh BN, Dwivedi P, Sarma BK, Singh GS, Singh HB (2019) A novel function of N-signalling in plants with special reference to *Trichoderma* interaction influencing plant growth, nitrogen use efficiency, and cross talk with plant hormones. Biotech 9(3):1–13. https://doi.org/10.1007/s13205-019-1638-3

Sofo A, Scopa A, Manfra M, Nisco MD, Tenore G, Troisi J, Fiori RD, Novellino E (2011) *Trichoderma harzianum* strain T-22 induces changes in phytohormone levels in cherry rootstocks *(Prunus cerasus* x *P. canescens)*. Plant Growth Regul 65:421–425. https://doi.org/10.1007/s10725-011-9610-1

Sood M, Kapoor D, Kumar V, Sheteiwy MS, Ramakrishnan M, Landi M, Araniti F, Sharma A (2020) *Trichoderma:* the "secrets" of a multitalented biocontrol agent. Plan Theory 9(762). https://doi.org/10.3390/plants9060762

Stewart A, Hill R (2014) Applications of *Trichoderma* in plant growth promotion. In: Biotechnology and biology of *Trichoderma*. Elsevier, Amsterdam, pp 413–428. https://doi.org/10.1016/B978-0-444-59576-8.00031-X

Sunpapao A, Chairin T, Ito S (2018) The biocontrol by *Streptomyces* and *Trichoderma* of leaf spot disease caused by *Curvularia oryzae* in oil palm seedlings. Biol Control 123:36–42. https://doi.org/10.1016/j.biocontrol.2018.04.017

Tewari RK, Paek KY (2011) Salicylic acid-induced nitric oxide and ROS generation stimulate ginsenoside accumulation in *Panax ginseng* roots. J Plant Growth Regul 30:396–404. https://doi.org/10.1007/s00344-011-9202-3

Vera-sirera F, Gomez MD, Perez-amador MA (2016) DELLA proteins, a group of GRAS transcription regulators that mediate gibberellin signalling. In: Plant transcription factors: evolutionary, structural and functional aspects. Academic Press, San Diego, pp 313–324. https://doi.org/10.1016/B978-0-12-800854-6/00020-8

Vieira PM, Santos MP, Andrade CM, Souza-neto OA, Ulhoa CJ, Aragão FJL (2017) Overexpression of an aquaglyceroporin gene from *Trichoderma harzianum* improves water-use efficiency and drought tolerance in *Nicotiana tabacum*. Plant Physiol Biochem. https://doi.org/10.1016/j.plaphy.2017.10.012

Vitti A, Pellegrini E, Nali C, Lovelli S, Sofo A, Valerio M, Scopa A, Nuzzaci M (2016) *Trichoderma harzianum* T-22 induces systemic resistance in tomato infected by *Cucumber mosaic virus*. Front Plant Sci 7(1520):1–11. https://doi.org/10.3389/fpls.2016.01520

Wani SH, Kumar V, Shriram V, Kumar S (2016) Phytohormones and their metabolic engineering for abiotic stress tolerance in crop plants. Crop J 4:162–176. https://doi.org/10.1016/j.cj.2016.01.010

Yang W, Zhu C, Ma X, Li G, Gan L, Ng D, Xia K (2013) Hydrogen peroxide is a second messenger in the salicylic acid-triggered adventitious rooting process in mung bean seedlings. PLoS One 8(12):1–14. https://doi.org/10.1371/journal.pone.0084580

Yasmeen R, Siddiqui ZS (2017) Physiological responses of crop plants against *Trichoderma harzianum* in saline environment. Acta Bot Croat. https://doi.org/10.1515/botcro-2016-0054

Yuan M, Huang Y, Ge W, Jia Z, Song S, Zhang L, Huang Y (2019) Involvement of jasmonic acid, ethylene and salicylic acid signalling pathways behind the systemic resistance induced by *Trichoderma longibrachiatum* H9 in cucumber. BMC Genomics 20(144):1–13

Zainap AEH, Nur Ain Izzati MZ, Asma A, Mohd Hafiz I, Mohd Termizi Y (2020) Biocontrol efficacy of Trichoderma asperellum-enriched coconut fiber against Fusarium wilts of cherry tomato. J App Microbiol 129:991–1003

Zhang F, Yuan J, Yang X, Cui Y, Chen L, Ran W, Qirong S (2013) Putative *Trichoderma harzianum* mutant promotes cucumber growth by enhanced production of indole acetic acid and plant colonization. Plant Soil 368:433–444. https://doi.org/10.1007/s11104-012-1519-6

Zhang S, Gan Y, Xu B (2019) Mechanisms of the IAA and ACC- deaminase producing strain of *Trichoderma longibrachiatum* T6 in enhancing wheat seedling tolerance to NaCl stress. BMC Plant Biol 19(22):1–18

Metabolomic Approaches to Study *Trichoderma*-Plant Interactions

Nishtha Mishra, Priyanka Chauhan, Pratibha Verma, S. P. Singh, and Aradhana Mishra

Contents

1 Introduction

The process of industrialization has forced significant and essential enhanced agricultural production worldwide for the availability of food to the growing population during the past decades. The increased demand for agricultural and its allied activities has led to serious environmental as well as social problems. The various methods have been opted for minimizing the global concern of environmentally

N. Mishra · P. Chauhan · P. Verma · A. Mishra (✉)
Division of Microbial Technology, CSIR-National Botanical Research Institute, Lucknow, India

Academy of Scientific and Industrial Research (AcSIR), Ghaziabad, India

S. P. Singh
Acharya Narendra Deva University of Agriculture and Technology, Ayodhya, India

N. Amaresan et al. (eds.), *Advances in* Trichoderma *Biology for Agricultural Applications*, Fungal Biology, https://doi.org/10.1007/978-3-030-91650-3_10

sustainable agriculture production. In this context, one of the potential uses of plant growth-promoting microbes (PGPM) has come forward with the least negative impact on the environment. These beneficial microbes alter various plant metabolic pathways by direct and indirect mechanisms such as phytohormone production, enhanced nutrient uptake, and changes in gene expression (Dutta and Thakur 2017; Singh and Gaur 2017).

Secondary metabolites are a group of natural compounds, heterogeneous in nature which helps for the survivability and basic functions of an organism. The functions comprise competition, symbiosis, metal transport, differentiation, etc., (Demain and Fang 2000). SMs comprised antibiotics which are natural products, produced by various microbes in the process of sporulation and development. *Trichoderma* count first in the segment of fungi as a biocontrol agent globally (Whipps and Lumsden 2001) as well as known for the production of secondary metabolites (Ghisalberti and Sivasithamparam 1991).

Secondary metabolism is the process of biomass production to complete metabolite biosynthesis which provides competitive benefits to the producers (Ruiz et al. 2010). Secondary metabolites secreted by microbes are the products of primary metabolites in specialized pathways which often can be grouped in the genome and can be affected by some of the global regulators (Herbert 1989). Genes of fungi are associated with the biosynthetic process of SMs, which are not transcribed in undergoing process during in vitro conditions. It can be triggered by imitating a complex of the microbial communities (Vinale et al. 2017).

Trichoderma is one of the most prevalent culturable fungi ubiquitously present in approximately all types of soil characterized by opportunistic avirulent plant symbiosis. The various strains of *Trichoderma* have been widely reported for plant growth promotion via direct and indirect mechanisms. The various mechanisms proposed for explanation of plant growth promotion include control of minor pathogens, enhanced nutrient uptake, increment in carbohydrate metabolism, photosynthesis, and phytohormone production (Stewart et al. 2014). *Trichoderma* has been proved for a strong influence on the production of indole acetic acid (IAA), gibberellic acid, and ethylene. There are various previous studies for understanding the underlying mechanisms of *Trichoderma* and its interaction with plants with the different strains such as *T. harzianum* strains -T22,[1] T39, and A6 and strain of *T. atroviride* P1 in several edible crops like *Pisum sativum* (pea), *Lycopersicum esculentum* (tomato), and *Brassica napus* (canola). The secondary metabolites showed pivotal role in the suppression of plant pathogens by exhibiting the antagonistic activity (Vinale et al. 2008). Metabolome of an organism can be well defined as the set of low molecular weight compounds which can be considered as phenotypic indication of organism. Nowadays, for understanding the complete biological events, metabolomics study has significant importance.

[1] Species identities are cited as initially published, and the current taxonomic status of each species requires verification.

2 The Interactive Strategy of *Trichoderma* and Plants During Biotic and Abiotic Stress

Trichoderma spp. possess several modes of actions in plants during stress condition which are as follows:

(a) Control of phytopathogen by the production of antibiotics in combination with extracellular cell wall degrading enzymes
(b) Struggle in nutrients uptake (i.e. iron and other micronutrients such as nitrogen or carbon sources)
(c) Competitions for colonization
(d) Signalling for the development of plant resistance (Harman et al. 2004; Vinale et al. 2008; Whipps and Lumsden 2001)

Several studies have investigated mechanism of different plant-pathogen interactions. These approaches are provided to changes in plant's metabolism during interaction with pathogens and identify specific metabolites secreted by plants which help to enhance plant's immunity. The secondary metabolites secreted by the *Trichoderma* spp. have been proven in agriculture for all the beneficial effects exerted (Vinale et al. 2008) acting as antibiotics as well as in synergistic form with other compounds. The several strains of genus *Trichoderma* exhibit the property of controlling various phytopathogens such as *Rhizoctonia solani*, *Alternaria alternata*, *Fusarium* spp., and *Pythium* spp. (Lorito et al. 2010). These groups of microbes are demonstrated as bio-fertilizer and exhibit positive effect on crops as well as provide disease resistance towards both biotic and abiotic stresses which has been shown in Fig. 1.

2.1 Secondary Metabolites in Diverse Species of Trichoderma

These group of fungi are productive in biosynthesis of secondary metabolites such as volatile and non-volatile terpenes; NRPs, pyrones, peptaibols siderophores, and nitrogen containing compounds as well as 373 different molecules are also identified, in which specific activity of these molecules has still to be identified (Reino et al. 2008; Mukherjee et al. 2012a, b, Crutcher et al. 2013). These are also observed as intermediates of chemical exchange among inhabitants of soil in numerous ecological niches with the minimal use like micrograms per litre for facilitating symbiosis with microorganisms, insects, pests, and higher animals.

Plethora, a metabolite secreted by *Trichoderma* into their close proximity, requires minimum nutritional needs, which generally can be utilized for many agricultural, industrial, and medical benefits. The diverse metabolites of *Trichoderma* and its interaction with several plants have been studies in many previous studies presented in Table 1.

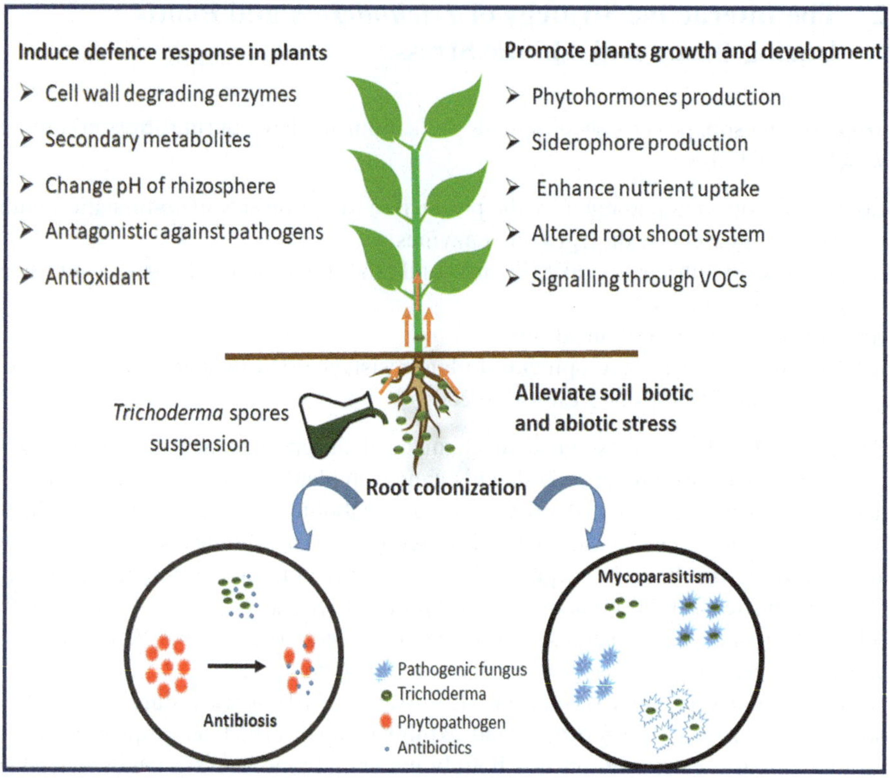

Fig. 1 Role of *Trichoderma* in biotic and abiotic stress amelioration

3 Role of Secondary Metabolites in Plants

3.1 Plant Growth Regulators

Trichoderma metabolites act as signalling compounds and influence the plant growth and productivity (Benítez et al. 2004). The purified metabolites of microbe showed similar results on the host and pathogen similarly as the living microbes (Vinale et al. 2008). Various secondary metabolites as well as their biological effects have been studied and shown in Table 2. The abundant diversity of chemical secondary metabolites secreted by *Trichoderma* spp. acts with different natures of growth promotion as well as inhibition (Vinale et al. 2008; Luo et al. 2010). The metabolite production showed its activity in concentration-dependent manner in the crops. The inhibitory effect of trichocaranes A and B has been found active up to 40% at 10^{-4} M concentration, whereas trichocaranes C showed its activity up to 86% at 10^{-3} M concentration (Macías et al. 2000).

The isolation, identification, and biological activity of secondary metabolites produced by *T. koningii* and *T. harzianium* (6-pentyl-α-pyrone) have been studied. The wheat coleoptile assay was carried out to study the concentration-dependent

Table 1 *Trichoderma* metabolites and interaction with plants during biotic and abiotic stress

S. No.	*Trichoderma* species	Metabolites	Interaction with plants	Reference
1.	*Trichoderma viride*	Gliotoxins	Showed biocontrol activity against some plant pathogenic fungi	Vey et al. (2001)
2.	*Trichoderma koningii*	Peptaibols, trichokonins VI and VIII	Showed broad-spectrum antimicrobial activity against a range of important plant pathogens, such as *R. solani*, *Fusarium oxysporum*, *Verticillium dahliae*, and *Botrytis cinerea*	Yang et al. (2016)
3.	*Trichoderma harzianum*	Peptaibols trichorzianine A1	Inhibit the spore germination, as well as hyphal elongation, of plant pathogenic fungi	Goulard et al. (1995), Lee et al. (1999)
4.	*T. koningii* and *T. harzianum*	Pyrone 6-PP	Reduces growth of *F. oxysporum* and *R. solani*	Claydon et al. (1987), Simon et al. (1988)
5.	*T. longibrachiatum*	5-Hydroxyvertinolide	Antagonistic to the fungus *Mycena citricolor*, cause American leaf spot disease of coffee	Andrade et al. (1992)
6.	*T. viride*	Vridepyronone	Showed 90% growth inhibition of *S. rolfsii* at a minimum inhibitory concentration (MIC) of 196 mg/ml	Evidente et al. (2003)
7.	*T. harzianum*	Harzianopyridone	Activity against *Phytophthora cinnamomi*, *B. cinerea*, and *Leptosphaeria maculans* and 90% of the growth of *R. solani*, *F. oxysporum*, and *S. rolfsii*	Sivasithamparam and Ghisalberti (1998), Ahluwalia et al. (2015)
8.	*T. koningii*	Koninginins A and B	Exhibit activity against *G. graminis* var. *tritici*	Cutler et al. (1989, 1991)
9.	*T. harzianum* and *T. koningii*	Stigmasterol	Showed antifungal activities against *R. solani*, *S. rolfsii*, *M. phaseolina*, and *F. oxysporum*	Ahluwalia et al. (2015), Ahluwalia et al. (2014)

(continued)

Table 1 (continued)

S. No.	*Trichoderma* species	Metabolites	Interaction with plants	Reference
10.	*T. harzianum*	Anthraquinones	Exhibit antifungal active against *R. solani*, *S. rolfsii*, *M. phaseolina*, and *F. oxysporum*	Ahluwalia et al. (2015)
11.	*T. cremeum*	Lactone cremenolide	Promotion of tomato seedling growth and showed antifungal activities against *R. solani*, *B. cinerea*, and *F. oxysporum*	Vinale et al. (2016)
12.	*T. arundinaceum*	Aspinolide C	Antibiotic effect against *B. cinerea* and *Fusarium sporotrichioides*, also played an important role in the induction of plant resistance against phytopathogenic fungi	Malmierca et al. (2015)
13.	*T. cerinum*	Cerinolactone	Inhibit growth of *Rosellinia necatrix*	Vinale et al. (2012a, b)
14.	*T. brevicompactum*	Trichothecenes	Inhibitory activity on *R. solani*, *B. cinerea*, and *Colletotrichum lindemuthianum*	Shentu et al. (2014)
15.	*T. harzianum* and *T. viride*	Diterpene harziandione	Antifungal activity against *S. rolfsii*	Mannina et al. (1997)
16.	*Trichoderma koningiopsis*	Koninginin D	Exhibit antifungal activity against several phytopathogens, such as *F. oxysporum*, *Bipolaris sorokiniana*, *P. cinnamomi*, and *Pythium middletonii*	Dunlop et al. (1989)
17.	*T. harzianum*	Harzianic acid	Plant growth promoter and antimicrobial agents against different plant pathogenic fungi, such as *Pythium irregulare*, *Sclerotinia sclerotiorum*, and *R. solani*	Vinale et al. (2012a, b)
18.	*T. atroviride*	Trichodermester A	Antifungal agents against *Phaeosphaerella theae*	Tang et al. (2020)

(continued)

Table 1 (continued)

S. No.	*Trichoderma* species	Metabolites	Interaction with plants	Reference
19.	*T. pseudokoningii*	Trichokonin VI	Showed antifungal activity by inducing extensive apoptotic programmed cell	Su et al. (2012)
20.	*T. brevicompactum*	Trichodermin	Fungitoxic metabolite against *Candida* spp.	Shentu et al. (2014), Tijerino et al. (2011)
21.	*T. virens*	Abscisic acid (ABA)	Regulates stomatal aperture in *A. thaliana*	Contreras-Cornejo et al. (2015b)
22.	*T. viride*	Alamethicin	Induces plant defence in lima bean and pathogen resistance	Kenerley (2012b)
23.	*T. virens*	β-Myrcene	Regulates the expression of abiotic and biotic stress-related genes related	Crutcher et al. (2013)
24.	*T. brevicompactum*	Trichodermin	Phytotoxic effect	Malmierca et al. (2015)
25.	*Trichoderma* spp.	Chitinases	Hydrolytic enzymes of the fungal cell wall	Gruber and Seidl-Seiboth (2012), Tzelepis et al. (2015)
26.	*Trichoderma* spp.	Coprogen B	Solubilizes iron unavailable to the plant	Vinale et al. (2012a, b)
27.	*T. brevicompactum*	Trichodermin	Phytotoxic effect	Malmierca et al. (2015)
28.	*T. harzianum*	Pachybasin	Increases the number of coils of the biocontrol agent against *R. solani*	Lin et al. (2012)
29.	*T. virens*	β-Caryophyllene	Attracts nematodes that prey on insect larvae	Lopez-Bucio (2014)
30.	*T. atroviride* and *T. virens*	Indole-3-carboxaldehyde	Induces adventitious root formation in *A. thaliana*	Contreras-Cornejo et al. (2011)

activity of metabolites in which activity of phytotoxicity detected at 10^{-3} M but not at 10^{-4} M. Cerinolactone one of the SMs has been characterized from filtrate of *T. cerinum* and found in induction of tomato seedlings growth after 3 days of treatment (Vinale et al. 2012,). The living strains of *Trichoderma* effect the growth, yield, and nutrient uptake of soybean in combination with harzianic acid, 6-pentyl-α-pyrone (Marra et al. 2020). SMs also enhanced fatty acid production such as oleic, linolenic, 11-eicosenoic, and stearic acid in harvested seeds.

According to Kim et al. (2006) ectotrophic fungus is known which produced plant growth-promoting metabolite (5-hydroxy-1-(3-methyl-3-buten-1-ynyl)-7-oxabicyclo [4.1.0]-hept-3-en-2-one) in liquid cultures of the fungus.

Table 2 Beneficial-microbe metabolome -bacteria, fungus, rhizospheric microbes, AMF

S.no.	Metabolites	Microbe	Biological effect	Reference
1.	L-Glutamic acid, L-aspartic acid, L-phenylalanine, L-lysine, L-methionine, L-threonine, and L-tryptophan	*Corynebacterium glutamicum*	Additive to animal feed	Sun et al. (2015)
2.	Phenol, flavonoids, and tannins	*Glomus* spp.	The concentration of phenols, flavonoids, and total tannins was favoured by mycorrhizal inoculation even at the highest levels of P	Pedone-Bonfim et al. (2013)
3.	Artemisin	*Rhizophagus intraradices*	Increases isoprenoids by induction of the MEP pathway	Mandal et al. (2015)
4.	Bacoside	*Bacillus megaterium*, *Rhizophagus intraradices*	Significantly enhance fresh biomass, essential oil content, nutrient uptake, and reduced root-knot infestation. Augmented of the phenolic, flavonoid, free radical scavenging activity, and total antioxidant	Gupta and Pandey (2015)
5.	Linolene, 1,8-cineole, linalool, carvone	*Claroideoglomus etunicatum*, *Claroideoglomus lalmellosum*	Allow plant growth in low fertility soils, reduce fertilizer inputs, and increase aromatic plant production of essential oils	Karagiannidis et al. (2011)
6.	Calendoflavosid, isorhamnetin, malonyl, and glucoside	*Claroideoglomus etunicatum*, *Claroideoglomus claroideum*, *Rhizophagus intraradices*	Increased the biomass of marjoram (1.5-fold), the number of marigold flowers (1.2-fold), and the yield of rosmarinic acid and lithospermic acid isomers of marjoram (1.5-fold) and lemon balm (1.2-fold)	Engel et al. (2016)

(continued)

Table 2 (continued)

S.no.	Metabolites	Microbe	Biological effect	Reference
7.	Phenolic content, flavonoids, and curcumin	*Glomus*, *Gigaspora*, *Acaulospora* sp.	Showed strong antioxidant activities	Dutta and Neog (2016)
8.	Ajmalicine	*Pseudomonas fluorescens*	Increase the ascorbic acid content in plant under salinity stress condition	Jaleel et al. (2007)
9.	Ajmalicine and serpentine	AMF species	Change the content of alkaloid and expression pattern of the genes	Andrade et al. (2013)
10.	Lipoxygenase and essential oil	*Rhizophagus clarus*, *Claroideoglomus etunicatum*, *P. aduncum*	Induce the metabolic activity of plant and increase the content of several monoterpenes and sesquiterpenes in leaves	de Oliveira et al. (2019)
11.	Flavonoids	*Gigaspora albida*	Increase the production of pharmacologically important foliar biomolecules, mainly flavonoids	Oliveira et al. (2015)
12.	2,4-Diacetylphloroglucinol	*Pseudomonas fluorescens*	Biological control against *Xanthomonas oryzae* pv. *oryzae*	Velusamy and Gnanamanickam (2008)
13.	Phenazine-1-carboxylic acid	*Pseudomonas putida* P15	Antifungal	Pathma et al. (2010)
14.	Rhizoxin analogs	*Pseudomonas fluorescens* Pf-5	Antifungal	Loper et al. (2008)
15.	Thuricin CD 19	*Bacillus thuringiensis* DPC 6431,	Bactericidal, bacteriolytic	Rea et al. (2010) B. anthracis
16.	3,4-DHB 17	*Bacillus thuringiensis*, *Bacillus cereus*, and *Bacillus anthracis*	Iron chelation	Zawadzka et al. (2009)
17.	Bleomycin	*S. verticillus*, *S. mobaraensis* ATCC 15003	Antibiotic	Radwan et al. (2011)

(continued)

Table 2 (continued)

S.no.	Metabolites	Microbe	Biological effect	Reference
18.	HCN	*Pseudomonas fluorescens*	*Alternaria alternata* OTA36; *Alternaria brassicola* OCA1; *Alternaria brassiceae* OCA3; *Collectotrichum gleosporidose* OGC1 revealed broad-spectrum anti-fungal activity	Ramyasmruthi et al. (2012)
19.	Taxol	*Taxomyces andreanae*	Chemotherapeutic agent	Strobel et al. (1993)
20.	Rohitukine	*Fusarium* sp.	Anticancer drug	Mohana Kumara et al. (2012)
21.	Hypericin	*Chaetomium*	*Antiviral drug*	Kusari et al. (2008)

Antifungal activities along with plant growth promotion of harzianolide have been studied against different phytopathogens (Vinale et al. 2008).

3.2 Antimicrobial Agents

Non-pathogenic microorganisms have been involved in inducing disease resistance against several pathogens in crops. PGPR helps in protection of aboveground plant part from various phytopathogens through colonization of roots in crops as well as root exudation process (Doornbos et al. 2012). SMs produced by *Trichoderma* have antagonistic activity against pathogenic fungus. In the previous years, organic chemists were only interested in SMs and were concerned primarily with the isolation, identification, as well as biosynthesis of these metabolites rather than with aspects of fungal metabolism and ecological interactions. There are various types of antifungal SMs such as epipolythiodioxopiperazines, peptaibols, pyrones, butenolides, pyridones, azaphilones, koninginins, steroids, anthraquinones, lactones, and trichothecenes produced by different strains of *Trichoderma* and showed antagonistic effect against pathogens such as *S. rolfsii*, *P. irregulare*, *S. sclerotiorum*, and *R. solani*. PGPF has been proved in eliciting the induction of systemic response (ISR) in plants. *Trichoderma* spp. (Harman et al. 2004; Vinale et al. 2008; Keswani et al. 2013; Singh et al. 2019), along with other *Penicillium* sp. GP 16-2 (Hossain et al. 2008), have been studied in several crops for plant growth promotion.

3.3 Root Colonization

Plant's root contains complex morphological structure with various physiological response. The initiation of lateral roots involves an auxin-dependent signalling cascade which activates pericycle cells followed by primordium development and regulated by IAA (Péret et al. 2009). *T. virens* has been reported for promoting shoot growth and lateral root development in *A. thaliana* (Contreras-Cornejo et al. 2009). The isolation of *Trichoderma* has been done from nearly all climatic zones, from various root ecosystems, and its growth in rhizosphere is enabled by secretion of polysaccharides in roots vicinity. Gravel et al. (2007), Contreras-Cornejo et al. (2009), and Vargas et al. (2009) investigated the importance of sucrose in the process of root colonization through *Trichoderma* spp. The signalling between *Trichoderma* and plant's root was found to be dependent on root-derived exudates (Bais et al. 2006). The newer advanced techniques are flourishing for exploring the mechanisms of direct interaction between host's root and *Trichoderma*. Shoresh et al. (2010) found an alteration in the systemic response in proteome, transcriptome, and MAMP relationships in leaves after *Trichoderma* colonization in the roots. *Trichoderma* invasion in plants leads to activation of rapid ion fluxes, oxidative burst (ROS formation), and callose deposition which is followed by polyphenol biosynthesis (Shoresh et al. 2010).

In this context, direct interaction of *Trichoderma* spp. has also utilized a wide range of proteins such as HYD1 protein from *T. asperellum* (Viterbo and Chet 2006) in assisting root colonization. Some proteins such as Swollenin SWO protein and the endopolygalacturonase PG1 have been reported for facilitating the root penetration from *T. harzianium* (Brotman et al. 2008; Morán-Diez et al. 2009).

3.4 ISR Response

ISR response in plants starts from microbial cell surface known as pathogen or microbe-associated molecular patterns (PAMPs or MAMPs) with the recognition of specific components (Schwessinger and Zipfel 2008). PAMP-triggered immunity (PTI) refers to the activation of defence response in the host through interaction between PAMP and corresponding plant receptor (Jones and Dangl 2006). As similar to PAMPs, numerous varied MAMPs from many microbes have been connected with ISR (Bakker et al. 2020;). ISR starts with the production of several reactive oxygen species (ROS) such as nitric oxide, ethylene (ET), and biosynthesis of antimicrobial substances and later involves the accumulation of callose.

T. harzianum has been reported for the promotion of chlorophyll biosynthesis individually as well as inoculation in drought-stressed plants. According to Azarmi et al. (2011), *T. harzianium* improves the photosynthetic capacities due to enhanced production of phytohormones (Resende et al. 2014), along with the biosynthesis of photosynthetic pigments in tomato.

Trichoderma spp. are involved in activation of systemic defence response of plants in response to peptaibol secondary metabolites (Szekeres et al. 2005; Viterbo et al. 2007; Luo et al. 2010; Druzhinina et al. 2011). Peptaibols are linear peptide composed of non-protein genic amino acids (i.e. isovaline and α-amino isobutyric acid), in which N-terminal group is acetylated and C-terminus group contains an amino alcohol (i.e. phenylalaninol, valinol, leucinol, isoleucinol, or tryptophanol).

The enzyme mixture was recognized in maize recently as PKS/NRPS which was performing the important role in defence-related responses (Mukherjee et al. 2012a, b). Sm1/Ep11 elicitor has well studied from *Trichoderma* spp. for activation of ISR response (Djonovic 2006), and deletion of this gene caused inappropriate ISR response in maize (Djonović et al. 2007).

4 Biosynthesis and Regulation of Secondary Metabolites in *Trichoderma*

The biosynthetic pathway of fungal secondary metabolites follows unique and intricate biochemical pathways. These are the products of primary metabolism derived from only few precursors such as acetyl-CoA, mevalonate, and amino acids which build up their building block backbones (Demain and Fang 2000; Keller et al. 2005).

The secondary metabolite derived through *Trichoderma* spp. includes non-ribosomal peptides such as peptaibiotics, siderophores, and diketopiperazines-like gliotoxin, gliovirin, polyketides, terpenes, pyrones, and isocyane metabolites. The production of these substances depends on species as well as on the strain of the fungus, so biosynthesis of whole range is hard to be performed by the single fungus under laboratory conditions. The production of metabolites needs specific prompt stimuli for its biosynthesis during laboratory conditions.

The majority of genes involved in the process are part of large biosynthetic gene clusters comprising core enzymes such as non-ribosomal peptide synthetases (NRPSs), polyketide synthases (PKSs), or terpene synthases/cyclases, accessory enzymes (like cytochrome P450s, oxidoreductases, methyl transferases), etc. and, in some cases, genes for transporters and transcription factors (Mukherjee et al. 2012a, b; Bansal and Mukherjee 2016; Mukherjee et al. 2013).

The secondary metabolism process in fungal species involves a tightly regulated cellular process influenced by the environmental conditions and regulatory factors which helps in understanding regulation of SMs. The complex proteins, pH signalling, and other microorganisms influence the expression of secondary metabolism-related genes in *Trichoderma* spp. as similar to other fungal spp. (Atanasova et al. 2013; Bazafkan et al. 2015; Fekete et al. 2014; Karimi-Aghcheh et al. 2013; Malmierca et al. 2015; Mukherjee and Kenerley 2010). Mukherjee et al. (2012a) studied the role of *T. virens* PacC orthologue which directs the biosynthesis of secondary metabolite and iron transport. The expression of genes encoding the NRPS Tex15, a neighbouring cytochrome P450, as well as siderophore-related biogenesis enzymes and transporters were altered in ΔpacC mutants in which various biosynthetic pathway could not be activated.

5 Strategies for Metabolite Production in *Trichoderma*

Trichoderma have various strategies to produce metabolites for plant diseases resistance. The rhizosphere shows extensive network of communication occurring between plants and their associated microbes through the exchange and insight communication of signals.

For the past decades, several approaches are used for the detection of SMs, and accurate analysis of several sensitive compounds is also established. During the past few years, several studies have demonstrated the analysis of secondary metabolites in plant-fungus interaction. These approaches help to identify the specific metabolites which play an important role to enhance the plant's immunity. For example, LC-MS analyse glucosinolate metabolite which was shown to mediate broad-spectrum antifungal and antibacterial defence. GC-MS analysis reveals the metabolic profile of nodulated alfalfa plants, which indicated different stages of nodule organogenesis, which is conducted by global physiological adaptation. Moreover, spectrophotometric and simple chromatographic methods are also used for analysis of secondary metabolites (Lisec et al. 2006). Metabolites that are detected and quantified by mass spectrophotometry are divided into three main subgroups based on physicochemical and molecular mass: (i) polar low molecular mass metabolites (primary metabolites), detected by gas chromatography GC-MS; (ii) polar high molecular mass metabolites (secondary metabolites), detected by liquid chromatography (LC-MS); and (iii) nonpolar (lipid) metabolites, detected by LC-MS or GC-MS analysis (Brotman 2013). There is not any direct relationship between metabolites and secondary metabolite synthesis genes. Secondary metabolites are produced as the result of many genes and their enzymes (Smedsgaard and Nielsen 2005). The sequencing of fungal genomes revealed that gene groups associated with SMs exceed the number of SMs from a certain fungus and several gene clusters from the estimated ones remain silent (Khan et al. 2020). There are different molecular approaches involved in the regulation of these silent genes. There are two approaches known in metabolomics for the identification of molecular weight of SMs produced by organisms which consist of untargeted as well as targeted. Targeted approaches are the method to identify known compounds, while untargeted approaches are techniques for searching all known and unknown compounds. Several chromatographic techniques, such as gas and liquid chromatography as well as mass spectrometry, are useful for the analysis of metabolites in a complex sample. These techniques are helpful for the detection of a large number of metabolites. LC-MS allows for the detection of mid to nonpolar metabolites, and GC-MS is used for the study of both volatile and polar small substances (Kluger 2015). Peptaibiotics, which is extracted from fungal culture, detected by LC-MS, whereby the specific amino acid, Aib, for

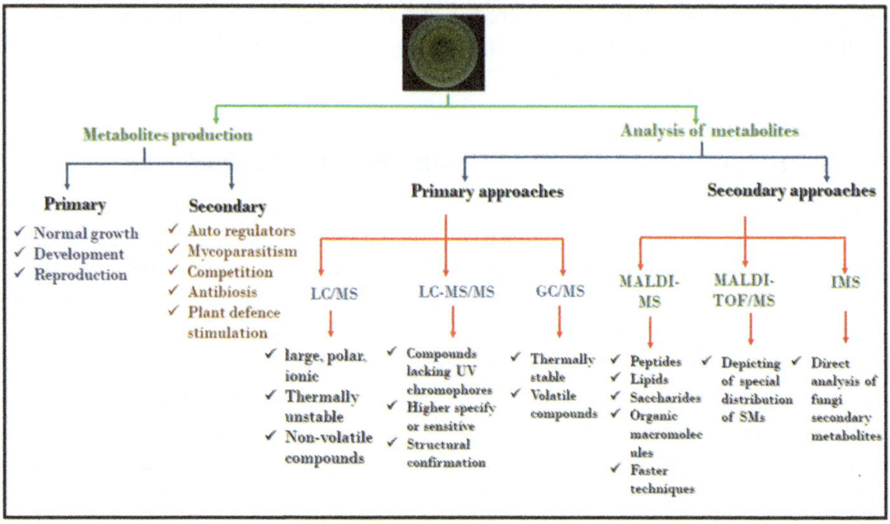

Fig. 2 Methods of analysis and production of secondary metabolites by *Trichoderma*

peptaibiotics can be indicated by mass differences of D m/z 85. The matrix-assisted laser desorption/ionization time-of-flight mass spectrometry (MALDI-TOF-MS) helps to discover the new bioactive SMs in fungus and provides an advanced approach that is much faster than the other traditional bioactive screening techniques. This technique was used for the detection of peptaibol production profiles from 28 different *Trichoderma* species (Neuhof et al. 2007). VOCs are volatile SMs that can be determined by GC-MS without chemical derivatization from liquid culture extracts (Kluger 2015). For example, extraction with organic solvents has been applied for the investigation of VOC production in *T. harzianum* and *T. viride*[2] cultures (Claydon et al. 1987). Imaging mass spectrometry (IMS) is an advanced technique that allows the direct analysis of living fungi for SMs. IMS produces images depicting the spatial distribution of natural products. There are many methods for production and analysis of metabolites which are shown in Fig. 2. MALDI-IMS has been used for the analysis of metabolites in living bacterial communities. Recently, MALDI-IMS was used to visualize the SMs in the mycoparasitic interaction of *R. solani* and *T. atroviride*. There is a minute or even no sample preparation required which makes the MALDI techniques well suited to the analysis of co-cultivations (Fang and Dorrestein 2014).

[2] Species identities are cited as initially published, and the current taxonomic status of each species requires verification.

5.1 Mechanism of SMs Produced by Trichoderma

Numerous SMs are produced by *Trichoderma* spp., such as harzianolides, peptaibols, and certain volatile compounds which have antifungal activity along with plant growth promotion which helps to increase plant immunity against biotic and abiotic stresses.6-pp (6-pentyl-α-pyrone or 6PP) one of the metabolites secreted by *Trichoderma* spp. reduces the mycelial growth of *F. oxysporum*, *B. cinerea*, and *R. solani*, for promotion of plant growth, and induces systemic resistance, which also acts as an auxin-like compound (Khan et al. 2020). Tomato plants, after the treatment with 6-PP produced γ-aminobutyric acid and acetylcholine, which helps in developing resistance against pathogens. The antifungal activities of peptaibols are due to their capacity to form ion channels in membranes and inhibit the enzymes responsible for the synthesis of cell walls.

Various studies have been conducted to investigate the mechanism of *Trichoderma* SMs are bioactive compounds against several pathogens. Besides that, SMs also have application as aroma compounds in the food industry. 6-Pentyl-α-pyrone (6PP) and harzianic acid (HA), a metabolite of *Trichoderma*, were used for the treatments on olive trees. According to Vinale (2009), 6PP was isolated from the liquid culture of *T. atroviride* strain P1, whereas HA was extracted from the liquid culture of *T. harzianum* strain M10. Metabolite solutions were prepared by diluting the compound with distilled water up to the final concentration used for the treatments. For both HA and 6PP, 0.01% ethyl acetate was added to facilitate resuspension and was successively evaporated under cabinet flow (Marra et al. 2020). *T. reesei* is known for production of beneficial SMs, which facilitate in production of metabolites with antibiotic activity including polyketides, pyrones, terpenes, and polypeptides.

6 *Trichoderma* Metabolomics Approaches in the Improvement of Agriculture

There are several SMs secreted by *Trichoderma* strains which are useful in the agriculture sector. Agricultural production is directly affected by soil fertility and their properties, which are dependent on metabolites, biomass, and soil microorganisms (Lehman et al. 2015). As part of metabolome, microorganism metabolites play an important role in maintenance of ecosystem resistance to biotic and abiotic stress. Agricultural modification by the use of PGP microbes is used to enhance soil fertility and wellbeing, which also provides soil disease suppression effects on plant pathogens. In this context, *Trichoderma* spp. have achieved a special status by producing large amounts of extracellular enzymes for mineralization of organic nutrients, used by plants as nutritional materials (Lorito et al. 2010; Contreras-Cornejo et al. 2009). The abundance of *Trichoderma* spp. in soil under different climatic condition has the ability to degrade organic substance in soil and competitive saprophytic and metabolic versatility (Mbarki et al. 2017). Yadav et al. (2009) reported that *T. viride* have great potential to restore soil fertility and promote sugarcane growth. Furthermore, Barakat (2008) reported that the *T. harzianum* Jn14 was added to organically amended soil for the suppression of *R. solani* causative agent of damping off disease. Additionally, Trichokonin a secondary metabolite has showed activity in inducing defence responses against tobacco mosaic virus (TMV) in *Nicotiana tabacum*. 100 nM concentration of trichokonin through foliar treatment leads to 54% lesion inhibition, 57% reduction in average lesion diameter, and 30% reduction in average lesion area in systemic tissue (Luo et al. 2010). *Trichoderma* spp. enhance the rate of decomposition process and act as a good natural decomposition agent. *Trichoderma* are also known for increasing the rate of palm oil effluent and empty fruit bunches decomposition rate from 4–6 months to 21–45 days (Amira et al. 2011).

Besides that, the volatile and non-volatile compounds of *Trichoderma* spp. also use for enhancing the flavour of the food products. According to Claydon et al. (1987) *Trichoderma* have been reported for the production of 6-pentyl-2H-pyran-2-one a volatile compound possesing coconut aroma in the viable cultures which subsequently can be used as a coconut flavouring agent in bakery products. Therefore, *Trichoderma* spp. could be one of the sustainable opportunities in current agricultural technologies.

7 Conclusion and Future Perspective

Trichoderma spp. are one of the rhizospheric inhabitants well known for its various plant growth properties. Omics studies such as transcriptomics, proteomics, and metabolomics provide new insights for better understanding of the biological events occurring inside the organisms. Due to diverse secondary metabolites in microbes, it's very important and interesting to know its various biosynthetic pathways as well as mode of interaction among various plant species during biotic and abiotic stress conditions. The multifaceted role of *Trichoderma* for plant growth and development forces researchers to study the metabolic profiling of several plant species with different accessions. Metabolite profiling can help in investigation of the complex genome organization as well as its downstream activation and signalling. The secondary metabolites secreted by microbes act as communication signal and defence molecules as well as in the variety of applications. The complete knowledge of fungal species with its associated secondary metabolites will lead to development of potent biological agents as well as new strains which helps in the ample production of pharmaceutically and biotechnologically compatible secondary metabolites.

Acknowledgements Authors are thankful to the director of CSIR-National Botanical Research Institute, Lucknow, India, for providing all the necessary facilities. Nishtha Mishra acknowledges Department of Science and Technology for the fellowship awarded to her.

References

Ahluwalia V, Walia S, Sati OP, Kumar J, Kundu A, Shankar J, Paul YS (2014) Isolation, characterisation of major secondary metabolites of the Himalayan *Trichoderma koningii* and their antifungal activity. Arch Phytopathol Plant Protect 47(9):1063–1071

Ahluwalia V, Kumar J, Rana VS, Sati OP, Walia S (2015) Comparative evaluation of two *Trichoderma harzianum* strains for major secondary metabolite production and antifungal activity. Nat Prod Res 29(10):914–920

Amira RD, Roshanida AR, Rosli MI, Zahrah MSF, Anuar JM, Adha CN (2011) Bioconversion of empty fruit bunches (EFB) and palm oil mill effluent (POME) into compost using Trichoderma virens. Afr J Biotechnol 10(81):18775–18780

Andrade R, Ayer WA, Mebe PP (1992) The metabolites of *Trichoderma longibrachiatum*. Part 1. Isolation of the metabolites and the structure of trichodimerol. Can J Chem 70(10):2526–2535

Andrade SAL, Malik S, Sawaya ACHF, Bottcher A, Mazzafera P (2013) Association with arbuscular mycorrhizal fungi influences alkaloid synthesis and accumulation in *Catharanthus roseus* and *Nicotiana tabacum* plants. Actaphysiologiaeplantarum 35(3):867–880

Atanasova, L., Crom, S.L., Gruber, S., Coulpier, F., Seidl-Seiboth, V., Kubicek, C.P. and Druzhinina, I.S., 2013. Comparative transcriptomics reveals different strategies of Trichodermamycoparasitism. BMC genomics, 14(1), pp.1–15

Azarmi R, Hajieghrari B, Giglou A (2011) Effect of Trichoderma isolates on tomato seedling growth response and nutrient uptake. Afr J Biotechnol 10(31):5850–5855

Bais HP, Weir TL, Perry LG, Gilroy S, Vivanco JM (2006) The role of root exudates in rhizosphere interactions with plants and other organisms. Annu Rev Plant Biol 57:233–266

Bakker PA, Berendsen RL, Van Pelt JA, Vismans G, Yu K, Li E, Van Bentum S, Poppeliers SW, Gil JJS, Zhang H, Goossens P (2020) The soil-borne identity and microbiome-assisted agriculture: looking back to the future. Mol Plant 13(10):1394–1401

Bansal R, Mukherjee PK (2016) Identification of novel gene clusters for secondary metabolism in Trichoderma genomes. Microbiology 85(2):185–190

Barakat RM (2008) The effect of Trichoderma harzianum in combination with organic amendment on soil suppressiveness to Rhizoctonia solani. Phytopathol Mediterr 47(1):11–19

Benítez T, Rincón AM, Limón MC, Codon AC (2004) Biocontrol mechanisms of Trichoderma strains. Int Microbiol 7(4):249–260

Brotman Y (2013) 12 The use of metabolomic approaches to study Trichoderma–plant interactions. In: Trichoderma: biology and applications, p 209

Brotman Y, Briff E, Viterbo A, Chet I (2008) Role of swollenin, an expansin-like protein from Trichoderma, in plant root colonization. Plant Physiol 147(2):779–789

Cardoza RE, Hermosa MR, Vizcaíno JA, Sanz L, Monte E, Gutiérrez S (2005) Secondary metabolites produced by *Trichoderma* and their importance in the biocontrol process. Microorganism for industrial enzymes and biocontrol, pp.1–22

Claydon N, Allan M, Hanson JR, Avent AG (1987) Antifungal alkyl pyrones of *Trichoderma harzianum*. Trans Br Mycol Soc

Collins RP, Halim AF (1972) Characterization of the major aroma constituent of the fungus *Trichoderma viride*. J Agric Food Chem 20:437–438. https://doi.org/10.1021/jf60180a010

Contreras-Cornejo HA, Macías-Rodríguez L, Cortés-Penagos C, López-Bucio J (2009) *Trichoderma virens*, a plant beneficial fungus, enhances biomass production and promotes lateral root growth through an auxin-dependent mechanism in Arabidopsis. Plant Physiol 149(3):1579–1592

Contreras-Cornejo HA, Macías-Rodríguez L, Vergara AG, López-Bucio J (2015) Trichoderma modulates stomatal aperture and leaf transpiration through an abscisic acid-dependent mechanism in Arabidopsis. Journal of Plant Growth Regulation, 34(2), pp.425–432

Crutcher FK, Parich A, Schuhmacher R, Mukherjee PK, Zeilinger S, Kenerley CM (2013) A putative terpene cyclase, vir4, is responsible for the biosynthesis of volatile terpene compounds in the biocontrol fungus Trichoderma virens. Fungal Genet Biol 56:67–77

Cutler HG, Himmelsbach DS, Arrendale RF, Cole PD, Cox RH (1989) Koninginin A: a novel plant growth regulator from *Trichoderma koningii*. Agric Biol Chem 53(10):2605–2611

Cutler HG, Himmelsbach DS, Yagen B, Arrendale RF, Jacyno JM, Cole PD, Cox RH (1991) Koninginin B: a biologically active congener of koninginin A from *Trichoderma koningii*. J Agric Food Chem 39(5):977–980

de Oliveira JSF, Xavier LP, Lins A, Andrade EHA, Maia JGS, de Mello AH, Setzer WN, Ramos AR, da Silva JKR (2019) Effects of inoculation by arbuscular mycorrhizal fungi on the composition of the essential oil, plant growth, and lipoxygenase activity of *Piper aduncum* L. AMB Express 9(1):29

Degenkolb T, Gräfenhan T, Nirenberg HI, Gams W, Brückner H, (2006) Trichodermabrevicompactum complex: rich source of novel and recurrent plant-protective polypeptide antibiotics (peptaibiotics). Journal of agricultural and food chemistry, 54(19), pp.7047–7061

Demain AL, Fang A (2000) The natural functions of secondary metabolites. In: History of modern biotechnology, vol I, pp 1–39

Djonovic S (2006) Howell CRand Kenerley CM, Sm1, a proteinaceous elicitor secreted by the biocontrol fungus Trichoderma virens induces plant defense responses and systemic resistance. Mol Plant-Microbe Interact 19:838–853

Djonović S, Vargas WA, Kolomiets MV, Horndeski M, Wiest A, Kenerley CM (2007) A proteinaceous elicitor Sm1 from the beneficial fungus Trichoderma virens is required for induced systemic resistance in maize. Plant Physiol 145(3):875–889

Doornbos RF, van Loon LC, Bakker PA (2012) Impact of root exudates and plant defense signaling on bacterial communities in the rhizosphere. A review. Agron Sustain Dev 32(1):227–243

Druzhinina IS, Seidl-Seiboth V, Herrera-Estrella A, Horwitz BA, Kenerley CM, Monte E, Mukherjee PK, Zeilinger S, Grigoriev IV, Kubicek CP (2011) Trichoderma: the genomics of opportunistic success. Nat Rev Microbiol 9(10):749–759

Dunlop RW, Simon A, Sivasithamparam K, Ghisalberti EL (1989) An antibiotic from *Trichoderma koningii* active against soilborne plant pathogens. J Nat Prod 52(1):67–74

Dutta SC, Neog B (2016) Accumulation of secondary metabolites in response to antioxidant activity of turmeric rhizomes co-inoculated with native arbuscular mycorrhizal fungi and plant growth promoting rhizobacteria. Sci Hortic 204:179–184

Dutta J, Thakur D (2017) Evaluation of multifarious plant growth promoting traits, antagonistic potential and phylogenetic affiliation of rhizobacteria associated with commercial tea plants grown in Darjeeling, India. PLoS One 12(8):e0182302

Evidente A, Cabras A, Maddau L, Serra S, Andolfi A, Motta A (2003) Viridepyronone, a new antifungal 6-substituted 2 h-pyran-2-one produced by *Trichoderma viride*. J Agric Food Chem 51(24):6957–6960

Engel R, Szabo K, Abranko L, Rendes K, Füzy A, Takács T (2016) Effect of arbuscular mycorrhizal fungi on the growth and polyphenol profile of marjoram, lemon balm, and marigold. Journal of agricultural and food chemistry, 64(19), pp.3733–3742

Fang J, Dorrestein PC (2014) Emerging mass spectrometry techniques for the direct analysis of microbial colonies. Curr Opin Microbiol 19:120–129

Ghisalberti EL, Sivasithamparam K (1991) Antifungal antibiotics produced by Trichoderma spp. Soil Biol Biochem 23(11):1011–1020

Ghisalberti EL, Hockless DC, Rowland C, White AH (1992) Harziandione, a new class of diterpene from Trichoderma harzianum. Journal of natural products, 55(11), pp.1690–1694

Goulard C, Hlimi S, Rebuffat S, Bodo B (1995) Trichorzins HA and MA, antibiotic peptides from *Trichoderma harzianum* I. Fermentation, isolation and biological properties. J Antibiot 48(11):1248–1253

Gravel V, Antoun H, Tweddell RJ (2007) Growth stimulation and fruit yield improvement of greenhouse tomato plants by inoculation with Pseudomonas putida or *Trichoderma* atroviride: possible role of indole acetic acid (IAA). Soil Biol Biochem 39(8):1968–1977

Gruber S, Seidl-Seiboth V (2012) Self versus non-self: fungal cell wall degradation in Trichoderma. Microbiology, 158(1), pp.26–34

Gupta R, Tiwari S, Saikia SK, Shukla V, Singh R, Singh SP, Kumar PV, Pandey R (2015) Exploitation of microbes for enhancing bacoside content and reduction of Meloidogyne incognita infestation in Bacopa monnieri L. Protoplasma, 252(1), pp.53–61

Harman GE, Howell CR, Viterbo A, Chet I, Lorito M (2004) Trichoderma species—opportunistic, avirulent plant symbionts. Nat Rev Microbiol 2(1):43–56

Herbert RB (1989) The biosynthesis of secondary metabolites. Springer Science & Business Media

Hossain MM, Sultana F, Kubota M, Hyakumachi M (2008) Differential inducible defense mechanisms against bacterial speck pathogen in Arabidopsis thaliana by plant-growth-promoting-fungus Penicillium sp. GP16-2 and its cell free filtrate. Plant Soil 304(1):227–239.56

Jaleel CA, Gopi R, Manivannan P, Panneerselvam R (2007) Responses of antioxidant defense system of *Catharanthus roseus* (L.) G. Don. to paclobutrazol treatment under salinity. ActaPhysiologiaePlantarum 29(3):205–209

Jones JD, Dangl JL (2006) The plant immune system. Nature 444(7117):323–329

Karimi-Aghcheh, R., Bok, J.W., Phatale, P.A., Smith, K.M., Baker, S.E., Lichius, A., Omann, M., Zeilinger, S., Seiboth, B., Rhee, C. and Keller, N.P., 2013. Functional analyses of Trichoderma reesei LAE1 reveal conserved and contrasting roles of this regulator. G3: Genes| Genomes| Genetics, 3(2), pp.369–378

Keller NP, Turner G, Bennett JW (2005) Fungal secondary metabolism—from biochemistry to genomics. Nat Rev Microbiol 3(12):937–947

Keswani C, Singh SP, Singh HB (2013) A superstar in biocontrol enterprise: *Trichoderma* spp. Biotech Today 3(2):27–30

Khan RAA, Najeeb S, Hussain S, Xie B, Li Y (2020) Bioactive secondary metabolites from *Trichoderma* spp. against phytopathogenic fungi. Microorganisms 8(6):817

Kim HJ, Vinale F, Ghisalberti EL, Worth CM, Sivasithamparam K, Skelton BW, White AH (2006) An antifungal and plant growth promoting metabolite from a sterile dark ectotrophic fungus. Phytochemistry 67(20):2277–2280

Kluger B (2015) Biosynthesis and molecular genetics of fungal secondary metabolites, vol II. Springer, New York

Kusari S, Lamshöft M, Zühlke S, Spiteller M (2008) An endophytic fungus from Hypericum perforatum that produces hypericin. Journal of Natural Products, 71(2), pp.159–162

Lee SJ, Yeo WH, Yun BS, Yoo ID (1999) Isolation and sequence analysis of new peptaibol, boletusin, from Boletus spp. J Peptide Sci 5(8):374–378

Lehman RM, Acosta-Martinez V, Buyer JS, Cambardella CA, Collins HP, Ducey TF, Halvorson JJ, Jin VL, Johnson JM, Kremer RJ, Lundgren JG (2015) Soil biology for resilient, healthy soil. J Soil Water Conserv 70(1):12A–18A

Lisec J, Schauer N, Kopka J, Willmitzer L, Fernie AR (2006) Gas chromatography mass spectrometry–based metabolite profiling in plants. Nat Protoc 1(1):387

Lorito M, Woo SL, Harman GE, Monte E (2010) Translational research on Trichoderma: from 'omics to the field. Annu Rev Phytopathol 48:395–417

López-Bucio JS, Dubrovsky JG, Raya-González J, Ugartechea-Chirino Y, López-Bucio J, de Luna-Valdez LA, Ramos-Vega M, León P, Guevara-García A (2014) Arabidopsis thaliana mitogen-activated protein kinase 6 is involved in seed formation and modulation of primary and lateral root development. Journal of experimental botany, 65(1), pp.169–183

Luo Z, Wang E, Sun OJ (2010) Can no-tillage stimulate carbon sequestration in agricultural soils? A meta-analysis of paired experiments. Agric Ecosyst Environ 139(1-2):224–231

Macías FA, Castellano D, Molinillo JM (2000) Search for a standard phytotoxic bioassay for allelochemicals. Selection of standard target species. J Agric Food Chem 48(6):2512–2521

Malmierca MG, Barua J, McCormick SP, Izquierdo-Bueno I, Cardoza RE, Alexander NJ, Hermosa R, Collado IG, Monte E, Gutiérrez S (2015) Novel aspinolide production by *Trichoderma arundinaceum* with a potential role in *Botrytis cinerea* antagonistic activity and plant defence priming. Environ Microbiol 17(4):1103–1118

Mannina L, Segre AL, Ritieni A, Fogliano V, Vinale F, Randazzo G, Maddau L, Bottalico A (1997) A new fungal growth inhibitor from *Trichoderma viride*. Tetrahedron 53(9):3135–3144

Mandal S, Upadhyay S, Wajid S, Ram M, Jain DC, Singh VP, Abdin MZ, Kapoor R (2015) Arbuscular mycorrhiza increase artemisinin accumulation in Artemisia annua by higher expression of key biosynthesis genes via enhanced jasmonic acid levels. Mycorrhiza, 25(5), pp.345–357

Marra R, Coppola M, Pironti A, Grasso F, Lombardi N, d'Errico G, Sicari A, Bolletti Censi S, Woo SL, Rao R, Vinale F (2020) The application of Trichoderma strains or metabolites alters the olive leaf metabolome and the expression of defense-related genes. J Fungi 6(4):369

Mbarki S, Cerdà A, Brestic M, Mahendra R, Abdelly C, Pascual JA (2017) Vineyard compost supplemented with Trichoderma harzianum T78 improve saline soil quality. Land Degrad Dev 28(3):1028–1037

Mishra A, Singh SP, Mahfooz S, Shukla R, Mishra N, Pandey S, Dwivedi S, Pandey V, Shirke PA, Nautiyal CS, (2019) External Supplement of Impulsive Micromanager *Trichoderma* Helps in Combating CO_2 Stress in Rice Grown Under FACE. Plant Molecular Biology Reporter, 37(1), pp.1–13

Morán-Diez E, Hermosa R, Ambrosino P, Cardoza RE, Gutiérrez S, Lorito M, Monte E (2009) The ThPG1 endopolygalacturonase is required for the *Trichoderma* harzianum–plant beneficial interaction. Mol Plant-Microbe Interact 22(8):1021–1031

Mohana Kumara P, Zuehlke S, Priti V, Ramesha BT, Shweta S, Ravikanth G, Vasudeva R, Santhoshkumar TR, Spiteller M, Uma Shaanker R (2012) Fusarium proliferatum, an endophytic fungus from Dysoxylum binectariferum Hook. f, produces rohitukine, a chromane alkaloid possessing anti-cancer activity. Antonie Van Leeuwenhoek, 101(2), pp.323–329

Mukherjee PK, Kenerley CM (2010) Regulation of morphogenesis and biocontrol properties in Trichoderma virens by a VELVET protein, Vel1. Appl Environ Microbiol 76(7):2345–2352

Mukherjee PK, Buensanteai N, Moran-Diez ME, Druzhinina IS, Kenerley CM (2012a) Functional analysis of non-ribosomal peptide synthetases (NRPSs) in Trichoderma virens reveals a polyketide synthase (PKS)/NRPS hybrid enzyme involved in the induced systemic resistance response in maize. Microbiology 158(1):155–165

Mukherjee PK, Buensanteai N, Moran-Diez ME, Druzhinina IS, Kenerley CM (2012b) Functional analysis of non-ribosomal peptide synthetases (NRPSs) in Trichoderma virens reveals a polyketide synthase (PKS)/NRPS hybrid enzyme involved in the induced systemic resistance response in maize. Microbiology 158(1):155–165

Mukherjee PK, Horwitz BA, Herrera-Estrella A, Schmoll M, Kenerley CM (2013) Trichoderma research in the genome era. Annu Rev Phytopathol 51:105–129

Neuhof T, Dieckmann R, Druzhinina IS, Kubicek CP, von Döhren H (2007) Intact-cell MALDI-TOF mass spectrometry analysis of peptaibol formation by the genus Trichoderma/Hypocrea: can molecular phylogeny of species predict peptaibol structures? Microbiology 153(10):3417–3437

Oliveira MS, Campos MA, Silva FS (2015) Arbuscular mycorrhizal fungi and vermicompost to maximize the production of foliar biomolecules in *Passiflora alata* Curtis seedlings. J Sci Food Agric 95(3):522–528

Pathma J, Ayyadurai N, Sakthivel N (2010) Assessment of genetic and functional relationship of antagonistic fluorescent pseudomonads of rice rhizosphere by repetitive sequence, protein coding sequence and functional gene analyses. The Journal of Microbiology, 48(6), pp.715–727

Péret B, De Rybel B, Casimiro I, Benková E, Swarup R, Laplaze L, Beeckman T, Bennett MJ (2009) Arabidopsis lateral root development: an emerging story. Trends Plant Sci 14(7):399–408

Ramyasmruthi S, Pallavi O, Pallavi S, Tilak K, Srividya S (2012) Chitinolytic and secondary metabolite producing *Pseudomonas fluorescens* isolated from Solanaceaerhizosphere effective against broad spectrum fungal phytopathogens. Asian J Plant Sci Res 2(1):16–24

Radwan HH, Moussa IM, Alsarra IA (2011) Optimization of a fed-batch fermentation process for production of bleomycin by Streptomyces mobaraensis ATCC 15003. African Journal of Biotechnology, 10(9), pp.1690–1965

Reino JL, Guerrero RF, Hernández-Galán R, Collado IG (2008) Secondary metabolites from species of the biocontrol agent Trichoderma. Phytochem Rev 7(1):89–123

Resende MIP, Jakoby ICMC, dos Santos LCR, Soares MA, Pereira FADI, Souchie EL, Silva FG (2014) Phosphate solubilization and phytohormone production by endophytic and rhizosphere Trichoderma isolates of guanandi (Calophyllum brasiliense Cambess). Afr J Microbiol Res 8(27):2616–2623

Ruiz B, Chávez A, Forero A, García-Huante Y, Romero A, Sánchez M, Rocha D, Sánchez B, Rodríguez-Sanoja R, Sánchez S, Langley E (2010) Production of microbial secondary metabolites: regulation by the carbon source. Critical reviews in microbiology, 36(2), pp.146–167

Schwessinger B, Zipfel C (2008) News from the frontline: recent insights into PAMP-triggered immunity in plants. Curr Opin Plant Biol 11(4):389–395

Shentu X, Zhan X, Ma Z, Yu X, Zhang C (2014) Antifungal activity of metabolites of the endophytic fungus *Trichoderma brevicompactum* from garlic. Braz J Microbiol 45(1):248–254

Shoresh M, Harman GE, Mastouri F (2010) Induced systemic resistance and plant responses to fungal biocontrol agents. Annu Rev Phytopathol 48:21–43

Simon A, Dunlop RW, Ghisalberti EL, Sivasithamparam K (1988) *Trichoderma koningii* produces a pyrone compound with antibiotic properties. Soil Biol Biochem 20(2):263–264

Singh SP, Gaur R (2017) Endophytic Streptomyces spp. underscore induction of defense regulatory genes and confers resistance against Sclerotium rolfsii in chickpea. Biol Control 104:44–56

Singh SP, Pandey S, Mishra N, Giri VP, Mahfooz S, Bhattacharya A, Kumari M, Chauhan P, Verma P, Nautiyal CS, Mishra A (2019) Supplementation of *Trichoderma* improves the alteration of nutrient allocation and transporter genes expression in rice under nutrient deficiencies. Plant Physiol Biochem 143:351–363

Sivasithamparam K, Ghisalberti EL (1998) Secondary metabolism in *Trichoderma* and *Gliocladium*. *Trichoderma* and *Gliocladium*. In: Kubicek CP, Harman GE (eds) Basic biology, taxonomy and genetics

Smedsgaard J, Nielsen J (2005) Metabolite profiling of fungi and yeast: from phenotype to metabolome by MS and informatics. J Exp Bot 56:273–286. https://doi.org/10.1093/jxb/eri068

Strobel GA, Stierle A, Hess WM (1993) Taxol formation in yew—Taxus. Plant Science, 92(1), pp.1–12

Stoppacher N, Kluger B, Zeilinger S, Krska R, Schuhmacher R, (2010) Identification and profiling of volatile metabolites of the biocontrol fungus Trichoderma atroviride by HS-SPME-GC-MS. Journal of Microbiological Methods, 81(2), pp.187–193

Su HN, Chen ZH, Song XY, Chen XL, Shi M, Zhou BC, Zhao X, Zhang YZ (2012) Antimicrobial peptide trichokonin VI-induced alterations in the morphological and nanomechanical properties of *Bacillus subtilis*. PLoS One 7(9):e45818

Sun X, Shen X, Jain R, Lin Y, Wang J, Sun J, Wang J, Yan Y, Yuan Q (2015) Synthesis of chemicals by metabolic engineering of microbes. Chem Soc Rev 44(11):3760–3785

Szekeres A, Leitgeb B, Kredics L, Antal Z, Hatvani L, Manczinger L, Vágvölgyi C (2005) Peptaibols and related peptaibiotics of Trichoderma. Acta Microbiol Immunol Hung 52(2):137–168

Tang XX, Liu SZ, Sun YY, He FM, Xu GX, Fang MJ, Zhen W, Qiu YK (2020) New cyclopentenoneacrylic acid derivatives from a marine-derived fungus *Trichoderma atroviride* H548. Nat Prod Res:1–8

Tijerino A, Cardoza RE, Moraga J, Malmierca MG, Vicente F, Aleu J, Collado IG, Gutiérrez S, Monte E, Hermosa R (2011) Overexpression of the trichodiene synthase gene tri5 increases trichodermin production and antimicrobial activity in Trichoderma brevicompactum. Fungal Genetics and Biology, 48(3), pp.285–296

Vargas WA, Mandawe JC, Kenerley CM (2009) Plant-derived sucrose is a key element in the symbiotic association between *Trichoderma* virens and maize plants. Plant Physiol 151(2):792–808

Velusamy P, Gnanamanickam SS (2008) The effect of bacterial secondary metabolites on bacterial and fungal pathogens of rice. In: Secondary metabolites in soil ecology. Springer, Berlin/Heidelberg, pp 93–106

Vey A, Hoagland RE, Butt TM (2001) Toxic metabolites of fungal biocontrol agents. In: Fungi as biocontrol agents: progress, problems and potential, vol 1, pp 311–346
Vinale F, Sivasithamparam K, Ghisalberti EL, Marra R, Woo SL, Lorito M (2008) *Trichoderma*–plant–pathogen interactions. Soil Biol Biochem 40(1):1–10
Vinale, F., Flematti, G., Sivasithamparam, K., Lorito, M., Marra, R., Skelton, B.W. and Ghisalberti, E.L., (2009) Harzianic acid, an antifungal and plant growth promoting metabolite from Trichoderma harzianum. Journal of Natural Products, 72(11), pp.2032–2035
Vinale F, Flematti G, Sivasithamparam K, Lorito M, Marra R, Skelton BW, Ghisalberti EL (2009a) Harzianic acid, an antifungal and plant growth promoting metabolite from Trichoderma harzianum. J Nat Prod 72(11):2032–2035
Vinale F, Flematti G, Sivasithamparam K, Lorito M, Marra R, Skelton BW, Ghisalberti EL (2009) Harzianic acid, an antifungal and plant growth promoting metabolite from Trichoderma harzianum. J Nat Prod 72:2032–2035
Vinale F, Girona IA, Nigro M, Mazzei P, Piccolo A, Ruocco M, Woo S, Rosa DR, Herrera CL, Lorito M (2012a) Cerinolactone, a hydroxy-lactone derivative from *Trichoderma cerinum*. J Nat Prod 75(1):103–106
Vinale F, Sivasithamparam K, Ghisalberti EL, Ruocco M, Woo S, Lorito M (2012b) Trichoderma secondary metabolites that affect plant metabolism. Nat Prod Commun 7(11):1934578X1200701133
Vinale F, Strakowska J, Mazzei P, Piccolo A, Marra R, Lombardi N, Manganiello G, Pascale A, Woo SL, Lorito M (2016) Cremenolide, a new antifungal, 10-member lactone from *Trichoderma cremeum* with plant growth promotion activity. Nat Prod Res 30(22):2575–2581
Vinale F, Nicoletti R, Lacatena F, Marra R, Sacco A, Lombardi N, d'Errico G, Digilio MC, Lorito M, Woo SL (2017) Secondary metabolites from the endophytic fungus Talaromyces pinophilus. Nat Prod Res 31(15):1778–1785
Viterbo ADA, Chet I (2006) TasHyd1, a new hydrophobin gene from the biocontrol agent *Trichoderma asperellum,* is involved in plant root colonization. Mol Plant Pathol 7(4):249–258
Viterbo ADA, Wiest ARIC, Brotman Y, Chet ILAN, Kenerley C (2007) The 18mer peptaibols from Trichoderma virens elicit plant defence responses. Molecular Plant Pathology, 8(6), pp.737–746
Whipps JM, Lumsden RD (2001) Commercial use of fungi as plant disease biological control agents: status and prospects. In: Fungal biocontrol agents: progress, problems and potential, pp 9–22
Xiao-Yan S, Qing-Tao S, Shu-Tao X, Xiu-Lan C, Cai-Yun S, Yu-Zhong Z (2006) Broad-spectrum antimicrobial activity and high stability of Trichokonins from Trichodermakoningii SMF2 against plant pathogens. FEMS Microbiology Letters, 260(1), pp.119–125
Yadav RL, Shukla SK, Suman A, Singh PN (2009) Trichoderma inoculation and trash management effects on soil microbial biomass, soil respiration, nutrient uptake and yield of ratoon sugarcane under subtropical conditions. Biol Fertil Soils 45(5):461–468
Zawadzka AM, Abergel RJ, Nichiporuk R, Andersen UN, Raymond KN (2009) Siderophore-mediated iron acquisition systems in Bacillus cereus: identification of receptors for anthrax virulence-associated petrobactin. Biochemistry, 48(16), pp.3645–3657

Management of Salinity Stress by the Application of *Trichoderma*

Nagamani Adusumilli and Sarojini Chakravarthy Kolli

Contents

1 Introduction

Salinization affects about 30% of the irrigated land of the world, increasing this area approximately 1–2% per year due to salt-affected land surfaces (FAO 2004). Recently it has been estimated that approximately 1125 million hectares of land are salt-affected, of which approximately 76 million hectares have been salinized by human-induced activities (Sanower 2019). The main reason for degradation of soil is salinization which is converting the fertile lands into unsuitable agricultural lands (FAO 2004). Globally, 20% and 33% of total cultivated and irrigated agricultural

N. Adusumilli (✉)
Department of Botany, UCS Saifabad, Osmania University, Hyderabad, Telangana State, India

S. C. Kolli
Government Degree College, Hyderabad, Telangana State, India

N. Amaresan et al. (eds.), *Advances in* Trichoderma *Biology for Agricultural Applications*, Fungal Biology, https://doi.org/10.1007/978-3-030-91650-3_11

lands, respectively, are stricken by high salinity (Shrivastava and Kumar 2015). As a result, 0.8 to 3.6% soils became saline globally, and now about 1125 million hectares of land under the threat of salinity, and 1.5 million hectares of lands are becoming useless for agricultural production (Hossain 2019). By 2050, 50% of arable land may face the problems of salinity (Wang et al. 2003). The accumulation of salt in soil adversely effects the plant anatomy and physiology resulting in both positive and negative impacts (Abuqamar et al. 2009; Zelm et al. 2020). The root system is first and foremost organ of the plant that faces the stress due to high salt accumulation in the soil. The effect of this abiotic stress causes immediate as well as long-term changes in the plant (Hernández 2019; Isayenkov and Maathuis 2019; Lamers et al. 2020). This stress immediately induces osmotic stress resulting in reduced water availability consequentially hampering the growth of the plant. Later, its continuing effect induces ion toxicity resulting in nutrient imbalance in the cytosol (Munns 2005). Hence, salinity is one of the crucial problems among many abiotic stresses that affects the productivity of plants globally (Ma et al. 2020). The plants under stress conditions generally limit the growth and productivity that reduces the yield of the crop up to 90 percent (Mushtaq et al. 2020). The high salinity affects whole plant health, and considerable changes are noted morphologically, biochemically, metabolically, and physiologically and expression of plant gene properties (Zhao et al. 2020).

The plants like halophytes sustain themselves by adopting strategies when exposed to higher salinity levels (i.e., excesses of Na^+, Cl^-, SO_4^{2-}). Generally, most crop species are salt sensitive (called glycophytes), so it is a strategy in the recent years to breed for resistance or to induce tolerance against salt stress to increase crop yield produced on salinized agricultural lands. To mitigate the salinity stress in plants, breeding for increased tolerance through gene transfer is tried, but this appears to be unsuccessful in genetically engineered transgenic plants (Katori et al. 2010; Munns et al. 2000). Another approach to increase/enhance the tolerance to salinity problem is by employing plant growth-promoting microbes (PGPM) like mycorrhizae and bacterial organisms (Hidangmayum and Dwivedi 2018; Hontzeas et al. 2006; Guo et al. 2018; Gupta et al. 2020; Manaf and Zayed 2015; Mastouri et al. 2010, Shrivastava and Kumar 2015; Talaat and Shawky 2011). Among PGPM microbes, avirulent symbiotic endophytic common soil fungus *Trichoderma* proven very effective in inducing tolerance in plants against the high salinity effects (Brotman et al. 2013; Chet 2016; Contreras-Cornejo et al. 2014; Kumar et al. 2020; López-Bucio et al. 2015; Mishra et al. 2020). *Trichoderma* is having successful history in combating plant diseases through multitude of mechanisms (Alfiky and Weisskopf 2021; Ferreira and Musumeci 2021; Harman et al. 2004; Hermosa et al. 2012; Nagamani and Sarojini 2012; Sood et al. 2020).

The association of *Trichoderma* with plants causes changes in the vicinity of rhizosphere which directly effects the plant functions. The well-known character of this fungus is extensively documented as plant growth enhancer, disease resistance inducer, and rhizosphere colonizer, increasing nutrient uptake in plants (Akter et al. 2019; Schuster and Schmoll 2010) through direct and indirect interactions. *Trichoderma* releases varied number of enzymes and metabolites that facilitate the plant to gain tolerance or resistance against biotic and/or abiotic stresses (Vinale et al. 2014; Ramírez-Valdespino et al. 2019; Sarojini and Nagamani 2020; Waghunde et al. 2016). The studies focused on the role of *Trichoderma* to ameliorate plant tolerance against the salinity stress revealed several insights explaining the mechanisms responsible to combat this abiotic stress and those insights are reviewed in this article.

2 *Trichoderma* Species Used in Management of Salinity Stress in Plants

Earlier findings on *Trichoderma* research are remarkable in managing health, productivity, and diseases of plants. The recent investigations using *Trichoderma* are concentrated toward the management of salinity stress on plants as it has emerged as critical abiotic constraint threatening the yield of crops and hampering the modern agricultural development (Zhao et al. 2020).

Trichoderma taxonomy web site (https://trichokey.com/index.php/*Trichoderma*-taxonomy-2020) shows a total of 464 identified species of which only 375 species are with valid names as on July 2020 (Cai and Druzhinina 2021). Among these, few species received attention to manage salinity stress. The most widely employed or used species against salinity stress is *T. harzianum* (Table 1). Other species used for this purpose belongs to *T. afroharzianum*, *T. aggressivum* f. sp. *europaeum*, *T. asperellum*, *T. asperelloides*, *T. atroviride*, *T. citrinoviride*, *T. hamatum* *T. longibrachiatum*, *T. parareesei*, *T. saturnisporum* *T. virens*, *T. viride*, and *T. yunnanense*. A variety of crops were tested to overcome soil salt stress using abovementioned species of *Trichoderma* (Table 1). The strains able to control plant diseases are employed against tolerance to salinity problems (Anwer et al. 2020; Guo et al. 2018; Poosapati et al. 2014; Regragui and Lahlou 2005; Rubio et al. 2014; Zhang et al. 2016). The endophytic *Trichoderma* strains enhanced plant growth are also able to induce resistance toward salt, osmotic stresses, and other abiotic factors (Zaidi et al. 2013). The

3 In Vitro Studies of the Salinity Stress Mitigation Using *Trichoderma* spp.

Among a dozen of *Trichoderma* strains tested, the *T. asperellum* Tvb1 was most tolerant to saline and high pH when grown on high salt and pH medium (Anwer et al. 2020) in comparison to wild strain (WT). This strain was isolated from hot springs; all others are from arable soils. The growth of salinity-tolerant *T. asperellum* mutant constructed with *Agrobacterium tumefaciens*-mediated transformation system was faster when subjected to different NaCl concentrations, and the growth showed no variation but spore producing ability was inhibited under NaCl stress.

Table 1 *Trichoderma* species used in the management of salinity stress

S. no	In vitro/in vivo studies	*Trichoderma* species	Reference
1.	Maize (*Zea- mays* L.) var. NT6621 and rice (*Oryza sativa* L.) var. Kernel	*T. harzianum* (Th-6)	Yasmeen and Siddiqui (2017)
2.	*Arabidopsis thaliana*	*T. asperelloides* T203	Brotman et al. (2013)
		T. virens	Contreras-Cornejo et al. (2014)
		T. atroviride	
3.	Bean (*Phaseolus vulgaris* L.)	*T. atroviride*, *T. harzianum* sensu lato species complex, *T. longibrachiatum* – *T. orientale* species complex	Gal-Hemed et al. (2011)
		T. asperelloides	
4.	Black pepper (*Piper nigrum* L.)	*T. harzianum*	Sri Vithya et al. (2018)
5.	Black pine (*Pinus thunbergia)*	*T. harzianum*	Min et al. (2014)
		T. hamatum	
6.	Broad bean (cv. *Vicia faba* L.)	*T. viride*	Abdel Kareem et al. (2016)
7.	Chilli (*Capsicum frutescens* L.)	*Trichoderma* spp.	Muthukumar et al. (2011)
8.	Cucumber (*Cucumis sativus* L.)	*T. harzianum*	Zhang et al. (2018)
9.	Egg plant	*T. harzianum*	Velmurugan et al. (2015)
10.	Faba beans (*Vicia faba* L.)	*T. harzianum*	El-Baki and Mostafa (2014)
11.	Groundnut	*Trichoderma* spp.	Taufiq and Yusnawan (2020)
12.	Indian mustard (*Brassica juncea* L)	*T. harzianum*	Ahmad et al. (2015)
13.	In vitro studies	*T. asperellum*	Anwer et al. (2020)
		T. asperellum T59 mutants	Guo et al. (2018)
		T. asperellum, *T. hamatum*, and *T. longibrachiatum*	Poosapati et al. (2014)
		T. aggressivum f. sp. *europaeum*, *T. saturnisporum*, and *T. longibrachiatum*	Sánchez-Montesinos et al. (2019)
		T. harzianum	Mohamed and Haggag (2006)
14.	Maize (*Zea mays* L.)	*T. citrinoviride* (T11C)	Yeşilyurt et al. (2018)

(continued)

Table 1 (continued)

S. no	In vitro/in vivo studies	*Trichoderma* species	Reference
15.	Maize (*Zea mays* L.), rice (*Oryza sativa* L.)	*T. harzianum*	Yasmeen and Siddiqui (2017)
16.	*Ochradenus baccatus* (Del.)	*T. hamatum*	Hashem et al. (2014)
17.	Pea (*Pisum sativum* L.)	*T. asperellum* (T42)	Singh and Dwivedi (2018)
18.	Rapeseed *Brassica napus* cv. Jura	*T. parareesei*	Poveda (2020)
19.	Rice (*Oryza sativa* L.)	*Trichoderma* spp.	Rawat et al. (2016)
20.	Soyabean (*Glycine max*(L.) Merr.)	*T. harzianum*	Khomari and Davari (2017)
21.	Sweet pea (*Lathyrus odoratus* L.)	*T. asperellum* (T42)	Singh and Dwivedi (2018)
22.	Tomato (*Solanum lycopersicum* L.)	*Trichoderma* spp.	Kashyap et al. (2020)
		T. afroharzianum T-22	Mastouri et al. (2010)
		T. parareesei, *T. reesei*	Rubio et al. (2014)
		T. harzianum mutants Th50M6 and Th50M11	Mohamed and Haggag (2006)
23.	Transgenic tobacco lines (*Nicotiana tabacum* L.)	*T. harzianum*	Dana et al. (2006), Sun et al. (2020)
24.	Wheat (*Triticum aestivum* L.)	*T. yunnanense*	Oljira et al. 2020
		T. afroharzianum	
		T. reesei	Ikram et al. (2019)
		T. longibrachiatum T6	Zhang et al. (2016, 2019)
25.	Wheat (*Triticum aestivum* L.) in-vitro studies	*T. aureoviride* and *T. harzianum*	Ripa et al. (2019)
26.	*Populus davidiana* X *P. alba* var. *pyramidalis Louche*	*T. asperellum* mutant T59	Guo et al. (2018)
27.	Melon (*Cucumis melo* var. *Piñonet*)	*T. aggressivum* f. sp. *europaeum*, *T. saturnisporum*, and *T. longibrachiatum*	Sánchez-Montesinos et al. (2019)
28.	Vineyard (*Vitis venifera* L.)	*T. harzianum* T78	Mbarki et al. (2016)

The NaCl tolerance of the T59 strain was higher than that of the WT strain (Guo et al. 2018). *T. asperellum*, TaDOR673 strain, is thermotolerant and salt tolerant and effectively controls the collar rot disease in groundnut (Poosapati et al. 2014), while some other strains in their study showed more salt tolerant but less thermotolerant and poor in disease control. The screening studies on *T. aggressivum* f. sp. *europaeum*, *T. saturnisporum*, and *T. longibrachiatum* revealed the salinity tolerance on high concentration of NaCl containing medium, and *Cucumis melo* var. *Piñonet* (melon variety Piel de sapo) seedling disease control is highest with *T. longibrachiatum* (Sánchez-Montesinos et al. 2019). The mutants of *T. harzianum* generated by irradiation were tolerant to high NaCl concentration in growth medium and showed proper linear growth; some altered morphology; high production of gliotoxin, gliovirin, trichodermin, and phenols; extracellular hydrolytic enzymes; and proline accumulation in cells in comparison to wild type (Mohamed and Haggag 2006). The growth inhibition of *Verticillium dahliae* with *T. harzianum* decreased in presence of high salt concentration, but still the inhibition was significant where the antagonistic action of metabolites decreased with an increase in salt concentration (Regragui and Lahlou 2005). It is interesting to note that the capability of marine isolates to tolerate increasing osmotic pressure (halotolerance) is a strain- or clade-specific novelty rather than a characteristic of a species (Gal-Hemed et al. 2011). The mutants with overexpressed genes are proved to be efficient to overcome the salinity stress. These studies indicate that all *Trichoderma* strains are not salinity tolerant, and strains showing tolerance to other abiotic factors mostly sustain salt concentration in the substrate simultaneously able to control plant pathogens in vitro as well as in vivo.

4 *Trichoderma* Formulations and Application Methods for Managing Salinity Stress in Crops

Trichoderma inoculum is prepared and applied to plants by following different procedures. It is applied as dry powders, spore suspensions sometimes mixed with basic materials for binding.

Trichoderma agar discs from actively growing culture are inoculated on rice (*Oryza sativa*) grains with sterilized distilled water. After sufficient incubation, the *Oryza sativa* grains dried and grained to fine powder then mixed with talcum powder (mesh 350 with 95% whiteness) and 1% carboxy methyl cellulose (CMC). This powdered formulation is coated on sterilized broad bean (cv. *Vicia faba* L.) (Abdel Kareem et al. 2016). Oljira et al. (2020) also used powered formulations but followed different method. *Trichoderma* was grown on boiled wheat (*Triticum aestivum* L.) grain for 3 weeks, and the conidial masses were separated by centrifugation. The pelleted conidial form was collected, later dried for 3 days under aerated and aseptic conditions. The dried mass was powdered and used for seed coating of wheat (*T. aestivum*) seeds for the study. Ahmad et al. (2015) used lyophilized culture powder for inoculations on seeds. The mycelium of *Trichoderma* grown on potato dextrose broth was collected and lyophilized under vacuum to get powder. Later the culture powder was mixed with talc and carboxy-methyl cellulose. The final concentration of the carrier material per gram was 2×10^9 cfu, and it was applied to the pot at the rate of $10 gkg^{-1}$ soil before sowing.

Others used spore suspension directly to the soil or seeds as biopriming. The pregerminated *Cucumis melo* var. *Piñonet* (melon) seeds sowed in pots were inoculated with 5 mL spore suspension of "*Trichoderma* sp." in each pot at 50 X 10^6 propagules/plant (Sánchez-Montesinos et al. 2019). *Solanum lycopersicum* (tomato) seeds were soaked in the *Trichoderma* suspension at a rate of 10^5 spores / mL (Mohamed and Haggag 2006). *Zea mays* L. cv. *samada* 07 seeds were bioprimed with "*Trichoderma* sp." conidial suspension at the rate of 1 × 107 spores /mL (Yeşilyurt et al. 2018) where *Trichoderma* inoculum was mixed with 2% carboxymethyl cellulose and 1% Tween 20. *Lathyrus odoratus* (Pea) seeds were directly soaked in *Trichoderma* spore suspension for the pot experiments (Singh and Dwivedi 2018). Poosapati et al. (2014) also used spore suspension of *Trichoderma* to soak tomato seeds for field experiments. The maize (*Zea mays* L. var. NT6621) and rice (*Oryza sativa* L var. *Kernel*) seeds were treated with *T. harzianum* conidia (colony forming unit 67.3 10^{-3}) mixed with 2% gum arabic as sticker (Yasmeen and Siddiqui 2017). The roots of tomato seedlings were submerged in a suspension of 2 × 10^6 conidia / ml before transplanting into plastic containers with sterile sand-peat (1:1) (Martínez et al. 2015). The conidial suspension of *T. virens* (10^6 spores/mL) prepared from 7-day-old culture on potato dextrose agar (PDA) was used as a seed coating by dipping the soyabean seeds into the spore suspension prior to the planting (Yusnawan et al. 2019). Mastouri et al. (2010) used encapsulated conidia for these studies. *Trichoderma* conidia were coated onto cellulose and encapsulated with tapioca dextran, and conidial suspension in sterile type I water was prepared for treating tomato seeds. Their studies used the conidial suspension at the rate of 20 μl g–1 to deposit 2 × 10^7 cfu g^{-1} of seed. *T. asperellum* spore suspension (1 × 10^9 spores/L) was prepared and 200 mL of the suspension diluted to each liter of soil to get appropriate concentration of 1 × 10^3 (A1); 1 × 10^6 (A2); and 1 × 10^9 (A3) spores/L (FU et al. 2017). *Trichoderma* inoculum was added to the *Cucumis sativus* (cucumber) and *Arabidopsis* root system in hydroponic solution, at the rate of 10^5 germinated spores mL^{-1}, and a final concentration of 10^6spores/g soil was mixed with the soil for pot experiments (Brotman et al. 2012). The spores were collected from the agar plates previously grown with *Trichoderma* by washing with sterile distilled water and adjusted to a concentration of 10^6 mL^{-1} to use as seed coating of *Arabidopsis thaliana*, chilli, cucumber, *Cucurbita pepo*, *Nicotiana benthamiana*, tomato, *Lotus japonicum*, and *Cynara cardunculus* under growth chamber studies (Ruocco et al. 2015). *Trichoderma* spores in a 5 μL aliquot of water were inoculated 5 cm below the growing root tips of *Arabidopsis thaliana* in the Petri pates and resealed after inoculation (Nieto-Jacobo et al. 2017).

Trichoderma inoculum collected from previously grown on agar plates was used as starter inoculum for solid-state fermentation on sterilized *Oryza sativa* (rice). Later the spores were collected and adjusted to 1 × 10^7 spores mL^{-1} and then used to treat lettuce seedlings by using a root dip method for greenhouse trials (Fiorentino et al. 2018). Three different dosages of *T. viride*, a commercial product used (2.0, 4.0 and 6.0 $g.kg^{-1}$ seed) as seed dresser, were applied (Leo Daniel et al. 2011).

While studying the *Arabidopsis thaliana* seedlings under salt stress, the *Trichoderma* fungal spore densities of 10^6 spores were inoculated on 0.2× MS medium with 100 mM NaCl or without salt by placing the spores at 5 cm in the opposite ends of agar plates containing 4-day-old germinated *Arabidopsis* seedlings

(Contreras-Cornejo et al. 2014). "*Trichoderma* sp." was added to the soil as a wheat (*Triticum aestivum*) bran-peat mixture (Gal-Hemed et al. 2011). After pre-soaking of seeds in water for 12 h, the soybean seeds were presoaked in water for 12 h then bioprimed with *T. harzianum* for 10 g/kg of seeds (Khomari and Davari 2017). After germination of *Trichoderma*-coated seeds, some inoculum was again applied below the soil surface by syringe to ensure infection (Soliman et al. 2020). Rice (*Oryza sativa*) straw was used to prepare biostraw with *Trichoderma* inoculum isolated from saline soils and used to treat cowpea seeds (Hamed et al. 2015). Sterilized tomato seeds were coated with an aqueous suspension containing *Trichoderma* spores 2×10^8 mL^{-1} (1 ml of spore suspension per 30 seeds) and then air-dried overnight in an open Petri dish under a laminar flow hood and sown in pot soil (Rubio et al. 2014). Guo et al. (2018) used *Trichoderma* spore suspension to treat *Populus davidiana* × *P. alba* var. *pyramidalis* (PdPap poplar) seedlings at the concentration of 10^6spores/mL. Vineyard composts were supplemented with the high salt-tolerant *T. harzianum* T78 to rehabilitate the saline soils (Mbarki et al. 2016).

The seed coating method using *Trichoderma* is a popular method in agriculture also in experimental trials. This may be owing to convenience, environmental safety, accuracy, and cost-effectiveness (Ma et al. 2020).

4.1 *Effect of* Trichoderma *on Management of Salinity Stress in Plants*

Many investigations extensively recorded the soil salinity effects on plants (Bhattarai et al. 2020; Cheeseman 1988; Hernández 2019; Isayenkov and Maathuis 2019; Ma et al. 2020; Mushtaq et al. 2020; Volkmar et al. 1998; Zelm et al. 2020; Zhao et al. 2020). The higher salt concentration (salinity) in the soil causes ionic and osmotic stress on plants. This triggers morphological, biochemical, and physiological responses in plants leading to less crop productivity. High salt content in soil causes decrease in soil water potential (osmotic phase). A rapid increase of salt in the cell walls or cytoplasm occurs when the vacuoles can no longer sequester incoming salts (ionic phase). Afterward salt concentration builds up in older leaves which hastens their death. Because of this injury, supply of carbohydrates and/or growth hormones to the meristematic regions of the plant reduces those results in growth inhibition of plant (Mazher et al. 2007). In reality, the excessive inflow of salts under salinized conditions manipulates the production of specific metabolites of the plant which limits the photosynthetic rate that is visible phenotypically as reduced growth (Shannon et al. 1994).

The interaction of *Trichoderma* with plants under salinity conditions ameliorated these effects (Brotman et al. 2013). The microbes colonizing roots in the rhizosphere usually have anatomical as well as physiological interaction with plants. The rhizosphere imparts very important niche that accommodates the rhizosphere microbes by providing both space and nutrients for their growth and activity. This facility is well adopted by the species of *Trichoderma* which is an outstanding feature of this genus that acquired title "rhizosphere competent fungus." This character is an important aspect which provides tolerance/resistance to plants against innumerable stresses including salinity stress (Brotman et al. 2013). The rhizosphere colonization enhanced the salt-tolerant metabolites in the *Arabidopsis thaliana* roots inoculated with *Trichoderma asperelloides* T203. It is assumed that colonization of roots with *Trichoderma* triggers a signaling cascade that activates a variety of defense mechanisms in plants to overcome stresses (Shoresh et al. 2010).

The management of salinity stress on cereals and non-cereals by means of different *Trichoderma* species was assessed by various workers (Table 1). Two broad bean (cv. *Vicia faba* L.) varieties treated with *T. viride* enhanced shoot growth when grown with or without 250 mM NaCl + *Trichoderma* (Mahmood et al. 2019). While managing salt stress in Indian mustard, *T. harzianum* increased biomass along with increased oil content (Ahmad et al. 2015). The same effect is found in salinity stressed tomato plants with *T. parareesei* treatment (Poveda 2020). *Arabidopsis* and cucumber (*Cucumis sativus* L.) plants treated with *T. asperelloides* T203, prior to imposition of salt stress, showed significant improvement in seed germination (Brotman et al. 2013) and yield of rapeseed after treating with *T. parareesei* (+T6) and *T. parareesei* transformation control (+Tp-TC) (Poveda 2020). *Arabidopsis* plant growth was promoted under normal as well as saline conditions with the treatment of *T. virens* (Tv29.8) and *T. atroviride* IMI 206040 (Contreras-Cornejo et al. 2014). The salt stress of the plants causes reduction in photosynthetic pigments due to decrease in enzymatic activity (Ahmad et al. 2015) which was ameliorated with the treatment of *Trichoderma* species in *Zea mays* (maize) and *Oryza sativa* (rice) (Yasmeen and Siddiqui 2018) and rapeseed (Poveda 2020).

Studying effect of *T. harzianum* T78 on saline soils growing vineyard did not show any promising result but when it amended with compost in saline soil improved the soil microbiological quality (Mbarki et al. 2016). Moreover, the colony-forming units did not decrease even thou the salt concentration is increased which was negative in natural soil where T78 was inoculated without compost. This study indicates the supplementation of organic matter which has increased the activity of T78 to overcome the salinity effect.

The glycophytes accumulate excessive production of ROS causing progressive oxidative damage and ultimately cell death (Gupta and Huang 2014; Zelm et al. 2020). Plants under stress conditions respond and develop some complex regulatory mechanisms to protect themselves against the ill effects caused by any kind of stresses. The salt-tolerant plants also develop several physiological and biochemical modifications by compartmentalization of excessive salts and activation of antioxidants (Schachtman and Munns 1992) and also synthesize many types of compounds like soluble proteins and sugars to stabilize the cellular structure, to maintain cell turgor, and to regulate osmoticum (Bartels and Sunkar 2005). High levels of ROS formation due to stress response cause damage of the biomolecules in plants. The low or reasonable amounts of ROS act as second messengers that help the plants in signaling at cellular level. This aids in modifying, scavenging, or reducing the ROS chemicals by the formation of enzymatic or non-enzymatic antioxidants in the plant

(Bhattarai et al. 2020). The enzymatic antioxidants comprise superoxide dismutase (SOD), catalase (CAT), guaiacol peroxidase (GPX), enzymes of ascorbate glutathione (AsA-GSH) cycle such as ascorbate peroxidase (APX), monodehydroascorbate reductase (MDHAR), dehydroascorbate reductase (DHAR), and glutathione reductase (GR) (Noctor and Foyer 1998). The nonenzymatic antioxidants present inside the cell are the Ascorbate (AsA), reduced glutathione (GSH), carotenoids, tocopherols, and phenolics. Maintenance of these antioxidants' ability benefits the plant to scavenge or detoxify the harmful ROS that is correlated to enhanced levels of tolerance in plants against the stresses (Bahmani et al. 2015). Hence, to identify the tolerance of plants against the salt stress, the formation of ROS and antioxidant formation is analyzed by measuring them with several techniques. These kinds of modifications were noticed when *Trichoderma* was applied to plants under salt stress conditions. *Trichoderma yunnanense*, *T. afroharzianum*, and *Bacillus licheniformis* improved net photosynthesis rate; water use efficiency and; root and shoot biomass production in *Triticum aestivum* (Oljira et al. 2020). Enhanced IAA production and ACC-deaminase activity with NaCl stress in *T. longibrachiatum* regulated the plant genes' expression of IAA, and ethylene synthesis in *T. aestivum* seedling roots subjected to salt stress reduced the toxicity of ions in the *T. aestivum* plant cells. The causes for the improvement of growth of *T. aestivum* (wheat) plants treated with *T. longibrachiatum* under salt stress are 1. reduction in the production of ethylene; 2. an increase in IAA; 3. reduced accumulation of Na+ ions; 4. improvement in K+ uptake; 5. K+/Na+ ratio; and 6. transcriptional level of Na+/H+ antiporter gene expression in both roots and shoots (Zhang et al. 2019). An antioxidative defense system enhancement and scavenging of excessive ROS produced by plants under salt stress was achieved in *T. aestivum* (wheat) seedlings treated with *T. longibrachiatum* T6 (Zhang et al. 2016). *T. harzianum* improved the uptake of beneficial elements, stimulated compatible solute accumulation, and elevated the level of antioxidant enzymes in *Brassica juncea* L. under salt stress (Ahmad et al. 2015). After treating the rapeseed plants with *T. harzianum* (+T34), the oxidative stress was lowered significantly under salt stress conditions while without the treatment the oxidative stress in the plants increased significantly. The application of *T. parareseesei* (+T6) and *T. parareesei* transformation control (+Tp-Tc) was beneficial in lowering the oxidative stress in rapeseed plants (Poveda 2020). The exogenous foliar application of Salicylic acid (SA) alone and along with *Trichoderma* was reported to result in the detoxification of ROS (H_2O_2 and $O^{\cdot -}$), lowering of lipid peroxidation, enrichment of osmo-protectants like proline coupled with stimulation of antioxidative enzymes activity (superoxide dismutase and ascorbate peroxidase) in pea (*Pisum sativum* L.) plants (Singh and Dwivedi 2018). A similar kind of observation is reported in *Ochradenus baccatus* (Del.) when treated with *T. hamatum* (Bonord.) Bainier (Abeer Hashem et al. 2014). The growth of the plant, lateral roots, and root hair induction in *Arabidopsis* was noticed when inoculated with *Trichoderma* spp. under saline conditions (Contreras-Cornejo et al. 2014). This enhancement is associated with increased levels of abscissic acid, L-proline, and ascorbic acid and increased elimination of Na^+ through root exudates. It is also attributed that it is due to increased IAA levels and osmoprotectiveness of the plant.

The studies on managing salinity stress on cereals – *Triticum aestivum*, Zea *mays*, and *Oryza sativa* – has shown remarkable findings. The maize seedlings treated with *T. asperellum* were larger than untreated plants under salinity conditions. The increased concentration of *T. asperellum* spores in the suspension used to treat the maize plants improved its plant height, root thickness, root length, root weight and water content in the cells also (Fu et al. 2017). The *Triticum aestivum* (wheat) plants developed tolerance against salinity which was evident from the experimental results showing a higher water use efficiency (WUF), lower intercellular CO_2, stomatal conductance, and transpiration in *Triticum aestivum* plants grown after treating seeds with *Trichoderma* under salinity stress (Oljira et al. 2020). The inoculated *Trichoderma* induced these photosynthetic parameters in *Triticum aestivum* plants where the effects were visible clearly in growth and biomass. This enhancement is greater with fungi rather than with bacteria (Oljira et al. 2020). The siderophores producing *T. asperellum* Q1 strain has a real potential to enhance the growth of other crops like cucumber by inducing physiological protection under saline stress and its siderophores have the potentiality of alleviating negative effect of salinity and iron deficiency (Qi and Zhao 2013).

Rawat et al. (2011) showed that salinity-tolerant isolates of *T. harzianum* have efficiency to reduce the severity of the effects due to salinity by strengthening the plant stand. The treated plants showed lower accumulation of MDA (malondialdehyde) content, whereas proline content and phenolics were higher under both non-saline and saline conditions. The treated plants showed lower accumulation of MDA content, whereas proline content and phenolics were higher under both non-saline and saline conditions.

The *Trichoderma* strains helping plants to overcome salinity effects also able to control plant pathogens. This was observed when three isolates of *Trichoderma* (*T. longibrachiatum, T. aggressivum* f. sp. *europaeum* and *T. saturnisporum*) regardless of their origin reduced the salinity stress resulting in larger melon plants and an increase in percentage of dry weight more than 80% for *T. longibrachiatum* or an increase in root dry weight close to 50% (Sánchez-Montesinos et al. 2019). *T. hamatum* treatment increased seed germination of *Ochradenus baccatus* grown in NaCl salt stress (Hashem et al. 2014). With the application of *Trichoderma* under salt stress conditions decreased MDA content and H_2O_2 in all tested broad bean genotypes (Abdel Kareem et al. 2016). The *Trichoderma* interaction with plant were more distinct in enhancing overall growth, decreasing the transportation of Na^+ to shoot from root. This gives tolerance to plants to sustain the salinity effect by stabilizing the cytoplasm from the toxic effect of Na^+ ions accumulation and accelerating the defense related antioxidant enzyme activities in the broad bean genotypes.

The studies on transgenic plants helpful in increasing tolerance to salt stress revealed their role in management of salinity in plants. The *APX*-gene (derived from *Pisum sativum* L.) in transgenic tomato plants (*Lycopersicon esculentum* L.) tolerated the salt injury (Chen et al. 2010). The gene tApx overexpressed either in tobacco or in *Arabidopsis* tolerated oxidative stress (Ueda et al. 2017). Other genes like *MDHAR* gene in tobacco (Zhang 2015) and *AtDHAR1* gene in potato plants (Song et al. 2012) enhanced tolerance to salt stress. Even though certain identified genes in transgenic plants overexpressed to improve tolerance against oxidative stress (due to generation of ROS), one component of antioxidative defense system may not change the other pathways (Cheeseman 1988; Mian et al. 2011). Rubio et al. (2014) observed a maximum upregulation of the salt tolerance *SOS1* gene after 24 h of treatment with "*Trichoderma* sp." (T6) in tomato (*Lycopersicon esculentum* L.) plants. The downregulation or activation of certain genes in *L. esculentum* plants is induced by the interaction of *T. parareesei* in their studies. Their studies indicated that plants responded to the interaction of T6 which activated the responsive element binding protein 2 (AREB2) and responsive element binding protein 2 (*SOS1*) genes in *Lycopersicon esculentum* L. plants which are associated with responses of ABA and salt tolerance, respectively. But ascorbate peroxidase (APX1) enzyme was not responded significantly which acts as negative regulator for ABA biosynthesis in early stages of plant growth period. When *EIN2* is not regulated in *Lycopersicon esculentum* L. plants, *AREB2* is activated as ABA synthesis is not hampered due to downregulation of EIN2 during salt stress condition in *L. esculentum* (tomato) plants. During salt stress condition in *L. esculentum* plants, ABA synthesis is unaffected because of downregulation of *EIN2*.

Viterbo and co-workers (2010) demonstrated that *Arabidopsis* and cucumber plants treated with *Trichoderma* prior to salt stress showed significant improvement in seed germination through expression of several gene related to osmoprotection. The defense-related gene expression during the time course of 0–6 days was reported after applying T6 in *Lycopersicon esculentum* plants to sustain the stress caused by biotic and abiotic factors. The genes related to jasmonic acid (JA)/ethylene (ET)-related *LOX1* and *EIN2* showed maximum upregulation after 24 h while salt tolerance *SOS1* gene of salicylic acid (SA) after 48 h of treated with T6. All these are known to enhance the tolerance to overcome salt stress in plants. Even expression levels of *AREB2* and *SOS1* genes in tomato seedlings were improved in presence of T6 which are related to ABA responses and salt tolerance, respectively.

5 Conclusions

The salt stress on plants is constantly increasing globally. The adaptations of crop plants against this stress are not evolving as fast as increase in soil salinity. Hence, alternative methodologies are being developed to maintain the soil/plant health and productivity of plants to meet the increasing population demand. These technologies implementation is always aimed at ecologically balanced systems. Thus, using microbial inoculants received importance particularly the wonder fungal organism *Trichoderma*. Several studies indicated the role of *Trichoderma* in managing the salt stress in plants. However, the application of efficient species on diversity of crops has not reached the field level. The inoculations of *Trichoderma* in plants enhanced the overexpression of certain genes in transgenic plants, but gene effect on multiple components for complete health of the plant is always recommended because a single component of gene may not change other genes expressions or pathways useful for management of overall effects caused by salinity stress. The utilization of *Trichoderma*'s ability to ameliorate the soil salinity problem will certainly help to achieve the objective of management of adverse impacts of salt stress on plants and improve plant/crop productivity and production.

References

Abdel Kareem SH, Mahmoud MG, Mahmoud E-MA, Sharaf M (2016) Interactive effect of *Trichoderma viride* on broad bean (cv. *Vicia faba* L.) genotypes grown under different salinity stress conditions. Inter J Ecotoxicol Ecobiol 1(3):141–151. https://doi.org/10.11648/j.ijee.20160103.21

Abeer Hashem EF, Abd_Allah AA, Alqarawi Asma A, Al Huqail DE (2014) Alleviation of abiotic salt stress in *Ochradenus baccatus* (Del.) by *Trichoderma hamatum* (Bonord.) Bainier. J Plant Interact 9(1):857–868

Abuqamar S, Luo HL, Laluk K, Mickelbart MV, Mengiste T (2009) Crosstalk between biotic and abiotic stress responses in tomato is mediated by the AIM1 transcription factor. Plant J 58:347–360. https://doi.org/10.1111/j.1365-313X.2008.03783.x

Ahmad P, Hashem A, Elsayed FA, Alqarawi AA, John R, Egamberdieva D, Gucel S (2015) Role of *Trichoderma harzianum* in mitigating NaCl stress in Indian mustard (*Brassica juncea* L.) through antioxidative defense system. Front Plant Sci 6:1–15. https://doi.org/10.3389/fpls.2015.00868

Akter F, Ahmed GU, Alam MF (2019) *Trichoderma*: a complete tool box for climate smart agriculture. Madridge J Agric Environ Sci 2(1):40–43. https://doi.org/10.18689/mjaes-1000107

Alfiky A, Weisskopf L (2021) Deciphering *Trichoderma* –plant –pathogen interactions for better development of biocontrol applications. J Fungi 7:61. https://doi.org/10.3390/jof7010061

Anwer MA, Singh K, Prasa BD, Yadav AK, Kumari P (2020) Abiotic stress tolerant Trichoderma asperellum Tvb1from hot spring and its antagonistic potential against soil borne phytopathogens. Int Arch App Sci Technol 11(3):70–90

Bahmani K, Noori SAS, Darbandi AI, Akbari A (2015) Molecular mechanisms of plant salinity tolerance: a review. AJCS 9(4):321–336

Bartels D, Sunkar R (2005) Drought and salt tolerance in plants. Crit Rev Plant Sci 24:23–58. https://doi.org/10.1080/07352680590910410

Bhattarai S, Biswas D, Fu Y, Biligetu B (2020) Morphological, physiological and genetic responses to salt stress in *Alfalfa*: a review. Agronomy 10:577. https://doi.org/10.3390/agronomy10040577

Brotman Y, Lisec J, Meret M, Chet I, Willmitzer L et al (2012) Transcript and metabolite analysis of the *Trichoderma* -induced systemic resistance response to *Pseudomonas syringae* in *Arabidopsis thaliana*. Microbiology 158:139–146

Brotman Y, Landau U, Cuadros-Inostroza Á, Takayuki T, Fernie AR, Chet I, Viterbo A, Willmitzer L (2013) *Trichoderma* -plant root colonization: escaping early plant defense responses and activation of the antioxidant machinery for saline stress tolerance. PLoS Pathog 9:e1003221. https://doi.org/10.1371/J.ppat.1003221

Cai F, Druzhinina IS (2021) In honor of John Bissett: authoritative guidelines on molecular identification of *Trichoderma*. Fungal Divers 107:1–69. https://doi.org/10.1007/s13225-020-00464-4

Cheeseman JM (1988) Mechanisms of salinity tolerance in plants. Plant Physiol 87:547–550. 0032-0889/88/87/0547/04/$01.00/0

Chen LT, Luo M, Wang YY, Wu K (2010) Involvement of *Arabidopsis* histone deacetylase HDA6 in ABA and salt stress response. J Exp Bot 61:3345–3353

Chet I (2016) *Trichoderma*: a biocontrol agent that induces resistance to both biotic and abiotic stress. In: Abstracts of the 2nd Global Summit on Plant science, London, 06–08 October 2016. doi:https://doi.org/10.4172/2329-955X.C1.006
Contreras-Cornejo HA, Macías-Rodríguez L, Alfaro-Cuevas R, López-Bucio J (2014) *Trichoderma* spp. improve growth of Arabidopsis seedlings under salt stress through enhanced root development, osmolite production, and Na+ elimination through root exudates. MPMI 27(6):503–514. https://doi.org/10.1094/MPMI-09-13-0265-R
Dana MM, Pintor-Toro JA, Cubero B (2006) Transgenic tobacco plants overexpressing chitinases of fungal origin show enhanced resistance to biotic and abiotic stress agents. Plant Physiol 142:722–730. https://doi.org/10.32604/phyton.2020.08795
El-Baki GA, Mostafa D (2014) The potentiality of *Trichoderma* harzianum in alleviation the adverse effects of salinity in faba bean plants. Acta Biol Hung 65(4):451–468. https://doi.org/10.1556/abiol.65.2014.4.9
FAO (2004) World Agricultural Center, FAOSTAT Agricultural statistic Data-Base Gateway
Ferreira FV, Musumeci MA (2021) Trichoderma as biological control agent: scope and prospects to improve efficacy. World J Microbiol Biotechnol 37:90. https://doi.org/10.1007/s11274-021-03058-7
Fiorentino N, Ventorino V, Woo SL, Pepe O, De Rosa A, Gioia L, Romano I, Lombardi N, Napolitano M, Colla G, Rouphael Y (2018) *Trichoderma* -based biostimulants modulate rhizosphere microbial populations and improve n uptake efficiency, yield, and nutritional quality of leafy vegetables. Front Plant Sci 9:743. https://doi.org/10.3389/fpls.2018.00743
Fu J, Liu Z, Li Z, Wang Y, Yang K (2017) Alleviation of the effects of saline-alkaline tress on maize seedlings by regulation of active oxygen metabolism by *Trichoderma asperellum*. PLoS One 12(6):e0179617. https://doi.org/10.1371/journal.pone.0179617
Gal-Hemed I, Atanasova L, Komon-Zelazowska M, Druzhinina IS, Viterba A, Yarde O (2011) Marine isolates of *Trichoderma* spp. as potential halotolerant agents of biological control for arid-zone agriculture. Appl Environ Microbiol 77:5100–5109
Guo R, Wang Z, Huang Y, Fan H, Liu Z (2018) Biocontrol potential of saline- or alkaline-tolerant *Trichoderma asperellum* mutants against three pathogenic fungi under saline or alkaline stress conditions. Braz J Microbiol 49(Suppl 1):236–245. https://doi.org/10.1016/j.bjm.2018.02.008. Epub 2018 Apr 13. PMID: 29691191; PMCID: PMC6328716
Gupta B, Huang B (2014) Mechanism of salinity tolerance in plants: physiological, biochemical, and molecular characterization. Int J Genomics Article ID 701596, 18 pages
Gupta S, Schillaci M, Walker R et al (2020) Alleviation of salinity stress in plants by endophytic plant-fungal symbiosis: current knowledge, perspectives and future directions. Plant Soil. https://doi.org/10.1007/s11104-020-04618-w
Hamed ER, Awad HM, Eman AG, Shehata HS (2015) *Trichoderma* asperellum isolated from salinity soil using rice straw waste as biocontrol agent for cowpea plant pathogens. J App Pharma Sci 2:091–098. https://doi.org/10.7324/JAPS.2015.58.S15
Harman GE, Howell CR, Viterbo A, Chet I, Lorito M (2004) *Trichoderma* species – opportunistic, avirulent plant symbionts. Nat Rev Microbiol 2:43–56
Hashem A, Abd_Allah EF, Alqarawi AA, Al Huqail AA, Egamberdieva D (2014) Alleviation of abiotic salt stress in *Ochradenus baccatus* (Del.) by *Trichoderma hamatum* (Bonord.) Bainier. J Plant Interact 9(1):857–868. https://doi.org/10.1080/17429145.2014.983568
Hermosa R, Viterbo A, Chet I, Monte E (2012) Plant beneficial effects of *Trichoderma* and of its genes. Microbiology 158:17–25. https://doi.org/10.1099/mic.0.052274-0
Hernández JA (2019) Salinity tolerance in plants: trends and perspectives. Int J Mol Sci 20:2408. https://doi.org/10.3390/ijms20102408
Hidangmayum A, Dwivedi P (2018) Plant responses to *Trichoderma* spp. and their tolerance to abiotic stresses: a review. J Pharmacogn Phytochem 7(1):758–766
Hontzeas N, Catherine H, Bernard G (2006) Reaction mechanisms of the bacterial enzyme 1-aminocyclopropane-1-carboxylate deaminase. Biotechnology advances 24:420–426. https://doi.org/10.1016/j.biotechadv.2006.01.006

Hossain SM (2019) Present scenario of global salt affected soils, its management and importance of salinity research. Int Res J Bio Sci 1(1):1–3. https://doi.org/10.1073/pnas.190309197

Ikram M, Ali N, Jan G, Iqbal A, Hamayun M, Jan FG, Hussain A, Lee IJ (2019) *Trichoderma* reesei improved the nutrition status of wheat crop under salt stress. J Plant Interact 14(1):590–560. https://doi.org/10.1080/17429145.2019.1684582

Isayenkov SV, Maathuis FJM (2019) Plant salinity stress: many unanswered questions remain. Front Plant Sci 10:80. https://doi.org/10.3389/fpls.2019.00080

Kashyap PL, Solanki MK, Kushwaha P, Srivastava AK, Kumar S (2020) Biocontrol potential of salt-tolerant *Trichoderma* and *Hypocrea* isolates for the management of tomato root rot under saline environment. J Soil Sci Plant Nutr 20:160–176. https://doi.org/10.1007/s42729-019-00114-y

Katori T, Ikeda A, Iuchi S, Kobayashi M, Shinozaki K, Maehashi K, Sakata Y, Tanaka S, Taji T (2010) Dissecting the genetic control of natural variation in salt tolerance of Arabidopsis thaliana accessions. J Exp Bot 61:1125–1138

Khomari S, Davari M (2017) *Trichoderma* -induced enhancement of soybean seedling performance in response to salt stress. J Plant Physiol Breed 7(1):27–39

Kumar S, Pandey S, Rathore Singh R, Utkarsh KK (2020) Multi-trait *Trichoderma* for improving plant health of grain legumes. In: Microbial mitigation of stress response of food legumes, 2nd edn, Chapter: 7. CRC Press, Taylor & Francis, pp 85–102. https://doi.org/10.1201/9781003028413-8

Lamers J, van der Meer T, Testerink C (2020) How plants sense and respond to stressful environments. Plant Physiol 182:1624–1635. https://doi.org/10.1104/pp.19.01464

Leo Daniel AE, Praveen Kumar G, Desai S, Mir Hassan ASK (2011) In vitro characterization of *Trichoderma viride* for abiotic stress tolerance and field evaluation against root rot disease in *Vigna mungo* L. J Biofertil Biopest 2:111. https://doi.org/10.4172/2155-6202.1000111

López-Bucio J, Pelagio-Flores R, Herrera-Estrella A (2015) *Trichoderma* as biostimulant: exploiting the multilevel properties of a plant beneficial fungus. Sci Hortic 196:109–123

Mahmood R, Ijaz M, Qamar S, Bukhari SA, Malik K (2019) Abiotic stress signaling in rice crop. In: Advances in rice research for abiotic stress tolerance. Elsevier, Amsterdam, pp 551–569

Ma Y, Dias MC, Freitas H (2020) Drought and salinity stress responses and microbe-induced tolerance in plants. Front Plant Sci 11:591911. https://doi.org/10.3389/fpls.2020.591911

Manaf HH, Zayed MS (2015) Productivity of cowpea as affected by salt stress in presence of endomycorrhizae and Pseudomonas fluorescens. Ann Agric Sci 60(2):219–226. https://doi.org/10.1016/j.aoas.2015.10.013. https://www.sciencedirect.com/science/article/pii/S0570178315000500

Martínez RS, Chavera GS, Mamani WH, Molina MR (2015) Native strains of *Trichoderma* from northern Chile: adaptive tolerance in boric saline soils. Interciencia 40(4):263–269

Mastouri F, Björkman T, Harman GE (2010) Seed treatment with *Trichoderma harzianum* alleviates biotic, abiotic, and physiological stresses in germinating seeds and seedlings. Phytopathology 100:1213–1221

Mazher AM, El-Quesni EMF, Farahat MM (2007) Responses of ornamental plants and woody trees to salinity. World J Agric Sci 3(3):386–395

Mbarki S, Cerdà A, Brestic M, Mahendra R, Abdelly Ch, Pascua JA (2016) Vineyard compost supplemented with *Trichoderma harzianum* T78 improve saline soil quality. Land Degrad Dev. https://onlinelibrary.wiley.com/. Accessed 22 May 2016 doi:https://doi.org/10.1002/ldr.2554

Mian A, Oomen RJFJ, Isayenkov S, Sentenac H, Maathuis FJM, Véry AA (2011) Over-expression of an Na+ – and K+ –permeable HKT transporter in barley improves salt tolerance. Plant J 68:468–479

Min YJ, Park MS, Fong JJ, QuanY JS, Lim YW (2014) Diversity and saline resistance of endophytic fungi associated with *Pinus thunbergii* in coastal shelterbelts of Korea. J Microbiol Biotechnol 24(3):324–333. https://doi.org/10.4014/jmb.1310.10041

Mishra BK, Saxena SN, Kant K (2020) Abiotic stress tolerance in horticultural crops by phyto-beneficial microbial inoculants: a review. MOJ Food Process Technol 8:126–130

Mohamed HAA, Haggag WM (2006) Biocontrol potential of salinity tolerant mutants of *Trichoderma harzianum* against *Fusarium oxysporum*. Braz J Microbiol 37(2):181–191. https://doi.org/10.1590/S1517-83822006000200016

Munns R (2005) Genes and salt tolerance: bringing them together. Plant Physiol 167:645–663

Munns R, Guo J, Cramer PJB, G.R. (2000) Leaf water status controls day-time but not daily rates of leaf expansion in salt-treated barley. Aust J Plant Physiol 27:949–957

Mushtaq Z, Faizan S, Gulzar B (2020) Salt stress, its impacts on plants and the strategies plants are employing against it: a review. J App Biol Biotechnol 8(03):81–91. https://doi.org/10.7324/JABB.2020.80315

Muthukumar A, Eswaran A, Sanjeevkumas K (2011) Exploitation of *Trichoderma* species on the growth of *Pythium aphanidermatum* in chilli. Braz J Microbiol 42:1598–1607

Nagamani A, Sarojini CK (2012) Green fungus for benefit of green plants. In: Ram Reddy S, Singara Charya MA, Girisham S (eds) Microbial diversity exploration & bioprospecting. Scientific Publishers, Jodhpur, pp 238–260

Nieto-Jacobo MF, Steyaert JM, Salazar-Badillo FB, Nguyen DV, Rostás M, Braithwaite M, De Souza JT, Jimenez-Bremont JF, Ohkura M, Stewart A, Mendoza-Mendoza A (2017) Environmental growth conditions of *Trichoderma* spp. affects indole acetic acid derivatives, volatile organic compounds, and plant growth promotion. Front Plant Sci 8:102. https://doi.org/10.3389/fpls.2017.00102. PMID: 28232840; PMCID: PMC5299017

Noctor G, Foyer CH (1998) Ascorbate and glutathione: keeping active oxygen under control. Ann Rev Plant Biol 49:249–279

Oljira AM, Hussain T, Waghmode TR, Zhao H, Sun H, Liu X, Wang X, Liu B (2020) *Trichoderma* enhances net photosynthesis, water use efficiency, and growth of wheat (*Triticum aestivum* L.) under salt stress. Microorganisms 8:1565. https://doi.org/10.3390/microorganisms8101565

Poosapati S, Ravulapalli PD, Tippirishetty N, Vishwanathaswamy DK, Chunduri S (2014) Selection of high temperature and salinity tolerant *Trichoderma* isolates with antagonistic activity against *Sclerotium rolfsii*. Springer Plus 3:641. http://www.springerplus.com/content/3/1/641

Poveda J (2020) *Trichoderma parareesei* favors the tolerance of Rapeseed (*Brassica napus* L.) to salinity and drought due to a chorismate mutase. Agronomy 10:118. https://doi.org/10.3390/agronomy10010118

Qi WZ, Zhao L (2013) Study of the siderophore-producing *Trichoderma* asperellum Q1 on cucumber growth promotion under salt stress. J Basic Microbiol 53:355–364. https://doi.org/10.1002/jobm.201200031

Ramírez-Valdespino CA, Casas-Flores S, Olmedo-Monfil V (2019) *Trichoderma* as a model to study effector-like molecules. Front Microbiol 10:1030. https://doi.org/10.3389/fmicb.2019.01030

Rawat L, Singh Y, Shukla N et al (2011) Alleviation of the adverse effects of salinity stress in wheat (*Triticum aestivum* L.) by seed biopriming with salinity tolerant isolates of *Trichoderma harzianum*. Plant Soil 347:387. https://doi.org/10.1007/s11104-011-0858-z

Rawat L, Bisht TS, Upadhayay RG, Kukreti A (2016) *Trichoderma harzianum* enhancing plant growth parameters and reducing deleterious effects of natural saline-sodic soil in rice. Int J Trop Agric 34(6):1855–1867

Regragui A, Lahlou H (2005) Effect of salinity on *in vitro Trichoderma harzianum* antagonism against *Verticillium dahliae*. Pak J Biol Sci 8:872–876

Ripa FR, Cao W, Tong S, Sun J (2019) Assessment of plant growth promoting and abiotic stress tolerance properties of wheat endophytic fungi. Biomed Res Int Article ID 6105865, 12 pages. doi:https://doi.org/10.1155/2019/6105865

Rubio B, Quijada NM, Pérez E, Domínguez S, Monte E, Hermosa R (2014) Identifying beneficial qualities of *Trichoderma parareesei* for plants. Appl Environ Microbi 80:1864–1873. https://doi.org/10.1128/AEM.03375-13

Ruocco M, Lanzuise S, Lombardi N, Woo SL, Vinale F, Marra R, Varlese R, Manganiello G, Pascale A, Scala V, Turrà D, Scala F, Lorito M (2015) Multiple roles and effects of a novel *Trichoderma* hydrophobin. Mol Plant Microbe Interact 28(2):167–179. https://doi.org/10.1094/MPMI-07-14-0194-R. PMID: 25317667

Sánchez-Montesinos B, Diánez F, Moreno-Gavira A, Gea FJ, Santos M (2019) Plant growth promotion and biocontrol of *Pythium ultimum* by saline tolerant *Trichoderma* isolates under salinity stress. Int J Environ Res Public Health 16:2053. https://doi.org/10.3390/ijerph16112053

Sanower HM (2019) Present scenario of global salt affected soils, its management and importance of salinity research. Int Res J Biol Sci 1(1):1–3

Sarojini CK, Nagamani A (2020) *Trichoderma* -its paramount role in agriculture. In: New and future developments in microbial biotechnology and bioengineering. Recent advances in application of fungi and fungal metabolites: environmental and industrial aspects, pp 69–83. https://doi.org/10.1016/B978-0-12-821007-9.00007-3

Schachtman DP, Munns R (1992) Sodium accumulation in leaves of *Triticum* species that differ in salt tolerance. Aust J Plant Physiol 19:331–340

Schuster A, Schmoll M (2010) Biology and biotechnology of *Trichoderma*. Appl Microbiol Biotechnol 87(3):787–799. https://doi.org/10.1007/s00253-010-2632-1. Epub 2010 May 12. PMID: 20461510; PMCID: PMC2886115

Shannon MC, Grieve CM, Francois LE (1994) Whole plant response to salinity. Dekker, New York

Shoresh M, Harman GE, Mastouri F (2010) Induced systemic resistance and plant responses to fungal biocontrol agents. Annu Rev Phytopathol 48:21–43

Shrivastava P, Kumar R (2015) Soil salinity: a serious environmental issue and plant growth promoting bacteria as one of the tools for its alleviation. Saudi J Bio Sci 22:123–131

Singh AK, Dwivedi P (2018) Salicylic acid and *Trichoderma* ameliorate salt stress responses in pea (*Pisum sativum* L.). IJAEB 11(2):387–395. https://doi.org/10.30954/0974-1712.04.2018.21

Soliman MH, Alnusaire TS, Abdelbaky NF, Alayafi AAM, Hasanuzzaman M, Rowezak MM, El-Esawi M, Elkelish A (2020) *Trichoderma*-induced improvement in growth, photosynthetic pigments, proline, and glutathione levels in Cucurbita pepo seedlings under salt stress. Phyton Int J Exp Bot 89(3):473–486. https://doi.org/10.32604/phyton.2020.08795

Song Y, Ji D, Li S, Wang P, Li Q, Xiang F (2012) The dynamic changes of DNA methylation and histone modifications of salt responsive transcription factor genes in soybean. PLoS One 7:1–11

Sood M, Kapoor D, Kumar V, Sheteiwy MS, Ramakrishnan M, Landi M, Araniti F, Sharma A (2020) *Trichoderma:* the "secrets" of a multitalented biocontrol agent. Plan Theory 9:762. https://doi.org/10.3390/plants9060762

Sri Vithya R, Gopal KS, Sujatha VS, Devi KMD, Girija D (2018) Abiotic stress tolerant *Trichoderma harzianum* strain for growth promotion and foot rot management in black pepper. J Plant Crops 46(1) https://doi.org/10.25081/jpc.2018.v46.i1.3535

Sun L, Lei P, Wang Q, Ma J, Zhan Y, Jiang K, Xu Z, Xu H (2020) The endophyte *Pantoea alhagi* NX-11 alleviates salt stress damage to rice seedlings by secreting exopolysaccharides. Front. Microbiol. 10:3112. https://doi.org/10.3389/fmicb.2019.03112

Sun H, Sun, X. Wang, H, Ma X (2020) Advances in salt tolerance molecular mechanism in tobacco plants. Hereditas 157, 5. https://doi.org/10.1186/s41065-020-00118-0

Talaat NB, Shawky BT (2011) Influence of arbuscular mycorrhizae on yield, nutrients, organic solutes, and antioxidant enzymes of two wheat cultivars under salt stress. J Plant Nutr Soil Sci 174:283–291

Taufiq A, Yusnawan E (2020) Influence of *Trichoderma* as a seed treatment on the growth and yield of groundnut under saline environment. J Degrade Min Land Manage 8(1):2401–2409. https://doi.org/10.15243/jdmlm.2020.081.2401

Ueda M, Matsui A, Tanaka M, Nakamura T, Abe T, Sako K, Sasaki T, Kim JM, Ito A, Nishino N et al (2017) The distinct roles of class I and II RPD3-like histone deacetylases in salinity stress response. Plant Physiol 175:1760–1773

Velmurugan A, Swarnam TP, Kumar K, Sakthivel K, Rachael, Sarojini R (2015) Evaluation of *Trichoderma* isolates for induced soil salinity stress Management in egg Plant. Trends Biosci 8:1595–1601

Vinale F, Sivasithamparam K, Ghisalberti E, Woo S, Marco N, Marra NL, Alberto P, Ruocco M, Stefania L, Gelsomina M, Lorito M (2014) *Trichoderma* secondary metabolites active on plants and fungal pathogens. Open Mycol J 8. https://doi.org/10.2174/1874437001408010127

Viterbo A, Landau U, Kim S, Chernin L, Chet I (2010) Characterization of ACC deaminase from the biocontrol and plant growth-promoting agent *Trichoderma asperellum* T203. FEMS Microbial Lett 305(1):42–48. https://doi.org/10.1111/j.1574-6968.2010.01910.x

Volkmar KM, Hu Y, Steppuhn H (1998) Physiological responses of plants to salinity: a review. Can J Plant Sci 78:19–27

Waghunde RR, Shelake RM, Sabalpara AN (2016) *Trichoderma:* a significant fungus for agriculture and environment Afr. J Agric Res 11:1952–1965

Wang W, Vinocur B, Altman A (2003) Plant responses to drought, salinity, and extreme temperatures: toward genetic engineering for stress tolerance. Planta 218:1–14

Yasmeen R, Siddiqui ZR (2017) Physiological responses of crop plants against *Trichoderma harzianum* in saline environment. Acta Bot Croa 76(2):154–162. https://doi.org/10.15243/jdmlm.2017.044.945

Yasmeen R, Siddiqui ZS (2018) Ameliorative effects of *Trichoderma harzianum* on monocot crops under hydroponic saline environment. Acta Physiol Plant 40:4. https://doi.org/10.1007/s11738-017-2579-2

Yeşilyurt A, Pehlivan N, Durmuş NA, Karaoğlu S (2018) *Trichoderma citrinoviride*: a potent biopriming agent for the alleviation of salt stress in maize. HJBC 46(1):101–111. https://dergipark.org.tr/en/pub/hjbc/issue/40706/469871

Yusnawan E, Inayati A, Baliadi1 YC (2019) Effect of soybean seed treatment with *Trichoderma virens* on its growth and total phenolic content. AIP Conf Proc 2120, 020003. https://doi.org/10.1063/1.5115604

Zaidi N, Manzoor D, Singh S, Singh U (2013) *Trichoderma* species as abiotic stress relievers in plants. In: Gupta VK, Schmoll M, Herrera-Esterella A, Druzhinina I, Upadhyay RS, Tuohy M (eds) *Trichoderma*, their uses and application in bioengineering and biotechnology. Elsevier, The Netherlands. https://doi.org/10.1016/B978-0-444-59576-8.00038-2

Zelm E, Zhang Y, Testerink C (2020) Salt tolerance mechanisms of plants. Ann Rev Plant Biol 71:403–433

Zhang B (2015) MicroRNA: a new target for improving plant tolerance to abiotic stress. J Exp Bot 66:1749–1761

Zhang S, Gan Y, Xu B (2016) Application of plant-growth-promoting fungi *Trichoderma longibrachiatum* T6 enhances tolerance of wheat to salt stress through improvement of antioxidative defense system and gene expression. Front Plant Sci 7:1–11

Zhang F, Wang Y, Liu C, Chen F, Ge H, Tian F, Yang T, Ma K, Zhang Y (2018) *Trichoderma harzianum* mitigates salt stress in cucumber via multiple responses. Ecotoxicol Environ Safety (IF 4872). https://doi.org/10.1016/j.ecoenv.2018.11.084

Zhang S, Xu B, Gan Y (2019) Seed treatment with *Trichoderma longibrachiatum* T6 promotes wheat seedling growth under NaCl stress through activating the enzymatic and non-enzymatic antioxidant defense systems. Int J Mol Sci 20:3729

Zhao C, Zhang H, Song C, Zhu J-K, Shabala S (2020) Mechanisms of plant responses and adaptation to soil salinity. Innovation 1(1):1–41. https://doi.org/10.1016/j.xinn.2020.100017

Tolerance to and Alleviation of Abiotic Stresses in Plants Mediated by *Trichoderma* spp.

Lorena Brito Pimentel Rodrigues dos Santos, Naiane Oliveira-Santos, Jacqueline Viana Fernandes, Juan Carlos Jaimes-Martinez, Jorge Teodoro De Souza, Valter Cruz-Magalhães, and Leandro Lopes Loguercio

Contents

L. B. P. R. dos Santos · N. Oliveira-Santos · J. V. Fernandes · J. C. Jaimes-Martinez · L. L. Loguercio (✉)
Department of Biological Sciences (DCB), State University of Santa Cruz (UESC), Ilhéus, BA, Brazil
e-mail: leandro@uesc.br

J. T. De Souza (✉) · V. Cruz-Magalhães
Department of Phytopathology (DFP), Federal University of Lavras (UFLA), Lavras, MG, Brazil
e-mail: jorge.souza@ufla.br

N. Amaresan et al. (eds.), *Advances in* Trichoderma *Biology for Agricultural Applications*, Fungal Biology, https://doi.org/10.1007/978-3-030-91650-3_12

1 Introduction

Climate change, as an ongoing scenario, is already markedly interfering with agricultural productivity and food security in the world, with a gloomy scenario in the near future (Zari 2014; Daryanto et al. 2016; Naumann et al. 2018). According to a FAO report (FAO 2007), less than 5% of the entire land area of the planet has not yet been altered by human activity. Food production issues associated with abiotic stresses in plants will remain in evidence in the near future, partially driven by (i) the consequences of COVID-19's huge impact in global economy, (ii) the necessity of reducing carbon footprint on Earth (Stern 2016), and (iii) the loss of biodiversity that will affect ecosystems and economy in an interdependent manner (Dasgupta 2008; Trisos et al. 2020; Rousseau and Deschacht 2020).

Throughout evolution, plants have accumulated mechanisms of response to various environmental factors that cause stress, allowing their adaptation to a variety of environments (Devi et al. 2017). These responses involve individually or jointly regulated signaling pathways, involving molecules such as ions, metabolites, cofactors, phytohormones, reactive oxygen species (ROS), and mitogen-activated phosphorylation cascades (MAP kinases activities) for induction of adaptive responses (Lata et al. 2018). In addition, plants make symbiotic/mutualistic interactions with rhizospheric and phyllospheric microorganisms that often result in an integrated relationship in which the roles of endophytes help induce abiotic stress tolerance for the holobiont (Lewis 1985; Marasco et al. 2012; Hardoim et al. 2015). To cope with harmful effects of abiotic stresses, microorganisms can directly synthesize anti-stress protective compounds (e.g., amino acids, glycine betaines, polyamines, amides, etc.) or act indirectly, through interfering with plant gene expression and synthesis of enzymes, hormones, and signaling proteins/molecules that activate the plant's stress-response system soon after exposure (Schulz et al. 2002; Chakraborty et al. 2015). In addition, microbes can promote growth, which aids in the prevention of losses in plant vitality (Harman and Uphoff 2019).

The fungal genus *Trichoderma* stands out in the context of microbial-induced beneficial effects to plants, as it is the basis for a variety of commercially available biopesticides, biofungicides, biofertilizers, and soil conditioners (Harman et al. 2004; Vinale et al. 2008; López-Bucio et al. 2015). This is possible because this genus has several species with multiple abilities, such as antagonism to a diversity of plant pathogens, enhancement of plant defense mechanisms, and improvement of plant growth and development (Loguercio et al. 2009a; El_Komy et al. 2015; Jalali et al. 2017; Ghorbanpour et al. 2018; Lombardi et al. 2018). Another interesting characteristic that has been described for *Trichoderma* species is related to their contribution to the relief of abiotic stresses in plants (e.g., Xiang et al. 2012; Calvo et al. 2014). Species of this fungus display a genetic arsenal that allow the production of an array of metabolites with antifungal and antibiotic activity, as well as with bioactivities of potential pharmaceutical use (Duran et al. 2010). Some of these metabolites can also help plant hosts to cope with harmful effects of abiotic stresses (Meena et al. 2017). In this context, the multifunctional properties of *Trichoderma* are highly advantageous for the development of environmentally sustainable strategies for agriculture (Harman 2011a, b; Glare et al. 2012; Berg et al. 2013; Chakraborty et al. 2015; Chojnacka 2015; Kumar and Verma 2018; Lata et al. 2018).

In this chapter, we adopted a systematic/integrative evaluation of the literature to describe the use of *Trichoderma* spp. in the alleviation of the negative effects of abiotic stresses in plants. The methods were based on qualitative and quantitative assessments of the scientific literature according to the methods described below. The main topics analyzed were (i) temporal and geographic aspects of the selected studies, (ii) species of *Trichoderma* and the plants/crops involved, (iii) types of abiotic stresses, (iv) mechanisms that *Trichoderma* use to minimize the negative effects of abiotic stresses in plants, and (v) plant genes possibly involved in interactive mechanisms with *Trichoderma* that ameliorate the stresses. Application potentialities and future research directions on this issue were also addressed.

2 Overview of the Methods and Procedures

The elaboration of the questions and the research protocol (Table 1) for the systematic part of the review research in this chapter were based on five components of the method described by Kitchenham (2004): (i) *population*, plant species that suffer the effects of abiotic stress; (ii) *intervention*, *Trichoderma* species/isolates applications; (iii) *comparison*, stress-bearing plants with vs without effects from interactions with *Trichoderma* species; (iv) *hypothesis*, *Trichoderma* spp., on average, which reduce the negative effects caused by abiotic stresses on plants; and (v) *type of study*, scientific articles containing primary studies. Quantitative and qualitative data collection during the research was based, therefore, on the question of what is the magnitude of the *Trichoderma* species' potential to ameliorate the negative effects caused by abiotic stresses on plants. The overall sequence of systematic steps of this chapter is described briefly as follows: the literature research was carried out using two keywords, "*Trichoderma*" and "abiotic stress," with the aid of the "Publish or Perish" version 6.2 program (P&P) (Harzing 2007) for the publications until February 2018; the Google Scholar was chosen as the main database used through the P&P program because it provides access to studies from virtually all databases and publishers available on the web. For the period of 2018–2020, the same two keywords were searched directly in the Web of Science, Scopus, and PubMed databases; this procedure aimed to provide a more representative sampling of the literature, with emphasis on the more recent research covered in the three mostly accessed databases. The initial search by P&P was performed using the two keywords above located anywhere in the full text of the publications. Since the P&P's criterion for ranking the retrievable studies is their number of citations, which we considered as an important parameter for quality and representativeness of our sampling (Harzing 2007), the research was limited to the first 1000 studies in English retrievable by P&P. The first approach to all these retrieved studies was based on the detailed reading of the corresponding titles and abstracts to retain only

Table 1 Research protocol for the systematic review of web-based scientific literature

General information	
Description	The genus *Trichoderma* is widely known to have isolates that are used as biocontrol agents of plant diseases and promoters of plant growth. In addition, other studies suggest that isolates of this genus may also provide plant tolerance to a variety of abiotic stresses. This study aims to describe the current situation of knowledge about *Trichoderma* species/isolates that showed this improvement effect in plants submitted to different types of abiotic stress
Objectives	Check which *Trichoderma* species can increase the plant's tolerance to stressful abiotic factors
	Use data collection to investigate information on plant/crop amplitude researched on stress relief
	To analyze the types of abiotic stresses, the species of the genus *Trichoderma* can decrease in plants
	Evaluate the methodologies used in the studies regarding the mechanisms of action that are observed in the *plant-Trichoderma* interaction to reduce the effects caused in the plant due to the abiotic stress suffered
	Gather information on the alteration of the expression of plant genes involved in the mechanisms of action against the negative effects of abiotic stress in the presence of species of the genus *Trichoderma*
Aspects of research	
Question	What is the magnitude of the potential of *Trichoderma* fungal species to control or decrease the negative effects caused by abiotic stresses on plants?
Population	Plant species that suffer from abiotic stress and interact with *Trichoderma* spp.
Intervention	Decreased effects of abiotic stress on plant interaction with *Trichoderma* species
Comparison	Measurable effects of plants with stress and no interaction with *Trichoderma* vs plants with stress and interaction with *Trichoderma* species/isolates
Hypothesis	*Trichoderma* spp. decreases the negative effects caused by abiotic stresses on plants
Expected result	The systematic analysis of the related literature will allow to verify the hypothesis formulated in relation to the mitigation of stresses caused by abiotic factors in plants as a result of its interaction with *Trichoderma* spp.
Type of studies	Primary studies in the form of scientific articles
Identification of studies	
Keywords	"*Trichoderma*," "abiotic stress"
Search string	"*Trichoderma*" **and** "abiotic stress"
Font selection criteria for search	Peer-reviewed editors/journals and editorial boards
	Available on the Internet
List of search sources	PubMed
	Scopus
	Web of Science
	Google academic (Publish or Perish)

(continued)

Table 1 (continued)

Online search strategy	*Google Scholar*-based "Publish or Perish" v. 6.2 (Harzing 2007) until 2018 (Research in 07/Feb2018)
	Research in *PubMed*, *Scopus* and *web of science* databases for 2018–2020 (research in 08/Sep/2020)
Selection and evaluation of studies	
Inclusion and exclusion criteria for studies	*Inclusion*:
	Written in English
	Primary studies/articles (including special editions)
	Articles focused on abiotic stress
	Exclusion:
	Not aligned with the object of study
	Simple or expanded abstract, review, chapter/book, dissertation, and thesis
	Article that has no plant experiments
	Article focusing on biocontrol and/or other biotic stressors (e.g., phytopathogens)
	Failure or inconsistency between methodology and results/conclusions
Strategy for the initial selection of studies	Detailed reading of:
	Title
	Summary
	Keywords
Strategy for the final selection of studies	Detailed reading of the full text of the article
	Presence of all inclusion criteria
	Absence of all exclusion criteria
Evaluation of the quality of the study	*Research* online:
	"Publish or Perish" quality criteria (based on the number of citations per year)
	Selected studies:
	Inclusion and exclusion criteria
	Subjective judgment of agreement between hypotheses, experimental procedures, results, and conclusions

(continued)

those specifically dealing with the central theme of this research. From this procedure, 134 papers were selected (Fig. 1), including 71 primary studies, 30 reviews, 28 book chapters, 3 theses/dissertations, and 2 open letters. To confirm the quality and consistency of these studies, the criterion of displaying a recognized peer review system and editorial board was observed. Based on this experience, for the studies retrieved directly from the three databases indicated above for the years 2018–2020, the focus was directly on the "title" and "abstract" sections to select studies specifically related to our investigation. With this procedure, 46 primary studies were initially retrieved, and, after the analysis of the abstract contents, 19 papers were retained and added to the local database under assessment (Fig. 1).

Table 1 (continued)

Data synthesis and presentation of results	
Data extraction strategy	*Items to collect/evaluate:*
	Objective (Abstract)
	Conclusions (abstract)
	Keywords;
	Country where the study was done
	Trichoderma species involved in stress mitigation
	Trichoderma isolates
	Type of stress
	Plant/crop used for the experiments
	Genes involved in the interaction plant-*Trichoderma*
	Variables/effects measures: increase of biomass (plant size, fresh and dry weight); higher gene expression (plant or fungus); physiological parameters etc.
	Mean and standard deviation or error of measures of effect
	New ideas raised in the evaluated study (*Discussion Section*)
Data summarization strategy	Tables, graphic, images, description in text
Publishing strategy	Scientific journal with scope of agricultural sciences, plant biology, applied microbiology, and biotechnology

In the next step, we applied a series of inclusion/exclusion criteria established in the protocol, so that those studies presented in the form of proceedings' abstracts, theses, dissertations, reviews, book chapters, and open letters were removed; only the 71 research articles containing primary studies remained. For the three databases direct search, two papers were removed, leaving us with 17 studies for the next steps (Fig. 1). With all these initially selected studies, the reading of the full text was performed for data extraction and qualitative/quantitative assessments. During this process, three articles from the P&P search and five from the three databases direct searches were further removed based on lack of key information required for our research analyses (i.e., they did not meet inclusion criteria), leaving us with a final number of 68 + 12 articles (Fig. 1), in a total of 80 primary studies articles that composed the literature database used for the systematic part of this chapter (Table 2). Further validation and integration of the systematized knowledge collected were achieved by assessing related publications, through regular (classical) database search, according to specific aspects of interest suggested by the up-to-date literature obtained in the systematic review.

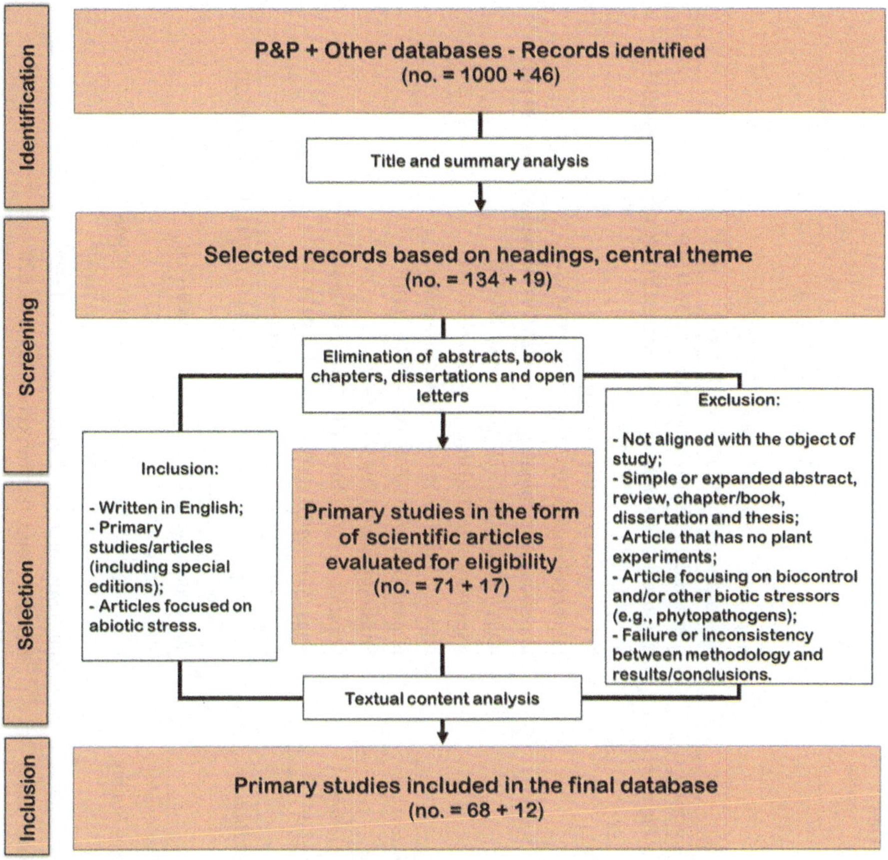

Fig. 1 Flow diagram of the search strategy, selection of studies, and data management procedure on the role of the genus *Trichoderma* in generating plant tolerance to a variety of abiotic stresses

3 When, Where, and How *Trichoderma* Has Been Tested for Abiotic Stress Alleviation?

3.1 Language and Timing of the Science on Trichoderma*-Plant-Abiotic Stress Interactions*

Considering the most recurring words in the titles of the 1046 initial studies retrieved and in the 80 finally selected papers, and after taking the searching keywords "*Trichoderma*" and "abiotic stress" away from the analyses, the words "plant(s)," "*harzianum*," and "growth" (Fig. 2) were highlighted. After selection of the studies according to the established criteria (see review methods), the words related to the

Table 2 Final database of articles included in the systematic review

Authors	Scientific journal	DOI access
Abd El-Baki and Mostafa 2014	*Acta Biologica Hungarica*	https://doi.org/10.1556/ABiol.65.2014.4.9
Ahmad et al. 2015	*Frontiers in Plant Science*	https://doi.org/10.3389/fpls.2015.00868
Azarmi et al. 2011	*African Journal of Biotechnology*	https://doi.org/10.5897/AJB10.1600
Babu et al. 2014	*Journal of Environmental Management*	https://doi.org/10.1016/j.jenvman.2013.10.009
Badar et al. 2015	*Journal of Pharmacognosy and Phytochemistry*	https://www.phytojournal.com/archives/2015/vol3issue6/PartB/3-6-38.1-841.pdf/3-6-38.1.pdf
Bae et al. 2009	*Journal of Experimental Botany*	https://doi.org/10.1093/jxb/erp165
Bakhshandeh et al. 2020	*Plant Growth Regulation*	https://doi.org/10.1007/s10725-019-00556-5
Becquer et al. 2018	*Cuban Journal of Agricultural Science*	http://scielo.sld.cu/pdf/cjas/v51n4/2079-3480-cjas-51-04-489.pdf
Brotman et al. 2013	*PLoS Pathogens*	https://doi.org/10.1371/journal.ppat.1003221
Buysens et al. 2016	*Applied Soil Ecology*	https://doi.org/10.1016/j.apsoil.2016.04.011
Caporale et al. 2014	*Journal of Plant Physiology*	https://doi.org/10.1016/j.jplph.2014.05.011
Chepsergon et al., 2016	*British Microbiology Research Journal*	https://doi.org/10.9734/BMRJ/2016/26015
Chepsergon et al. 2018	*Plant Pathology & Quarantine*	https://doi.org/10.5943/ppq/8/1/5
Contreras-Cornejo et al. 2014	*Molecular Plant-Microbe Interactions*	https://doi.org/10.1094/MPMI-09-13-0265-R
Contreras-Cornejo et al. 2015	*Journal of Plant Growth Regulation*	https://doi.org/10.1007/s00344-014-9471-8
Dana et al. 2006	*Plant Physiology*	https://doi.org/10.1104/pp.106.086140
Devi et al. 2017	*Indian Journal of Experimental Biology*	http://nopr.niscair.res.in/bitstream/123456789/41180/1/IJEB%2055%284%29%20235-241.pdf
Dixit et al. 2011	*PLOS ONE*	https://doi.org/10.1371/journal.pone.0016360
Donoso et al. 2008	*Applied and Environmental Microbiology*	https://doi.org/10.1128/AEM.02013-07
Elkelish et al. 2020	*Environmental and Experimental Botany*	https://doi.org/10.1016/j.envexpbot.2019.103946
Fu et al. 2017	*PLOS ONE*	https://doi.org/10.1371/journal.pone.0179617
Ghorbanpour et al. 2018	*Scientia Horticulturae*	https://doi.org/10.1016/j.scienta.2017.11.028
Govarthanan et al. 2018	*Ecotoxicology and Environmental Safety*	https://doi.org/10.1016/j.ecoenv.2018.01.020

Guler et al. 2016	*Acta Physiologiae Plantarum*	https://doi.org/10.1007/s11738-016-2153-3
Gusain et al. 2014	*African Journal of Agricultural Research*	https://doi.org/10.5897/AJAR2014.8575
Hanci et al. 2014	*Tarım Bilimleri Araştırma Dergisi*	http://ijans.org/index.php/ijans/article/view/287/280
Hashem et al. 2014	*Journal of Plant Interactions*	https://doi.org/10.1080/17429145.2014.983568
Hermosa et al. 2011	*Journal of Plant Physiology*	https://doi.org/10.1016/j.jplph.2011.01.027
Jalali et al. 2017	*Fungal Ecology*	https://doi.org/10.1016/j.funeco.2017.06.007
Khomari and Davari 2017	*Journal of Plant Physiology and Breeding*	https://breeding.tabrizu.ac.ir/article_6350_6427f52ffa3652b6761f0e890a3c2ef1.pdf
Khomari et al. 2017	*New Zealand Journal of Crop and Horticultural Science*	https://doi.org/10.1080/01140671.2017.1352520
Khoshmanzar et al. 2020	*International Journal of Environmental Science and Technology*	https://doi.org/10.1007/s13762-019-02405-4
Kumar et al. 2016	*Journal of Basic Microbiology*	https://doi.org/10.1002/jobm.201600369
Ma et al. 2020	*Pakistan Journal of Botany*	https://doi.org/10.30848/PJB2020-3(25)
Mastouri et al. 2010	*Phytopathology*	https://doi.org/10.1094/PHYTO-03-10-0091
Mastouri et al. 2012	*Molecular Plant-Microbe Interactions*	https://doi.org/10.1094/MPMI-09-11-0240
Mishra et al. 2016	*World Journal of Microbiology and Biotechnology*	https://doi.org/10.1007/s11274-016-2086-4
Mona et al. 2017	*Journal of Integrative Agriculture*	https://doi.org/10.1016/S2095-3119(17)61695-2
Montero-Barrientos et al. 2010	*Journal of Plant Physiology*	https://doi.org/10.1016/j.jplph.2009.11.012
Nongmaithem and Bhattacharya 2017	*International Journal of Current Microbiology and Applied Sciences*	https://doi.org/10.20546/ijcmas.2017.606.116
Nzioki and Mutisya 2016	*International Journal of Agriculture and Environmental Research*	http://ijaer.in/uploads/ijaer_02__55.pdf
Pandey et al. 2016	*Planta*	https://doi.org/10.1007/s00425-016-2482-x
Pehlivan et al. 2017	*Acta Physiologiae Plantarum*	https://doi.org/10.1007/s11738-017-2375-z

(continued)

Table 2 (continued)

Authors	Scientific journal	DOI access
Poosapati et al. 2014	*SpringerPlus*	https://doi.org/10.1186/2193-1801-3-641
Poveda et al. 2019	*Frontiers in Plant Science*	https://doi.org/10.3389/fpls.2019.01478
Poveda 2020	*Agronomy*	https://doi.org/10.3390/agronomy10010118
Qi and Zhao 2012	*Journal of Basic Microbiology*	https://doi.org/10.1002/jobm.201200031
Rawat et al. 2011	*Plant and Soil*	https://doi.org/10.1007/s11104-011-0858-z
Rawat et al. 2012	*Journal of Plant Pathology*	https://doi.org/10.4454/JPP.FA.2012.026
Rawat et al. 2013	*Archives of Phytopathology and Plant Protection*	https://doi.org/10.1080/03235408.2013.769316
Rawat et al. 2016	*Molecular Soil Biology*	https://doi.org/10.5376/msb.2016.07.0004
Roatti et al. 2013	*Phytopathology*	https://doi.org/10.1094/PHYTO-02-13-0040-R
Rouphael et al. 2017	*Frontiers in Plant Science*	https://doi.org/10.3389/fpls.2017.00131
Rubio et al. 2017	*Frontiers in Plant Science*	https://doi.org/10.3389/fpls.2017.00294
Sánchez-Montesinos et al. 2019	*International Journal of Environmental Research and Public Health*	https://doi.org/10.3390/ijerph16112053
Sharma and Singh 2014	*Journal of Applied and Natural Science*	https://doi.org/10.31018/jans.v6i2.479
Shukla et al. 2012	*Plant Physiology and Biochemistry*	https://doi.org/10.1016/j.plaphy.2012.02.001
Shukla et al. 2014	*Annals of Applied Biology*	https://doi.org/10.1111/aab.12160
Singh and Dwivedi 2018	*Journal of Pharmacognosy and Phytochemistry*	http://www.phytojournal.com/archives/2018/vol7issue2/PartAJ/7-2-300-161.pdf
Singh et al. 2018	*International Journal of Current Microbiology and Applied Sciences*	https://doi.org/10.20546/ijcmas.2018.705.174
Singh et al. 2019	*Plant Physiology and Biochemistry*	https://doi.org/10.1016/j.plaphy.2019.09.015
Singh et al. 2020b	*Scientific Reports*	https://doi.org/10.1038/s41598-020-61,140-w
Soliman et al. 2020	*Phyton*	https://doi.org/10.32604/phyton.2020.08795
Song et al. 2014	*Applied Soil Ecology*	https://doi.org/10.1016/j.apsoil.2014.09.007
Su et al. 2017	*Chemosphere*	https://doi.org/10.1016/j.chemosphere.2017.02.048
Tripathi et al. 2013	*Ecotoxicology and Environmental Safety*	https://doi.org/10.1016/j.ecoenv.2012.10.017

Tripathi et al. 2017	*Environmental Pollution*	https://doi.org/10.1016/j.envpol.2016.12.073
Vargas et al. 2017	*Environmental and Experimental Botany*	https://doi.org/10.1016/j.envexpbot.2017.01.009
Vieira et al. 2017	*Plant Physiology and Biochemistry*	https://doi.org/10.1016/j.plaphy.2017.10.012
Vithya et al. 2018	*Journal of Plantation Crops*	https://doi.org/10.25081/jpc.2018.v46.i1.3535
Yasmeen and Siddiqui 2017	*Acta Botanica Croatica*	https://doi.org/10.1515/botcro-2016-0054
Zaidi et al. 2017	*Field Crops Research*	https://doi.org/10.1016/j.fcr.2017.05.003
Zhang et al. 2015	*Plant Physiology and Biochemistry*	https://doi.org/10.1016/j.plaphy.2015.05.007
Zhang et al. 2016a	*Canadian Journal of Plant Science*	https://doi.org/10.1139/cjps-2014-0265
Zhang et al. 2016b	*Frontiers in Plant Science*	https://doi.org/10.3389/fpls.2016.01405
Zhang et al. 2018	*Ecotoxicology and Environmental Safety*	https://doi.org/10.1016/j.ecoenv.2018.03.047
Zhang et al. 2019a	*BMC Plant Biology*	https://doi.org/10.1186/s12870-018-1618-5
Zhang et al. 2019b	*International Journal of Molecular Sciences*	https://doi.org/10.3390/ijms20153729
Zhao and Zhang 2015	*Journal of Integrative Agriculture*	https://doi.org/10.1016/S2095-3119(14)60966-7
Zhao et al. 2014	*Journal of Basic Microbiology*	https://doi.org/10.1002/jobm.201400148

main types of stresses investigated became highlighted. Other frequent title words found for the 1046 studies were "tolerance," "resistance," "induced," "response," "stresses," "gene," and "expression," thus referring to the interaction between *Trichoderma* and plants (Fig. 2a). When observing the wording of the 80 finally selected primary studies, the following groups of terms acquired more relevance: (i) "growth," as the main response variable for plant studies of this nature; (ii) "*harzianum*" and "*asperellum*," related to the most common *Trichoderma* species used in the studies; (iii) "rice" and "maize", indicating the most tested plant species; (iv) "drought," "salt," and "salinity," as well as "cadmium" and "arsenic' (representing metal-polluting elements), which refer to the most studied stresses; and (v) "seed(ling)" as the main part of the plant for inoculation/assessments (Fig. 2b). The word "gene" is relatively recurrent in both word clouds. These results, at a first glance, point to the trend that the research specifically dealing with abiotic stresses is focusing on those two *Trichoderma* species, three crops, and three types of stress, with a preferable form of inoculation.

From a temporal standpoint, research on *Trichoderma*, beyond their use as biocontrol agents against phytopathogens, began to grow exponentially from 2006 onward (Fig. 2), likely due to, at least in part, an increased consciousness of the negative consequences of global warming and climate changes for sustainable agriculture. It is worth to highlight the years of 2014, followed by 2017, in which more studies were published. The number of articles on *Trichoderma* in general began to increase in the 2000s, roughly coinciding with the raise in the number of commercially available bioproducts (Waghunde et al. 2016). Since 2014, there are already more than 250 registered bioproducts in the world that are based on *Trichoderma* species, either individually or in combinations (Woo et al. 2014), which correspond to around 60% of the world's biofungicide market. *Trichoderma harzianum* comprises ~83% of these products (Topolovec-Pintarić 2019) and also corresponds to one of the most recurrent words in the recovered studies (Fig. 2b). Bioproducts represent a small share of the plant-protection market, mainly due to their slow activity and dependence on environmental factors, which has been seen as a constraint to their effectiveness in the field (Singh et al. 2018); further issues related to difficulties and costs of registration add to this context (Topolovec-Pintarić 2019). However, the reported increase in their utilization likely reflects the current demand for healthier foods, free from chemical residues (Gomiero 2018). The use of *Trichoderma* as biofertilizers to improve plant growth has facilitated registration, thereby increasing its availability in the market (Topolovec-Pintarić 2019). It is noteworthy that the potential of offering bioproducts at lower costs for smallholders to deal with their production necessities can assist with food security globally (Harman 2011b).

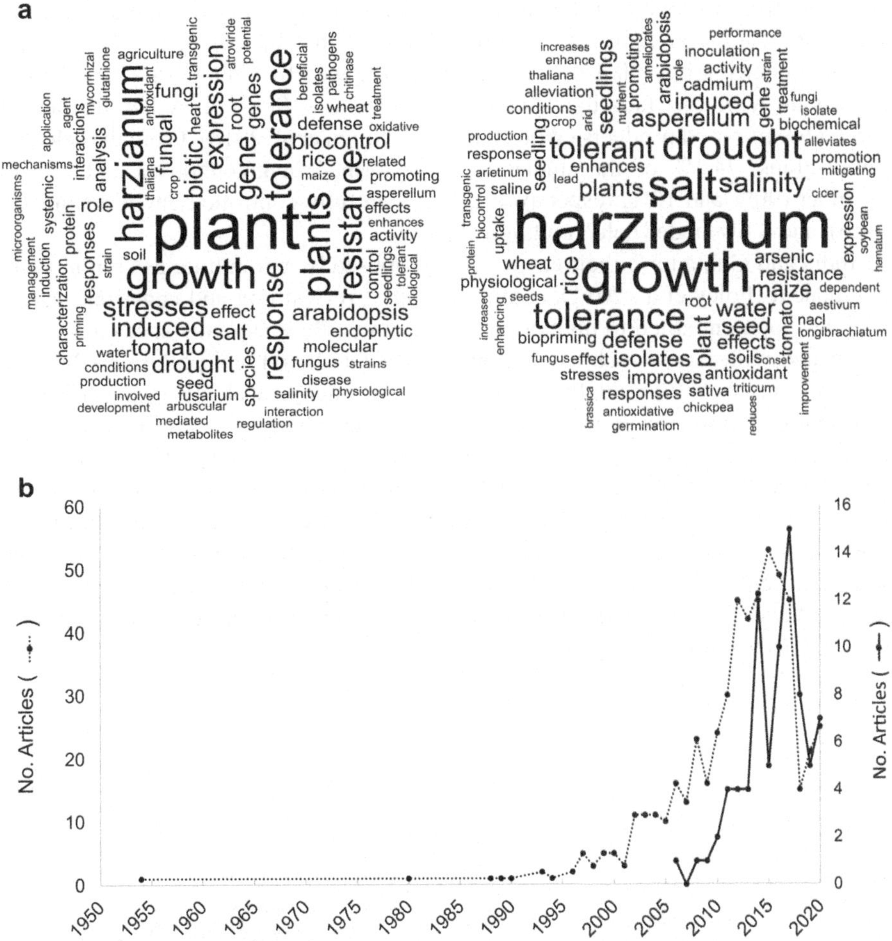

Fig. 2 Temporal distribution of scientific publications involving "*Trichoderma*" and "abiotic stress." The word clouds were assembled with the most repeated words in the titles of the 1046 of the total search (**a**) and with the 80 articles systematically selected (**b**) (the size of the words indicating the frequency of appearance). The time curves represent the number of publications selected; the dotted-line curve (Y-axis on the left) corresponds to the 1046 initial studies retrieved and the solid-line curve (Y-axis on the right) to the final database of 80 articles

3.2 Geographic Distribution of Studies with Trichoderma *and Abiotic Stresses in Plants*

To assess whether there was any geographical tendency for studies of this nature, the articles systematically retrieved were distributed as indicated on the map (Fig. 3), and the type of experiment performed was also registered. The 80 selected studies were performed in 19 countries, where 67.5% were from Asia, followed by countries from Europe and the Americas (15 and 11.25% respectively), and from Africa (6.25%). The significant number of studies in Asia was due to India's outstanding contribution (30% of total articles), followed by China (16.25%). This is consistent with the fact that India contributes to ~90% of Asian market of *Trichoderma*-based products (Woo et al. 2014; Singh et al. 2018). From the selected publications, 74.1% of the studies were performed in greenhouses and growth chamber (90% and 10%, respectively), 17.3% in fully controlled environments (in vitro), and 8.6% under field conditions (Fig. 3).

India and China have their economies composed by agriculture as an important component (Foley et al. 2011), combined with a very strong and consistent industrial development allied to high population counts. These circumstances tend to be associated with issues such as environmental degradation and pollution (Ballescá 2016; Chopra 2016), especially by heavy metals (Sodango et al. 2018). Furthermore, human population growth, urbanization, and climate changes are further challenges to be faced, in order to cope with food production in an environmentally sustainable way (Foley et al. 2011; Du et al. 2018). All these issues must be dealt properly to assure global food security (Godfray et al. 2010; He et al. 2013). It has been proposed that investments in agronomic research and development toward sustainable strategies and products (e.g., *Trichoderma*-based bioproducts) can not only help solving those challenges but also stimulate agricultural productivity on a long-term scale (Heisey and Fuglie 2018).

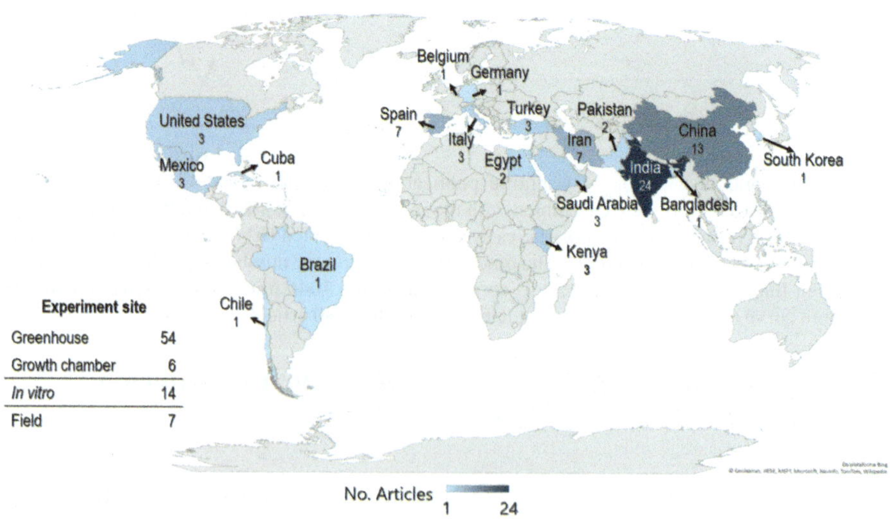

Fig. 3 Geographic distribution and types of experiments for the systematically selected primary studies of our search

3.3 Trichoderma *Species, Their Origin, Targeted Plants, and Inoculation Methods*

A total of 175 *Trichoderma* isolates were associated with abiotic stresses in the selected studies, with 78.9% distributed in 16 species and the remaining not identified to the species level (Fig. 4a). As indicated by the word clouds (Fig. 2b), *T. harzianum* and *T. asperellum* were the most abundant species, with 75 and 21 isolates, respectively, within the 138 isolates that were identified up to the species level. The other species all together occurred in a frequency of 30.4%: *T. longibrachiatum* with eight; *T. atroviride* with six; *T. afroharzianum* and *T. britannicum* with five isolates each; *T. virens* with four; *T. parareesei* with three; *T. asperelloides*, *T. hamatum*, and *T. reesei* with two isolates each; and *T. aggressivum*, *T. koningiopsis*, *T. simmonsii*, *T. saturnisporum*, and *T. viride* with one representative each (Fig. 4a). The data obtained on the sources of these isolates indicated that most came from collections of the study-affiliated or collaborating institutions (38.3%) or from rhizospheric soil (28.6%, Fig. 4a). Considering only isolates from collections' material, 55.2% were *T. harzianum*. When collection isolates were not taken into account, 72.2% of the isolates were from soil. Isolates from contaminated environments (mining tailings and contaminated soil) were specifically tested against stresses caused by heavy metals. Only in four studies, *Trichoderma* species (*T. harzianum*, *T. asperellum*, and *T. atroviride*) were used as formulated bioproducts, which were then tested for their effects on plant responses to abiotic stresses (Fig. 4a). In general, data suggest that the observed roles of *Trichoderma* in abiotic stress relief come from research primarily aimed at complementing the current knowledge on activities, applications, and bioproducts' development of particular isolates, most of them already known as biocontrol agents (e.g., Woo et al. 2014; Waghunde et al. 2016; Anam et al. 2019). Since only a small proportion of different *Trichoderma* species/isolates have been studied as mitigators of abiotic stresses (Fig. 4), there is still much exploration to be done, given the large diversity found in this genus worldwide (De Souza et al. 2006; Loguercio et al. 2009a; Kubicek et al. 2011; Feitosa et al. 2019).

The most frequent plant species found in the selected studies were maize (*Zea mays*, 12.8%), rice (*Oryza sativa*, 11.6%), tomato (*Solanum lycopersicum*, 10.5%), *Arabidopsis thaliana* (9.3%), and wheat (*Triticum aestivum*, 8.1%) (Fig. 4b; also see Fig. 2). Among the methods used to inoculate *Trichoderma*, seed biopriming alone (i.e., soaking seeds with suspensions of fungal spores to allow seed germination before planting) was the most used (Fig. 4b). *Trichoderma* spore suspensions directly applied into the soil (liquid or powder) or on the roots (by spraying) corresponded to 44.2% of the studies. Other inoculation methods, including mycelium discs for volatile compounds experiments and in vitro techniques, as well as inoculation of flowers and leaf tissues comprised the remaining 17.5% (Fig. 4b). These three predominant inoculation methods correspond to those usually planned for and used in large-scale crop applications, mainly for the most studied plant species (Fig. 4b), which combine ease of product manipulation and delivery with lower costs (Parnell et al. 2016; Rocha et al. 2019).

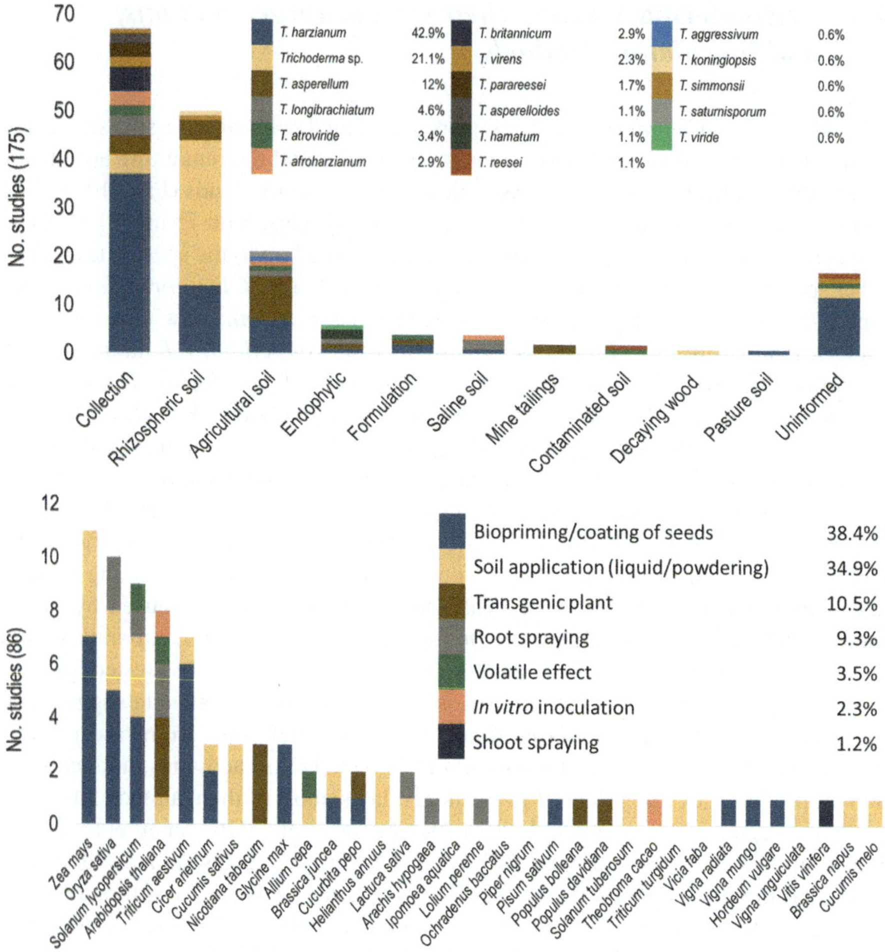

Fig. 4 Distribution of *Trichoderma* species and sources, inoculation methods, and host plants. (**a**) The sources of *Trichoderma* isolates for the reported studies are shown in the X-axis; the taxonomic definition found for the experimental isolates within the 80 articles is appearing in the center of the graph. All isolates in which their species were not defined are collectively represented by "*Trichoderma* spp." (**b**) Plant species used in the experiments as targets for the applied stresses and inoculated with the *Trichoderma* isolates; distribution of the modes of inoculation appear in the center of the graph. The total number of studies considered (86) exceeded the 80 systematically selected articles, as in some of them, there was more than a single type of study/experiment being reported

Trichoderma is one of the most abundant and widespread fungal genus in the world and has characteristics that justify the amount and depth of studies on them (e.g., reviews by Harman et al. 2004; Vinale et al. 2008; Schuster and Schmoll 2010; López-Bucio et al. 2015). *Trichoderma* spp. can adapt to a diversity of environments, not only due to their ability to sporulate in response to a complex and intertwined variety of environmental factors (Loguercio et al. 2009b; Steyaert et al. 2010a, b, c) but also due to a phylogenetic and genome-printed high opportunism (Druzhinina et al. 2011) that allow the occupation of a broad array of niches and environmental gradients (Mukherjee et al. 2013; Egidi et al. 2019; Jiao and Lu 2020). The production of a variety of hydrolytic enzymes (e.g., reviewed by Schuster and Schmoll 2010; Mukherjee et al. 2013; Waghunde et al. 2016), a great ability to control cell-wall synthesis and repair in themselves and in their hosts (Gruber and Seidl-Seiboth 2012; Kappel et al. 2020), and some tolerance of certain isolates to higher temperatures (>32 °C) during growth (Chang et al. 1997) certainly contribute to this wide niche occupancy (including a great variety of plant hosts). Some species have an endophytic lifestyle, colonizing plants by penetrating root cells and remaining throughout the plant life cycle (Harman et al. 2004, 2019; Contreras-Cornejo et al. 2018). *Trichoderma harzianum* is the most used species in bioproducts and in experiments to control plant pathogens and the one most commonly found in soil environments (Vinale et al. 2008; Mukherjee et al. 2013; Woo et al. 2014; Waghunde et al. 2016), which explains why it is the species most frequently found in this review (Fig. 4a). Since *T. harzianum* is a species complex, with multiple cryptic species, i.e., a complex group of morphologically indistinguishable species (Chaverri et al. 2015), this is likely another reason for its higher frequency in the systematically retrieved studies dealing with abiotic stress relief in plants. *Trichoderma* spp. are predominantly saprophytic fungi in soil, litter, organic matter, and rhizospheric ecosystem of all climatic zones, and their diverse metabolic capacity allows them to colonize soils of different habitats (Vinale et al. 2008; Druzhinina et al. 2011; Mukherjee et al. 2013). It is such a strong competitive nature of these fungal species that provide rapid rhizospheric establishment, root colonization (including interaction with arbuscular mycorrhizal fungi; Mehta and Sirari 2019), pathogenic microflora control, and plant-growth promotion (Hidangmayum et al. 2018). These characteristics, therefore, allow to explain their frequent interaction with the surface of plant roots, so that strategies of isolation (or inoculation) of these isolates tend to be often related to forest or agricultural soils and seed coating/biopriming (Topolovec-Pintarić 2019; Rocha et al. 2019) (Fig. 4).

The experimental plants used for the interactive experiments with *Trichoderma* tend to be mainly crop species that are mostly recognized as displaying short life cycles, small sizes, easy propagation, and considerable economic importance, being well-established model plants for a great variety of research in plant biology and agricultural sciences (Fig. 4b). Moreover, the high frequency of inoculation methods involving seeds and/or soil (~3/4) suggests a natural overlap between basic/applied research and technological development of methods/products for agricultural applications. The biopriming of seeds with *Trichoderma* spp. has been used to improve seedling vigor, which can be triggered by the release and/or production of enzymes and phytohormones involved in seed viability and germination rates and speed (Kumar et al. 2014; Babychan and Simon 2017), as well as in resistance against pathogens (Mastouri et al. 2010; Singh et al. 2019, 2020). With *Trichoderma* inoculation in roots/soil, additional features occur such as alteration of soil microflora and increase of nutrients availability, due to degradation of many complex substrates. Currently, the use of changing microbial communities of cultivated soils and improvement of the performance and vigor have been widely used in agricultural production (Harman and Uphoff 2019).

4 Types of Abiotic Stresses in Plants Alleviated by *Trichoderma*

From the 80 final articles selected, 105 abiotic stresses were identified, which were classified into 13 groups (Fig. 5). The highest proportion of the studied stresses were saline stress (36.2%), agreeing with the word clouds (Fig. 2); within this fraction, 92.7% corresponded to the effects of the salts as a single factor, with the remaining three studies (7.3%) assessing this factor in combination with high temperature and osmotic and alkaline stresses (one study each) (Fig. 5). Drought stress was the second most represented (27.6%), with water deficit being analyzed in combination with heat stress in one study (Fig. 5). Stresses caused by heavy metals comprised 21% of the studies retrieved. These 22 studies included 7 chemical elements and were distributed as follows: arsenic (six), cadmium (five), lead (four), copper (three), zinc (two), and chromium and nickel (one each) (Fig. 5). Taken together, these three types of abiotic stress comprised 83.8% of the experiments involving *Trichoderma* isolates and plants.

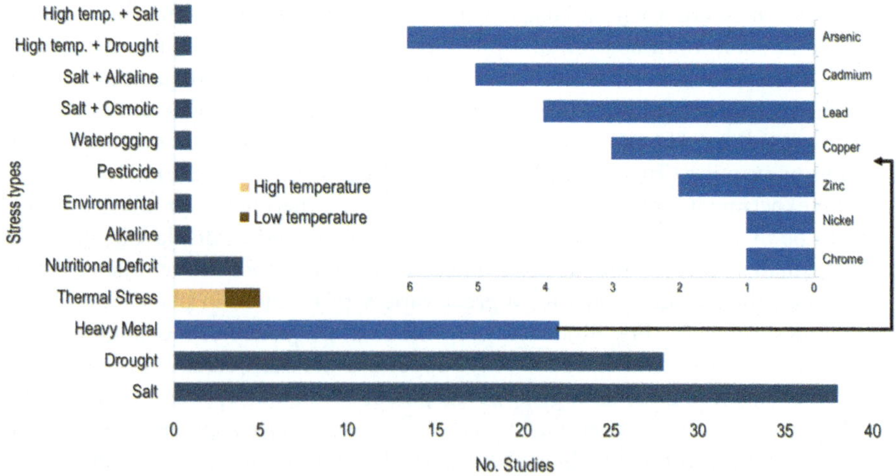

Fig. 5 Types of abiotic stresses in plants alleviated by *Trichoderma*. Twelve different types of abiotic stresses (single or in combination) were found in the selected studies (left-side graph). The number of studies reporting stresses caused by heavy metals were discriminated by each metal (right-side graph). Thermal stress was further divided (proportionally) into high and low temperatures

Under the circumstances of climate change, salinity and drought can be viewed as the most relevant types of abiotic stress that can affect crop production (Munns and Gilliham 2015); moreover, they are interconnected not only due to their direct relationship with water availability (Nuccio et al. 2018) but also through their effects in the osmotic balance and regulation in plant cells (Mastouri et al. 2010; Ikram et al. 2019; Poveda 2020). From this standpoint, our analysis indicated that ~2/3 of the current science on *Trichoderma*-mediated abiotic stress relief deals with the physiological and/or biochemical responses of plants toward osmoregulation and water use efficiency (Munns and Gilliham 2015; Ikram et al. 2019; Khoshmanzar et al. 2020), which are major issues expected to affect plant survival, growth, and productivity in a climate change context (Daryanto et al. 2016; Naumann et al. 2018; Khoshmanzar et al. 2020).

By disrupting osmotic equilibrium, saline stress alters membrane stability, increases the toxicity of ions within the plant cells, and affects photosynthetic rates (Khomari and Davari 2017; Meena et al. 2017; Mona et al. 2017; Ikram et al. 2019). Due to the lower availability of water created by a higher osmotic pressure (an effect similar to that caused by drought), the plants tend to respond physiologically to these stresses as if they were in a process of acclimation (Farooq et al. 2009; Filippou et al. 2013). The stress induces changes in membrane function, which tends to disrupt the ionic phase, so that cell toxicity results from accumulation of ions, which causes oxidative stress and biochemical imbalances (Begum et al. 2019); depending on their intensity, duration, and speed, these changes can lead to either acclimation or apoptosis (Filippou et al. 2013; Yang and Guo 2018). Furthermore, plants under drought conditions suffer from water supply limitations both by the root system and from the transpiration losses (Tardieu et al. 2018), although a decrease in transpiration rates is a major plant response to this stress (Farooq et al. 2009). The consequent decrease in water potential interferes with the photosynthetic process, by affecting the stomatal opening/conductance, much as a result of responsive-hormones synthesis, as well as of changes in the chlorophyll and carotenoid contents (Mona et al. 2017; Begum et al. 2019). In terms of cellular processes, these water-deficit stresses affect cell division, cell-wall dynamics, primary and secondary metabolism, regulation of hormones and synthesis, and accumulation of reactive oxygen species (ROS) (Bray 2007; Takahashi et al. 2018; Tardieu et al. 2018; Zhang et al. 2019a, b). Reduction in size of leaves and seeds, root growth suppression, and flowering/fruiting delays are additional stressing effects at morphological and physiological levels (Mastouri et al. 2012; Osakabe et al. 2014). Since all of these effects ultimately lead to decrease in plant growth and productivity, *Trichoderma* treatments appear as a relevant option (Mona et al. 2017; Ikram et al. 2019; Zhang et al. 2019a; Poveda 2020) for the development of salt- and drought-tolerance to cope with those additional types of stresses (Farooq et al. 2009; Filippou et al. 2013).

The next most recurrent stress in the studies was caused by heavy metals (Fig. 5). *Trichoderma* spp. applications have shown to be promising alternatives for amelioration of this stress, either alone or combined with salinity. Interestingly, such conditions allow improved phytoremediation activities for plants in metal-polluted soils (Anam et al. 2019; Li et al. 2019). In general, the presence of these metals in soil can affect plants in a variety of forms, such as reducing seed germination, chlorophyll contents, photosynthesis, and ATP synthesis; altering water balance, nutrient absorption by roots, mitochondrial and chloroplast activities, cell signaling, and enzymatic activities; and increasing membrane lipid peroxidation, levels of ROS, etc. (Ghori et al. 2019; Arif et al. 2019). Usually, all these disturbances can lead to a net effect of decreasing and/or halting plant growth and to necrosis of parts or the whole plant (Groppa et al. 2007). Soil, water, air, and trophic chain pollution is mainly caused by anthropic actions of industrial (power and heat, metallurgy, steelmaking, leather, paper, textile, electroplating, electronics, petrochemistry, waste and landfills, etc.), agricultural (chemical fertilizers and pesticides, sewage irrigation), mining (coal, crude oil, iron, and other metals), and urban life (He et al. 2013; Hu et al. 2014; Etesami 2018). For instance, due to the large and strong industrial, urban, and rural development of the last decades in densely populated regions, India and China have shown one of the highest levels of soils, water, and air contamination by heavy metals in the world (Hu et al. 2014; Paul 2017; Mukherjee et al. 2020), especially in rural areas, which have been generating much concern about food security and human health (He et al. 2013; Huang et al. 2018; Yang et al. 2018). Hence, these circumstances also help explaining the highest proportion of studies found for these two countries (Fig. 3).

5 Parameters Evaluated in the Studies of *Trichoderma*-Plant-Abiotic Stresses

The most assessed variables in studies with plant-*Trichoderma*-abiotic stress interactions can be classified as indirect or direct responses: in the former group, the final phenotypic effects (i.e., plants growth and development) are evaluated, whereas in the latter, biochemical/cellular pathways and compounds related to physiological and photosynthetic processes are gauged (Table 3). To act on recovery and/or amelioration of the adverse effects that the abiotic stresses cause in plants, *Trichoderma* spp. interfere in the physiology, biochemistry, and morphology of the host through the diverse genetic and metabolic arsenal available in this fungal genus. The quantification of relief effects of abiotic stresses in plants by *Trichoderma* has been studied by an array of response variables, which, in some cases, can link to possible mechanisms of action. These parameters are related to physiological, morphological, physical, and (bio)chemical aspects, which could be classified into four main categories by conceptual affinity (Table 3; Fig. 6).

Table 3 Parameters used to study mechanisms possibly involved in the alleviation of abiotic stresses in plants by *Trichoderma*[a]

Response variables	No. articles	Δ ("trat" – "ctrl")[b] Min (%)	Max (%)	References[c]
1. Growth/development (173)[d]				
Grain yield	5	−12.6	1160.0	Becquer et al. 2018; Tripathi et al. 2017
Number leaves	6	−39.0	77.7	Azarmi et al. 2011
Leaf area	7	−96.8	993.0	Azarmi et al. 2011; Singh and Dwivedi 2018
Shoot fresh wght	15	−75.2	744.1	Azarmi et al. 2011
Root fresh wght	17	−73.8	374.4	Azarmi et al. 2011; Abd El-Baki et al. 2014
Germination	18	−8.3	516.3	Montero-Barrientos et al. 2010; Nzioki and Mutisya 2016
Root dry weight	24	−74.7	4457.7	Mastouri et al. 2012; Abd El-Baki et al. 2014
Shoot dry weight	25	−87.7	416.7	Azarmi et al. 2011; Hashem et al. 2014
Shoot length	27	−28.6	199.6	Abd El-Baki et al. 2014; Shukla et al. 2014
Root length	29	−20.7	290.9	Mishra et al. 2016; Vieira et al. 2017
2. Physiology/photosynthesis (93)				
Transpiration	3	−17.8	82.2	Vieira et al. 2017
Intercell CO_2*	3	3.8	*−24.8*	Vieira et al. 2017
Chl fluoridation	7	−46.9	132.9	Azarmi et al. 2011; Rawat et al. 2012
Net photosynth.	7	−15.9	412.5	Vieira et al. 2017
Stomat. conduct.	8	−59.6	243.6	Azarmi et al. 2011; Shukla et al. 2012
Relate H_2O content	13	−1.1	170.0	Vieira et al. 2017; Shukla et al. 2014
Photosynthetic pigments				
Chlorophyll α	12	−48.0	123.5	Singh and Dwivedi 2018; Badar et al. 2015
Chlorophyll b	12	−23.5	428.0	Singh and Dwivedi 2018; Hashem et al. 2014
Total chlorophyll	18	−25.0	525.0	Jalali et al. 2017
Total carotenoid	10	−39.4	122.9	Singh and Dwivedi 2018; Elkelish et al. 2020
3. Stress-related activities (117)				
Transloc. factor	2	−23.8	300.0	Vargas et al. 2017
Lipid peroxid.*	3	*−6.9*	*−58.1*	Dixit et al. 2011; Nongmaithem and Bhattacharya 2017
Electrolytic leak*	5	*2.4*	*−58.5*	Poveda 2020
Membrane stability index	11	−57.4	101.6	Tripathi et al. 2013; Hashem et al. 2014
Malondialdehyde (mda)*	20	*−99.9*	*−137.3*	Abd El-Baki and Mostafa 2014; Kumar et al. 2016

(continued)

Table 3 (continued)

Response variables	No. articles	Δ ("trat" – "ctrl")[b] Min (%)	Max (%)	References[c]
Antioxidant enzyme activity				
GPX	7	−10.3	148.7	Singh et al. 2019; Dixit et al. 2011
GR	9	−15.5	200.0	Pehlivan et at. 2017; Tripathi et al. 2017
APX	10	10.0	764.7	Guler et al. 2016; <u>Singh et al. 2019</u>
POD	10	−44.8	136.8	Fu et al. 2017; Devi et al. 2017
CAT	19	−30.2	563.2	Fu et al. 2017; Chepsergon et al. 2016
SOD	21	−40.3	359.9	<u>Singh et al. 2019</u>; Chepsergon et al. 2016
4. Compounds' levels/content (102)				
Flavonoids	2	409.8	502.8	Mona et al. 2017; <u>Elkelish et al. 2020</u>
Superox. dismutase (O^{2-})*	2	*−25.7*	*−63.5*	Fu et al. 2017
Siderophores	2	0	151.7	Zhao et al. 2014
Amino acids	3	–	–	–
Soluble sugar	5	9.8	47.6	Fu et al. 2017
Phytohormones	8	−78.1	1154.5	Singh et al. 2019
Heavy metal conc.*	9	*75.0*	*−84.2*	Song et al. 2014; Vargas et al. 2017
Protein	10	−48.5	68.5	Abd El-Baki and Mostafa, 2014
Ions	10	−67.6	2335.3	Azarmi et al. 2011
Total phenol	11	−4.4	192.2	Kumar et al. 2016; Rawat et al. 2013
H Perox (H_2O_2)*	16	108.5	*−77.9*	Shukla et al. 2014; Pehlivan et at. 2017
Proline	24	−81.7	350.0	Abd El-Baki and Mostafa 2014; Rawat et al. 2012

[a]The four categories were defined according to a conceptual affinity among their response variables
[b]Differences (in %; control = 100%) between the values obtained for each variable, considering the treatments with application of *Trichoderma* ("trat") in relation to the treatments with only the abiotic stress(es)("ctrl"). The "Min" and the "Max" columns correspond to the lowest and highest D values (differences between "trat" and "ctrl") found for a given parameter in the set of articles containing it (see "No. articles" column). "*" indicates those variables that describe damaging stress effects to the plants; *italicized* "min" and "max" values in the table are those in which *negative* D values indicate *positive/favorable* effects of *Trichoderma* to the plant host in the amelioration of the stress
[c]The references in this column belong to the final selected database of articles used in the systematic review treated in this chapter (see Table 2)
[d]The numbers within parentheses for each of the four main categories indicate the total number of *experiments* (i.e., *variables per article*) and correspond to the sum of values for the "No. articles" column within each category; this column, therefore, indicates the amount of articles systematically selected to compose our final database, in which a given response variable (indicated on the left) was found. Hence, a single article can be counted more than once, in case it has reported various response variables at the same time
[**<u>note</u>**: the underlined citations refer to studies also related to molecular analyses discussed for Fig. 6, Sect. 6]

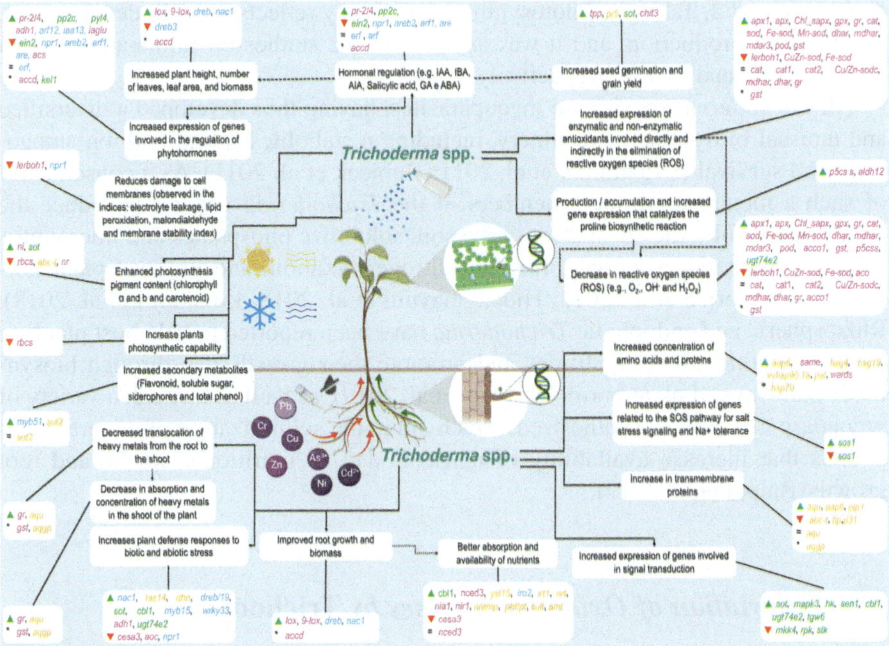

Fig. 6 Possible mechanisms involved in the interaction between *Trichoderma* and plants in response to abiotic stresses. The information retrieved on genetic products in this figure refers to 13 articles in which studies of gene expression were found. Different colors represent four main groups of activities/functions identified for the plant genes involved in stress mitigation: blue, transcription factors; pink, metabolic pathways; green, signal transduction; orange, structural proteins and protective compounds. The symbols located on the left inside the balloons represent gene expression modulation in relation to treatment *Trichoderma*-plant-stress vs plant-stress only: ▲green, upregulation of gene expression; ▼red, downregulation in gene expression; '=' expression level without significant difference; '•' transgenic plants expressing *Trichoderma* genes

5.1 *Influence of* Trichoderma *on Plant Growth and Development*

Out of the 485 experiments found in the 80 systematically selected articles whose data was collected (Table 3), the most frequent group of variables were growth and development parameters (35.7%), mainly root and shoot biomasses (fresh and dry length and weight), which are measures of plant vitality as evidence of their recovery from stresses. Taking the results of *Trichoderma* application into account, the overall positive effects on plant growth and development could be observed, with increases in relation to control treatments varying from 77.7% improvement in the number of leaves to 4457.7% raise in root dry weight (Table 3).

Another relevant group of parameters evaluated in addressing *Trichoderma* effects on plant stresses was more specifically related to plant physiology, mostly focusing on photosynthesis and represented 19.2% of the variables evaluated in this study (group # 2, Table 3). Photosynthetic efficiency reflects growth, development, and biomass production, and it was assessed in the studies on saline and drought stresses (56% and 25.8%, respectively).

All *Trichoderma* species are mycoparasites, having thus developed a diversified and unusual biosynthetic machinery, including metabolites acting both on antagonism and survival (Druzhinina et al. 2011; Kubicek et al. 2011). As a consequence of such a metabolic variety, members of the *Trichoderma* genus can reduce the concentration of toxic substances in the soil, solubilize phosphates and micronutrients, synthesize siderophores, increase nitrogen fixation, and produce plant hormones (Mukherjee et al. 2013; Hidangmayum et al. 2018; Lombardi et al. 2018). Rhizospheric and endophytic *Trichoderma* have been reported to help host plants to adapt to abiotic stress conditions and promote their growth also through biosynthetic pathways of plant hormones (Yan et al. 2019), as well as through a variety of secondary metabolites synthesized, which aid in the solubilization of mineral compounds that increase availability of nutrients and so nutritional uptake and root growth (Rajput et al. 2019).

5.2 Alleviation of Oxidative Stresses by Trichoderma

The other two categories of variables addressed in experiments with *Trichoderma* were represented in our dataset as follows: 24.1% for variables related to enzyme activities and cellular functions directly affected by the stresses and 21% for levels and rates of compounds synthesized as responses to the stresses (Table 3). Within the group of variables gauging activities directly related to stress responses (group # 3, Table 3), almost the totality of the retrieved studies deals with either antioxidant activities (65%) or membrane/lipid effects (33.3%). Within the antioxidant enzyme activities related to oxidative stress response/regulation, superoxide dismutase (SOD), catalase (CAT), peroxidase (POD), ascorbate peroxidase (APX), and glutathione reductase (GR) and peroxidase (GPX) were the most prominent found (Table 3). In terms of membrane-related studies, the most prevalent specific activities were lipid peroxidation (with special interest in the use of malondialdehyde), which made up 51.2% of the analyses in this subgroup (Table 3). The composition and stability of the plasma membrane, which was used to test stress damage levels to plant cells, made up 28.2% of the studies, and the remaining four assessments dealt with electrolytic leak and translocation factors (two studies each).

The parameters relating to the content of certain substances synthesized by plants (Table 3, group # 4) were grouped as such because they are indicators of, or relate to stress states, or yet belong in metabolisms or processes that assist in physiological recovery from the action of abiotic stresses. As a result from the higher concentration of studies in saline/drought stresses (Fig. 5), the highest frequency of studies in this category # 4 (Table 3) were related to proline levels (22.5%), which corresponded 62.5% of the studies with saline stress and 20.8% of the studies on drought tolerance (Fig. 5). Among the compounds identified in the systematically retrieved studies, there was a focus on secondary metabolites related to stress responses, such as ROS (17.6%), phenolics (10.7%), and phytohormones (7.8%). Ions and heavy metal contents also appeared well represented (9.8 and 8.8%, respectively), as some of the research was focused on this type of abiotic stress (Fig. 5).

As much as biotic factors (e.g., fungal diseases, herbivory, etc.), stressing factors of abiotic nature also cause the overproduction of reactive oxygen species (ROS) in plants, which lead to metabolic toxicity, damage to the membranes, inhibition of photosynthetic apparatus and steps, and changes in hormonal levels, among others (Selvakumar et al. 2012). The major ROS species formed (superoxide, O2–, hydroxyl, OH–, and hydrogen peroxide, H_2O_2) react chemically with virtually all metabolites of the plants, including proteins, lipids, and nucleic acids (Nath et al. 2013; Harman et al. 2019). As in low concentrations, ROS act as signaling molecules, with specific signatures of their steady-state levels, depending on the type of cell of the plant (Choudhury et al. 2017). The regulation of ROS levels is very precise in plant cells, being related to a fine-tuned balance between their perception and detoxification, and the redox state of the cell, with a particular relevance for chloroplasts in this metabolism (Farooq et al. 2009; Meyer et al. 2020). In this context, antioxidant compounds and enzymes act coordinately on the fine modulation of these mechanisms (Mittler 2002). *Trichoderma* spp. have shown to also depend on ROS signaling for a variety of their own cellular processes and responses to environmental cues (Cruz-Magalhães et al. 2019), thereby having a clear modulatory interference in plants, when interacting with them.

The majority of the studies on abiotic stresses involving plants and *Trichoderma* have shown to focus on drought/salinity (Fig. 5). Knowledge generated in this aspect indicates that major protection of plant cells against these stresses occurs by the promotion of osmolites' synthesis or accumulation, which increases both the water absorption and retention capacity of the cells and the activities of enzymatic and non-enzymatic antioxidants (Hameed et al. 2014; Waghunde et al. 2016; Pachauri et al. 2019). A recurrent mechanism of action found in the studies with *Trichoderma* spp. was the production and accumulation of proline, an amino acid that acts as cellular osmoprotector (Harman et al. 2019), mostly in three ways: (i) by protecting intracellular macromolecules against reactive oxygen species (ROS) attack, (ii) by serving as a source of carbon and nitrogen for the cell as a result of its oxidative metabolism, and (iii), as discussed above, by acting as a modulator of the osmotic balance of the cell (Christgen and Becker 2019). Some underlying mechanisms of exogenous phytohormones production by *Trichoderma*, such as similar forms of abscisic acid (ABA), can also protect the plant from oxidative damage (Bano et al. 2012; Khan et al. 2015), as well as modulate other stress-response metabolisms. Members of the *Trichoderma* genus are outstanding producers of secondary metabolites with functions already known (Table 3, Fig. 6), although many of such compounds are still unknown. There are more than 2000 natural products, such as peptaibols, non-ribosomal peptides, polypeptide, terpenes, and steroids produced by *Trichoderma* ssp., which play important roles in their interaction with plants (Mukherjee et al. 2012). As mentioned above, *Trichoderma* species throughout evolution have developed the ability to produce a large amount of extracellular enzymes and secondary metabolites (Mukherjee et al. 2012; Kubicek et al. 2019), as well as very effective systems of resistance and repair of cellular and molecular damages (Duran et al. 2010; Ghorbanpour et al. 2018), a capability that can extend the protection to their hosts (Harman et al. 2019).

6 Plant Genes Influenced by *Trichoderma* in Response to Abiotic Stresses

6.1 Outline of the Studies

The analytical review of this chapter allowed us to provide a glance on the current status of research on genes and their products that can be related to the beneficial fungus-host interaction in response to abiotic stresses. Out of the systematically assembled database, about 26.3% of its articles were identified as comprising studies of this nature, dealing with in vivo biochemical and molecular methods; all the data we found on gene expression patterns related to *Trichoderma*-plant-stress interaction were related to plant genes (Table 3; Fig. 6), and all of them were previously known to be involved in plant stress responses and in the transport of macro and micronutrients. The largest amounts of these studies were on drought (33.3%) and saline stresses (28.6%); of the remainder, 9.5% by heavy metals evaluated stress by high temperature, low temperature, nutritional deficit, and waterlogging with 4.8% each, and 9.5% gauged combined stresses (drought + high temperature and salinity + osmotic stress). The species of *Trichoderma* used in these molecular genetics' studies were *T. harzianum* (seven), *T. parareesei* (six) *T. britannicum* (three), *T. asperelloides*, and *T. longibrachiatum* (two each); *T. afroharzianum*, *T. asperellum*, *T. hamatum*, *T. virens*, and *T. reesei* (one each); and a study in which there was no identification at the species level. The plant species investigated in these studies were *Arabidopsis thaliana*, *Brassica napus*, *Solanum lycopersicum*, *Nicotiana tabacum*, *Cicer arietinum*, *Oryza sativa*, *Populus bolleana*, *Triticum aestivum*, *Theobroma cacao*, *Zea mays*, and *Vitis vinifera*. The studies on stress-responsive genetic expression reported the majority of the genes (77.2%) as being upregulated as the result of stress, both in the above- and belowground parts of the plants (Fig. 6). It is important to mention that 23.8% of these studies were performed with transgenic plants, in which overexpression of *Trichoderma*-derived transgenes (supposedly induced in the fungus as a response to some abiotic stress) were investigated on their effects in modulating plant gene expression in response to the abiotic stresses (e.g., Meena and Swapnil 2019; Mota et al. 2019).

The studies selected concerning the molecular aspects of the stress alleviation mechanisms of plants by *Trichoderma* were sufficiently consistent with the physiological characteristics of the assessed plants under abiotic stress conditions (Table 3). Essentially, there were four major groups of activities identified for the plant genes involved in stress mitigation: (i) transcription factors (TFs) directly involved in stress-response gene expression modulation, (ii) genes responsive to metabolism and oxidative stresses, (iii) signal-transduction pathways, and (iv) synthesis of structural/protective proteins and compounds (Table 4; Fig. 6).

6.2 Transcription Factors

With regard to the genes encoding TFs, studies related to their expression altered in response to the *Trichoderma*-plant-stress interaction showed a tendency of them to refer mostly to hormonal and pathogen-related signaling pathways and dehydration-responsive genes; they were *nac1/nac6* (Ghorbanpour et al. 2018; Singh et al. 2019); *dreb* (dehydration-responsive element binding proteins, Brotman et al. 2013; Pandey et al. 2016; Rubio et al. 2017; Singh et al. 2020b); *zfp* and *p13* (zinc-finger domain factors related to transcriptional repression, Bae et al. 2009); *erf* (ethylene-responsive factor, Roatti et al. 2013; Elkelish et al. 2020; Poveda 2020); *npr1, are, areb2, arf* (TFs related to salicylic acid, ABA, and auxin signaling pathways, Rubio et al. 2017; Singh et al. 2019; Elkelish et al. 2020); *iaa13, myb15, myb51, wrky33* (TFs related to secondary metabolites synthesis, auxin, jasmonate/salicylate signaling pathways, Brotman et al. 2013); *iro2* (iron-regulated transcription factor, Singh et al. 2019). Interestingly, recent full-genome comparisons have shown the class of transcription factors genes as one of the most abundant in the core genome of *Trichoderma* (Kubicek et al. 2019).

6.3 Plant Genes Responsive to Oxidative Stresses

Another relevant biological function identified for the studied plant genes was associated with pathways of direct response to stresses and to metabolic changes resulting from the stress effects (Fig. 6). The genes within this category included *p5cs* (encoding pyrrolin-5-carboxylate synthetase enzyme, which catalyzes a rate-limiting step reaction of proline synthesis, Ghorbanpour et al. 2018); methyltransferase and alcohol dehydrogenase (Brotman et al. 2013; Ma et al. 2020; Elkelish et al. 2020); *chit3* and *pr-2* (acid endochitinase and pathogenesis-related type 2 protein, i.e., beta-1,3-glucanase) (Roatti et al. 2013); *acc deaminase* and *oxidase* (Zhang et al. 2016a; Zhang et al. 2019a; Elkelish et al. 2020; Poveda 2020;); small subunit of Rubisco complex (catalyzes the limiting step of CO_2 fixation), cellulose synthase, lipoxygenase (oxylipin synthesis), phosphatase involved in the last step of trehalose synthesis, invertase involved sucrose hydrolysis, and nitrate/ferredoxin-nitrite reductase (Bae et al. 2009; Roatti et al. 2013; Singh et al. 2019); and genes/enzymes involved in ROS metabolism, such as *nadph* oxidase 1, dehydroascorbate reductases, *gst* (glutathione transferase), and all those genes encoding the antioxidant enzymes indicated in Table 3 (Montero-Barrientos et al. 2010; Dixit et al.

Table 4 Identification of plant genes from the systematically retrieved studies
Different colors represent the categories indicated in Fig. 6. Blue, transcription factors; pink, metabolic pathways; green, signal transduction; orange, structural proteins and protective compounds

Abbreviation	Gene function
are	ABA-responsive element binding protein 2
arf	Auxin response factor like
dreb	Dehydration responsive element bindings protein
erf	Ethylene-Responsive transcription factor
iaa13	Auxin-responsive protein IAA13-like
iro2	Protein iron-related transcription factor 2
nac	NAC domain-containing protein
npr1	Regulatory protein NPR1
myb	MyB-Domain Protein
wrky33	Member of the 'WRKY' family of transcription factors
zfp, p13	Protein with 'zinc finger' domain
accd	1-AminoCyclopropane-1-Carboxylate deaminase
acco/aco	1-aminocyclopropane-1-carboxylate oxidase 1
acs	1-aminocyclopropane-1-carboxylic acid synthase
adh	Alcohol Dehydrogenase
aldh	Delta-1-pirrolina-5-carboxilato desidrogenase
aoc	Cyclam se of allene oxide
apx	Ascorbate peroxidase enzyme
cat	Catalase enzyme
cesa3	Putative protein with cellulose synthase activity
chit3	Acid Endochitinase 3
dhar	Enzyme Desidroascorbate Reductase
gpx	Enzyme Glutathione Peroxidase
gr	Enzyme Glutathione Reductase
gst	Glutathione Transferase enzyme
iaglu	Indole-3-acetate beta-glucosyltransferase
lerboh1	NADPH oxidase 1
lox	Lipoxygenase enzyme
mdhar	Enzyme Monodesidroascorbato Reductase
nced3	9-cis-epoxycarotenoid dioxygenase
ni	Putative alkaline/neutral Invertase
nia1	Nitrate reductase [NADH] 1-like
nir1	Ferredoxin-nitrite reductase
nr	Nitrate reductase
p5cs s	Delta 1-Pyrrolin-5-Carboxylato Synthetase
pal	Phenylalanine ammonia-lyase activity
pod	Peroxidase
pr-2	Pathogenesis Related prot. no.2 (beta-1, 3-glucanase)
rbcs	Small subunit of rubisco complex
same	S-adenosyl-L-methionine-Dependent Methyltransferase
sod	Enzyme Superoxide Dismutase
tpp	Trehalose-6-phosphate phosphatase
wards	Alanine aminotransferase
cbl1	Calcineurin B-Like protein 1

(continued)

ein2	Protein insensitive to ethylene 2
hk	Enzyme Histidine kinase
kel1	Protein with 5 repeated Kelch-like domains
mapk3	MAP kinase 3
mkk4	MAP kinase kinase 4
pp2c	Phosphatase protein 2C
pyl4	Receptor for abscisic acid (ABA)
rpk	Putative receptor protein kinase
sen1	Protein associated withsenescence
sos1	Salt overly sensitive 1
sot	Sorbitol transporter
stk	Serine/threonine protein kinase
sult	Sulfotransferase gene 1
ugt74e2	UDP-glycosyltransferase 74E2
aap6	Amino Acid Permease 6
abc-t	Transmembrane protein that binds to ATP
amt	Ammonium transporter
aqgp	Transmembrane protein 'Aquaglyceroporin' (family of aquaporins)
aqu	Transmembrane protein 'Aquaporin'
dhn	Protein 'Dehydrin'
hsg4	Heat-Shock Gene 4
hsg19	Heat-Shock Gene 19
hsp70	Heat-Shock Protein 70
irt1	Iron regulated metal transporter
nramp	Nramp metal transporter Mn uptake
nrt	Nitrate transporter gene
osm-1	Osmotic stress
pip1	Aquaporin PIP1
pht/pt	Phosphate transporter
pr-4	Pathogenesis Related prot. no. 4 (thaumatin-like PR-protein)
pr-5	Pathogenesis Related prot. no 5
sut2	Sucrose transporter protein involve in flowering and grain development
tas14	Dehydratorine (Group 2 LEA Proteins)
tip, p31	Intrinsic protein of the tonoplasto
vvhsp90.1a	Heat-Shock Protein 90
ysl15	Iron (III)-deoxymugineic acid transporter

Different colors represent the categories indicated in Figure 6. blue: transcription factors; pink: metabolic pathways; green: signal transduction; orange: structural proteins and protective compounds.

2011; Mastouri et al. 2012; Brotman et al. 2013; Rubio et al. 2017; Tripathi et al. 2017; Zhang et al. 2019a, b; Elkelish et al. 2020; Singh et al. 2020).

6.4 *Signal Transduction Pathways*

The third group of genes identified as having their expression altered as a function of abiotic stress effects was related to signaling proteins involved in stress-response physiology of plants (Table 4; Fig. 6). This group comprises the following genes/proteins: *sos1* (signal protein of the *salt overly sensitive* pathway, Montero-Barrientos et al. 2010; Rubio et al. 2017; Zhang et al. 2019a); *hk*, *rpk*, *mapk*3, and 4, *stk* (histidine, receptor protein, MAP, and serine/threonine kinases), *sen1* (senescence associated), *pp2c* (phosphatase protein 2C, possibly related to ABA pathway), and *sot* (sorbitol transporter) (Bae et al. 2009); *ein2* (ethylene-insensitive protein, central to this hormone signaling pathway, Rubio et al. 2017); *pyl4* (abscisic acid receptor – required for ABA-mediated responses, Poveda 2020); *cbl1* (*calcineurin B-like 1* protein, sensor of calcium levels, interacting/regulating a family of kinases located in endomembranes) and *ugt*74e2 (UDP-glycosyltransferase 74E2, related to signaling of drought stress and auxin homeostase, Brotman et al. 2013); *kel1* (encodes a protein with 5 repeated Kelch-like domains, characteristic of gene families involved in cell morphology and protein-protein interactions, Hermosa et al. 2011); and *tgw6* (Trafficking protein particle complex subunit, Zhang et al. 2019a).

6.5 Genes Involved in Transport and Protection Against Abiotic Stresses

Finally, the last group of stress-responsive plant genes are those encoding proteins with either a directly protective activity to ameliorate the effects of the stressing agent or a transporting activity for molecules and substances used for this protection (Table 4; Fig. 6). In the former subgroup, we found *tas14* (dehydratorine of group 2 late embryogenesis abundant proteins, Ghorbanpour et al. 2018) and *dhn* (dehydrin, cellular protection against dehydration, also a LEA protein, Pandey et al. 2016; Singh et al. 2020b); *hsp70, -4, -19 -90* (heat shock chaperones, Monteiro-Barrientos et al. 2010; Roatti et al. 2013); and *osm1* (osmotic stress-sensitive mutant, belonging in the superfamily of SNARE proteins involved in vesicle/membrane fusion, Roatti et al. 2013). In the second subgroup of genes/proteins with transporting activity, there were *Aqgp*, *aqu*, *tip*, *p31*, *pip1* (aquaporin-type transmembrane proteins, Bae et al. 2009; Pandey et al. 2016; Vieira et al. 2017; Elkelish et al. 2020; Singh et al. 2020b); *abc* and *pr-5* (ATP-binding transporters) and pathogenesis-related type 5, i.e., osmotin-like membrane located protein (Bae et al. 2009); *aap6* (amino acid permease 6, transmembrane transporters, Brotman et al. 2013); and *Ysl15*, *irt1*, *nrt*, *nramp*, *sut2*, *pht/pt*, *amt* (macro and micronutrient transporters, Singh et al. 2019).

The assessment of expression modulation of genes investigated in the *Trichoderma*-plant-abiotic stress interaction suggest an interplay of a variety of cellular and physiological mechanisms, many with a cross talk among signaling and metabolic pathways responsive to both biotic and abiotic stresses (Choudhury et al. 2017; Mendoza-Mendoza et al. 2018; Meyer et al. 2020). Plants have to deal simultaneously with multiple environmental stress-related cues, thus displaying a complex integration of stimuli and defense signals. Prioritizing certain physiological responses is a fine-tuned regulation resulting from plant-microbe interactions, whose understanding will be advantageous for crop improvements (Schenk et al. 2012). Further studies supported by the multi-omics, high-throughput, and comparative genomics approaches can unravel structural and functional aspects of these complex regulatory networks with unprecedent detail (Zeilinger et al. 2016; Meena et al. 2017; Kubicek et al. 2019; Arif et al. 2019), thereby providing additional opportunities for biotechnological development of *Trichoderma*-based bioproducts directed toward mitigation of plant stresses caused by abiotic factors (Waghunde et al. 2017; Szczałba et al. 2019; Topolovec-Pintarić 2019).

7 Conclusions and Perspectives

Environmental degradation imperils quality of life on Earth, and bioeconomy is a recent view that may properly handle the threatening circumstances. Bioeconomy has been developed on the basis of three visions – bio-ecology, bio-resources, and bio-technology (Bugge et al. 2016). In this regard, a vast array of studies on *Trichoderma* spp. have been widely reported in the literature, mostly due to their efficiency as biological control agents of plant pathogens, direct plant growth promotion, and the synthesis of a diverse of compounds with industrial applications (Vinale et al. 2008; Schuster and Schmoll 2010; Mukherjee et al. 2013). Nevertheless, from the beginning of this century, researchers have been pointing out that this fungal genus is even more multifaceted and so with an increasing potential for a wider diversity of applications akin to the bioeconomy view.

In this chapter, we systematically reviewed and discussed the use of *Trichoderma* to mitigate the negative effects of abiotic stresses on plants and discussed the consequences and potential applications of these findings, including areas of knowledge with strengths and gaps in this theme. An up-to-date sampling of articles containing primary studies reported in journals relevant to the areas of biological control, mycology, bioprospection, biotechnology development, and bioproducts were gathered, with their data being collected and assessed in details in this chapter. With specific tools, it was possible to prioritize the retrieval of more highly cited publications, which helped us to work with an amount of representative research of quality in this field (Figs. 1 and 2). Our survey was able to retrieve interesting information on the current status of research with *Trichoderma*, their interaction with plants, and the mitigation of abiotic stresses (Figs. 3, 4, 5 and 6).

Trichoderma of various species are commonly used as biocontrol agents and/or growth promoters, making up about 3/5 of the biofungicide market in the world (Topolovec-Pintarić 2019); since a multifunctional characteristic can add value to bioproducts, *Trichoderma* isolates with additional phenotypes of abiotic stress-relief for plants (Zhang et al. 2016a; Anam et al. 2019; Szczałba et al. 2019; Poveda 2020) can provide a very advantageous benefits/costs relationship for environmentally sustainable food production strategies (Harman 2011a). In a region with various environmental degradation issues to solve (Chopra 2016), India is an example of a country taking robust steps in this direction, with a significant contribution not only on *Trichoderma* spp. science but also on their bioproducts' market (Woo et al. 2014). Despite having more than 300 species already described in this genus (Kubicek et al. 2019), and also a great fraction of functional isolates not yet characterized up to species level, more than 3/4 of the studies involving *Trichoderma*-plant-abiotic stress interactions have appeared to be restricted to 16 main species, with a highlight for *T. harzianum* and *T. asperellum*. The two main sources of isolates for these studies (~1/3 each) are institutional collections and rhizospheric soils. Therefore, there is yet plenty of opportunity for bioprospection and basic studies, not only to unravel novel *Trichoderma* isolates/species bearing abiotic stress-relief effects on plants but also to further understand the underpinning mechanisms of this desirable phenotype.

A tendency was found for the studies to focus on model plants with agronomic/economic relevance, most with short reproductive cycles. Moreover, the competitive ability of *Trichoderma* species that allows rapid rhizospheric establishment and roots colonization likely explains the preference for experimental inoculations based on soil application and seed coating and biopriming; the ease of later field applications for bioproducts developed in this manner adds to this circumstance. With regard to the abiotic stresses studied in the context of plant-*Trichoderma* interactions, the focus has been essentially on those caused by excess of salt, drought, and heavy metals, which can be considered as coherent with the major environmental issues affecting the security and sustainability of food production worldwide. As a consequence, the parameters assessed are all directly and indirectly related to shoot and root weight and length, as well as major physiological processes, such as photosynthesis, general cellular redox state and oxidative-stress control/response, and protective compounds synthesis. Not unexpectedly, the molecular and genetic mechanisms studied in this regard have been strictly related with those response variables and could be conceptually classified as transcription factors, metabolic/oxidative stress and signaling pathways, and direct stress-protective molecules. Metanalytical approaches are currently underway to advance the dissection of current information on the *Trichoderma* effects in the improvement of plant growth and development under abiotic stresses. Taking all this information together, and assessing knowledge-integration studies and reviews, it became evident the astonishing complexity of regulatory mechanisms and networks already unveiled in the *Trichoderma*-plant interaction, as well as the universe yet to be researched in this field.

We reviewed here the state-of-the-art of knowledge on the use of *Trichoderma* spp. in aiding plants to cope with a variety of stresses caused by climatic and edaphic abiotic factors; despite some trends and gaps observed in the pertinent investigated literature, the great potential of this fungal genus for developing alternative applications of biotechnological interest (agriculture, industry, environment, and health) is clear. Augmentation of salinity and pollution by an array of xenobiotics (Rosegrant et al. 2009; Munns and Gilliham 2015), as well as global warming effects such as high temperatures, alterations in rainfall cycles, and longer droughts (Godfray et al. 2010; Foley et al. 2011), are relevant examples of these environmental impacts. More than 4/5 of global ecosystems functioning and processes that underpin support services for people are already affected by climate change (Ogar et al. 2020), with a significant role for abiotic factors. The current issues on food production and security will require robust and coordinated actions on scientific and political arenas to bring forth environmentally sustainable solutions for the global economy, hugely impacted by the COVID-19's pandemic. Those solutions depend upon consistent reductions in both our carbon footprint on the planet (Stern 2016) and in the biodiversity losses, which interdependently affect ecosystems and economy (Dasgupta 2008; Trisos et al. 2020). The large spectrum of compounds and bioproducts that can be extracted from and formulated on the basis of species/isolates of the *Trichoderma* genus can certainly contribute to broaden the strategies and options for agricultural production with environmental sustainability and safety. Considering current unresolved issues related to the bioproduct registration system and the scope of its applicability, e.g. (Berg et al. 2013; Chojnacka 2015), it may be considered advantageous for a biotechnological product to bear multiple simultaneous functions, a possibility that is clearly present in *Trichoderma* (Vinale et al. 2008; Schuster and Schmoll 2010; Mukherjee et al. 2013; Hidangmayum et al. 2018). The various biological activities of *Trichoderma* with beneficial effects for their host plants have a high potential for adding economical and sustainability values to existing and yet to be developed bioproducts and derivatives.

Acknowledgments The authors thank the Fundação de Amparo à Pesquisa do Estado da Bahia (FAPESB), Coordenação de Aperfeiçoamento de Pessoal de Nível Superior (CAPES), and Conselho Nacional de Desenvolvimento Científico e Tecnológico (CNPq) for financial assistance and scholarships support. We are grateful to Ronilma Fernandes Alves for valuable input in the beginning of this study.

References[1]

Anam GB, Reddy MS, Ahn YH (2019) Characterization of *Trichoderma asperellum* RM-28 for its sodic/saline-alkali tolerance and plant growth promoting activities to alleviate toxicity of red mud. Sci Total Environ 662:462–469. https://doi.org/10.1016/j.scitotenv.2019.01.279

Arif N, Sharma NC, Yadav V et al (2019) Understanding heavy metal stress in a rice crop: toxicity, tolerance mechanisms, and amelioration strategies. J Plant Biol 62(4):239–253. https://doi.org/10.1007/s12374-019-0112-4

Babychan M, Simon S (2017) Efficacy of *Trichoderma* spp. against *Fusarium oxysporum* f. sp. *lycopersici*. (FOL) infecting pre-and post-seedling of tomato. J Pharmacogn Phytochem 6(4):616–619

Ballescá AB (2016) China's economic development and its effects on air pollution: accountability, policies and health risks. Universitat Autònoma de Barcelona

[1] Please, note that the references included in the systematic review are listed in Table 2 only, regardless if they are cited in the text or not.

Bano A, Ullah F, Nosheen A (2012) Role of abscisic acid and drought stress on the activities of antioxidant enzymes in wheat. Plant Soil Environ 58(4):181–185. https://doi.org/10.17221/210/2011-pse

Begum N, Ahanger MA, Su Y et al (2019) Improved drought tolerance by AMF inoculation in maize (*Zea mays*) involves physiological and biochemical implications. Plan Theory 8(12):579. https://doi.org/10.3390/plants8120579

Berg G, Zachow C, Müller H et al (2013) Next-generation bio-products sowing the seeds of success for sustainable agriculture. Agronomy 3:648–656. https://doi.org/10.3390/agronomy3040648

Bray EA (2007) Plant response to water-deficit stress. In: Encyclopedia of life sciences. John Wiley & Sons, Ltd, Chichester. https://doi.org/10.1038/npg.els.0001298

Bugge M, Hansen T, Klitkou A (2016) What is the bioeconomy? A review of the literature. Sustainability 8(7):691. https://doi.org/10.3390/su8070691

Calvo P, Nelson L, Kloepper JW (2014) Agricultural uses of plant biostimulants. Plant Soil 383(1):3–41. https://doi.org/10.1007/s11104-014-2131-8

Chakraborty U, Chakraborty B, Dey P, Chakraborty A (2015) Role of microorganisms in alleviation of abiotic stresses for sustainable agriculture. In: Chakraborty U, Chakraborty B (eds) Abiotic stresses in crop plants. CABI, Wallingford/Boston. https://doi.org/10.1079/9781780643731.0232

Chang P-FL, Xu Y, Narasimhan ML et al (1997) Induction of pathogen resistance and pathogenesis-related genes in tobacco by a heat-stable *Trichoderma* mycelial extract and plant signal messengers. Physiol Plant 100(2):341–352. https://doi.org/10.1111/j.1399-3054.1997.tb04792.x

Chaverri P, Branco-Rocha F, Jaklitsch W et al (2015) Systematics of the *Trichoderma harzianum* species complex and the re-identification of commercial biocontrol strains. Mycologia 107(3):558–590. https://doi.org/10.3852/14-147

Chojnacka K (2015) Innovative bio-products for agriculture. Open Chem 13(1):932–937. https://doi.org/10.1515/chem-2015-0111

Chopra R (2016) Environmental degradation in India: Causes and consequences. Int J Appl Environ Sci 11(6):1593–1601

Choudhury FK, Rivero RM, Blumwald E, Mittler R (2017) Reactive oxygen species, abiotic stress and stress combination. Plant J 90(5):856–867. https://doi.org/10.1111/tpj.13299

Christgen SL, Becker DF (2019) Role of proline in pathogen and host interactions. Antioxid Redox Signal 30(4):683–709. https://doi.org/10.1089/ars.2017.7335

Contreras-Cornejo HA, Macías-Rodríguez L, del-Val E, Larsen J (2018) The root endophytic fungus *Trichoderma atroviride* induces foliar herbivory resistance in maize plants. Appl Soil Ecol 124:45–53. https://doi.org/10.1016/j.apsoil.2017.10.004

Cruz-Magalhães V, Nieto-Jacobo MF, van Zijll de Jong E et al (2019) The NADPH oxidases Nox1 and Nox2 differentially regulate volatile organic compounds, fungistatic activity, plant growth promotion and nutrient assimilation in *Trichoderma atroviride*. Front Microbiol 9:3271. https://doi.org/10.3389/fmicb.2018.03271

Daryanto S, Wang L, Jacinthe PA (2016) Global synthesis of drought effects on maize and wheat production. PLoS One 11(5):e0156362. https://doi.org/10.1371/journal.pone.0156362

Dasgupta P (2008) Creative accounting. Nature 456:44. https://doi.org/10.1038/twas08.44a

De Souza JT, Pomella AWV, Bowers JH et al (2006) Genetic and biological diversity of *Trichoderma stromaticum*, a mycoparasite of the cacao witches'-broom pathogen. Phytopathology 96(1):61–67. https://doi.org/10.1094/PHYTO-96-0061

Druzhinina IS, Seidl-Seiboth V, Herrera-Estrella A et al (2011) *Trichoderma*: the genomics of opportunistic success. Nat Rev Microbiol 9(10):749–759. https://doi.org/10.1038/nrmicro2637

Du JB, Han TF, Gai JY et al (2018) Maize-soybean strip intercropping: achieved a balance between high productivity and sustainability. J Integr Agric 17(4):747–754. https://doi.org/10.1016/S2095-3119(17)61789-1

Duran R, Cary JW, Calvo AM (2010) Role of the osmotic stress regulatory pathway in morphogenesis and secondary metabolism in filamentous fungi. Toxins 2(4):367–381. https://doi.org/10.3390/toxins2040367

Egidi E, Delgado-Baquerizo M, Plett JM et al (2019) A few Ascomycota taxa dominate soil fungal communities worldwide. Nat Commun 10(1):1–9. https://doi.org/10.1038/s41467-019-10373-z

El_Komy MH, Saleh AA, Eranthodi A, Molan YY (2015) The plant pathology journal characterization of novel *Trichoderma asperellum* isolates to select effective biocontrol agents against tomato *Fusarium* Wilt. Plant Pathol J 31(1):50–60. https://doi.org/10.5423/PPJ.OA.09.2014.0087

Etesami H (2018) Bacterial mediated alleviation of heavy metal stress and decreased accumulation of metals in plant tissues: mechanisms and future prospects. Ecotoxicol Environ Saf 147:175–191. https://doi.org/10.1016/j.ecoenv.2017.08.032

FAO (2007) Mapping biophysical factors that influence agricultural production and rural vulnerability. http://www.fao.org/tempref/docrep/fao/010/a1075e/a1075e00.pdf. Accessed 12 Nov 2019

Farooq M, Wahid A, Kobayashi N et al (2009) Plant drought stress: effects, mechanisms and management. In: Sustainable agriculture. Springer Netherlands, pp 153–188. https://doi.org/10.1007/978-90-481-2666-8_12

Feitosa YB, Cruz-Magalhães V, Argolo-Filho RC et al (2019) Characterization of genetic diversity on tropical *Trichoderma* germplasm by sequencing of rRNA internal transcribed spacers. BMC Res Notes 12(1):1–6. https://doi.org/10.1186/s13104-019-4694-1

Filippou P, Tanou G, Molassiotis A, Fotopoulos V (2013) Plant acclimation to environmental stress using priming agents. In: Plant acclimation to environmental stress. Springer, New York, pp 1–27. https://doi.org/10.1007/978-1-4614-5001-6_1

Foley JA, Ramankutty N, Brauman KA et al (2011) Solutions for a cultivated planet. Nature 478:337–342. https://doi.org/10.1038/nature10452

Ghori NH, Ghori T, Hayat MQ et al (2019) Heavy metal stress and responses in plants. Int J Environ Sci Technol 16(3):1807–1828. https://doi.org/10.1007/s13762-019-02215-8

Glare T, Caradus J, Gelernter W et al (2012) Have biopesticides come of age? Trends Biotechnol 30(5):250–258. https://doi.org/10.1016/J.TIBTECH.2012.01.003

Godfray HCJ, Beddington JR, Crute IR et al (2010) Food security: the challenge of feeding 9 billion people. Science 327(5967):812–818. https://doi.org/10.1126/science.1185383

Gomiero T (2018) Food quality assessment in organic vs. conventional agricultural produce: findings and issues. Appl Soil Ecol 123:714–728. https://doi.org/10.1016/j.apsoil.2017.10.014

Groppa MD, Tomaro ML, Benavides MP (2007) Polyamines and heavy metal stress: the antioxidant behavior of spermine in cadmium- and copper-treated wheat leaves. Biometals 20(2):185–195. https://doi.org/10.1007/s10534-006-9026-y

Gruber SG, Seidl-Seiboth V (2012) Self versus non-self: fungal cell wall degradation in *Trichoderma*. Microbiology 158(1):26–34. https://doi.org/10.1099/mic.0.052613-0

Hameed A, Dilfuza E, Abd-Allah EF et al (2014) Salinity stress and arbuscular mycorrhizal symbiosis in plants. In: Miransari M (ed) Use of microbes for the alleviation of soil stresses, vol 1. Springer, New York, pp 139–159. https://doi.org/10.1007/978-1-4614-9466-9_7

Hardoim PR, van Overbeek LS, Berg G et al (2015) The hidden world within plants: ecological and evolutionary considerations for defining functioning of microbial endophytes. Microbiol Mol Biol Rev 79(3):293–320. https://doi.org/10.1128/mmbr.00050-14

Harman G (2011a) *Trichoderma*-not just for biocontrol anymore. Phytoparasitica 39(2):103–108. https://doi.org/10.1007/s12600-011-0151-y

Harman GE (2011b) Multifunctional fungal plant symbionts: new tools to enhance plant growth and productivity. New Phytol 189(3):647–649. https://doi.org/10.1111/j.1469-8137.2010.03614.x

Harman GE, Uphoff N (2019) Symbiotic root-endophytic soil microbes improve crop productivity and provide environmental benefits. Scientifica. https://doi.org/10.1155/2019/9106395

Harman GE, Howell CR, Viterbo A et al (2004) *Trichoderma* species — opportunistic, avirulent plant symbionts. Nat Rev Microbiol 2(1):43–56. https://doi.org/10.1038/nrmicro797

Harman GE, Doni F, Khadka RB, Uphoff N (2019) Endophytic strains of *Trichoderma* increase plants' photosynthetic capability. J Appl Microbiol 130(2):529–546. https://doi.org/10.1111/jam.14368

Harzing AW (2007) Publish or Perish. Computer Program. https://harzing.com/resources/publish-or-perish

He B, Yun ZJ, Shi JB, Bin JG (2013) Research progress of heavy metal pollution in China: sources, analytical methods, status, and toxicity. Chin Sci Bull 58(2):134–140. https://doi.org/10.1007/s11434-012-5541-0

Heisey PW, Fuglie KO (2018) Agricultural research investment and policy reform in high-income countries. ERR-249, U.S. Department of Agriculture, Economic Research Service. doi:https://doi.org/10.22004/AG.ECON.276235

Hidangmayum A, Dwivedi P, Padmanabh Dwivedi C (2018) Plant responses to *Trichoderma* spp. and their tolerance to abiotic stresses: a review. J Pharmacogn Phytochem 7:758–766

Hu H, Jin Q, Kavan P (2014) A study of heavy metal pollution in China: current status, pollution-control policies and countermeasures. Sustainability 6(9):5820–5838. https://doi.org/10.3390/su6095820

Huang Y, Chen Q, Deng M et al (2018) Heavy metal pollution and health risk assessment of agricultural soils in a typical peri-urban area in Southeast China. J Environ Manag 207:159–168. https://doi.org/10.1016/j.jenvman.2017.10.072

Ikram M, Ali N, Jan G et al (2019) *Trichoderma reesei* improved the nutrition status of wheat crop under salt stress. J Plant Interact 14(1):590–602. https://doi.org/10.1080/17429145.2019.1684582

Jiao S, Lu Y (2020) Abundant fungi adapt to broader environmental gradients than rare fungi in agricultural fields. Glob Chang Biol 26(8):4506–4520. https://doi.org/10.1111/gcb.15130

Kappel L, Münsterkötter M, Sipos G et al (2020) Chitin and chitosan remodeling defines vegetative development and *Trichoderma* biocontrol. PLoS Pathog 16(2):e1008320. https://doi.org/10.1371/journal.ppat.1008320

Khan AL, Hussain J, Al-Harrasi A et al (2015) Endophytic fungi: resource for gibberellins and crop abiotic stress resistance. Crit Rev Biotechnol 35(1):62–74. https://doi.org/10.3109/07388551.2013.800018

Kitchenham B (2004) Procedures for performing systematic reviews. [Online] Available at: https://pdfs.semanticscholar.org/d36f/e9d7839596d58fa008121db57dc7cadda338.pdf

Kubicek CP, Herrera-Estrella A, Seidl-Seiboth V et al (2011) Comparative genome sequence analysis underscores mycoparasitism as the ancestral life style of *Trichoderma*. Genome Biol 12(4):1–15. https://doi.org/10.1186/gb-2011-12-4-r40

Kubicek CP, Steindorff AS, Chenthamara K et al (2019) Evolution and comparative genomics of the most common *Trichoderma* species. BMC Genomics 20(1):1–24. https://doi.org/10.1186/s12864-019-5680-7

Kumar A, Verma JP (2018) Does plant—microbe interaction confer stress tolerance in plants: a review? Microbiol Res 207:41–52. https://doi.org/10.1016/j.micres.2017.11.004

Kumar S, Thakur M, Rani A (2014) *Trichoderma*: mass production, formulation, quality control, delivery and its scope in commercialization in India for the management of plant diseases. Afr J Agric Res 9(53):3838–3852. https://doi.org/10.5897/AJAR2014

Lata R, Chowdhury S, Gond SK, White JF (2018) Induction of abiotic stress tolerance in plants by endophytic microbes. Lett Appl Microbiol 66(4):268–276. https://doi.org/10.1111/lam.12855

Lewis D (1985) Symbiosis and mutualism: crisp concepts and soggy semantics. In: Boucher DH (ed) The biology of mutualism. Croom-Helm, London, pp 29–39

Li X, Zhang X, Wang X et al (2019) Bioaugmentation-assisted phytoremediation of lead and salinity co-contaminated soil by *Suaeda salsa* and *Trichoderma asperellum*. Chemosphere 224:716–725. https://doi.org/10.1016/j.chemosphere.2019.02.184

Loguercio LL, de Carvalho AC, Niella GR et al (2009a) Selection of *Trichoderma stromaticum* isolates for efficient biological control of witches' broom disease in cacao. Biol Control 51(1):130–139. https://doi.org/10.1016/J.BIOCONTROL.2009.06.005

Loguercio LL, Santos LS, Niella GR et al (2009b) Canopy-microclimate effects on the antagonism between *Trichoderma stromaticum* and *Moniliophthora perniciosa* in shaded cacao. Plant Pathol 58(6):1104–1115. https://doi.org/10.1111/j.1365-3059.2009.02152.x

Lombardi N, Vitale S, Turrà D et al (2018) Root exudates of stressed plants stimulate and attract *Trichoderma* soil fungi. Mol Plant-Microbe Interact 31(10):982–994. https://doi.org/10.1094/MPMI-12-17-0310-R

López-Bucio J, Pelagio-Flores R, Herrera-Estrella A (2015) *Trichoderma* as biostimulant: exploiting the multilevel properties of a plant beneficial fungus. Sci Hortic 196:109–123. https://doi.org/10.1016/j.scienta.2015.08.043

Marasco R, Rolli E, Ettoumi B et al (2012) A drought resistance-promoting microbiome is selected by root system under desert farming. PLoS One 7(10):e48479. https://doi.org/10.1371/journal.pone.0048479

Mastouri F, Björkman T, Harman GE (2010) Seed treatment with *Trichoderma harzianum* alleviates biotic, abiotic, and physiological stresses in germinating seeds and seedlings. Phytopathology 100(11):1213–1221. https://doi.org/10.1094/PHYTO-03-10-0091

Meena M, Swapnil P (2019) Regulation of *wrky* genes in plant defence with beneficial fungus *Trichoderma* : current perspectives and future prospects. Arch Phytopathol Plant Prot 52(1–2):1–17. https://doi.org/10.1080/03235408.2019.1606490

Meena KK, Sorty AM, Bitla UM et al (2017) Abiotic stress responses and microbe-mediated mitigation in plants: the omics strategies. Front Plant Sci 8:172. https://doi.org/10.3389/fpls.2017.00172

Mehta CM, Sirari K (2019) A review on mycorrhiza-mycoparasite interactions in the rhizosphere. Plant Arch 19(2):508–511

Mendoza-Mendoza A, Zaid R, Lawry R et al (2018) Molecular dialogues between *Trichoderma* and roots: role of the fungal secretome. Fungal Biol Rev 32(2):62–85. https://doi.org/10.1016/j.fbr.2017.12.001

Meyer AJ, Dreyer A, Ugalde JM et al (2020) Shifting paradigms and novel players in Cys-based redox regulation and ROS signaling in plants - and where to go next. Biol Chem 402(3):399–423. https://doi.org/10.1515/hsz-2020-0291

Mittler R (2002) Oxidative stress, antioxidants and stress tolerance. Trends Plant Sci 7(9):405–410. https://doi.org/10.1016/S1360-1385(02)02312-9

Mota TM, Oshiquiri LH, Lopes ÉCV et al (2019) Hsp genes are differentially expressed during *Trichoderma asperellum* self-recognition, mycoparasitism and thermal stress. Microbiol Res 227:126296. https://doi.org/10.1016/j.micres.2019.126296

Mukherjee PK, Horwitz BA, Kenerley CM (2012) Secondary metabolism in *Trichoderma* - a genomic perspective. Microbiology 158(1):35–45. https://doi.org/10.1099/mic.0.053629-0

Mukherjee PK, Horwitz BA, Singh US et al (2013) *Trichoderma* in agriculture, industry and medicine: an overview. In: Mukherjee PK, Horwitz BA, Singh US et al (eds) *Trichoderma*: biology and applications. CABI, Wallingford, pp 1–9

Mukherjee I, Singh UK, Singh RP et al (2020) Characterization of heavy metal pollution in an anthropogenically and geologically influenced semi-arid region of East India and assessment of ecological and human health risks. Sci Total Environ 705:135801. https://doi.org/10.1016/j.scitotenv.2019.135801

Munns R, Gilliham M (2015) Salinity tolerance of crops - what is the cost? New Phytol 208(3):668–673. https://doi.org/10.1111/nph.13519

Nath K, Jajoo A, Poudyal RS et al (2013) Towards a critical understanding of the photosystem II repair mechanism and its regulation during stress conditions. FEBS Lett 587(21):3372–3381. https://doi.org/10.1016/j.febslet.2013.09.015

Naumann G, Alfieri L, Wyser K et al (2018) Global changes in drought conditions under different levels of warming. Geophys Res Lett 45(7):3285–3296. https://doi.org/10.1002/2017GL076521

Nuccio ML, Paul M, Bate NJ et al (2018) Where are the drought tolerant crops? An assessment of more than two decades of plant biotechnology effort in crop improvement. Plant Sci 273:110–119. https://doi.org/10.1016/j.plantsci.2018.01.020

Ogar E, Pecl G, Mustonen T (2020) Science must embrace traditional and indigenous knowledge to solve our biodiversity crisis. One Earth 3(2):162–165. https://doi.org/10.1016/j.oneear.2020.07.006

Osakabe Y, Osakabe K, Shinozaki K, Tran LSP (2014) Response of plants to water stress. Front Plant Sci 5:86. https://doi.org/10.3389/fpls.2014.00086

Pachauri S, Sherkhane PD, Mukherjee PK (2019) Secondary metabolism in *Trichoderma*: chemo- and geno-diversity. In: Satyanarayana T, Das S, Johri B (eds) Microbial diversity in ecosystem sustainability and biotechnological applications. Springer, Singapore, pp 441–456. https://doi.org/10.1007/978-981-13-8487-5_17

Parnell JJ, Berka R, Young HA et al (2016) From the lab to the farm: an industrial perspective of plant beneficial microorganisms. Front Plant Sci 7:1110. https://doi.org/10.3389/fpls.2016.01110

Paul D (2017) Research on heavy metal pollution of river ganga: a review. Ann Agrar Sci 15(2):278–286. https://doi.org/10.1016/j.aasci.2017.04.001

Rajput RS, Ram RM, Vaishnav A, Singh HB (2019) Microbe-based novel biostimulants for sustainable crop production. In: Microbial diversity in ecosystem sustainability and biotechnological applications. Springer, Singapore, pp 109–144

Rocha I, Ma Y, Souza-Alonso P et al (2019) Seed coating: a tool for delivering beneficial microbes to agricultural crops. Front Plant Sci 10:1357. https://doi.org/10.3389/fpls.2019.01357

Rosegrant MW, Ringler C, Zhu T (2009) Water for agriculture: maintaining food security under growing scarcity. Annu Rev Environ Resour 34:205–222. https://doi.org/10.1146/annurev.environ.030308.090351

Rousseau S, Deschacht N (2020) Public awareness of nature and the environment during the COVID-19 crisis. Environ Resour Econ 76(4):1149–1159. https://doi.org/10.1007/s10640-020-00445-w

Schenk PM, Carvalhais LC, Kazan K (2012) Unraveling plant-microbe interactions: can multi-species transcriptomics help? Trends Biotechnol 30(3):177–184. https://doi.org/10.1016/j.tibtech.2011.11.002

Schulz B, Boyle C, Draeger S et al (2002) Endophytic fungi: a source of novel biologically active secondary metabolites. Mycol Res 106(9):996–1004. https://doi.org/10.1017/S0953756202006342

Schuster A, Schmoll M (2010) Biology and biotechnology of *Trichoderma*. Appl Microbiol Biotechnol 87(3):787–799. https://doi.org/10.1007/s00253-010-2632-1

Selvakumar G, Panneerselvam P, Ganeshamurthy AN (2012) Bacterial mediated alleviation of abiotic stress in crops. In: Maheshwari D (ed) Bacteria in agrobiology: stress management. Springer-Verlag, Berlin/Heidelberg, pp 205–224. https://doi.org/10.1007/978-3-642-23465-1_10

Singh P, Singh J, Ray S et al (2020) Seed biopriming with antagonistic microbes and ascorbic acid induce resistance in tomato against *Fusarium* wilt. Microbiol Res 237:126482. https://doi.org/10.1016/j.micres.2020.126482

Sodango TH, Li X, Sha J, Bao Z (2018) Review of the spatial distribution, source and extent of heavy metal pollution of soil in China: impacts and mitigation approaches. J Heal Pollut 8(17):53–70. https://doi.org/10.5696/2156-9614-8.17.53

Stern N (2016) Economics: current climate models are grossly misleading. Nature 530(7591):407–409. https://doi.org/10.1038/530407a

Steyaert JM, Weld RJ, Loguercio LL, Stewart A (2010a) Rhythmic conidiation in the blue-light fungus *Trichoderma pleuroticola*. Fungal Biol 114(2–3):219–223. https://doi.org/10.1016/J.FUNBIO.2010.01.001

Steyaert JM, Weld RJ, Mendoza-Mendoza A, Stewart A (2010b) Reproduction without sex: conidiation in the filamentous fungus *Trichoderma*. Microbiology 156(10):2887–2900. https://doi.org/10.1099/mic.0.041715-0

Steyaert JM, Weld RJ, Stewart A (2010c) Ambient pH intrinsically influences *Trichoderma* conidiation and colony morphology. Fungal Biol 114(2–3):198–208. https://doi.org/10.1016/J.FUNBIO.2009.12.004

Szczałba M, Kopta T, Gąstoł M, Sękara A (2019) Comprehensive insight into arbuscular mycorrhizal fungi, *Trichoderma* spp. and plant multilevel interactions with emphasis on biostimulation of horticultural crops. J Appl Microbiol 127(3):630–647. https://doi.org/10.1111/jam.14247

Takahashi F, Kuromori T, Sato H, Shinozaki K (2018) Regulatory gene networks in drought stress responses and resistance in plants. In: Iwaya-Inoue M, Sakurai M, Uemura M (eds) Advances in experimental medicine and biology. Springer, New York, pp 189–214. https://doi.org/10.1007/978-981-13-1244-1_11

Tardieu F, Simonneau T, Muller B (2018) The physiological basis of drought tolerance in crop plants: a scenario-dependent probabilistic approach. Annu Rev Plant Biol 69:733–759. https://doi.org/10.1146/annurev-arplant-042817-040218

Topolovec-Pintarić S (2019) *Trichoderma*: invisible partner for visible impact on agriculture. In: *Trichoderma* - the most widely used fungicide. IntechOpen. https://doi.org/10.5772/intechopen.83363

Trisos CH, Merow C, Pigot AL (2020) The projected timing of abrupt ecological disruption from climate change. Nature 580(7804):496–501. https://doi.org/10.1038/s41586-020-2189-9

Vinale F, Sivasithamparam K, Ghisalberti EL et al (2008) *Trichoderma*-plant-pathogen interactions. Soil Biol Biochem 40(1):1–10. https://doi.org/10.1016/j.soilbio.2007.07.002

Waghunde RR, Shelake RM, Sabalpara AN (2016) *Trichoderma*: a significant fungus for agriculture and environment. Afr J Agric Res 11(22):1952–1965. https://doi.org/10.5897/AJAR2015.10584

Woo SL, Ruocco M, Vinale F et al (2014) *Trichoderma*-based products and their widespread use in agriculture. Open Mycol J 8(1):71–126. https://doi.org/10.2174/1874437001408010071

Xiang W, Zhao L, Xu X et al (2012) Mutual information flow between beneficial microorganisms and the roots of host plants determined the bio-functions of biofertilizers. Am J Plant Sci 3(8):1115–1120. https://doi.org/10.4236/ajps.2012.38134

Yan L, Zhu J, Zhao X et al (2019) Beneficial effects of endophytic fungi colonization on plants. Appl Microbiol Biotechnol 103(8):3327–3340. https://doi.org/10.1007/s00253-019-09713-2

Yang Y, Guo Y (2018) Elucidating the molecular mechanisms mediating plant salt-stress responses. New Phytol 217(2):523–539. https://doi.org/10.1111/nph.14920

Yang Q, Li Z, Lu X et al (2018) A review of soil heavy metal pollution from industrial and agricultural regions in China: pollution and risk assessment. Sci Total Environ 642:690–700. https://doi.org/10.1016/j.scitotenv.2018.06.068

Zari M (2014) Ecosystem services analysis in response to biodiversity loss caused by the built environment. SAPIENS 7:1–14

Zeilinger S, Gruber S, Bansal R, Mukherjee PK (2016) Secondary metabolism in *Trichoderma* - chemistry meets genomics. Fungal Biol Rev 30(2):74–90. https://doi.org/10.1016/j.fbr.2016.05.001

Trichoderma Genes for Abiotic Stress Tolerance in Plants

Nurhidayu Al-Saari, Nur Sabrina Ahmad Azmi, and Nurul Hidayah Samsulrizal

Contents

1 Introduction

The food demand is expected to increase globally by 2050 due to the increase of the worldwide population from 7.7 billion to 9.7 billion in 2050. However, there are several limiting factors that affect crop yield production globally such as climate change, the occurrence of pests and diseases, limited soil availability, and many more. Agricultural production is heavily dependent on rainfall frequency, temperature, atmospheric carbon dioxide content, and other devastating incidents such as typhoons, drought, heavy metal contamination, flooding, and other extreme events. These abiotic stresses not only cause a great reduction in producing sustainable crop yields but also influence the distribution and behavioral patterns of biotic stresses.

N. Al-Saari
International Institute for Halal Research and Training, International Islamic University Malaysia, Kuala Lumpur, Malaysia

N. S. A. Azmi (✉) · N. H. Samsulrizal
Department of Plant Science, Kulliyyah of Science, International Islamic University Malaysia, Pahang, Malaysia
e-mail: sabrinaazmi@iium.edu.my

N. Amaresan et al. (eds.), *Advances in* Trichoderma *Biology for Agricultural Applications*, Fungal Biology, https://doi.org/10.1007/978-3-030-91650-3_13

Therefore, to be able to supply enough food production, we need to come up with a sustainable and eco-friendly solution that can aid in plant performance under current environmental patterns.

Species from the genus *Trichoderma* are ubiquitous soil fungi that are recognized for their role as a biocontrol agent against plant pathogens. Besides that, *Trichoderma* is well known for their roles in improving plant growth and development. The accessibility of *Trichoderma* spp. genomes provides us with insights into the identification and characterization of useful *Trichoderma* genes in alleviating abiotic stress. Briefly, this chapter will describe the current knowledge of the *Trichoderma* gene's ability in mitigating abiotic stress in plants.

2 Heat Shock Proteins

Many biotic/abiotic stresses including extreme temperatures trigger changes in transcription and translation machinery of all organisms to activate the synthesis of protein groups called "heat shock proteins" (HSPs), "stress-induced proteins," or "stress proteins" (Lindquist and Craig 1988). HSPs that act as molecular chaperones are also crucial in protein folding homeostasis, preventing stress-induced aggregation of partially denatured proteins, and later assist them to restore their native three-dimensional conformations when conducive environments are reinstated (Parsell and Lindquist 1993; Sitia and Braakman 2003; Huttner and Strasser 2012). Due to their ability to aggregate upon heat-induced denaturation and overexpression, many chaperons are often called HSPs (Ansari and Mande 2018). HSPs are categorized based on their molecular weights, which vary from 10 to more than 100 kDa in molecular size and are found ubiquitously in different cellular compartments. There are currently five main HSP families in animals and plants, namely, HSP100, HSP90, HSP70, HSP60, and small HSP (sHSPs) (Sarkar et al., 2009; Waters 2013).

Previous research has shown that HSPs can impart thermotolerance in a variety of species (Sung and Guy 2003; Montero-Barrientos et al. 2007), and their synthesis has been extensively studied in yeasts (Sanchez and Lindquist 1990), filamentous fungi (Stephanou and Demopoulos 1986; Rezaie et al. 2000), plants (Vierling 1991; Parsell and Lindquist 1993; Li et al. 2009; Nekrasov et al. 2009; Liu and Howell 2010), and animals (Sun and MacRae 2005). Even though HSPs have been studied for more than a decade, little is known about this protein class in *Trichoderma* species. At present, 63 *Trichoderma* genome sequences are available in public databases (https://www.ncbi.nlm.nih.gov/genome/?term=trichoderma) with only twelve genomes annotated (Kubicek et al. 2019). Based on published literature and experimental findings, the following section aims to provide up-to-date information on the mode of responses of intracellular HSPs of *Trichoderma* species to extracellular stresses. Furthermore, molecular characters and possible function of HSPs in the living body with specific references to *Trichoderma*-plant interaction will be explored.

TrichoEST project (Rey et al. 2004), a functional genomics study conducted on eight *Trichoderma* spp., is still a good reference in understanding the HSPs genes in *Trichoderma*. The researchers discovered several ESTs with close identity to *hsp23* in *T. virens* T59 and *hsp70* in *T. harzianum* T34 (Rey et al. 2004). Separate subsequent studies reported the cloning, characterization, and expression of *hsp23* (Montero-Barrientos et al. 2007) and *hsp70* (Montero-Barrientos et al. 2008) in the biocontrol model strain of *T. harzianum* T34 upon thermal shock assays. Both HSP23 and HSP70 families have heavily conserved regions that can be used for their identification at a molecular level. The 214 amino acids of *hsp23* in *T. virens* T59 consists of 59 amino acids of an α-crystallin domain that are highly conserved in sHSPs. Similar sizes domains were also observed in sHSPs of other filamentous fungi and located along a highly conserved C-terminal extension preceded by a poorly conserved N-terminal region when they aligned together (Montero-Barrientos et al. 2007). The *hsp70* of *T. harzianum* T34 has a domain that contains a 44 kDa N-terminal ATP-binding region and a 25–30 kDa C-terminal substrate-binding region that is described as signatures to the HSP70 family. The end of the C-terminal tail of T34-*hsp70* also includes the highly conserved EEVD terminal sequence, which is considered as a signature feature of cytosolic HSP70 proteins in all organisms (Montero-Barrientos et al. 2008).

HSPs are reported to be induced by exposure of cells to thermal and other abiotic stress conditions. In their study, Montero-Barrientos et al. (2007) have observed an increased transcript level of the *hsp23* gene in response to thermal shock (4, 10 and 41 °C), oxidative conditions with exposure to 10% ethanol and 1.2 mM paraquat, and osmotic stress conditions with 10% NaCl in a growth medium. In contrast, the *hsp70*gene of *T. harzianum* T34 is not induced by low temperature (Montero-Barrientos et al. 2008). An attempt to grow the fungus at six different temperatures discovered a rise in the transcript level of *hsp70* in response to heat shock at 37 °C and 41 °C, oxidative condition (3 mM hydrogen peroxide and 10% ethanol), and osmotic stress at 1 M mannitol (Montero-Barrientos et al. 2008). The family of HSPs are generally acknowledged to play an important role in cross-tolerance to environmental perturbations (Dubeau et al. 1998; Todgham et al. 2005). Following an overexpression technique in *T. harzianum* T34, many researches were performed to elucidate the possible action of both *hsp23* and *hsp70*. It is found that overexpression of *hsp23* and *hsp70* is not only capable to confer thermotolerance to *T. harzianum* T34 but also increases their tolerance to other abiotic stresses (Montero-Barrientos et al. 2007, 2008).

Trichoderma spp. as biostimulants is capable of promoting plant growth. Strawberry (*Fragaria x ananassa*) plants treated with three selected *Trichoderma* strains (*T. afroharzianum* T22 and TH1 and *T. virens* GV41) have been shown to successfully promote plant growth, improved fruit production and preferred anthocyanin, and other antioxidant accumulation in red ripened fruits (Lombardi et al. 2020). Proteomic analysis of fruits harvested from the treated plants demonstrated that the microbial inoculants had a significant impact on the representation of proteins involved in responses toward stress or external stimuli and other physiological processes. Such proteins include the HSP70 isoforms, HSP91, HSP20, and various chaperons (Lombardi et al. 2020). In another study, *hsp70* genes of *T. harzianum* are genetically engineered to *Arabidopsis* for thermotolerance and increased resistance to salt, osmotic, and oxidative stresses (Montero-Barrientos et al. 2010). Despite no alteration of *Arabidopsis* phenotype was observed, an in vivo assay confirms tolerance to heat and the presence of cross-talk between different stress-response pathways in the plant. Their findings also indicated that the fungal HSP70 protein functions as a negative regulator of the heat shock factor (HSF) transcriptional activity. Thus, it prevents the synthesis of new HSPs and their accumulation in the transgenic plant causing thermotolerance. However, the upregulation of stress marker genes involved in salt and oxidative stress responses found in transgenic lines following heat stress suggested that proteins other than HSF could also involve in the regulation of these genes (Montero-Barrientos et al. 2010).

The roles of HSPs in fungal biology are variable, and the expression of these proteins can occur both in response to stress and during basal metabolism. Studies demonstrated that HSPs are involved in morphogenic change, adaptation to stress, and antifungal resistance to *Trichoderma* (Tereshina 2005; Lamoth et al. 2015; Mota et al. 2019). Advancement in sequencing technology, genetic, and proteomic research has led to the isolation and the study of HSPs from many organisms. However, information is still lacking for *Trichoderma* species. Considering the importance of this class of proteins for cellular homeostasis, more data mining of the available *Trichoderma* genome, and/or overexpression, studies will provide insights and a better understanding of their roles in *Trichoderma* growth and interaction with biotic/abiotic stresses.

3 Glucosidase with Kelch-Repeat Domains

T. harzianum kel1 gene carries 338 amino acids and encodes for a putative Kelch-repeat domain protein. It shares similar homology to *Arabidopsis* epithiospecifier proteins (ESP) (Hermosa et al. 2011) and can be found in many organisms, especially in eukaryotes. In *Arabidopsis*, ESP has been identified to be involved in glucosinolate hydrolysis through the formation of nitriles and epithionitriles (Wittstock and Burow 2007). Thus, it is speculated that the *kel1* gene may also involve in glucosinolate hydrolysis. A study by Hermosa et al. (2011) found that deletion of the *kel1* gene mutants in *Trichoderma* showed a significantly lower glucosidase activity compared to the wild type of *T. harzianum*T34 strain. This suggests the involvement of *kel1* in the increased production of glucosidase activity in salt and osmotic conditions. On the other hand, they also found higher germination percentages with significantly lower abscisic acid (ABA) levels in the *kel1* overexpressing plants in osmotic and salt conditions. The lower levels of ABA may contribute to the plant's ability to germinate and develop cotyledons (Gonzalez-Guzman et al. 2002). Therefore, these findings suggest that *kel1* may be able to contribute toward the higher resistance toward salt and osmotic stress in plants.

4 Aquaglyceroporin Gene

The main intrinsic protein (MIPs) is a membrane channel family found in mammals, plants, insects, fungi, and bacteria that is required for osmotic cell stabilization. Aquaporins (AQPs), glycerol facilitators (GlpFs), and aquaglyceroporins are the three main subgroups of MIPs (Froger et al. 2001). Aquaglyceroporins seem to be of special importance in the study of the molecular basis for both water and solutes in identifying whether mixed channels have a distinct molecular mechanism (Ben Amira et al. 2018; Bienert et al. 2008; Maurel et al. 2009, 2016; Tanaka et al. 2008). A gene encoding an aquaglyceroporin (*aqp*) in *T. harzianum* has been previously shown to be upregulated during biocontrol of the plant pathogen *Fusarium solani* (Vieira et al. 2013, 2017).

T. harzianum aqp gene has been shown to play a role in controlling physiological functions and responses in periods of water stress (Ben Amira et al. 2018; Vieira et al. 2013). Overexpression of *T. harzianum aqp* gene in tobacco (*Nicotiana tabacum*) showed that it has a lot of promise to be developed as drought-resistant transgenic plants (Vieira et al. 2017, 2018). Besides, overexpressed *aqp* in *Phaseolus vulgaris* (French bean) has demonstrated excellent growth-promoting properties. Plants that came through the interaction with the *T. harzianum aqp*-overexpressing transformant often had larger leaves and a higher dry weight by showing increased root and shoot volumes, as well as water quality and drought resistance (Brandão et al. 2019). Therefore, genetic modification of aquaglyceroporin from the *Trichoderma* genome can improve plant performance for agricultural applications such as the development of stress-tolerant plants and application in plant molecular breeding.

5 Glutathione S-Transferase Gene

Glutathione S-transferase (GST) is a large enzyme superfamily that is known for their function in detoxification by eliminating membrane lipid peroxides through the glutathione conjugation, thus protecting the plants from an oxidative burst. Several studies have suggested the role of glutathione S-transferase in protecting the plants by alleviating several abiotic stresses such as heavy metal stress (Zhang et al. 2013), radiation, and ultraviolet damage (Liu and Li 2002). *T. virens GST* gene consists of 252 amino acids and it contains N-terminal thioredoxin-fold domain that plays a role in protecting the cells against oxidant and heavy metal ion toxicity (Dixit et al. 2011a, b). Plant exposure to heavy metals such as cadmium (Cd) can lead to the accumulation of reactive oxygen species (ROS) that can inhibit DNA and protein synthesis. Therefore, excess ROS in plants needs to be scavenged to protect them from any damage induced by ROS. Tobacco (*Nicotiana tabacum*)-expressing *GST* showed better plant growth with enhanced tolerance toward Cd without increasing the Cd accumulation in the plants when exposed to different concentrations of Cd (Dixit et al. 2011b). Thus, the development of the transgenic tobacco expressing *T. virens GST* not only can help plants to tolerate Cd but also can help in limiting Cd accumulation in plants when grown in Cd-contaminated soil.

Plants can metabolize xenobiotic pollutants such as polycyclic aromatic hydrocarbons. However, plants lack complete metabolic pathways compared to bacteria and fungi. Several fungi have been reported to be able to degrade polycyclic aromatic hydrocarbons (Luch 2005) in which *T. virens* is one of them. A recent study showed that transgenic tobacco-expressing *T. virens GST* is able to degrade anthracene to naphthalene derivatives (Dixit et al. 2011a). This demonstrates a promising potential in the enhancement of the anthracene tolerance in plants. Therefore, the GST gene from *T. virens* showed a potential application in developing transgenic plants that have the capability to tolerate heavy metal and PAH stress.

6 Conclusion

The availability of *Trichoderma* spp. genomes has provided us with a great opportunity to identify and explore the functions of their genes in mitigating abiotic stresses. It has been proven that *Trichoderma* spp. consist of many useful genes that can help them to adapt to different and harsh environmental conditions. Therefore, current knowledge in genomic and molecular can help us in developing transgenic plants expressing *Trichoderma* genes, which can be a sustainable option in producing plants that are able to acclimatize in adverse conditions.

References

Ansari MY, Mande SC (2018) A glimpse into the structure and function of atypical type I chaperonins. Front Mole Biosci 5:1–8

Ben Amira M, Mom R, Lopez D, Chaar H, Khouaja A, Pujade-Renaud V, Fumanal B, Gousset-Dupont A, Bronner G, Label P, Julien JL (2018) MIP diversity from *Trichoderma*: structural considerations and transcriptional modulation during mycoparasitic association with *Fusarium solani* olive trees. PLoS One 13(3):e0193760

Bienert GP, Thorsen M, Schüssler MD, Nilsson HR, Wagner A, Tamás MJ, Jahn TP (2008) A subgroup of plant aquaporins facilitate the bi-directional diffusion of As (OH) 3 and Sb (OH) 3 across membranes. BMC Biol 6(1):1–15

Brandão RS, Qualhato TF, Valdisser PAMR, de CB Côrtes MV, Vieira PM, Silva RN, Junior ML, Ulhoa CJ (2019) Evaluation of *Trichoderma harzianum* mutant lines in the resistance induction against white mold and growth promotion of common bean. BioRxiv:713776

Dixit P, Mukherjee PK, Sherkhane PD, Kale SP, Eapen S (2011a) Enhanced tolerance and remediation of anthracene by transgenic tobacco plants expressing a fungal glutathione transferase gene. J Hazard Mater 192(1):270–276

Dixit P, Mukherjee PK, Ramachandran V, Eapen S (2011b) Glutathione transferase from *Trichoderma virens* enhances cadmium tolerance without enhancing its accumulation in transgenic *Nicotiana tabacum*. PLoS One 6(1):e16360

Dubeau SF, Pan F, Tremblay GC, Bradley TM (1998) Thermal shock of salmonin vivo induces that heat shock protein hsp70 and confers protection against osmotic shock. Aquaculture 11:311–323

Froger A, Rolland JP, Bron P, Lagrée V, Le Cahérec F, Deschamps S, Hubert JF, Pellerin I, Thomas D, Delamarche C (2001) Functional characterization of a microbial aquaglyceroporin. Microbiology 147(5):1129–1135

Gonzalez-Guzman M, Apostolova N, Belles JM, Barrero JM, Piqueras P, Ponce MR, Micol JL, Serrano R, Rodriguez PL (2002) The short-chain alcohol dehydrogenase ABA2 catalyzes the conversion of xanthoxin to abscisic aldehyde. Plant Cell 14:1833–1846

Hermosa R, Botella L, Keck E, Jiménez JA, Montero-Barrientos M, Arbona V, Gomez-Cadenas A, Monte E, Nicolas C (2011) The overexpression in *Arabidopsis thaliana* of a *Trichoderma harzianum* gene that modulates glucosidase activity and enhances tolerance to salt and osmotic stresses. J Plant Physiol 168(11):1295–1302

Huttner S, Strasser R (2012) Endoplasmic reticulum-associated degradation of glycoproteins in plants. Front Plant Sci 3(67)

Kubicek CP, Steindorff AS, Chenthamara K, Manganiello G, Henrissat B, Zhang J, Cai F, Kopchinskiy AG, Kubicek EM, Kuo A, Baroncelli R, Sarrocco S, Noronha EF, Vannacci G, Shen Q, Grigoriev IV, Druzhinina IS (2019) Evolution and comparative genomics of the most common *Trichoderma* species. BMC Genomics 20(1):485. 1–485.24

Lamoth F, Juvvadi PR, Steinbach WJ (2015) Histone deacetylase inhibition as an alternative strategy against invasive aspergillosis. Front Microbiol 16(6):96

Li J, Zhao-Hui C, Batoux M, Nekrasov V, Roux M, Chinchilla D, Zipfel C, Jones JD (2009) Specific ER quality control components required for biogenesis of the plant innate immune receptor EFR. Proc Natl Acad Sci U S A 106:15973–15978

Lindquist S, Craig EA (1988) The heat-shock proteins. Annu Rev Genetic 22:631–677

Liu JX, Howell SH (2010) Endoplasmic reticulum protein quality control and its relationship to environmental stress responses in plants. Plant Cell 22:2930–2942

Liu XF, Li JY (2002) Characterization of an ultra-violet inducible gene that encodes glutathione S-transferase in *Arabidopsis thaliana*. Acta Genet Sin 29(5):458–460

Lombardi N, Caira S, Troise AD, Scaloni A, Vitaglione P, Vinale F et al (2020) *Trichoderma* applications on strawberry plants modulate the physiological processes positively affecting fruit production and quality. Front Microbiol 11:1364

Luch A (2005) The carcinogenic effects of polycyclic aromatic hydrocarbons. Imperial College Press, London. ISBN 1-86094-417-5

Maurel C, Santoni V, Luu DT, Wudick MM, Verdoucq L (2009) The cellular dynamics of plant aquaporin expression and functions. Curr Opin Plant Biol 12(6):690–698

Maurel C, Verdoucq L, Rodrigues O (2016) Aquaporins and plant transpiration. Plant Cell Environ 39(11):2580–2587

Montero-Barrientos M, Cardoza RE, Gutiérrez S, Monte E, Hermosa R (2007) The heterologous overexpression of hsp23, a small heat-shock protein gene from *Trichoderma virens*, confers thermotolerance to *T. harzianum*. Curr Genet 52(1):45–53

Montero-Barrientos M, Hermosa R, Nicolás C, Cardoza RE, Gutiérrez S, Monte E (2008) Overexpression of a *Trichoderma* HSP70 gene increases fungal resistance to heat and other abiotic stresses. Fungal Genet Biol 45(11):1506–1513

Montero-Barrientos M, Hermosa MR, Cardoza RE, Gutiérrez S, Nicolás C, Monte E (2010) Transgenic expression of the *Trichoderma harzianum* hsp70 gene increases *Arabidopsis* resistance to heat and other abiotic stresses. J Plant Physiol 167:659–665

Mota TM, Oshiquiri LH, Lopes ÉCV, Barbosa Filho JR, Ulhoa CJ, Georg RC (2019) Hsp genes are differentially expressed during *Trichoderma asperellum* self-recognition, mycoparasitism and thermal stress. Microbiol Res 227:126–296

Nekrasov V, Li J, Batoux M, Roux M, Chu ZH, Lacombe S, Rougon A, Bittel P, Kiss-Papp M, Chinchilla D, van Esse HP, Jorda L, Schwessinger B, Nicaise V, Thomma BP, Molina A, Jones JD, Zipfel C (2009) Control of the pattern-recognition receptor EFR by an ER protein complex in plant immunity. EMBO J 28(211):3428–3438

Parsell DA, Lindquist S (1993) The function of heat-shock proteins in stress tolerance: degradation and reactivation of damaged proteins. Annu Rev Genet 27:437–496

Rezaie S, Ban J, Mildner M, Poitschek C, Brna C, Tschachler E (2000) Characterization of a cDNA clone encoding a 70 kDA heat-shock protein from the dermatophyte pathogen Trichophyton rubrum. Gene 241:27–33

Rey M, Llobell A, Monte E, Scala F, Lorito M (2004) Genomics of Trichoderma in Fungal Genomics, Vol. 4, Elsevier Science, Amsterdam, 225–248

Sanchez Y, Lindquist S (1990) HSP104 required for induced thermotolerance. Science 248(4959):1112–1115

Sarkar NK, Kim YK, Grover A (2009) Rice sHSPs: genomic organization and expression profiling under stress and development. BMC Genomics 10:393. https://doi.org/10.1186/1471-2164-10-393

Sitia R, Braakman I (2003) Quality control in the endoplasmic reticulum protein factory. Nature 426:891–894

Stephanou G, Demopoulos NA (1986) Heat shock phenomena in *Aspergillus nidulans*. The effect of heat on mycelial protein synthesis. Curr Genet 10:791–796

Sun Y, MacRae TH (2005) The small heat shock proteins and their role in human disease. FEBS J 272(11):2613–2627

Sung DY, Guy CL (2003) Physiological and molecular assessment of altered expression of hsc70-1 in *Arabidopsis*. Evidence for pleiotropic consequences. Plant Physiol 132:979–987

Tanaka M, Wallace IS, Takano J, Roberts DM, Fujiwara T (2008) NIP6; 1 is a boric acid channel for preferential transport of boron to growing shoot tissues in *Arabidopsis*. Plant Cell 20(10):2860–2875

Tereshina VM (2005) Thermotolerance in fungi: the role of heat shock proteins and trehalose. Microbiology 74:247–257

Todgham AE, Schulte PM, Iwama GK (2005) Cross-tolerance in the tidepool sculpin: the role of heat shock proteins. Physiol Biochem Zool 78(2):133–144

Vieira PM, Coelho ASG, Steindorff AS, de Siqueira SJL, do Nascimento Silva, R. and Ulhoa, C.J. (2013) Identification of differentially expressed genes from *Trichoderma harzianum* during growth on cell wall of *Fusarium solani* as a tool for biotechnological application. BMC Genomics 14(1):1–11

Vieira PM, Santos MP, Andrade CM, Souza-Neto OA, Ulhoa CJ, Aragão FJL (2017) Overexpression of an aquaglyceroporin gene from *Trichoderma harzianum* improves water-use efficiency and drought tolerance in *Nicotiana tabacum*. Plant Physiol Biochem 121:38–47

Vieira PM, Zeilinger S, Brandao RS, Vianna GR, Georg RC, Gruber S, Aragao FJL, Ulhoa CJ (2018) Overexpression of an aquaglyceroporin gene in the fungal biocontrol agent *Trichoderma harzianum* affects stress tolerance, pathogen antagonism and *Phaseolus vulgaris* development. Biol Control 126:185–191

Vierling E (1991) The roles of heat shock proteins in plants. Annu Rev Plant Physiology Plant Mol Biol 42:579–620

Waters ER (2013) The evolution, structure, and expression of the plant sHSPs. J Exp Bot 64:391–403

Wittstock U. And Burow M. 2007. Tipping the scales - specifier proteins in glucosinolate hydrolysis. IUBMB Life, 598(1), pp. 744–751

Zhang Y, Liu J, Zhou Y, Gong T, Wang J, Ge Y (2013) Enhanced phytoremediation of mixed heavy metal (mercury)–organic pollutants (trichloroethylene) with transgenic alfalfa co-expressing glutathione S-transferase and human. J Hazard Mater 260:1100–1107

Part IV
Practical Aspects of *Trichoderma* Commercialization in Agriculture

Development, Production, and Storage of *Trichoderma* Formulations for Agricultural Applications

Ravulapalli Durga Prasad, Kella S. V. Poorna Chandrika, Suseelendra Desai, Kothur Greeshma, and Sriramappa Vijaykumar

Contents

*Many of the *Trichoderma* species names reported from research work carried out before 2005 in the chapter are not characterized based on DNA barcode (Druzhinina et al. 2005, Fungal Genetics and Biology 42 (10), 813-828). Species reported as *T. viride* from Indian literature are mostly *T. asperellum* as per DNA barcoding (Sriram et al., 2013, Current Science 104 (10), 1332-1340).

R. D. Prasad (✉) · K. S. V. Poorna Chandrika · K. Greeshma · S. Vijaykumar
ICAR-Indian Institute of Oilseeds Research, Hyderabad, India

S. Desai
ICAR-Central Research Institute for Dryland Agriculture, Hyderabad, India

N. Amaresan et al. (eds.), *Advances in* Trichoderma *Biology for Agricultural Applications*, Fungal Biology, https://doi.org/10.1007/978-3-030-91650-3_14

1 Production

The major concern in commercial production of biocontrol systems is to obtain adequate growth and sporulation of the biocontrol agent. The biomass production of the antagonist is not easy owing to the specific requirement of nutritional and environmental conditions for the growth of an organism. *Trichoderma* biomass must be produced in a cost-effective way, and it should be viable at each processing step, such as harvesting, drying, formulation, storage, and delivery (Ramanujam et al. 2010). Much work has been done on the production of *Trichoderma* spp. by liquid- and solid-state fermentation methods. Production of these antagonists can be done easily using cheaper substrates.

1.1 Solid Fermentation

In solid-state fermentation (SSF), microorganisms are grown on solid materials with optimum moisture content. In this type of fermentative process, the quantity of water should not exceed the saturation capacity of the solid bed in which the organism is being cultivated. Water is essential for the growth and sporulation of microorganisms. In SSF, water present is in thin layers and sometimes gets absorbed into the substrate (Kumar et al. 2014). Solid fermentation involves interactions of microbial biomass with the wetted solid substrate and the microorganism can grow on and within the substrate. Microbial biomass within the matrix consumes the substrate and secretes metabolites and enzymes (Padmasari 2005). In general, no prior sophisticated formulation procedures are used for the product formed from solid- or semisolid-state fermentation. Based on the solid phase used, there are two types of SSF systems. The first system involves cultivation on natural substrates such as agricultural by-products. The second system involves cultivation on an inert support impregnated with a liquid medium (Kumar et al. 2014). SSF has gained a lot of interest because of its product recovery and reduced energy requirements.

Several researchers have optimized growth requirements for the production of *Trichoderma* on agricultural by-products and wastes. Several media have been used including bran of cereal crops, different agro-waste materials, and industrial by-products for the mass multiplication of *Trichoderma* species (Sangeetha et al. 1993; Zaidi and Singh 2004; Singh and Joshi 2007; Bhagat and Pan 2007; Sangle et al. 2002; Saju et al. 2002; Tiwari et al. 2004; Gangadharan and Jeyarajan 1990; Rini and Sulochana 2007; Gupta et al. 2016; Guzmán et al. 2014; Ahuja and Bhatt 2018; Naeimi et al. 2020; De la Cruz-Quiroz et al. 2017). Among the by-products, wheat bran supported good growth of biomass and suitable for mass multiplication of *T.* cf. *harzianum* (Heins et al. 1978; Martin et al. 1984). Similarly, the combination of wheat bran and sawdust was used for *Trichoderma* mass multiplication (Elad et al. 1980; Mukhopadhyay et al. 1986). A mixture of wheat bran and maize bran was also found to be a good medium for *Trichoderma* (Kapoor and Kumar 2004).

Growth medium comprising of pulse bran with sawdust supported high biomass and spores of *Trichoderma* compared to wheat bran (Dubey and Patel 2002). Corn in a bag bioreactor (Lewis and Papavizas 1980; De la Cruz-Quiroz et al. 2017), barley grains (Abd-El Moity and Shatala 1981), and sorghum (Padmanabhan and Alexander 1984; Upadhyay and Mukhopadhyay 1986) have also been used as substrate for *Trichoderma* production.

Among the agro-waste materials, sugarcane bagasse, corn cobs, rice straw, and groundnut shell were found to be good media. Dubey and Patel (2002) used wheat straw, groundnut shells, and mushroom bedstraw for mass multiplication of *Trichoderma*. *Trichoderma* or *Gliocladium* were grown on peat-bran substrate to yield between 5×10^7 and 3×10^{10} colony-forming units (cfu) g^{-1} substrate after 14 days of growth (Maplestone et al. 1991). Thangavelu et al. (2004) tested five different organic substrates, viz., rice bran, rice chaffy grain, farm yard manure, banana pseudostem, and banana leaf, for mass multiplication of *T.* cf. *harzianum*. Singh and Singh (2007) found that tea waste served as the best substrate for mass multiplication of the bioagents and the mass-multiplied cultures could be stored for 3 months without much reduction in the population of the biocontrol agents. Clay saturated with 10% molasses was found to support maximum spore production of *Trichoderma* (Blackman and Kabana 1975).

1.2 Liquid Fermentation

The common difficulties faced in the SSF are high volume of substrate, risk of contamination, large space for processing, sterilization, inoculation, incubation, and storage. To overcome these issues, liquid-state fermentation (LSF) was developed through which large quantities of biomass can be produced within a few days under axenic conditions. Hence, the industry has widely adapted to this method of mass multiplication. The LSF term is applied for the processes in which water-soluble materials are used for the microbial growth. Water is essential for microbial growth. In liquid fermentation, water is present in thick layers and often gets absorbed by the substrates.

In many countries, deep tank fermentation using inexpensive media is being followed. Deep tank fermentation system was employed in liquid formulation which makes it as a more preferred approach for biomass production in Europe and North America (Churchill 1982). Inexpensive growth media such as molasses and brewer's yeast are used for production in liquid formulation (Papavizas et al. 1984; Sankar and Jeyarajan 1996). Similarly, liquid media, viz., molasses-soy powder and molasses and jaggery, were found to support good growth of *Trichoderma* (Prasad and Rangeshwaran 2000a; Prasad et al. 2002a, b). Molasses-soy medium (MSM) standardized for mass production of *Trichoderma* by liquid fermentation yielded maximum biomass, viable propagules, and spores as compared to standard molasses-yeast medium (MYM). The MSM medium could serve as a better alternative to MYM for the commercial production of *Trichoderma* species (Prasad and

Rangeshwaran 2000b). However, as the availability of molasses is limited to areas near to sugar factories, attempts were made to utilize jaggery as an alternative. Jaggery-soy medium gave more biomass and viable propagules compared to MYM but it was not superior to MSM (Prasad et al. 2002a, b). Potato dextrose broth, glucose nitrate broth, maltose peptone broth, sabouraud dextrose broth, and molasses-yeast extract broth were tested for mass production of *T. viride* (*T.* cf. *viride*). Among them, molasses-yeast extract broth was found to be suitable for the maximum growth of this fungus (Khan et al. 2011).

Optimized fermentation conditions could result in maximum biomass of *Trichoderma* spp. in short-time by using appropriate medium in a fermenter with aeration, agitation, temperature, pH, and antifoam controls than in shake-flask cultures and such technology is industry-friendly for mass production of *Trichoderma* spp. Studies have revealed that maximum amount of biomass and viable propagules of *T. harzianum* (*T.* cf. *harzianum*) can be obtained within 96 h of fermentation in a fermenter with aeration, agitation, and temperature controls (Prasad et al. 1997).

1.3 Biphasic Production of Trichoderma

A biphasic production system with initial liquid-state fermentation for 4 days in any suitable medium followed by tray culturing of the biomass for 3–4 days is being followed at ICAR-Indian Institute of Oilseeds Research, Hyderabad, India, to obtain maximum conidial biomass. The biomass is then air-dried for different durations to impart desiccation tolerance to conidia (Indian Patent No patent No: 316651 dated 23.07.2019).

2 Influence of Cultural Conditions on *Trichoderma* Growth

An important factor that can favor the processes of sporulation in *Trichoderma* is C/N ratio. Serna-Díaz et al. (2020) reported that barley straw with C/N ratios of 160:1 improved spore production of *T. viride* (*T.* cf. *viride*). A combination of sucrose at 30,000 ppm as carbon source and ammonium nitrate at 3000 ppm as nitrogen source significantly enhanced the mycelial growth and conidial production by *T. harzianum* in wheat straw, rice husk, and millet grains. However, the addition of carbon and nitrogen sources to sorghum grains and rice grains resulted in negative effect on sporulation of *T. harzianum* (Rajput et al. 2014). Sui Ming (2019) optimized solid fermentation medium that contains grass powder/wheat bran/rice bran at 3:2:1 ratio supplemented with glucose 1%, peptone 0.05%, and dipotassium hydrogen phosphate 0.01% for enhanced sporulation of *T. harzianum*. Jayaswal et al. (2003) studied the influence of physiological and environmental factors on antagonistic strain of *T. viride* RSR7. Both mycelial growth and sporulation of *T. viride* were observed when sucrose, peptone, and trehalose were supplemented to the medium as sole carbon sources. Both growth and sporulation were favored by ammoniacal forms of nitrogen compared to nitrite or nitrate forms.

The pH is a key parameter that impacts both growth and sporulation, while carbon concentration and C:N ratio strongly affected spore production time. At fixed pH, the C:N ratio had a limited influence on spore yield, but was critical for spore shelf life. The highest spore longevity was found in a medium with a pH of 7.0 (Agosin et al. 1997).

The moisture of the substrate is very important for the growth of the fungus. Optimal moisture content facilitates the development of mycelium and later, sporulation. However, the moisture requirements for optimum development differed with the choice of the microorganism (Serna-Díaz et al. 2020). Flodman and Noureddini (2013) cultivated *T. reesei* on spent maize grains from distilleries with an initial humidity of 50% and obtained 7.5×10^8 spores/gdm after 136 h of cultivation on solid substrate with mechanical agitation. Moisture contents of 68.41 ± 0.08% for wheat bran and 34.33 ± 0.91% for rice were found to be optimum for the growth of *T. harzianum*, *T. viride*, and *T. koningii*. The increase of moisture to 73.13 ± 0.31% decreased the quantity of spores produced significantly (Rosane et al. 2008).

Besides moisture and water activity, solid-state fermentation is also affected by the solid composition and structure of the substrate and the type of the microbial strain being multiplied. The availability and accessibility of the nutrients in the solid matrix depend on the solid porosity and structure of the material which in turn is influenced by the moisture and sterilization. Hydration of the solid media can be done by soaking in water for specified time until optimal moisture level is attained. Soaking of hard grains overnight and later drying under shade by spreading as a thin layer for 2–3 h could help in maintaining optimum moisture required for growth of *Trichoderma*.

3 Formulations of *Trichoderma*

Several types of formulations of the biocontrol agent have been developed world over, and some of them have obtained regulatory approval for field applications as per the specific guidelines of the country concerned. Efficient species and strains of *Trichoderma* spp. can be used to develop different formulations as per local requirement and also to suit various delivery methods (Romão-Dumaresq et al. 2012). Secondary metabolites, especially antibiotics and lytic enzymes produced by efficient strains of *Trichoderma*, have been marketed by industries in the form of formulations for agricultural applications (Woo et al. 2014; Błaszczyk et al. 2014).

3.1 Liquid Formulations

The deep tank fermentation system is being followed for liquid formulation in Europe and North America regions for more than four decades (Churchill 1982). Other cheaper liquid components such as yeast, molasses, soy flour broths, and agars are used for production as well as in formulation (Papavizas et al. 1984). The advantage of using liquid growth media for production and formulations allows longer shelf life by providing nutrients, pH, temperature, etc. and also reduces contamination issues while handling (Whipps 1997). Other ingredients such as mineral oils, vegetable oils, etc. are also used in formulations which will check the prolific growth of *Trichoderma* resulting in longer shelf life and less hindrance with other environmental conditions (Herrera et al. 2020; Navaneetha et al. 2015).

3.2 Solid Formulations

In solid formulations, solid media like agricultural residues such as wheat and rice straw, sugarcane bagasse, groundnut shells, corn cobs, sawdust, and rice bran in combination or alone are used as food base or substrate for *Trichoderma* (Cumagun and Lapis 1993; Papavizas et al. 1984; Nelson et al. 1988). Talc and different clays are used as bulking materials for solid formulations (Prasad et al. 2002a, b). To obtain stable product drying is an important step for prolonged shelf life of *Trichoderma* in both solid and liquid formulations (Jin et al. 1992). The solid formulations involve minimum cost but require larger space for production, processing, and storage. However, the transportation losses are minimal in these types of formulations as compared to liquid formulations.

3.3 *Encapsulation of* Trichoderma

Many researchers have attempted microencapsulation (Vemmer and Patel 2013; Ma et al. 2015; Rathore et al. 2013) and nanoencapsulation (Guilger et al. 2017; Ahluwalia et al. 2014) techniques for *Trichoderma* entrapment in carriers like polymers, composites, nanocarriers, etc. These techniques of encapsulation or entrapment use a carrier around the active principle as a physical barrier which protects from external stress due to ultraviolet radiation, sunlight, oxidation, high temperatures, etc. (Sris et al. 2012; McLoughlin 1994; Paulo and Santos 2017). Due to protection from external stresses, the microbes can survive for longer duration with good metabolic activity even under ambient conditions. The encapsulation around microbes also functions as a physical barrier and helps in enhanced persistence in natural microenvironment and controlled or sustained release of active principles targeting insect-pest (Cassidy et al. 1996).

3.3.1 *Microencapsulation of* Trichoderma

Physical (spray drying), chemical (polymerization), and physicochemical (coacervation and ionic gelation) methods have been reported for microencapsulation of *Trichoderma*. While spray drying is a cheaper method of microencapsulation, the high temperatures applied for microencapsulation could result in loss of viability of the *Trichoderma* propagules. Hence, this method could be helpful for the strains that produce heat-resistant spores. Polymerization process utilizes monomers, initiator, etc. As the monomers used for microencapsulation are mostly toxic and can inhibit the growth of *Trichoderma*, nowadays greener alternatives for such molecules are being searched. Ionic gelation for *Trichoderma* is being followed largely. Coacervation process can also be attempted to deliver *Trichoderma* formulations. Encapsulation of *Trichoderma* spores in novel lignin-based polyelectrolyte shells was attempted. The encapsulation was done through layer-by-layer technique resulting in self-stabilizing spore dispersion. The lignin shell can protect the *Trichoderma* spores from the external stresses by keeping the spores in resting state. These spores were applied by trunk injection method to manage grapevine diseases. After injection, the encapsulated spores will be released due to lignin-degrading enzymes secreted by the pathogen. This is popularly known as the "Trojan horse concept." The spores will germinate and attack the pathogen (Peil et al. 2020). An active principle, i.e., naringinase enzyme produced by one of the *Trichoderma* sp., was entrapped in chitosan nanocapsules and alginate nanocapsules. These carriers protect the naringinase enzyme from environmental stresses like temperature, pH, etc. Chitosan-PEG blend plasticized solutions were used for entrapment of *Trichoderma*, and these were applied around the seeds as seed coating. The films formed uniformly around the seed exhibited enhanced compatibility and protection during storage to *Trichoderma* spp. In this method, the rejuvenation of spores under favorable conditions was observed (Chandrika et al. 2019; Prasad et al. 2020). Encapsulation of *Trichoderma* conidia in different polysaccharide-based polymer matrices through different methods has resulted in increased shelf life of formulations (Muñoz-Celaya et al. 2012; Jin and Custis 2011).

3.3.2 Nanoencapsulations of *Trichoderma*

Trichoderma spores when used in conjunction with nanoparticles or nanosystems can show enhanced potential. Organic and inorganic nanoparticles or nanosystems of less than 100 nm particle size have been reported as potential sources for *Trichoderma* carriers for formulations (Kim et al. 2006). The large surface area of nanoparticles offers better holding and distribution of *Trichoderma* spores and improved diffusion in plants. The nanomaterials can be synthesized by different methods like chemical, physical, and biogenic ways. In biogenic synthesis, reduction of precursors to nanomaterials requires organic reducing molecules like proteins, amino acids, sugars, enzymes, etc. It is a cost-effective method and environmentally friendly in nature (Mohanpuria et al. 2008). These organic reducing molecules are being produced by *Trichoderma* as metabolites. Those metabolites from *Trichoderma* help in the production of nanomaterials (Lloyd 2003; Siddiqi and Husen 2017). The remnants of *Trichoderma* spores and nanomaterials after conversion are being used in disease management in crops. Among different nanomaterials reported to be synthesized by *Trichoderma* species, silver nanoparticles are of notable importance due to its synergistic activity. The synergistic activity of silver nanoparticles is due to coating of nanoparticles with metabolites produced from *Trichoderma* and is an added advantage in stability and augmenting the biological control ability (Fraceto et al. 2018; Rodrigues et al. 2013). Few species of *Trichoderma* produced selenium nanoparticles of varied size and surface charge and exhibited good control of pearl millet downy mildew (Nandini et al. 2017). Several studies with nanomaterials and *Trichoderma* combination for management of various crop stresses are being conducted at laboratory level. These products need to be tested for their stability, toxicity to the bioagents, and residual toxicity in the environment. They should also comply with the regulatory needs before field-level exploitation.

4 Agricultural Application of *Trichoderma*

Delivery strategies reported earlier are application of live spores through foliar spraying, seed treatment, and soil application of talc-based formulations, wettable powders, suspension concentrates, etc. available in the market. Most commonly used formulations are wettable powders (WP) whereas the other formulations such as granules, liquid, and solids are relatively less adopted by the industry. The substrates such as a coco mat or peat moss, cereal grains such as rice, or broken corn that support the growth of a *Trichoderma* culture until sporulation are being used directly for soil application. The liquid formulations include emulsions and concentrated liquid suspensions. All liquid formulations and few solid formulations like WP and granules of *Trichoderma* are used for spray on foliar and aboveground plant parts, root dipping, soil drenching, seed treatment, and mixing with during irrigation water in the form of fertigation and in hydroponics at recommended doses of application. Other solid formulations like pellets and dry flowables are directly applied to the soil at time of seeding or transplanting. For management of few phytopathogens, frequent application of *Trichoderma* spp. is recommended. However, as frequent application of biocontrol agents is not only expensive but also labor intensive, a few inexpensive strategies were reported such as using honeybees for continuous dissemination of *Trichoderma* inoculum from hive to flowers for management of floral diseases in crops like strawberry and raspberry and *Botrytis* grey mold of apple and fire blight of pear (Kevan et al. 2003; Maccagnani et al. 2006; Delaplane and Mayer 2000; Kovach et al. 2000). The success of this method of dissemination depends on attraction of the formulation by honeybees, efficiency of the formulation, and inoculum load that could be carried during flight of the bees. In grapevines, coconut, and some other palm trees, the application of *Trichoderma* using dowel impregnation for insertion into holes drilled into the tree trunk provided systemic protection against the pathogens for 4–5 years (Woo et al. 2014). Similar novel delivery approaches that are economically competitive with enhanced efficacy are the need of the hour. Ditta (2012) and Mishra et al. (2017) suggested that the development of micro- and nanotechnological interventions to deliver and to enhance the activity of *Trichoderma* to be more environmentally friendly can be a highly promising area of research.

Most frequently *Trichoderma* grown on different bioproducts is directly applied indiscriminately to plants without proper knowledge of its persistence and mechanism of action. Although, these are popular practices of application of *Trichoderma* spp., there are certain drawbacks that can reduce its effectiveness including nonuniform application; less persistence due to sensitivity to unsuitable soil microenvironment like moisture, temperature, etc.; low viability; and varied performance under different agroclimatic conditions (Sris et al. 2012; Vemmer and Patel 2013). Several adjuvants are being added to existing *Trichoderma* formulations to address those drawbacks. Further, due to improper screening under various laboratory- and field-level conditions, end users are facing problems such as reliability, reproducibility of results, and inadequate quality of formulations. To derive maximum benefits *Trichoderma* spp. under field conditions, the neglected issues such as stability during storage and transport, prolonged shelf life, and persistence under soil microenvironment need to be addressed urgently. The formulations must be cost-effective and ease in application method (Parnell et al. 2016).

5 Storage and Shelf Life

The major challenges to biopesticide development and formulations include the shorter shelf life compared to conventional pesticides. Conidia of *Trichoderma* derived from solid-state fermentation are highly tolerant of abiotic stresses compared with propagules or biomass derived from liquid fermentation (Watanabe et al. 2006). Liquid fermentation can facilitate abundant production of conidial biomass at a shorter period (Harman et al. 1991). Hence, there is a need to improve the shelf life of *Trichoderma* formulations derived from liquid fermentation. Many factors like medium and inoculum type (Elzein et al. 2004), method of drying, the addition of protectants (Friesen et al. 2006), and environmental conditions during storage (Connick et al. 1996) affect the viability of the formulation.

Different interventions can be made during fermentation or at post-fermentation stages to extend the shelf life of formulations. Addition of chitin in the production medium or in the formulation of the bioagent *T. harzianum* enhanced the shelf life. Addition of colloidal chitin at 0.2% to liquid-based production medium and pure chitin in talc formulations of *T. harzianum* enhanced the shelf life by additional 2 months compared to normal shelf life of 4 to 5 months (Sriram et al. 2010). The osmoticum of the production medium can be adjusted by the addition of polyethylene glycol or glycerol that can induce trehalose production and provide the desiccation tolerance (Jin et al. 1991, 1996). Since the biomass of a microbial biocontrol agent is dried after mixing with the carrier material to avoid possible contamination during shelf life, the conidial biomass has to be desiccation-tolerant, besides having high spore viability. Compared with polyethylene glycol (PEG), the addition of glycerol was found to be more beneficial since PEG addition resulted in reduced biomass though it provided desiccation tolerance. The use of glycerol as the osmoticant is particularly effective in initiating micro-cycle conidiation. Accumulation of trehalose that is responsible for stabilizing membranes of cells during desiccation in conidia of *T. harzianum* was correlated with desiccation tolerance (Jin et al. 1996). Addition of glycerol at 3 and 6% extended the shelf life (with viability of >2 × 10^6 CFU g^{-1}) to 7 and 12 months, respectively, compared to 4–5 months of shelf life in formulations derived without the addition of glycerol. In bio-efficacy tests, even after storage for 12 months, formulations derived with the addition of glycerol at 3 or 6% in the production medium were effective (Sriram et al. 2011). Navaneetha et al. (2015) developed suspension concentrate (SC) formulation using conidial biomass of *Trichoderma* resulting in improved yield of biomass and desiccation-tolerant viable propagules at storage temperature of 4 °C. The colony-forming units (CFU) were found to be highest in all the SC formulations of *T. harzianum* Th4 SC and *T. asperellum* Tv5 SC for 18 months period. Storage at 30 ± 2 °C of the SC formulation also retained shelf life of the product until 14 months, but thereafter population declined significantly. A novel biopolymer chitosan-based liquid formulation of *T. harzianum* (Th4d) developed (Prasad et al. 2020) was able to maintain viable counts of log 10.0 and log 10.2 over a period of 6 months at storage temperatures of 30 °C and 4 °C, respectively. However, the antagonistic activity remained unaffected against three plant pathogens, viz., *Macrophomina phaseolina*, *Fusarium oxysporum* f. sp. *ricini*, and *Aspergillus niger* over a period of 6 months of storage.

6 Conclusion

Trichoderma is one of the most researched biocontrol agents for the management of biotic stresses in crops. Bio-inputs play a vital role on commercial agriculture as they not only reduce cost of cultivation and thus enhance profitability but also are considered eco-friendly. However, often the end users have difficulties in the field-level use of this technology. The major bottlenecks are standardized protocols for mass multiplication, formulation, application, and enhanced shelf life of the *Trichoderma* formulations. Research and development are in progress in developing industry-friendly robust protocols to resolve these issues. The demand for *Trichoderma* will be growing in the years to come as there will be more stringent regulations for the use of chemical molecules for biotic stress management. The novel technological inventions in other fields such as molecular biology, nanotechnology, and material sciences should be dovetailed to develop eco-friendly and user-friendly formulations of *Trichoderma*. Over and above, climate change and climatic variability are also known to impact crop-pathogen-biocontrol agent interactions, and there is a need for thorough understanding of the future crop disease scenarios vis-a-vis biocontrol strategies. The genetic diversity among *Trichoderma* strains across different agro-ecologies is yet to be systematically characterized and cataloged to pick up the candidate strains for various crop production systems. Similarly, research is also required to critically understand the frequency of the application of *Trichoderma* within and over crop seasons, and such information will be very valuable for the farmers to budget their crop protection strategies.

References

Abd-El Moity TH, Shatala MN (1981) Biological control of white rot disease of onion *(Sclerotium cepivorum)* by *Trichoderma harzianum*. Phytopathology 100:29–39

Agosin E, Volpe D, Mun~oz G, San Martin R, Crawford A (1997) Effect of culture conditions on spore shelf life of the biocontrol agent *Trichoderma harzianum*. World J Microbiol Biotechnol 13:225–232

Ahluwalia V, Kumar J, Sisodia R, Shakil NA, Walia S (2014) Green synthesis of silver nanoparticles by *Trichoderma harzianum* and their bio-efficacy evaluation against *Staphylococcus aureus* and *Klebsiella pneumonia*. Ind Crop Prod 55:202–206

Ahuja V, Bhatt AK (2018) *Trichoderma viride* (MTCC 800): a potential candidate for agri-horti waste utilization by solid state fermentation. Int J Environ Sci Technol 15:2679–2684

Bhagat S, Pan S (2007) Mass multiplication of *Trichoderma harzianum* on agricultural byproducts and their evaluation against seedling blight *(Rhizoctonia solani)* of mungbean and collar rot (*Sclerotium rolfsii*) of groundnut. Indian J Agric Sci 77(9):33–38

Blackman PA, Kabana RR (1975) A new system for the growth and delivery of biological control agent of the soil. Phytopathology 65:819–821

Błaszczyk L, Siwulski M, Sobieralski K, Lisiecka J, Jędryczka M (2014) *Trichoderma* spp. – application and prospects for use in organic farming and industry. J Plant Prot Res 54(4):309–317

Cassidy MB, Lee H, Trevors JT (1996) Environmental applications of immobilized microbial cells: a review. J Ind Microbiol 16(2):79–101

Chandrika KSVP, Prasad RD, Varsha G (2019) Development of chitosan-PEG blended films using *Trichoderma*: enhancement of antimicrobial activity and seed quality. Int J Biol Macromol 126:282–290

Churchill BW (1982) Mass production of microorganisms for biological control. In: Charudattan R, Walker HL (eds) Biological control of weeds with plant pathogens. Wiley, New York, pp 139–156

Connick WJ Jr, Daigle DJ, Boyette CD, Williams KS, Vinyard BT, Quimby PC Jr (1996) Water activity and other factors that affect the viability of *Colletotrichum truncatum* conidia in wheat flour-kaolin granules (pesta). Biocontrol Sci Tech 6:277–284

Cumagun CJR, Lapis DB (1993) Note: practical approach in mass production of *Trichoderma* spp. as a means of biological control against sheath blight of rice. Philipp Agric 76:251–257

De la Cruz-Quiroz R, Roussos S, Hernandez-Castillo D, Rodríguez-Herrera R, López-López LI, Castillo F, Aguilar CN (2017) Solid-state fermentation in a bag bioreactor: effect of corn cob mixed with Phytopathogen biomass on spore and Cellulase production by Trichoderma asperellum. Fermentation Processes, Angela Faustino Jozala, IntechOpen. https://doi.org/10.5772/64643

Delaplane K, Mayer D (2000) Crop pollination by bees. CABI Publishing, New York

Ditta A (2012) How helpful is nanotechnology in agriculture? Adv Nat Sci Nanosci Nanotechnol 3(3):033002

Dubey SC, Patel B (2002) Mass multiplication of antagonists and standardization of effective dose for management of web blight of urd and mung bean. Indian Phytopathol 55(3):338–341

Elad Y, Katan J, Chet I (1980) Physical, biological and chemical control integrated for soil borne diseases of potato. Phytopathology 70:418–522

Elzein A, Kroshel J, Mueller-Stoever D (2004) Effects of inoculum type and propagules, concentration on shelf life of pasta formulations containing *Fusarium oxysporum* Foxy2, a potential myco-herbicide agent for Striga spp. Biol Control 30:203–211

Flodman HR, Noureddini H (2013) Effects of intermittent mechanical mixing on solid-state fermentation of wet corn distillers grain with *Trichoderma reesei*. Biochem Eng J 81:24–28

Fraceto LF, Maruyama CR, Guilger M, Mishra S, Keswani C, Singh HB, de Lima R (2018) *Trichoderma harzianum* -based novel formulations: potential applications for management of Next-Gen agricultural challenges. J Chem Technol Biotechnol 93(8):2056–2063

Friesen TJ, Holloway G, Hill GA, Pugsley TS (2006) Effect of conditions and protectants on the survival of *Penicillium bilaiae* during storage. Biocontrol Sci Tech 16:89–98

Gangadharan K, Jeyarajan R (1990) Mass multiplication of *Trichoderma* spp. J Biol Control 4:70–71

Guilger M, Pasquoto-Stigliani T, Bilesky-Jose N, Grillo R, Abhilash PC, Fraceto LF (2017) Biogenic silver nanoparticles based on *Trichoderma harzianum*: synthesis, characterization, toxicity evaluation and biological activity. Sci Rep 7

Gupta M, Mahajan S, Sharma D, Gupta S, Mallick SA (2016) Agrowastes as substrates for hydrolytic enzymes and biomass production of native isolates of *Trichoderma* spp. Indian Phytopathol 69:558–564

Guzmán N, De la Cruz R, Aguilar NC (2014) Use of citrus waste for production of *Trichoderma harzianum* by solid state fermentation. https://doi.org/10.13140/2.1.5081.1688

Harman GE, Jin X, Stasz TE, Peruzzotti G, Leopold AC, Taylor AG (1991) Production of conidial biomass of *Trichoderma harzianum* for biological control. Biol Control 1:23–28

Heins Y, Gaffar A, Bakaer KF (1978) Integrated control of Rhizoctonia solani damping off of radish – effect of successive plantings, PCNB and *Trichoderma harzianum* on pathogen and disease. Phytopathology 68:900–907

Herrera W, Valbuena O, Pavone-Maniscalco D (2020) Formulation of *Trichoderma asperellum* TV190 for biological control of *Rhizoctonia solani* on corn seedlings. Egypt J Biol Pest Contr 30:44

Jayaswal RK, Singh R, Lee YS (2003) Influence of physiological and environmental factors on growth and sporulation of an antagonistic strain of *Trichoderma viride* RSR 7. Mycobiology 31(1):36–41

Jin X, Custis D (2011) Microencapsulating aerial conidia of *Trichoderma harzianum* through spray drying at elevated temperatures. Biol Control 56(2):202–208

Jin X, Harman GE, Taylor AG (1991) Conidial biomass and desiccation tolerance of *Trichoderma harzianum* produced at different water potentials. Biol Control 1:237–243

Jin X, Hayes KC, Harman GE (1992) Principles in the development of biological control systems employing *Trichoderma* species against soil-born plant pathogenic fungi. In: Lantham GF (ed) Frontiers in industrial mycology. Chapman & Hall Inc., New York, pp 174–195

Jin X, Taylor AG, Harman GE (1996) Development of media and automated liquid fermentation methods to produce desiccation tolerant propagules of *Trichoderma harzianum*. Biol Control 7:267–274

Kapoor AS, Kumar P (2004) Evaluation of different organic substrates, variability of *Trichoderma harzianum* and management of white rot of pea. Pl Dis Res 19(1):55–56

Kevan PG, Al-Mazra'awi MS, Sutton JC, Tam L, Boland G, Broadbent B, Thompson SV, Brewer GJ (2003) Using pollinators to deliver biological control agents against crop pests. In: Downer RA, Mueninghoff JC, Volgas GC (eds) Pesticide formulations and delivery systems: meeting the challenges of the current crop protection industry, vol 1430. American Society for Testing and Materials International, West Conshohocken, pp 148–153

Khan S, Bagwan N, Mohammed AI, Tamboli R (2011) Mass multiplication and shelf life of liquid fermented final product of *Trichoderma viride* in different formulations. Adv Biores 2:178–182

Kim J, Grate JW, Wang P (2006) Nanostructures for enzyme stabilization. Chem Eng Sci 61:1017–1026

Kovach J, Petzoldt R, Harman GE (2000) Use of honey bees and bumble bees to disseminate *Trichoderma harzianum* 1295-22 to strawberries for botrytis control. Biol Control 18:235–242

Kumar S, Thakur M, Rani A (2014) *Trichoderma*: mass production, formulation, quality control, delivery and its scope in commercialization in India for the management of plant diseases. Afr J Agric Res 9(53):3838–3852

Lewis JA, Papavizas GC (1980) Integrated control of Rhizoctonia fruit rot of cucumber. Phytopathology 70:85–89

Lloyd JR (2003) Microbial reduction of metals and radionuclides. FEMS Microbiol Rev 27(2):411–425

Ma X, Wang X, Cheng J, Nie X, Yu X, Zhao Y (2015) Microencapsulation of *Bacillus subtilis* B99-2 and its biocontrol efficiency against *Rhizoctonia solani* in tomato. Biol Control 90:34–41

Maccagnani B, Bazzi C, Biondi E, Tesoriero D, Maini S (2006) Potential of *Osmia cornuta* as a carrier of antagonistic bacteria in biological control of fire blight: a comparison with *Apis mellifera*. Acta Hortic 704:379–386

Maplestone PA, Whipps JM, Lynch JM (1991) Effect of peat-bran inoculum of *Trichoderma* species on biological control of *Rhizoctonia solani* in lettuce. Plant Soil 136:257–263

Martin SB, Abawi GS, Hoch HC (1984) Influence of the antagonists Laetisaria arvalis on infection of table beets by Phoma betae. Phytopathology 74:1092–1096

McLoughlin AJ (1994) Controlled release of immobilized cells as a strategy to regulate ecological competence of inocula. In: Biotechnics/wastewater [Internet]. Springer, Berlin/Heidelberg, pp 1–45

Mishra S, Keswani C, Abhilash PC, Fraceto LF, Singh HB (2017) Integrated approach of agri-nanotechnology: challenges and future trends. Front Plant Sci 8

Mohanpuria P, Rana NK, Yadav SK (2008) Biosynthesis of nanoparticles: technological concepts and future applications. J Nanopart Res 10(3):507–517

Mukhopadhyay AN, Patel GH, Brahmbhatt A (1986) *Trichoderma harzianum*- a potential biocontrol agent for tobacco damping off. Tob Res 12:26–35

Muñoz-Celaya AL, Ortiz-García M, Vernon-Carter EJ, Jauregui-Rincón J, Galindo E, Serrano-Carreón L (2012) Spray-drying microencapsulation of *Trichoderma harzianum* conidias in carbohydrate polymers matrices. Carbohydr Polym 88(4):1141–1148

Naeimi S, Khosravi V, Varga A, Vagvolgyi C, Kredics L (2020) Screening of organic substrates for solid-state fermentation, viability and bioefficacy of *Trichoderma harzianum* AS12-2, a biocontrol strain against Rice sheath blight disease. Agronomy 10:1258

Nandini B, Hariprasad P, Prakash HS, Shetty HS, Geetha N (2017) Trichogenic selenium nanoparticles enhance disease suppressive ability of *Trichoderma* against downy mildew disease caused by *Sclerospora graminicola* in pearl millet. Sci Rep 7(1):1–6

Navaneetha T, Prasad RD, Venkateswara Rao L (2015) Liquid formulation of *Trichoderma* species for management of gray mold in castor *(Ricinus communis L.)* and Alternariaster leaf blight in sunflower (*Helianthus annuus L.).* J Biofertil Biopestici 6:149

Nelson EB, Harman GE, Nash GT (1988) Enhancement of *Trichoderma*- induced biological control of Pythium seed rot and pre-emergence damping-off of peas. Soil Biol Biochem 20:145–150

Padmanabhan P, Alexander KC (1984) Biological control of cane seedling root rot. Sugarcane 4:11–12

Padmasari Y (2005) Fungal mats in solid – state fermentation. PhD Thesis, Wageningen University, Wageningen

Papavizas GC, Dunn MT, Lewis JA, Beagle-Ristaino J (1984) Liquid fermentation technology for experimental production of biocontrol fungi. Phytopathology 74:1171–1175

Parnell JJ, Berka R, Young HA, Sturino JM, Kang Y, Barnhart DM (2016) From the lab to the farm: an industrial perspective of plant beneficial microorganisms. Front Plant Sci

Paulo F, Santos L (2017) Design of experiments for microencapsulation applications: a review. Mater Sci Eng C:1327–1340

Peil S, Beckers SJ, Fischer J, Wurma FR (2020) Biodegradable, lignin-based encapsulation enables delivery of *Trichoderma reesei* with programmed enzymatic release against grapevine trunk diseases. Mater Today Bio 7:100061

Prasad RD, Rangeshwaran R (2000a) A modified liquid medium for mass production of *Trichoderma* by fermentation process. Plant Disease Res 15:209–211

Prasad RD, Rangeshwaran R (2000b). An improved medium for mass production of the biocontrol fungus *Trichoderma harzainum*. J Mycol Pl Pathol 30: 233–235

Prasad RD, Rangeshwaran R, Kumar SP (1997) Mass production of *Trichoderma harzianum* and its shelf-life in different formulations. In: Proceedings of "International Conference on Integrated Plant Disease Management for Sustainable Agriculture", vol I, pp 358–360

Prasad RD, Rangeshwaran R, Anuroop CP, Phanikumar PR (2002a) Bioefficacy and shelf life of conidial and chlamydospore formulation of *Trichoderma harzianum*. J Biol Control 16:145–148

Prasad RD, Rangeshwaran R, Sunanda CR (2002b) Jaggery-an easily available alternative to molasses for mass production of *Trichoderma harzianum*. Pl Dis Res 17:363–365

Prasad RD, Poorna Chandrika KSV, Varsha G (2020) A novel chitwosan biopolymer based *Trichoderma* delivery system: storage stability, persistence and bio efficacy against seed and soil borne diseases of oilseed crops. Microbiol Res 237

Rajput AQ, Khanzada MA, Shahzad S (2014) Effect of different organic substrates and carbon and nitrogen sources on growth and shelf life of *Trichoderma harzianum*. J Agr Sci Technol 16:731–745

Ramanujam B, Prasad RD, Sriram S, Rangeswaran R (2010) Mass production, formulation, quality control and delivery of *Trichoderma* for plant disease management. J Pl Prot Sci:1–8

Rathore S, Desai PM, Liew CV, Chan LW, Heng PWS (2013) Microencapsulation of microbial cells. J Food Eng 116(2):369–381

Rini CR, Sulochana KK (2007) Substrate evaluation for multiplication of *Trichoderma* spp. J Trop Agric 45(1–2):58–60

Rodrigues AG, Ping LY, Marcato PD, Alves OL, Silva MCP, Ruiz RC (2013) Biogenic antimicrobial silver nanoparticles produced by fungi. Appl Microbiol Biotechnol 97(2):775–782

Romão-Dumaresq AS, Araújo WL, Talbot NJ, Thornton CR (2012) RNA interference of endochitinases in the sugarcane endophyte *Trichoderma virens* reduces its fitness as a biocontrol agent of pineapple disease. PLoS One 7(10)

Rosane S, Cavalcante HLS, Lima GAS, Pinto CAT, Gava SR (2008) Effect of moisture on *Trichoderma* conidia production on corn and wheat bran by solid state fermentation. Food Bioprocess Technol 1:100–104

Saju KA, Anandraj M, Sarma YR (2002) On farm production of *Trichoderma harzianum* using organic matter. Indian Phytopathol 55(3):255–259

Sangeetha P, Jeyarajan R, Panicker S (1993) Mass multiplication of biocontrol agent *Trichoderma* spp. Indian J Mycol Plant Pathol 23:328–330

Sangle UR, Wuike RV, Gade RM (2002) Influence of original moisture present in the substrates on the growth and sporulation of *Trichoderma* spp. Pestology XXVI(9):46–48

Sankar P, Jeyarajan R (1996) Biological control of sesamum root rot by seed treatment with *Trichoderma* spp. and *Bacillus subtilis*. Indian J Mycol Plant Pathol 26:147–153

Serna-Díaz MG, Mercado Y, Jimenez A, Anducho-Reyes MA, Medina J, Tuoh-Mora JC, Tellez A (2020) Use of barley straw as a support for the production of conidiospores of *Trichoderma harzianum*. Biotechnol Rep 26

Siddiqi KS, Husen A (2017) Fabrication of metal nanoparticles from fungi and metal salts: scope and application. Nanoscale Res Lett 11(1)

Singh V, Joshi BB (2007) Mass multiplication of *Trichoderma harzianum* on sugarcane press mud. Indian Phytopathol 60(4):530–531

Singh J, Singh PN (2007) Effect of abiotic factors on growth, sporulation and population of antagonists. Ann Plant Soil Res 9:156–158

Sriram S, Palanna KB, Ramanujam B (2010) Effect of chitin on the shelf-life of *Trichoderma harzianum* in talc formulation. Indian J Agric Sci 80:930–932

Sriram S, Roopa KP, Savitha MJ (2011) Extended shelf-life of liquid fermentation derived talc formulations of *Trichoderma harzianum* with the addition of glycerol in the production medium. Crop Prot 30(10):1334–1339

Sris J, Seethadevi A, Suria Prabha K, Muthuprasanna P, Pavitra P (2012) Microencapsulation: a review. Int J Pharma Bio Sci 3:509–531

Sui Ming (2019) The solid fermentation state's optimization of *Trichoderma harzianum* . IOP Conf. Ser.: Mater Sci Eng 612 022111

Thangavelu R, Palaniswamy A, Velazhahan R (2004) Mass production of *Trichoderma harzianum* for managing Fusarium wilt of banana. Agric Ecosyst Environ 103(1):259–263

Tiwari AK, Krishna K, Razdan VK, Rather TR (2004) Mass production of *Trichoderma viride* on indigenous substrates. Ann Plant Protect Sci 12(1):71–74

Upadhyay JP, Mukhopadhyay AN (1986) Biological control of *Sclerotium rolfsii* by *Trichoderma harzianum* in sugarbeet. Trop Pest Manage 32:215–220

Vemmer M, Patel AV (2013) Review of encapsulation methods suitable for microbial biological control agents. Biol Control 67(3):380–389

Watanabe S, Kato H, Kumakura K, Ishibashi E, Nagayama K (2006) Properties and biological control activities of aerial and submerged spores in *Trichoderma asperellum* SKT-1. J Pestic Sci 31:375–379

Whipps JM (1997) Developments in the biological control of soil borne plant pathogens. In: Callow JA (ed) Advances in botanical research incorporating advances in plant pathology, vol 26, pp 1–123

Woo SL, Ruocco M, Vinale F, Nigro M, Marra R, Lombardi N (2014) *Trichoderma*-based products and their widespread use in agriculture. Open Mycol J 8(1):71–126

Zaidi NW, Singh NS (2004) Mass multiplication of *Trichoderma harzianum* on cow dung. Indian Phytopathol 57(2):189–192

Regulatory Issues in Commercialization of *Trichoderma*-Based Products in the USA

Gary E. Harman

Contents

1 Federal Regulations

The Federal Insecticide, Fungicide, and Rodenticide Act (FIFRA[1]) is defined as "with certain exceptions, a pesticide is any substance or mixture of substances intended for preventing, destroying, repelling, or mitigating any pest, or intended for use as a plant regulator, defoliant, or desiccant, or desiccant, or any nitrogen stabilizer." Any use outside this definition does not require registration under FIFRA.

[1] https://www.epa.gov/enforcement/federal-insecticide-fungicide-and-rodenticide-act-fifra-and-federal-facilities

G. E. Harman (✉)
Cornell University, Ithaca, NY, USA
e-mail: geh3@cornell.edu

N. Amaresan et al. (eds.), *Advances in* Trichoderma *Biology for Agricultural Applications*, Fungal Biology, https://doi.org/10.1007/978-3-030-91650-3_15

FIFRA enforcement is focused on the sale, distribution, and use (which can include disposal) of pesticides. Generally, before a pesticide may be sold or distributed in the USA, it must be registered (licensed) with the US Environmental Protection Agency (EPA, www.epa.gov). Before EPA may register a pesticide under FIFRA, the applicant must show, among other things, that using the pesticide according to specifications will not generally cause unreasonable adverse effects on the environment. Taking into account the economic, social, and environmental costs and benefits of the use of any pesticide, FIFRA defines the term unreasonable adverse effects on the environment to mean the following (www.epa.gov):

- "Any unreasonable risk to man or the environment, taking into account the economic, social, and environmental costs and benefits of the use of any pesticide"
- "Any human dietary risk from residues that result from use of a pesticide in or on any food inconsistent with the standard under section 408[2] of the Federal Food, Drug, and Cosmetic Act"

The version of FIFRA, enacted in 1947, required that persons registering pesticides distributed in interstate commerce with the US Department of Agriculture (USDA, www.usda.gov) and established a rudimentary set of labeling provisions. Concerns regarding the toxic effects of pesticides and residues on applicators, non-target species, the environment, and food prompted significant changes in the original FIFRA legislation. Some subsequent amendments occurred as a result of the current statute. In 1972, the Federal Environmental Pesticide Control Act amended FIFRA, as did the Pesticide Registration Improvement Act of 2004 (PRIA1).

The primary objective of FIFRA is to ensure that, when applied as instructed, pesticides will not generally cause unreasonable risk to human health or the environment.

2 Types of Documents Required

2.1 Documentation Required

- Manufacturing process
- Safety information
- Product identity
- Good laboratory compliance statement
- Analysis of samples
- Microbial contamination
- Certification of limits

Any of the documents may be confidential. This must be stated on the page or section where this is required. Sections used in a recent application for our

[2] https://www.epa.gov/laws-regulations/summary-federal-food-drug-and-cosmetic-act

Trichoderma strains details and their preservation, information on growth media for production, scale-up procedures and harvest, and preparation of formulations. Many of the tests and data may require third-party tests to avoid conflict of interest. In our applications, the product identity, contamination testing, and safety testing were done by outside laboratories. In addition, private laboratories and companies are required in preparation of dossiers to be submitted to EPA and in taxonomic identity studies.

Chemical pesticides are expressed by units of weight. This is largely meaningless for biological agents. Instead, the most useful measure is colony forming units/gram (CFU/g). It is possible to produce different formulations that have tenfold differences in CFU/g, at the same weight. Such measurements need to be submitted to EPA, since they are not part of the usual regulatory framework.

2.2 Manufacturer's Use Permit

This dossier includes information on taxonomic data. Taxonomy in *Trichoderma* and most other organisms requires genetic sequences. Older systems, based on morphology, are no longer adequate. Others include the history of the strain, the life cycle of the organism, characteristics of the stain, pH, temperature optima, modes of action, and the pathogen or pest to be controlled.

2.3 Safety Information

Safety information for human health for microbial pesticides include (see Data Requirements for Pesticide Registration[3]):

- Acute oral toxicity
- Acute dermal toxicity
- Acute inhalation toxicity
- Primary eye irritation
- Primary dermal irritation
- Dermal sensitization

In addition, ecological and non-target effects may be required including:

- Avian inhalation testing
- Avian oral testing
- Estuarine and marine animal testing
- Freshwater aquatic invertebrate testing
- Freshwater fish testing

---

[3] https://www.epa.gov/pesticide-registration/data-requirements-pesticide-registration#dh

- Honey bee testing
- Non-target insect testing
- Non-target plant testing

Unless foliar application will be used, most of the tests in the list above are unnecessary, except for the last one (non-target plant testing) and harm to beneficial organisms testing. Before testing is initiated, it is useful to discuss protocols and testing procedures with EPA personnel. An in-person pre-submission conference is important. In my experience, EPA personnel are quite helpful, but any discussions are not binding and will depend on the interpretation of the regulations based on what is found in the application.

Regulation of *Trichoderma* stain is considered to be microbial pesticides. Regulations for this category can be found in Series No. 885 Microbial Pesticide Test Guidelines.[4] One required dossier is deposition of cultures in a nationally recognized type culture collection, such as American Type Culture Collection (www.atcc.org) or the USDA's Northern Regional Type Culture Collection (https://nrrl.ncaur.usda.gov/). Another required dossier is on storage stability. Typically, cultures are required to remain viable for one year, although different time periods may be proposed. Any special storage conditions, such as refrigeration, need to be provided.

Another permit that may be required is residue analysis in plants or other organisms. However, it is possible to apply for an exemption to residue analysis. *T. virens* strain G-41 received such an exemption, and this is provided in Federal Register portal (search https://www.federalregister.gov/ for [*Trichoderma virens* 2012]).[5] The summary is provided.

> This regulation establishes an exemption from the requirement of a tolerance for residues of *Trichoderma* virens strain G-41 in or on all food commodities when applied as a fungicide and used in accordance with good agricultural practices. BioWork, Inc. submitted a petition to EPA under the Federal Food, Drug and Cosmetic Act (FFDCA) requesting an exemption from the requirement of a tolerance. This regulation eliminated the need to establish a maximum permissible level of *Trichoderma* virens strain G-41 under the FFDCA.

It may be useful to include other information in the application, such as resistance to abiotic and biotic stresses, crop yield enhancement, and photosynthetic improvement. Such claims are permitted and may serve as validation of a specific product.

Most biological agents are not harmful. More than 50 different organisms are listed in Gwynn (2014), and all are at low levels of toxicity. Some *Trichoderma* strains are problematic, for example, *T. longibrachiatum* grows at human body temperature and is an opportunistic pathogen of immunocompromised people. *T. aggressivum* is a pathogen of mushrooms (Harman et al. 2010).

A number of *Trichoderma* strains are registered. A partial list is provided in Table 1.

---

[4] https://www.epa.gov/test-guidelines-pesticides-and-toxic-substances/series-885-microbial-pesticide-test-guidelines

[5] https://www.federalregister.gov/documents/2012/02/01/2012-2216/trichoderma#:~:text=This%20regulation%20establishes%20an%20exemption%20from%20the%20requirement,and%20used%20in%20accordance%20with%20good%20agricultural%20practices

Table 1 List of *Trichoderma*-based products approved for pesticide use. A general reference is Gwynn (2014)

Strain	Where registered	Safety	Reference
Trichoderma asperellum TV1	France, Spain, Italy, and other countries	Not toxic to animals or non-target organisms	https://sitem.herts.ac.uk/aeru/bpdb/Reports/2047.htm
T. atroviride I 1237	France	Not toxic to a range of organisms	https://sitem.herts.ac.uk/aeru/bpdb/Reports/2047.htm
T. asperellum ICC012	USA	Not toxic, exemption from requirement of tolerance	https://www.law.cornell.edu/cfr/text/40/180.1294
T. asperellum T34	USA, Spain, Belgium, France, Italy, Holland, and other countries	Not toxic or harmful	www.federalregister.gov/documents/2020/09/25/… https://sitem.herts.ac.uk/aeru/bpdb/Reports/2043.htm
T. atroviride IMI 206040	Sweden, Italy	Not toxic or harmful	https://sitem.herts.ac.uk/aeru/bpdb/Reports/2044.htm
T. atroviride T-11	Sweden, Italy	Not toxic or harmful	https://sitem.herts.ac.uk/aeru/bpdb/Reports/2045.htm
T. gamsii ICC 080	USA	Not toxic, exemption from requirement of tolerance	https://www.law.cornell.edu/cfr/text/40/180.1293 https://nevegetable.org/table-19-fungicides-and-bactericides-labeled-vegetable-transplants
T. virens GL-21 (formerly *Gliocladium virens)*	USA	Not toxic, exemption from requirement of tolerance	https://nevegetable.org/table-19-fungicides-and-bactericides-labeled-vegetable-transplants https://www.federalregister.gov/documents/2012/02/01/2012-2216/trichoderma
T. hamatum TH382	USA	Not toxic, exemption from requirement of tolerance	https://www3.epa.gov/pesticides/chem_search/cleared_reviews/csr_PC-119205_18-May-10.pdf
T. virens G-41 (formerly *G. virens)*	USA	Not toxic	https://www3.epa.gov/pesticides/chem_search/reg_actions/registration/decision_176604_06-Feb-12.pdf

(continued)

Table 1 (continued)

Strain	Where registered	Safety	Reference
T. afroharzianum T22 (formerly *T. harzianum*)	USA, Holland	Not toxic, exemption from requirement of tolerance	https://archive.epa.gov/pesticides/biopesticides/web/html/frnotices_119202.html
T. atroviride K5	Pending USA	Not toxic	Harman et al. (2018)
T. atroviride SC1	USA	Not toxic, exemption from requirement of tolerance	https://www.federalregister.gov/documents/2020/07/31/2020-15695/trichoderma-atroviride-strain-sc1-exemption-from-the-requirement-of-a-tolerance

2.4 Organic Certifications

In the USA, and elsewhere, organic certification is administrated by Organic Farming Associations in each of the 50 states, e.g., https://nofany.org/. These associations use federal funding. There is also the non-profit organization the Organic Materials Research Institute (OMRI https://www.omri.org/).

Differences exist between the programs, but in general they are based on methods of manufacture and whether or not components occur naturally. No genetically altered materials can be used nor can any material be synthesized or manufactured. The websites above list procedures, applications, and lists of approved materials.

In some cases, one source of a material may be organically approved, while another is not. For example, sodium nitrate derived from bat guano (Chilean nitrate) is approved, with limitations, while the same compound that is manufactured is prohibited. The statement for material is as follows:

"This product contains highly soluble nitrogen and must be applied in a manner that does contribute to the contamination of crops, soil, or water. Its use must be part of an organic system plant that maintains or improves natural resources of the operation, including soil and water quality, and that complies with crop nutrient and soil fertility requirements."

2.5 Inert Ingredients

Registration packets to EPA require information on inert ingredients. The following is a quote from EPA Inert Ingredients Regulation:[6]

> Most pesticide products contain substances in addition to the active ingredient(s) that are referred to as inert ingredients or sometimes as "other ingredients. An inert ingredient generally is any substance (or group of similar substances) other than an active ingredient that is intentionally included in a pesticide product. Examples of inert ingredients include emulsifiers, solvents, carriers, aerosol propellants, fragrances and dyes.

Safety information is required. The EPA provides a list (25b) that lists materials approved for use as inert ingredients[7] in Inert Ingredients Eligible for FIFRA 25(b) pesticide products (Revised November 2016).

Safety information for specific compounds is provided at the website of the Centers for Disease Control and Prevention (CDC, https://www.cdc.gov/) and the National Institute for Occupational Safety and Health (NIOSH) Pocket Guide to Chemical Hazards at https://www.cdc.gov/niosh/npg/.

[6] https://www.epa.gov/pesticide-registration/inert-ingredients-regulation

[7] https://www.epa.gov/minimum-risk-pesticides/inert-ingredients-approved-use-minimum-risk-pesticide-products

3 Conclusions

Registrations are expensive and usually cost more than $100,000, and those costs may double or triple depending on the complexity and requirements for registration. In the author's experience, at least a year is required. This is a very good reason to hire a consultant who is very familiar with regulatory requirements. This adds to expenses initially but minimizes total expenses and time required.

References

Gwynn RL (ed) (2014) The Manual of Biocontrol Agents, 5th edn. British Crop Protection Council, Alton

Harman GE, Obregón MA, Samuels GJ, Lorito M (2010) Changing models of biocontrol in the developing and developed world. Plant Dis 94:928–939

Harman GE, Cadle-Davidson M, Nosir W (2018) Patent application WO2017192117A1. Highly effective and multifunctional microbial compositions and uses. In: USPTO (ed) https://patents-googlecom/patent/WO2017192117A1/en?q=multifunctional&inventor=Harman. Advanced Biological Marketing, p 12

Part V
New Industrial Applications of *Trichoderma*

Industrially Important Genes from *Trichoderma*

Şeyma Çolakoğlu Özkaya, Hüseyin Okan Soykam, and Günseli Bayram Akçapınar

Contents

Authors Şeyma Çolakoğlu Özkaya and Hüseyin Okan Soykam have equally contributed to this chapter.

Ş. Çolakoğlu Özkaya
Department of Medical Biology and Genetics, Institute of Health Sciences, Marmara University, Maltepe, Istanbul, Turkey

H. O. Soykam
Department of Biostatistics and Bioinformatics, Institute of Health Sciences, Acibadem University, Istanbul, Turkey

G. Bayram Akçapınar (✉)
Department of Medical Biotechnology, Institute of Health Sciences, Acibadem University, Istanbul, Turkey
e-mail: gunseli.akcapinar@acibadem.edu.tr

N. Amaresan et al. (eds.), *Advances in* Trichoderma *Biology for Agricultural Applications*, Fungal Biology, https://doi.org/10.1007/978-3-030-91650-3_16

1 *Trichoderma* as a Cosmopolitan Genus with Colossal Industrial Potential

Being a cosmopolitan genus, *Trichoderma* spp. occupy diverse habitats, including but not limited to soil, decaying wood, and rhizosphere (Druzhinina and Kubicek 2013). Diversity of these habitats not only reflects their adaptability and opportunistic potential but also diversity of their genes and gene families to fit these ecological niches. Their rapid growth in soil allows for their easier isolation. *Trichoderma* genus currently hosts at least 460 species as listed in the International Commission on *Trichoderma* Taxonomy webpages (Druzhinina and Cai 2020) as of August 2020. Many species in this genus are important producers of cellulolytic enzymes (Payne et al. 2015; Bischof et al. 2016; Kubicek et al. 2019). The most important industrial cellulase producer is *Trichoderma reesei*, which was isolated in Solomon Islands during the World War II (Mandels and Reese 1960). It was identified as *Trichoderma viride* at first and named as QM6a strain and then renamed as *Trichoderma reesei* due to its marked characteristics than *Trichoderma viride* (E.G. Simmons 1977). Although the industrial interest with various *Trichoderma* species started with cellulases and their immense potential for lignocellulosic biomass transformations (Bischof et al. 2016), it has rapidly spread to other carbohydrate active enzyme (CAZy) families (Lombard et al. 2014) and genes and gene families involved in secondary metabolite production (Mukherjee et al. 2012b; Atanasova et al. 2013b; Zeilinger et al. 2016). Recently, gene families encoding for small secreted cysteine-rich proteins (SSCPs) have been standing out due to their surface-modifying activities (Nakari-Setälä et al. 1996, 1997; Linder et al. 2001; Przylucka et al. 2017a) and their involvement in a plethora of biotrophic interactions (Seidl et al. 2006; Atanasova et al. 2013a; Guzmán-Guzmán et al. 2017; Przylucka et al. 2017b; Cai et al. 2020).

2 Available *Trichoderma* Genomes

Since many *Trichoderma* species are utilized as biofertilizers and in biocontrol, a precise understanding of their biology and evolution and genes involved in their interactions is required. The first *Trichoderma* (*Hypocreales*, *Ascomycota*) genome was published in 2008 (Martinez et al. 2008). Since then, until 2021, a total of 24 genomes from this genus are published in the JGI database (https://mycocosm.jgi.doe.gov/Trichoderma/Trichoderma.info.html, accession date: 21.01.2021). These belong to *Trichoderma arundinaceum* IBT 40837; *T. asperelloides* TR356; *T. asperellum* CBS 433.97; *T. atrobrunneum* ITEM 908; *T. atroviride* B10, F7, and P1; *T. atroviride* v2.0; *T. brevicompactum* IBT40841; *T. citrinoviride* TUCIM 6016; *T. gamsii* T6085; *T.* spp. NJAU 4742; *T. hamatum* GD12; *T. harzianum* CBS 226.95; *T.* spp. M10; *T. afroharzianum* T22; *T. harzianum* TR274; *T. longibrachiatum* ATCC 18648; *T. longibrachiatum* MK1; *T. parareesei* CBS 125925; *T.* spp.; *T. reesei*

QM6a; *T. reesei* RUT C-30; *and T. virens* Gv29–8 (Cai and Druzhinina 2021). A comparative genomic analysis of the first three whole genomes of *Trichoderma* underlined the innate mycoparasitic lifestyle of these species; further comparison of the genomes of 12 common species of *Trichoderma* has further supported mycoparasitism as the ancestral lifestyle (Kubicek et al. 2011).

Comparative genomic analysis of these species would reveal novel genes and gene families that might be useful for industrial applications.

3 Industrially Important Genes and Gene Families from *Trichoderma* Species

3.1 Carbohydrate Active Enzymes (CAZymes)

Trichoderma species are well-known producers of diverse enzyme families with the ability to degrade and modify biopolymers, such as lignocellulose, as shown in Fig. 1. Harzianum clade hosting *T. harzianum and T. guizhuense* is enriched in the number of CAZymes (Lombard et al. 2014) in parallel to the total gene number, which is a mere reflection of their adaptability and biotrophic interactions.

(i) **Glycoside Hydrolases (GH)**

As lignocellulose, being the most abundant and outspread biopolymer on earth, *Trichoderma* species are enriched in glycoside hydrolase (GH) families, which have the ability to degrade biopolymers such as lignocellulose. A comparative genome analysis of *Trichoderma* species as shown in Fig. 1 indicates that almost half of the CAZymes belong to GH family, ranging from ~48% to 60%. GH18 chitinases, GH3 beta-glycosidases, GH16 beta-1,3/1,4-glucanases, and GH5 family members are abundantly represented in these 24 *Trichoderma* genomes as also shown via comparative genomic analysis of 12 representative species by Kubicek et al. (2019). The average number of GH family genes is 237, whereas the average numbers of GH18, GH3, GH16, and GH5 family genes within the 24 genomes are 26, 15, 15, and 10, respectively. While some *Trichoderma* species such as the endophytic *T. brevicompactum* with biocontrol potential possesses 247 GH enzyme-encoding genes, 34 of which belong to GH18 family within their published genome (Proctor et al. 2018); *T. reesei* QM6a and RutC30 strains have a less number of GH family genes with respect to other species.

Cellulases are the most famous enzymes of *Trichoderma* spp. with industrial potential. They are cellulose-hydrolyzing enzymes and belong to the GH family according to CAZy classification of carbohydrate active enzymes (Lombard et al. 2014). *T. reesei*, being the key industrial species for cellulase production, has been known to express a number of endoglucanases (EG), enzymes degrading the cellulose chain from the inside; cellobiohydrolases (CBH) or exoglucanases degrading the cellulose chain from the reducing and nonreducing ends; and, finally,

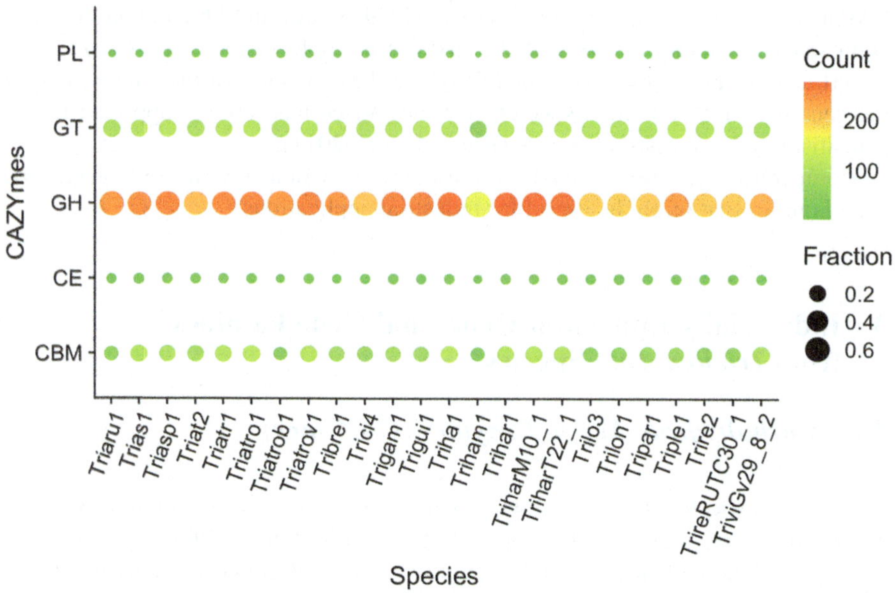

Fig. 1 Numbers and distribution of industrially important CAZy families in 24 *Trichoderma* genomes based on the genomic analysis data in Joint Genome Institute (JGI) https://mycocosm.jgi.doe.gov/Trichoderma/Trichoderma.info.html, accession date: 21.01.2021). PL, polysaccharide lyase; GT, glycosyltransferase; GH, glycoside hydrolase; CE, carbohydrate esterase; CBM, carbohydrate binding module. Coloring indicates the number of genes in each family represented in the genomes, and fraction indicates the fraction of each family with respect to the total number of PL, GT, GH, CE, and CBMs within each species genome. [*Trichoderma arundinaceum* IBT 40837 (Triaru1), *Trichoderma asperelloides* TR356 (Trias1), *Trichoderma asperellum* CBS 433.97 (Triasp1), *Trichoderma atrobrunneum* ITEM 908 (Triat2), *Trichoderma atroviride* B10 (Triatr1), F7 (Triatro1) and P1 (Triatrob1), *Trichoderma atroviride* v2.0 (Triatrov1), *Trichoderma brevicompactum* IBT40841 (Tribre1), *Trichoderma citrinoviride* TUCIM 6016 (Trici4), *Trichoderma gamsii* T6085 (Trigam1), *Trichoderma* spp. NJAU 4742 (Trigui1), *Trichoderma hamatum* GD12 (Triham1), *Trichoderma harzianum* CBS 226.95 (Triha1), *Trichoderma* spp. M10 (TriharM10_1), *Trichoderma afroharzianum* T22 (TriharT22_1), *Trichoderma harzianum* TR274 (Trihar1), *Trichoderma longibrachiatum* ATCC 18648 (Trilo3), *Trichoderma longibrachiatum* MK1 (Trilon1), *Trichoderma parareesei* CBS 125925 (Tripar1), *Trichoderma* spp. TPhu1 (Triple1), *Trichoderma reesei* QM6a (Trire2), *Trichoderma reesei* RUT C-30 (TrireRUTC30_1), *Trichoderma virens* Gv29–8 (TriviGv29_8_2) (Cai and Druzhinina 2021)]

beta-glucosidases (BGL) with the capability to hydrolyze cellooligosacharides such as cellobiose into its glucose monomers. Although the genome mining of their genome indicates the presence of eight endoglucanases, two cellobiohydrolases, and up to six beta-glucosidase genes, an extensively studied *T. reesei* cellulase system is composed of at least five EGs (EGI to EGV) and two CBHs (CBHI and CBHII) with two additional BGLs (BGLI and BGLII) (Häkkinen et al. 2012). These enzymes act in synergy for the degradation of cellulosic polymers (Woodward 1991; Akcapinar et al. 2011). In addition to these cellulases, *T. reesei* genome harbors hemicellulase enzymes mainly composed of four endoxylanases (XYNI to

XYN IV), one beta-xyloside (BXLI), a mannanase (MANI), an acetyl xylan esterase (AXEI), an α-glucuronidase (GLRI), an α-L-arabinofuranosidase (ABFI), three α-galactosidases (AGLI to AGLIII), and an acetyl esterase (AESI) (Häkkinen et al. 2012). Characterized and putative cellulases and hemicellulases of *T. reesei* with the GH family groups are listed in Table 1. Moreover, *T. reesei* genome is also enriched with chitinases from GH18 family. There are 19 putative and one characterized chitinase in the genome (Häkkinen et al. 2012). Ike et al. (2006) identified and characterized Chi46 chitinase from *T. reesei* PC-3-7 strain by heterologous expression in *E. coli*. The recombinant enzyme exhibited an exochitinase activity toward colloidal chitin and endochitinase activity toward chitosan 7B and N-acetylchitooligosaccharides. Chitinase-encoding genes, such as *chit33*, *chit37*, and *chi42* from *T. harzianum* and *chit36* and *chi42* from *T. asperellum*, were also among the characterized ones (de la Cruz et al. 1992; Viterbo et al. 2002; Steyaert et al. 2004; Boer et al. 2007). An antifungal chitinase, CHIT46 from *T. harzianum* GIM3.442, heterologously expressed in *P. pastoris,* was reported for use in colloidal chitin conversion (Deng et al. 2019). Some of the chitinases from GH18 were shown to exhibit antifungal activity and could be used as biotechnological agents for biocontrol against pathogenic fungi (Lienemann et al. 2009; Wu et al. 2013). All of the aforementioned enzymes and their orthologs in other *Trichoderma* species could be good candidates for industrial applications in said areas.

(ii) **Glycosyltransferases (GT)**

Glycosyltransferases are a family of enzymes with the ability of transferring glycosyl residues from a specific donor to an acceptor (Lairson et al. 2008). Therefore, they are involved in anabolism and catabolism of a variety of biological molecules, such as carbohydrates, antibiotics, and glycan-containing structures such as glycolipids, glycoproteins, and proteo- and peptidoglycans. They also play important roles in the biosynthesis and modeling of the fungal cell walls (Klutts et al. 2006). GTs are the second largest family of CAZymes represented in the *Trichoderma* genomes, forming almost 20% of the CAZymes (Fig. 1). *T. longibrachiatum* genomes contain a higher number of GTs than other species. *T. reesei* genome was shown to possess 103 GTs, which is very close to the average number observed among *Sordariomycetes* (Martinez et al. 2008). A study of the genomic regions of the *T. harzianum* IOC-3844 strain based on assembled BAC sequences and RNA-Seq analysis revealed three GTs – α-mannosyltransferase (GT71), GnT-III/β-1,4-N-acetylglucosaminyltransferase III (GT17), and a candidate β-xylosyltransferase (GT90) – involved in the biosynthesis of fungal cell wall. GT71 was also reported to be co-induced with a cellulase gene (Crucello et al. 2015).

Although there are few studies on the characterization of glycosyltransferases that belong to *Trichoderma* spp., recent in silico comparative genomic studies underline their numbers and diversity. In a recent study (Nauom et al. 2019), a six-hairpin glycosidase-like glycosyltransferase enzyme has been identified from the enriched secretome of *T. harzianum* after interacting with the cell walls of *Sclerotinia sclerotiorum* and *Fusarium oxysporum.* Currently, due to the limited research and literature, this family of enzymes is being underestimated. However, their further

Table 1 Cellulases and hemicellulases of *Trichoderma reesei* (Häkkinen et al. 2012)

Enzyme	CAZy family	References
EGI-CEL7B	GH7	Mitsuishi et al. (1990), Saloheimo et al. (1997)
EGII-CEL5A	GH5	Medve et al. (1998)
EGIII-CEL12A	GH12	Saloheimo et al. (1988)
EGIV-CEL61A	GH61	Saloheimo et al. (1997)
EGV-CEL45A	GH45	Saloheimo et al. (1994)
CEL5B[a]	GH5	Foreman et al. (2003)
CEL61B[a]	GH61	
CEL74A[a]	GH74	
CBHI-CEL7A	GH7	Shoemaker et al. (1983; Mitsuishi et al. (1990)
CBHII-CEL6A	GH6	Penttilä et al. (1988), Koivula et al. (1996)
BGLI-CEL3A	GH3	Fowler and Brown (1992)
BGLII-CEL1A	GH1	Takashima et al. (1999)
CEL1B[a]	GH1	Foreman et al. (2003)
CEL3B[a]	GH3	
CEL3C[a]		
CEL3D[a]		
CEL3E[a]		
CEL3G[a]		Zou et al. (2018)
XYNI	GH11	Biely et al. (1994), Törrönen and Rouvinen (1995)
XYNII	GH11	Törrönen and Rouvinen (1995), Jänis et al. (2001)
XYNIII	GH10	Ogasawara et al. (2006)
XYNIV	GH30	Tenkanen et al. (2013)
XYNV[a]	GH11	Metz et al. (2011)
BXLI	GH3	Drouet et al. (1994)
MANI	GH5	Stålbrand et al. (1995)
AXEI	CE5	Zhang et al. (2011a)
AXEII[a]	CE5	Foreman et al. (2003)
GLRI	GH67	Margolles-Clark et al. (1996a, b)
ABFI	GH54	Margolles-Clark et al. (1996c)
ABFII[a]	GH62	(Foreman et al. 2003)
ABFIII[a]	GH54	Herpoël-Gimbert et al. (2008)
AGLI	GH27	Margolles-Clark et al. (1996b)
AGLII	GH36	
AGLIII	GH27	
AESI	CE16	Li et al. (2008), Puchart et al. (2016)
AES[a]	CE16	Häkkinen et al. (2012)

[a]Putative

characterization and analysis may allow for their effective application in biocontrol against pathogenic fungi.

(iii) **Polysaccharide Lyases (PL)**

Polysaccharide lyases are a family of enzymes that use a β-elimination mechanism to cleave polysaccharides rather than hydrolysis (Lombard et al. 2010). Anionic polysaccharides are cleaved by lyase enzymes. Polyuronates, such as alginate, pectin, glucuronan, xanthan, and hyaluronan, are degraded by corresponding lyase enzymes. Polyuronates form important components of the cell wall and extracellular polysaccharides. This group of enzymes has an important function in microbial plant invasion or degradation. Catalytic action of these enzymes results in plant tissue maceration, cellular lysis, and cell wall modifications, thereby increasing the accessibility of the plant biomass for other enzymes (Lombard et al. 2010; Atanasova et al. 2018). *Trichoderma* spp. harbors four to eight polysaccharide lyases. Within the CAZymes, PLs are the least represented enzymes. There are eight PLs in the *T. atroviride* genome (Fig. 1).

There are a limited number of PLs of *Trichoderma* species that have been functionally characterized. Glucuronan lyase enzyme from *Trichoderma* sp. isolated from compost was previously shown to be induced by fermentation on cellouronate (β-(1→4)-polyglucuronate) as the sole carbon source (Delattre et al. 2006). A similar PL from *T. reesei,* glucuronan lyase belongs to PL20 family and was heterologously expressed in the methylotrophic yeast *Pichia pastoris* (Konno et al. 2009a). The enzyme represented a novel subclass of PLs and exhibited Ca^{2+}-dependent enzyme activity similar to other pectate lyases. The enzyme performed the endolytic depolymerization of the cellouronate substrate through β-elimination. Crystal structure of this enzyme was also resolved to 1.8 A°, representing the first resolved PL structure from *Trichoderma* (Konno et al. 2009b). Other identified *T. reesei* putative PLs include two PL7 alginate lyases, one PL8 chondroitin lyase, and a PL20 endo-β-1,4-glucuronan lyase (Häkkinen et al. 2012).

(iv) **Carbohydrate Esterases (CE)**

Carbohydrate esterase families make up at most 4% of the CAZymes in published *Trichoderma* genomes (Fig. 1). As indicated in Fig. 1, *T. asperelloides* TR356 and *T. harzianum* possess the highest number of CE genes (21) in their genomes, whereas *T. atroviride* P1 only has eight CEs. Häkkinen and coworkers performed re-annotation of the genes encoding for CAZymes of *T. reesei* by performing transcription in the presence of lignocellulosic substrates. Their analysis revealed a total of 22 carbohydrate esterases belonging to CE1, CE3, CE4, CE5, CE9, CE14, CE15, and CE16 families with S-formylglutathione hydrolase, esterase, acetyl xylan esterase, esterase/suberinase, chitin deacetylase, imidase, cutinase, acetyl xylan esterase, N-acetyl-glucosamine-6-phosphate deacetylase, N-acetylglucosaminylphosphatidylinositol de-N-acetylase, glucuronoyl esterase, and acetyl esterase activities. Only two of the CEs were studied and characterized. These were AXEI (Zhang et al. 2011a) and AESI (Li et al. 2008; Puchart et al. 2016), which function synergistically during the degradation of lignocellulose with the GH family cellulases and other hemicellulases (Table 1). *T. reesei* AXEI is also known to possess a carbohydrate-binding module. However, the crystal structure was determined at 1.9 A° resolution only for the catalytic domain (Hakulinen et al. 2000). A more recent study by Ferreira Filho et al. (2017) revealed the presence of 22 CEs in the *T. harzianum* genome by RNA-Seq analysis. These families of carbohydrate esterases could find potential industrial use in the degradation and modification of recalcitrant materials.

(v) **Carbohydrate-Binding Modules (CBM)**

Although CBMs are not enzymes by themselves, they are found linked to the catalytic domains of a diversity of CAZymes and function in substrate binding and aid catalytic domains during catalysis. CBMs of *Trichoderma* cellulases belonging to CBM1 family aid in binding of the enzyme to the cellulosic substrate (Reinikainen et al. 1992). CBM1 family are generally known to exhibit a preference for crystalline cellulose and increase the hydrolytic activity of their corresponding enzymes on less soluble cellulosic substrates (Bayer et al. 1998; Seiboth et al. 2011). The generally accepted paradigm for CBMs postulates that CBMs act by increasing the effective concentration of the enzyme on the polysaccharide substrate surface, thereby helping the catalytic domain to come close proximity with the substrate (Várnai et al. 2013).

T. atroviride and *T. harzianum* genomes are enriched in CBMs as seen in Fig. 1. They have more than a hundred CBMs. *T. reesei* GHs possess 25 CBM-containing enzymes induced by lignocellulose that belong to CBM1, CBM18, CBM24, CBM42, and CBM43 families (Häkkinen et al. 2012). The predicted number of CBMs is around 55 in the *T. reesei* published genomes. In a recent study, 46 CBMs were identified in *T. harzianum* via RNA-Seq analysis (Ferreira Filho et al. 2017).

CBMs, due to their high affinity for cellulosic substrates, have a potential to be used as cost-effective and efficient affinity purification tags. For example, a bacterial CBM (CBM2a) was used effectively for this purpose in fusion to protein A from *Staphylococcus aureus* for purification on Avicel PH101 (Rodriguez et al. 2004). The same CBM2a was also successfully used to purify well-known deglycosylation enzymes through the construction of CBM-EndoF1 and CBM-PNGaseF enzyme fusions (Kwan et al. 2005). Similarly, Sugimoto and coworkers constructed various fungal CBM1-red fluorescent protein (RFP) fusions and exhibited that CBM1-RFP fusion using CBM of CBHI (CEL7A) of *T. reesei* was expressed with high efficiency and finally the recovery rate of the purification on cellulose column was more than 80% (Sugimoto et al. 2012). Bayram Akcapinar and coworkers expressed codon-optimized and non-optimized EGI (CEL7B) from *T. reesei* in *P. pastoris* and exploited the presence of the natural CBM for purification on regenerated amorphous cellulose (Akcapinar et al. 2011).

These studies show that various CBMs, based on their attachment to diverse substrates, could serve as cost-effective, efficient, and highly scalable industrial purification tags.

3.2 Small Secreted Cysteine-Rich Protein Family

Fungi have a diversity of small secreted cysteine-rich protein (SSCP)-encoding genes, characterized by distinct cysteine motifs and cysteine content in their genomes. This family of proteins is postulated to be involved in a series of processes that govern the communication between the fungi and its immediate environment and also with other organisms. Three prominent subfamilies standing out in this group are hydrophobins (HFBs) (PF06766), cysteine-rich secreted proteins (CSPs) (PF00188), and cerato-platanins (CPs) (PF07249). An analysis of the 24 *Trichoderma* genomes based on the PFAM codes of these protein families reveals that hydrophobins are present and enriched in most of the *Trichoderma* species (Fig. 2). *T. atroviride* strains have the highest number of HFBs. Kubicek et al. (2019) performed further analysis of 12 commonly found *Trichoderma* spp. genomes for the presence of small secreted cysteine-rich proteins and revealed that 27 to 125 different SSCPs are present in these genomes. Each of these *Trichoderma* species genomes harbored three cerato-platanins, seven to 12 class II HFBs, and two to three pseudo-class I HFBs. They have also indicated that the variation in the numbers and diversity of this family is species specific.

Fungal hydrophobins are surface-active proteins belonging to small secreted cysteine-rich protein family. They are classified into at least two main classes (class I and class II) based on the solubility of their aggregates, cysteine spacing patterns, and hydropathy profiles (Nakari-Setälä et al. 1996; Linder et al. 2001; Przylucka et al. 2017a, b). Recent findings suggested that there are hydrophobins which do not fit into either classes (Jensen et al. 2010; Littlejohn et al. 2012). They have eight conserved cysteine residues forming four disulfide bridges, thereby giving these proteins their character and unusual stability. They exhibit affinity toward both hydrophobic and hydrophilic surfaces since they have an amphiphilic structure and reverse the surface properties. Hydrophobins coat the fungal mycelium and fungal spores. These proteins are thought to play important functions during the fungal life and in a range of host-pathogen interactions (Whiteford and Spanu 2002; Bayry et al. 2012; Guzmán-Guzmán et al. 2017; Cai et al. 2020). Class I hydrophobin, RodA of the human opportunistic pathogen *Aspergillus fumigatus*, is the main protein present on the fungal spores and acts as a Trojan horse by shielding the sugars on the spore surface from recognition by the human immune system, thereby aiding the pathogenesis (de Carrion et al. 2013).

Exploiting these features, recent studies reported application of hydrophobins in a variety of areas, such as in the coating of biomedical implants (Devine et al. 2019), in drug formulations (Haas Jimoh Akanbi et al. 2010), as coating layers for fibroblast activation (Janssen et al. 2004), and as coating layers for protein immobilization (Qin et al. 2007). Anticancer activity of a class I hydrophobin, SC3 from *Schizophyllum commune*, was shown in sarcoma and melanoma mouse models. Fungal hydrophobin caused a significant decrease in the size and weight of the melanoma. A microscopic analysis of the tumors indicated a strong antitumor effect

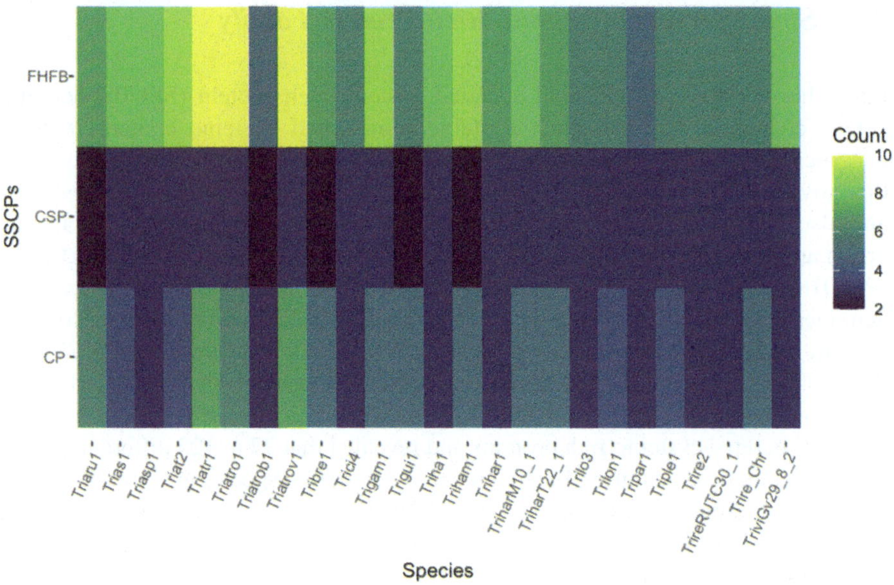

Fig. 2 Diversity and number of small secreted cysteine-rich protein (SSCP) families in 24 *Trichoderma* genomes from the genomic analysis data in Joint Genome Institute (JGI) (https://mycocosm.jgi.doe.gov/Trichoderma/Trichoderma.info.html), accession date: 21.01.2021). FHFB, fungal hydrophobins (PF06766); CSP, cysteine-rich secreted proteins (PF00188); CP, cerato-platanins (PF07249). Coloring indicates the number of genes in each family represented in each genome. [*Trichoderma arundinaceum* IBT 40837 (Triaru1), *Trichoderma asperelloides* TR356 (Trias1), *Trichoderma asperellum* CBS 433.97 (Triasp1), *Trichoderma atrobrunneum* ITEM 908 (Triat2), *Trichoderma atroviride* B10 (Triatr1), F7 (Triatro1) and P1 (Triatrob1), *Trichoderma atroviride* v2.0 (Triatrov1), *Trichoderma brevicompactum* IBT40841 (Tribre1), *Trichoderma citrinoviride* TUCIM 6016 (Trici4), *Trichoderma gamsii* T6085 (Trigam1), *Trichoderma* spp. NJAU 4742 (Trigui1), *Trichoderma hamatum* GD12 (Triham1), *Trichoderma harzianum* CBS 226.95 (Triha1), *Trichoderma* spp. M10 (TriharM10_1), *Trichoderma afroharzianum* T22 (TriharT22_1), *Trichoderma harzianum* TR274 (Trihar1), *Trichoderma longibrachiatum* ATCC 18648 (Trilo3), *Trichoderma longibrachiatum* MK1 (Trilon1), *Trichoderma parareesei* CBS 125925 (Tripar1), *Trichoderma* spp. TPhu1 (Triple1), *Trichoderma reesei* QM6a (Trire2), *Trichoderma reesei* RUT C-30 (TrireRUTC30_1), *Trichoderma virens* Gv29–8 (TriviGv29_8_2) (Cai and Druzhinina 2021)]

on both tumors, possibly through an immunomodulation mechanism (Akanbi et al. 2013).

A recent study reported the use of a fungal HFB (HGFI) for the modification of a fat-soluble drug, menaquinone-7 (Tang et al. 2021). HFB-modified menaquinone-7 was shown to significantly promote osteoblast differentiation. Osteoclast differentiation was shown to be inhibited. Zhao et al. (2016) improved the serum stability and in vivo half-life of glucagon-like peptide-1 (GLP-1) by using recombinant HGFI and its mutant produced in *P. pastoris* as drug carriers. The mutant hydrophobin was designed as a controlled pH drug-release system for GLP-1 as a drug candidate for type II diabetes. HFBI-coated niosomes were prepared as an alternative to PEG coating for the delivery of doxorubicin to cancer cell lines. Niosomes coated with HFB were shown to exhibit better size distribution, higher entrapment efficiency, more sustained release profile, enhanced biocompatibility, and improved anticancer effects in comparison to PEG coating (Barani et al. 2020).

HFBs were found to enhance cutinase activity (Espino-Rammer et al. 2013). To this end, recombinantly produced HFB4 and HFB7 from *T. virens* were successfully used to modify PET and glass surfaces (Przylucka et al. 2017a). Moreover, a HFB-cutinase fusion protein was constructed to improve the depolymerization and recycling of PET (Ribitsch et al. 2015). One of the well-studied fungal hydrophobins of *T. reesei* HFBs was used as fusion tags for purifying recombinant proteins by aqueous two-phase separation (Linder et al. 2004).

Cerato-platanins are important fungal effector proteins. They are believed to function in the fungal interactions. They are involved in interaction with plants by eliciting plant resistance reactions (Seidl et al. 2006). EPL1 from the industrious biocontrol fungi *T. harzianum* was shown to be involved in mycoparasitism against a phytopathogenic fungi, plant resistance induction, and self-cell wall protection (Gomes et al. 2015).

Recently, a new family of small secreted cysteine-rich proteins, called hyphosphere (HFS) proteins, encoded by *hfs1, hfs2,* and *hfs3* genes and two novel hydrophobins encoded by *hfb11* and *hfb12* genes were identified by genome mining of *T. guizhouense* (Zhao et al. 2021). Similar to other SSCPs, these larger-sized HFS proteins exhibited their unique pattern of eight single cysteine residues (C-CXXXC-C-C-C-C). Of these families, HFS1 and HFB12 proteins were heterologously produced in *Pichia pastoris* and shown to be surface active. Moreover, addition of the recombinantly produced HFS1 to the glass wool improved the attachment of the *T. guizhouense* strain. However, addition of HFS1 to the roots of *Solanum lycopersicum* (tomato) seedlings exhibited a reverse pattern, indicating a possible role in root colonization.

These unusual and remarkable abovementioned properties render the small secreted cysteine-rich protein family as an ideal candidate for a diversity of industrial, agricultural, and therapeutic applications.

3.3 Genes Involved in Bioactive Secondary Metabolite Synthesis

Fungi are notorious producers of a large repertoire of secondary metabolites (SMs), which are also termed natural products, that are not directly required for growth yet have critical roles in signaling, development, and interaction with other organisms, and in some cases they are even vital for survival (Mukherjee et al. 2012b; Brakhage 2013; Zeilinger et al. 2016). Fungal SMs, nonpolar small molecules with low molecular mass, are important for humans due to their potential to be applied as novel and innovative therapeutic agents (Degenkolb et al. 2008).

As a result of their huge potential to produce an array of SMs, *Trichoderma* species are well-known as the most potent biocontrol agents in use today (Khan et al. 2020). Not only SMs produced by *Trichoderma* have beneficial effects on crop plants, but they also have antagonistic effects against numerous bacteria, yeast, and fungi. *Trichoderma*-derived SMs are used as commercial biofungicides. In addition, some secondary metabolites demonstrate good therapeutic agent properties, and therefore, they are potential drug candidates (Reino et al. 2008; Khan et al. 2020).

The biosynthesis of fungal SMs is performed by unique biochemical pathways using the primary metabolite pool, such as acetyl-CoA, mevalonate, and amino acids (Keller 2019). Unlike the genes required for the synthesis of primary metabolites, the genes encoding the enzymes required for the synthesis of SMs are located in clusters that are not expressed ubiquitously under standard laboratory conditions (Brakhage and Schroeckh 2011; Zeilinger et al. 2016). These gene clusters consist of polyketide synthases (PKSs), non-ribosomal peptide synthases (NRPSs), terpene synthases (TCs), and PKS-NPRS that generate hybrid metabolites (Kubicek et al. 2011). Numbers of core secondary metabolism-related genes in the *Trichoderma* genomes are presented in Fig. 3. (*Trichoderma* genome cluster information was taken from JGI and National Center for Biotechnology Information (NCBI) databases) (Date of access 21.01.2021).

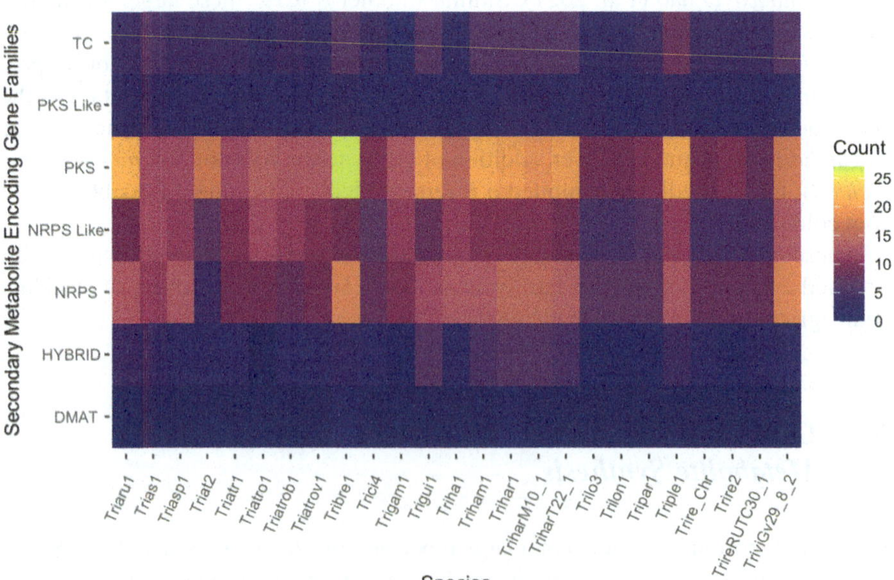

Fig. 3 Number and distribution of secondary metabolite encoding gene families over 24 *Trichoderma* genomes based on the genomic analysis data in Joint Genome Institute (JGI) (https://mycocosm.jgi.doe.gov/Trichoderma/Trichoderma.info.html, accession date: 21.01.2021). *DMAT* tryptophan dimethylallyltransferase, *NRPS*, non-ribosomal peptide synthetase, *PKS* polyketide synthase, *TC* terpene cyclase. Coloring indicates the number of genes in each family represented in each genome

(i) **Polyketide Synthase**

Polyketides constitute a large group of important secondary metabolites that exhibit significant multiplicity with both their structures and functions (Risdian et al. 2019). They exert a diverse range of bioactivities, such as antibacterial (e.g., rapamycin, tetracycline), antifungal (e.g., aflatoxin B1, fusaric acid), anticancer (e.g., doxorubicin), and anticholesterol (e.g., lovastatin, compactin) (Zeilinger et al. 2016; Risdian et al. 2019). These exceptional properties make them clinically important. Moreover, some organisms such as bacteria, fungi, plants, and insects use polyketides as protective compounds, while some insects use it for pheromonal communication (Pankewitz and Hilker 2008; Khosla 2009; Mukherjee et al. 2012b).

Biosynthesis of polyketides is a very complex process because they are synthesized from simple units, such as acetyl-CoA and malonyl-CoA, by multifunctional enzymes called polyketide synthases (PKSs). The process requires a multitude of enzymatic reactions performed by an acyl transferase (AT), a ketoacyl synthase (KS), and a phosphopantetheine attachment site domain (Keller et al. 2005; Mukherjee et al. 2012b).

Although *Trichoderma* genomes are rich in PKS-encoding genes, there are limited genomic studies reported. When the first genome sequence of the symbiotic fungus *T. reesei* was published, it was seen to contain two NRPS-PKS hybrid coding genes and 11 PKS genes. *T. atroviride* and *T. virens* each encode 18 PKS genes. Phylogenetic analysis of the PKS genes of *T. reesei, T. virens*, and *T. atroviride* revealed that nine of the *T. reesei* PKS-encoding genes are orthologous to *T. virens'* and most of the PKS genes are reported to belong to lovastatin/citrine or fumonisins (Martinez et al. 2008; Kubicek et al. 2011; Baker et al. 2012). Phylogenetic analysis of PKS-encoding genes also revealed that these genes are responsible for *Trichoderma* yellow-green pigment. Most of the PKS-encoding genes found in *T. reesei* were also found in *T. virens* and *T. atroviride*, but approximately half of these genes are reported as recent additions to these two species from evolutionary perspective. It is suggested that these genes were passed on through recombination. The contribution of horizontal gene transfer (HGT) to the evolution of PKS-encoding gene is still under debate. Although there is evidence of HGT of modular PKS genes among bacteria, there is little evidence for HGT with fungi PKS genes (Ridley et al. 2008; Baker et al. 2012).

Tajima's D test was used to understand whether the PKS-encoding genes are products of HGT or another evolutionary mechanism. It was reported that PKS-encoding genes of *Trichoderma* evolve under purifying selection and divided into five clades: (a) nonreducing PKS clades I and II, (b) nonreducing PKS clade III, (c) reducing PKS clade I and II (lovastatin type), (d) reducing PKS clade III (fumonisin type), and (e) a small clade that occupied a basal position to the reducing PKS clades (Tajima 1989; Baker et al. 2012).

A few fungal SMs are produced by PKS-NPRS/NPRS-PKS hybrid enzymes. Hybrid enzyme-derived chimeric metabolites, referred to as lipopeptides, are generally known to be bioactive (Fisch 2013; Keller 2019). These hybrids, neglected by phylogenetic analysis studies, are thought to be generated by means of gene duplications, gene loss, or HGT events (Theobald et al. 2019). The number and distribution of hybrid genes in the *Trichoderma* genomes are presented in Fig. 3. According to the functional analyzed results of *T. virens* NPRS-PKS hybrid enzyme TEX13, it has been reported that TEX13 causes the induction of the *pal* gene that was involved in defense during *Trichoderma*-plant interactions (Mukherjee et al. 2012a).

(ii) **Non-ribosomal Peptide Synthase**

Non-ribosomal peptide synthases (NRPSs) are multimodal enzymes that synthesize non-ribosomal peptides, which are economically and ecologically important secondary metabolites of bacterial or fungal origin. Unlike peptides that are ribosomally synthesized and have posttranslational modification, NRPs contain modified versions of both proteinogenic (e.g., methylated or hydroxylated amino acids and D-forms) and non-proteinogenic (e.g., isovaline) amino acids. Produced peptides could be either linear or cyclic. These metabolites also exhibit a wide range of biological properties, such as antimicrobial (e.g., penicillins), antifungal (e.g., echinocandin), anticancer (e.g., terrequinone), and immunosuppressive (e.g., cyclosporine A) (Keller et al. 2005; Strieker et al. 2010). The main groups of NRPs from *Trichoderma* spp. are peptaibiotics, epidithiodioxopiperazines (ETPs), and siderophores (Zeilinger et al. 2016).

Peptaibiotics

Peptaibiotics are mostly linear peptides composed of four to 21 residues and molecular weight ranging from 500 to 2100 Da. According to their chemical structures, they are divided into six subgroups: peptaibols, lipopeptaibols, lipoaminopeptides, cyclic peptaibiotics, other peptaibiotics, and all Aib-replaced peptides (Stoppacher et al. 2013; Zeilinger et al. 2016).

Peptaibols, major class of peptaibiotics, are linear peptides with five to 20 amino acid residues produced by peptaibol synthetases, which are characterized by the presence of high levels of nonstandard amino acids (hydroxylated C-terminus and the N-terminal acetylated amino acids) (Tamandegani et al. 2020). After the isolation of the first peptaibols, suzukacillin and alamethicin, interest in these bioactive substances continues to increase and studies continue to develop (Ooka et al. 1966; Meyer and Reusser 1967; Zeilinger et al. 2016). A number of sequences and structures continue to increase in the peptaibol database hosted by the School of Crystallography, Birkbeck College, University of London, UK http://peptaibol.cryst.bbk.ac.uk (Whitmore and Wallace 2004). Although peptaibols isolated from *Acremonium, Tylopilus, Boletus, Bionectria, Paecilomyces, Emericellopsis, Cephalosporium, Stilbella, Gliocladium,* and *Tolypocladium* species have been reported, *Trichoderma* species are known as major producers of peptaibols (Stoppacher et al. 2013). Peptaibols with antimicrobial activities against bacteria, fungi, and viruses have been reported and their mode of action is largely based on their physical, chemical, and biological features. For example, by creating voltage-dependent ion channels in plasma membranes, they increase membrane permeability and consequently cause cell death by cytoplasmic leakage (Mueller and Rudin 1968; Chugh and Wallace 2001; Tamandegani et al. 2020).

NPRSs have modules that recognize, activate, and modify a single residue and add specific monomers step by step to produce final peptide. There are many peptaibol synthetases (7-, 11-, 14-, 18-, 19-, 20-modules) in *Trichoderma* genomes (Mukherjee et al. 2011; Degenkolb et al. 2012). The first peptaibol from *Trichoderma* species was peptaibol synthetase gene *tex1*, identified from *T. virens* GV29-8 in 2002 (Wiest et al. 2002; Tamandegani et al. 2020). Mutagenesis studies showed that this gene was responsible for the production of only 18 amino acid peptaibols because it eliminated all peptaibol isoforms. This gene contains a 62.8-kb continuous open reading frame (ORF) and encodes a mature protein of approximately 2.3 MDa (Wiest et al. 2002). Studies revealed that the predicted protein structure includes 18 peptide synthetase modules and multiple NRPS genes are responsible for peptaibol synthesis in *T. virens*. Further studies with *T. virens* exhibited that 14-modules of NPRS were required for the synthesis of shorter peptaibols (11- and 14-residues) (Viterbo et al. 2007; Mukherjee et al. 2011). However, as a result of this study, it was postulated that the 14-module genes should skip three modules in order to be responsible for synthesizing both 11-residue and 14-residue peptaibols, whereas module skipping has not yet been demonstrated (Wenzel et al. 2005, 2006; Degenkolb et al. 2012).

Epipolythiodioxopiperazines

Epipolythiodioxopiperazines (ETPs) are a poorly studied class of secondary metabolite toxins characterized by the presence of an internal sulfur-bridged dioxopiperazine ring. The toxicity of ETPs arises from the disulfide bridges, inhibiting proteins via cross-linking from cysteine residues and generating reactive oxygen species (ROS), such as superoxide or hydrogen peroxide (Gardiner et al. 2005; Patron et al. 2007).

There are many ETPs, almost all of which are produced by ascomycetes. Gliotoxin, the best known ETP, gets its name from the fungus *Gliocladium fimbriatum*, from which it was originally identified. Gliotoxin is also produced by *Aspergillus fumigatus*, *Aspergillus terreus*, *Aspergillus flavus*, *Aspergillus niger*, *Penicillium terlikowskii*, and *T. virens* (Gardiner et al. 2005; Patron et al. 2007). Gliotoxins display bioactivity against the human pathogenic fungus *Aspergillus fumigatus*. Gliotoxins produced by *T. virens* also play an important role in the biocontrol of plant pathogenic fungi (Howell 2006; Scharf et al. 2016). The strain of *T. virens* called the "Q strain" produces gliotoxin. The "P strains" of *T. virens* produce another ETP, gliovirin. Gliovirin produced by "P" strains acts as a protoplasm coagulation factor against oomycetes, such as *Pythium ultimum*, and is also reported to inhibit growth of *Phytophthora palmivora* and *Phytophthora megakarya* (Howell and Stipanovic 1983; Mukherjee et al. 2012b; Pakora et al. 2018). Since its discovery in 1982, 83 molecules of the gliovirin family have been reported (Howell and Stipanovic 1983; Zhu et al. 2020).

Although it was discovered owing to its antibacterial/antifungal effect, its powerful effect on cell lines also attracted the attention of researchers. Experiments with mammalian cell lines have revealed that gliotoxin causes both apoptotic and necrotic cell death by causing mitochondrial membrane damage through the production of ROS (Vigushin et al. 2004) . In another study, it has been shown that gliotoxin (>50 mM) causes necrosis through the redox-sensitive calcium channel in the plasma membrane of murine thymocytes (Hurne et al. 2002). Similar to gliotoxin, the effects of gliovirin in mammalian cells have been studied. In a study with mammalian cell lines, it was revealed that gliovirin inhibited expression of the proinflammatory cytokines tumor necrosis factor-α (TNF-α) and interleukin-2 (IL-2) with a number of signal transduction pathways, leading to nuclear factor kappa B (NF-kB) activation. So, gliovirin is a candidate anticancer drug (Rether et al. 2007).

When *Aspergillus fumigatus'* entire genome was sequenced in 2005, the genes responsible for gliotoxin biosynthesis were also identified (Gardiner and Howlett 2005; Nierman et al. 2005). Like other secondary metabolites, the genes responsible for gliotoxin biosynthesis are found in clusters. This cluster includes 12 genes, such as core enzyme GliP (NRPS dioxopiperazine synthetase), helper biosynthetic enzymes, and zinc finger transcription factor *gliZ* (Gardiner et al. 2005; Scharf et al. 2016). The gene cluster of *T. virens* consists of only eight genes, closely related to the gene cluster of *Aspergillus fumigatus*. By creating mutations in the *gliP* locus of *T. virens*, this cluster has been shown to be responsible for the production of gliotoxin. These mutants grow faster, and they are more sensitive to oxidative stress. It has also been reported that these mutants that did not produce gliotoxin have shown less activity against *Pythium ultimum* (Vargas et al. 2014; Scharf et al. 2016). Interestingly, *T. reesei* does not produce gliotoxin, although it harbors six genes of the GliP and the Gli cluster. The smaller cluster of the *T. reesei* missing other genes may explain why this species does not produce gliotoxin (Mukherjee et al. 2012b). In addition to the GliP cluster, the *T. virens* genome also has the SirP gene cluster similar to the gene responsible for the biosynthesis of the phytotoxin sirodesmin PL in the phytopathogen *Leptosphaeria maculans*. Metabolites associated with this cluster remain unclear because they cannot be expressed under standard laboratory conditions (Patron et al. 2007).

Siderophores

Iron is an essential element necessary for the survival of organisms. This element is critical for cellular enzyme activities (e.g., DNA replication enzymes). Microorganisms, including viruses, bacteria, and fungi, have different iron uptake/use mechanisms to support their ability to uptake/use iron (Wallner et al. 2009; Chhabra et al. 2020). In fungi, the two most common mechanisms for iron uptake including (1) reductive iron assimilation (RIA) and (2) non-reductive (siderophore-mediated) iron uptake have been described (Howard 2004). The RIA mechanism involves reduction of ferric iron to the ferrous form via ferrireductases. Most fungi use siderophore-mediated iron uptake mechanism that transports iron-bound siderophore complexes via transmembrane transporters (Philpott 2006; Bairwa et al. 2017). Siderophores are low-molecular-mass (Mr < 1500) ferric iron-chelating secondary metabolites. They are important in biotrophic interactions with plants and with other microbes (Renshaw et al. 2002; Wallner et al. 2009). Most of the fungal siderophores are categorized in two groups: (1) hydroxamates (e.g., coprogens, ferrichromes, and fusarinines) and (2) polycarboxylates (Howard 1999; Renshaw et al. 2002). Siderophores are synthesized by NRPS, and they consist of L-ornithine, a non-proteinogenic amino acid (Lehner et al. 2013).

Trichoderma spp., known as siderophore producers, are used in the biocontrol of plant pathogenic fungi (Benítez et al. 2004). Anke et al. used nine *Trichoderma* strains and identified siderophore coprogen, coprogen B, ferricrosin and fusgene type and coprogen, ferricrosin and palmitoylcoprogen from micelle (Anke et al. 1991). Other studies have reported the presence of cis- and trans-fusarinine, dimerum acid, ferrichrome C, fusarinine B, fusigen (fusarinine C), and Nα-dimethylisoneocoprogen II in *Trichoderma* spp. (Lehner et al. 2013). *T. reesei, T. virens, and T. atrovirideTrichoderma atroviride (T. atroviride)* whole genome sequencing results revealed that all three *Trichoderma* spp. have a single ferricrosin synthesis gene belonging to the secondary metabolite gene cluster (Kubicek et al. 2011). Studies suggest that this gene is indeed involved in ferricrosin synthesis and protection against oxidative stress in *T. virensTrichoderma virens (T. virens)* (Mukherjee et al. 2012b). *T. virens* and *T. reesei* orthologs (SidD and NPS6) have two putative gene clusters that contain an NRPS as the core member known to play a role in siderophore synthesis, while the *T. atroviride* has only the NPS6 orthologs (Kubicek et al. 2011). Importance of the siderophores in microbial competition and biocontrol may continue to be revealed as the number of genetic studies increases.

(iii) **Terpenoids**

Terpenoids are various volatile and nonvolatile secondary metabolites synthesized from the five-carbon isopentyl units, isopentenyl diphosphate (IPP), and its isomer dimethylallyl diphosphates (DMAPP). Despite their remarkable variety, terpenoids are synthesized by terpene synthases (TC)/terpene cyclases (TS) from only a few precursor molecules (Christianson 2008; Stoppacher et al. 2010). These molecules are ecologically and economically important due to their interspecies and intraspecific communication and defense properties.

Trichoderma spp. have a wide variety of terpenoids that play important roles in the physiology and interaction with other organisms, acting as chemical messengers, structural components of membranes, and inducers of plant defense responses (Bansal and Mukherjee 2016; Vicente et al. 2020). Numerous terpenoids have been identified in *Trichoderma* spp., such as volatile terpenes, tetracyclic diterpene harziandione, and sesquiterpenes. Although there are a large number of terpenes isolated from *Trichoderma* species, extensive analyses of TC genes and biosynthetic pathways have not yet been reported. Only a few of the TC members have been functionally characterized. A gene cluster responsible for the biosynthesis of 24 volatile sesquiterpenes in *T. virens* was identified using mutants. Volatile terpenes from *T. reesei, T. atroviride,* and *T. virens* were identified in the same study (Mukherjee et al. 2006; Crutcher et al. 2013; Bansal and Mukherjee 2016). The Vir cluster consisting of a terpene cyclase, four cytochrome P450s, a monooxygenase, and a major facilitator superfamily (MFS) transporter, which enables the production of unique terpenoids, has been reported to exist only in *T. virens,* not in *T. reesei* and

T. atroviride (Mukherjee et al. 2006; Crutcher et al. 2013). In another study using *T. britannicum hmgR*-silenced mutants, this gene encodes hydroxy-methylglutaryl-CoA reductase (HMGR), and a decrease in antifungal activity was reported against *Rhizoctonia solani* and *Fusarium oxysporum* (Cardoza et al. 2007). After a comparative analysis of TSs of 21 strains of *Trichoderma* spp. and using *T. gamsii* T6085 for expression studies, the researchers reported that they identified 15 TS groups. Besides, they demonstrated the presence of clade-specific enzymes (Vicente et al. 2020). Although genome mining studies have the potential to analyze and reveal the entire TC/TS gene family in *Trichoderma* spp., current studies are mostly limited to certain species.

(iv) **6-Pentyl Pyrone**

6-Pentyl pyrone (6-PP), the "coconut aroma," compound produced by *Trichoderma* spp. has antifungal activity and acts as a plant growth regulator (Vinale et al. 2008). Classified as volatile organic compounds (VOCs), 6-PPs include alcohols, ketones, alkanes, and alkenes (Korpi et al. 2009). The synthesis of fungal VOCs is affected by ambient conditions, such as pH, temperature, and light, and is species specific (Wilkins et al. 2003).

While the enzymes and genes involved in the biosynthesis of 6-PP are well-known in plants, in *Trichoderma* it is still uncertain. However, as a result of comparative genome analysis, a *lipoxygenase* gene (ID 33350) was identified in *T. atroviride*, which is not found in *T. virens* and *T. reesei*. This gene has been reported to be upregulated during the interaction with *R. solani* (Kubicek et al. 2011; Atanasova et al. 2013a). Rubio *et al.* disrupted a transcription factor, *ctf1* gene, by homologous recombination and showed that *ctf1* plays a role in the production of 6-PP and the antifungal activity of *T. harzianum* (Rubio et al. 2009). VOCs produced by *T. asperelloides* PSU-P1 have antifungal activity, promote plant growth, and induce defensive responses in *Arabidopsis thaliana*. Researchers report that cell wall-degrading enzyme, chitinase (CHI), and β-1,3-glucanase (GLU) genes were upregulated in the *A. thaliana* (Phoka et al. 2020). Few studies have been reported on these interesting compounds. Further elucidation of the biosynthetic pathways and functional genomic analysis of this group would decipher their distinctive properties similar to other secondary metabolites and allow their practical application.

Trichoderma has an enormous potential to produce secondary metabolites, and although it has been known since the 1930s, secondary metabolite pathways, associated genes, and functional studies are still limited. In addition to whole genome sequencing studies, the use of new tools of mining the fungal secondary metabolome (SMURF, MIDDAS, FunGeneClusterS, SeMPI) and metabolomic studies will enable us to reach much more for this species (Keller 2019).

4 Computational Approaches to Genome Mining and Industrial Gene Discovery

Since the revolutionary breakthrough of genome sequencing (Sanger et al. 1977), especially next-generation sequencing (NGS) (Schuster 2008), the cost per sequencing a genome has been in a drastically reducing trend over years (van Nimwegen et al. 2016). This enabled researchers and similar R&D-focused entities to produce relatively large amount of data in a relatively smaller time frame with comparably low cost and effort as evident by NCBI Sequence Read Archive (SRA) (Leinonen et al. 2011) database growth (Kodama et al. 2012). Due to the high-throughput nature of the sequencing process and worldwide high demand in sequencing (van Dijk et al. 2014), the speed and progress at which manual annotating and biological association of such datum has significantly stalled behind the global rate of NGS data yield (Baker 2010). This inevitably led researchers to investigate computational approaches to assign biological context to such data by means of annotation, curation, and systemization. Several manually and automatically curated biological databases (Tateno et al. 2002; Pruitt et al. 2007; Benson et al. 2013; Siva 2008; Grigoriev et al. 2012; Howe et al. 2020; Clough and Barrett 2016; Parkinson et al. 2007; Bateman et al. 2017; Hunter et al. 2009; Griffiths-Jones et al. 2003; Winnenburg et al. 2006; Bateman et al. 2004; Berman et al. 2000; Murzin et al. 1995; Orengo et al. 1997; Cuomo and Birren 2010) were created as outcomes of such attempts.

After initial sequencing of complete human genome, several offshoot projects, such as 1000 Genomes (Siva 2008), have been launched in order to systematically utilize NGS and its benefits in several domains. Two particular examples of such projects in this context is the 1000 Fungal Genomes (Grigoriev et al. 2014) and the Fungal Genome Initiative (FGI) (Cuomo and Birren 2010). The Fungal Genome Initiative has been initiated by Broad Institute (BROAD) and supported by National Genome Research Institute, the National Science Foundation, the National Institute of Allergy and Infectious Disease, and the US Department of Agriculture. FGI doesn't have any genome data that belongs to *Trichoderma* spp. to this date yet. 1000 Fungal Genomes project has been launched and conducted by JGI and its collaborators. Their results are published under MycoCosm portal (https://genome.jgi.doe.gov/portal/), which is developed and maintained by JGI. Several fungal species and their sub-strains have been actively researched, and their genomes are sequenced and assembled in order to provide reference genome assemblies to be used for further studies. These efforts provided the sequence content basis on which further annotation and biological association studies could be conducted by researchers. There are currently genome assemblies and their respective gene annotations in MycoCosm for 24 *Trichoderma* strains as described in the previous sections.

Raw sequence information contained in genomes, however, does not constitute any useful knowledge in and of themselves, and therefore, these sequences have to be annotated to assign biologically meaningful information to them. This annotation effort, called as genome annotation, can be classified into three levels:

(a) Nucleotide-level annotation in which genetic features, such as gene, open reading frame (ORF), exon, coding region, noncoding region, etc., are labeled.
(b) Protein-level annotation aims to assign function to the products of genome produced by aforementioned labeled genetic features, such as genes, ORFs, etc.
(c) Process-level annotation's main goal is to put genes and their functions into biological context with respect to cellular and organismal physiology and environmental interactions (Stein 2001).

As the chief aim of nucleotide-level annotation is to discover genes hidden in genome by mining, we will focus on this annotation level throughout this section. That is, we will first briefly survey how fungal genome annotation is performed and then discuss how genome annotation processes are used operatively to discover novel genes (nucleotide level), to assign functions to genes (protein level), and to provide biological context to these functions regarding industrial applications (process level).

Several institutes such as JGI and BROAD have put decades of effort into finding genes and annotating proteins in fungal genomes. Out of their experience in doing so, Haridas et al. (2018) and Haas et al.(2011) published manuscripts that conceptualize and summarize whole genome annotation process and provide retrospective overviews into the field. Haridas et al. (2018) offer a simpler summary and divide fungal genome annotation into three main steps. First is noncoding features in the genome identified by identifying repeats, transposable elements (TE), transfer RNA (tRNA), small nucleolar RNA (snoRNA), ribosomal RNA (rRNA), and other noncoding RNA (ncRNA). Following this identification, gene prediction approaches are applied to determine gene structures and mine genes, aka gene modeling or gene discovery.

There are three main approaches to gene prediction: (i) ab initio gene prediction, which relies on statistical methods without leveraging any prior information; (ii) homology-based gene prediction, which maps genes from previously known organisms based on sequence similarity; and (iii) transcriptome-based gene prediction, which uses transcripts to predict genes via either mapping RNA-Seq reads to genome or mapping RNA-Seq assemblies to genome and build gene models from those aligned transcripts. In practice, a variety of gene predictors from these three categories are applied, and each gene predictor's gene model is regarded as a distinct line of evidence. It is best to inspect these gene models visually and select best models by combining multiple evidence models. A scoring filtering procedure, where every model is evaluated by transcriptome and homology support and the use of tools that select the best model for each locus, is suggested. This step may provide both nucleotide-level and protein-level annotations by modeling gene structures and annotating their proteins. In practice, this gene prediction step constitutes most of the annotation efforts. Lastly, functional annotation of proteins and noncoding genomic features are performed, mostly concurrently. This is performed by means of three general approaches: (i) characterization of protein sequence regions, such as domains, families, etc., (ii) similarity-based search against existing proteins with known functions, and (iii) mapping to functional annotation schemes predefined as gene and function classification schemes, such as gene ontologies and molecular pathways. Annotating to these classification schemes helps toward providing biological context into functional annotations and hence completing annotation to process level. As an exception, mitochondrial gene annotation should be theoretically performed by starting predicting tRNAs since mitochondrial genes transcribed polycistronically and cleaved by tRNAs. However, tools developed for mitochondrial annotation are still in infancy compared to other genomic features, and there still remains a probability of substantial errors due to the methodological insufficiencies and relatively small size of the mitochondrial genome, high levels of length variability of rRNAs, and errors occurring in circular assembly of the mitochondrial genome. On a final note, they provide concrete advices on how to check for correctness and evaluate for accuracy of the annotated genomes, and they provide detailed examples and usage of tools with accompanying codes under a procedural narrative. In their earlier work (Grigoriev et al. 2006), Grigoriev and coworkers surveyed the fungal genome annotation field, discussed achievements and challenges in the field, described plethora of methods along with their advantages and disadvantages, expanded on fungal gene structure and its challenges in gene prediction, and elaborated on themes, such as pseudogene annotation, annotation with experimental data, validation of gene models, and manual curation. They additionally provided a conceptualized annotation pipeline as a workflow diagram.

5 Bioinformatics Resources and Tools

Briefly mentioned in the beginning of this section, a myriad of databases emerged out as by-products of computational tool development studies and annotation efforts. Moreover, several fungal and *Trichoderma* specific tools were developed, and resources have been compiled. These databases, resources, and tools are discussed in this section in relation to their application to fungal genomes and *Trichoderma* spp.

Sequence reads resulting from sequencing have to be assembled into comprehensive genomic units, such as contigs, scaffolds, and chromosomes. This is achieved by merging reads in between them in such consecutive order that their starts and ends overlap with each other in order to constitute cohesive units while considering sequencing errors. This so-called fragment assembly problem is aimed to be solved by assemblers that can be categorized as (a) de novo assemblers that assemble fragments from scratch without priory information and guidance and (b) reference-based assemblers that assemble fragments by using a reference assembly as a template. Since the performance comparison between various assemblers are out of scope of this section, Zhang et al. (2011b) can be referred for such NGS-focused comparison. In addition, a de novo assembly guide with a specific focus on fungal genomes can be found in Haridas et al. (2011). Most widely used assemblers include Abyss (Simpson et al. 2009), SOAPdenovo2 (Luo et al. 2012), SPAdes (Bankevich et al. 2012), and Velvet (Zerbino and Birney 2008) for NGS. Canu (Koren et al.

2017) is a worthful and widely used tool to mention, developed to assemble Oxford Nanopore sequencing (ONS) long reads as opposed to short reads of NGS. ONS is a third-generation sequencing (TGS) that is rapidly being adopted, and there are studies reflecting on its advantages over short-read sequencing technologies and challenges contained therein (Branton et al. 2009). A survey on genome assembly and genome analyses for TGS can be found in Wee et al. (2019).

Since different approaches produce different assembly results for the same sequence sets, resulting assemblies must be put through a quality control process to decide which ones are better or to decide whether they are feasible or not. Booker et al. (2005) provide an iterative framework within which to produce high-quality assemblies and to identify errors and issues during the process which starts from sequencing, whereas Darling et al. (2011) defined a system to measure assembly quality under several scoring metrics, and O'Neil and Emrich (2013) investigated which metrics reflect real assembly quality for de novo transcriptome assembly. There are also multiple genome assembly assessment and visualization tools and pipelines available to researchers. Some of the widely used and recently developed tools for evaluation include but not limited to QUAST (Gurevich et al. 2013), REAPR (Hunt et al. 2013), GenomeQC (Manchanda et al. 2020), SolexaQA (Cox et al. 2010), dnAQET (Yavas et al. 2019), Referee (Thomas and Hahn 2019), and SQUAT (Yang et al. 2019), which can be used for both pre- and post-assembly quality assessment. Aside from assessment under various quality scoring schemes, BUSCO (Simão et al. 2015) is able to assess both genome assembly and annotation completeness based on evaluation of gene content in the genome. Additionally, GMASS (Kwon et al. 2019) is developed to measure similarity between genome assemblies in a structured way as there are multiple assemblies even for same strains or samples. Mapleson et al. (2015) developed RAMPART, a configurable workflow management system for de novo genome assemblies that guides users on parameters such as assembler selection. Icarus (Mikheenko et al. 2016) was developed to visualize assemblies; meanwhile, tools such as QUAST (Gurevich et al. 2013) and SQUAT (Yang et al. 2019) offer visualization capabilities in addition to assessment. Though these tools can be applied to any kind of organism, FGMP (Cissé and Stajich 2019) has been recently developed particularly to assess fungal genome completeness, which claims to be an especially accurate tool for fungal genomes and could be also useful for non-model organisms without reference genomes.

Gene prediction performed on assembled genomes consists of several steps, and myriads of tools are used at each step, mostly concurrently as outlined previously. RepeatMasker (Tarailo-Graovac and Chen 2009) is used to identify known repeat sequence elements, whereas RepeatScout (Price et al. 2005) and RepeatModeler2 (Flynn et al. 2020) are used to identify novel repeat sequence elements in genomes. tRNAscan-SE (Lowe and Eddy 1997) is used to predict tRNAs. snoSeeker (Yang et al. 2006) is used to identify snoRNAs. While there are specific tools developed to predict specific types of ncRNAs, Infernal (Nawrocki et al. 2009; Nawrocki and Eddy 2013) can be used as a general method to predict all ncRNAs, which have corresponding covariance models available in RFAM (Griffiths-Jones et al. 2003) database.

Widely used *ab initio* gene prediction tools include but not limited to GeneMark. hmm (Lukashin and Borodovsky 1998), FGENESH, Augustus (Stanke et al. 2004, 2008), SNAP (Korf 2004), TigrScan, and GlimmerHMM (Majoros et al. 2004).

Homology-based gene prediction tools utilize preexisting databases, such as UniRef90 (Bateman et al. 2017), Uniprot/Swiss-Prot (Bateman et al. 2017), nonredundant database (nr) of RefSeq (Pruitt et al. 2007), etc., to map sequences to and produce gene models based on homology. GeneWise (Birney et al. 2004) is such a tool developed to predict gene structures by using other known homolog protein sequences.

In order to utilize transcriptomes to perform gene predictions, RNA-Seq reads can be mapped to genome to predict transcripts by using a standard transcript alignment suites such as Cufflinks (Trapnell et al. 2012). Alternatively, PASA (Haas et al. 2003) can be used to map RNA-Seq assemblies to the genome itself and build gene models from these alignments.

IGV (Thorvaldsdóttir et al. 2013) and similar visualization tools can be used to view predicted gene models in order to evaluate and verify them manually. After the optional visual inspection, EVidenceModeler (Haas et al. 2008) can be utilized to select the best gene model for each locus. EUGENE (Schiex et al. 2001) and GeneComber (Shah et al. 2003) can be used to combine different evidence sources and even outputs of different gene prediction software.

Functional annotation of predicted gene structures is performed with three general approaches. Direct characterization of protein domains and families can be performed running HMMER (Finn et al. 2011) against PFAM (Bateman et al. 2004) to reveal pfam domains in the genome. SignalP (Almagro Armenteros et al. 2019) can be used to identify signal peptides, suggesting protein secretions in gene structures, and also to validate predicted gene models. TMMHMM (Krogh et al. 2001) can predict transmembrane proteins, whereas PSORT (NAKAI 1999) can be used to predict cellular localization of predicted proteins. In addition to PFAM (Bateman et al. 2004), InterProScan (Hunter et al. 2009) offers a useful collection of functional and structural protein domains that can be used for functional annotation. In addition to specialized tools, as a second way, manual BLAST (Johnson et al. 2008) search against databases such as nr Ref-Seq and UniProt/Swiss-Prot or to any domain-specific database is commonly used by researchers. To this end, TRICHOBLAST (Kopchinskiy et al. 2005) has been developed to aid *Trichoderma* spp. research by providing a rich database of *Trichoderma* spp.-related sequences. Lastly, gene classification systems, such as EuKaryotic Orthologous Groups (KOG) (Koonin et al. 2004), Gene Ontology (GO) (Ashburner et al. 2000; The Gene Ontology Consortium 2021), and Kyoto Encyclopedia of Genes and Genomes (KEGG) (Kanehisa et al. 2006) etc., can be used either as an alternative way or to further enrich the annotation level. On a special remark, CAZymes Analysis Toolkit (CAT) (Park et al. 2010) can be used to search for and analyze carbohydrate-active enzymes using CAZymes database (Garron and Henrissat 2019).

Last but not least, particularly developed for fungal domain, FunGAP (Min et al. 2017) offers a streamlined pipeline that utilizes BLAST (Johnson et al. 2008), Pfam (Bateman et al. 2004), and BUSCO (Simão et al. 2015) to perform gene prediction and annotation in fungal genomes.

6 Conclusions

With the available *Trichoderma* spp. genomes reaching around 24, comparative genomics allowed researchers to find and study genes and gene families that could be important for industrial applications in addition to genes involved in secondary metabolism and lignocellulose conversion. Genes and/or gene families encoding for other CAZymes in addition to lignocellulolytic ones, small secreted cysteine-rich protein families (mainly hydrophobins), and secondary metabolite clusters are gaining attention, owing to the diverse properties of the protein products for industrial, agricultural, or therapeutic applications. A thorough and deeper understanding and comparative analysis of the available and novel *Trichoderma* genomes, efficient application of existing computational tools, and development of specific computational tools for fungal genome mining would reveal novel genes and/or gene families that could have industrial, agricultural, and therapeutic potential.

References

Akanbi MHJ, Post E, van Putten SM et al (2013) The antitumor activity of hydrophobin SC3, a fungal protein. Appl Microbiol Biotechnol 97:4385–4392. https://doi.org/10.1007/s00253-012-4311-x

Akcapinar GB, Gul O, Sezerman U (2011) Effect of codon optimization on the expression of *Trichoderma reesei* endoglucanase 1 in *Pichia pastoris*. Biotechnol Prog 27:1257–1263. https://doi.org/10.1002/btpr.663

Almagro Armenteros JJ, Tsirigos KD, Sønderby CK et al (2019) SignalP 5.0 improves signal peptide predictions using deep neural networks. Nat Biotechnol 37:420–423. https://doi.org/10.1038/s41587-019-0036-z

Anke H, Kinn J, Bergquist K-E, Sterner O (1991) Production of siderophores by strains of the genus *Trichoderma*. Biol Metals 4:176–180. https://doi.org/10.1007/BF01141311

Ashburner M, Ball CA, Blake JA et al (2000) Gene ontology: tool for the unification of biology. Nat Genet 25:25–29. https://doi.org/10.1038/75556

Atanasova L, Crom SL, Gruber S et al (2013a) Comparative transcriptomics reveals different strategies of *Trichoderma* mycoparasitism. BMC Genomics 14:121. https://doi.org/10.1186/1471-2164-14-121

Atanasova L, Knox BP, Kubicek CP et al (2013b) The polyketide synthase gene pks4 of *Trichoderma reesei* provides pigmentation and stress resistance. Eukaryot Cell 12:1499–1508. https://doi.org/10.1128/EC.00103-13

Atanasova L, Dubey M, Grujić M et al (2018) Evolution and functional characterization of pectate lyase PEL12, a member of a highly expanded Clonostachys rosea polysaccharide lyase 1 family. BMC Microbiol 18. https://doi.org/10.1186/s12866-018-1310-9

Bairwa G, Hee Jung W, Kronstad JW (2017) Iron acquisition in fungal pathogens of humans. Metallomics 9:215–227. https://doi.org/10.1039/c6mt00301j

Baker M (2010) Next-generation sequencing: adjusting to data overload. Nat Methods 7:495–499. https://doi.org/10.1038/nmeth0710-495

Baker SE, Perrone G, Richardson NM et al (2012) Phylogenomic analysis of polyketide synthase-encoding genes in *Trichoderma*. Microbiology (Reading) 158:147–154. https://doi.org/10.1099/mic.0.053462-0

Bankevich A, Nurk S, Antipov D et al (2012) SPAdes: a new genome assembly algorithm and its applications to single-cell sequencing. J Comput Biol 19:455–477. https://doi.org/10.1089/cmb.2012.0021

Bansal R, Mukherjee PK (2016) The Terpenoid biosynthesis toolkit of *Trichoderma*. Nat Prod Commun 11:431–434

Barani M, Mirzaei M, Torkzadeh-Mahani M et al (2020) A new formulation of hydrophobin-coated niosome as a drug carrier to cancer cells. Mater Sci Eng C 113:110975. https://doi.org/10.1016/j.msec.2020.110975

Bateman A, Coin L, Durbin R et al (2004) The Pfam protein families database. Nucleic Acids Res 32:D138–D141. https://doi.org/10.1093/nar/gkh121

Bateman A, Martin MJ, O'Donovan C et al (2017) UniProt: the universal protein knowledgebase. Nucleic Acids Res 45:D158–D169. https://doi.org/10.1093/nar/gkw1099

Bayer EA, Chanzy H, Lamed R, Shoham Y (1998) Cellulose, cellulases and cellulosomes. Curr Opin Struct Biol 8:548–557. https://doi.org/10.1016/s0959-440x(98)80143-7

Bayry J, Aimanianda V, Guijarro JI et al (2012) Hydrophobins–unique fungal proteins. PLoS Pathog 8:e1002700. https://doi.org/10.1371/journal.ppat.1002700

Benítez T, Rincón AM, Limón MC, Codón AC (2004) Biocontrol mechanisms of *Trichoderma* strains. Int Microbiol 7:249–260

Benson DA, Cavanaugh M, Clark K et al (2013) GenBank. Nucleic Acids Res 41:D36–D42. https://doi.org/10.1093/nar/gks1195

Berman HM, Westbrook J, Feng Z et al (2000) The protein data bank. Nucleic Acids Res 28:235–242. https://doi.org/10.1093/nar/28.1.235

Biely P, Kremnický L, Alföldi J, Tenkanen M (1994) Stereochemistry of the hydrolysis of glycosidic linkage by endo-beta-1,4-xylanases of *Trichoderma reesei*. FEBS Lett 356:137–140. https://doi.org/10.1016/0014-5793(94)01248-2

Birney E, Clamp M, Durbin R (2004) GeneWise and Genomewise. Genome Res 14:988–995. https://doi.org/10.1101/gr.1865504

Bischof RH, Ramoni J, Seiboth B (2016) Cellulases and beyond: the first 70 years of the enzyme producer *Trichoderma reesei*. Microb Cell Factories 15:106. https://doi.org/10.1186/s12934-016-0507-6

Boer H, Simolin H, Cottaz S et al (2007) Heterologous expression and site-directed mutagenesis studies of two *Trichoderma harzianum* chitinases, Chit33 and Chit42, in *Escherichia coli*. Protein Expr Purif 51:216–226. https://doi.org/10.1016/j.pep.2006.07.020

Booker JD, Swift KG, Brown NJ (2005) Designing for assembly quality: strategies, guidelines and techniques. J Eng Des 16:279–295. https://doi.org/10.1080/09544820500126672

Brakhage AA (2013) Regulation of fungal secondary metabolism. Nat Rev Microbiol 11:21–32. https://doi.org/10.1038/nrmicro2916

Brakhage AA, Schroeckh V (2011) Fungal secondary metabolites - strategies to activate silent gene clusters. Fungal Genet Biol 48:15–22. https://doi.org/10.1016/j.fgb.2010.04.004

Branton D, Deamer DW, Marziali A et al (2009) The potential and challenges of nanopore sequencing. In: Nanoscience and technology. Co-Published with Macmillan Publishers Ltd, pp 261–268

Cai F, Druzhinina IS (2021) In honor of John Bissett: authoritative guidelines on molecular identification of *Trichoderma*. Fungal Divers. https://doi.org/10.1007/s13225-020-00464-4

Cai F, Gao R, Zhao Z et al (2020) Evolutionary compromises in fungal fitness: hydrophobins can hinder the adverse dispersal of conidiospores and challenge their survival. ISME J 14:2610–2624. https://doi.org/10.1038/s41396-020-0709-0

Cardoza RE, Hermosa MR, Vizcaíno JA et al (2007) Partial silencing of a hydroxy-methylglutaryl-CoA reductase-encoding gene in *Trichoderma harzianum* CECT 2413 results in a lower level of resistance to lovastatin and lower antifungal activity. Fungal Genet Biol 44:269–283. https://doi.org/10.1016/j.fgb.2006.11.013

Chhabra R, Saha A, Chamani A et al (2020) Iron pathways and iron chelation approaches in viral, microbial, and fungal infections. Pharmaceuticals (Basel) 13. https://doi.org/10.3390/ph13100275

Christianson DW (2008) Unearthing the roots of the terpenome. Curr Opin Chem Biol 12:141–150. https://doi.org/10.1016/j.cbpa.2007.12.008

Chugh JK, Wallace BA (2001) Peptaibols: models for ion channels. Biochem Soc Trans 29:565–570. https://doi.org/10.1042/bst0290565

Cissé OH, Stajich JE (2019) FGMP: assessing fungal genome completeness. BMC Bioinform 20:184. https://doi.org/10.1186/s12859-019-2782-9

Clough E, Barrett T (2016) The gene expression omnibus database. In: Mathé E, Davis S (eds) Statistical genomics: methods and protocols. Springer, New York, pp 93–110

Cox MP, Peterson DA, Biggs PJ (2010) SolexaQA: At-a-glance quality assessment of Illumina second-generation sequencing data. BMC Bioinform 11:485. https://doi.org/10.1186/1471-2105-11-485

Crucello A, Sforça DA, Horta MAC et al (2015) Analysis of genomic regions of *Trichoderma harzianum* IOC-3844 related to biomass degradation. PLoS One 10:e0122122. https://doi.org/10.1371/journal.pone.0122122

Crutcher FK, Parich A, Schuhmacher R et al (2013) A putative terpene cyclase, *vir4*, is responsible for the biosynthesis of volatile terpene compounds in the biocontrol fungus *Trichoderma virens*. Fungal Genet Biol 56:67–77. https://doi.org/10.1016/j.fgb.2013.05.003

Cuomo CA, Birren BW (2010) Chapter 34: The fungal genome initiative and lessons learned from genome sequencing. In: Methods in enzymology. Academic Press, pp 833–855

Darling AE, Tritt A, Eisen JA, Facciotti MT (2011) Mauve assembly metrics. Bioinformatics 27:2756–2757. https://doi.org/10.1093/bioinformatics/btr451

de Carrion SJ, Leal SM, Ghannoum MA et al (2013) The RodA Hydrophobin on *aspergillus fumigatus* spores masks Dectin-1– and Dectin-2–dependent responses and enhances fungal survival in vivo. J Immunol 191:2581. https://doi.org/10.4049/jimmunol.1300748

de la Cruz J, Hidalgo-Gallego A, Lora JM et al (1992) Isolation and characterization of three chitinases from *Trichoderma harzianum*. Eur J Biochem 206:859–867. https://doi.org/10.1111/j.1432-1033.1992.tb16994.x

Degenkolb T, von Döhren H, Nielsen KF et al (2008) Recent advances and future prospects in peptaibiotics, hydrophobin, and mycotoxin research, and their importance for chemotaxonomy of *Trichoderma* and *Hypocrea*. Chem Biodivers 5:671–680. https://doi.org/10.1002/cbdv.200890064

Degenkolb T, Karimi Aghcheh R, Dieckmann R et al (2012) The production of multiple small peptaibol families by single 14-module Peptide synthetases in *Trichoderma/Hypocrea*. Chem Biodivers 9:499–535. https://doi.org/10.1002/cbdv.201100212

Delattre C, Michaud P, Keller C et al (2006) Purification and characterization of a novel glucuronan lyase from *Trichoderma* sp. GL2. Appl Microbiol Biotechnol 70:437–443. https://doi.org/10.1007/s00253-005-0077-8

Deng J-J, Shi D, Mao H-H et al (2019) Heterologous expression and characterization of an antifungal chitinase (Chit46) from *Trichoderma harzianum* GIM 3.442 and its application in colloidal chitin conversion. Int J Biol Macromol 134:113–121. https://doi.org/10.1016/j.ijbiomac.2019.04.177

Devine R, Singha P, Handa H (2019) Versatile biomimetic medical device surface: hydrophobin coated, nitric oxide-releasing polymer for antimicrobial and hemocompatible applications. Biomater Sci 7:3438–3449. https://doi.org/10.1039/c9bm00469f

Drouet P, Zhang M, Legoy MD (1994) Enzymatic synthesis of alkyl beta-D-xylosides by transxylosylation and reverse hydrolysis. Biotechnol Bioeng 43:1075–1080. https://doi.org/10.1002/bit.260431110

Druzhinina IS, Cai F (2020) Taxonomy 2020 – ICTT. https://trichoderma.info/trichoderma-taxonomy-2020/. Accessed 24 Jan 2021

Druzhinina IS, Kubicek CP (2013) Ecological genomics of *Trichoderma*. In: The ecological genomics of fungi. Wiley, pp 89–116

E.G. Simmons (1977) *Trichoderma reesei* Tampa, p 618

Espino-Rammer L, Ribitsch D, Przylucka A et al (2013) Two novel class II Hydrophobins from *Trichoderma* spp. stimulate enzymatic hydrolysis of poly(ethylene terephthalate) when expressed as fusion proteins. Appl Environ Microbiol 79:4230–4238. https://doi.org/10.1128/aem.01132-13

Ferreira Filho JA, Horta MAC, Beloti LL et al (2017) Carbohydrate-active enzymes in *Trichoderma harzianum*: a bioinformatic analysis bioprospecting for key enzymes for the biofuels industry. BMC Genomics 18. https://doi.org/10.1186/s12864-017-4181-9

Finn RD, Clements J, Eddy SR (2011) HMMER web server: interactive sequence similarity searching. Nucleic Acids Res 39:W29–W37. https://doi.org/10.1093/nar/gkr367

Fisch KM (2013) Biosynthesis of natural products by microbial iterative hybrid PKS–NRPS. RSC Adv 3:18228–18247. https://doi.org/10.1039/C3RA42661K

Flynn JM, Hubley R, Goubert C et al (2020) RepeatModeler2 for automated genomic discovery of transposable element families. PNAS 117:9451–9457. https://doi.org/10.1073/pnas.1921046117

Foreman PK, Brown D, Dankmeyer L et al (2003) Transcriptional regulation of biomass-degrading enzymes in the filamentous fungus *Trichoderma reesei*. J Biol Chem 278:31988–31997. https://doi.org/10.1074/jbc.M304750200

Fowler T, Brown RD (1992) The *bgl1* gene encoding extracellular beta-glucosidase from *Trichoderma reesei* is required for rapid induction of the cellulase complex. Mol Microbiol 6:3225–3235. https://doi.org/10.1111/j.1365-2958.1992.tb01777.x

Gardiner DM, Howlett BJ (2005) Bioinformatic and expression analysis of the putative gliotoxin biosynthetic gene cluster of *Aspergillus fumigatus*. FEMS Microbiol Lett 248:241–248. https://doi.org/10.1016/j.femsle.2005.05.046

Gardiner DM, Waring P, Howlett BJ (2005) The epipolythiodioxopiperazine (ETP) class of fungal toxins: distribution, mode of action, functions and biosynthesis. Microbiology (Reading) 151:1021–1032. https://doi.org/10.1099/mic.0.27847-0

Garron M-L, Henrissat B (2019) The continuing expansion of CAZymes and their families. Curr Opin Chem Biol 53:82–87. https://doi.org/10.1016/j.cbpa.2019.08.004

Gomes EV, do Costa MN, de Paula RG et al (2015) The Cerato-Platanin protein EPL1 from *Trichoderma harzianum* is involved in mycoparasitism, plant resistance induction and self cell wall protection. Sci Rep 5:17998. https://doi.org/10.1038/srep17998

Griffiths-Jones S, Bateman A, Marshall M et al (2003) Rfam: an RNA family database. Nucleic Acids Res 31:439–441. https://doi.org/10.1093/nar/gkg006

Grigoriev IV, Martinez DA, Salamov AA (2006) Fungal genomic annotation. In: Arora DK, Berka RM, Singh GB (eds) Applied mycology and biotechnology. Elsevier, pp 123–142

Grigoriev IV, Nordberg H, Shabalov I et al (2012) The genome portal of the Department of Energy Joint Genome Institute. Nucleic Acids Res 40:D26–D32. https://doi.org/10.1093/nar/gkr947

Grigoriev IV, Nikitin R, Haridas S et al (2014) MycoCosm portal: gearing up for 1000 fungal genomes. Nucleic Acids Res 42:D699–D704. https://doi.org/10.1093/nar/gkt1183

Gurevich A, Saveliev V, Vyahhi N, Tesler G (2013) QUAST: quality assessment tool for genome assemblies. Bioinformatics 29:1072–1075. https://doi.org/10.1093/bioinformatics/btt086

Guzmán-Guzmán P, Alemán-Duarte MI, Delaye L et al (2017) Identification of effector-like proteins in *Trichoderma* spp. and role of a hydrophobin in the plant-fungus interaction and mycoparasitism. BMC Genet 18. https://doi.org/10.1186/s12863-017-0481-y

Haas Jimoh Akanbi M, Post E, Meter-Arkema A et al (2010) Use of hydrophobins in formulation of water insoluble drugs for oral administration. Colloids Surf B Biointerfaces 75:526–531. https://doi.org/10.1016/j.colsurfb.2009.09.030

Haas BJ, Delcher AL, Mount SM et al (2003) Improving the *Arabidopsis* genome annotation using maximal transcript alignment assemblies. Nucleic Acids Res 31:5654–5666. https://doi.org/10.1093/nar/gkg770

Haas BJ, Salzberg SL, Zhu W et al (2008) Automated eukaryotic gene structure annotation using EVidenceModeler and the program to assemble spliced alignments. Genome Biol 9:R7. https://doi.org/10.1186/gb-2008-9-1-r7

Haas BJ, Zeng Q, Pearson MD et al (2011) Approaches to fungal genome annotation. Mycology 2:118–141. https://doi.org/10.1080/21501203.2011.606851

Häkkinen M, Arvas M, Oja M et al (2012) Re-annotation of the CAZy genes of *Trichoderma reesei* and transcription in the presence of lignocellulosic substrates. Microb Cell Factories 11:134. https://doi.org/10.1186/1475-2859-11-134

Hakulinen N, Tenkanen M, Rouvinen J (2000) Three-dimensional structure of the catalytic core of acetylxylan esterase from *Trichoderma reesei*: insights into the deacetylation mechanism. J Struct Biol 132:180–190. https://doi.org/10.1006/jsbi.2000.4318

Haridas S, Breuill C, Bohlmann J, Hsiang T (2011) A biologist's guide to de novo genome assembly using next-generation sequence data: a test with fungal genomes. J Microbiol Methods 86:368–375. https://doi.org/10.1016/j.mimet.2011.06.019

Haridas S, Salamov A, Grigoriev IV (2018) Fungal Genome Annotation. In: de Vries RP, Tsang A, Grigoriev IV (eds) Fungal genomics: methods and protocols. Springer, New York, pp 171–184

Herpoël-Gimbert I, Margeot A, Dolla A et al (2008) Comparative secretome analyses of two *Trichoderma reesei* RUT-C30 and CL847 hypersecretory strains. Biotechnol Biofuels 1:18. https://doi.org/10.1186/1754-6834-1-18

Howard DH (1999) Acquisition, transport, and storage of iron by pathogenic fungi. Clin Microbiol Rev 12:394–404

Howard DH (2004) Iron gathering by zoopathogenic fungi. FEMS Immunol Med Microbiol 40:95–100. https://doi.org/10.1016/S0928-8244(03)00301-8

Howe KL, Contreras-Moreira B, De Silva N et al (2020) Ensembl genomes 2020—enabling non-vertebrate genomic research. Nucleic Acids Res 48:D689–D695. https://doi.org/10.1093/nar/gkz890

Howell CR (2006) Understanding the mechanisms employed by *Trichoderma virens* to effect biological control of cotton diseases. Phytopathology 96:178–180. https://doi.org/10.1094/PHYTO-96-0178

Howell CR, Stipanovic RD (1983) Gliovirin, a new antibiotic from *Gliocladium virens*, and its role in the biological control of *Pythium ultimum*. Can J Microbiol 29:321–324. https://doi.org/10.1139/m83-053

Hunt M, Kikuchi T, Sanders M et al (2013) REAPR: a universal tool for genome assembly evaluation. Genome Biol 14:R47. https://doi.org/10.1186/gb-2013-14-5-r47

Hunter S, Apweiler R, Attwood TK et al (2009) InterPro: the integrative protein signature database. Nucleic Acids Res 37:D211–D215. https://doi.org/10.1093/nar/gkn785

Hurne AM, Chai CLL, Moerman K, Waring P (2002) Influx of calcium through a redox-sensitive plasma membrane channel in thymocytes causes early necrotic cell death induced by the epipolythiodioxopiperazine toxins. J Biol Chem 277:31631–31638. https://doi.org/10.1074/jbc.M201699200

Ike M, Nagamatsu K, Shioya A et al (2006) Purification, characterization, and gene cloning of 46 kDa chitinase (Chi46) from *Trichoderma reesei* PC-3-7 and its expression in *Escherichia coli*. Appl Microbiol Biotechnol 71:294–303. https://doi.org/10.1007/s00253-005-0171-y

Jänis J, Rouvinen J, Leisola M et al (2001) Thermostability of endo-1,4-beta-xylanase II from *Trichoderma reesei* studied by electrospray ionization Fourier-transform ion cyclotron resonance MS, hydrogen/deuterium-exchange reactions and dynamic light scattering. Biochem J 356:453–460. https://doi.org/10.1042/0264-6021:3560453

Janssen MI, van Leeuwen MB, van Kooten TG et al (2004) Promotion of fibroblast activity by coating with hydrophobins in the beta-sheet end state. Biomaterials 25:2731–2739. https://doi.org/10.1016/j.biomaterials.2003.09.060

Jensen BG, Andersen MR, Pedersen MH et al (2010) Hydrophobins from *Aspergillus* species cannot be clearly divided into two classes. BMC Res Notes 3:344. https://doi.org/10.1186/1756-0500-3-344

Johnson M, Zaretskaya I, Raytselis Y et al (2008) NCBI BLAST: a better web interface. Nucleic Acids Res 36:W5–W9. https://doi.org/10.1093/nar/gkn201

Kanehisa M, Goto S, Hattori M et al (2006) From genomics to chemical genomics: new developments in KEGG. Nucleic Acids Res 34:D354–D357. https://doi.org/10.1093/nar/gkj102

Keller NP (2019) Fungal secondary metabolism: regulation, function and drug discovery. Nat Rev Microbiol 17:167–180. https://doi.org/10.1038/s41579-018-0121-1

Keller NP, Turner G, Bennett JW (2005) Fungal secondary metabolism - from biochemistry to genomics. Nat Rev Microbiol 3:937–947. https://doi.org/10.1038/nrmicro1286

Khan RAA, Najeeb S, Hussain S et al (2020) Bioactive secondary metabolites from *Trichoderma* spp. against Phytopathogenic fungi. Microorganisms 8:817. https://doi.org/10.3390/microorganisms8060817

Khosla C (2009) Structures and mechanisms of polyketide synthases. J Org Chem 74:6416–6420. https://doi.org/10.1021/jo9012089

Klutts JS, Yoneda A, Reilly MC et al (2006) Glycosyltransferases and their products: cryptococcal variations on fungal themes. FEMS Yeast Res 6:499–512. https://doi.org/10.1111/j.1567-1364.2006.00054.x

Kodama Y, Shumway M, Leinonen R, on behalf of the International Nucleotide Sequence Database Collaboration (2012) The sequence read archive: explosive growth of sequencing data. Nucleic Acids Res 40:D54–D56. https://doi.org/10.1093/nar/gkr854

Koivula A, Lappalainen A, Virtanen S et al (1996) Immunoaffinity chromatographic purification of cellobiohydrolase II mutants from recombinant *Trichoderma reesei* strains devoid of major endoglucanase genes. Protein Expr Purif 8:399–400

Konno N, Igarashi K, Habu N et al (2009a) Cloning of the *Trichoderma reesei* cDNA encoding a Glucuronan Lyase belonging to a novel polysaccharide Lyase family. Appl Environ Microbiol 75:101–107. https://doi.org/10.1128/AEM.01749-08

Konno N, Ishida T, Igarashi K et al (2009b) Crystal structure of polysaccharide lyase family 20 endo-beta-1,4-glucuronan lyase from the filamentous fungus *Trichoderma reesei*. FEBS Lett 583:1323–1326. https://doi.org/10.1016/j.febslet.2009.03.034

Koonin EV, Fedorova ND, Jackson JD et al (2004) A comprehensive evolutionary classification of proteins encoded in complete eukaryotic genomes. Genome Biol 5:R7. https://doi.org/10.1186/gb-2004-5-2-r7

Kopchinskiy A, Komoń M, Kubicek CP, Druzhinina IS (2005) TRICHOBLAST: a multilocus database for *Trichoderma* and *Hypocrea* identifications. Mycol Res 109:658–660. https://doi.org/10.1017/S0953756205233397

Koren S, Walenz BP, Berlin K et al (2017) Canu: scalable and accurate long-read assembly via adaptive k-mer weighting and repeat separation. Genome Res 27:722–736. https://doi.org/10.1101/gr.215087.116

Korf I (2004) Gene finding in novel genomes. BMC Bioinform 5:59. https://doi.org/10.1186/1471-2105-5-59

Korpi A, Järnberg J, Pasanen A-L (2009) Microbial volatile organic compounds. Crit Rev Toxicol 39:139–193. https://doi.org/10.1080/10408440802291497

Krogh A, Larsson B, von Heijne G, Sonnhammer ELL (2001) Predicting transmembrane protein topology with a hidden markov model: application to complete genomes11Edited by F. Cohen. J Mol Biol 305:567–580. https://doi.org/10.1006/jmbi.2000.4315

Kubicek CP, Herrera-Estrella A, Seidl-Seiboth V et al (2011) Comparative genome sequence analysis underscores mycoparasitism as the ancestral life style of *Trichoderma*. Genome Biol 12:R40. https://doi.org/10.1186/gb-2011-12-4-r40

Kubicek CP, Steindorff AS, Chenthamara K et al (2019) Evolution and comparative genomics of the most common *Trichoderma* species. BMC Genomics 20:485. https://doi.org/10.1186/s12864-019-5680-7

Kwan EM, Boraston AB, McLean BW et al (2005) N-glycosidase-carbohydrate-binding module fusion proteins as immobilized enzymes for protein deglycosylation. Protein Eng Des Sel 18:497–501. https://doi.org/10.1093/protein/gzi055

Kwon D, Lee J, Kim J (2019) GMASS: a novel measure for genome assembly structural similarity. BMC Bioinform 20:147. https://doi.org/10.1186/s12859-019-2710-z

Lairson L, Henrissat B, Davies GJ, Withers SG (2008) Glycosyltransferases: structures, functions, and mechanisms. Annu Rev Biochem 77:521–555. https://doi.org/10.1146/annurev.biochem.76.061005.092322

Lehner SM, Atanasova L, Neumann NKN et al (2013) Isotope-assisted screening for iron-containing metabolites reveals a high degree of diversity among known and unknown siderophores produced by *Trichoderma* spp. Appl Environ Microbiol 79:18–31. https://doi.org/10.1128/AEM.02339-12

Leinonen R, Sugawara H, Shumway M, Collaboration on behalf of the INSD (2011) The sequence read archive. Nucleic Acids Res 39:D19–D21. https://doi.org/10.1093/nar/gkq1019

Li X-L, Skory CD, Cotta MA et al (2008) Novel family of carbohydrate Esterases, based on identification of the *Hypocrea jecorina* acetyl esterase gene. Appl Environ Microbiol 74:7482. https://doi.org/10.1128/AEM.00807-08

Lienemann M, Boer H, Paananen A et al (2009) Toward understanding of carbohydrate binding and substrate specificity of a glycosyl hydrolase 18 family (GH-18) chitinase from *Trichoderma harzianum*. Glycobiology 19:1116–1126. https://doi.org/10.1093/glycob/cwp102

Linder M, Selber K, Nakari-Setälä T et al (2001) The hydrophobins HFBI and HFBII from *Trichoderma reesei* showing efficient interactions with nonionic surfactants in aqueous two-phase systems. Biomacromolecules 2:511–517. https://doi.org/10.1021/bm0001493

Linder MB, Qiao M, Laumen F et al (2004) Efficient purification of recombinant proteins using hydrophobins as tags in surfactant-based two-phase systems. Biochemistry 43:11873–11882. https://doi.org/10.1021/bi0488202

Littlejohn KA, Hooley P, Cox PW (2012) Bioinformatics predicts diverse *Aspergillus* hydrophobins with novel properties. Food Hydrocoll 27:503–516. https://doi.org/10.1016/j.foodhyd.2011.08.018

Lombard V, Bernard T, Rancurel C et al (2010) A hierarchical classification of polysaccharide lyases for glycogenomics. Biochem J 432:437–444. https://doi.org/10.1042/BJ20101185

Lombard V, Golaconda Ramulu H, Drula E et al (2014) The carbohydrate-active enzymes database (CAZy) in 2013. Nucleic Acids Res 42:D490–D495. https://doi.org/10.1093/nar/gkt1178

Lowe TM, Eddy SR (1997) tRNAscan-SE: a program for improved detection of transfer RNA genes in genomic sequence. Nucleic Acids Res 25:955–964. https://doi.org/10.1093/nar/25.5.955

Lukashin AV, Borodovsky M (1998) GeneMark.hmm: new solutions for gene finding. Nucleic Acids Res 26:1107–1115. https://doi.org/10.1093/nar/26.4.1107

Luo R, Liu B, Xie Y et al (2012) SOAPdenovo2: an empirically improved memory-efficient short-read de novo assembler. GigaScience 1. https://doi.org/10.1186/2047-217X-1-18

Majoros WH, Pertea M, Salzberg SL (2004) TigrScan and GlimmerHMM: two open source ab initio eukaryotic gene-finders. Bioinformatics 20:2878–2879. https://doi.org/10.1093/bioinformatics/bth315

Manchanda N, Portwood JL, Woodhouse MR et al (2020) GenomeQC: a quality assessment tool for genome assemblies and gene structure annotations. BMC Genomics 21:193. https://doi.org/10.1186/s12864-020-6568-2

Mandels M, Reese ET (1960) Induction of cellulase in fungi by cellobiose. J Bacteriol 79:816–826

Mapleson D, Drou N, Swarbreck D (2015) RAMPART: a workflow management system for *de novo* genome assembly. Bioinformatics 31:1824–1826. https://doi.org/10.1093/bioinformatics/btv056

Margolles-Clark E, Saloheimo M, Siika-aho M, Penttilä M (1996a) The alpha-glucuronidase-encoding gene of *Trichoderma reesei*. Gene 172:171–172. https://doi.org/10.1016/0378-1119(96)00167-9

Margolles-Clark E, Tenkanen M, Luonteri E, Penttilä M (1996b) Three alpha-galactosidase genes of *Trichoderma reesei* cloned by expression in yeast. Eur J Biochem 240:104–111. https://doi.org/10.1111/j.1432-1033.1996.0104h.x

Margolles-Clark E, Tenkanen M, Nakari-Setälä T, Penttilä M (1996c) Cloning of genes encoding alpha-L-arabinofuranosidase and beta-xylosidase from *Trichoderma reesei* by expression in *Saccharomyces cerevisiae*. Appl Environ Microbiol 62:3840

Martinez D, Berka RM, Henrissat B et al (2008) Genome sequencing and analysis of the biomass-degrading fungus *Trichoderma reesei* (syn. *Hypocrea jecorina*). Nat Biotechnol 26:553–560. https://doi.org/10.1038/nbt1403

Medve J, Karlsson J, Lee D, Tjerneld F (1998) Hydrolysis of microcrystalline cellulose by cellobiohydrolase I and endoglucanase II from *Trichoderma reesei*: adsorption, sugar production pattern, and synergism of the enzymes. Biotechnol Bioeng 59:621–634

Metz B, Seidl-Seiboth V, Haarmann T et al (2011) Expression of biomass-degrading enzymes is a major event during Conidium development in *Trichoderma reesei*. Eukaryot Cell 10:1527. https://doi.org/10.1128/EC.05014-11

Meyer CE, Reusser F (1967) A polypeptide antibacterial agent isolated from *Trichoderma viride*. Experientia 23:85–86. https://doi.org/10.1007/BF02135929

Mikheenko A, Valin G, Prjibelski A et al (2016) Icarus: visualizer for *de novo* assembly evaluation. Bioinformatics 32:3321–3323. https://doi.org/10.1093/bioinformatics/btw379

Min B, Grigoriev IV, Choi I-G (2017) FunGAP: Fungal Genome Annotation Pipeline using evidence-based gene model evaluation. Bioinformatics 33:2936–2937. https://doi.org/10.1093/bioinformatics/btx353

Mitsuishi Y, Nitisinprasert S, Saloheimo M et al (1990) Site-directed mutagenesis of the putative catalytic residues of *Trichoderma reesei* cellobiohydrolase I and endoglucanase I. FEBS Lett 275:135–138. https://doi.org/10.1016/0014-5793(90)81457-y

Mueller P, Rudin DO (1968) Action potentials induced in biomolecular lipid membranes. Nature 217:713–719. https://doi.org/10.1038/217713a0

Mukherjee M, Horwitz BA, Sherkhane PD et al (2006) A secondary metabolite biosynthesis cluster in *Trichoderma virens*: evidence from analysis of genes underexpressed in a mutant defective in morphogenesis and antibiotic production. Curr Genet 50:193–202. https://doi.org/10.1007/s00294-006-0075-0

Mukherjee PK, Wiest A, Ruiz N et al (2011) Two classes of new Peptaibols are synthesized by a single non-ribosomal peptide Synthetase of *Trichoderma virens*. J Biol Chem 286:4544–4554. https://doi.org/10.1074/jbc.M110.159723

Mukherjee PK, Buensanteai N, Moran-Diez ME et al (2012a) Functional analysis of non-ribosomal peptide synthetases (NRPSs) in *Trichoderma virens* reveals a polyketide synthase (PKS)/NRPS hybrid enzyme involved in the induced systemic resistance response in maize. Microbiology (Reading) 158:155–165. https://doi.org/10.1099/mic.0.052159-0

Mukherjee PK, Horwitz BA, Kenerley CM (2012b) Secondary metabolism in *Trichoderma* – a genomic perspective. Microbiology 158:35–45. https://doi.org/10.1099/mic.0.053629-0

Murzin AG, Brenner SE, Hubbard T, Chothia C (1995) SCOP: a structural classification of proteins database for the investigation of sequences and structures. J Mol Biol 247:536–540. https://doi.org/10.1016/S0022-2836(05)80134-2

Nakai K (1999) PSORT : a program for detecting the sorting signals of proteins and predicting their subcellular localization. Trends Biochem Sci 24:34–35

Nakari-Setälä T, Aro N, Kalkkinen N et al (1996) Genetic and biochemical characterization of the *Trichoderma reesei* hydrophobin HFBI. Eur J Biochem 235:248–255. https://doi.org/10.1111/j.1432-1033.1996.00248.x

Nakari-Setälä T, Aro N, Ilmén M et al (1997) Differential expression of the vegetative and spore-bound hydrophobins of *Trichoderma reesei*–cloning and characterization of the *hfb2* gene. Eur J Biochem 248:415–423. https://doi.org/10.1111/j.1432-1033.1997.00415.x

Nauom S, da Silva Neto BR, Ribeiro MS et al (2019) Biochemical and molecular study of *Trichoderma harzianum* enriched Secretome protein profiles using lectin affinity chromatography. Appl Biochem Biotechnol 187:1–13. https://doi.org/10.1007/s12010-018-2795-2

Nawrocki EP, Eddy SR (2013) Infernal 1.1: 100-fold faster RNA homology searches. Bioinformatics 29:2933–2935. https://doi.org/10.1093/bioinformatics/btt509

Nawrocki EP, Kolbe DL, Eddy SR (2009) Infernal 1.0: inference of RNA alignments. Bioinformatics 25:1335–1337. https://doi.org/10.1093/bioinformatics/btp157

Nierman WC, Pain A, Anderson MJ et al (2005) Genomic sequence of the pathogenic and allergenic filamentous fungus *Aspergillus fumigatus*. Nature 438:1151–1156. https://doi.org/10.1038/nature04332

O'Neil ST, Emrich SJ (2013) Assessing *de novo* transcriptome assembly metrics for consistency and utility. BMC Genomics 14:465. https://doi.org/10.1186/1471-2164-14-465

Ogasawara W, Shida Y, Furukawa T et al (2006) Cloning, functional expression and promoter analysis of xylanase III gene from *Trichoderma reesei*. Appl Microbiol Biotechnol 72:995–1003. https://doi.org/10.1007/s00253-006-0365-y

Ooka T, Shimojima Y, Akimoto T et al (1966) A new antibacterial peptide "Suzukacillin". Agric Biol Chem 30:700–702. https://doi.org/10.1080/00021369.1966.10858667

Orengo C, Michie A, Jones S et al (1997) CATH – a hierarchic classification of protein domain structures. Structure 5:1093–1109. https://doi.org/10.1016/S0969-2126(97)00260-8

Pakora G-A, Mpika J, Kone D et al (2018) Inhibition of *Phytophthora* species, agents of cocoa black pod disease, by secondary metabolites of *Trichoderma* species. Environ Sci Pollut Res 25:29901–29909. https://doi.org/10.1007/s11356-017-0283-9

Pankewitz F, Hilker M (2008) Polyketides in insects: ecological role of these widespread chemicals and evolutionary aspects of their biogenesis. Biol Rev 83:209–226. https://doi.org/10.1111/j.1469-185X.2008.00040.x

Park BH, Karpinets TV, Syed MH et al (2010) CAZymes analysis toolkit (CAT): web service for searching and analyzing carbohydrate-active enzymes in a newly sequenced organism using CAZy database. Glycobiology 20:1574–1584. https://doi.org/10.1093/glycob/cwq106

Parkinson H, Kapushesky M, Shojatalab M et al (2007) ArrayExpress—a public database of microarray experiments and gene expression profiles. Nucleic Acids Res 35:D747–D750. https://doi.org/10.1093/nar/gkl995

Patron NJ, Waller RF, Cozijnsen AJ et al (2007) Origin and distribution of epipolythiodioxopiperazine (ETP) gene clusters in filamentous ascomycetes. BMC Evol Biol 7:174. https://doi.org/10.1186/1471-2148-7-174

Payne CM, Knott BC, Mayes HB et al (2015) Fungal cellulases. Chem Rev 115:1308–1448. https://doi.org/10.1021/cr500351c

Penttilä ME, André L, Lehtovaara P et al (1988) Efficient secretion of two fungal cellobiohydrolases by *Saccharomyces cerevisiae*. Gene 63:103–112. https://doi.org/10.1016/0378-1119(88)90549-5

Philpott CC (2006) Iron uptake in fungi: a system for every source. Biochim Biophys Acta 1763:636–645. https://doi.org/10.1016/j.bbamcr.2006.05.008

Phoka N, Suwannarach N, Lumyong S et al (2020) Role of volatiles from the endophytic fungus *Trichoderma asperelloides* PSU-P1 in biocontrol potential and in promoting the plant growth of *Arabidopsis thaliana*. J Fungi (Basel) 6. https://doi.org/10.3390/jof6040341

Price AL, Jones NC, Pevzner PA (2005) De novo identification of repeat families in large genomes. Bioinformatics 21:i351–i358. https://doi.org/10.1093/bioinformatics/bti1018

Proctor RH, McCormick SP, Kim H-S et al (2018) Evolution of structural diversity of trichothecenes, a family of toxins produced by plant pathogenic and entomopathogenic fungi. PLoS Pathog 14:e1006946. https://doi.org/10.1371/journal.ppat.1006946

Pruitt KD, Tatusova T, Maglott DR (2007) NCBI reference sequences (RefSeq): a curated non-redundant sequence database of genomes, transcripts and proteins. Nucleic Acids Res 35:D61–D65. https://doi.org/10.1093/nar/gkl842

Przylucka A, Akcapinar GB, Bonazza K et al (2017a) Comparative physiochemical analysis of hydrophobins produced in *Escherichia coli* and *Pichia pastoris*. Colloids Surf B: Biointerfaces 159:913–923. https://doi.org/10.1016/j.colsurfb.2017.08.058

Przylucka A, Akcapinar GB, Chenthamara K et al (2017b) HFB7 – a novel orphan hydrophobin of the Harzianum and Virens clades of *Trichoderma*, is involved in response to biotic and abiotic stresses. Fungal Genet Biol 102:63–76. https://doi.org/10.1016/j.fgb.2017.01.002

Puchart V, Agger JW, Berrin J-G et al (2016) Comparison of fungal carbohydrate esterases of family CE16 on artificial and natural substrates. J Biotechnol 233:228–236. https://doi.org/10.1016/j.jbiotec.2016.07.003

Qin M, Wang LK, Feng XZ et al (2007) Bioactive surface modification of mica and poly(dimethylsiloxane) with hydrophobins for protein immobilization. Langmuir 23:4465–4471. https://doi.org/10.1021/la062744h

Reinikainen T, Ruohonen L, Nevanen T et al (1992) Investigation of the function of mutated cellulose-binding domains of *Trichoderma reesei* cellobiohydrolase I. Proteins 14:475–482. https://doi.org/10.1002/prot.340140408

Reino J, Guerrero R, Hernández-Galán R, Collado I (2008) Secondary metabolites from species of the biocontrol agent *Trichoderma*. Phytochem Rev 7:89–123. https://doi.org/10.1007/s11101-006-9032-2

Renshaw JC, Robson GD, Trinci APJ et al (2002) Fungal siderophores: structures, functions and applications. Mycol Res 106:1123–1142. https://doi.org/10.1017/S0953756202006548

Rether J, Serwe A, Anke T, Erkel G (2007) Inhibition of inducible tumor necrosis factor-alpha expression by the fungal epipolythiodiketopiperazine gliovirin. Biol Chem 388:627–637. https://doi.org/10.1515/BC.2007.066

Ribitsch D, Herrero Acero E, Przylucka A et al (2015) Enhanced cutinase-catalyzed hydrolysis of polyethylene terephthalate by covalent fusion to hydrophobins. Appl Environ Microbiol 81:3586–3592. https://doi.org/10.1128/AEM.04111-14

Ridley CP, Lee HY, Khosla C (2008) Evolution of polyketide synthases in bacteria. Proc Natl Acad Sci U S A 105:4595–4600. https://doi.org/10.1073/pnas.0710107105

Risdian C, Mozef T, Wink J (2019) Biosynthesis of polyketides in *Streptomyces*. Microorganisms 7. https://doi.org/10.3390/microorganisms7050124

Rodriguez B, Kavoosi M, Koska J et al (2004) Inexpensive and generic affinity purification of recombinant proteins using a family 2a CBM fusion tag. Biotechnol Prog 20:1479–1489. https://doi.org/10.1021/bp0341904

Rubio MB, Hermosa R, Reino JL et al (2009) Thctf1 transcription factor of *Trichoderma harzianum* is involved in 6-pentyl-2H-pyran-2-one production and antifungal activity. Fungal Genet Biol 46:17–27. https://doi.org/10.1016/j.fgb.2008.10.008

Saloheimo M, Lehtovaara P, Penttilä M et al (1988) EGIII, a new endoglucanase from *Trichoderma reesei*: the characterization of both gene and enzyme. Gene 63:11–22. https://doi.org/10.1016/0378-1119(88)90541-0

Saloheimo A, Henrissat B, Hoffrén AM et al (1994) A novel, small endoglucanase gene, *egl5*, from *Trichoderma reesei* isolated by expression in yeast. Mol Microbiol 13:219–228. https://doi.org/10.1111/j.1365-2958.1994.tb00417.x

Saloheimo M, Nakari-Setälä T, Tenkanen M, Penttilä M (1997) cDNA cloning of a *Trichoderma reesei* cellulase and demonstration of endoglucanase activity by expression in yeast. Eur J Biochem 249:584–591. https://doi.org/10.1111/j.1432-1033.1997.00584.x

Sanger F, Nicklen S, Coulson AR (1977) DNA sequencing with chain-terminating inhibitors. Proc Natl Acad Sci 74:5463–5467. https://doi.org/10.1073/pnas.74.12.5463

Scharf DH, Brakhage AA, Mukherjee PK (2016) Gliotoxin–bane or boon? Environ Microbiol 18:1096–1109. https://doi.org/10.1111/1462-2920.13080

Schiex T, Moisan A, Rouzé P (2001) Eugène: an eukaryotic gene finder that combines several sources of evidence. In: Gascuel O, Sagot M-F (eds) Computational biology. Springer, Berlin/Heidelberg, pp 111–125

Schuster SC (2008) Next-generation sequencing transforms today's biology. Nat Methods 5:16–18. https://doi.org/10.1038/nmeth1156

Seiboth B, Ivanova C, Seidl-Seiboth V (2011) *Trichoderma reesei*: a fungal enzyme producer for cellulosic biofuels. In: Biofuel production-recent developments and prospects. IntechOpen

Seidl V, Marchetti M, Schandl R et al (2006) EPL1, the major secreted protein of *Hypocrea atroviridis* on glucose, is a member of a strongly conserved protein family comprising plant defense response elicitors. FEBS J 273:4346–4359. https://doi.org/10.1111/j.1742-4658.2006.05435.x

Shah SP, McVicker GP, Mackworth AK et al (2003) GeneComber: combining outputs of gene prediction programs for improved results. Bioinformatics 19:1296–1297. https://doi.org/10.1093/bioinformatics/btg139

Shoemaker S, Schweickart V, Ladner M et al (1983) Molecular cloning of Exo–cellobiohydrolase I derived from *Trichoderma reesei* strain L27. Bio/Technology 1:691–696. https://doi.org/10.1038/nbt1083-691

Simão FA, Waterhouse RM, Ioannidis P et al (2015) BUSCO: assessing genome assembly and annotation completeness with single-copy orthologs. Bioinformatics 31:3210–3212. https://doi.org/10.1093/bioinformatics/btv351

Simpson JT, Wong K, Jackman SD et al (2009) ABySS: a parallel assembler for short read sequence data. Genome Res 19:1117–1123. https://doi.org/10.1101/gr.089532.108

Siva N (2008) 1000 Genomes project. Nat Biotechnol 26:256–256. https://doi.org/10.1038/nbt0308-256b

Stålbrand H, Saloheimo A, Vehmaanperä J et al (1995) Cloning and expression in *Saccharomyces cerevisiae* of a *Trichoderma reesei* beta-mannanase gene containing a cellulose binding domain. Appl Environ Microbiol 61:1090–1097. https://doi.org/10.1128/AEM.61.3.1090-1097.1995

Stanke M, Steinkamp R, Waack S, Morgenstern B (2004) AUGUSTUS: a web server for gene finding in eukaryotes. Nucleic Acids Res 32:W309–W312. https://doi.org/10.1093/nar/gkh379

Stanke M, Diekhans M, Baertsch R, Haussler D (2008) Using native and syntenically mapped cDNA alignments to improve *de novo* gene finding. Bioinformatics 24:637–644. https://doi.org/10.1093/bioinformatics/btn013

Stein L (2001) Genome annotation: from sequence to biology. Nat Rev Genet 2:493–503. https://doi.org/10.1038/35080529

Steyaert JM, Stewart A, Jaspers MV et al (2004) Co-expression of two genes, a chitinase (*chit42*) and proteinase (*prb1*), implicated in mycoparasitism by *Trichoderma hamatum*. Mycologia 96:1245–1252

Stoppacher N, Kluger B, Zeilinger S et al (2010) Identification and profiling of volatile metabolites of the biocontrol fungus *Trichoderma atroviride* by HS-SPME-GC-MS. J Microbiol Methods 81:187–193. https://doi.org/10.1016/j.mimet.2010.03.011

Stoppacher N, Neumann NKN, Burgstaller L et al (2013) The comprehensive peptaibiotics database. Chem Biodivers 10:734–743. https://doi.org/10.1002/cbdv.201200427

Strieker M, Tanović A, Marahiel MA (2010) Nonribosomal peptide synthetases: structures and dynamics. Curr Opin Struct Biol 20:234–240. https://doi.org/10.1016/j.sbi.2010.01.009

Sugimoto N, Igarashi K, Samejima M (2012) Cellulose affinity purification of fusion proteins tagged with fungal family 1 cellulose-binding domain. Protein Expr Purif 82:290–296. https://doi.org/10.1016/j.pep.2012.01.007

Tajima F (1989) Statistical method for testing the neutral mutation hypothesis by DNA polymorphism. Genetics 123:585–595

Takashima S, Nakamura A, Hidaka M et al (1999) Molecular cloning and expression of the novel fungal beta-glucosidase genes from *Humicola grisea* and *Trichoderma reesei*. J Biochem 125:728–736. https://doi.org/10.1093/oxfordjournals.jbchem.a022343

Tamandegani PR, Marik T, Zafari D et al (2020) Changes in Peptaibol production of *Trichoderma* species during in vitro antagonistic interactions with fungal plant pathogens. Biomol Ther 10. https://doi.org/10.3390/biom10050730

Tang H, Zhu Z, Zheng Z et al (2021) A study of hydrophobins-modified menaquinone-7 on osteoblastic cells differentiation. Mol Cell Biochem. https://doi.org/10.1007/s11010-021-04062-z

Tarailo-Graovac M, Chen N (2009) Using RepeatMasker to identify repetitive elements in genomic sequences. Curr Protoc Bioinformatics 25:4.10.1–4.10.14. https://doi.org/10.1002/0471250953.bi0410s25

Tateno Y, Imanishi T, Miyazaki S et al (2002) DNA data Bank of Japan (DDBJ) for genome scale research in life science. Nucleic Acids Res 30:27–30. https://doi.org/10.1093/nar/30.1.27

Tenkanen M, Vršanská M, Siika-aho M et al (2013) Xylanase XYN IV from *Trichoderma reesei* showing exo- and endo-xylanase activity. FEBS J 280:285–301. https://doi.org/10.1111/febs.12069

The Gene Ontology Consortium (2021) The gene ontology resource: enriching a GOld mine. Nucleic Acids Res 49:D325–D334. https://doi.org/10.1093/nar/gkaa1113

Theobald S, Vesth TC, Andersen MR (2019) Genus level analysis of PKS-NRPS and NRPS-PKS hybrids reveals their origin in *Aspergilli*. BMC Genomics 20:847. https://doi.org/10.1186/s12864-019-6114-2

Thomas GWC, Hahn MW (2019) Referee: reference assembly quality scores. Genome Biol Evol 11:1483–1486. https://doi.org/10.1093/gbe/evz088

Thorvaldsdóttir H, Robinson JT, Mesirov JP (2013) Integrative Genomics Viewer (IGV): high-performance genomics data visualization and exploration. Brief Bioinform 14:178–192. https://doi.org/10.1093/bib/bbs017

Törrönen A, Rouvinen J (1995) Structural comparison of two major endo-1,4-xylanases from *Trichoderma reesei*. Biochemistry 34:847–856. https://doi.org/10.1021/bi00003a019

Trapnell C, Roberts A, Goff L et al (2012) Differential gene and transcript expression analysis of RNA-seq experiments with TopHat and Cufflinks. Nat Protoc 7:562–578. https://doi.org/10.1038/nprot.2012.016

van Dijk EL, Auger H, Jaszczyszyn Y, Thermes C (2014) Ten years of next-generation sequencing technology. Trends Genet 30:418–426. https://doi.org/10.1016/j.tig.2014.07.001

van Nimwegen KJM, van Soest RA, Veltman JA et al (2016) Is the $1000 genome as near as we think? A cost analysis of next-generation sequencing. Clin Chem 62:1458–1464. https://doi.org/10.1373/clinchem.2016.258632

Vargas WA, Mukherjee PK, Laughlin D et al (2014) Role of gliotoxin in the symbiotic and pathogenic interactions of *Trichoderma virens*. Microbiology (Reading) 160:2319–2330. https://doi.org/10.1099/mic.0.079210-0

Várnai A, Siika-aho M, Viikari L (2013) Carbohydrate-binding modules (CBMs) revisited: reduced amount of water counterbalances the need for CBMs. Biotechnol Biofuels 6:30. https://doi.org/10.1186/1754-6834-6-30

Vicente I, Baroncelli R, Morán-Diez ME et al (2020) Combined comparative genomics and gene expression analyses provide insights into the terpene synthases inventory in *Trichoderma*. Microorganisms 8:1603. https://doi.org/10.3390/microorganisms8101603

Vigushin DM, Mirsaidi N, Brooke G et al (2004) Gliotoxin is a dual inhibitor of farnesyltransferase and geranylgeranyltransferase I with antitumor activity against breast cancer in vivo. Med Oncol 21:21–30. https://doi.org/10.1385/MO:21:1:21

Vinale F, Sivasithamparam K, Ghisalberti EL et al (2008) A novel role for *Trichoderma* secondary metabolites in the interactions with plants. Physiol Mol Plant Pathol 72:80–86. https://doi.org/10.1016/j.pmpp.2008.05.005

Viterbo A, Montero M, Ramot O et al (2002) Expression regulation of the endochitinase chit36 from Tr*ichoderma asperellum* (*T. harzianum* T-203). Curr Genet 42:114–122. https://doi.org/10.1007/s00294-002-0345-4

Viterbo A, Wiest A, Brotman Y et al (2007) The 18mer peptaibols from *Trichoderma virens* elicit plant defence responses. Mol Plant Pathol 8:737–746. https://doi.org/10.1111/j.1364-3703.2007.00430.x

Wallner A, Blatzer M, Schrettl M et al (2009) Ferricrocin, a siderophore involved in intra- and transcellular iron distribution in *Aspergillus fumigatus*. Appl Environ Microbiol 75:4194–4196. https://doi.org/10.1128/AEM.00479-09

Wee Y, Bhyan SB, Liu Y et al (2019) The bioinformatics tools for the genome assembly and analysis based on third-generation sequencing. Brief Funct Genomics 18:1–12. https://doi.org/10.1093/bfgp/ely037

Wenzel SC, Kunze B, Höfle G et al (2005) Structure and biosynthesis of myxochromides S1-3 in *Stigmatella aurantiaca*: evidence for an iterative bacterial type I polyketide synthase and for module skipping in nonribosomal peptide biosynthesis. Chembiochem 6:375–385. https://doi.org/10.1002/cbic.200400282

Wenzel SC, Meiser P, Binz TM et al (2006) Nonribosomal peptide biosynthesis: point mutations and module skipping lead to chemical diversity. Angew Chem Int Ed Engl 45:2296–2301. https://doi.org/10.1002/anie.200503737

Whiteford JR, Spanu PD (2002) Hydrophobins and the interactions between fungi and plants. Mol Plant Pathol 3:391–400. https://doi.org/10.1046/j.1364-3703.2002.00129.x

Whitmore L, Wallace BA (2004) Analysis of peptaibol sequence composition: implications for in vivo synthesis and channel formation. Eur Biophys J 33:233–237. https://doi.org/10.1007/s00249-003-0348-1

Wiest A, Grzegorski D, Xu B-W et al (2002) Identification of peptaibols from *Trichoderma virens* and cloning of a peptaibol synthetase. J Biol Chem 277:20862–20868. https://doi.org/10.1074/jbc.M201654200

Wilkins K, Larsen K, Simkus M (2003) Volatile metabolites from indoor molds grown on media containing wood constituents. Environ Sci Pollut Res 10:206–208. https://doi.org/10.1065/espr2002.05.119

Winnenburg R, Baldwin TK, Urban M et al (2006) PHI-base: a new database for pathogen host interactions. Nucleic Acids Res 34:D459–D464. https://doi.org/10.1093/nar/gkj047

Woodward J (1991) Synergism in cellulase systems. Bioresour Technol 36:67–75. https://doi.org/10.1016/0960-8524(91)90100-X

Wu Q, Bai L, Liu W et al (2013) Construction of a *Streptomyces lydicus* A01 transformant with a *chit42* gene from *Trichoderma harzianum* P1 and evaluation of its biocontrol activity against *Botrytis cinerea*. J Microbiol 51:166–173. https://doi.org/10.1007/s12275-013-2321-8

Yang J-H, Zhang X-C, Huang Z-P et al (2006) snoSeeker: an advanced computational package for screening of guide and orphan snoRNA genes in the human genome. Nucleic Acids Res 34:5112–5123. https://doi.org/10.1093/nar/gkl672

Yang L-A, Chang Y-J, Chen S-H et al (2019) SQUAT: a sequencing quality assessment tool for data quality assessments of genome assemblies. BMC Genomics 19:238. https://doi.org/10.1186/s12864-019-5445-3

Yavas G, Hong H, Xiao W (2019) dnAQET: a framework to compute a consolidated metric for benchmarking quality of *de novo* assemblies. BMC Genomics 20:706. https://doi.org/10.1186/s12864-019-6070-x

Zeilinger S, Gruber S, Bansal R, Mukherjee PK (2016) Secondary metabolism in *Trichoderma* – chemistry meets genomics. Fungal Biol Rev 30:74–90. https://doi.org/10.1016/j.fbr.2016.05.001

Zerbino DR, Birney E (2008) Velvet: algorithms for *de novo* short read assembly using de Bruijn graphs. Genome Res 18:821–829. https://doi.org/10.1101/gr.074492.107

Zhang J, Siika-Aho M, Tenkanen M, Viikari L (2011a) The role of acetyl xylan esterase in the solubilization of xylan and enzymatic hydrolysis of wheat straw and giant reed. Biotechnol Biofuels 4:60. https://doi.org/10.1186/1754-6834-4-60

Zhang W, Chen J, Yang Y et al (2011b) A practical comparison of *de novo* genome assembly software tools for next-generation sequencing technologies. PLoS One 6:e17915. https://doi.org/10.1371/journal.pone.0017915

Zhao L, Xu H, Li Y et al (2016) Novel application of hydrophobin in medical science: a drug carrier for improving serum stability. Sci Rep 6:26461. https://doi.org/10.1038/srep26461

Zhao Z, Cai F, Gao R et al (2021) At least three families of hyphosphere small secreted cysteine-rich proteins can optimize surface properties to a moderately hydrophilic state suitable for fungal attachment. Environ Microbiol. https://doi.org/10.1111/1462-2920.15413

Zhu M, Zhang X, Huang X et al (2020) Irregularly bridged Epipolythiodioxopiperazines and related analogues: sources, structures, and biological activities. J Nat Prod 83:2045–2053. https://doi.org/10.1021/acs.jnatprod.9b01283

Zou G, Jiang Y, Liu R et al (2018) The putative β-glucosidase BGL3I regulates cellulase induction in *Trichoderma reesei*. Biotechnol Biofuels 11:314. https://doi.org/10.1186/s13068-018-1314-6

Biosynthesis of Metal-Based Nanoparticles by *Trichoderma* and Its Potential Applications

Ernestina Castro-Longoria

Contents

1 Introduction

During the last decade, nanotechnology has become one of the most thrust areas of research. The synthesis of nanoparticles (NPs), which range from 1 to 100 nm, is of major importance due to the unique properties the materials display in the nanometric size, differing from the properties these materials have in bulk. The specific properties of each nanomaterial will depend on its own particular characteristics,

E. Castro-Longoria (✉)
Departamento de Microbiología, Centro de Investigación Científica y de Educación Superior de Ensenada (CICESE), Ensenada, Baja California, México
e-mail: ecastro@cicese.mx

N. Amaresan et al. (eds.), *Advances in* Trichoderma *Biology for Agricultural Applications*, Fungal Biology, https://doi.org/10.1007/978-3-030-91650-3_17

such as size, shape, and composition, and, to some extent, their surrounding medium. Due to their unique properties, NPs can be used in a wide range of applications, for example, food and personal care products (Weir et al. 2012; Musial et al. 2020), antimicrobial agents (Sánchez-López et al. 2020), cancer therapy and diagnosis (Brigger et al. 2012), sensor technology (Naama et al. 2015), catalysis (Zibareva et al. 2019), and biological optical imaging (Wu et al. 2019), among others. Hence, the development of clean, nontoxic, and environmentally friendly processes for the synthesis of nanomaterials is gaining importance. Among the bio-resources used for the biosynthesis of nanoparticles, fungi are considered excellent candidates due to their easy cultivation, fast growth rate, and the high amount of enzymes and secondary metabolites production, which replace chemicals used as reducing, capping, and stabilizing agents (Castro-Longoria 2016).

Since the production of NPs using fungi has been fully demonstrated, it is desirable to work with non-pathogenic species, especially if the resulting NPs are intended for medical applications (Castro-Longoria 2016). Currently, some species like *T. harzianum*, *T. atroviride*, and *T. asperellum* are used as bio-remediation agents against other plant pathogenic fungi and to promote plant and root growth of economically important crops (Ghazanfar et al. 2018). However, it is important to mention that an increasing number of fungal species may cause life-threatening infections in immunocompromised patients. In the case of *Trichoderma*, some species like *T. harzianum*, *T. koningii*, *T. longibrachiatum*, *T. pseudokoningii*, and *T. viride* have been identified as etiologic agents of infections in immunocompromised hosts (Richter et al. 1999; Chouaki et al. 2002; De Miguel et al. 2005); also see Kredics et al., in this book (chapter "Trichodermosis: Human Infections Caused by *Trichoderma* Species"). Therefore, any fungal species must always be managed with care and following all safety protocols.

Still, *Trichoderma* species are considered good candidates for nanoparticle production since they are easy to manage and can be easily cultivated in various agricultural and domestic wastes such as corn flour, sorghum grain, wheat bran, farm yard manure, tea waste, wheat straw, rice bran, and vegetable waste (Srivastava et al. 2007). Therefore, this group of species represents a feasible alternative for the development of new nanomaterials with enhanced properties. In recent years, the number of reports using *Trichoderma* species for the production of metallic nanoparticles has increased, and results include promising potential applications. However, there is still much work to be done in order to clearly demonstrate the advantage of using *Trichoderma* as reducing agents and the efficacy of the produced nanomaterials. In this chapter, the work on the production of metal-based nanoparticles using *Trichoderma* species is resumed; also the suggested applications of the biosynthesized material are described.

2 General Protocol for the Synthesis and Characterization of *Trichoderma* Nanoparticles

Synthesis of nanoparticles using species of *Trichoderma* is achieved using cell-free filtrates or supernatants. For this, the fungus is cultivated in liquid medium, generally under agitation, although static cultures may be used (Omran et al. 2019). After cultivation for 5–7 days, biomass is removed by filtration, and the filtrate is used as reducing agent. In the case of using the supernatant, the fungus is also cultivated in liquid medium, and then the biomass is obtained and washed with distilled water. Afterward, biomass is placed in deionized water for the release of metabolites by the fungus, usually for 2–3 days. Then, biomass is removed by filtration, and the resulting supernatant is ready to be used for biosynthesis. The supernatant or cell-free filtrate is then mixed with metallic salt solutions, usually prepared at 1–10 mM concentration. Some parameters are adjusted during the synthesis process such as amount of biomass used, temperature, pH, and time of incubation. By making these adjustments, differences in particle size are obtained, although the control over size is not fully achieved.

Generally, the first indicative of NPs formation is the color change of the reaction, from colorless (or the color of the filtrates) to brown, for silver nanoparticles (Fig. 1a, inset). In the case of gold nanoparticles, the reaction turns pink or red (Fig. 1b, inset). A UV–Vis spectroscopy analysis is commonly carried out to identify the formation of nanoparticles, since these are known to exhibit a UV–Vis absorption maximum because of surface plasmon resonance. Silver nanoparticles exhibit absorption range at 400–500 nm (Fig. 1a), while gold nanoparticles in the range of 500–600 nm (Fig. 1b) (Saravanakumar et al. 2016). Transmission electron microscopy (TEM) analysis of the synthesized nanoparticles will provide the information about size range and shape. Other analyses such as Fourier transform infrared spectroscopy (FTIR) and energy dispersive X-ray (EDX) are frequently carried

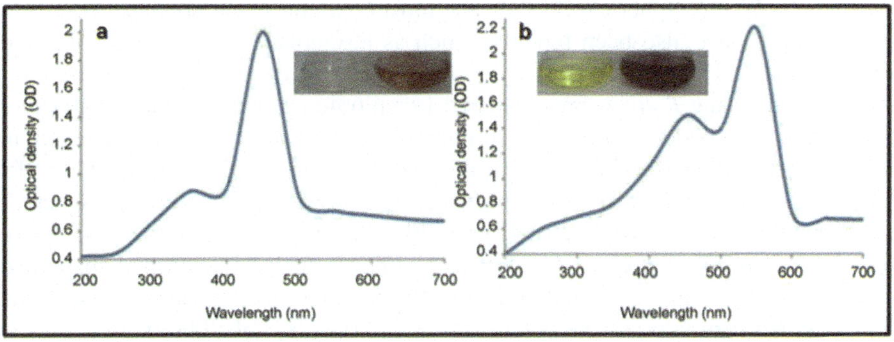

Fig. 1 UV–Vis spectrophotometry analysis of nanoparticles synthesized by *Trichoderma hamatum*. (**a**) Silver nanoparticles (AgNPs), (**b**) gold nanoparticles (AuNPs). Synthesis showed maximum absorbance at 450 nm for AgNPs and 540 nm for AuNPs. (Reproduced and modified from Saravanakumar et al. 2016 with permission from Elsevier)

out to determine the nature of the biomolecules involved in the synthesis process and the elemental composition of a sample, respectively.

Spherical or quasi-spherical nanoparticles are mostly obtained; however, different shapes and sizes can also be formed; those are developed by evolution of morphology from pseudospherical nanoparticles to more complex structures, i.e., pseudospheres fuse to form nanoprisms and nanoplates (Mukherjee et al. 2012). Isolation of nanoparticles by size from the reacting masses is possible by differential centrifugation (Mukherjee et al. 2012; Maliszewska 2013).

Biosynthesized nanoparticles using *Trichoderma* are reported with no precipitation or aggregation to up to 3–6 months of storage at ambient temperature (Maliszewska 2013; Ponmurugan 2016). Therefore, it is assumed that biomolecules secreted by *Trichoderma* act as capping agents conferring them stability, i.e., preventing them from aggregation and preserving their characteristics such as composition, shape, and crystallinity. Stability is a very important characteristic since nanoparticles with good stability preserve their intrinsic properties. Therefore, using *Trichoderma* species could increase the shelf storage of biosynthesized nanoparticles. In fact, the presence of proteins as stabilizing agents of nanoparticles was demonstrated when using *T. koningiopsis* as a reducing agent (Salvadori et al. 2014).

3 Biosynthesis of *Trichoderma* Silver Nanoparticles

The biosynthesis of silver nanoparticles (AgNPs) has received considerable attention since it is well documented by the potent antimicrobial activity they display. In this respect, *Trichoderma* has attracted the attention in nanotechnology, since it is considered a non-pathogenic fungus and is currently used as a biocontrol agent. Several species of *Trichoderma* have been successfully used for the production of AgNPs. In general, small range sizes are reported, with particles typically in the range of 1–60 nm. Spherical NPs are the most common shape obtained; however, other shapes have also been reported, such as hexagonal, triangular, cuboid, and rod-like (Table 1).

In a study using *T. viride*, in which cell-free filtrate was put in contact with silver nitrate, the first evidence of AgNPs formation was the color change of the reaction, from colorless to brown (Fig. 2a). The resulting solution was then analyzed for NPs characterization, and the UV–Vis spectroscopy analysis confirmed the presence of AgNPs with an absorption peak at 421 nm (Fig. 2b), suggesting that small NPs were obtained. Transmission electron microscopy (TEM) analysis confirmed shape and size distribution; spherical and occasionally rod-like silver nanoparticles in the range of 5–40 nm in size were obtained (Fig. 2c) (Fayaz et al. 2009a).

Most studies use a solution of silver nitrate ($AgNO_3$) at 1 mM for the biosynthesis of AgNPs; however, other concentrations are also used, obtaining differences in the curves of absorbance (Fig. 3a). As mentioned in the previous section, by FTIR analysis, the nature of the biomolecules involved in the synthesis can be identified.

Table 1 *Trichoderma* species used to produce silver nanoparticles and their suggested applications

Fungi	Size (nm)	Shape[a]	Application	Source or reference
T. asperellum	13–18	Quasispherical[b]Round	NA	Mukherjee et al. (2008)
	8–60	Round	NA	Devi et al. (2013)
T. atroviride	14.01–21.02	Hexagonal	Larvicidal	Singh and Prakash (2015)
	100–200	Spherical	Antifungal	Ponmurugan (2016)
	15–25	Anisotropic	Antibacterial, antioxidant, cytotoxic	Saravanakumar and Wang (2018)
	10–15	Spherical	Antimicrobial	Abdel-Azeem et al. (2020)
T. hamatum	1–150	Spherical	Bioelectricity production	Saravanakumar et al. (2016)
T. harzianum	30–50	Spherical and rods	NA	Singh and Raja (2011)
	8–60	Round	NA	Devi et al. (2013)
	4.66	Spherical	Anti-Fasciolasis	Gherbawy et al. (2013)
	51.1	Spherical	Antibacterial	Ahluwalia et al. (2014)
	10–20	NR	Larvicidal & pupicidal	Sundaravadivelan and Padmanabhan (2014)
	19–63	Sperical and ellipsoid	Seed germination	Shelar and Chavan (2015)
	12.7 ± 0.8	Spherical	Antifungal	El-Moslamy et al. (2017)
	5.33–29.46	Round and oval	Antimicrobial	El-Waseif et al. (2017)
	20–30	Spherical	Antifungal	Guilger et al. (2017)
	3–20	Round	Antibacterial	Noshad et al. (2019)
	5–18	Spherical	Antifungal	Consolo et al. (2020)
	5–50	Variable	Antifungal	Kalia et al. (2020)
T. koningii	8–24	Spherical	Antibacterial	Tripathi et al. (2013)
T. longibrachiatum	8–60	Round	NA	Devi et al. (2013)

(continued)

Table 1 (continued)

Fungi	Size (nm)	Shape[a]	Application	Source or reference
	5–25	Spherical	Antifungal	Elamawi et al. (2018)
	5–11	Spherical, triangular, cuboid and hexagonal	Antibacterial	Omran et al. (2019)
T. pseudokoningii	8–60	Round	NA	Devi et al. (2013)
T. reesei	5–50	Variable	NA	Vahabi et al. (2011)
	3–4, 15–17	Spherical	Waste removal from water	Gemishev et al. (2019a)
	1–4, 15–25	Quasispherical[b]	NA	Gemishev et al. (2019b)
T. sp	Polydispersed	Round	Antibacterial	Ramos et al. (2020)
T. virens	8–60	Round	NA	Devi et al. (2013)
	5–5 0	Sherical, oval	Antifungal	Tomah et al. (2020)
T. viride	5–40	Spherical, rodlike	Antibacterial	Fayaz et al. (2009a)
	2–100	Spherical, rodlike, nanoplates	NA	Fayaz et al. (2009b)
	5–40	Spherical, rodlike	Antibacterial	Fayaz et al. (2010a)
	2–4	Spherical	NA	Fayaz et al. (2010b)
	28–59.17	Bowl like	Antibacterial	Chitra and Annadurai (2013)
	1–50	Globular	Antibacterial	Elgorban et al. (2016)
	9–60	Spherical and ellipsoidal	Antibacterial	Shelar (2016)
	10–20	Spherical	Antibacterial	Kumari et al. (2017a)
	2–5, 50–100	Spherical, pentagonal & hexagonal	Antibacterial	Kumari et al. (2017b)
	10–20	Spherical	Antifungal	Kumari et al. (2019)
	0.1–10	Spherical	Anticancer & immunomodulatory	Adebayo-Tayo et al. (2019)
	100–250	Spherical & irregular	Antifungal	Manikandaselvi et al. (2020)
T. viz.	8–60	Round	NA	Devi et al. (2013)
T. interfusant	59.66 ± 4.18	Spherical	Antifungal	Hirpara and Gajera (2020)

[a]Shape is named as originally reported, [b]not reported but visible in image, *NA* no suggested application

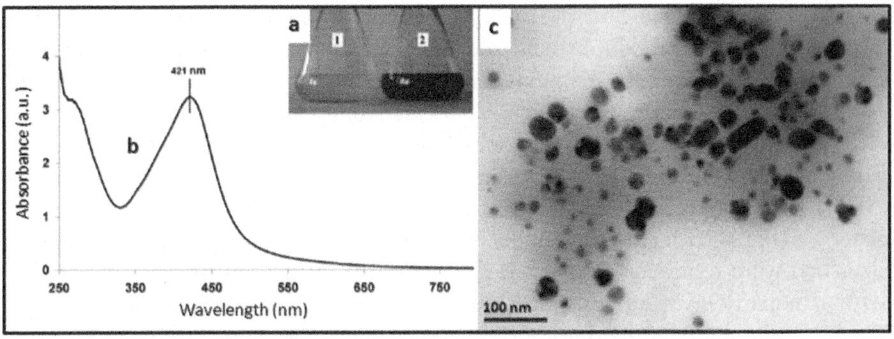

Fig. 2 Biosynthesis of silver nanoparticles using *Trichoderma viride*. (**a**) Picture of flask containing the solution of cell filtrate with 10^{-3} M of silver nitrate, before reaction (flask 1) and after 24 h of reaction (flask 2), (**b**) UV–Vis absorption spectra of silver nanoparticles after 24 h of reaction, (**c**) bright field TEM micrograph of synthesized silver nanoparticles. (Reproduced and modified from Fayaz et al. 2009a, with permission from the American Chemical Society (ACS))

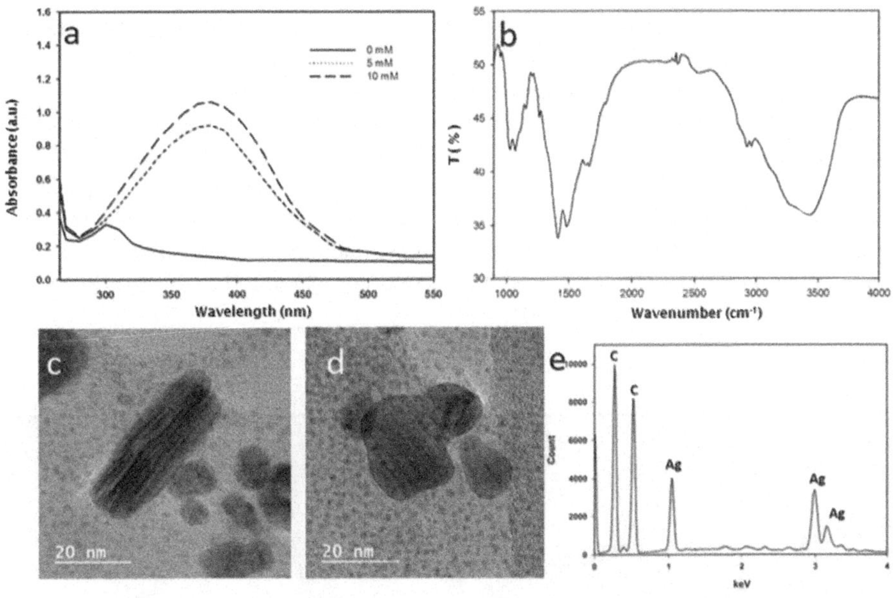

Fig. 3 Characterization of silver nanoparticles biosynthesized using *Trichoderma atroviride*. (**a**) UV–Vis spectra, (**b**) FTIR spectrum, (**c–d**) TEM micrograph, (**e**) EDX analysis. (Reproduced from Saravanakumar and Wang (2018), with permission from Elsevier)

In the work by Saravanakumar and Wang (2018), FTIR spectrum shows alkaline, amine, proteins, and aromatic peptides at the bands of 1115.4 and 3450 cm^{-1}, assigned to the metallic and O stretching vibrations of the metallic oxides, respectively (Fig. 3b). Although the resulting formation of small spherical nanoparticles is predominantly documented, some studies frequently report the formation of anisotropic particles (Fig. 3c–d). EDX analysis is carried out to confirm the elemental composition of a sample, and the analysis of nanoparticles synthesized by *T. atroviride* displayed the elemental signal (Ag) at high percentage, indicating the affluent synthesis of AgNPs (Fig. 3e) (Saravanakumar and Wang 2018).

In a study using *T. viride* (Chitra and Annadurai 2013), the biosynthesized silver nanoparticles were air-dried and examined under a scanning electron microscope (SEM), and the formation of silver nanoparticles with bowl-like shapes was observed (Fig. 4a, b). However, under TEM analysis, nanoparticles revealed smaller sizes and were reported to have a range size of 28–59.17 nm; indeed, the synthesized nanoparticles were described as polydispersed (Chitra and Annadurai 2013). Then, it becomes clear that in order to have certainty about size range, TEM analysis must be also carried out since smaller nanoparticles are not visualized under SEM.

The stability of nanoparticles is an important characteristic since it would increase their shelf storage without losing their properties. Stable nanoparticles are produced when using fungal extracts or fungal supernatants as reducing agents. In fact, by FTIR analysis, several studies detect proteins and other biomolecules around the AgNPs (Omran et al. 2019), which would act as reducing as well as stabilizing agents during synthesis. Not all studies report stability of nanoparticles, but nanoparticles obtained with *T. asperellum* were reported to be stable after storage for over 6 months without significant aggregation (Mukherjee et al. 2008).

The optimization of AgNPs production using *T. harzianum* was investigated by El-Moslamy et al. (2017). The authors used the Taguchi experimental design and found that it was possible to develop a formulation using minimum raw materials to have the most significant effect on the dry mass weight of nano-Ag. This method was used for the first time for mycosynthesis production of AgNPs, and it was concluded that the production increased three times compared to basal conditions.

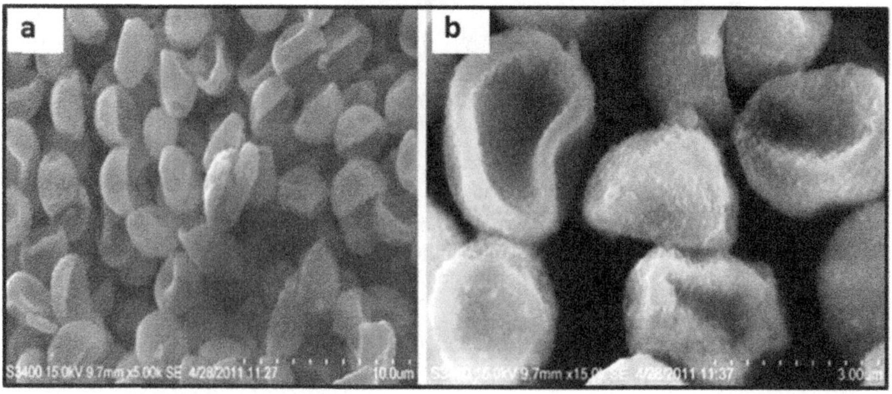

Fig. 4 SEM analysis of silver nanoparticles biosynthesized using *Trichoderma viride*. (**a**) Image showing the bowl shape of nanoparticles, (**b**) closer view. (Reproduced from Chitra and Annadurai 2013)

4 Biosynthesis of *Trichoderma* Gold Nanoparticles

Biosynthesis of gold nanoparticles (AuNPs) has been successful using cell-free filtrates and supernatants of *Trichoderma* species. Different size ranges and shapes are reported; size of particles range from 1 to 150 nm, and shapes include spherical, pseudospherical, nanotriangle, nanoprisms, triangular, cubical, hexagonal, rod-like, and flower-like (Table 2). The synthesis protocol in most cases is adjusted depending on the aim of the study, but in general cell-free filtrate or supernatants are put in contact with tetrachloroauric acid ($HAuCl_4$) where the change in color of the reaction is the first evidence of AuNPs formation. Depending on the fungal species and the parameters of the reaction, the color may vary from light pink to dark red or violet (Saravanakumar et al. 2016; Abdel-Kareem and Zohri 2018). The supernatant of *T. hamatum* was used for the synthesis of AuNPs, and the color of the reaction mixture changed from light yellow to ruby-red (Fig. 1b). UV–Vis spectroscopy analysis confirmed the formation of AuNPs with maximum absorption at 530–560 nm (Fig. 1b). TEM analysis revealed cubical shape nanoparticles ranging from 5 to 150 nm with mean size of 82 ± 2.8 nm (Saravanakumar et al. (2016).

As mentioned earlier, the synthesis reaction can be adjusted varying parameters such as pH, temperature, amount of biomass, concentration of metallic solution, time of incubation, etc. In a study, AuNPs were synthesized with a newly isolated *Trichoderma* strain (WL-Go), and the optimal conditions for AuNPs synthesis were as follows: $HAuCl_4$ 1.0 mmol l^{-1}, biomass 0.5 g, and pH 7–11. The maximum of UV–Vis spectra was at 550 nm (Fig. 5a) indicating formation of nanoparticles. TEM analysis revealed that the shape of AuNPs were mostly spheres, but triangles and hexagons were also present (Fig. 5b) (Qu et al. 2017). AuNPs of diverse shapes such as spheres, pentagons, and hexagons of 5–30 nm in size were also obtained with optimized protocol using *T. hamatum* (Abdel-Kareem and Zohri 2018). Similar results were obtained using *T. koningii*, obtaining AuNPs ranging from polydisperse small spheres to large triangles and hexagons of 30–40 nm and thickness of 5–10 nm (Maliszewska et al. 2009). However, using the same *T. koningii* species and a modified protocol, small spherical shaped AuNPs with average size of 14 ± 4 nm were obtained. The author reported that it was possible to separate AuNPs by size using sucrose gradients; spheres from 10 to 14 nm were concentrated in the 30% fraction and spheres from 12 to 17 nm in the 40% fraction (Maliszewska 2013).

Gold nanoprisms and nanoplates of different shapes and sizes are developed by evolution of morphology from pseudospherical nanoparticles to more complex forms. This process was documented and studied in AuNPs synthesized with *T. asperellum*. TEM analyses revealed that several pseudospheres fuse to form nanoprisms and nanotriangles. It was concluded that the wide spectrum of morphologies was due to a slow rate of reduction of $HAuCl_4$ by the constituents of the cell-free fungal extract (Mukherjee et al. 2012).

As mentioned in the previous section, stability of nanoparticles is important in order to increase their shelf storage. However, not many studies report the time at

Table 2 *Trichoderma* species used to produce gold and other metal-based nanoparticles and their suggested applications

Fungi	Type of NP produced	Size (nm)	Shape[a]	Application	Source or reference
T. asperellum	Au	15 30 100	Pseudospherical nanotriangles nanoprisms	NA	Mukherjee et al. (2012)
	Se	49.5, 61.3, 130.2	Irregular and spherical	Anti-mildew and zoosporicidal	Nandini et al. (2017)
	CuO	10–190	Spherical	Anticancer	Saravanakumar et al. (2019)
T. atroviride	Au	50–75, 10–50	Triangular nanopaltes and spherical	Antifungal	Ponmurugan (2016)
	Se	157.9, 168.4, 67.9	Irregular	Anti-mildew and zoosporicidal	Nandini et al. (2017)
	Se	60.48–123.16	Spherical	Antifungal	Joshi et al. (2019)
	Cu	5–25	Irregular spherical	Antifungal	Natesan et al. (2020)
T. brevicompactum	Se	99.6, 109.2, 199.6	Irregular	Anti-mildew and zoosporicidal	Nandini et al. (2017)
T. citrinoviridae	TiO_2	10–400	Triangular, pentagonal, spherical and rod	Antibacterial	Arya et al. (2020)
T. hamatum	Au	5–150	Cubical	Bioelectricity production	Saravanakumar et al. (2016)
	Au	5–30	Spherical, pentagonal and hexagonal	Antibacterial	Abdel-Kareem and Zohri (2018)
T. harzianum	CdS	3–8	Spherical	Photocatalysis	Bhadwal et al. (2014)
	Au	26–34	Spherical	Detection of Hg^{2+}	Tripathi et al. (2014)
	Au-Ag	10–25	Triangles, spheres, rods, squares, hexagonal	Catalytic	Tripathi et al. (2015)
	Se	60.8, 140.4, 103.5	Spherical	Anti-mildew and zoosporicidal	Nandini et al. (2017)
	Au	26–34	Spherical	Antibacterial, catalytic	Tripathi et al. (2018)

(continued)

Table 2 (continued)

Fungi	Type of NP produced	Size (nm)	Shape[a]	Application	Source or reference
	Te	Variable	Variable	NA	Liang et al. (2019)
	Se	Variable	Variable	NA	Liang et al. (2019)
	Se	60	Irregular	Antifungal	Hu et al. (2019)
	T-β-D-glu-ZnO	30–186	Spherical	Anticancer and antibacterial	Saravanakumar et al. (2020)
	CuO	38–77 width, 135–320 length	Elongated fibers	Antifungal	Consolo et al. (2020)
	ZnO	27–40 width, 134–200 length	Fan and bouquet structure	NA	Consolo et al. (2020)
	FeO	20–60	Variable	Antifungal	Kalia et al. (2020)
	ZnO	10–40	Variable	Antifungal	Kalia et al. (2020)
	ZnO	12–35	Variable	Antibacterial	Shobha et al. (2020)
	Au	15	Spherical	NA	do Nascimento et al. (2020)
T. koningii	Au	30–40	Spheres, triangles, hexagons	NA	Maliszewska et al. (2009)
	Au	14±4	Spherical	Anticancer	Maliszewska (2013)
T. koningiopsis	Cu	87.5	Spherical	NA	Salvadori et al. (2014)
T. longibrachiatum	Se	87.5, 256.1, 158.4	Spherical & irregular	Anti-mildew and zoosporicidal	Nandini et al. (2017)
	Au	102.93–123.99	Flower like	Biomedical	Elegbede et al. (2020)
T. reesei	ZnO	12–35	Variable	Antibacterial	Shobha et al. (2020)
Trichoderma sp.	Se	20–220	Spherical and pseudo-spherical	NA	Diko et al. (2020a)
	Se	80-180	Spherical	NA	Diko et al. (2020a)
	PbSe	10–30	Cubic	Photocatalysis	Diko et al. (2020b)

(continued)

Table 2 (continued)

Fungi	Type of NP produced	Size (nm)	Shape[a]	Application	Source or reference
T. virens	Se	96.2, 158.8, 312.5	Spherical & irregular	Anti-mildew and zoosporicidal	Nandini et al. (2017)
T. viride	Au	4–15	Spherical	Antibacterial	Fayaz et al. (2011)
	TiO_2	60–86.67	Spherical	Bio-pesticidal	Chinnaperumal et al. (2018)
T. sp	Au	NR	Spheres, triangles, hexagons	Dye decoloration	Qu et al. (2017)
	Au	1–24	Spherical and pseudo-spherical	Degradation of pollutants	Qu et al. (2018)
	Au	0–120	Spherical	Biocatalisys & antibacterial	Mishra et al. (2014)

[a]Shape is named as originally reported, *NR* not reported, *NA* no suggested application

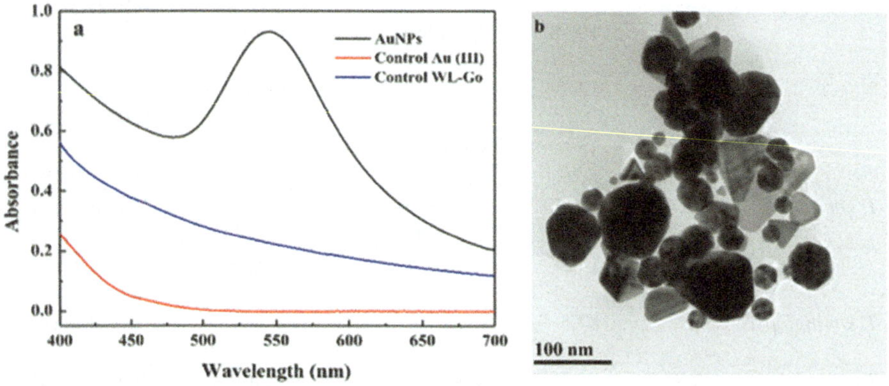

Fig. 5 Gold nanoparticles synthesized using *Trichoderma* sp. strain WL-Go. (**a**) UV–Vis spectra of the dispersed AuNPs after 40 h incubation, $HAuCl_4$ solution, and control strain WL-Go, (**b**) TEM image of AuNPs synthesized by strain WL-Go. (Reproduced from Qu et al. 2017, with permission from Elsevier)

which AuNPs remain stable. Some studies report that AuNPs synthesized by *Trichoderma* species remain stable for 3 months (Fayaz et al. 2011; Tripathi et al. 2014; Ponmurugan 2016) and up to 6 months of storage at ambient temperature (Maliszewska 2013).

5 Biosynthesis of Other *Trichoderma* Metal-Based Nanoparticles

Up to the present time, most studies on the biosynthesis of metallic nanoparticles using *Trichoderma* species are on silver and gold nanoparticles (Tables 1 and 2). Nevertheless, recent studies have produced other *Trichoderma* metal-based nanoparticles such as Se, Cu, CdS, Te, PbSe, ZnO, and TiO_2. Most of these reports are on the production of selenium nanoparticles (SeNPs), while studies on the biosynthesis of other meta-based nanoparticles are scarce (Table 2).

The biosynthesis of SeNPs has been carried out using a solution of Na_2SeO_3 (Nandini et al. 2017; Joshi et al. 2019; Hu et al. 2019) or SeO_2 (Diko et al. 2020a). The first evidence on the formation of nanoparticles by the reduction of selenite ions is visualized as a color change in the solution, from pale yellow to orange-red (Nandini et al. 2017; Joshi et al. 2019; Hu et al. 2019; Diko et al. 2020a). Interestingly, *Trichoderma harzianum* was able to grow on selenium and tellurium-containing media at concentrations of 1 mM. The fungal surface turned red in color in the case of selenium and black in the case of tellurium; scanning electron microscopy analysis confirmed the formation of SeNPs and tellurium nanoparticles (TeNPs) (Liang et al. 2019). In a study by Diko et al. (2020b) using *Trichoderma* sp. (strain WL-Go), the formation of spherical SeNPs was reported (Fig. 6a–b). Remarkably, with an optimized synthesis protocol at pH 8 with 0.5 g biomass of strain WL-Go and (1:1) mM of SeO_2: $Pb(NO_3)_2$, the formation of 10–30 nm cubic faced centered PbSeNPs was achieved (Fig. 6c–d). The authors stated that the form of the biomaterial was influenced by the synergy between Se and Pb ions in the presence of secreted proteins by strain WL-Go.

Another interesting contribution is the synthesis of copper oxide nanoparticles (CuONPs), which were synthesized using cell-free extract of *T. asperellum*. Synthesis of CuONPs was evidenced through color change from light-yellow to dark-brown. Analysis by UV-Vis spectroscopy indicated the surface plasmon resonance peak between 285 and 295 nm. Ultra high-resolution scanning electron microscopy analysis revealed agglomerated spherical CuONPs (Fig. 7a). The presence of Cu and O was confirmed by EDS mapping (Fig. 7b, c) and by the EDS spectrum (Fig. 7d). By field emission transmission electron microscopy analysis, the spherical shape of synthesized CuONPs was confirmed (Fig. 7e, f), which were in the size range of 10–190 nm (Saravanakumar et al. 2019).

6 Antibacterial Activity of *Trichoderma* Nanoparticles

Among the studies for the synthesis of nanoparticles using *Trichoderma*, there are reports in which nanoparticles have been tested for antibacterial capacity (Tables 1 and 2). Silver nanoparticles (AgNPs) are the most frequent nanomaterial that is proposed as antibacterial agent; analyses in vitro have demonstrated the

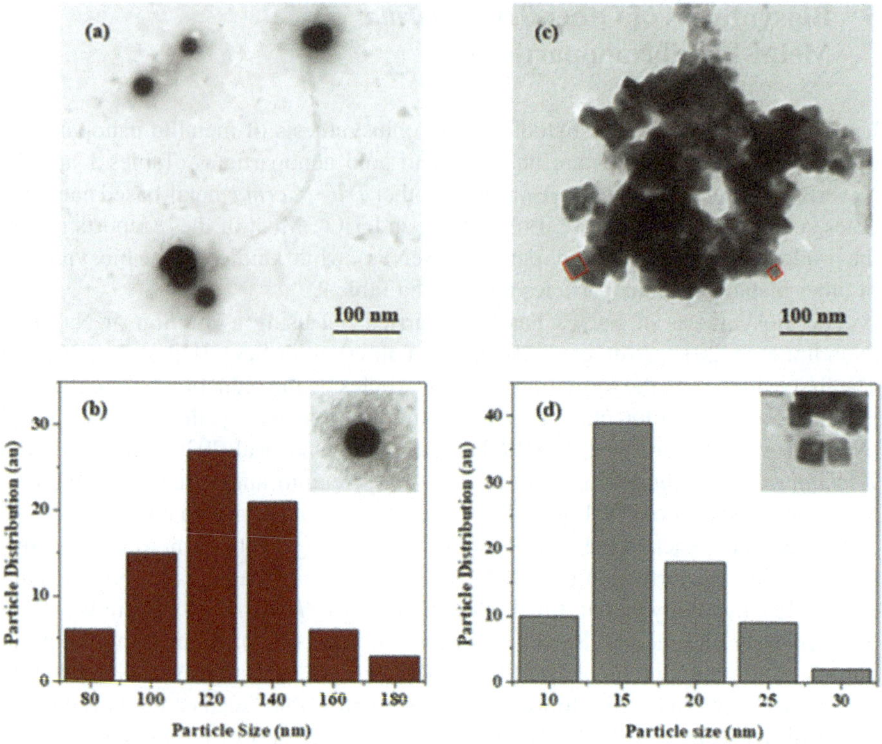

Fig. 6 Selenium (SeNPs) and lead selenide (PbSeNPs) nanoparticles synthesized using *Trichoderma* sp. (**a**) TEM micrograph of SeNPs, (**b**) corresponding size distribution analysis, inset shows spherical morphology, (**c**) TEM micrograph of PbSeNPs, (**d**) corresponding size distribution analysis, inset shows cubic morphology. (Reproduced from Diko et al. 2020b, with permission from Elsevier)

antimicrobial capacity against harmful bacteria such as *Escherichia coli*, *Pseudomonas aeruginosa*, *Klebsiella pneumoniae*, *K. aeruginosa*, *Shigella flexneri*, *S. sonnei*, *Serratia marcescens*, *Salmonella typhimurium*, and *Staphylococcus aureus*. Furthermore, it has been proved that the use of cell filtrates of *Trichoderma* enhances the antimicrobial efficacy of the synthesized nanoparticles, since the filtrate possess antimicrobial metabolites (Kumari et al. 2017a). Also, another advantage in using *Trichoderma* species was detected when the antibacterial capacity of biosynthesized AgNPs using cell-free filtrate of *T. viride* were compared to citrate stabilized nanoparticles. Biosynthesized nanoparticles were internalized inside the bacterial cell more efficiently than the ones chemically synthesized. Thus, it was concluded that biologically synthesized AgNPs coated with antimicrobial metabolites of *T. viride* were more potent in killing bacteria than chemical nanoparticles (Kumari et al. 2017a).

The well diffusion or disk diffusion methods are good assessments to preliminarily detect antibacterial activity of nanoparticles. The agar plate surface is inoculated

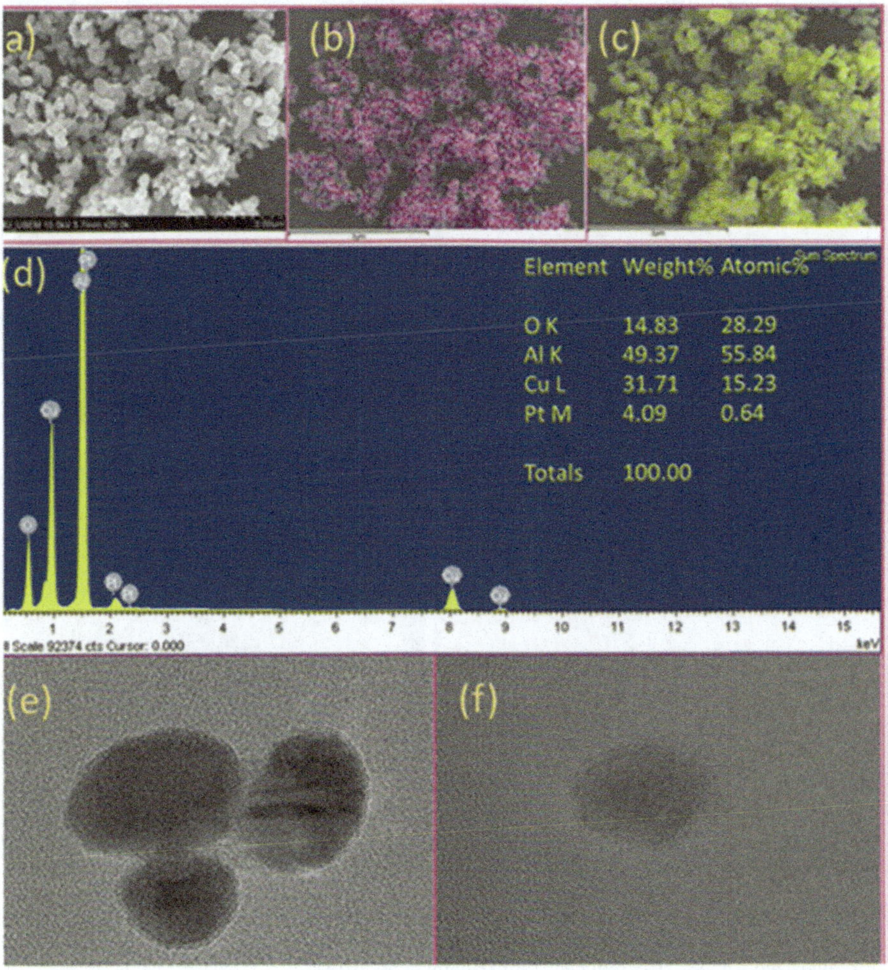

Fig. 7 Characterization of copper oxide nanoparticles (TA-CuONPs) synthesized by *Trichoderma asperellum*. (**a**) Ultra high-resolution scanning electron microscopy analysis (**b**) EDS mapping of Cu, (**c**) EDS mapping of O, (**d**) EDS spectrum of CU, and (**e, f**) field emission transmission electron microscopy images of TA-CuONPs. (Reproduced from Saravanakumar et al. 2019, with permission from Elsevier)

by spreading a volume of the bacterial inoculum over the entire agar surface. Then, wells or disks are loaded with biosynthesized nanoparticles, and after 24 h of incubation, inhibition of bacteria is detected. Antibacterial activity is evaluated by measuring the diameter of inhibition zone. After preliminary assays, the minimum inhibition concentration (MIC) and/or minimum bactericidal concentration (MBC) has to be determined, since the concentrations needed to inhibit bacterial growth or to kill bacteria are significantly different with respect to the concentrations used in the well or disk diffusion methods. In a study using AgNPs synthesized by *T.*

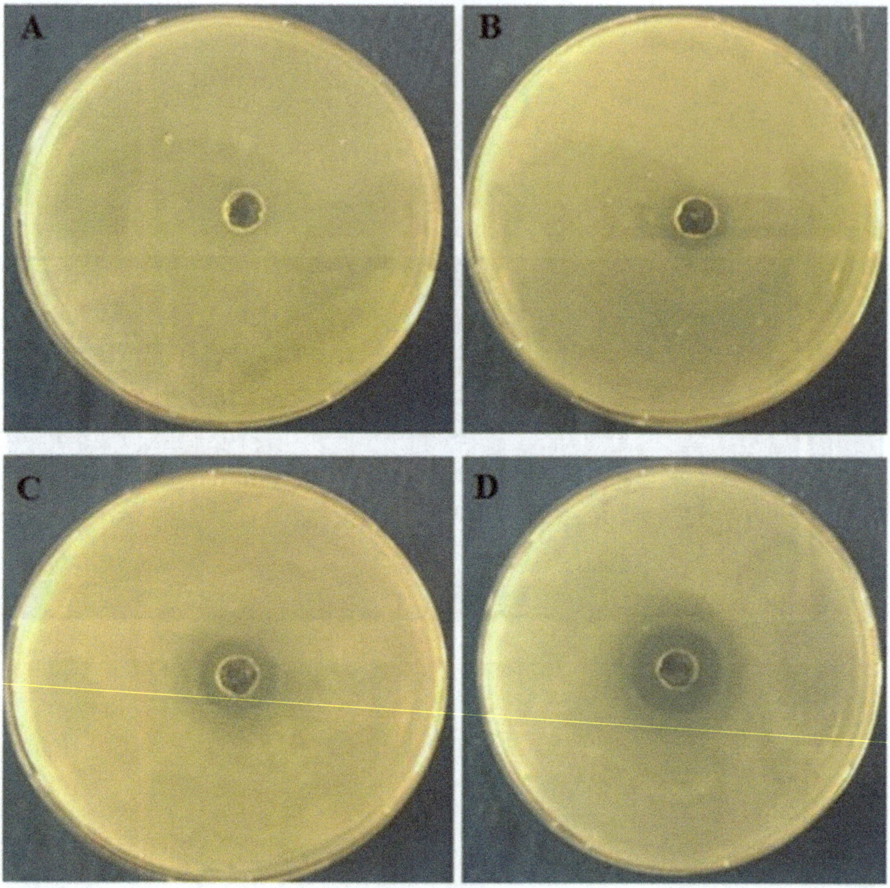

Fig. 8 In vitro antibacterial activity of AgNPs synthesized using *Trichoderma koningii*. Zone of inhibition against *Salmonella typhimurium* loaded with (**a**) 0 μg, (**b**) 2 μg, (**c**) 5 μg, and (**d**) 7 μg of biosynthesized AgNPs. (Reproduced from Tripathi et al. 2013)

koningii, this was clearly shown, and it was found that the diameter of inhibition zone of *Salmonella typhimurium* increased with increasing concentration of AgNPs in a well diffusion assay using 2, 5, and 7 μg of AgNPs (Fig. 8). However, the necessary concentration for bacterial inhibition (MIC) and bacterial death (MBC) was 25 μg ml^{-1} and 45 μg ml^{-1}, respectively (Tripathi et al. 2013).

Antimicrobial activity of AgNPs against human pathogens is shape and size dependent (Osonga et al. 2020). This was demonstrated in a study using AgNPs of different shapes and sizes synthesized by *T. viride*. Maximum inhibition was found with spherical nanoparticles (2–5 nm) showing 40, 51, 43, 53.9, and 55.8% against *S. sonnei*, *E. coli*, *S. marcescens*, *S. aureus*, and *P. aeruginosa*, respectively, while pentagonal and hexagonal nanoparticles (50–100 nm) demonstrated lower inhibition, displaying 32, 41, 31, 42.84, and 42.80% as compared to control at the same concentration of nanoparticles. Remarkably, when using NPs of similar size, penta- and hexagonal NPs showed 15–18% more antagonistic effects on tested pathogens in comparison with spherical NPs. Thus, it was concluded that shape and size played a major role in enhancing antimicrobial capacity of AgNPs, both singly and synergistically combined with antibiotics (Kumari et al. 2017b).

It is well known that the concentrations of AgNPs needed for the inhibition of Gram-negative and Gram-positive bacteria are different; chemically synthesized AgNPs were tested against both types of bacteria, and lower concentrations were needed for the inhibition of Gram-negative bacteria (Shrivastava et al. 2007). This was also found when using AgNPs synthesized with *Trichoderma* spp. isolated from the *Bertholletia excelsa* (Brazil nut) seeds. The minimal inhibitory concentration (MIC) and the minimal bactericide concentration (MBC) of AgNPs needed to inhibit Gram-negative bacteria were lower and were attributed to the thin peptidoglycan layer (2–3 nm) between the cytoplasmic membrane and the outer membrane (Ramos et al. 2020).

Searching for antimicrobial alternatives, a few reports have evaluated gold nanoparticles (AuNPs) synthesized by *Trichoderma* species against pathogenic bacteria (Table 2). Similarly, as reported with AgNPs using the well diffusion method, the diameter of inhibition zone increases with increasing concentration of AuNPs. Biosynthesized AuNPs by *T. harzianum* were used against *E. coli*, and the diameter of inhibition zones were 3, 5, and 9 mm with 4, 6, and 12 μg concentrations of AuNPs, respectively. The minimum inhibitory concentration (MIC) of AuNPs was 20 μg/ml. Thus, it was concluded that AuNPs show effective antibacterial activity (Tripathi et al. 2018). Also, dose-dependent antimicrobial activity of biosynthesized AuNPs by *Trichoderma* sp. was recorded against *P. syringae*, *E. coli,* and *S. sonnei* (Mishra et al. 2014).

In another study, vancomycin was bound to AuNPs synthesized using *T. viride* (VBGNPs). The VBGNPs exhibited notable antibacterial activity against *E. coli*, while vancomycin alone had no significant influence on the bacterial growth. Analysis by TEM revealed that VBGNPs were accumulated in the outer membrane of *E. coli*, while some of them successfully penetrated into cells (Fayaz et al. 2011).

Recently, other metallic nanoparticles synthesized by *Trichoderma* species have been evaluated as antibacterial agents; for instance, titanium nanoparticles (TiO_2NPs) synthesized with the extract of *T. citrinoviride* showed antibacterial activity (100 μg/mL) against planktonic cells of extremely drug-resistant (XDR) *P. aeruginosa* clinical isolates (Arya et al. 2020). In a study carried out by Shobha et al. (2020), the authors used *T. harzianum* and *T. reesei* for the synthesis of ZnONPs and reported that nanoparticles were able to inhibit the growth of *Xanthomonas oryzae* pv. *oryzae* in in vitro assays, in a dose-dependent manner. Also, *T. harzianum* was used for the synthesis of zinc oxide nanoparticles conjugated with β-D-glucan from barley (T-β-D-glu-ZnO NPs), and their antibacterial capacity was investigated. The authors found excellent antibacterial activity of T-β-D-glu-ZnO NPs against *S. typhi*, MRSA, and *E. coli* which was evidenced by in vitro and in vivo antibacterial experiments (Saravanakumar et al. 2020).

7 Antifungal Activity of *Trichoderma* Nanoparticles

Silver nanoparticles are the most widely nanomaterial tested for antimicrobial capacity, mostly against bacteria, but an increasing number of reports have also analyzed the antifungal properties of nano-silver and other metal-based nanoparticles. Searching for ecofriendly methods, several *Trichoderma* species have been used for the production of nanoparticles, and their antifungal capacity has been evaluated (Tables 1 and 2). It has been reported that mycogenic AgNPs display antifungal activity with efficiencies comparable to the positive controls, especially toward clotrimazole and nystatin (Abdel-Azeem et al. 2020).

Furthermore, AgNPs synthesized by *Trichoderma* species are reported to be more effective than chemically synthesized particles of comparable size. This was demonstrated by Kumari et al. (2019) who used AgNPs synthesized by *T. viride* and found enhanced antifungal activity of biosynthesized NPs against *Fusarium oxysporum* and *Alternaria brassicicola*, in comparison to chemically synthesized AgNPs (CSNP) of similar shape and size. Scanning electron microscopy (SEM) analysis also revealed that the architecture of *A. brassicicola* was partially damaged by CSNP (Fig. 9b) and completely disintegrated after the treatment with biosynthesized NPs (Fig. 9c, d). It was reported that biosynthesized AgNPs possess enhanced properties because of the cell-free extract of *T. viride*, which have multiple modes of actions including protein degradation, and complete disruption of cell by cellular lyses and disruption of osmotic balance and cell wall leakage. Then, it is impossible for the pathogen to recover after injury resulting in a complete inhibition (Kumari et al. 2019). In a recent study, it was also demonstrated in in vitro assays the prominent antifungal capacity of AgNPs synthesized using *T. virens* against the soilborne pathogen *Sclerotinia sclerotiorum*. The biosynthesized AgNPs showed a high percentage inhibition against hyphal growth, sclerotial formation, and myceliogenic germination of sclerotia. SEM analysis revealed a direct interaction between nanoparticles and fungal cells, including AgNPs' contact, accumulation, lamellar fragment production, and micropore or fissure formation on fungal cell walls (Tomah et al. 2020).

Although AgNPs synthesized by *Trichoderma* species are effective against pathogenic fungi, not all pathogens respond equally. In a study, AgNPs synthesized using *T. longibrachiatum* were tested against nine fungal pathogens: *Alternaria alternata, Fusarium verticillioides, F. moniliforme, Aspergillus flavus, A. heteromorphus, Penicillium glabrum, P. brevicompactum, Pyricularia grisea*, and *Helminthosporium oryzae*. The results obtained showed that AgNPs were most effective against *P. grisea, F. verticillioides, H. oryzae, F. moniliforme*, and *A. alternata* with inhibition percentages of 98.9, 96.4, 95.1, 93.6, and 93.0%, respectively (Elamawi et al. 2018).

Also, various reports have documented that some *Fusarium* species present no response or lower percentage of inhibition when exposed to NPs. This was evidenced when using AgNPs synthesized using two strains of *T. harzianum* (EMCC 540 and SYA.F4). By the agar well diffusion method, nanoparticles from both

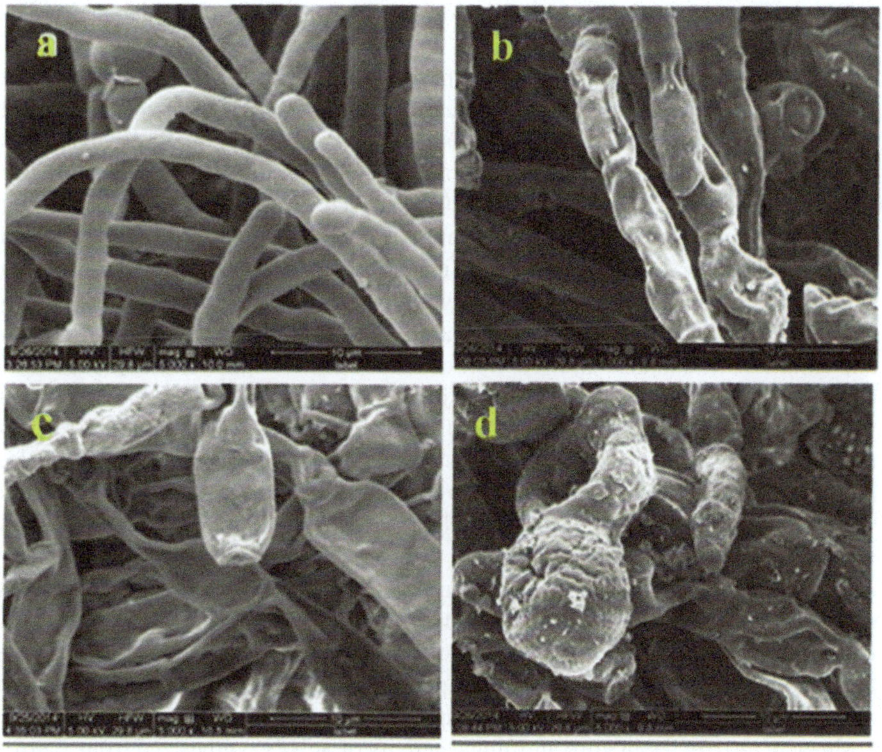

Fig. 9 SEM micrograph of *Alternaria brassicicola*. (**a**) Healthy mycelia network without any treatment; (**b**) sample treated with chemically synthesized silver nanoparticles, showing partially damaged mycelia; (**c**, **d**) sample treated with silver nanoparticles biosynthesized using *T. viride*, showing completely damaged mycelia and spores. (Reproduced from Kumari et al. 2019 with permission from Elsevier)

strains were screened against *Fusarium proliferatum*, *Fusarium* sp., *Botrytis cinerea*, *Rhizoctonia solani*, and *F. oxysporum*. Results showed that AgNPs produced using *T. harzianum* SYA.F4 were effective against most fungal pathogens; however, no response against *F. proliferatum* was detected (El-Moslamy et al. 2017). Similarly, mycogenic AgNPs using the endophytic *T. atroviride* showed significant ($P < 0.05$) antifungal activity against *Candida albicans* and *Aspergillus* sp. However, *Fusarium oxysporum* f. sp. *lycopersici* showed a complete resistance toward both the mycogenic and chemical AgNPs (Abdel-Azeem et al. 2020). In another study, AgNPs synthesized by *T. longibrachiatum* were effective against *P. grisea*, *F. verticillioides*, *H. oryzae*, *F. moniliforme*, and *A. alternata*. However, lower efficiency was observed against *F. oxysporum* (Elamawi et al. 2018). Other *Fusarium* species appear to be more susceptible to AgNPs since a study reported AgNPs synthesized by *T. viride* as a suitable agent against *Fusarium moniliforme* (Manikandaselvi et al. 2020). Kalia et al. (2020) also reported AgNPs and FeONPs synthesized by *T.*

harzianum as effective for the inhibition of *F. moniliforme*, in a concentration-dependent manner.

Not only nano-silver has proved to have antifungal properties, the efficacy of both silver and gold nanoparticles synthesized by *T. atroviride* were also successful against the tea pathogenic fungus *Phomopsis theae*. In vitro analyses revealed a considerable suppression on the growth of *P. theae* using both types of nanoparticles (Ponmurugan 2016). Natesan et al. (2020) also used *T. atroviride* CuNPs to inhibit the tea pathogens *Poria hypolateritia* and *P. theae* which causes red root-rot and Phomopsis canker diseases. Selenium nanoparticles derived from *T. harzianum* (TSeNPs) were analyzed and compared with traditional SeNPs for antifungal properties. The synthesized TSeNPs showed antifungal properties against *F. verticillium* and *A. alternata* and also showed control functionalities against *Alternaria* toxins (83% of tenuazonic acid (TeA) and 79% of alternariol (AOH) reduction), fumonisin B1 (63% of FB1 reduction), and deoxynivalenol (76% of DON reduction), respectively. Also, it was reported that expression of synthetic genes (FUM1, PA, TRI5, and TRI6) and mycotoxins production were substantially decreased.

In a recent study, the synthesis of AgNPs, CuONPs, and ZnONPs was carried out with *T. harzianum*. The resulting NPs were used to conduct antifungal activity assays against *A. alternata*, *P. oryzae*, and *S. sclerotiorum*. AgNPs and CuONPs reduced significantly the mycelial growth of *A. alternata* and *P. oryzae* in a dose-dependent manner. However, although treatment with ZnONPs displayed a tendency to reduce mycelial growth of the plant pathogens, no significant differences were found to conclude an antifungal capacity (Consolo et al. 2020). In a study by Joshi et al. (2019), the in vitro antifungal activity of SeNPs synthesized from *T. atroviride* against *Pyricularia grisea* was reported. It was also found that fungal inhibition increased with increased concentration of particles (Fig. 10).

8 Application of *Trichoderma* Nanoparticles in Agriculture

Trichoderma species are successfully used for the efficient control of fungal phytopathogens. However, short shelf life, low on-field stability, and irregular performance in different agro-climatic regions are some problems associated with these commercial formulations (Fraceto et al. 2018). Therefore, new strategies are urgently required for the efficient management of pathogens. Nanotechnology offers promising applications in the agricultural area. New technologies such as the microencapsulation of fungi and the biogenic synthesis of nanoparticles are strategies that could be used for disease control and thus contribute to sustainable agricultural practices (Guilger et al. 2017; Fraceto et al. 2018; Manikandaselvi et al. 2020).

In this regard, *Trichoderma* species are currently considered as promising agents for nanoparticles production, and the efficacy of the resultant nanomaterials are evaluated as antimicrobial agents for plant pathogens. For instance, as specified above, in a study conducted by Ponmurugan (2016), AgNPs and AuNPs synthesized with *T. atroviride* were evaluated against the tea pathogenic fungus *P. theae*. In vitro

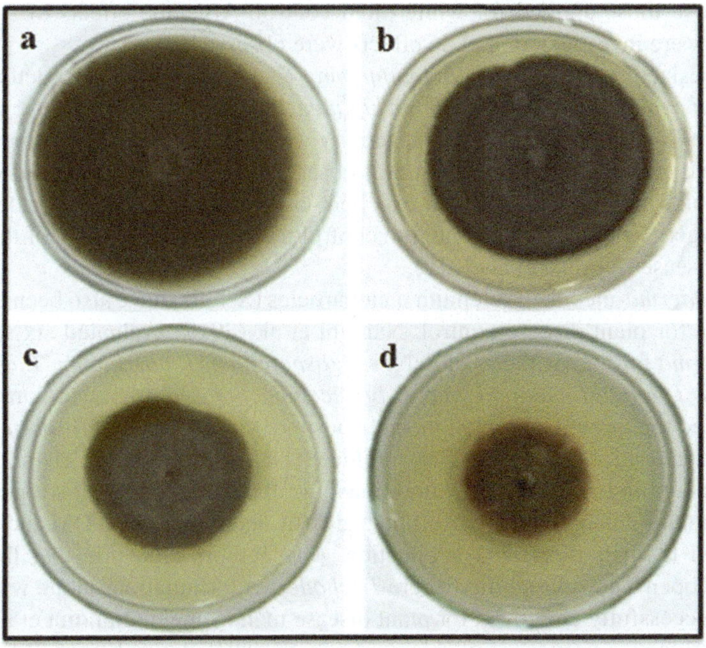

Fig. 10 In vitro antifungal activity of SeNPs synthesized using *Trichoderma atroviride*. Growth of *Pyricularia grisea* on PDA plates at different concentrations, (**a**) 0 ppm, (**b**) 50 ppm, (**c**) 100 ppm, and (**d**) 200 ppm of biosynthesized SeNPs. (Reproduced and modified from Joshi et al. 2019)

antifungal studies revealed a considerable suppression on the growth of *P. theae*. Field experiments were also conducted with soil application and wound dressing. A significant reduction in canker size was observed in plants treated with gold and silver nanoparticles. Also, improvement in leaf yield was noted in response to these treatments. The author concludes that metallic nanoparticles could improve the efficacy in management of stem disease in tea plantations.

Biosynthesized AgNPs were also suggested for agricultural purposes to increase the viability of seeds. Disease-free and healthy-looking seeds of sunflower (*Helianthus annuus*) and soybean (*Glycine max*) were obtained by per-soaked in AgNPs solution of *T. harzianum*. Also, the percentage of seed germination was enhanced and increased, with increased soaking time in the AgNPs solution (Shelar and Chavan 2015). Guilger et al. (2017) also reported AgNPs by *T. harzianum* as a new alternative in agriculture for the white mold control. The fungus *Sclerotinia sclerotiorum* is the causal agent of the white mold disease, and although there are different ways of controlling this organism, none of these inhibit sclerotia, which are its resistance structures. By using the mycogenic AgNPs against *S. sclerotiorum*, it was possible to inhibit sclerotia germination and mycelial growth. Although the nanoparticles showed cytotoxicity and genotoxicity, the most toxic concentrations were above those applied for white mold control. Also, the effects of AgNPs on soybean were investigated, and no effects were observed.

Synthesized AgNPs by *T. harzianum* were also evaluated against actinomycete *Clavibacter michiganensis* subsp. *michiganensis,* which is the causative pathogen of tomato canker disease. Antibiotic activity was detected at low concentrations of 1 mM, and increased inhibition zone was observed at 2.5 mM. The authors conclude that the biosynthesis process is an excellent candidate for industrial scale production of AgNPs and has the potential to control the bacterial pathogen of the tomato canker disease (Noshad et al. 2019).

Trichoderma-mediated selenium nanoparticles (SeNPs) have also been evaluated as agents for plant disease control. Nandini et al. (2017) evaluated six species of *Trichoderma* for the synthesis of SeNPs: *T. asperellum, T. harzianum, T. atroviride, T. virens, T. longibrachiatum*, and *T. brevicompactum*. All obtained nanoparticles suppressed the growth, sporulation, and zoospore viability of *Sclerospora graminicola*, the causative agent of downy mildew (DM) disease in pearl millet (PM). Furthermore, under greenhouse conditions, the application of SeNPs together with *T. asperellum* enhanced the early plant growth and suppressed DM incidence as compared to their individual application. The authors conclude that the results obtained open a new possibility where *Trichoderma* formulations along with SeNPs can be successfully employed for plant disease management (Nandini et al. 2017).

The protective properties of selenium nanoparticles derived from *T. harzianum* (TSeNPs) were also assessed in in vitro assays in maize and pear. It was found that between the traditional SeNPs and TSeNPs-treated groups, TSeNPs had better protective effect in both maize and pear. Thus, the authors concluded that the biogenic nanoparticles are valuable functional materials with great potential for practical plant protection and food safety prevention (Hu et al. 2019).

Selenium nanoparticles (SeNPs) synthesized from *T. atroviride* were also tested against the plant pathogen *Colletotrichum capsici*, which causes anthracnose disease in chili, and against *Alternaria solani*, which causes early blight of tomato. Assays of plant pathogens inhibition were carried out on healthy chili and tomato leaves. It was found that SeNPs inhibited the infection of *C. capsici* on chili leaves (Fig. 11a) and *A. solani* on tomato leaves (Fig. 11b). The authors concluded that SeNPs could be useful to manage plant diseases in an eco-friendly manner, due to their mycogenic synthesis and antifungal activity against different phytopathogens (Joshi et al. 2019).

Mycogenic nanoparticles using *Trichoderma* seems to be a feasible option to inhibit plant pathogens and thus using them for agricultural purposes. However, further research should be carried out in order to determine the effects of nanomaterials in the environment before using them for agricultural applications.

Fig. 11 In vitro antifungal leaflet assay of SeNPs synthesized using *Trichoderma atroviride*. (**a**) Leaflet assay of 1-month-old chili and (**b**) tomato leaves primed with various concentrations of mycogenic SeNPs (0, 10, 25, 50, and 100 ppm) and then artificially inoculated with *Colletotrichum capsici* and *Alternaria solani*, respectively. (Reproduced and modified from Joshi et al. 2019)

9 Possible Biomedical Applications of *Trichoderma* Nanoparticles

Currently, the main biomedical application that has been suggested for *Trichoderma* nanoparticles is their use as antimicrobial agents. As described in previous sections, nanoparticles synthesized by *Trichoderma* species display excellent antibacterial and antifungal capacity, thus representing a feasible option to treat infections caused by multi-resistant microbes. Although the application of these nanoparticles in the clinic is not yet a reality, there is also an increasing interest in searching eco-friendly nanomaterials for their use in other biomedical applications. Hence, some species of *Trichoderma* could be used for the biosynthesis of nanoparticles for their potential use in cancer therapy, bioimaging, biosensors, hyperthermia, photoablation therapy, and targeted drug delivery.

In fact, recent reports have suggested the potential use of *Trichoderma*-mediated nanoparticles for cancer treatment. Adebayo-Tayo et al. (2019) reported that SeNPs synthesized by *T. viride* exhibited cytotoxicity against Hep-2C cell line and RD cell line in a dose-dependent manner and had immune-stimulation potential by increasing the production of IgA and IgM. Thus, the authors concluded that the anticancer and immunomodulatory potential of the SeNPs synthesized by *T. viride* justifies its biomedical application and showcases the biotechnological relevance of the fungus (Adebayo-Tayo et al. 2019).

Similarly, AgNPs synthesized using *T. atroviride* were reported to trigger the cancer cell death at a concentration-dependent manner. The study reported AgNPs with antioxidant and cytotoxicity activities; thus, the authors proposed further investigations for their biomedical applications (Saravanakumar and Wang 2018). In another study, copper oxide nanoparticles (CuONPs) were synthesized using the extract of *T. asperellum*, and it was found that *Trichoderma*-mediated CuONPs possessed anticancer properties, since they induced photothermolysis of A549 cancer cells by reactive oxygen species generation, nucleus damage, mitochondrial membrane potential ($\Delta\Psi m$), and regulatory protein expression (Saravanakumar et al. 2019).

Also, Saravanakumar et al. (2020) reported the anticancer activity of β-D-glucan-zinc oxide nanoparticles (β-D-glu-ZnO NPs). The authors first synthesized ZnONPs using the fungal mycelia water extract derived from *T. harzianum* and then successfully conjugated ZnONPs with β-D-glucan (β-D-glu-ZnO NPs), and the conjugation was confirmed by PACE (polysaccharide analysis by carbohydrate gel electrophoresis) and FTIR. Nanoparticles exhibited a dose-dependent inhibitory effect to human pulmonary carcinoma A549 cells. These results indicated that ZnONPs and β-D-glu-ZnO NPs induced the cancer cell death through necrosis and apoptosis pathway, respectively.

Recently, AuNPs synthesized with fungal xylanases from *T. longibrachiatum* were reported to show excellent anticoagulant and thrombolytic activities on human blood. The authors suggested the biomedical application of these AuNPs in the potential management of blood coagulation disorders and thrombotic diseases (Elegbede et al. 2020).

10 Other Potential Applications of *Trichoderma* Nanoparticles

Most studies on *Trichoderma*-mediated nanoparticles have suggested the resulting nanomaterials as antimicrobial agents for the control of human and plants pathogens. However, other interesting applications including their use for the treatment of wastewaters, mosquito control, fruit preservation, and generation of bioelectricity, among others, have been reported. For instance, gold nanoparticles (AuNPs) and gold–silver (Au–Ag) alloy nanoparticles synthesized by biomass of *T. harzianum* were suggested as an effective candidate for catalytic degradation of toxic pollutants. AuNPs were analyzed for catalytic activity against methylene blue (MB) as a model pollutant in water. MB was degraded 39% in 30 min in the presence of AuNPs and sodium borohydrate, and the rate constant (k) was found to be 0.2×10^{-3} s^{-1} (Tripathi et al. 2018). Alloy NPs were found to have enhanced catalytic activity toward the reduction of MB by $NaBH_4$ in aqueous media, having a catalytic rate of 0.88×10^{-3} s^{-1} (Tripathi et al. 2015).

T. viride was used for the biosynthesis of AuNPs, and the resulting nanoparticles also served as an efficient biocatalyst, which reduced 4-nitrophenol to 4-aminophenol in the presence of NaBH4 (Mishra et al. 2014). AuNPs synthesized using *Trichoderma* sp. (strain WL-Go) also exhibited efficient catalytic capability for degradation of aromatic pollutants, and it was found that AuNPs could efficiently catalyze the decolorization of various azo dyes with efficiency from 82.2% to 97.5% (Qu et al. 2017; Qu et al. 2018). Using the same *Trichoderma* sp. WL-Go strain, lead selenide nanoparticles (PbSeNPs) were synthesized and used as catalyst for investigating the antioxidant activity using 2,2-diphenyl-1-picrylhydrazyl (DPPH) and photodegradation ability of rhodamine B dye. The results showed that the PbSeNPs could eliminate up to 88.60% of free radicals and could photodegrade 82% of rhodamine B in 30 min, thus suggesting that PbSeNPs can be used to efficiently eliminate free radicals and for the treatment of persistent organic pollutants in wastewaters (Diko et al. 2020b). Silver nanoparticles synthesized using *T. reesei* were also suggested for the treatment of wastewaters since they were able to immobilize potassium amyl xanthate from model wastewater (Gemishev et al. 2019a).

T. harzianum was suggested for selenium and tellurium biorecovery since selenium oxide and tellurium oxide as well as the formation of elemental selenium and tellurium were found after growing *T. harzianum* with 1 mM selenite and tellurite. The authors found that the hyphal matrix provided nucleation sites for metalloid deposition with extracellular protein and extracellular polymeric substances localizing the resultant Se or Te nanoparticles (Liang et al. 2019).

Another application of nanoparticles synthesized by *Trichoderma* includes detection of mercury(II) ions. A simple, cost-effective, and selective method for the colorimetric detection of mercury(II) ions was reported using AuNPs by *T. harzianum*. The minimum concentration detected was 2.6 nM. The assay showed high specificity toward Hg^{2+} in a complex environment (Tripathi et al. 2014).

Silver nanoparticles using *T. viride* were synthesized and incorporated into sodium alginate and tested for vegetable and fruit preservation. The results suggested that AgNPs incorporated into sodium alginate for coating vegetables and fruits are suitable for preservation. It was found that the use of this coating increased the shelf life of carrot and pear when compared to control with respect to weight loss and soluble protein content (Fayaz et al. 2009a).

The mosquito *Aedes aegypti* is the vector for transmitting dengue, chikungunya, and yellow fever. Thus, AgNPs synthesized using *T. atroviride* were tested for the control of the insect. It was found that AgNPs showed significant efficacies against first, second, third, and fourth instar larvae of *A. aegypti*. The synthesized AgNPs were reported as a new promising candidate for application in mosquito control (Sundaravadivelan and Padmanabhan 2014; Singh and Prakash 2015).

Finally, the generation of bioelectricity is another interesting application of silver and gold NPs synthesized by *T. hamatum*. Bioelectricity was generated by using sulfate-reducing bacteria and different concentrations (2–40 mg l^{-1}) of AgNPs and AuNPs as a biocatalyst/microbicide and artificial sewage water as a substrate in a microbial fuel chamber. The authors reported this novel methodology as an alternative to generate bioelectricity and, most importantly, as economically cheap, green, and eco-friendly technique (Saravanakumar et al. 2016).

11 Summary and Conclusions

The current use of *Trichoderma* species as biocontrol agents opens the possibility of using them in other applications such as the fabrication of nanomaterials. Thus, *Trichoderma* species represent a feasible option for the synthesis of metallic nanoparticles with a high potential to be successfully used in our daily lives. Although some species of this group of organisms have been identified as etiologic agents of infections in immunocompromised hosts, the synthesis of nanomaterials is mostly carried out with fungal extracts or supernatants, without the use of any living material such as spores or fragmented mycelia. Therefore, their use in the biosynthesis protocol may not represent any danger if the fungus is always managed with care, following all safety protocols. At present, there is enough evidence that metallic nanoparticles mediated by *Trichoderma* have excellent antibacterial properties, and their possible use in the clinic seems promising. The same is true for their potential use in agricultural practices, since the nanoparticles of various materials such as silver, copper, and selenium possess antifungal properties, being an alternative for the control of pathogens affecting commercially important plants and crops. However, a deep analysis of the potential cytotoxic effects in human cells and their impact in the environment is still needed.

References

Abdel-Azeem A, Nada AA, O'Donovan A, Thakur VK, Elkelish A (2020) Mycogenic silver nanoparticles from endophytic *Trichoderma atroviride* with antimicrobial activity. J Renew Mater 8(2):171–185. https://doi.org/10.32604/jrm.2020.08960

Abdel-Kareem MM, Zohri AA (2018) Extracellular mycosynthesis of gold nanoparticles using *Trichoderma hamatum*: optimization, characterization and antimicrobial activity. Lett Appl Microbiol 67(5):465–475. https://doi.org/10.1111/lam.13055

Adebayo-Tayo BC, Ogunleye GE, Ogbole O (2019) Biomedical application of greenly synthesized silver nanoparticles using the filtrate of *Trichoderma viride*: Anticancer and immunomodulatory potentials. Polim Med 49(2):57–62. https://doi.org/10.17219/pim/116086

Ahluwalia V, Kumar J, Sisodia R, Shakil NA, Walia S (2014) Green synthesis of silver nanoparticles by *Trichoderma harzianum* and their bio-efficacy evaluation against Staphylococcus aureus and Klebsiella pneumonia. Ind Crop Prod 55:202–206. https://doi.org/10.1016/j.indcrop.2014.01.026

Arya S, Sonawane H, Math S, Tambade P, Chaskar M, Shinde D (2020) Biogenic titanium nanoparticles (TiO2 NPs) from *Tricoderma citrinoviride* extract: synthesis, characterization and antibacterial activity against extremely drug-resistant *Pseudomonas aeruginosa*. Int Nano Lett:1–8

Bhadwal AS, Tripathi RM, Gupta RK, Kumar N, Singh RP, Shrivastav A (2014) Biogenic synthesis and photocatalytic activity of CdS nanoparticles. RSC Adv 4(19):9484–9490. https://doi.org/10.1039/c3ra46221h

Brigger I, Dubernet C, Couvreur P (2012) Nanoparticles in cancer therapy and diagnosis. Adv Drug Deliv Rev 64:24–36

Castro-Longoria E (2016) Fungal biosynthesis of nanoparticles, a cleaner alternative, pp 323–351. https://doi.org/10.1007/978-3-319-42852-9_13

Chinnaperumal K, Govindasamy B, Paramasivam D, Dilipkumar A, Dhayalan A, Vadivel A, Sengodan K, Pachiappan P (2018) Bio-pesticidal effects of *Trichoderma viride* formulated titanium dioxide nanoparticle and their physiological and biochemical changes on Helicoverpa armigera (Hub.). Pestic Biochem Physiol 149:26–36. https://doi.org/10.1016/j.pestbp.2018.05.005

Chitra K, Annadurai G (2013) Bioengineered silver nanobowls using *Trichoderma viride* and its antibacterial activity against gram-positive and gram-negative bacteria. J Nanostruct Chem 3(1):1–7. https://doi.org/10.1186/2193-8865-3-9

Chouaki T, Lavarde V, Lachaud L, Raccurt CP, Hennequin C (2002) Invasive infections due to *Trichoderma* species: report of 2 cases, findings of in vitro susceptibility testing, and review of the literature. Clin Infect Dis 35(11):1360–1367. https://doi.org/10.1086/344270

Consolo VF, Torres-Nicolini A, Alvarez VA (2020) Mycosinthetized Ag, CuO and ZnO nanoparticles from a promising *Trichoderma harzianum* strain and their antifungal potential against important phytopathogens. Sci Rep 10(1):1–9. https://doi.org/10.1038/s41598-020-77294-6

De Miguel D, Gómez P, González R, García-Suárez J, Cuadros JA, Bañas MH, Romanyk J, Burgaleta C (2005) Nonfatal pulmonary *Trichoderma viride* infection in an adult patient with acute myeloid leukemia: report of one case and review of the literature. Diagn Microbiol Infect Dis 53(1):33–37. https://doi.org/10.1016/j.diagmicrobio.2005.04.009

Devi TP, Kulanthaivel S, Kamil D, Borah JL, Prabhakaran N, Srinivasa N (2013) Biosynthesis of silver nanoparticles from *Trichoderma* species. Indian J Exp Biol 51(7):543–547

Diko CS, Zhang H, Lian S, Fan S, Li Z, Qu Y (2020a) Optimal synthesis conditions and characterization of selenium nanoparticles in *Trichoderma* sp. WL-Go culture broth. Mater Chem Phys 246(August 2019):122583. https://doi.org/10.1016/j.matchemphys.2019.122583

Diko CS, Qu Y, Henglin Z, Li Z, Ahmed Nahyoon N, Fan S (2020b) Biosynthesis and characterization of lead selenide semiconductor nanoparticles (PbSe NPs) and its antioxidant and photocatalytic activity. Arab J Chem. https://doi.org/10.1016/j.arabjc.2020.06.005

do Nascimento JM, Cruz ND, de Oliveira GR, Sá WS, de Oliveira JD, Ribeiro PRS, Leite SGF (2020) Evaluation of the kinetics of gold biosorption processes and consequent biogenic synthesis of AuNPs mediated by the fungus *Trichoderma harzianum*. Environ Technol Innov 101238. https://doi.org/10.1016/j.eti.2020.101238

Elamawi RM, Al-Harbi RE, Hendi AA (2018) Biosynthesis and characterization of silver nanoparticles using *Trichoderma longibrachiatum* and their effect on phytopathogenic fungi. Egypt J Biol Pest Control 28(1):1–11. https://doi.org/10.1186/s41938-018-0028-1

Elegbede JA, Lateef A, Azeez MA, Asafa TB, Yekeen TA, Oladipo IC, Aina DA, Beukes LS, Gueguim-Kana EB (2020) Biofabrication of gold nanoparticles using xylanases through valorization of corncob by *Aspergillus niger* and *Trichoderma longibrachiatum*: antimicrobial, antioxidant, anticoagulant and thrombolytic activities. Waste Biomass Valoriz 11(3):781–791. https://doi.org/10.1007/s12649-018-0540-2

Elgorban AM, Al-Rahmah AN, Sayed SR, Hirad A, Mostafa AAF, Bahkali AH (2016) Antimicrobial activity and green synthesis of silver nanoparticles using *Trichoderma viride*. Biotechnol Biotechnol Equip 30(2):299–304. https://doi.org/10.1080/13102818.2015.1133255

El-Moslamy SH, Elkady MF, Rezk AH, Abdel-Fattah YR (2017) Applying Taguchi design and large-scale strategy for mycosynthesis of nano-silver from endophytic *Trichoderma harzianum* SYA.F4 and its application against phytopathogens. Sci Rep 7(October 2016):1–22. https://doi.org/10.1038/srep45297

El-Waseif AAM, Ibrahim NA, Ahmed SAEA, Abdallah N (2017) Bioactivity enhancement with microbial silver nanoparticles produced and characterized from *Streptomyces* and *Trichoderma*. Egypt J Exp Biol 13(2):439–445. https://doi.org/10.5455/egyjebb.20171108083304

Fayaz MA, Balaji K, Girilal M, Kalaichelvan PT, Venkatesan R (2009a) Mycobased synthesis of silver nanoparticles and their incorporation into sodium alginate films for vegetable and fruit preservation. J Agric Food Chem 57(14):6246–6252. https://doi.org/10.1021/jf900337h

Fayaz AM, Balaji K, Kalaichelvan PT, Venkatesan R (2009b) Fungal based synthesis of silver nanoparticles-an effect of temperature on the size of particles. Colloids Surf B: Biointerfaces 74(1):123–126. https://doi.org/10.1016/j.colsurfb.2009.07.002

Fayaz AM, Balaji K, Girilal M, Yadav R, Kalaichelvan PT, Venketesan R (2010a) Biogenic synthesis of silver nanoparticles and their synergistic effect with antibiotics: a study against gram-positive and gram-negative bacteria. Nanomedicine 6(1):103–109. https://doi.org/10.1016/j.nano.2009.04.006

Fayaz M, Tiwary CS, Kalaichelvan PT, Venkatesan R (2010b) Blue orange light emission from biogenic synthesized silver nanoparticles using *Trichoderma viride*. Colloids Surf B: Biointerfaces 75(1):175–178. https://doi.org/10.1016/j.colsurfb.2009.08.028

Fayaz AM, Girilal M, Mahdy SA, Somsundar SS, Venkatesan R, Kalaichelvan PT (2011) Vancomycin bound biogenic gold nanoparticles: a different perspective for development of anti VRSA agents. Process Biochem 46(3):636–641. https://doi.org/10.1016/j.procbio.2010.11.001

Fraceto LF, Maruyama CR, Guilger M, Mishra S, Keswani C, Singh HB, de Lima R (2018) *Trichoderma harzianum*-based novel formulations: potential applications for management of Next-Gen agricultural challenges. J Chem Technol Biotechnol 93(8):2056–2063. https://doi.org/10.1002/jctb.5613

Gemishev OT, Panayotova MI, Panayotov VT, Zaharieva ES (2019a) Extracellular biosynthesis of silver nanoparticles by *Trichoderma reesei* and their application to remove xanthogenate from wastewater. Mater Methods Technol 13(February):230–240. https://doi.org/10.1037//0033-2909.I26.1.78

Gemishev OT, Panayotova MI, Mintcheva NN, Djerahov LP, Tyuliev GT, Gicheva GD (2019b) A green approach for silver nanoparticles preparation by cell-free extract from *Trichoderma reesei* fungi and their characterization. Mater Res Express 6(9). https://doi.org/10.1088/2053-1591/ab2e6a

Ghazanfar MU, Raza M, Raza W, Qamar MI (2018) *Trichoderma* as potential biocontrol agent, its explotation in agriulture: a review. Plant Prot 2(3):109–135. http://esciencepress.net/journals/PP

Gherbawy YA, Shalaby IM, El-Sadek MSA, Elhariry HM, Abdelilah BA (2013) The anti-fasciolasis properties of silver nanoparticles produced by *Trichoderma harzianum* and their improvement of the anti-fasciolasis drug Triclabendazole. Int J Mol Sci 14(11):21887–21898. https://doi.org/10.3390/ijms141121887

Guilger M, Pasquoto-Stigliani T, Bilesky-Jose N, Grillo R, Abhilash PC, Fraceto LF, De Lima R (2017) Biogenic silver nanoparticles based on *Trichoderma harzianum*: synthesis, characterization, toxicity evaluation and biological activity. Sci Rep 7(October 2016):1–13. https://doi.org/10.1038/srep44421

Hirpara DG, Gajera HP (2020) Green synthesis and antifungal mechanism of silver nanoparticles derived from chitin- induced exometabolites of *Trichoderma* interfusant. Appl Organomet Chem 34(3):1–15. https://doi.org/10.1002/aoc.5407

Hu D, Yu S, Yu D, Liu N, Tang Y, Fan Y, Wang C, Wu A (2019) Biogenic *Trichoderma harzianum*-derived selenium nanoparticles with control functionalities originating from diverse recognition metabolites against phytopathogens and mycotoxins. Food Control 106(January):106748. https://doi.org/10.1016/j.foodcont.2019.106748

Joshi SM, De Britto S, Jogaiah S, Ito SI (2019) Mycogenic selenium nanoparticles as potential new generation broad spectrum antifungal molecules. Biomolecules 9(9). https://doi.org/10.3390/biom9090419

Kalia A, Kaur J, Kaur A, Singh N (2020) Antimycotic activity of biogenically synthesised metal and metal oxide nanoparticles against plant pathogenic fungus *Fusarium moniliforme* (*F. fujikuroi*). Indian J Exp Biol 58(April):263–270

Kumari M, Shukla S, Pandey S, Giri VP, Bhatia A, Tripathi T, Kakkar P, Nautiyal CS, Mishra A (2017a) Enhanced cellular internalization: a bactericidal mechanism more relative to biogenic nanoparticles than chemical counterparts. ACS Appl Mater Interfaces 9(5):4519–4533

Kumari M, Pandey S, Giri VP, Bhattacharya A, Shukla R, Mishra A, Nautiyal CS (2017b) Tailoring shape and size of biogenic silver nanoparticles to enhance antimicrobial efficacy against MDR bacteria. Microb Pathog 105:346–355. https://doi.org/10.1016/j.micpath.2016.11.012

Kumari M, Giri VP, Pandey S, Kumar M, Katiyar R, Nautiyal CS, Mishra A (2019) An insight into the mechanism of antifungal activity of biogenic nanoparticles than their chemical counterparts. Pestic Biochem Physiol 157(March):45–52. https://doi.org/10.1016/j.pestbp.2019.03.005

Liang X, Perez MAMJ, Nwoko KC, Egbers P, Feldmann J, Csetenyi L, Gadd GM (2019) Fungal formation of selenium and tellurium nanoparticles. Appl Microbiol Biotechnol 103(17):7241–7259. https://doi.org/10.1007/s00253-019-09995-6

Maliszewska I (2013) Microbial mediated synthesis of gold nanoparticles: preparation, characterization and cytotoxicity studies. Digest J Nanomater Biostruct 8(3):1123–1131

Maliszewska I, Aniszkiewicz, Sadowski Z (2009) Biological synthesis of gold nanostructures using the extract of *Trichoderma koningii*. Acta Phys Pol A 116(SUPPL):163–165. https://doi.org/10.12693/aphyspola.116.s-163

Manikandaselvi S, Sathya V, Vadivel V, Sampath N, Brindha P (2020) Evaluation of bio control potential of AgNPs synthesized from *Trichoderma viride*. Adv Nat Sci Nanosci Nanotechnol 11(3):035004. https://doi.org/10.1088/2043-6254/ab9d16

Mishra A, Kumari M, Pandey S, Chaudhry V, Gupta KC, Nautiyal CS (2014) Biocatalytic and antimicrobial activities of gold nanoparticles synthesized by *Trichoderma* sp. Bioresour Technol 166(February 2019):235–242. https://doi.org/10.1016/j.biortech.2014.04.085

Mukherjee P, Roy M, Mandal BP, Dey GK, Mukherjee PK, Ghatak J, Tyagi AK, Kale SP (2008) Green synthesis of highly stabilized nanocrystalline silver particles by a non-pathogenic and agriculturally important fungus *T. asperellum*. Nanotechnology 19(7). https://doi.org/10.1088/0957-4484/19/7/075103

Mukherjee P, Roy M, Mandal BP, Choudhury S, Tewari R, Tyagi AK, Kale SP (2012) Synthesis of uniform gold nanoparticles using non-pathogenic bio-control agent: evolution of morphology from nano-spheres to triangular nanoprisms. J Colloid Interface Sci 367(1):148–152. https://doi.org/10.1016/j.jcis.2011.08.085

Musial J, Krakowiak R, Mlynarczyk DT, Goslinski T, Stanisz BJ (2020) Titanium dioxide nanoparticles in food and personal care products – what do we know about their safety? Nanomaterials 10(6):1–23. https://doi.org/10.3390/nano10061110

Naama S, Hadjersi T, Keffous A, Nezzal G (2015) CO_2 gas sensor based on silicon nanowires modified with metal nanoparticles. Mater Sci Semicond Process 38(February):367–372. https://doi.org/10.1016/j.mssp.2015.01.027

Nandini B, Hariprasad P, Prakash HS, Shetty HS, Geetha N (2017) Trichogenic-selenium nanoparticles enhance disease suppressive ability of *Trichoderma* against downy mildew disease caused by *Sclerospora graminicola* in pearl millet. Sci Rep 7(1):1–11. https://doi.org/10.1038/s41598-017-02737-6

Natesan K, Ponmurugan P, Gnanamangai BM, Manigandan V, Joy SPJ, Jayakumar C, Amsaveni G (2020) Biosynthesis of silica and copper nanoparticles from *Trichoderma*, *Streptomyces* and *Pseudomonas* spp. evaluated against collar canker and red root-rot disease of tea plants. Arch Phytopathol Plant Protect:1–30. https://doi.org/10.1080/03235408.2020.1817258

Noshad A, Iqbal M, Folkers L, Hetherington C, Khan A, Numan M, Ullah S (2019) Antibacterial effect of silver nanoparticles (AgNPs) synthesized from *Trichoderma harzianum* against *Clavibacter michiganensis*. J Nano Res 58(July):10–19. https://doi.org/10.4028/www.scientific.net/jnanor.58.10

Omran BA, Nassar HN, Younis SA, Fatthallah NA, Hamdy A, El-Shatoury EH, El-Gendy NS (2019) Physiochemical properties of *Trichoderma longibrachiatum* DSMZ 16517-synthesized silver nanoparticles for the mitigation of halotolerant sulphate-reducing bacteria. J Appl Microbiol 126(1):138–154. https://doi.org/10.1111/jam.14102

Osonga FJ, Akgul A, Yazgan I, Akgul A, Eshun GB, Sakhaee L, Sadik OA (2020) Size and shape-dependent antimicrobial activities of silver and gold nanoparticles: a model study as potential fungicides. Molecules (Basel, Switzerland) 25(11). https://doi.org/10.3390/molecules25112682

Ponmurugan P (2016) Biosynthesis of silver and gold nanoparticles using *Trichoderma atroviride* for the biological control of *Phomopsis* canker disease in tea plants. IET Nanobiotechnol 11(3):261–267

Qu Y, Shen W, Pei X, Ma F, You S, Li S, Wang J, Zhou J (2017) Biosynthesis of gold nanoparticles by *Trichoderma* sp. WL-Go for azo dyes decolorization. J Environ Sci 56:79–86

Qu Y, Li X, Lian S, Dai C, Jv Z, Zhao B, Zhou H (2018) Biosynthesis of gold nanoparticles using fungus *Trichoderma* sp. WL-Go and their catalysis in degradation of aromatic pollutants. IET Nanobiotechnol 13(1):12–17. https://doi.org/10.1049/iet-nbt.2018.5177

Ramos MM, dos Morais ES, da Sena IS, Lima AL, de Oliveira FR, de Freitas CM, Fernandes CP, de Carvalho JCT, Ferreira IM (2020) Silver nanoparticle from whole cells of the fungi *Trichoderma* spp. isolated from Brazilian Amazon. Biotechnol Lett 42(5):833–843. https://doi.org/10.1007/s10529-020-02819-y

Richter S, Cormican MG, Pfaller MA, Lee CK, Gingrich R, Rinaldi MG, Sutton DA (1999) Fatal disseminated *Trichoderma longibrachiatum* infection in an adult bone marrow transplant patient: species identification and review of the literature. J Clin Microbiol 37(4):1154–1160. https://doi.org/10.1128/jcm.37.4.1154-1160.1999

Salvadori MR, Ando RA, Do Nascimento O, C.A. and Correa, B. (2014) Bioremediation from wastewater and extracellular synthesis of copper nanoparticles by the fungus *Trichoderma koningiopsis*. J Environ Sci Health Part A Toxic Hazard Subst Environ Eng 49(11):1286–1295. https://doi.org/10.1080/10934529.2014.910067

Sánchez-López E, Gomes D, Esteruelas G, Bonilla L, Lopez-Machado AL, Galindo R, Cano A, Espina M, Ettcheto M, Camins A, Silva AM, Durazzo A, Santini A, Garcia ML, Souto EB (2020) Metal-based nanoparticles as antimicrobial agents: an overview. Nanomaterials 10(2):1–39. https://doi.org/10.3390/nano10020292

Saravanakumar K, Wang MH (2018) *Trichoderma* based synthesis of anti-pathogenic silver nanoparticles and their characterization, antioxidant and cytotoxicity properties. Microb Pathog 114:269–273. https://doi.org/10.1016/j.micpath.2017.12.005

Saravanakumar K, Mubarakali D, Kathiresan K, Thajuddin N, Alharbi NS, Chen J (2016) Biogenic metallic nanoparticles as catalyst for bioelectricity production: a novel approach in microbial fuel cells. Mater Sci Eng B Solid-State Mater Adv Technol 203:27–34. https://doi.org/10.1016/j.mseb.2015.10.006

Saravanakumar K, Shanmugam S, Varukattu NB, MubarakAli D, Kathiresan K, Wang MH (2019) Biosynthesis and characterization of copper oxide nanoparticles from indigenous fungi and its effect of photothermolysis on human lung carcinoma. J Photochem Photobiol B Biol 190:103–109. https://doi.org/10.1016/j.jphotobiol.2018.11.017

Saravanakumar K, Jeevithan E, Hu X, Chelliah R, Oh DH, Wang MH (2020) Enhanced anti-lung carcinoma and anti-biofilm activity of fungal molecules mediated biogenic zinc oxide nanoparticles conjugated with β-D-glucan from barley. J Photochem Photobiol B Biol 203:111728. https://doi.org/10.1016/j.jphotobiol.2019.111728

Shelar GB (2016) Resemblances between the biosynthesized silver nanoparticles of *Aspergillus niger* and *Trichoderma viride* and their antioxidant activity. Eur J Biomed Pharm Sci 3(11):195–199

Shelar GB, Chavan AM (2015) Myco-synthesis of silver nanoparticles from *Trichoderma harzianum* and its impact on germination status of oil seed. Biolife 3(1):109–113

Shobha B, Lakshmeesha TR, Ansari MA, Almatroudi A, Alzohairy MA, Basavaraju S, Alurappa R, Niranjana SR, Chowdappa S (2020) Mycosynthesis of ZnO nanoparticles using *Trichoderma* spp. Isolated from rhizosphere soils and its synergistic antibacterial effect against *Xanthomonas oryzae* pv. *oryzae*. J Fungi 6(3):1–19. https://doi.org/10.3390/jof6030181

Shrivastava S, Bera T, Roy A, Singh G, Ramachandrarao P, Dash D (2007) Characterization of enhanced antibacterial effects of novel silver nanoparticles. Nanotechnology 18(22). https://doi.org/10.1088/0957-4484/18/22/225103

Singh G, Prakash S (2015) Virulency of novel nanolarvicide from *Trichoderma atroviride* against *Aedes aegypti* (Linn.): a CLSM analysis. Environ Sci Pollut Res 22(16):12559–12565. https://doi.org/10.1007/s11356-015-4531-6

Singh P, Raja RB (2011) Biological synthesis and characterization of silver nanoparticles using the fungus *Trichoderma harzianum*. Asian J Exp Biol Sci 2(4):600–605

Sundaravadivelan C, Padmanabhan MN (2014) Effect of mycosynthesized silver nanoparticles from filtrate of *Trichoderma harzianum* against larvae and pupa of dengue vector *Aedes aegypti* L. Environ Sci Pollut Res 21(6):4624–4633

Tomah AA, Alamer ISA, Li B, Zhang JZ (2020) Mycosynthesis of silver nanoparticles using screened *Trichoderma* isolates and their antifungal activity against *Sclerotinia sclerotiorum*. Nanomaterials 10(10):1–15. https://doi.org/10.3390/nano10101955

Tripathi RM, Gupta RK, Shrivastav A, Singh MP, Shrivastav BR, Singh P (2013) *Trichoderma koningii* assisted biogenic synthesis of silver nanoparticles and evaluation of their antibacterial activity. Adv Nat Sci Nanosci Nanotechnol 4(3). https://doi.org/10.1088/2043-6262/4/3/035005

Tripathi RM, Gupta RK, Singh P, Bhadwal AS, Shrivastav A, Kumar N, Shrivastav BR (2014) Ultra-sensitive detection of mercury(II) ions in water sample using gold nanoparticles synthesized by *Trichoderma harzianum* and their mechanistic approach. Sensors Actuators B Chem 204:637–646. https://doi.org/10.1016/j.snb.2014.08.015

Tripathi RM, Gupta RK, Bhadwal AS, Singh P, Shrivastav A, Shrivastav BR (2015) Fungal biomolecules assisted biosynthesis of Au-Ag alloy nanoparticles and evaluation of their catalytic property. IET Nanobiotechnol 9(4):178–183. https://doi.org/10.1049/iet-nbt.2014.0043

Tripathi RM, Shrivastav BR, Shrivastav A (2018) Antibacterial and catalytic activity of biogenic gold nanoparticles synthesised by *Trichoderma harzianum*. IET Nanobiotechnol 12(4):509–513. https://doi.org/10.1049/iet-nbt.2017.0105

Vahabi K, Mansoori GA, Karimi S (2011) Biosynthesis of silver nanoparticles by fungus *Trichoderma reesei* (A route for large-scale production of AgNPs). Insci J 1(1):65–79. https://doi.org/10.5640/insc.010165

Weir A, Westerhoff P, Fabricius L, von G. N. (2012) Titanium dioxide nanoparticles in food and personal care products. Environ Sci Technol 46(4):2242–2250. https://doi.org/10.1021/es204168d.Titanium

Wu Y, Ali MRK, Chen K, Fang N, El-Sayed MA (2019) Gold nanoparticles in biological optical imaging. Nano Today 24(February):120–140. https://doi.org/10.1016/j.nantod.2018.12.006

Zibareva IV, Ilina LY, Vedyagin AA (2019) Catalysis by nanoparticles: the main features and trends. React Kinet Mech Catal 127(1):19–24

Trichoderma Role in Anthropogenic Pollutions Mycoremediation: Pesticides and Heavy Metals

Jorge Poveda

Contents

1 Anthropogenic Pollution and Soils

When talking about soil, we refer to the most superficial layers of the earth's crust, which are capable of supporting plant growth, being the result of the action of environmental conditions on natural bodies and remains of living organisms, therefore representing a dynamic body made up of liquids, gases, mineral and organic solids, and living organisms. In its study, it must always be taken into account that it represents a very dynamic body formed by an abiotic component and a biotic one that is closely related to each other. This is very important in understanding its formation process and all those that take place continuously (nutrient cycle, decomposition of organic matter, degradation of pollutants, etc.) (Cachada et al. 2018). In this sense,

J. Poveda (✉)
Institute for Multidisciplinary Research in Applied Biology, Universidad Pública de Navarra, Pamplona, Spain
e-mail: jorge.poveda@unavarra.es

N. Amaresan et al. (eds.), *Advances in* Trichoderma *Biology for Agricultural Applications*, Fungal Biology, https://doi.org/10.1007/978-3-030-91650-3_18

the soil not only represents the main substrate for food production but is also a natural habitat of biodiversity, a store of nutrients, and a regulator of the water cycle (Cachada et al. 2018; Koul and Taak 2018).

Anthropogenic activity can seriously affect the functions of soils, mainly through the agricultural and forestry sectors, construction, tourism, and industrial activities, increasing in conjunction with the increase in the world population. This leads to an increase in the amount and intensity of soil use, causing compaction, erosion, salinization, pollution, acidification, and loss of organic matter and biodiversity (Cachada et al. 2018).

With regard to soil contamination, it represents a serious problem in world expansion after the Industrial Revolution, due to the massive use of agrochemicals, the burning of fossil fuels, and the expansion of industrial activity. A soil is considered contaminated when it does not carry out its own processes or cannot be used for its estimated use due to the presence of contaminants (Cachada et al. 2018). The main sources of anthropogenic soil contamination include solid wastes (domestic waste, market wastes, hospital wastes, kitchen wastes, slaughterhouse wastes, industrial wastes, livestock and poultry wastes, ceramic wastes, glass, and metals), agrochemicals (pesticides, fertilizers, hormones, and animal manure), radioactive wastes, chemical wastes (hydrocarbons, solvents, and measured metals), and mining and smelting (heavy metals) (Shankar 2017; Cachada et al. 2018; Koul and Taak 2018).

Soil pollution not only causes serious environmental effects but is also against health. Contaminants present in the soil can affect human health directly when inhaled or in contact with the skin, but the most common is that they enter through ingestion, by contaminating food and water (such as aquifers). The effects that they can have on health are very diverse depending on each specific pollutant; for example, exposure to heavy metals such as chromium, solvents, hydrocarbons, or pesticides can lead to the appearance of cancer, neuromuscular disorders, and/or congenital disorders. Similarly, soil pollutants affect ecosystems, reducing the reproductive capacity of organisms, their ability to feed themselves, and their growth and development, causing serious changes in their populations and communities. Therefore, soil pollution affects ecosystem functions, by modifying its living component (Shankar 2017; Koul and Taak 2018).

The main economic sector affected by soil pollution is agriculture, seriously affecting both productivity and the quality of crops. The proportion of contaminated soil continues to increase throughout the world, estimating current losses in crop yields by about 15–25% as a consequence of soil and water contamination. Pollutants can prevent the absorption of nutrients by the roots of crops, by interacting with them, modifying the soil pH and electrical conductivity, and causing the loss of soil fertility. In addition, many of these pollutants can be absorbed by plants and stored in their tissues, contaminating food, such as heavy metals. On the other hand, soil pollution can significantly affect the water supply to plants, by increasing the salinity of soils and preventing its infiltration/percolation (Saha et al. 2017; Koul and Taak 2018; Elbana et al. 2019). In this sense, it is important to highlight that agriculture is also an important source of soil and air pollution due to the use of agrochemicals, such as pesticides (Bauer et al. 2016).

1.1 Pesticides

Pesticides are chemical substances used on agricultural land and public and private areas in order to eliminate, avert, deter, control, and/or kill populations of biological agents that cause harm to human interest (Mahmood et al. 2016; Ozkara et al. 2016). Its use in the agricultural sector is above 5 billion pounds worldwide (Mahmood et al. 2016), although it is constantly increasing, due to the need for higher food productivity to feed the growing world population (Ozkara et al. 2016). At present, it is considered that 40% of agricultural production is lost as a result of pests, pathogens, and weeds, a percentage that would be higher without the use of pesticides, an important sign of their need today (Mahmood et al. 2016). Therefore, since the nineteenth century, the use of chemical pesticides in pest control has caused a widespread release of these xenobiotics into the environment. Specifically, more than 500 different pesticide formulations are currently in use, which affect not only their target organisms but many non-target organisms, including humans. In addition, many of these pesticides are hardly degraded and can persist up to 30 years in water or soils, such as organochlorine insecticides, which influences their easy probability of entering the food chain (Ozkara et al. 2016).

In this sense, the risks derived from the use of chemical pesticides considerably exceed the benefits obtained, having drastic effects in aquatics and terrestrial ecosystems, affecting animal and plant biodiversity by acting on non-target species. Among the different groups, insecticides are the most harmful pesticides for the environment, followed by fungicides and herbicides, due to their toxicity (Mahmood et al. 2016). The pesticides most frequently detected in soils are contaminated with organophosphorus pesticides (OPs), having been detected in more than 90% of soils in China. Organochlorine pesticides (OCs) include many products banned around the world, but due to their persistence, they are easily detected in soils today, as is the case with 2,2-bis(4-chlorophenyl)-1,1,1-trichloroethane pesticides (DDTs). On the other hand, the pesticides most implicated in poisoning problems on the planet are the anti-cholinesterase pesticides, which include both organophosphates and carbamates. Almost 70% of pesticides used in agriculture continuously contaminate soil and water through their residues. About 40% are herbicides, 30% insecticides, and the remainder include all pesticides used against plant pathogens (Sun et al. 2018).

The uncontrolled use of pesticides causes serious damage to biodiversity, both directly and by accumulating in the food chains. The most common is to report a reduction in the amount and variety of weeds, shrubs, and insects in the ecosystem, but the populations of higher animals such as birds are also reduced. As regards human health, the World Health Organization has indicated that 3 million cases of pesticide poisoning and more than 20 thousand deaths are registered annually. Its effects on health are highly variable and dependent on various factors, although they can immediately cause headaches, respiratory tract irritation, digestive problems, or signs on the skin such as rash and blisters, while its effects are chronically reflected in damage to the immune system, neurological, cancer, or reproductive problems, among others (Mahmood et al. 2016).

1.2 Heavy Metals

The term heavy metals includes all those elements with a density greater than 4 g cm^{-3}, which includes both metals and metalloids (such as arsenic). Although some of them represent essential elements in many biological processes, in high concentrations they can be very harmful to the environment and health. This is because they are not degradable and easily accumulate in organisms. For this reason, in Europe, heavy metals are considered the main pollutants of soil and water (Vareda et al. 2019). The origin of environmental pollution by heavy metals can be found in natural processes, such as erosion, weathering, or volcanoes, although its main source is human activity, which includes textile and paint industries, mining, smelting, wastewater, or use of agrochemicals (Mishra and Nautiyal 2009).

The effect of heavy metals in the environment can be very serious, as a consequence of their persistence and ubiquity. Its toxicity affects all the components of the ecosystem, since they are accumulated in the tissues and pass easily to the different steps of the trophic chain. In soils and waters, high concentrations of heavy metals cause a decrease in soil microbial biomass, diversity, and activities (Abdu et al. 2017). As far as human health is concerned, exposure can occur through ingestion, inhalation, or contact with the skin, causing serious damage to the central nervous system and various vital organs or cancer (Varhdan et al. 2019). As examples, excessive exposure to chromium (Cr) is related to cancer, to mercury (Hg) is related to immune and nervous diseases, to lead (Pb) is related to cardiovascular and neurological diseases, or to cadmium (Cd) is related to cancer or endocrine damage (Mishra and Nautiyal 2009).

In agriculture, the entry of heavy metals is due to irrigation with wastewater, fertilization with livestock manure, and the use of agrochemicals (Rai et al. 2019). The main route of entry of heavy metals into plant tissues is through the roots by absorption, being easily transported by the vascular bundles to the entire plant. Their toxic effect can be highly variable depending on the ability to tolerate their presence, but they generally inhibit germination, growth, and development, by deactivating different enzymes and causing stress responses, such as the accumulation of reactive oxygen species (ROS), which causes serious losses of productivity in crops (Rai et al. 2016; Bhardwaj et al. 2020).

2 Mycoremediation

Bioremediation is defined as the use of different organisms, usually microorganisms or plants, to remove or neutralize the pollutants present in the environment. In the case of microorganisms, their main mechanism of action is based on the production and release of enzymes that interact with pollutants and degrade them completely or convert them into less harmful products (Dangi et al. 2019).

Mycoremediation is based on the use of fungi for the elimination of pollutants from the environment or, at least, their adverse effects (Gupta et al. 2017). Through the secretion of enzymes and other chemical compounds that modify the chemical bioavailability of heavy metals, organic chemicals, and radionuclides, fungi are able to degrade these pollutants. In this way, fungi metabolize and immobilize contaminants in the mycosphere or store them in their own cells (Singh et al. 2020). The main groups of enzymes produced by fungi and involved in the degradation of pollutants include the extracellular oxidoreductases (such as tyrosinases, laccases, manganese peroxidases, lignin peroxidases, etc.), involved in giving fungi the ability to grow on recalcitrant substrates; cell-bound enzymes (such as cytochrome P450s), involved in the formation of intracellular metabolites; and different transferase enzymes (such as nitroreductases, quinone reductases, etc.), involved in the conjugation of pollutants to form nontoxic compounds that are released into the environment (Singh et al. 2020).

Fungi are capable of surviving in a wide diversity of different habitats, even in massively contaminated places, from where there are several groups that are isolated as possible bioremediation agents. Lignocellulosic materials are mainly biodegraded by white-rot fungi, which has been reported with the ability to bioremediate environments contaminated with endocrine disrupting chemicals, such as pesticides, as a consequence of the action of their ligninolytic enzymes on contaminants. In the bioremediation of heavy metals, marine fungi stand out, capable of inactivating their toxic ions with strategies similar to those used to tolerate the high salinity of their habitat of origin. Finally, the other large group of fungi used in bioremediation is encompassed by those that are isolated from those extreme environments where the pollutant is present, for example, wastewater from mining, known as extremophilic fungi (Deshmukh et al. 2016; Singh et al. 2020).

Fungal species capable of biodegrading almost all biodegradable pollutants have been described. As far as toxic recalcitrant compounds are concerned, we are talking about organic compounds that are very persistent in the environment and that have, in a remarkable way, carcinogenic capacity. These pollutants are mainly biodegraded by fungi such as *Curvularia*, *Aspergillus*, *Mucor*, or *Penicillium*, thanks to the high production of lipases, as in the case of hydrocarbons. Regarding heavy metals, fungi have the highest tolerance and bioremediation capacity against Cd, Cu, and Ni, and they are also capable of mycoremediating various pollutants that present them, such as dyes or pesticides. Regarding municipal solid wastes, their fermentation for the production of biogas and compost applicable as organic fertilizer in agriculture is being considered. In this sense, greater efficiency is required in the process, thanks to the hydrolytic enzymatic machinery of different fungal species, which include cellulases, proteases, amylases, and lipases (Deshmukh et al. 2016; Singh et al. 2020).

Although the ability of various unique species to biodegrade pollutants in soils and waters has been described, the process can be very slow, without completely eliminating pollutants from the site. Furthermore, fungal inoculants require an adaptation time to the contaminated environment to be able to develop and act effectively. In this sense, the accessibility and bioavailability of the contaminant can also significantly reduce the efficiency of the mycoremediation under field conditions process. Furthermore, the partial degradation of certain organic compounds, such as pesticides, can lead to the formation of new pollutants that are more toxic to the environment. For this reason, many authors highlight the importance of using bioremediation consortia formed by fungi and bacteria, greatly increasing their effectiveness and reducing all possible associated limitations. Also, there is even the possibility of transforming endogenous microorganisms in the contaminated environment with genes that allow them to biodegrade the pollutants (Gupta et al. 2017).

2.1 Pesticides

Bioremediation of pesticides in the environment can be carried out by both plants and microorganisms. The main site of pesticide phytoremediation is the rhizosphere, where there is also great microbial bioremediation activity. Therefore, plants are capable of directly degrading the pesticides present in the soil, as well as indirectly, by providing nutrients to their rhizospheric microbiota. In addition, plants can assimilate pesticides and store them in their tissues, degrade them internally through their own enzymatic machinery or that of their endophytic microbiota, and/or transform them into volatile forms that they release into the atmosphere (Eevers et al. 2017).

Regarding the bioremediation of pesticides through the use of microorganisms, numerous investigations have been carried out in recent years, and even complete books have been dedicated, such as the one edited by Singh, in 2016. In this sense, they have described very diverse microbial communities capable of mineralizing, transforming, or degrading pesticides, being the bacteria the group with the largest number of species described so far. As bacterial examples, various species of the genus *Pseudomonas* are capable of degrading pesticides such as the herbicide 2,4-dichlorophenoxyacetic acid (2,4-D), organochlorine pesticide like endosulfan and lindane, or organophosphorus insecticide chlorpyrifos. In the group of cyanobacteria, different *Microcystis* and *Anabaena* have been described with the ability to biodegrade organophosphorus and organochlorine insecticides or the glyphosate herbicide. As regards fungi, for example, various species of white-rot fungi have been described with the ability to degrade various pesticides like atrazine, aldrin, diuron, DDT, chlordane, gamma-hexachlorocyclohexane (γ-HCH), dieldrin, lindane, heptachlor, metalaxyl, mirex, or terbuthylazine (Prabha et al. 2017; Parween et al. 2018).

The mechanisms used by fungi to biodegrade pesticides present in soils and waters include their polar hydroxylation and demethylation, esterification, dehydrogenation, hydroxylation, and dioxygenation, for which it is essential to have a significant enzymatic capacity (Maqbool et al. 2016; Spina et al. 2018).

2.2 *Heavy Metals*

Regarding heavy metals, without entering into metallic pollutants, such as hydrocarbons or dyes, there are also contaminated environments that can be bioremediated through the use of plants and/or microorganisms. Plants remove heavy metals from soils and waters after absorbing them by the roots through their phytoaccumulation in different tissues and organs, their phytodegradation through different enzymes and metabolic pathways, and their phytovolatilization through their transformation to volatile forms that they release into the atmosphere, thanks to its phytostabilization and thanks to its transformation into nontoxic forms by root exudates (Muthusaravanan et al. 2018).

Today, heavy metal removal by microorganisms represents a series of advantages over other strategies, such as its simple-to-use, low cost, high adsorption capacity and large availability (Yin et al. 2019). In the case of bacteria, they are capable of acting against the toxic effects of heavy metals in the environment by producing siderophores that chelate them, as different *Pseudomonas* species do with Pb, or by producing metal-binding proteins, called metallothioneins, by different species of *Bacillus* in the presence of Pb (Choudhary et al. 2017). As regards fungi, the main fungal species involved in the mycoremediation of heavy metals are included within the genera *Aspergillus*, *Trichoderma*, and *Penicillium*. For this, they use strategies based on extracellular sequestration by extracellular polymeric substances, such as chitosan, or on intracellular sequestration by storing them in vacuoles (Choudhary et al. 2017; Ul Hassan et al. 2017).

2.3 *Other Pollutants*

Many other pollutants related to pesticides and heavy metals can be present in the environment and are susceptible to being eliminated by the action of fungi, such as hydrocarbons, aromatic amines, or radioactive wastes.

Hydrocarbons are classified as aliphatic or aromatic according to their chemical structure, both types being present in oil and natural gas, as sources of origin. These compounds are not only used as fuel but also represent the raw material for many substances in the chemical industry, such as dyes, solvents, varnishes, etc. Although resulting from biogenic and geological processes, petroleum hydrocarbons become severe pollutants when dispersed in the environment. Polycyclic aromatic hydrocarbons (PAHs) are benzene rings fused pollutants widely present in nature and with carcinogenic activity, while aliphatic hydrocarbons are related to affectations of the nervous system. The way in which fungi degrade hydrocarbons intra- and extracellularly is through enzymes that oxidize them, to form water and nontoxic or less toxic residues (Conejo-Saucedo et al. 2019; Daccò et al. 2020a; Li et al. 2020).

Aromatic amines (AAs), and their derivative compounds, are organic pollutants from very diverse industries specialized in the production of dyes, refined oils, cosmetics, agrochemicals, adhesives, medicine, etc. Due to their origin, they are widely present in the environment, posing a serious danger, as they are carcinogenic compounds. This group of chemicals also includes several groups of pesticides, due to their chemical structure. The main way by which fungi are able to eliminate them from the environment is based on their N-acetylation by enzymes N-acetyltransferases (de Lima et al. 2018). However, yeasts capable of supporting their toxicity and reducing radioactivity have already been used, through the release of carboxylic acids and the formation of biofilms (Tkavc et al. 2018).

3 *Trichoderma* and Bioremediation

The genus *Trichoderma* includes a group of fungal species widely distributed throughout the world due to their rapid growth, their ability to use different substrates and to tolerate the presence of different contaminants (Sharma et al. 2019; Hu et al. 2020). Its main current economic interest is based on its use as a biocontrol agent in agriculture and as a producer of enzymes in different industries (Jangir et al. 2017), although in recent years its relevance in other sectors has been increasing, as a promoter of plant growth and tolerance to abiotic stresses (Poveda et al. 2019a; Poveda 2020), source of genes for use in biotechnology (Poveda et al. 2019b), or mycoremediator.

In its interaction with the plant, *Trichoderma* behaves as a root endophyte, colonizing only the outermost layers of the root, due to a plant defense response mediated by salicylic acid, which prevents the fungus from reaching the vascular bundles and behaving like a systemic pathogen (Alonso-Ramírez et al. 2014; Poveda et al. 2020a). In this way, *Trichoderma* is also capable of activating systemic plant defenses against the attack of pests and/or pathogens (Poveda et al. 2020b) and acts as a biofertilizer (Zhang et al. 2018a).

One of the characteristics that make *Trichoderma* a good alternative for its use in agriculture is its resistance to various fungicides, which allows its inclusion of integrated crop protection management programs. Some of the fungicides to which *Trichoderma* is resistant are azoxystrobin, metalaxyl, carbendazim, chlorothalonil, copper oxy chloride, mancozeb, boscalid, cyazofamid, myclobutanil, pentachloronitrobenzene, propamocarb, or trifloxystrobin, among others (Shashikumar et al. 2019; Widmer et al. 2019). Similarly, *Trichoderma* is able to tolerate the presence of many different contaminants, which, together with its ability to eliminate them or reduce their toxicity, make it an effective mycoremediator agent.

How generally *Trichoderma* strains are selected to be used for mycoremediation is based on isolation from contaminated environments, where it is certain that he is able to survive (Tripathi et al. 2013). This is because *Trichoderma* is capable of obtaining resources from a wide variety of different substrates, as well as surviving extreme conditions, which makes it a better alternative than many other microorganisms used in bioremediation (Solanki et al. 2019).

Following, the different studies carried out in the mycoremediation of pesticides, heavy metals, and other pollutants by using different strains of *Trichoderma* (which have been compiled in Table 1) will be explained. This process of bioremediation by *Trichoderma* is carried out by different mechanisms, depending on the chemical nature of the specific pollutant, including, mainly, biosorption/bioaccumulation, biovolatilization by enzymatic conversation, and phytobial remediation, or microbe-assisted phytoremediation (Tripathi et al. 2013).

4 *Trichoderma* and Pesticides

First studies that determine the ability of *Trichoderma* to degrade different pesticides began in the 1990s (Katayama and Matsumura 1993), but it is in the last 20 years when the mechanisms involved and the wide diversity of pesticides on which it is capable of act.

Fungicides currently represent the only effective control strategy for various plant diseases. Its massive, repeated, and uncontrolled use leads to its excessive accumulation in soils, with very negative effects on the environment (Baćmaga et al. 2019). Both in vitro and in field, the ability of *Trichoderma* to degrade this group of compounds by up to 85% in 5 days has been reported (Sharma et al. 2016; Podbielska et al. 2020). This is a consequence of the action of the cytochrome P450 enzyme, implicated in the degradation of the fungicide climbazole (Manasfi et al. 2020).

With regard to insecticides, *T. atroviride* is capable of degrading in vitro up to 96% of compounds such as the organophosphate insecticide dichlorvos or the neonicotinoid insecticide imidacloprid (Tang et al. 2009; He et al. 2014). Dichlorvos (O, O-dimethyl-2,2-dichlorovinyl phosphate) is an insecticide that causes serious damage to aquatic ecosystems due to its water solubility, being efficiently degraded by the enzymes hygromycin B phosphotransferase and a paraoxonase-like enzyme of *T. atroviride* (Tang et al. 2009; Sun et al. 2019). On the other hand, the capacity of *T. asperellum* to favor a degradation of up to 75% in 5 days of the organophosphate insecticide phoxim has been described, by favoring an increase in glutathione S-transferase, peroxidase, and polyphenol peroxidase activity in tomato roots (Chen et al. 2020).

Herbicides are a group of pesticides widely used on all crops. Its presence in soils and waters causes serious environmental damages, mainly against the microorganisms present and when entering the food chain (Singh and Singh 2016). Glyphosate has been the most widely used herbicide in the last 30 years, as a

Table 1 *Trichoderma* as a mycoremediation agent for pesticides, heavy metals, and other pollutants, indicating the study carried out and its mechanism of action

Pollutant						
Group		Name	*Trichoderma* species[a]	Experiment	Mechanism of action	References
Pesticides	Antibiotics	Fluoroquinolone	*T. harzianum* *T. asperellum*	In vitro	Degradation by cytochrome P450 enzyme	Manasfi et al. (2020)
	Fungicides	Benzimidazole	*T. harzianum* *T. viride* *T. atroviride*	In vitro	Not indicated	Sharma et al. (2016)
		Climbazole	*T. harzianum* *T. asperellum*	In vitro	Degradation by cytochrome P450 enzyme	Manasfi et al. (2020)
		Penthiopyrad	*T. harzianum*	In vitro In field	Not indicated	Podbielska et al. (2020)
	Insecticides	Dichlorvos	*T. atroviride*	In vitro	Degradation by hygromycin B phosphotransferase enzyme	Tang et al. (2009)
			T. atroviride	In vitro	Degradation by paraoxonase-like enzyme	Sun et al. (2019)
		Imidacloprid	*T. atroviride*	In vitro	Not indicated	He et al. (2014)
		Phoxim	*T. asperellum*	In soil	Increased glutathione S-transferase, peroxidase and polyphenol peroxidase activity in tomato roots	Chen et al. (2020)

		T. koningii *T. citrinoviride* *T. harzianum* *T. viride* *T. virens*	*In vitro* In soil	Dechlorination and hydroxylation	Nykiel-Szymańska et al. (2020)
	Glyphosate	*T. viride*	*In vitro*	Use as a phosphorus resource	Arfarita et al. (2013)
		T. viride	In soil	Not indicated	Arfarita et al. (2016)
		T. inhamatum	*In vitro* In field	Urease activity	Kunanbayev et al. (2019)
	Imazethapyr	*T. brevicompactum*	In field	Not indicated	Zhigou et al. (2018)
	Metolachlor	*T. koningii* *T. citrinoviride* *T. harzianum* *T. viride* *T. virens*	*In vitro* In soil	Dechlorination and hydroxylation	Nykiel-Szymańska et al. (2020)
	Sulfosulfuron	*Trichoderma* sp.	In soil	Decarboxylation on the sulphonyl urea bridge and the hydrolytic cleavage of the sulfonylamide linkage	Yadav and Choudhury (2014)
General broad spectrum pesticides	Pentachlorophenol	*T. harzianum*	In vitro In soil	Methylation of phenolic compounds	Rigot and Matsumura (2002)
		T. harzanium *T. piluliferum* *T. aureoviride*	In vitro	Not indicated	Sing et al. (2014)
		T. harzianum	In vitro	Not indicated	Vacondio et al. (2015)
	3-Chloropropionic acid	*Trichoderma* sp.	In vitro	Dehalogenation	Edbeib (2020)

(continued)

Table 1 (continued)

Pollutant		Trichoderma species[a]	Experiment	Mechanism of action	References
Group	Name				
Heavy metals	Arsenic	*T. asperellum*	In vitro	Biosorption	Su et al. (2010)
		T. asperellum	In vitro	Biosorption, reduction and methylation	Su et al. (2011)
		T. asperellum	In soil	Methylation	Su et al. (2017)
		Trichoderma sp.	In soil	Not indicated	Tripathi et al. (2017)
		Trichoderma sp.	In vitro In soil	Phosphatase, dehydrogenase, cellulase, urease, amylase, and invertase activities	Govarthanan et al. (2018)
		Trichoderma sp.	In soil	Urease activity that forms metal carbonates	Govarthanan et al. (2019)
		T. harzianum	In soil	Biosorption	Pehlivan et al. (2020)
	Barium	*T. atroviride*	In vitro	Biosorption	Kacprzak and Malina (2005)

			T. atroviride	In soil	Not indicated	Cao et al. (2008)
			T. koningii	In soil	Not indicated	Wang et al. (2009)
			T. harzianum	In vitro	Not indicated	de Freitas Lima et al. (2011)
			T. viride	In vitro	Biosorption	Sahu et al. (2012)
			T. asperellum	In vitro	Not indicated	Mohsenzadeh and Shahrokhi (2014)
			T. harzianum *T. asperellum*	In vitro	Not indicated	Hoseinzadeh et al. (2017)
			Trichoderma sp.	In vitro	Not indicated	Nongmaithem et al. (2016)
			Trichoderma sp.	In soil	Not indicated	Herliana et al. (2018)
			T. asperellum	In soil	Not indicated	Zhang et al. (2018b)
			T. asperellum	In vitro	Biosorption	Maurya et al. (2019)
			T. simmonsii	In vitro	Biosorption	Yaghoubian et al. (2019)

(continued)

Table 1 (continued)

Pollutant					
Group	Name	*Trichoderma* species[a]	Experiment	Mechanism of action	References
	Chromium	*T. viride*	In vitro	Biosorption	Bishnoi et al. (2007)
		Trichoderma sp.	In vitro	Biosorption	Vankar and Bajpai (2008)
		Trichoderma sp.	In soil	Not indicated	Kacprzak et al. (2014)
		Trichoderma sp.	In vitro	Biosorption	Shukla and Vankar (2014)
		T. pseudokoningii	In vitro	Reduction and biosorption	Ray and Sur (2016)
		Trichoderma sp.	In vitro	Biosorption	Smily and Sumithra (2017)
		T. virens	In vitro	Biosorption	Tansengco et al. (2018)
		T. lixii	In vitro	Biosorption	Kumar and Dwivedi (2019)
		T. harzianum	In soil	Biosorption	Pehlivan et al. (2020)
		T. asperellum	In vitro	Reduction	Saranya et al. (2020)

		T. atroviride	In vitro	Biosorption	Yazdani et al. (2009)
		T. harzianum *T. aureoviride* *T. virens*	In vitro	Biosorption	Siddiquee et al. (2013)
		Trichoderma sp.	In soil	Not indicated	Kacprzak et al. (2014)
		T. koningiopsis	In vitro	Biosorption	Salvadori et al. (2014)
		T. harzianum	In soil	Not indicated	Adebiyi (2017)
		T. asperellum	In soil	Not indicated	Vargas et al. (2017)
		T. virens	In vitro	Biosorption	Tansengco et al. (2018)
		T. asperellum	In vitro	Biosorption	Ladi et al. (2020)
		T. harzianum	In soil	Biosorption	Pehlivan et al. (2020)
		T. brevicompactum	In vitro	Biosorption	Zhang et al. (2020)
		T. lixii	In vitro	Biosorption	Kumar and Dwivedi (2021)
	Iron	*T. atroviride*	In vitro	Biosorption	Kacprzak and Malina (2005)
		T. harzianum	In soil	Not indicated	Adebiyi (2017)

(continued)

Table 1 (continued)

Pollutant					
Group	Name	*Trichoderma* species[a]	Experiment	Mechanism of action	References
	Lead	*T. harzianum*	In soil	Not indicated	Adams et al. (2007)
		T. viride	In vitro	Biosorption	Sahu et al. (2012)
		T. harzianum *T. aureoviride* *T. virens*	In vitro	Biosorption	Siddiquee et al. (2013)
		T. harzianum	In soil	Not indicated	Adebiyi (2017)
		T. harzianum *T. asperellum*	In vitro	Not indicated	Hoseinzadeh et al. (2017)
		Trichoderma sp.	In vitro In soil	Phosphatase, dehydrogenase, cellulase, urease, amylase, and invertase activities	Govarthanan et al. (2018)
		T. virens	In vitro	Biosorption	Tansengco et al. (2018)
		T. asperellum	In soil	Not indicated	Zhang et al. (2018b)
		Trichoderma sp.	In soil	Urease activity that forms metal carbonates	Govarthanan et al. (2019)
		T. asperellum	In soil	Not indicated	Li et al. (2019)
		T. asperellum	In vitro	Biosorption	Maurya et al. (2019)
		T. viride	In vitro	Biosorption	Luo et al. (2020)
		T. harzianum	In soil	Biosorption	Pehlivan et al. (2020)
		T. asperellum	In vitro	Biosorption	Sun et al. (2020)
		T. brevicompactum	In vitro	Biosorption	Zhang et al. (2020)
	Manganese	*T. harzianum*	In soil	Not indicated	Adams et al. (2007)

		T. atroviride	In soil	Not indicated	Cao et al. (2008)
		T. harzianum *T. aureoviride* *T. virens*	In vitro	Biosorption	Siddiquee et al. (2013)
		Trichoderma sp.	In soil	Not indicated	Kacprzak et al. (2014)
		Trichoderma sp.	In vitro	Not indicated	Nongmaithem et al. (2016)
		T. harzianum *T. asperellum*	In vitro	Not indicated	Hoseinzadeh et al. (2017)
		T. virens	In vitro	Biosorption	Tansengco et al. (2018)
		T. asperellum	In soil	Not indicated	Rosmana et al. (2019)
	Uranium	*T. harzianum*	In vitro	Biosorption	Akhtar et al. (2007)
	Zinc	*T. atroviride*	In vitro	Biosorption	Kacprzak and Malina (2005)
		T. harzianum	In soil	Not indicated	Adams et al. (2007)
		T. atroviride	In vitro	Biosorption	Yazdani et al. (2010)
		T. harzianum *T. aureoviride* *T. virens*	In vitro	Biosorption	Siddiquee et al. (2013)
		T. harzianum	In soil	Not indicated	Adebiyi (2017)
		T. virens	In vitro	Biosorption	Tansengco et al. (2018)
		T. harzianum	In soil	Biosorption	Pehlivan et al. (2020)

(continued)

Table 1 (continued)

Pollutant						
Group		Name	*Trichoderma* species[a]	Experiment	Mechanism of action	References
Other pollutants	Dyes	Anthraquinone dyes	*T. lixii*	In vitro	Biosorption and enzymatic degradation	Adnan et al. (2017)
		Congo red	*T. harzianum*	In vitro	Degradation by laccase enzyme	Ranimol et al. (2018)
		Cresol Red	*T. harzianum*	In vitro	Degradation by manganese peroxidase, lignin peroxidase, laccase, 1,2- and 2,3-dioxygenase enzymes	Nor et al. (2015)
		Crystal violet	*T. harzianum*	In vitro	Degradation by laccase enzyme	Ranimol et al. (2018)
			T. asperellum	In vitro	Degradation by laccase enzyme	Shanmugam et al. (2017a)
		Malachite green	*T. asperellum*	In vitro	Degradation by laccase enzyme	Shanmugam et al. (2017b)
			T. harzianum	In vitro	Degradation by laccase enzyme	Ranimol et al. (2018)
		Methylene blue	*T. harzianum*	In vitro	Degradation by laccase enzyme	Ranimol et al. (2018)

				enzyme	(2009)
		T. longibrachiatum	In soil	Degradation by phenoloxidase enzyme	Andreolli et al. (2016)
		T. reesei	In vitro	Not indicated	Nazifa et al. (2018)
		T. harzianum	In soil	Not indicated	Elshafie et al. (2020)
	Polycyclic aromatic hydrocarbons (PAHs)	*T. longibrachiatum*	In soil	Not indicated	Cobas et al. (2013)
		T. viride	In soil	Not indicated	Szczepaniak et al. (2015)
		T. reesei	In soil	Degradation by dehydrogenase enzyme	Yao et al. (2015)
		T. asperellum	In soil	Degradation by catechol 1,2 dioxygenase, laccase, and peroxidase enzymes	Zafra et al. (2015)
		Trichoderma sp.	In vitro	Not indicated	Al Farraj et al. (2020)
		T. harzianum	In vitro	Not indicated	Miles et al. (2020)
		T. lixii	In soil	Use as a carbon resource	Venice et al. (2020)
	Used engine oil	*T. harzianum*	In vitro	Use as a carbon resource	Daccò et al. (2020b)
Wastewaters	Cassava wastewater	*T. harzianum* *T. viride*	In vitro	Cyanide removal	Saravanan et al. (2019)
	Synthetic wastewater	*T. pubescens*	In vitro	Not indicated	Narendran et al. (2019)
	Table olive processing water	*T. viride*	Sequencing Batch Reactor	Phenols and organic carbon removal	D'Urso et al. (2008)
		T. harzianum	In vitro	Phenols removal	Campaniello et al. (2020)

(continued)

Table 1 (continued)

Pollutant						
Group		Name	*Trichoderma* species[a]	Experiment	Mechanism of action	References
	Other industrial and urban wastes	Detergents	*T. harzianum*	In vitro	Degradation by invertase and protease enzymes	Jakovljević et al. (2015)
		Municipal solid waste leachates	*T. harzianum*	In vitro	Not indicated	Awasthi et al. (2017)
		Phenolic compounds	*T. atroviride*	In vitro	Degradation by laccase enzyme	Chakroun et al. (2010)
			T. viride	In vitro	Degradation by laccase enzyme	Divya et al. (2013)
			T. aureoviridae	In vitro	Degradation by laccase enzyme	Lawrance et al. (2019)
		Cyanide	*T. harzianum* *T. pseudokoningi*	In vitro	Cyanide hydratase and rhodanese activities	Ezzi and Lynch (2002)
			T. harzianum *T. pseudokoningi*	In soil	Not indicated	Ezzi and Lynch (2005a)
			T. harzianum *T. pseudokoningi*	In vitro	Use as a carbon and nitrogen resource	Ezzi and Lynch (2005b)
			T. koningii *T. harzianum* *T. atroviride*	In vitro	Cyanide hydratase and rhodanese activities	Zhou et al. (2007)
		Ethoxylated oleyl-cetyl alcohol	*T. viride*	In vitro	Degradation by protease enzyme	Jakovljević (2020)
		Waste plastics	*T. atroviride*	In vitro	Degradation by laccase enzyme	Balcázar-López et al. (2016)
		2,4,6-trinitrotoluene (TNT)	*T. viride*	In vitro	Not indicated	Alothman et al. (2020)

[a]Species identities are cited as initially published, and the current taxonomic status of each species requires verification

consequence of the development of specifically resistant transgenic crops. Its toxicity is described as particularly harmful to animals as it occurs in the food chain, although its carcinogenic capacity has not been fully demonstrated (Xu et al. 2019). In this sense, *T. viride* and *T. inhamatum* have been described with the ability to degrade glyphosate, both in vitro and in the field, up to 70%, due to its use as a phosphorus resource and the action of urease enzymes (Arfarita et al. 2013, 2016; Kunanbayev et al. 2019). Alachlor herbicide has also been described as possibly carcinogenic, but it is certainly an endocrine-disrupting compound. Its total degradation in 7 days has been reported by various species of *Trichoderma* through its dechlorination and hydroxylation, intervening cytochrome P450 and laccase enzymes; as it happens with metolachlor, another chloroacetanilide herbicide (Nykiel-Szymańska et al. 2018, 2020).

Finally, there is a group of pesticides that can be used against a wide variety of pathogens and pests, the so-called broad spectrum. Pentachlorophenol has been used as a general biocide for many different purposes, becoming very harmful to the environment and health, by forming an important reservoir source of dioxins and furans (Verbrugge et al. 2018). *T. harzianum* has been reported as a potent mycoremediation agent for this pesticide through the methylation of phenolic compounds, degrading it to 100% in 7 days, in vitro and in soil (Rigot and Matsumura 2002; Vacondio et al. 2015). Moreover, the bioremedial capacity of *Trichoderma* can be used to obtain compounds of interest in different industries. Through the dehalogenation of the broad spectrum pesticide 3-chloropropionic acid, *Trichoderma* is capable of forming propionic acid, an additive widely used in animal feed and in the manufacture of biodegradable polymers (Edbeib 2020).

5 *Trichoderma* and Heavy Metals

As in the case of pesticides, the first studies that demonstrated the ability of *Trichoderma* to eliminate heavy metals from the environment date from the 1990s (Krantz-Rülcker et al. 1996). The main mechanism used by *Trichoderma* to heavy metals mycoremediation is its biosorption.

Cadmium (Cd) is a non-essential trace metal, very toxic for the environment and health. In humans, Cd can cause lung cancer in long-term exposure or kidney and bone damages in high exposure (Liu et al. 2017). Through the biosorption of Cd, *Trichoderma* species, such as *T. asperellum* or *T. harzianum*, are capable of reducing its presence in vitro by up to 90% in 21 days (Mohsenzadeh and Shahrokhi 2014; Hoseinzadeh et al. 2017; Maurya et al. 2019). As a consequence, *Trichoderma* is capable of increasing the tolerance in Cd-contaminated soils of crack willow (*Salix fragilis*) (Adams et al. 2007), spinach (Herliana et al. 2018), or *Arabidopsis thaliana* (Zhang et al. 2018b), also increasing Cd phytoaccumulation in oilseed rapes (*Brassica napus* and *B. juncea*) (Cao et al. 2008; Wang et al. 2009).

Lead (Pb) is a toxic metal from waste batteries and paint, mining and smelting activities, and combustion of fossil fuels. It is a very harmful element for health, since it is a powerful neurotoxic that can lead to death (Arnemo et al. 2016). Several species of *Trichoderma* have been reported with the ability to reduce the amount of Pb in vitro above 95% in 21 days due to its biosorption (Siddiquee et al. 2013; Tansengco et al. 2018; Maurya et al. 2019), in which the functional groups of its polysaccharides, with a high affinity for metal ions, are involved (Sun et al. 2020). Through this mechanism, it has been described how earthworms are capable of eliminating the Pb present in the soil, by having *T. brevicompactum* in their intestine (Zhang et al. 2020). Another mechanism reported in *Trichoderma* has been the formation of metal carbonates by different enzymatic activities (such as phosphatase, dehydrogenase, cellulase, urease, amylase, and invertase), removing 70% of Pb in contaminated soils (Govarthanan et al. 2018, 2019). In this way, *T. harzianum* and *T. asperellum* are capable of improving the tolerance of *S. fragilis*, *A. thaliana*, and *Suaeda salsa* in soils contaminated with Pb, reducing oxidative stress in the plant (Adams et al. 2007; Zhang et al. 2018b; Li et al. 2019).

Copper (Cu) is an essential element for plants as it is involved in numerous physiological processes. However, high levels of Cu are very harmful for plant growth, being also toxic for animals (Rehman et al. 2019). In vitro, *Trichoderma* is able to remove the Cu present up to 85% in 120 h (Yazdani et al. 2009; Tansengco et al. 2018; Kumar and Dwivedi 2021), also observed in *T. brevicompactum* in intestinal earthworms (Zhang et al. 2020), although in soil its capacity is reduced to 20% removal (Pehlivan et al. 2020). By means of biosorption of Cu mediums, the dead biomass of *T. koningiopsis* has been used in the production of Cu nanoparticles (Salvadori et al. 2014). Moreover, *Trichoderma* is capable of increasing plant tolerance in soils contaminated with high amounts of Cu and increasing its phytoaccumulation (Kacprzak et al. 2014; Vargas et al. 2017).

Chromium (Cr) is a very useful metal to many industries. In nature, it is found as Cr(III), without being harmful, but when oxidized to its Cr(VI) form due to anthropogenic activity, it presents high toxicity. The main damage to the environment and health of Cr(VI) is due to its corrosive nature, causing serious injuries when in contact with internal epithelia (ingestion or inhalation) or external (Coetzee et al. 2020). In Cr(VI) bioremediation by *Trichoderma*, a reduction to Cr(III) is necessary followed by a biosorption (Ray and Sur 2016; Saranya et al. 2020). In this sense, *Trichoderma* is capable of eliminating almost 100% of Cr(IV) in vitro (Vankar and Bajpai 2008; Shukla and Vankar 2014) and 30% in soil (Pehlivan et al. 2020).

Nickel (Ni) is a heavy metal considered an essential microelement for many plant physiological processes involved in its correct growth and development. However, excessive amounts of Ni in soils or waters cause serious toxicity symptoms in plants, such as chlorosis and growth inhibition, since their photosynthetic, respiratory, and water and nutrient transport activity are reduced. Ni environmental pollution is mainly a consequence of the metallurgical and electroplating industries. In animals, Ni easily accumulates in tissues, causing serious embryo-toxic, teratogenic, and carcinogenic damages (Shahzad et al. 2018). The ability of several *Trichoderma* species to bioaccumulate Ni by biosorption has been reported, reducing its presence in the soil by up to 78% (Hoseinzadeh et al. 2017; Tansengco et al. 2018). Furthermore, in interaction with plants, *T. atroviride* and *T. asperellum*, as examples, are capable of increasing the tolerance and phytoaccumulation of *B. juncea* and cacao, respectively, in Ni-contaminated soils (Cao et al. 2008; Rosmana et al. 2019).

Zinc (Zn) is an essential element for many biological processes in all organisms, such as protein synthesis or cell division. Its main toxicity problem due to excessive pollution of the environment has been observed in aquatic ecosystems, where it can be very harmful to life (Andarani et al. 2020). Although *T. harzianum*, *T. atroviride*, and *T. virens* are capable of eliminating the compound by biosorption, their capacity is very low, with removal percentages of 50% in vitro (Yazdani et al. 2010; Siddiquee et al. 2013; Tansengco et al. 2018) and 10% in soil (Pehlivan et al. 2020).

Arsenic (As) is the most widely distributed metalloid on the planet. The contamination of aquifers by As is the main sequence of natural geochemical mechanisms, but there are also minor anthropogenic sources, such as agrochemicals. Being present in aquifers, it quickly enters the food chain, causing serious damage to the vascular, nervous, and skin systems, as well as cancer (Alka et al. 2020). The removal of As by *Trichoderma* is performed through its reduction and methylation before its biosorption, transforming it into the nontoxic forms As(V) and As(III) (Su et al. 2011; Su et al. 2017), thus such as the formation of metal carbonates by the action of urease enzymes (Govarthanan et al. 2019). In this way, up to 70% of As is eliminated in vitro (Govarthanan et al. 2018) and percentages close to 10% in soil (Pehlivan et al. 2020). Due to this, the tolerance of water spinach (*Ipomoea aquatic*) and chickpea is increased in soils contaminated with As by *Trichoderma* application (Su et al. 2017; Tripathi et al. 2017).

6 *Trichoderma* and Other Pollutants

In relation to pesticides and heavy metals, *Trichoderma* has also been reported as an efficient mycoremediation agent against a great variety of pollutants of very varied origin, through mechanisms of action such as those already described.

The main group of hydrocarbons polluting the environment are the polycyclic aromatic hydrocarbons (PAHs), whose adverse effects have already been described. The ability of *Trichoderma* to eliminate the toxicity of PAHs in different soils has been widely reported, due to its use as a carbon resource (Daccò et al. 2020b) by the action of various enzymes (dehydrogenase, catechol 1,2 dioxygenase, laccase, and peroxidase) (Yao et al. 2015; Zafra et al. 2015). In this way, it has been possible to eliminate up to 75% of the phenanthrene (Cobas et al. 2013; Zafra et al. 2015), 80% of the pyrene (Zafra et al. 2015; Al Farraj et al. 2020) and benzo[a]pyrene (Yao et al. 2015; Zafra et al. 2015), or 50% of naphthalene (Miles et al. 2020). In the same way, *Trichoderma* is capable of eliminating diesel present as a pollutant in different soils. In vitro, *T. reesei* eliminates up to 95% of the diesel in 40 days (Nazifa et al. 2018), while in soil the percentage is reduced to 70% by *T. harzianum* (Elshafie et al. 2020). Diesel degradation occurs through dehydrogenase and phenoloxidase enzymes (Mishra and Nautiyal 2009; Andreolli et al. 2016).

The main source of contamination by dyes comes from the widely distributed worldwide coloring industry, specifically from its wastewaters. The dyes present a great potential of damage to the environment, due to their mutagenic and carcinogenic capacity, and their direct damage to kidney, liver, brain, reproductive system, and central nervous system (Kaykhaii et al. 2018). The main mechanism of action of *Trichoderma* in the mycoremediation of dyes is through its enzymatic degradation. In vitro, *T. asperellum* and *T. harzianum* are capable of degrading by the action of laccase enzymes up to 98% of methylene blue (Ranimol et al. 2018) and malachite green (Shanmugam et al. 2017b; Ranimol et al. 2018), 96% of Congo red (Ranimol et al. 2018), or 60% of crystal violet (Shanmugam et al. 2017a; Ranimol et al. 2018). Although in the degradation of up to 88% of creson red in 30 days by *T. harzianum*, the activity of the enzymes manganese peroxidase, lignin peroxidase, and 1,2- and 2,3-dioxygenase has also been reported (Nor et al. 2015).

There are many other pollutants against which *Trichoderma*'s ability as a mycoremediation agent has been reported, which are listed in Table 1. Some of them include the degradation of detergents by invertase and protease enzymes from *T. harzianum* (Jakovljević et al. 2015), phenolic compounds or plastics by laccase enzymes (Balcázar-López et al. 2016; Lawrance et al. 2019), cyanide by cyanide hydratase and rhodanese enzymes (Ezzi and Lynch 2002; Zhou et al. 2007), and even 2,4,6-trinitrotoluene (TNT) (Alothman et al. 2020).

7 Conclusions

Soils and waters around the world present, to a greater or lesser extent, some pollutant that is seriously harmful to the environment and health. Due to their presence in soils and waters and their toxicity, pesticides and heavy metals represent the main pollutants in the agricultural system. In this sense, bioremediation is an effective strategy for the elimination of these contaminants, highlighting the role played by fungal enzymes in mycoremediation, which allows the degradation and/or conjugation of these harmful elements.

The use of fungi in the bioremediation of pollutants in soils and waters presents a series of limitations and drawbacks. The process can be very slow and incomplete, since the fungi need a period of adaptation to the new environment once they are inoculated, and have access to the contaminant of interest. Moreover, the transformation of pollutants to more toxic forms by fungal action can occur. However, mycoremediation is an innovative, cost-effective, and ecologically beneficial technology in removing contaminants such as pesticides and heavy metals.

Due to their ability to survive in highly polluted extreme environments and the extensive enzymatic library they possess, there are numerous species of the genus *Trichoderma* capable of effectively bioremediating a wide range of different contaminants. Thanks to its enzymatic activity, *Trichoderma* is capable of degrading in percentages close to 100% such polluting pesticides as glyphosate or pentachlorophenol. In addition, its mycoremediation capacity can have derived benefits, as is the case with the broad spectrum pesticide 3-chloropropionic acid, transformed by *Trichoderma* into propionic acid.

As far as heavy metals are concerned, in vitro it has been proven that *Trichoderma* is capable of effectively eliminating almost all of the contaminant by biosorption, although in its application on soils the elimination percentages are even reduced to one-tenth. Despite the wide variety of heavy metals that *Trichoderma* is capable of bioaccumulating, its low efficiency in natural environments represents a difficulty for its widespread use.

Furthermore, various species of *Trichoderma* have been described with the ability to remove many other pollutants from soils and waters, thanks to mechanisms similar to those used against pesticides and heavy metals. These include hydrocarbons, dyes, detergents, phenolic compounds, or cyanide. Therefore, *Trichoderma* is a powerful mycoremediation agent for the main current environmental pollutants, although even more studies are necessary on its application in natural environments, in order to obtain efficient elimination processes.

References

Abdu N, Abdullahi AA, Abdulkadir A (2017) Heavy metals and soil microbes. Environ Chem Lett 15:65–84

Adams P, De-Leij FA, Lynch JM (2007) *Trichoderma harzianum* Rifai 1295-22 mediates growth promotion of crack willow (*Salix fragilis*) saplings in both clean and metal-contaminated soil. Microb Ecol 54:306–313

Adebiyi KA (2017) Bioremediation of heavy metals in the soil by *Pseudomonas aeruginosa* and *Trichoderma harzianum* using *Solanum lycopersicum* as test plant. Asian J Soil Sci Plant Nutr:1–13

Adnan LA, Sathishkumar P, Yusoff ARM, Hadibarata T, Ameen F (2017) Rapid bioremediation of Alizarin Red S and Quinizarine Green SS dyes using *Trichoderma lixii* F21 mediated by biosorption and enzymatic processes. Bioprocess Biosyst Eng 40:85–97

Akhtar K, Akhtar MW, Khalid AM (2007) Removal and recovery of uranium from aqueous solutions by *Trichoderma harzianum*. Water Res 41:1366–1378

Al Farraj DA, Hadibarata T, Elshikh MS, Al Khulaifi MM, Kristanti RA (2020) Biotransformation and degradation pathway of pyrene by filamentous soil fungus *Trichoderma* sp. F03. Water Air Soil Pollut 231:1–9

Alka S, Shahir S, Ibrahim N, Ndejiko MJ, Vo DVN, Abd Manan F (2020) Arsenic removal technologies and future trends: a mini review. J Clean Prod 123805

Alonso-Ramírez A, Poveda J, Martín I, Hermosa R, Monte E, Nicolás C (2014) Salicylic acid prevents *Trichoderma harzianum* from entering the vascular system of roots. Mol Plant Pathol 15:823–831

Alothman ZA, Bahkali AH, Elgorban AM, Al-Otaibi MS, Ghfar AA, Gabr SA et al (2020) Bioremediation of explosive TNT by *Trichoderma viride*. Molecules 25:1393

Andarani P, Yokota K, Saga M, Inoue T, Matsumoto Y (2020) Study of zinc pollution in river water: average mass balance based on irrigation schedule. River Res Appl 36:1286–1295

Andreolli M, Lampis S, Brignoli P, Vallini G (2016) *Trichoderma longibrachiatum* Evx1 is a fungal biocatalyst suitable for the remediation of soils contaminated with diesel fuel and polycyclic aromatic hydrocarbons. Environ Sci Pollut Res 23:9134–9143

Arfarita N, Imai T, Kanno A, Yarimizu T, Xiaofeng S, Jie W et al (2013) The potential use of *Trichoderma viride* strain FRP3 in biodegradation of the herbicide glyphosate. Biotechnol Biotechnol Equip 27:3518–3521

Arfarita N, Djuhari D, Prasetya B, Imai T (2016) The application of *Trichoderma viride* strain frp 3 for biodegradation of glyphosate herbicide in contaminated land. AGRIVITA J Agric Sci 38:275–281

Arnemo JM, Andersen O, Stokke S, Thomas VG, Krone O, Pain DJ, Mateo R (2016) Health and environmental risks from lead-based ammunition: science versus socio-politics. EcoHealth 13:618–622

Awasthi AK, Pandey AK, Khan J (2017) Potential of fungus *Trichoderma harzianum* for toxicity reduction in municipal solid waste leachate. Int J Environ Sci Technol 14:2015–2022

Baćmaga M, Wyszkowska J, Kucharski J (2019) The biochemical activity of soil contaminated with fungicides. J Environ Sci Health B 54:252–262

Balcázar-López E, Méndez-Lorenzo LH, Batista-García RA, Esquivel-Naranjo U, Ayala M, Kumar VV et al (2016) Xenobiotic compounds degradation by heterologous expression of a *Trametes sanguineus* laccase in *Trichoderma atroviride*. PLoS One 11:e0147997

Bauer SE, Tsigaridis K, Miller R (2016) Significant atmospheric aerosol pollution caused by world food cultivation. Geophys Res Lett 43:5394–5400

Bhardwaj R, Yadav SP, Singh RK, Tripathi VK (2020) Crop growth under heavy metals stress and its mitigation. In: New frontiers in stress management for durable agriculture. Springer, Singapore, pp 169–183

Bishnoi NR, Kumar R, Bishnoi K (2007) Biosorption of Cr (VI) with *Trichoderma viride* immobilized fungal biomass and cell free Ca-alginate beads. Indian J Exp Biol 45:657–664

Cachada A, Rocha-Santos T, Duarte AC (2018) Soil and pollution: an introduction to the main issues. In: Soil pollution. Academic, New York, pp 1–28

Cao L, Jiang M, Zeng Z, Du A, Tan H, Liu Y (2008). Trichoderma atroviride F6 improves phytoextraction efficiency of mustard (Brassica juncea (L.) Coss. var. foliosa Bailey) in Cd, Ni contaminated soils. Chemosphere, 71(9): 1769–1773

Campaniello D, Carlucci A, Speranza B, Raimondo ML, Cibelli F, Rosaria Corbo M, Bevilacqua A (2020) A comparative study on *Trichoderma harzianum* and a combination of *Candida/Bacillus* as tools for the bioremediation of table olive processing water. Microorganisms 8:878

Chakroun H, Mechichi T, Martinez MJ, Dhouib A, Sayadi S (2010) Purification and characterization of a novel laccase from the ascomycete *Trichoderma atroviride*: application on bioremediation of phenolic compounds. Process Biochem 45:507–513

Chen S, Yan Y, Wang Y, Wu M, Mao Q, Chen Y et al (2020) *Trichoderma asperellum* reduces phoxim residue in roots by promoting plant detoxification potential in *Solanum lycopersicum* L. Environ Pollut 259:113893

Choudhary M, Kumar R, Datta A, Nehra V, Garg N (2017) Bioremediation of heavy metals by microbes. In: Bioremediation of salt affected soils: an indian perspective. Springer, Cham, pp 233–255

Cobas M, Ferreira L, Tavares T, Sanromán MA, Pazos M (2013) Development of permeable reactive biobarrier for the removal of PAHs by *Trichoderma longibrachiatum*. Chemosphere 91:711–716

Coetzee JJ, Bansal N, Chirwa EM (2020) Chromium in environment, its toxic effect from chromite-mining and ferrochrome industries, and its possible bioremediation. Exposure Health 12:51–62

Conejo-Saucedo U, Olicón-Hernández DR, Robledo-Mahón T, Stein HP, Calvo C, Aranda E (2019) Bioremediation of polycyclic aromatic hydrocarbons (PAHs) contaminated soil through fungal communities. In: Recent advancement in white biotechnology through fungi. Springer, Cham, pp 217–236

D'Urso A, Gapes D, Bravi M (2008) Bioremediation of olive oil mill wastewaters by fungal (*Trichoderma viride*, strain 8/90) sequencing batch reactor. Chem Eng Trans 14:481–486

Daccò C, Girometta C, Asemoloye MD, Carpani G, Picco AM, Tosi S (2020a) Key fungal degradation patterns, enzymes and their applications for the removal of aliphatic hydrocarbons in polluted soils: a review. Int Biodeterior Biodegradation 147:104866

Daccò C, Nicola L, Temporiti MEE, Mannucci B, Corana F, Carpani G, Tosi S (2020b) *Trichoderma*: evaluation of its degrading abilities for the bioremediation of hydrocarbon complex mixtures. Appl Sci 10:3152

Dangi AK, Sharma B, Hill RT, Shukla P (2019) Bioremediation through microbes: systems biology and metabolic engineering approach. Crit Rev Biotechnol 39:79–98

de Freitas Lima A, Ferreira de Moura G, Barbosa de Lima MA, Mendes de Souza P, Alves da Silva CA, de Campos Takaki GM, Do Nascimento AE (2011) Role of the morphology and polyphosphate in *Trichoderma harzianum* related to cadmium removal. Molecules 16:2486–2500

de Lima DP, dos Santos EDA, Marques MR, Giannesi GC, Beatriz A, Yonekawa MK, Montanholi ADS (2018) Fungal bioremediation of pollutant aromatic amines. Curr Opin Green Sustain Chem 11:34–44

Deshmukh R, Khardenavis AA, Purohit HJ (2016) Diverse metabolic capacities of fungi for bioremediation. Indian J Microbiol 56:247–264

Divya LM, Prasanth GK, Sadasivan C (2013) Isolation of a salt tolerant laccase secreting strain of *Trichoderma* sp. NFCCI-2745 and optimization of culture conditions and assessing its effectiveness in treating saline phenolic effluents. J Environ Sci 25:2410–2416

Edbeib M (2020) 3-Chloropropionic acid (3cp) degradation and production of propionic acid by newly isolated fungus *Trichoderma* sp. Mf1. Int J Life Sci Biotechnol 3:41–50

Eevers N, White JC, Vangronsveld J, Weyens N (2017) Bio-and phytoremediation of pesticide-contaminated environments: a review. In: Advances in botanical research, vol 83. Academic, New York, pp 277–318

Elbana T, Gaber HM, Kishk FM (2019) Soil chemical pollution and sustainable agriculture. In: The soils of Egypt. Springer, Cham, pp 187–200

Elshafie HS, Camele I, Sofo A, Mazzone G, Caivano M, Masi S, Caniani D (2020) Mycoremediation effect of *Trichoderma harzianum* strain T22 combined with ozonation in diesel-contaminated sand. Chemosphere 126597

Ezzi MI, Lynch JM (2002) Cyanide catabolizing enzymes in *Trichoderma* spp. Enzym Microb Technol 31:1042–1047

Ezzi MI, Lynch JM (2005a) Plant microcosm studies demonstrating bioremediation of cyanide toxicity by *Trichoderma* and *Fusarium* spp. Biol Fertil Soils 42:40–44

Ezzi MI, Lynch JM (2005b) Biodegradation of cyanide by *Trichoderma* spp. and *Fusarium* spp. Enzym Microb Technol 36:849–854

Govarthanan M, Mythili R, Selvankumar T, Kamala-Kannan S, Kim H (2018) Myco-phytoremediation of arsenic-and lead-contaminated soils by Helianthus annuus and wood rot fungi, *Trichoderma* sp. isolated from decayed wood. Ecotoxicol Environ Saf 151:279–284

Govarthanan M, Mythili R, Kamala-Kannan S, Selvankumar T, Srinivasan P, Kim H (2019) *In-vitro* bio-mineralization of arsenic and lead from aqueous solution and soil by wood rot fungus, *Trichoderma* sp. Ecotoxicol Environ Saf 174:699–705

Gupta S, Wali A, Gupta M, Annepu SK (2017) Fungi: an effective tool for bioremediation. In: Plant-microbe interactions in agro-ecological perspectives. Springer, Singapore, pp 593–606

He X, Wubie AJ, Diao Q, Li W, Xue F, Guo Z et al (2014) Biodegradation of neonicotinoid insecticide, imidacloprid by restriction enzyme mediated integration (REMI) generated *Trichoderma* mutants. Chemosphere 112:526–530

Herliana O, Soesanto L, Mawadah E (2018) Phytobioremediation of cadmium-contaminated soil using combination of *Ipomoea reptans* Poir and *Trichoderma* sp. and its effect on spinach growth and yield. J Degraded Mining Lands Manag 6:1519

Hoseinzadeh S, Shahabivand S, Aliloo AA (2017) Toxic metals accumulation in *Trichoderma asperellum* and *T. harzianum*. Microbiology 86:728–736

Hu J, Zhou Y, Chen K, Li J, Wei Y, Wang Y et al (2020) Large-scale T*richoderma* diversity was associated with ecosystem, climate and geographic location. Environ Microbiol

Jakovljević VD (2020) Synergistic effect of *Fusarium lateritium* LP7 and *Trichoderma viride* LP5 promotes ethoxylated oleyl-cetyl alcohol biodegradation. J Environ Sci Health A 55:438–447

Jakovljević VD, Stojanović JD, Vrvić MM (2015) The potential application of fungus *Trichoderma harzianum* Rifai in biodegradation of detergent and industry. Chem Ind Chem Eng Q 21:131–139

Jangir M, Pathak R, Sharma S (2017) Trichoderma and its potential applications. In: Plant-microbe interactions in agro-ecological perspectives. Springer, Singapore, pp 323–339

Kacprzak M, Malina G (2005) The tolerance and Zn^{2+}, Ba^{2+} and Fe^{3+} accumulation by *Trichoderma atroviride* and *Mortierella exigua* isolated from contaminated soil. Can J Soil Sci 85:283–290

Kacprzak MJ, Rosikon K, Fijalkowski K, Grobelak A (2014) The effect of *Trichoderma* on heavy metal mobility and uptake by *Miscanthus giganteus*, Salix sp., *Phalaris arundinacea*, and *Panicum virgatum*. Appl Environ Soil Sci 2014

Katayama A, Matsumura F (1993) Degradation of organochlorine pesticides, particularly endosulfan, by *Trichoderma harzianum*. Environ Toxicol Chem 12:1059–1065

Kaykhaii M, Sasani M, Marghzari S (2018) Removal of dyes from the environment by adsorption process. Chem Mater Eng 6:31–35

Koul B, Taak P (2018) Soil pollution: causes and consequences. In: Biotechnological strategies for effective remediation of polluted soils. Springer, Singapore, pp 1–37

Krantz-Rülcker C, Allard B, Schnürer J (1996) Adsorption of IIB-metals by three common soil fungi-comparison and assessment of importance for metal distribution in natural soil systems. Soil Biol Biochem 28:967–975

Kumar V, Dwivedi SK (2019) Hexavalent chromium stress response, reduction capability and bioremediation potential of *Trichoderma* sp. isolated from electroplating wastewater. Ecotoxicol Environ Saf 185:109734

Kumar V, Dwivedi SK (2021) Bioremediation mechanism and potential of copper by actively growing fungus *Trichoderma lixii* CR700 isolated from electroplating wastewater. J Environ Manag 277:111370

Kunanbayev K, Churkina G, Rukavitsina I, Filippova N, Utebayev M (2019) Potential attractiveness of soil fungus *Trichoderma inhamatum* for biodegradation of the glyphosate herbicide. *Journal of*. Ecol Eng 20

Ladi E, Shukla N, Bohra Y, Tiwari AK, Kumar J (2020) Copper tolerant *Trichoderma asperellum* increases bio-efficacy of copper against *Phytophthora infestans* in dual combination. Phytoparasitica 48:357–370

Lawrance I, Sivaranjani V, Selvakumar AM, Khambhaty Y, Saravanan P (2019) Biodegradation of phenolic resin used in leather processing by laccase producing *Trichoderma aureoviridae*. Int J Environ Sci Technol 16:6857–6862

Li X, Zhang X, Wang X, Yang X, Cui Z (2019) Bioaugmentation-assisted phytoremediation of lead and salinity co-contaminated soil by *Suaeda salsa* and *Trichoderma asperellum*. Chemosphere 224:716–725

Li Q, Liu J, Gadd GM (2020) Fungal bioremediation of soil co-contaminated with petroleum hydrocarbons and toxic metals. Appl Microbiol Biotechnol:1–10

Liu Y, Xiao T, Perkins RB, Zhu J, Zhu Z, Xiong Y, Ning Z (2017) Geogenic cadmium pollution and potential health risks, with emphasis on black shale. J Geochem Explor 176:42–49

Luo D, Geng R, Wang W, Ding Z, Qiang S, Liang J et al (2020) *Trichoderma viride* involvement in the sorption of Pb (II) on muscovite, biotite and phlogopite: batch and spectroscopic studies. J Hazard Mater 401:123249

Mahmood I, Imadi SR, Shazadi K, Gul A, Hakeem KR (2016) Effects of pesticides on environment. In: Plant, soil and microbes. Springer, Cham, pp 253–269

Manasfi R, Chiron S, Montemurro N, Perez S, Brienza M (2020) Biodegradation of fluoroquinolone antibiotics and the climbazole fungicide by *Trichoderma* species. Environ Sci Pollut Res:1–11

Maqbool Z, Hussain S, Imran M, Mahmood F, Shahzad T, Ahmed Z et al (2016) Perspectives of using fungi as bioresource for bioremediation of pesticides in the environment: a critical review. Environ Sci Pollut Res 23:16904–16925

Maurya S, Naik SK, Choudhary JS, Kumar S (2019) Heavy metals scavenging potential of *Trichoderma asperellum* and *Hypocrea nigricans* isolated from acid soil of Jharkhand. Indian J Microbiol 59:27–38

Miles SM, Asiedu E, Balaberda AL, Ulrich AC (2020) Oil sands process affected water sourced *Trichoderma harzianum* demonstrates capacity for mycoremediation of naphthenic acid fraction compounds. Chemosphere 127281

Mishra A, Nautiyal CS (2009) Functional diversity of the microbial community in the rhizosphere of chickpea grown in diesel fuel-spiked soil amended with *Trichoderma ressei* using sole-carbon-source utilization profiles. World J Microbiol Biotechnol 25:1175–1180

Mohsenzadeh F, Shahrokhi F (2014) Biological removing of cadmium from contaminated media by fungal biomass of *Trichoderma* species. J Environ Health Sci Eng 12:102

Muthusaravanan S, Sivarajasekar N, Vivek JS, Paramasivan T, Naushad M, Prakashmaran J et al (2018) Phytoremediation of heavy metals: mechanisms, methods and enhancements. Environ Chem Lett 16:1339–1359

Narendran R, Kathiresan K, Sathishkumar RS, Kayalvizhi K, Sundaramanickam A (2019) Bioremoval of toxic substances in synthetic wastewater using *Trichoderma pubescens* (NPK2), isolated from mangrove soil. Biocatal Agric Biotechnol 19:101100

Nazifa TH, Ahmad MA, Hadibarata T, Aris A (2018) Bioremediation of diesel oil spill by filamentous fungus *Trichoderma reesei* H002 in aquatic environment. Int J Integr Eng 10

Nongmaithem N, Roy A, Bhattacharya PM (2016) Screening of *Trichoderma* isolates for their potential of biosorption of nickel and cadmium. Braz J Microbiol 47:305–313

Nor NM, Hadibarata T, Zubir MMFA, Lazim ZM, Adnan LA, Fulazzaky MA (2015) Mechanism of triphenylmethane Cresol Red degradation by *Trichoderma harzianum* M06. Bioprocess Biosyst Eng 38:2167–2175

Nykiel-Szymańska J, Bernat P, Słaba M (2018) Potential of *Trichoderma koningii* to eliminate alachlor in the presence of copper ions. Ecotoxicol Environ Saf 162:1–9

Nykiel-Szymańska J, Bernat P, Słaba M (2020) Biotransformation and detoxification of chloroacetanilide herbicides by *Trichoderma* spp. with plant growth-promoting activities. Pestic Biochem Physiol 163:216–226

Özkara A, Akyıl D, Konuk M (2016) Pesticides, environmental pollution, and health. In: Environmental health risk-hazardous factors to living species. IntechOpen

Parween T, Bhandari P, Sharma R, Jan S, Siddiqui ZH, Patanjali PK (2018) Bioremediation: a sustainable tool to prevent pesticide pollution. In: Modern age environmental problems and their remediation. Springer, Cham, pp 215–227

Pehlivan N, Gedik K, Eltem R, Terzi E (2020) Dynamic interactions of *Trichoderma harzianum* TS 143 from an old mining site in Turkey for potent metal (oid) s phytoextraction and bioenergy crop farming. J Hazard Mater 403:123609

Podbielska M, Kus-Liśkiewicz M, Jagusztyn B, Piechowicz B, Sadło S, Słowik-Borowiec M et al (2020) Influence of *Bacillus subtilis* and *Trichoderma harzianum* on penthiopyrad degradation under laboratory and field studies. Molecules 25:1421

Poveda J (2020) *Trichoderma parareesei* favors the tolerance of rapeseed (*Brassica napus* L.) to salinity and drought due to a chorismate mutase. Agronomy 10:118

Poveda J, Hermosa R, Monte E, Nicolás C (2019a) *Trichoderma harzianum* favours the access of arbuscular mycorrhizal fungi to non-host Brassicaceae roots and increases plant productivity. Sci Rep 9:1–11

Poveda J, Hermosa R, Monte E, Nicolás C (2019b) The *Trichoderma harzianum* Kelch protein ThKEL1 plays a key role in root colonization and the induction of systemic defense in Brassicaceae plants. Front Plant Sci 10:1478

Poveda J, Eugui D, Abril-Urias P (2020a) Could *Trichoderma* be a plant pathogen? Successful root colonization. In: Trichoderma. Springer, Singapore, pp 35–59

Poveda J, Abril-Urias P, Escobar C (2020b) Biological control of plant-parasitic nematodes by filamentous fungi inducers of resistance: *Trichoderma*, mycorrhizal and endophytic fungi. Front Microbiol 11

Prabha R, Singh DP, Verma MK (2017) Microbial interactions and perspectives for bioremediation of pesticides in the soils. In: Plant-microbe interactions in agro-ecological perspectives. Springer, Singapore, pp 649–671

Rai R, Agrawal M, Agrawal SB (2016) Impact of heavy metals on physiological processes of plants: with special reference to photosynthetic system. In: Plant responses to xenobiotics. Springer, Singapore, pp 127–140

Rai PK, Lee SS, Zhang M, Tsang YF, Kim KH (2019) Heavy metals in food crops: health risks, fate, mechanisms, and management. Environ Int 125:365–385

Ranimol G, Venugopal T, Gopalakrishnan S, Sunkar S (2018) Production of laccase from *Trichoderma harzianum* and its application in dye decolourisation. Biocatal Agric Biotechnol 16:400–404

Ray RR, Sur D (2016) Optimization of process parameters for effective bioremediation of chromium contaminated soil by *Trichoderma pseudokoningii*. Russ J Biol Res 8:69–79

Rehman M, Liu L, Wang Q, Saleem MH, Bashir S, Ullah S, Peng D (2019) Copper environmental toxicology, recent advances, and future outlook: a review. Environ Sci Pollut Res:1–14

Rigot J, Matsumura F (2002) assessment of the rhizosphere competency and pentachlorophenol-metabolizing activity of a pesticide-degrading strain of *Trichoderma harzianum* introduced into the root zone of corn seedlings. J Environ Sci Health B 37:201–210

Rosmana A, Sakrabani R, Sjam S, Nasaruddin N, Asman A, Pandin BYS (2019) Plant residue based-composts applied in combination with *Trichoderma asperellum* improve cacao seedling growth in soil derived from nickel mine area. Pak Agric Sci Forum 29:291–298

Saha JK, Selladurai R, Coumar MV, Dotaniya ML, Kundu S, Patra AK (2017) Impacts of soil pollution and their assessment. In: Soil pollution-an emerging threat to agriculture. Springer, Singapore, pp 37–73

Sahu A, Mandal A, Thakur J, Manna MC, Rao AS (2012) Exploring bioaccumulation efficacy of *Trichoderma viride*: an alternative bioremediation of cadmium and lead. Natl Acad Sci Lett 35(4):299–302

Salvadori MR, Ando RA, Oller Do Nascimento CA, Correa B (2014) Bioremediation from wastewater and extracellular synthesis of copper nanoparticles by the fungus *Trichoderma koningiopsis*. J Environ Sci Health A 49:1286–1295

Saranya N, Suganya E, Narayanasamy S, Sivaprakasam S, Sivasubramanian V, Pandian S, Selvaraj R (2020) 3-level Box–Behnkenoptimization of hexavalent chromium reduction by chromate resistant *Trichoderma asperellum* cells from simulated and industrial effluent. Environ Technol Innov 19:101024

Saravanan S, Kumari A, Sriram A (2019) Comparative study on treatment of cassava wastewater using free, immobilized and biofilm of *Trichoderma harzianum-Trichoderma viride* consortium. Water Desalin Treat 156:370–377

Shahzad B, Tanveer M, Rehman A, Cheema SA, Fahad S, Rehman S, Sharma A (2018) Nickel; whether toxic or essential for plants and environment-a review. Plant Physiol Biochem 132:641–651

Shankar S (2017) Management and remediation of problem soils, solid waste and soil pollution. In: Principles and applications of environmental biotechnology for a sustainable future. Springer, Singapore, pp 143–171

Shanmugam S, Ulaganathan P, Sivasubramanian S, Esakkimuthu S, Krishnaswamy S, Subramaniam S (2017a) *Trichoderma asperellum* laccase mediated crystal violet degradation-optimization of experimental conditions and characterization. J Environ Chem Eng 5:222–231

Shanmugam S, Ulaganathan P, Swaminathan K, Sadhasivam S, Wu YR (2017b) Enhanced biodegradation and detoxification of malachite green by *Trichoderma asperellum* laccase: degradation pathway and product analysis. Int Biodeterior Biodegradation 125:258–268

Sharma P, Sharma M, Raja M, Singh DV, Srivastava M (2016) Use of *Trichoderma* spp. in biodegradation of Carbendazim. Indian J Agric Sci 86:891–894

Sharma S, Kour D, Rana KL, Dhiman A, Thakur S, Thakur P et al (2019) *Trichoderma*: biodiversity, ecological significances, and industrial applications. In: Recent advancement in white biotechnology through fungi. Springer, Cham, pp 85–120

Shashikumar HM, Koulagi S, Navyashree SE (2019) Compatibility of *Trichoderma viride* and *Trichoderma harzianum* with fungicides against soil borne diseases of tomato and cabbage. Int J Curr Microbiol App Sci 8:1920–1928

Shukla D, Vankar PS (2014) Role of *Trichoderma* species in bioremediation process: biosorption studies on hexavalent chromium. In: Biotechnology and biology of trichoderma. Elsevier, pp 405–412

Siddiquee S, Aishah SN, Azad SA, Shafawati SN, Naher L (2013) Tolerance and biosorption capacity of Zn^{2+}, Pb^{2+}, Ni^{3+} and Cu^{2+} by filamentous fungi (*Trichoderma harzianum*, *T. aureoviride* and *T. virens*). Adv Biosci Biotechnol 4:570–583

Sing NN, Zulkharnain A, Roslan HA, Assim Z, Husaini A (2014) Bioremediation of PCP by *Trichoderma* and *Cunninghamella* strains isolated from sawdust. Braz Arch Biol Technol 57:811–820

Singh B, Singh K (2016) Microbial degradation of herbicides. Crit Rev Microbiol 42:245–261

Singh RK, Tripathi R, Ranjan A, Srivastava AK (2020) Fungi as potential candidates for bioremediation. In Abatement of environmental pollutants (pp. 177–191). Elsevier

Smily JRMB, Sumithra PA (2017) Optimization of chromium biosorption by fungal adsorbent, *Trichoderma* sp. BSCR02 and its desorption studies. HAYATI J Biosci 24:65–71

Solanki MK, Kashyap BK, Solanki AC, Malviya MK, Surapathrudu K (2019) Helpful linkages of *Trichodermas* in the process of mycoremediation and mycorestoration. In: Plant health under biotic stress. Springer, Singapore, pp 51–64

Spina F, Cecchi G, Landinez-Torres A, Pecoraro L, Russo F, Wu B et al (2018) Fungi as a toolbox for sustainable bioremediation of pesticides in soil and water. Plant Biosyst 152:474–488

Su S, Zeng X, Bai L, Jiang X, Li L (2010) Bioaccumulation and biovolatilisation of pentavalent arsenic by *Penicillin janthinellum*, *Fusarium oxysporum* and *Trichoderma asperellum* under laboratory conditions. Curr Microbiol 61:261–266

Su S, Zeng X, Bai L, Li L, Duan R (2011) Arsenic biotransformation by arsenic-resistant fungi *Trichoderma asperellum* SM-12F1, *Penicillium janthinellum* SM-12F4, and *Fusarium oxysporum* CZ-8F1. Sci Total Environ 409:5057–5062

Su S, Zeng X, Bai L, Williams PN, Wang Y, Zhang L, Wu C (2017) Inoculating chlamydospores of *Trichoderma asperellum* SM-12F1 changes arsenic availability and enzyme activity in soils and improves water spinach growth. Chemosphere 175:497–504

Sun S, Sidhu V, Rong Y, Zheng Y (2018) Pesticide pollution in agricultural soils and sustainable remediation methods: a review. Curr Pollut Rep *4*:240–250

Sun J, Yuan X, Li Y, Wang X, Chen J (2019) The pathway of 2, 2-dichlorovinyl dimethyl phosphate (DDVP) degradation by *Trichoderma atroviride* strain T23 and characterization of a paraoxonase-like enzyme. Appl Microbiol Biotechnol 103:8947–8962

Sun H, Meng M, Wu L, Zheng X, Zhu Z, Dai S (2020) Function and mechanism of polysaccharide on enhancing tolerance of *Trichoderma asperellum* under Pb^{2+} stress. Int J Biol Macromol 151:509–518

Szczepaniak Z, Cyplik P, Juzwa W, Czarny J, Staninska J, Piotrowska-Cyplik A (2015) Antibacterial effect of the *Trichoderma viride* fungi on soil microbiome during PAH's biodegradation. Int Biodeterior Biodegradation 104:170–177

Tang J, Liu L, Hu S, Chen Y, Chen J (2009) Improved degradation of organophosphate dichlorvos by *Trichoderma atroviride* transformants generated by restriction enzyme-mediated integration (REMI). Bioresour Technol 100:480–483

Tansengco M, Tejano J, Coronado F, Gacho C, Barcelo J (2018) Heavy metal tolerance and removal capacity of *Trichoderma* species isolated from mine tailings in Itogon, Benguet. Environ Nat Resour J 16:39–57

Ting ASY, Choong CC (2009) Bioaccumulation and biosorption efficacy of *Trichoderma* isolate SP2F1 in removing copper (Cu (II)) from aqueous solutions. World J Microbiol Biotechnol 25:1431–1437

Tkavc R, Matrosova VY, Grichenko OE, Gostinčar C, Volpe RP, Klimenkova P et al (2018) Prospects for fungal bioremediation of acidic radioactive waste sites: characterization and genome sequence of *Rhodotorula taiwanensis* MD1149. Front Microbiol 8:2528

Tripathi P, Singh PC, Mishra A, Chauhan PS, Dwivedi S, Bais RT, Tripathi RD (2013) *Trichoderma*: a potential bioremediator for environmental clean up. Clean Techn Environ Policy 15:541–550

Tripathi P, Singh PC, Mishra A, Srivastava S, Chauhan R, Awasthi S et al (2017) Arsenic tolerant *Trichoderma* sp. reduces arsenic induced stress in chickpea (*Cicer arietinum*). Environ Pollut 223:137–145

Ul Hassan Z, Ali S, Rizwan M, Ibrahim M, Nafees M, Waseem M (2017) Role of bioremediation agents (bacteria, fungi, and algae) in alleviating heavy metal toxicity. In: Probiotics in agroecosystem. Springer, Singapore, pp 517–537

Vacondio B, Birolli WG, Ferreira IM, Seleghim MH, Gonçalves S, Vasconcellos SP, Porto AL (2015) Biodegradation of pentachlorophenol by marine-derived fungus *Trichoderma harzianum* CBMAI 1677 isolated from ascidian *Didemnun ligulum*. Biocatal Agric Biotechnol 4:266–275

Vankar PS, Bajpai D (2008) Phyto-remediation of chrome-VI of tannery effluent by *Trichoderma* species. Desalination 222:255–262

Vardhan KH, Kumar PS, Panda RC (2019) A review on heavy metal pollution, toxicity and remedial measures: current trends and future perspectives. J Mol Liq 290:111197

Vareda JP, Valente AJ, Durães L (2019) Assessment of heavy metal pollution from anthropogenic activities and remediation strategies: a review. J Environ Manag 246:101–118

Vargas JT, Rodríguez-Monroy M, Meyer ML, Montes-Belmont R, Sepúlveda-Jiménez G (2017) *Trichoderma asperellum* ameliorates phytotoxic effects of copper in onion (*Allium cepa* L.). Environ Exp Bot 136:85–93

Venice F, Davolos D, Spina F, Poli A, Prigione VP, Varese GC, Ghignone S (2020) Genome sequence of *Trichoderma lixii* MUT3171, a promising strain for mycoremediation of PAH-contaminated sites. Microorganisms 8:1258

Verbrugge LA, Kahn L, Morton JM (2018) Pentachlorophenol, polychlorinated dibenzo-p-dioxins and polychlorinated dibenzo furans in surface soil surrounding pentachlorophenol-treated utility poles on the Kenai National Wildlife Refuge, Alaska USA. Environ Sci Pollut Res 25:19187–19195

Wang B, Liu L, Gao Y, Chen J (2009) Improved phytoremediation of oilseed rape (*Brassica napus*) by *Trichoderma* mutant constructed by restriction enzyme-mediated integration (REMI) in cadmium polluted soil. Chemosphere 74:1400–1403

Widmer, T. L. (2019). Compatibility of Trichoderma asperellum isolates to selected soil fungicides. Crop Protection, 120:91–96

Xu J, Smith S, Smith G, Wang W, Li Y (2019) Glyphosate contamination in grains and foods: An overview. Food Control 106:106710

Yadav U, Choudhury PP (2014) Biodegradation of sulfosulphuron in agricultural soil by *Trichoderma* sp. Lett Appl Microbiol 59:479–486

Yaghoubian Y, Siadat SA, Telavat MRM, Pirdashti H, Yaghoubian I (2019) Bio-removal of cadmium from aqueous solutions by filamentous fungi: *Trichoderma* spp. and *Piriformospora indica*. Environ Sci Pollut Res 26:7863–7872

Yao L, Teng Y, Luo Y, Christie P, Ma W, Liu F et al (2015) Biodegradation of polycyclic aromatic hydrocarbons (PAHs) by *Trichoderma reesei* FS10-C and effect of bioaugmentation on an aged PAH-contaminated soil. Bioremediation J 19:9–17

Yazdani M, Yap CK, Abdullah F, Tan SG (2009) *Trichoderma atroviride* as a bioremediator of Cu pollution: an *in vitro* study. Toxicol Environ Chem 91:1305–1314

Yazdani M, Yap CK, Abdullah F, Tan SG (2010) An *in vitro* study on the adsorption, absorption and uptake capacity of Zn by the bioremediator *Trichoderma atroviride*. Environ Asia 3:53–59

Yin K, Wang Q, Lv M, Chen L (2019) Microorganism remediation strategies towards heavy metals. Chem Eng J 360:1553–1563

Zafra G, Moreno-Montaño A, Absalón ÁE, Cortés-Espinosa DV (2015) Degradation of polycyclic aromatic hydrocarbons in soil by a tolerant strain of *Trichoderma asperellum*. Environ Sci Pollut Res 22:1034–1042

Zhang F, Huo Y, Cobb AB, Luo G, Zhou J, Yang G et al (2018a) *Trichoderma* biofertilizer links to altered soil chemistry, altered microbial communities, and improved grassland biomass. Front Microbiol 9:848

Zhang X, Li X, Yang H, Cui Z (2018b) Biochemical mechanism of phytoremediation process of lead and cadmium pollution with *Mucor circinelloides* and *Trichoderma asperellum*. Ecotoxicol Environ Saf 157:21–28

Zhang D, Yin C, Abbas N, Mao Z, Zhang Y (2020) Multiple heavy metal tolerance and removal by an earthworm gut fungus *Trichoderma brevicompactum* QYCD-6. Sci Rep 10:1–9

Zhiguo L, Luyang H, Yutong L, Qing L, Airong D (2018) Effects of *Trichoderma brevicompactum*, soybean, and imazethapyr on soil fungal diversity and pesticide residues. J Northeast Forest Univ 3:19

Zhou X, Xu S, Liu L, Chen J (2007) Degradation of cyanide by *Trichoderma* mutants constructed by restriction enzyme mediated integration (REMI). Bioresour Technol 98:2958–2962

Trichoderma and Its Products From Laboratory to Patient Bedside in Medical Science: An Emerging Aspect

Swapan Kumar Ghosh

Contents

1 Introduction

Trichoderma Pers. Fr. genus belongs to the family *Hypocreaceae*, order *Hypocreales*, and phylum *Ascomycota*. Although we have acquired huge knowledge of this genus, the taxonomy of *Trichoderma* is still rather incomplete, and the distinction of species in the genus *Trichoderma* remains problematic. The taxonomy and identification of *Trichoderma* were originally introduced by Christiaan Hendri Persoon in

S. K. Ghosh (✉)
Molecular Mycopathology Lab., Biocontrol and Cancer Research Unit, Ramakrishna Mission Vivekananda Centenary College (Autonomous), Kolkata, West Bengal, India

N. Amaresan et al. (eds.), *Advances in* Trichoderma *Biology for Agricultural Applications*, Fungal Biology, https://doi.org/10.1007/978-3-030-91650-3_19

1794 (Römer's Neues Mag. Bot. 1:92. 1794), but there have still been recent nomenclatural problems. As per Mukherjee et al. (2013), more than 200 well-defined species of *Trichoderma* exist, but phenotypic and phylogenetic analysis gave us information of about 260 species, which have been recognized and accepted (Qin and Zhuang 2016; Sun et al. 2016; Jaklitsch and Voglmayr 2015; Bissett et al. 2015). More recently, more than 300 species of the genus *Trichoderma* have been described (Zhang and Zhuang 2018; Bissett et al. 2015). According to Cai and Druzhinina (2021), validated spp. of *Trichoderma* are 375 species. The members of this genus produce bioactive compounds of clinical significance and enzymes with widespread industrial application (Mukherjee et al. 2013). Its role in ecology is well worth, that is, to decompose of plant and animal residues in the soil. Some *Trichoderma* species produce large amount of cellulase enzymes (Juhász et al. 2003). Several species of this genus have excellent antagonistic properties against plant pathogenic fungi (Sivan and Chet 1986; Naár and Kecskés 1995; Naseby et al. 2000). Therefore, they are frequently being applied in agrifields for the biological control of several plant diseases (Papavizas 1985; Ghosh et al. 2018). The discovery of gliotoxin in the early 1930s (Weindling 1934) encouraged scientists to search bioactive compounds or secondary metabolites within *Trichoderma*, and over the years research revealed that more than 1000 compounds have been estimated to be produced by this genus (Hermosa et al. 2012). Recently, a new species of *Trichoderma* – *T. hypoxylon* – has been reported to harbor enormous numbers of secondary metabolites (Sun et al. 2016). *Trichoderma*-derived secondary metabolites comprise of non-ribosomal peptides (NRPs) such as antibiotic peptides, known as peptaibols (peptaibiotics), siderophores and diketopiperazines-like gliotoxin and gliovirin, polyketides, terpenes, pyrones, and isocyanate metabolites; enzymes; fatty acids; etc. (Daniel et al. 2007; Reino et al. 2008; Zeilinger et al. 2016). Several secondary metabolites produced by *Trichoderma* species have been reported to have enormous pharmaceutical values such as antibacterial (Cheng et al. 2011; De Zotti et al. 2009), antiviral (Lu et al. 2002), antiprotozoal (Ciscotto et al. 2009), antifungal (Ande et al. 2008), anticancer activities (Shi et al. 2010; Liu et al. 2009), etc. So, it may be called as miniature of pharmaceutical factor.

Now due to fast growth of human population and climatic changes, humans are facing some fatal diseases like cancer, ailment of kidney, liver, diabetes, pandemic viral disease (COVID-19), etc. For the prevention and cure from them, scientists are tirelessly searching new natural compounds from different bio-sources, like plants, bacteria, fungi, animals, etc. As members of *Trichoderma* harbor hundreds of bioactives compounds, their proper detection, isolation, purification, and proper application in medical science are burning topics for research for human welfare. In this chapter, how much *Trichoderma* has drawn attention of scientists to apply them in medical science for managing acute or fatal diseases has been narrated.

2 Some Important Compounds Originated from *Trichoderma* and Their Application in Medical Sciences

The metabolites extracted and isolated from *Trichoderma* spp. include non-ribosomal peptides (NRPs) such as antibiotic peptides, known as peptaibols (peptaibiotics), siderophores and diketopiperazines-like gliotoxin and gliovirin, polyketides, terpenes, pyrones, and isocyanate metabolites, enzymes, fatty acids, etc. (Daniel et al. 2007; Reino et al. 2007; Zeilinger et al. 2016).

2.1 *Peptaibiotics or Peptaibols*

The discovery of peptide antibiotics (peptaibiotics) produced by fungi has attracted attention, because they are efficient weapon against pathogens. On the basis of chemical structures, peptaibiotics are generally categorized into different kinds of peptide such as the following: (a) peptaibols which bears at least an acylated N-terminus and an amide-bound amino alcohol at the C-terminus; (b) lipopeptaibols, where one or more fatty acid residues are incorporated in peptaibols; (c) lipoaminopeptides, also known as aminolipopeptides, which are Aib-containing peptides having one or more lipoaminoacid residues; (d) cyclic peptaibiotics, where the first and the last residues are bound to form an arch-shaped structure; (e) other peptaibiotics, including Aib-containing peptides, which do not fit into any of the other categories described above; and (f) the so-called all-Aib-replaced, a mixture of short-sequence peptaibols, in which the single marker Aib has been replaced by one of the standard amino acids (Neumann et al. 2015).

The name "peptaibol" is derived from *pep*tide, *Aib* (α-*a*mino *i*so***b***utyrate acid or α-methyl alanine), and amino alcoh*ol*, referring to these main features. Peptaibols are included in a class of compounds called peptaibiotics. They are defined as peptides derived from fungal secondary metabolism, consisting of approximately 4–21 amino acid residues. The 83% of all described peptaibiotics originated from *Trichoderma* (n = 738) or *Hypocrea* spp. (n = 92), the former name of *Trichoderma* teleomorph, now synonym. Among them, *T. viride, T. harzianum, T. virens, T. brevicompactum,* and *T. parceramosum* (name is not in use) are highly studied species (Stoppacher et al. 2013). One of the main features of peptaibols is the presence of non-proteinogenic amino acids, such as α-aminoisobutyrate acid (Aib) or IVA (isovaleric acid or α-ethyl alanine or ethylnorvaline (EtNor)), and many have a number of amino acids, either proline (Pro) or hydroxyproline (Hyp) (Stoppacher et al. 2013). Aib residue facilitates for the formation of helical structures due to the steric constraints imposed by the second methyl group on the Cα atom, while the amino acids promote the formation of bends or kinks in these structures (Chugh and Wallace 2001). At the N-terminus, there are modifications such as acyl or acetyl groups, and at the C-terminus there is the presence of an amino alcohol such phenylalaninol (Phe-OH), prolinol (Pro-OH), or valinol (Val-OH) (Daniel et al. 2007). They create voltage-dependent ion channels within the pathogen membrane (Milov et al. 2016). The first compound of this group was alamethicin, which was isolated from *T. viride* (Brewer et al. 1987; Meyer and Reusser 1967). The peptaibols can be divided into three groups on the basis of the chain lengths of the amino acid sequences: the long-sequence peptaibols with 18–20 amino acid residues, exemplified by the alamethicin or the suzukacillins from *T. viride* (Katz et al. 1985) or trichokonins from *T. koningii* (Huang et al. 1995) (20 amino acid residues) or trichorzianins *T. harzianum* (19 amino acid residues) or trichotoxins from *T. viride* (Brückner et al. 1985) (18 amino acid residues); the short-sequence peptaibols with 11–17 residues, exemplified by the harzianins (Lucaciu et al. 1997), the harzianins HA (14 residues) from *T. harzianum* (Rebuffat et al. 1995), the trikoningins KB (11 residues) from *T. koningii* (Auvin-Guette et al. 1993), and the trichorozins from *T. harzianum* (Iida et al. 1995); and the lipopeptaibols with 6 or 10 residues (Peggion et al. 2001) where the N-terminal amino acid is acylated by a short-chain fatty acid, e.g., trichogin A IV from *T. longibrachiatum* (Auvin-Guette et al. 1992) (11 residues). The species and strains of the genus *Trichoderma* are capable of producing all three groups of peptides, as exemplified by the trichoaureocins, isolated from *T. aureoviride*. Table 1 shows some examples of the peptides and peptaibols with their amino acid residues isolated from *Trichoderma* species. Several peptaibols from *Trichoderma* have already been loaded on "The Comprehensive Peptaibiotics Database," including information such as amino acid sequences, molecular formulae, monoisotopic masses, and groupings for 1043 peptaibiotics (Stoppacher et al. 2013; http://peptaibiotics-database.boku.ac.at). A graphical representation of peptides in peptide database up to 2014 has been displayed in Fig. 1a. The Comprehensive Peptaibiotics Database consists of 1297 peptaibiotic sequences. Up to June, 2014 (Neumann et al. 2015), the lipopeptide antibiotics (linear or cyclic) isolated from different fungi since 2000 from different sources have been reviewed by Zhao et al. (2019a, b). Fungi-derived lipopeptide antibiotics can be classified into four distinct categories: cyclic depsipeptides, peptaibiotics (e.g., peptaibols, lipoaminopeptides, and lipopeptaibols), non-depsipeptide cyclic lipopeptides (e.g., acetyl and anthranilic acid peptide derivatives, echinocandins, and aspochracins), and non-peptaibiotic linear lipopeptides (Zhao et al. 2019a, b). In cyclic depsipeptides, amide groups are replaced by corresponding lactone bonds due to the presence of a hydroxylated carboxylic acid or amino acid with a hydroxyl group in the core ring. Non-peptaibol peptaibiotics (lipoaminopeptides) like trichoderins A, B, and A1 and trichopolyn VI are isolated from *Trichoderma* sp. and *T. brevicompactum* (Zhao et al. 2019a, b; Suga et al. 2015). It is interesting to note the unique trichopolyn group from *T. polysporum* (Fig. 1b) as they have an *R*-2-methyldecanoyl group esterifying the N-terminal amino acid, a 2-amino-6-hydroxy-4-methyl-8-oxodecanoic acid residue at position 2, and the unusual C-terminal group (Fujita et al. 1981; Mihara et al. 1994). Zhao et al. (2019a, b) also compiled information on non-lipopeptide peptide antibiotics including 2,5-diketopiperazines (DKPs), as well as typical peptides and their analogs, originated from different fungi and different sources since 2000 in another review paper.

Table 1 A list of some peptaibols isolated from different species of *Trichoderma*

Trichoderma sp.	Peptaibol (amino acid number)	Reference
T. viride	Alamethicin (20)	Brewer et al. (1987) and Meyer and Reusser (1967)
	Trichotoxin A40 (acidic) (18)	Bruckner et al. (1985)
	Trichotoxin A50 (Neutral) (18)	Jaworski and Brückner (1999)
	Suzukacillin A (20)	Katz et al. (1985) and Krause et al. (2006)
	Trichovirins II (14)	Jaworski et al. (1999)
	Trichodecenins I, II (lipopeptaibols) (7)	Fujita et al. (1994)
T. polysporum	Polysporins A–D (20)	New et al. (1996)
	Trichosporins B–V (20)	Iida et al. (1993)
T. reesei	Paracelsin (20)	Bruckner et al. (1984)
T. saturnisporum	Paracelsin E (20)	Ritieni et al. (1995)
	Saturnisporins SA II, SA IV (20)	Rebuffat et al. (1993)
T. koningii	Trichokonins V–VIII (20)	Huang et al. (1995)
	Trikoningins KA, KB (19)	Auvin-Guette et al. (1993)
	Gliodeliquescins (20)	Huang et al. (1995)
T. longibrachiatum	Longibrachins A-II-b (neutral), B-II, and B-III (acidic) (20)	Mohamed-Benkada et al. (2006)
	Tricholongins B I, B II (19)	Rebuffat et al. (1991)
	Longibrachins LGBII, LGBIII (20)	Leclerc et al. (1998, 2001)
	Trichogin A IV (lip peptaibols) (11)	Auvin-Guette et al. (1992)
T. orientale	Hyporientalin A (20)	Touati et al. (2018)
T. harzianum	Trichorzianines A, B El (19)	Rebuffat et al. (1989) and Bodo et al. (1985)
	Trichokindins I–VII (18)	Iida et al. (1994)
	Harzianins HA (14)	Rebuffat et al. (1995) and Lucaciu et al. (1997)
	Trichorozins I–IV (11)	Iida et al. 1995
	Trichorzins (18)	Hlimi et al. (1995)
	Trichotoxin A50 (neutral)	Suwan et al. 2000
T. atroviride	Atroviridins A–C (20)	Oh et al. 2000
	Neoatroviridins A –D (18–20)	Oh et al. (2000)
	TA-17A-Ix to -IVx (17)	Carrouxa et al. (2013)
T. asperellum	Trichotoxin A-50 (18)	Chutrakul et al. (2008)
	Trichotoxin A-50 E (18) Trichotoxin A-50 F (18) Trichotoxin A-50 I (18) Trichotoxin A-50 J (18)	Stoppacher et al. (2013) and Tamandegani et al. (2020)
	Asperelines A–F (10)	Ren et al. (2009)
	Asperelines G and Hc (10)	Chen et al. (2013)
T. aureoviride	Trichoaureocins	Bruckner et al. (2002)
Trichoderma sp.	Trichofumins A–B (11), C–D (13)	Berg et al. (2003)

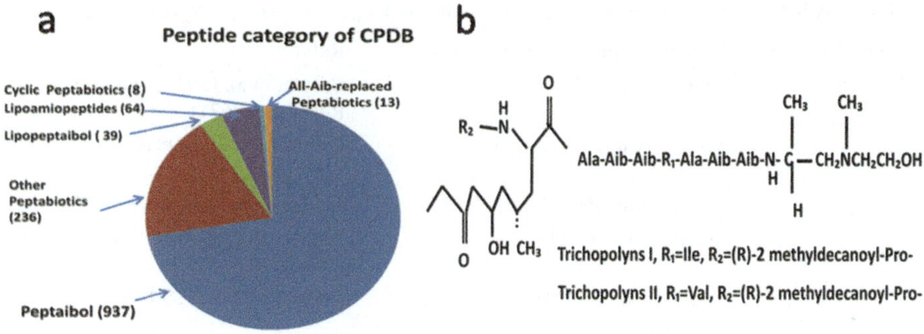

Fig.1 (**a**) Graphical presentation of the distribution of peptides including peptaibols in peptide database (Stoppacher et al. 2013). (Figure drawn on the basis of Stoppacher et al. 2013; Neumann et al. 2015 (total peptibiotics, 1297)). (**b**) Structure of trichopolyns I and II

Why Are Peptaibols Called as Non-ribosomal Peptides?
This is because these compounds do not result from gene transcription and subsequent translation but are formed from multienzyme complex called non-ribosomal peptide synthetases (NRPSs).

What Are the Steps of Synthesis?
In summary, these multienzyme complexes are formed by a set of modules, where each module has catalytic domains responsible for the synthesis steps (Schwarzer et al. 2003). In general, there are three main steps, the 1st step begins in domain A (adenylation), where the biosynthetic process begins with amino acid entry, which is activated by adenylation. In the 2nd step, the activated amino acid is attached to a PCP protein cofactor (HS-4′PP) that acts as a carrier between the catalytic centers, and the 3rd or the final step, in domain C (condensation), peptide bond formation occurs (Mootz et al. 2002; Schwarzer et al. 2003).

Molecular Mechanism of Action of Peptaibols
The mechanism of biological activity of peptaibols lies on the fact that they form voltage-based ion channels in the membrane of pathogens due to amphipathic nature (Chugh and Wallace 2001; Iida et al. 1995). Due to amphiphilic nature, peptaibols show “detergent-like” properties. It has been well-established that they perturb the permeability properties of phospholipid bilayers. The structural peculiarity of peptaibols confers an amphiphilic helix-favoring character that directs these peptides to assimilate in pathogen membrane and creates ion channels and permeabilizes the membrane (Iida et al. 1995) and consequently inducing cell death by cytoplasmic leakage (Chug et al. 2001). So, they are active against many dangerous pathogens, including cancer cells. The presence of rare amino acid α-aminoisobutyric acid (Aib) in their sequence confers α-helix structure empowering its bioactivity as well as resistance to the host or pathogen proteases (Ramachander Turaga 2020). The 20-residue alamethicin (Alm), which has been isolated from *Trichoderma viride*, has been intensively researched as a model molecule to study membrane channel behavior in defined lipid environments and also with regard to its antibiotic

effect on different pathogenic microorganisms. This compound creates multi-conductance channels in a voltage-dependent manner on inserting into membranes by combining to create barrel-shaped pores. To understand the multi-conductance type of Alm channels, the "barrel stave" model has been proposed (Fox and Richards 1982; Mathew and Balaram 1983b; Boheim et al. 1983) (Fig. 2), and here monomers of alamethicin form a helix bundle surrounding a central pore. Alamethicin forms mostly α-helical structure. Alamethicin molecules are 34 Ǎ long and sufficient to span lipid bilayers. The side chains of Gln-7, Glu-18, and Gln-19 are all situated on the same face of the helix and are seemed to create part of the lumen of the channel. Proline situating at 14 position (Pro-14) plays the major role for the insertion of alamethicin into the membrane, because it creates a bend point between two helical segments (Fox and Richards 1982). As helices have overall dipole moments along the direction of their helical axes, it was confirmed that the N-terminal helix must enter first into the membrane, leaving Pro-14 on the bilayer edge with the C-terminal helix lying along the membrane surface. Upon the application of a voltage, the C-terminus would reorient itself and insert fully into the membrane; then, a number of such helices would combine to form the channel (Chugh and Wallace 2001) (Fig. 2a). Several molecular models have been proposed to explain for voltage-dependent alamethicin pore formation (Sansom 1991; Latorre and Alvarez 1981). The electric field forces the rotation of the monomers from the membrane surface into its interior (Bauman and Mueller 1974; Schwarz et al. 1986; Rizzo et al. 1987). On the other hand, the second group of scientists suggested molecular model that seems the aggregation of peptide monomers at the membrane surface (Hall 1975; Boheim and Kolb 1978). The voltage pushes two or three monomers into the membrane simultaneously from a hexamer already preformed at the membrane solution interface (Boheim and Kolb 1978). The 3rd molecular model to explain the voltage-dependency of alamethicin pore formation is a flip-flop gating

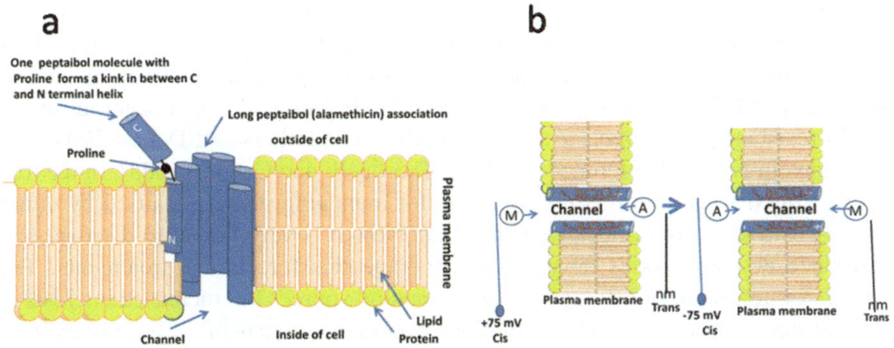

Fig. 2 Mechanism of channel formation in membrane by peptaibol. (**a**) "Barrel stave" model. (Drawn on the basis of information from Fox and Richards 1982; Chugh and Wallace 2001; Mathew and Balaram 1983b) showing long peptaibols associate each other to form a channel in the membrane. (**b**) Voltage dependent, a flip-flop gating mechanism. (Drawn on the basis of information of Beheim et al. 1983)

mechanism of single alamethicin molecules (Beheim et al. 1983) (Fig. 2b). This molecular model seems the presence of aggregates of antiparallel dipolar molecules situated perpendicularly to the membrane plane and connecting the hydrophobic layer at zero voltage. The energetically preferred aggregate structure is changed from antiparallel to parallel molecule direction by membrane voltage application. The application of an electric field forces one or more molecules into a parallel direction, leading to electrostatic repulsion and to the formation of water-filled pores. Each alamethicin channel is composed of between 6 and 12 monomers, but octamer is the most stable conducting forms (Fox and Richards I982). On the other hand, trichotoxin-A50E, an 18-residue peptaibol, induces a single-channel conductance (Duclohier et al. 2004). The peptide-membrane interactions are most likely dependent on membrane properties such as charge and lipid composition Therefore, unlike other antibiotics, the modes of all peptaibols are more or less similar to the arresting resistance created by the pathogen.

2.1.1 Peptaibols as Antimicrobial Compounds

In the golden era of antibiotics, non-ribosomal antibacterial peptides (NRAPs), such as penicillin, are very important as antibiotics in the clinical uses (Fleming 1929; Awan et al. 2017). In addition, vancomycin and colistin (Corona and Cattaneo 2017) are the last option to apply against gram-positive and Gram-negative pathogens, respectively. NRAPs are produced by non-ribosomal peptide synthetases (NRPSs) (Strieker et al. 2010). The quick appearance and spread of antibacterial resistance is global problem in medical science (Kupferschmidt 2016). Subsequently, the Centers for Disease Control and Prevention (CDCP) reported that greater than 2 million people are facing problems from antibiotic-resistant microbes and at least 23 000 people cannot survive from this problem per year in the USA alone (McKenna 2013). It has been recognized that the occurrence and distribution of multidrug-resistant (MDR) (plasmid-mediated resistance) Gram-negative bacteria (*Enterobacteriaceae*) to carbapenems (Walsh et al. 2011) and colistin (Liu et al. 2016; Wang et al. 2017), is jeopardizing healthcare practices globally. A similar threatening has been coming from Gram-positive bacteria, like the dangerous methicillin-resistant *Staphylococcus aureus* (MRSA) (Chambers and DeLeo 2009) and vancomycin-resistant enterococci (VRE) (Tacconelli and Cataldo 2008). In the true sense, effective antibiotic is not available for managing diseases caused by either Gram-positive or Gram-negative superbugs. Now it is very urgent to discover and introduce new antibiotics or alternative therapeutics for clinical treatments. The search for new antibiotic compounds from natural sources has increased interests in the last three decades. The production of membrane-active peptide antibiotics from different fungi has been recognized as intensive research. One of the first investigations of natural compounds isolated from *Trichoderma viride* exhibited that a new antibiotic (designated U-21963) was very effective against many pathogenic bacteria and fungi (Pyke and Dietz 1966). Moreover, several amphipathic peptides isolated from some *Trichoderma* spp. showed activity against human pathogenic

Mollicutes species. *Mollicutes* species were reported to be more or less sensitive to natural amphipathic peptides; it was dependent on both the species and the peptide. Peptaibols from *Trichoderma* have been reported to have potentiality to kill mycoplasma (Beven et al. 1998) and *Staphylococcus aureus* (Rebuffat et al. 1995). Both alamethicin and gramicidin S are very active against mycoplasmas (cell wall less mollicutes bacteria) (Nir-Paz et al. 2002). Trichorzins exhibited potent activities against *Mycoplasma* and *Spiroplasma* cells and also against *Staphylococcus aureus*. Different peptaibols have inhibitory effects on a wide range of bacteria: *Bacillus subtilis, B. cereus, Brucella bronchiseptica, Micrococcus luteus, Mycobacterium phlei, Staphylococcus aureus, Streptococcus faecalis, S. lactis,* and *S. thermophilus* (Szkeres et al. 2005). The synthesized and also hybrid peptides are nowadays screened for their antimicrobial specificity against pathogenic microorganisms.

A list of 35 novel peptaibols since 2000 from different fungi, including *Trichoderma*, has been presented (Zhao et al. 2019a, b). *T. atroviride* yields atroviridins A–C and neoatroviridins A–D, which are 20- and 18-mer peptaibols, inhibit the growth of Gram-positive bacteria (Oh et al. 2000, 2002). Atroviridins exhibited membrane-perturbing property, and their functions are dependent on their structural features (e.g., Aib and Iva adopt a helical conformation), and its action is like the neutral peptaibol alamethicin (Oh et al. 2002). Longibrachins A-II-b produced from *T. longibrachiatum* have activities against Gram-positive bacteria and fungus like *Aspergillus fumigatus* (Mohamed-Benkada et al. 2016). Both longibrachins B-II and B-III were applied against a lot of mycoplasmas, namely, *Spiroplasma apis, S. citri, S. floricola, Acholeplasma laidlawii, Mycoplasma gallisepticum, M. mycoides,* etc., and both showed effectiveness to kill these pathogens (Leclerc et al. 2001). In addition, scientists have noticed that longibrachins have the capacity to form membrane channels as they bear the Glu residue at the C-terminus of the peptide helix, which plays as an anchor at the cis-bilayer/water interface (Touati et al. 1999). Trichoderins A, A1, and B, and trichopolyn VI (non-peptaibol peptaibiotics) have a similar chemical structure characterized as protective 2-Me-decanoicacid FAs at the N-terminus, trichodiaminol at the C-terminus, and 2-amino-6-Hy-4-Me-8-oxodecanoic acid (AHMOD)/2-amino-4-Me-8-oxodeca-6- enoic acid in the side chain (Pruksakorn et al. 2010, 2011; Suga et al. 2015). Pruksakorn et al. (2011) also reported that trichoderins have potent antimycobacterial activity against *Mycobacterium smegmatis*, *M. bovis*, and *M. tuberculosis* under both aerobic condition and dormancy-inducing hypoxic condition. Interesting to note that the presence of AHMOD moiety in the structure of trichoderins displayed better antimycobacterial activity. MICs of trichoderins A against those three mycobacteria were 0.1, 0.02, and 0.12 μg/mL, respectively. Currently, very few compounds are available for antifungal therapies against a diverse array of pathogenic fungi, some of which are also sensitive to antibiotic peptides. Trichopolyns A and B are produced by *Trichoderma polysporum*. The MICs of trichopolyns A and B for *Candida albicans*, *C. utilis*, *Cryptococcus neoformans*, *Aspergillus fumigatus*, *A. niger*, *Penicillium chrysogenum*, and *T. mentagrophytes* were from 0.78 to 6.25 μg/ml (Fujita et al. 1881, 1981). Trichofumins A, B, C, and D and new 11 and 13mer peptaibols were isolated from *Trichoderma* sp. HKI 0276 and characterized and all showed antifungal activity (Berg et al. 2003). Recently, Touati et al. (2018) isolated hyporientalin A from a marine *Trichoderma orientale* and found that it was an anti-*Candida* peptaibol. So, hyporientalin A may be applied against candidiasis disease, which is resistant to other drugs.

2.1.2 Peptaibols as Anticancer Compounds

Although peptaibols exhibit antibiotic activities against bacteria and fungi, some studies recently showed that peptaibols have cytotoxicity toward some human cancer cell lines like lung epithelial and breast carcinoma cells. However, the mechanism behind peptaibol-induced cell death is still not clear. But there are few reports about the effects of peptaibols on human cancer cells (Wiest et al. 2002). Peltola et al. (2004) reported that the peptaibols from *T. harzianum* suppressed the growth of A549 cells and disturbed the mitochondrial membrane potential. Generally, three trichokonins (TKs), like TK-VI, TK-VII, and TK-VIII, have been recorded from *Trichoderma* spp. Trichokonin VI (TK-VI), which is composed of 20 amino acids, was isolated and characterized from *Trichoderma pseudokoningii* (this species is not found in the molecular database website). As we know that hepatocellular carcinoma (HCC) is a very common cancer in the world and is becoming highly resistant to currently available chemotherapeutic agents, Shi et al. (2010) tested this compound against this cancer cell line and got a satisfactory result in a dose-dependent manner. They revealed that TK-VI triggered two types of cell death, like calcium-calpain-Bax-mediated apoptosis and calcium-Bak-mediated autophagy in HepG2 cells. The trilongins BI, BII, BIII, and BIV, which are peptaibols containing 20 amino acid residues, were isolated and identified from *Trichoderma* sp. P8BDA1F1, an endophytic fungus from *Begonia venosa*. The setrilongins inhibited proteasome ChTL activity, with IC50 values of 6.5 ± 2.7, 4.7 ± 1.8, 6.3 ± 2.2, and 2.7 ± 0.5 μM, respectively. It was the first report of trilongins BI-BIV with proteasome target (Grigoletto Diana et al. 2020).

2.2 *Fatty Acids*

Two intracellular fungal metabolites, such as 16-methylheptadecanoic acid methyl ester (HDA) and 9,12-octadecadienoic acid (ODA), were isolated and purified from *Hypocrea lixii* TSK8 and *Hypocrea rufa* SKS2 (marine *Trichoderma*), respectively, and they were tested on oral cancer (KB) and skin carcinoma (A431) by using MTT assay. The inhibitory concentrations (IC_{50}) against KB oral cancer cells were found to be 18.75 ± 0.12 μg/mL for HDA and 75.50 ± 0.42 μg/mL for ODA, whereas IC_{50} values of HDA and ODA against A431 were recorded as 37.5 ± 0.42 μg/mL and 72.89 ± 0.15 μg/mL, respectively. The effect of HDA-triggered apoptosis *via* ROS-dependent inter-nucleosomal DNA fragmentation was confirmed by AGE analysis. Workers also recorded that HDA was a highly potent anticancer compound against the skin cancer of Swiss albino mice induced with skin cancer by 7,12-dimethylben z(a)anthracene (DMBA) and croton oil (CO) (Saravanakumar et al. 2015). Metabolite TM2 (4H-1,3-dioxin-4-one-2,3,6-trimethyl) isolated from *T. atroviride* induced the cell death and cytotoxicity, as revealed by cell viability test and Western blot analysis. According to microscopic, flow cytometer, and Western blot study, TM2-treated cells displayed higher ROS, cell death, and apoptosis-related protein expression than the control. This study confirmed that TM2 was a potential therapeutic agent for antiprostate cancer (Saravanakumar et al. 2019). You et al. (2010) isolated trichoderone, a novel cytotoxic cyclopentenone and cholera-7, 22-diene-3b, 5a, 6b-triol, with fair activities from the marine fungus *Trichoderma* sp.

2.3 Siderophores

Siderophores (Greek sidero meaning iron and phore meaning carrier) are low-molecular-weight (<10 KDa) iron-cheating organic compound, produced by microorganisms, like bacteria and fungi, and by some plants under iron-deficient condition. Plant growth-promoting microbes (PGPM) and biocontrol agents (BCA) produce siderophore, which is one of the important factors for plant growth promotion (Ghosh and Panja 2020a) and disease suppression (Ghosh and Panja 2020b; Ghosh et al. 2020). Recently, siderophores from mammalian cells also have been reported (Devireddy et al. 2010).

According to Schalk and Guillon (2013), siderophore can be classified as catecholate, hydroxamate, phenolate, carboxylate, and "mixed type" (those who have two or more functional groups) (Holden and Bachman 2015). (Fig. 3).

The maximum siderophores are oxygen donor, generally hexadentate ligands which create octahedral complexes with iron (Neu 2000). In case of hexadentate ligands, it coordinates Fe^{3+} ions to form Fe^{3+}–siderophore complex (Dhungana et al. 2007). Siderophores produced by fungi are both extracellular and intracellular, but bacterial siderophores are the only extracellular form (Raymond et al. 2003). Generally, fungi produce hydroxamate and carboxylate type of siderophores.

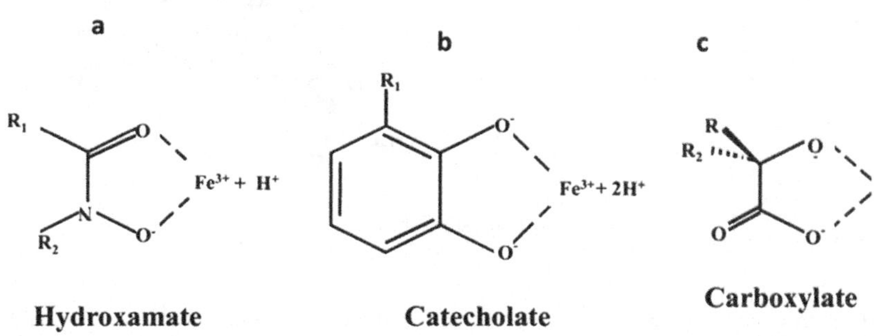

Fig. 3 Structures of siderophores with irons. (**a**) Hydroxamate, (**b**) catecholate, and (**c**) carboxylate

Trichoderma species produce coprogens (hydroxamate) (Zähner et al. 1963). *T. harzianum* has the capacity for the production of maximum hydroxamate and carboxylate type of siderophore, and it is better than other *Trichoderma* spp. like *T. viride, T. asperellum,* and *T. longibrachiatum* (Ghosh et al. 2017, 2020). Hussein and Joo (2012) recorded that *T. harzianum* produced 92.33% of siderophores. So, this species is a good source of commercial siderophore.

Microbial hydroxamate and carboxylate siderophores and their substituted derivatives are now being applied in medical sciences. *Trichoderma*-originated siderophores have not been widely used in medical purposes. So, we are introducing here a brief review of microbial siderophores on medical science to focus the need of siderophores in medical science. Some uses are as follows:

2.3.1 Specific Drug Delivery as "Trojan Horse" Approach (Siderophore-Antibiotic Conjugates)

The cellular uptake of antibiotics for drug-resistant pathogen is now the current problem. To solve this problem, scientists are applying siderophore as the delivery vehicle or "trojan horse" to deliver the antibiotics into pathogens by forming conjugate with the antibiotic (sideromycin) (Huang et al. 2013). The siderophore–antibiotic conjugates are of three types – natural, synthetic, and hybrid. Till now, catecholate and hydroxamate types of siderophores are applied as delivery vehicles in *Staphylococcus aureus* but carboxylate-type siderophores, such as staphyloferrin A, is better because this type exhibits better iron-binding activity in acidic environments than the former two (Milner et al. 2013). The classical example of this approach of drug delivery is to link ampicillin or amoxicillin with an artificial tris-catecholate siderophore (enterobactin) (Fig. 4a) (Ji et al. 2012; Mislin and

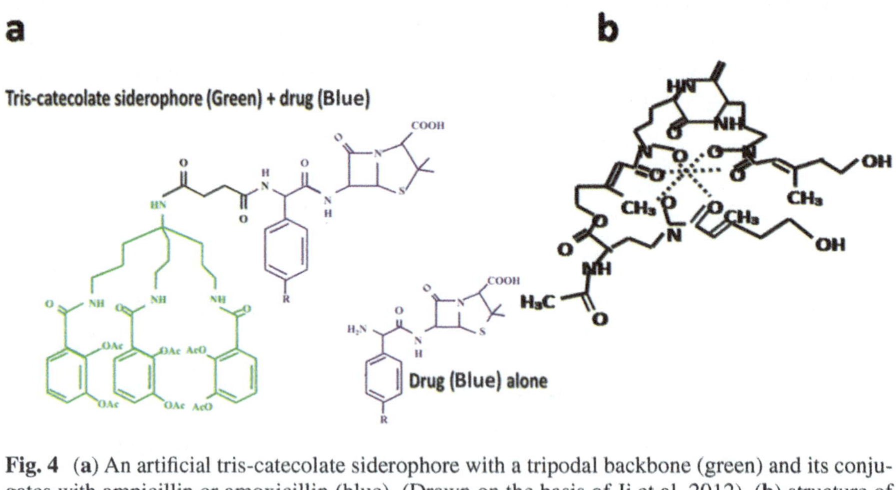

Fig. 4 (**a**) An artificial tris-catecolate siderophore with a tripodal backbone (green) and its conjugates with ampicillin or amoxicillin (blue). (Drawn on the basis of Ji et al. 2012), (**b**) structure of coprogen. (Drawn as per Pocsi et al. 2008)

Schalk 2014) against drug-resistant *P. aeruginosa*. The tested drug conjugates exhibited significant in vitro activities against different strains of *P. aeruginosa* with MICs ranging from 0.05 to 0.39 μM. (Ji et al. 2012). We can mention another two siderophore-conjugated β-lactam antibiotics, which are under human trials, and they are (i) S-649266, a catechol-substituted cephalosporin under trial phase III trial (Kohira et al. 2015), and (ii) BAL30072, a siderophore monosulfactam under phase I trial (Page 2013, 2019).

2.3.2 Treatment of Iron-Overload Diseases

The transfusional iron-overload diseases are sickle cell disease, Cooley's anemia, myelodysplasia, aplastic anemia, and Diamond-Blackfan anemia, and they are global iron load disease, but the focal iron-overload diseases are hemorrhagic stroke, Parkinson's disease, reperfusion damage, and macular degeneration (Bergeron et al. 2014). We know that the causes for iron-associated diseases are not same, but the mechanism of the iron-associated hamper is practically always more or less similar, i.e., generation of hydroxyl radicals by iron reaction with hydrogen peroxide which follows the Fenton reaction (Bergeron et al. 2014; Jomova and Valko 2011) (Eq. 1). The hydroxyl radicals are very reactive species, damaging from cell membrane components to DNA. In normal case, the presence of biological reducing agents, like glutathione, ascorbate, superoxide anion, and others, converts Fe(III) to Fe(II) (Eqs. 2 and 3) to protect the cellular molecular components. But due to blood transfusion, excess iron aggravates the problems. To solve this problem, iron-chelating siderophore strategy is the best option to apply.

Fenton reaction:

$$\mathrm{Fe}\left(\mathrm{II}\right)+\mathrm{H_2O_2}\rightarrow\mathrm{Fe}\left(\mathrm{III}\right)+\mathrm{HO}^{\bullet}+\mathrm{HO}^{-} \tag{1}$$

Reaction for the conversion (reduction of Fe(III) to Fe(II) (Bergeron et al. 2014):

(i) Via superoxide anion

$$\mathrm{Fe}\left(\mathrm{III}\right)+\mathrm{O}^{\bullet-}\rightarrow\mathrm{Fe}\left(\mathrm{II}\right)+\mathrm{O_2} \tag{2}$$

(ii) Via ascorbate

$$\mathrm{Fe}\left(\mathrm{III}\right)+\text{ascorbate}\rightarrow\mathrm{Fe}\left(\mathrm{II}\right)+\text{semi}-\text{dehydroascorbate} \tag{3}$$

As siderophore has a high iron chelating property, its reasonable application is to treat iron-overload diseases. Desferrioxamine B (DFO), a hydroxamate-based siderophore, is found in the WHO list of medicine (19th WHO list of essential medicine). The two synthetic iron chelators, like deferiprone (DFP) and deferasirox (DFX), are more improved modern drugs for clinical use for this purpose. The very recently introduced iron chelator siderophore-based drug, deferitazole, has drawn much attention as it possesses even wider therapeutic uses than others (Bergeron et al. 2014).

2.3.3 The Removal of Transuranic Elements

As we know, siderophores are also chelator of other mineral like aluminum and vanadium. So aluminum overload problem due to dialysis encephalopathy is solved by applying siderophore-based drugs. Similarly for the removal of vanadium from our body, siderophore-based drug like desferal has been come in medical science (Nagoba and Vedpathak 2011).

2.3.4 Siderophore for Heart and Cardiovascular Diseases

The most worth mentioned fungal (e.g., *Trichoderma* spp.) siderophores, which have an important use as medicine, specifically for heart and cardiovascular diseases, are coprogen (Fig. 4b) and ferrichrome (both are hydroxamate-type siderophores) (De Serrano 2017). "Iron hypothesis" says that excess iron influences heart and cardiovascular diseases (Sullivan 1981, 2009). Scavenging or removal of excess iron is necessary to prevent these diseases. As siderophores are efficient to chelate iron, their application is frequently used. Recent reports described the potential use of these two siderophores produced by fungi as anti-atherosclerotic agents. The researchers based their findings on the fact that iron accumulates in atherosclerotic lesions contributing to iron-dependent lipid oxidation. By utilizing siderophores as iron-chelating agents, the effects of lipid oxidation can be reduced, representing a potential improvement of the disease (Emri et al. 2013; Pocsi et al. 2008).

2.3.5 Siderophore in Cancer Therapy

Scientists have evaluated that siderophores, like desferrioxamines, O-trensox, desferriexochelins, and tachpyridine, are active against few cancer cell lines (Nakouti et al. 2013). For example, desferrioxamines have the capacity to decrease the growth of aggressive tumors in patients with neuroblastoma (NB) or leukemia (Buss et al. 2003; Lovejoy and Richardson 2003). As we know that cancerous cells require higher iron for rapid cell division, iron uptake and storage rates are also higher in cancer cells (Vaughn et al. 1987). So, iron chelators, like siderophores, are reasonable to use for cancer therapy (Wandersman and Delepelaire 2004).

2.3.6 Siderophores as Antimalarial Drugs

Plasmodium falciparum, causal protozoa of malaria, are sensitive to some siderophores (e.g., desferrioxamine B, produced by *Streptomyces pilosus*). Therefore, they are applicable to treat malaria (Nagoba and Vedpathak 2011).

2.3.7 Molecular Imaging Agents

Positron emission tomography (PET), computed tomography (CT), magnetic resonance imaging (MRI), ultrasonography (USG), etc. are now modern molecular or radiological imaging techniques, which are widely used for diagnostic of disease or infection. The microbial siderophores are recently used as radiopharmaceutical imaging agents. For example, siderophores are used in imaging pulmonary aspergillosis (Petrik et al. 2017; Doble et al. 2003).

2.3.8 Siderophore-Based Vaccine

Siderophore-based vaccine production is now an emerging area. The siderophore antigen conjugates to form vaccines for immunization against the urinary tract pathogen with uropathogenic *E. coli* murine model (Mike et al. 2016), and similarly immunization strategy by siderophore to inhibit the growth of enteric pathogens has been recorded (Sassone-Corsi et al. 2016; Bergeron et al. 2009).

2.4 Polyketides

2.4.1 Sorbicillinoids (Also Called Vertinoids)

A very important polyketide group found in *Trichoderma* is hexaketide, and on the carboxylate terminus, cyclization has happened. Sorbicillinoids (also called vertinoids) belong to (Harned and Volp 2011) this group. The term "sorbicillinoid" indicates the family as a whole and generally refers to any compound that consists of the carbon skeleton of sorbicillin carbon structure. Harned and Volp (2011) compiled the structures of 62 sorbicillinoids. Later on, several new species of this family were isolated (Lan et al. 2012; Fahad et al. 2014; Zhai et al. 2016). Bisorbicillinoids are hypothesized to be derived from sorbicillin (Fig. 5a), which is a natural compound (Abe et al. 1998a), or a closely related derivative such as sorbicillinol (Fig. 5b) (Abe et al. 2000a). Other vertinoids or bicillin derivatives, such as demethylsorbicillin (Fig. 5c), oxosorbicillinol (Fig. 5d) (Abe et al. 2000b), and epoxysorbicillinol, have also been obtained from several *Trichoderma* species. Some scientists have studied biosynthesis and chemical synthesis of sorbicillinoids (Harned and Volp 2011; Abe et al. 2000b, 2002; Sugaya et al. 2008). It has been hypothesized that sorbicillinol/sorbicillin acts as a precursor of most sorbicillinoids, and here the important enzyme involved for biosynthesis is polyketide synthases (PKs) (Abe et al. 2000b). Furthermore, the PKS gene cluster, having *SorbA, SorbB,* and *SorbC,* is involved in sorbicillin biosynthesis, and sorbicillinol plays like a key intermediate (Fahad et al. 2014). Nearly about 90 sorbicillinoids have been recorded mainly in terrestrial fungi and marine fungi, including *Trichoderma* spp. (Meng et al. 2017). Trichodimerol (Fig. 5e), which has been isolated from *T. longibrachiatum* (Andrade

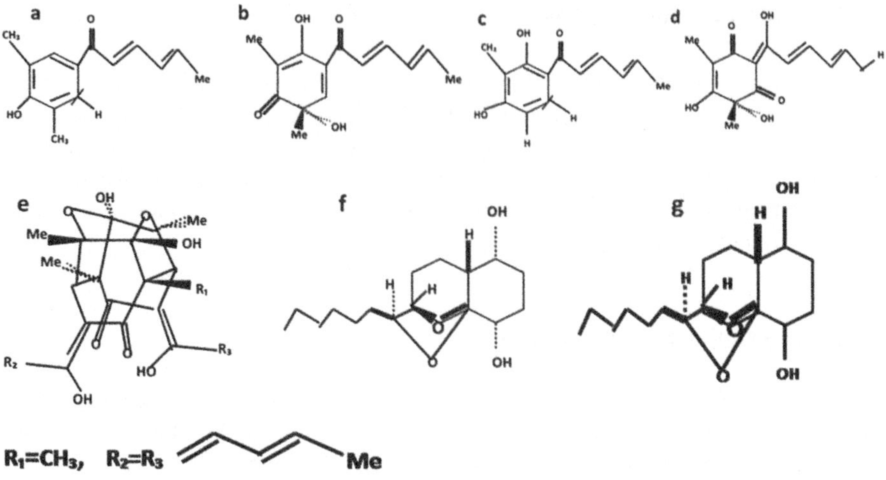

Fig. 5 Structures of polyketides: (**a**) sorbicillin, (**b**) sorbicillinol, (**c**) demethylsorbicillin, (**d**) oxo-sorbicillinol, (**e**) trichodimerol, (**f**) koninginin A, (**g**) koninginin B. (Figures drawn on the basis of Reino et al. 2007; Andrade et al. 1992)

et al. 1992), showed a good inhibitory activity against lipopolysaccharide-induced production of TNF-a (tumor necrosis factor a) in human monocytes, and so it raised a new hope for a potential treatment of septic shock (Mazzucco and Warr 1996). Fungal metabolites, named koninginins (Fig. 5f, g), and *Trichoderma*, ketone C, were isolated from solid fermentation products of *Trichoderma koningii*. Eight fungal polyketides were extracted, isolated, purified, and characterized from the *Trichoderma koningiopsis*, which is endophytic in *Panax notoginseng*. Out of them, four named koninginins N–Q were newly reported for the first time by the workers (Liu et al. 2016). All were tested for their antimicrobial activity, nitric oxide inhibition, and anticoagulant activity. More recently, seven fungal polyketides, namely, *ent*-koninginin A, 1,6-di-*epi*-koninginin A, 15-hydroxy- koninginin A, 10-deacetyl-koningiopisin D, koninginin T, koninginin L, and trichoketide A, were isolated and characterized from the culture extract of endophytic fungus *T. koningiopsis* QA-3. All compounds were applied separately against the human pathogen *E. coli*, and their MIC values ranged from 4 to 64 μg mL^{-1} (Shi et al. 2017) (Table 2).

According to the structural features, sorbicillinoids can be divided into four groups: monomeric sorbicillinoids, bisorbicillinoids, trisorbicillinoids, and hybrid sorbicillinoids.

Some polyketides were isolated from the different species of *Trichoderma*, and their anticancerous properties were evaluated by several scientists from time to time, and their findings are summarized and presented in the Table 3.

New polyketide derivatives, trichodermatides A–D, from the marine fungus *T. reesei* exhibited a strong cytotoxicity against A375-S2 human melanoma cell line (Sun et al. 2008).

Table 2 Some peptaibols with sequences of amino acids isolated from different species of *Trichoderma*

Peptaibol	Sequences of amino acids 1 2 3 4 5 6 7 8 9 10 11 12 13 14 15 16 17 18 19 20
Alamethicin	Ac-Aib-Pro-Aib-Ala-Aib-Ala-Gln-Aib-Val-Aib-Gly-Leu-Aib-Pro-Val-Aib-Aib-Glu-Gln-phoel-OH
Koningin	Ac-Aib-Gly-Ala-Aib-Ile-Gln-Aib-Aib-Aib-Ser-Leu-Aib-Pro-Val-Aib-Ile-Gln-Gln-Leuol
Trichorzin	Ac-Aib-Gly-Ala-Aib-Aib-Gln-Aib-Val-Aib-Gly–Leu-Aib-Pro-Leu-Aib-Aib-Gln-Leu
Harzianin	Ac-Aib-Asn-Leu-Aib-Pro-Ala-Ile-Aib- Pro-Iva-Leu-Aib-Pro-Leol
Trichogin A	Oc-Aib-Gly-Leu-Aib-Gly-Gly-Leu-Aib-Gly-Ile-Leol
Trichodecenins	1 2 3 4 5 6 7 Me(CH2)4CH=CH(CH2)3CO-Gly-Gly-Leu-Aib-Gly-Ile-leucinol (2,4 decenoyl-)

Abbreviations of uncommon amino acids or derivatives: *Aib* α-aminobutyric acid, *Iva* isovaline, *Ala* alanine, *Glu* glutamine, *Ser* serine, *Ile* isoleucine, *Pro* proline, *Leuol* leucinol, *Pheol* phenylalaninol, *Trpol* tryptophanol, *Val* valine, *Leu* leucine, *Ac* means acyl or acetyl groups, *Oc* stands for octanoyl

Table 3 Some polyketides isolated from different species of *Trichoderma* and their anticancerous properties in different human cancer cell lines with reference

Polyketide			
A. Sorbicillinoid			
Sorbicillinoid	Class	Fungi (Ref)	Anticancer activity (Ref)
Sorbicillin	Monomeric sorbicillinoids	*Trichoderma longibrachiatum* (Andrade et al. 1992) *Trichoderma* sp. (Lan et al. 2012) *Trichoderma* sp. f-13 (Du et al. 2009) *Trichoderma* sp. PR-35 (Wu et al. 2011) *Trichoderma* sp. (Abe et al. 1998a, b) *Trichothecium* sp.	HL-60 (Leukemia) cell line (IC50 = .12.7 μM) (Du et al. 2009) HeLa and HepG2 cells (IC50s = 1.6 and 27.2 μM, respectively (Ying et al. 2011) HL-60, U937 and T47D cell lines (IC50s = 6.55 to 28.55 μM. (Yao et al. 2015)
6-Demethylsorbicillin	Monomeric sorbicillinoids	*Trichoderma* sp. f-13 (Du et al. 2009)	HL-60 cell line (IC50 = 23.9 μM) (Du et al. 2009)
2 1,3 1-Dihydrosorbicillin	Monomeric sorbicillinoids	*Trichoderma* sp. (Lan et al. 2012) *Trichoderma* sp. f-13 (Du et al. 2009)	HeLa and HepG2 (IC50s = 7.4 and 44.4 μM respectively) (Ying et al. 2011) Many human cancer cell lines (IC50s = 9.19 to 21.93 μg/mL) (Lan et al. 2012)

(continued)

Table 3 (continued)

Polyketide			
(2*S*)-2,3-Dihydro-7-hydroxy-6,8-dimethyl-2-[(*E*)-prop-1-enyl]-chroman-4-one	Monomeric sorbicillinoids	*Trichoderma* sp. (Lan et al. 2012)	Human breast cancer cell line MCF-7 (IC50 = 9.51 μg/mL) (Lan et al. 2012)
(2*S*)-2,3-Dihydro-7-hydroxy-6-methyl-2-[(*E*)-prop-1-enyl]-chroman-4-one	Monomeric sorbicillinoids	*Trichoderma* sp. (Lan et al. 2012)	Human breast cancer cell line MCF-7 (IC50 = 7.82 μg/mL) (Lan et al. 2012)
(*E*)-6-(2,4-Dihydroxyl-5-methylphenyl)-6-oxo-2-hexenoic acid	Monomeric sorbicillinoids	*Trichoderma* sp. JH8 (Ma et al. 2011)	Human breast cancer cell line MCF-7 (IC50 = 9.51 μg/mL) (Lan et al. 2012)
Trichodimerol	Bisorbicillinoids	*Trichoderma longibrachiatum* UAMH 4159 (Andrade et al. 1992) *Trichoderma* sp. (Shirota et al. 1997) *Trichoderma* sp. (Neumann et al. 2007) *Trichoderma* sp. f-13 (Du et al. 2009) *Trichoderma* sp. JH8 (Ma et al. 2011) *Trichoderma* sp. USF-2690. (Abe et al. 1998b) *Trichothecium* sp. (Yao et al. 2015)	HL-60 cell line (IC50 = 7.8 μM) (Du et al. 2009) P388 and A549 cell lines (IC50s = 0.33 and 4.7 μM, respectively) ([Liu et al. 2005) HL-60, U937 and T47D cell lines (IC50s = 6.55 to 28.55 μM) (Yao et al. 2015)
Bislongiquinolide = bisorbibutenolide = trichotetronine	Bisorbicillinoids	*Trichoderma citrinoviride* ITEM 4484 (Evidente et al. 2009; Balde et al. 2010) *Trichoderma longibrachiatum* (Sperry et al. 1998) *Trichoderma longibrachiatum* UAMH 4159 (Andrade et al. 1992, 1997) *Trichoderma viride* (Abdel-Lateff et al. 2009) *Trichoderma* sp. (Shirota et al. 1997) *Trichoderma* sp. (Neumann et al. 2007) *Trichoderma* sp. f-13 (Du et al. 2009) *Trichoderma* sp. USF-2690 (Abe et al. 1998a, b)	U373, A549, SKMEL-28, OE21, Hs683, and B16F10 cell lines (IC50s of 4–22 μM) (Balde et al. 2010)

(continued)

Table 3 (continued)

Polyketide		
B. Anthraquinone		
Type	Fungi	Cancer cell line (Ref)
Chrysophanol	T. *harzianum* strain Th-R16 (Liu et al. 2007), *T. polysporum* (Donnelly and Sheridan 1986), *T. aureoviride*	A549 (non-small-cell lung), SK-OV-3 (ovary), SK-MEL-2 (melanoma), XF498 (central nervous system), and HCT-15 (colon) (IC50 values of 24.76, 7.28, 5.83, 30.0, and 30.0 μg/mL, respectively) (Lee et al. 2005); J5 human liver cancer cell line (IC50 = 120 μM) (Lu et al. 2010); human renal cell carcinoma Caki-2 cell (IC50 = 20 μM) (Choi et al 2016); SNU-C5 human; colon cancer cell (IC50 = 120 μM) (Lee et al. 2011)
C. Polyketide derivative		
Trichodermatides A–D	*T. reesei* (Sun et al. 2008)	A375-S2 human melanoma cell line (Sun et al. 2008)
Trichodenones A, B, and C	*T. harzianum* OUPSN115 (Thakur et al. 2003)	Leukemia P388 cell line (Thakur et al. 2003)

Trichodenones A, B, and C, extracted from the marine *T. harzianum* OUPSN115, showed a significant cytotoxicity against leukemia P388 cell line (Thakur et al. 2003). The group of trichodimerols showed antiviral and anti-inflammatory activities by inhibiting the prostaglandin H synthase 2 and tumor necrosis factor alpha (TFN^-) in human peripheral blood monocytes (Nicolaou et al. 1999).

2.4.2 Anthraquinones

Anthraquinones are a well-known polyketide group of metabolites of *Trichoderma* species. In 1967, a wild strain of *T. viride* isolated from soil produced pachybasin, chrysophanol, and emodin. Subsequently, *T. polysporum* when grown with *Fomes annosus*, also yielded these compounds (Donnelly and Sheridan 1986). In addition, chrysophanol was isolated from dry mycelium and culture filtrates of a *T. aureoviride* (De Stefano and Nicoletti 1999).

Chrysophanol

It is one polyketide (a tricyclic aromatic quinone) and is a 1,8-dihydroxy-3-methyl derivative of the 9,10-anthracenedione ring. Chrysophanol ($C_{15}H_{10}O_4$, the molecular weight is 254.2 g/mol) (Fig. 6a) is an anthracene derivative with two ketone groups attached to the central benzene ring. It is also known as chrysophanic acid. It is golden yellow or brown powder. It is synthesized in fungi through the PMA

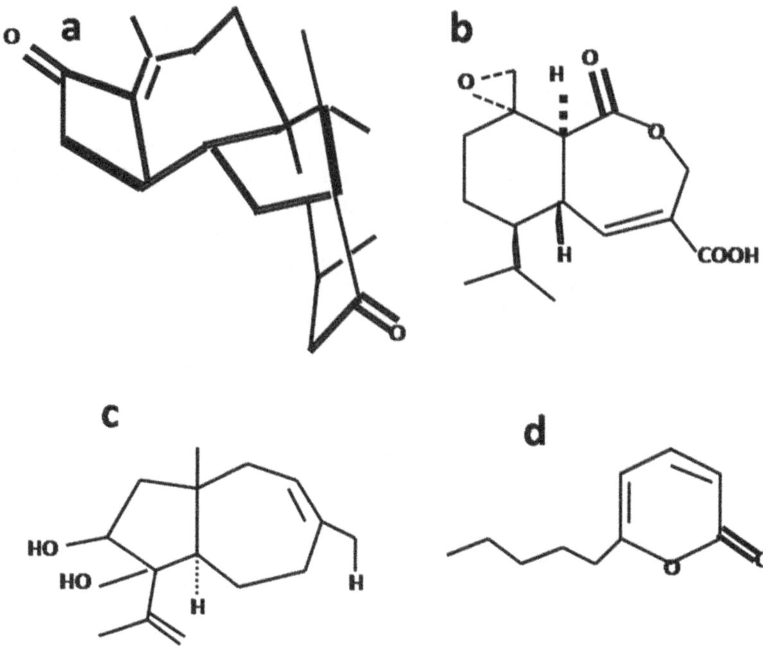

Fig. 6 Structures of anthraquinones: (**a**) chrysophanol, (**b**) emodin, (**c**) trichodermaol. (Drawn on the basis of Chukwujekwu et al. 2006; Reino et al. 2007)

(polymalonate) pathway, but in plants, its synthesis operates through both the shikimate and PMA pathways (Leistner and Zenk 1969). Although it is found in many organisms, including plants, microbes, and insects, it is present in huge amount in *Trichoderma harzianum* strain Th-R16 (Liu et al. 2007). *Trichoderma polysporum* can produce it when it is grown along *Fomes annosus* (Donnelly and Sheridan 1986) in a competitive survival mode. Its multi-faced application (antioxidant, anti-ulcer, anti-inflammatory, anticancer, neuroprotective, anti-aging, lung protective, and hepatoprotective properties) has been recorded in health science (Diaz-Muñoz et al. 2018).

Chrysophanol exhibited anticancer property against many human cancer cell lines, SK-OV-3 (ovary), SK- MEL-2 (melanoma), XF498 (central nervous system), HCT-15 (colon) (Lee et al. 2005a, b), MCF-7 and MDA-MB-231 (breast cancer), HL-60 and L1210 (human leukemia cells) (Kang et al. 2008; Ueno et al. 1995), J5 human liver cancer cells (Lu et al. 2010), Caki-2 (human renal), and A549 (human lung) (Choi 2016). Important human pathogenic fungi, like *Candida albicans*, *Cryptococcus neoformans*, *Trichophyton mentagrophytes*, and *Aspergillus fumigates*, are very sensitive to chrysophanol, and its MIC values are 50, 50, 25, and 50 μg/mL, respectively (Malik and Muller 2016; Agarwal et al. 2000). Its antiviral activity in vitro assay has been recorded against several viruses, like poliovirus type 2, herpes simplex virus type 1, human rhinovirus type 2, etc. (Ramana et al. 2017). JEV (Japanese encephalitis virus) was inhibited by 90% at 10 μg/mL of chrysophanol. The plaque reduction and virucidal activity assays showed the IC50 values of it were 15.82 and 0.75 μg/mL, respectively. It affects the viral replication in the early stage, and it is hypothesized that the CH_3 group attached to the C-3 position is the main factor for the antiviral activity. Furthermore, it can trigger a host innate immune response against JEV infection (Chang et al. 2014). Several researchers have attempted to chemically synthesize chrysophanol. Two methods, using Friedel-Craft and Diels-Alder reactions, have been particularly studied; in both these reactions, a common intermediate derivative, 4-methyl-6-methoxy-2-pyrone, is synthesized. This derivative, when heated with 5-hydroxy-1,4-naphthoquinone and hydrolyzed after oxidation, was found to yield chrysophanol at a rate of 62% (Jung et al. 1982). The structural study of chrysophanol suggested that the methyl group on the 3rd position and two hydroxyl groups on the 1st and 8th position of chrysophanol are responsible for its anticancer effects (Demirezer et al. 2016). Other biological activities might be due to the same factor. Interested readers are suggested to follow the comprehensive review on the natural sources, biosynthetic pathways, and pharmacology of chrysophanol presented by Prateeksha et al. (2019).

Emodin

Emodin (Fig. 6b) is also another anthraquinone polyketide, which inhibits the activity of both monoamine oxidase (Fujimoto et al. 1998) and tyrosine kinase. This compound acts also as an antimicrobial, antineoplastic, and cathartic agent (Wu et al. 2006; Huang et al. 2006; Ali et al. 2004) and exhibits a remarkable bacteriostatic effect on Gram-positive bacteria, especially against *Bacillus subtilis* and *Staphylococcus aureus* (Chukwujekwu et al. 2006).

2.4.3 Trichodermaol

Trichodermaol (Fig. 6c) is an anthraquinone derivative isolated from the combined culture of a strain of *Trichoderma* species and *Fusarium oxysporum* or *F. solani* proving active at a concentration of 50 μg/ml against *Bacillus subtilis* and *Streptococcus aureus* (Adachi et al. 1983).

2.5 *Terpenoids or Terpenes*

Terpenoids identified from *Trichoderma* spp. include volatile terpenes, the tetracyclicditerpene harziandione, sesquiterpenes such as the trichothecenes trichodermin and harzianum A, and the triterpene viridin (Stoppacher et al. 2010). Terpenes, which are frequently found as secondary metabolites in various fungi including *Trichoderma*, are chemically varied structural types. It is generally agreed that they all come from mevalonic acid through the intermediates isopentenyl and dimethyl

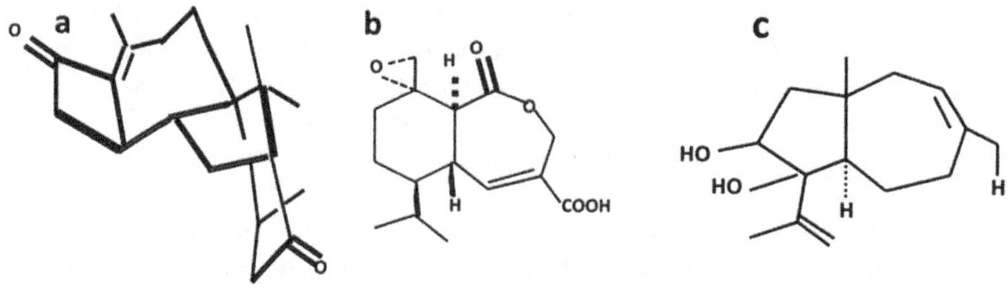

Fig. 7 Structures of compounds: (**a**) harziandione, (**b**) heptelidic acid, (**c**) daucane sesquiterpen. (Drawn on the basis of Reino et al. 2007)

allyl diphosphate, which together create geranyl diphosphate (a monoterpene). Gradually, further addition of isopentenyl diphosphate units creates farnesyl (sesquiterpene), geranylgeranyl diphosphate (diterpene), and sesquiterpene diphosphate (C25). The head-to-head addition of farnesyl diphosphate yields triterpene (C30, squalene) and tetraterpene (the C40). After that a myriad of cyclization occurs in each case. The diterpene harziandione (Fig. 7a) has been isolated from *T. harzianum*, and it has been reported to have an antifungal activity.

Triterpene, 3 beta-hydroxy-urs-12-en-28-oic acid, pentacyclic triterpenoid, which was isolated from *Trichoderma viride*, also endophytic in *Ziziphus mauritiana*, has a promising cytotoxicity effect against the HeLa cell line. IC50 value of the compound was 23.57μg/ml. Hence, the bioactive compounds of 3 beta-hydroxy urs-12-en-28-oic acid are a candidate agent for the treatment of cervical cancer (Sheeba et al. 2020). The heptelidic acid (koningic acid) (Fig. 7b) is a sesquiterpene lactone produced by *Trichoderma virens* and *Trichoderma koningii* (Reino et al. 2008), and it is an anticancerous compound. Interesting to note that *Trichoderma virens* has been categorized into P and Q strains on the basis of ability/inability to produce heptelidic acid. P-strains are able to produce heptelidic acid (Taylor et al. 2020).

Trichothecenes are a well-studied class of sesquiterpene-based mycotoxins (Useno 1983). Six new trichodermarins (A–F), together with the known trichothecenes (trichodermin, trichodermol, trichoderminol), were isolated from the soil-derived *Trichoderma brevicompactum* PSU-RSPG27. Compounds trichodermarin, trichodermin (Fig. 8a), trichodermol (Fig. 8b), and trichoderminol were tested for antifungal (*C. albicans, C. neoformans,* and *M. gypseum*), antimalarial (*P. falciparum*), and cytotoxic (KB and Vero cell lines) activities. The compound trichodermin displayed the most potent antifungal activity against *C. albicans, C. neoformans, and M. gypseum* with MIC values of 1, 4, and 2 μg/mL, respectively (Klaiklay et al. 2019). In addition, a culture of *T. harzianum* was found in 1994 to produce harzianum A (Fig. 8c) (Corley et al. 1994). This compound showed cytotoxicity to HT1080 and HeLa cell lines with IC50 values of 0.65 and 5.07 μg /ml, respectively (Lee et al. 2005a, b). Furthermore, a new compound, tricho-acorenol (Fig. 8d), has been isolated from *T. koningii*.

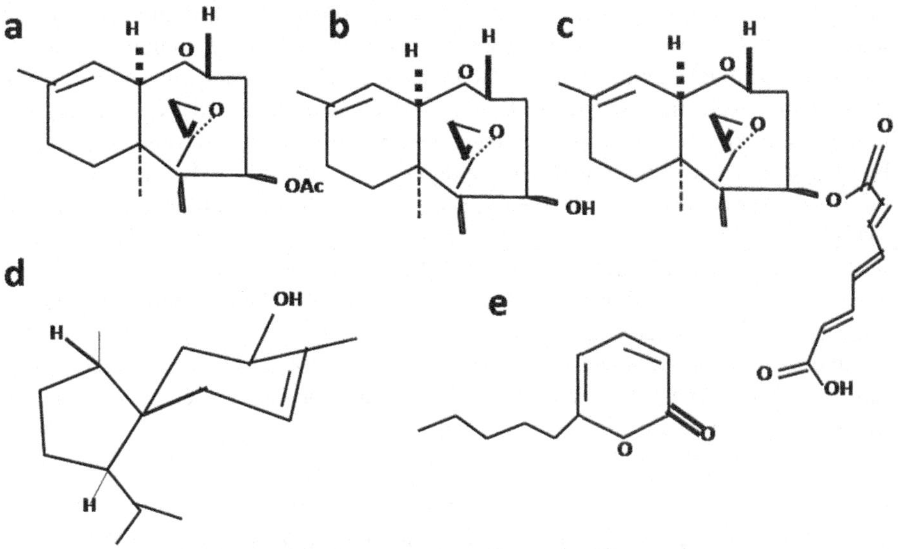

Fig. 8 Structures of some compounds: (**a**) trichodermin, (**b**) trichodermol, (**c**) harzianum A, (**d**) tricho-acorenol, (**e**) 6-pentyl-α-*pyrone* (6-PP). (Drawn on the basis of Rein et al. 2007)

2.6 Pyrones

Out of the members of pyrones, 6-pentyl-α-*pyrone* (6-PP) (Fig. 8e), an unsaturated lactone with coconut-like aroma, is very important in medical sciences. It was first isolated and characterized from *Trichoderma viride* and later on, it has been isolated from other many members of *Trichoderma* (El-Hasan et al. 2008; Oda et al. 2009). Several workers have stressed on *T. koningii* for the fermentative production of 6PP (Worasatit et al. 1994). The biosynthesis of 6PP and other lactones are not clear or confusable but Serrano-Carreon et al. (1993) proposed one hypothetical biosynthetic pathway of 6PP from linoleic acid. For lactone biosynthesis, a polypeptide pathway was also reported. One alternative biosynthetic pathway was glutamic acid metabolism. The chemical synthesis of 6PP is very cumbersome, and it requires a very high temperature (400 °C).

A screening was conducted among 60 endophytic fungal strains for their ability to produce 6-pentyl-*α*-pyrone (6PP). Of these isolated strains, four strains of *Trichoderma koningii* Oudemans were positive for 6PP production. Twenty-two strains of pathogens were treated with 6PP by the agar well diffusion assay, but it was interestingly noted that 6PP has only positive inhibitory activity against *Staphylococcus aureus* (MIC: 100 μg/mL) (Ismaiel and Ali 2017). The authors recorded that under scanning electron microscope (SEM), treated bacterial cells were distorted and lysed with bleb-like structure in the outer surface of some cells. Burow's solution-treated *S. aureus* cells showed the same kind of morphological changes (Hyo et al. 2012). Additionally, several extensive cellular damages were also observed in microbial cells, and the most frequent alteration noticed was the detachment of plasma membrane from the cell wall. Moreover, to demonstrate its mode of antifungal activity, *Aspergillus flavus*, *Penicillium expansum*, and *Fusarium acuminatum* were treated with 6PP at sub-MICs and examined by scanning and transmission electron microscopy. Several morphological alterations were caused by 6PP, such as the induction of surface depression with loosing hyphal linearity of the treated fungi. The inhibitory activity of 6PP was further demonstrated on aflatoxin B1 (AFB1) production by several strains of *Aspergillus flavus* and *Aspergillus parasiticus* grown in liquid medium, and the results showed that 6PP had a good efficacy in the suppression of AFB1 by 34.28–54.63%. These findings raised the hope of scientists to control the pathogenic organisms and their toxicity by this metabolite. Human pathogenic *Escherichia coli* was noted to be sensitive to 6PP at dilutions of 1:10 and 1:20, exhibiting an inhibition zone larger or equal to 12 mm in the disc diffusion assay (Cutler et al. 1986).

2.7 Enzymes as Anticancerous from Trichoderma

2.7.1 L-Lysine Oxidase

L-lysine oxidase (LOX) is an extracellular copper-dependent enzyme catalyzing lysine-derived cross-links in extracellular matrix proteins. Scientists paid attention to it due to its potential applications in biotechnology and tumor therapy. We know that cancerous cells are highly dependent on growth factors including amino acids, L-lysine alpha-oxidase reduces L-lysine; thus, the tumor cells die because of their inability to synthesize this amino acid. The L-lysine alpha-oxidase was isolated and purified from *Trichoderma harzianum* through ammonium sulfate precipitation, anion exchange, and gel filtration chromatography. The enzyme has two subunits (approximately 118 kDa and 58 kDa). L-lysine alpha-oxidase exhibited anticancer effect against the three carcinoma cell lines (Caco-2, HEp-2, and HepG2 cells). Treated cancer cell lines showed apoptotic effects, which were validated by cell and nuclear morphological changes, cell-cycle phase changes, and DNA fragmentation (El-Shanawany et al. 2018). This enzyme was also isolated from *Trichoderma viride*, and 10 units of L-lysine oxidase inhibited 82.5% of ovarian cancer cell line by cytotoxicity assay (Kalra et al. 2016). Kusakabe et al. (2014) observed the antitumor activity of this enzyme after isolating from *T. viride*.

2.7.2 L-Methioninase

It has been isolated from *Trichoderma harzianum*. The purified enzyme has a molecular mass of 48 kDa, and it inhibited the growth of human cell lines hepatocellular carcinoma (Hep-G2) and breast carcinoma (MCF-7) with IC_{50} values of 14.12 μg/ml and 20.07 μg/ml, respectively. The in vivo antitumor activity of l-methioninase was tested against DAL cell line-implanted Swiss albino mice. The enzyme effectively regressed the tumor volume and packed cell volume and decreased the viable cell count, and here the serum enzyme and lipid profile levels backed to normal levels in comparison to the control mice. These findings support that l-methioninase from *Trichoderma* is very effective against cancer cell lines in vitro and in vivo conditions (Salim et al. 2020).

2.8 Isocyano Metabolite

The first naturally occurring isocyano metabolite, xanthocillin, was reported from *Penicillium notatum* in 1956 (Scheuer 1992). The second microbial metabolite, dermadin (Fig. 9), was isolated 10 years later from *T. viride* (Pyke and Dietz 1966), and its antibiotic activity has been patented in 1971 (Coats et al. 1971). The same compound was also isolated from *T. koningii* together with trichoviridina. Baldwin et al. (1985) have been able to show that dermadin can be synthesized from the amino acid tyrosine (Fig. 9).

2.9 Others

2.9.1 A Non-peptide Compound (No Amide Group) as Anticancer

A non-peptide compound (no amide group) isolated from *T. harzianum* showed a high cytotoxic activity against a lot of cancer cell lines (human HaCaT – keratinocytes; THP-1 macrophage-like human cell line; A431 – human epidermoid carcinoma line; JurKat – human T-cells; K562 – chronic myeloid leukemia cell line; HEK293T – human embryonic kidney cells). The active principle of the compound is not a protein or peptide but refers to molecules without amide bonds (Patent No RU2465314C1 2012).

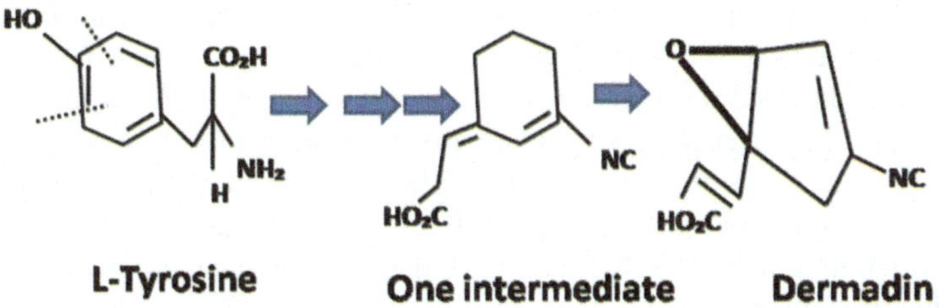

Fig. 9 Dermadin synthesis from L-tyrosine via intermediates. (Drawn on the basis of Sivasithamparam and Ghisalberti 2002)

2.9.2 Trichoderone

Trichoderone [(-)-(4R*, 5S*)-3-ethyl-4,5–dihydroxy-cyclopent-2-enone], a novel compound, first time isolated from a marine *Trichoderma,* was tested for cytotoxic property against six cancer cell lines, like DU-145 (prostate cancer), A549, NCI-H460 (non-small-cell lung cancer), MCF-7 (breast cancer), HeLa-229 (cervical cancer), and MDA-MB-435 (breast cancer) and normal human lung fibroblast cell line HLF. Trichoderone showed moderate cytotoxicity toward the six cancer cell lines, but this compound has no effect on the normal cell line HLF at concentrations up to 7.02 mM. The selectivity index of this compound was greater than 100 and even more than that of cisplatin (You et al. 2010).

2.9.3 TM1 and TM2

TM1 (two forms like 1, 3-dione-5, 5-dimethylcyclohexane and 2-enone-3hydroxy-5, 5-dimethylcylohex) and TM2 (4H-1,3-dioxin-4-one-2,3,6-trimethyl) were isolated and characterized from *T. atroviride*, but TM2 was very efficient for killing *Helicobacter pylori* and ***Shigella*** toxin-producing *Escherichia coli* (STEC) (Saravanakumar et al. 2019).

2.9.4 Trichodermamides A

Trichodermamides A (Fig. 10a) and B (Fig. 10b), modified dipeptides, were isolated from the cultures of *T. virens* isolated from marine environments. Trichodermamides **A** displayed a significant in vitro cytotoxicity against HCT-116 human colon carcinoma with an IC50 of 0.32 μg/ml.

2.9.5 Viridins

The steroidal antibiotics of the viridians show selective antifungal activity and specific inhibitory action at specific steps in the cell signaling process. These compounds carry an unusual furan ring fused between C-4 and C-6 of the steroid framework, some with an aromatic ring C. Viridin was identified as an antifungal metabolite in the fungus *Gliocladium virens* (*Trichoderma virens*). This compound has been also detected in other *Trichoderma* species such as *T. koningii*, *T. viride*, and *T. virens* (Singh et al. 2005). Derivatives of viridian (demethoxyviridin and demethoxyviridiol, wortmannolone (Fig. 10c), and virone (Fig. 10d) have been shown to be inhibitors of the phosphatidylinositol 3-kinase (Dodge et al. 1995). Such compounds can be used to treat PI 3-kinase-dependent conditions, particularly neoplasms, in human.

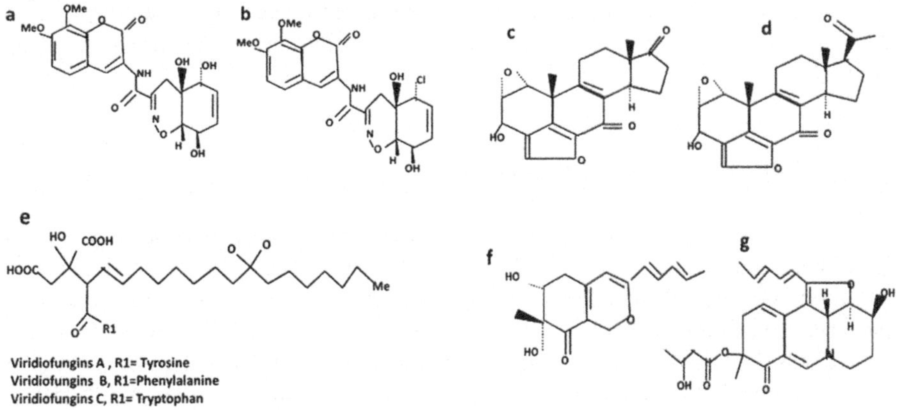

Fig. 10 Structures of compounds: (**a**) trichodermamides A, (**b**) trichodermamides B, (**c**) wortmannolone, (**d**) virone, (**e**) viridiofungins A–C, (**f**) harziphilone, and (**g**) fleephilone. (Drawn on the basis of Reino et al. 2007)

2.9.6 Viridiofungins

The structural element of citric acid is present in metabolites such as viridiofungins A–C (Fig. e), A1–4, B2, and Z2 obtained from the solid fermentation of *T. viride* (Mandala et al. 1997). The viridiofungins are potent broad spectrum fungicidal compounds with MFC (minimum fungicidal concentration) of 1–20 μg/ml against the *Candida, Cryptococcus,* and *Aspergillus* species. Later, it has been found that these compounds act as inhibitors of the farnesyl transferase and the farnesylation of the oncogenic RAS protein, indicating their potential to treat cancer.

2.9.7 Azaphilones

The azaphilones form a structurally diverse family of natural products containing a highly oxygenated bicyclic core and a chiral quaternary center. Two azaphilone-type compounds, harziphilone and fleephilone (Fig. 10f, g), were isolated from *T. harzianum*, and it was found that they had inhibitory activity against the binding REV (regulation of virion expression) protein to RRE (REV-responsive element) (Qian-Cutrone et al. 1996) of HIV. Furthermore, fleephilone demonstrated cytotoxicity at 38 μM/ml against the murine tumor cell line M-109 (Qian-Cutrone et al. 1996). Vinale et al. (2006) stressed on the identification of the major secondary metabolites, including T22 azaphilone produced by commercially used strains *T. afroharzianum* T22 (*T. harzianum* that time) and T39, their antifungal activity and involvement during the antagonistic interaction. Lebeau et al. (2017) used a new sustainable pigment extraction method (a six-stage pressurized liquid extraction protocol), for advanced mycelial pigment extraction, and they extracted a potential red pigment including *Monascus*-like azaphilone from four strains of fungi, including *Trichoderma atroviride*. The azaphilones (trigazaphilones) from *Trichoderma guizhouense* were recorded to have the antioxidant property (Pang et al. 2020).

2.9.8 Culture Filtrate and Fractions

The culture filtrate of *Trichoderma harzianum* exhibited anticancer activity against NCI-H292 lung cancer cells (Sinthujah et al. 2017). The culture supernatants of *T. harzianum* strain T9 and TS15 were applied against human pathogens and enterobacteria (seven Gram-positive bacteria, like *Bacillus subtilis* ATCC6633, *B. subtilis* TISTR008, *B. cereus* ATCC11778, *B. amyloliquefaciens* TISTR1045, *B. licheniformis* TISTR1010, *Staphylococcus aureus* ATCC1466, and *S. aureus* ATCC25923, and three Gram-negative bacteria, like *Salmonella typhi*, *Escherichia coli*, and *Vibrio cholera* isolate (clinical)), and it was found that culture supernatants were very effective to kill the human pathogens in vitro assay but *S. typhi* ATCC5784, which is a human-specific pathogen causing systemic febrile illness typhoid fever, showed the highest sensivity to the culture filtrate (Phupiewkham et al. 2015).

3 *Trichoderma* for Immunosuppressor Product or Drug

The organ transplantation is now an important medical surgery. For its success, one immunosuppressant drug is very much necessary to prevent acute rejection in organ transplantation (Konstantinovas et al. 2017). One widely used immunosuppressant drug is cyclosporin A (synonym: cyclosporine A or CsA). Cyclosporine A (empirical formula: $C_{62}H_{111}$ $N_{111}O_{12}$) (Fig. 11a) is a cyclic non-ribosomal peptide of 11 amino acids. Primarily, it was manufactured by Sandoz and approved for use by the FDA in 1983 and is now manufactured and marked in various products in markets by different drug companies in medical science (Fig. 11b). Furthermore, CsA has been applied against autoimmune diseases as an antifungal agent, an antirheumatic drug, an anti-asthmatic drug, a phosphoprotein phosphatase inhibitor, and an anti-coronaviral agent. It is also marketed as the ophthalmic solution and applied to increase tear production in patients suffering from keratoconjunctivitis sicca. Moreover, cyclosporine has been allowed for the treatment of nephrotic syndrome due to glomerular diseases such as focal and segmental glomerulosclerosis or membranous glomerulonephritis (https://go.drugbank.com/ drugs /DB00091). *Trichoderma polysporum* and *T. harzianum*, including other few fungi (*Beauveria bassiana, Fusarium oxysporum, Cyli7ndrocarpon lucidum*), produce CsA (Rodríguez et al. 2006; Azam et al. 2012). CsA interferes in cytokine signaling. Furthermore, immunosuppressive property is also recorded in gliotoxin (GT) and gliovirin (Konstantinovas et al 2017; Rocio Garcia-Rubio and Laura Alcazar-Fuoli 2018). They are produced by *Trichoderma virens* and *T. harzianum*, respectively, and are sulfur-containing fungal secondary metabolite (SM) of

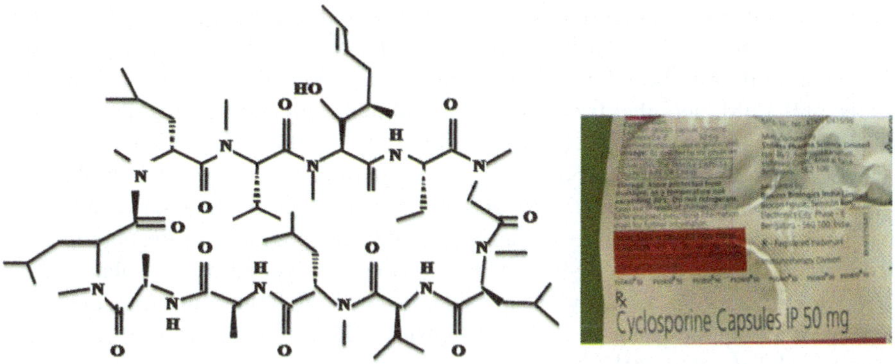

Fig. 11 (**a**) Structure of cyclosporine A. (Drawn on the basis of Thell et al. 2014, https://go.drugbank.com/drugs/DB00091). (**b**) Capsule of cyclosporine in market

epidithiodioxopiperazine (ETP) class of peptide, characterized by an internal disulfide bridge. The immunosuppressive molecules may comprise chemotherapy agents for autoimmunity and hypersensitivity reactions (Thell et al. 2014).

The mechanism behind the immunomodulatory function of CsA has been revealed. It, after binding to the receptor cyclophilin-1 within cells, produces a complex, cyclosporine-cyclophilin. This complex subsequently inhibits the function of calcineurin. As we know that calcineurin dephosphorylates and activates the transcription factor NFAT to stimulate expression of IL2, the dephosphorylation of NFAT is stopped or hampered, and consequently, the IL2-dependent T cell proliferation is reduced (Ge et al. 2012; Konstantinovas et al. 2017). Another important action of the mechanisms of CsA consists of inhibition of NO production by destabilization of the iNOS mRNA and interference with p38 and JNK signaling cascades. As its mode of action is multi-target pathway, the use of CAs is restricted, in particular during long-term treatment, as it shows some side effects, like hepatotoxicity, nephrotoxicity, neurotoxicity, and cytotoxicity (https://go.drugbank.com/drugs/DB00091).

4 *Trichoderma* as Anti-*Plasmodium* or Antimalaria

Malaria, a mosquito-borne infectious disease in humans and other animals, is caused by a parasitic protozoan genus *Plasmodium*. Previous research has established that most human deaths are caused by *P. falciparum*, but *P. vivax*, *P. ovale*, and *P. malariae* generally cause a milder form of malaria (Snounou et al. 1993; WHO 2010–2014; EPAR 1995–2009). The research works on the effect of *Trichoderma* as an antiprotozoal activity is very limited. An alkaloid compound hirsutellone, isolated from *Trichoderma* sp., showed antimalarial activity on *Plasmodium falciparum* (Isaka et al. 2006). Very recently, Klaiklay et al. (2019) isolated 13 trichothecenes compounds from *Trichoderma brevicompactum*. Out of them, six compounds were new.

The structures of all compounds were configured and tested against *Plasmodium falciparum* (K1 strain). Their finding showed that the trichothecene compound (trichodermin) was best for inhibiting this human malaria parasite with an IC_{50} value of 0.1 μM, but other trichothecenes exhibited comparatively less activity with IC_{50} values in the range of 7.1–9.6 μM. Similarly in 1999, Takashima and Wataya (1999) reported antimalarial activity of trichodermol derivative of trichothecenes (anthraquinones). Heptelidic acid and its derivative compound have also been shown to have in vitro activity toward the human malaria parasite *Plasmodium falciparum* (Tanaka et al. 1998). Chrysophanol (polyketide) is isolated from *Trichoderma polysporum* and is considered as a moderate antiprotozoal agent against chloroquine-resistant (W2) and sensitive (D6) strains of *Plasmodium falciparum* (Abdissa et al. 2017).

The activity of crude ethanolic extract of the fungus *Trichoderma stromaticum* on the growth of *P. falciparum* NF54 in infected human red blood cells (ihRBCs) was tested, and its antimalarial and anti-inflammatory activities in a mouse model of cerebral malaria were done (Cariaco et al. 2018). In their experiment, ethanolic extract was applied on *Plasmodium*-infected human RBCs (pihRBCs) for parasitemia study, and simultaneously in vivo study C57BL/6 mice were injected with *P. berghei* ANKA, treated daily with the ethanolic extract of *T. stomaticum*. Their results showed that ethanolic extract displayed a dose-dependent activity to inhibit *P. falciparum* in ihRBCs. Treated PbA-infected mice became more survival and decrease parasitemia at the beginning of infection in compare to control sets. Their neurological signs were also less. Simultaneously, systemically decreased levels of lipids and IFN-γ, ICAM-1, VCAM-1, and CCR5 cerebral expression were noticed in treated mice. Their finding surely encourages to find out the active compounds from the *Trichoderma* sp. as immune-modulatory and antimalarial drug development, which may improve the treatment of cerebral malaria (Cariaco et al. 2018).

5 *Trichoderma* in Malaria Vector Control

It has been reported that 13 spp., out of 46 spp., of *Anopheles* are malarial vectors. Indigenous strains of *Trichoderma asperellum* were isolated by us in our laboratory (Podder and Ghosh 2019) from soils of different districts of West Bengal and efficacy of larval killing by their crude ME (methanolic extract) and ME fractions and spores were evaluated against the larvae of *Anopheles* spp. The LD50 value of ME of this species was 0.073 mg/mL. Similarly, 12 methanolic fractions (MF1-MF12) were evaluated for activity against anopheline larvae. The MF8 exhibited the best larvicidal potentiality while MF1-MF4 had no activity. The LD50 value of MF8 of this species was 0.059 mg/mL while the LT50 value was 8.57 h. From this experiment, we noted that there was a proportional relationship among larval death, MF8 dose concentrations, and time. It was globally the first report to screen *Trichoderma asperellum* as a potential killer of mosquito larvae (Podder and Ghosh 2019) (Fig. 12).

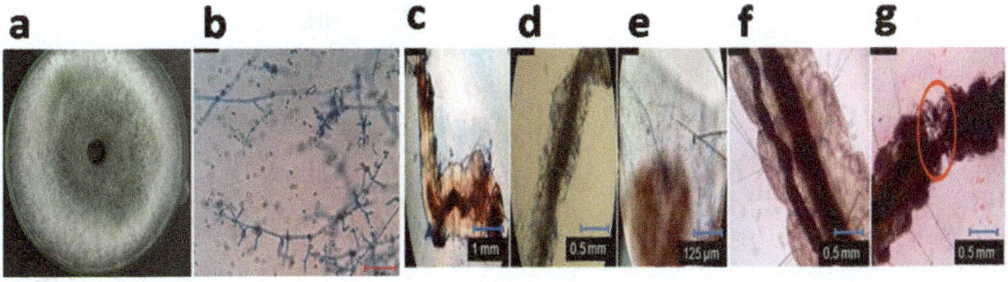

Fig. 12 *Trichoderma asperellum* as anopheline larvicide. (**a**) Cultural (in plate) and (**b**) microscopical characteristics of *T. asperellum*. (**c**) Spores (lactophenol cotton blue stained) attached on outer surface and plugging spiracles of treated larvae. (**d**, **e**) Hyphal outgrowth from the inner side of infected larvae. (**f**) Non-treated larvae stained with alizarine. (**g**) Tissue damage of ME-treated larvae (red circular mark area stained with alizarine. (Figures from Podder and Ghosh 2019)

6 *Trichoderma* as Anti-*Leishmania*

The culture filtrate of *Trichoderma asperelloides* was extracted in ethanol solution, and the extract was fractionized. Both the extract and its fractions were applied on promastigotes and amastigotes of *Leishmania amazonensis*, a major causative agent of cutaneous leishmaniasis in the New World (Lopes et al. 2020). The extract and fractions exhibited antileishmanial property on *L. amazonensis* parasites, and its pharmacological activity was associated with the low-molecular-weight fraction (LMWF) of ethanolic extract. Microscopical observation displayed that morphological alterations in the mitochondria and the flagellar pocket of promastigotes occurred. In addition, more lipid body and acidocalcisome formation, microtubule disorganization of the cytoplasm were noticed, and more vacuoles in the cytoplasm when amastigotes were present. Their finding suggested *Trichoderma* fungi as a good resource for developing chemotherapeutic leishmanicidal agents (Lopes et al. 2020). The trilongins BI, BII, BIII, and BIV, which are peptaibols containing 20 amino acid residues, were isolated and identified from *Trichoderma* sp. P8BDA1F1. These compounds were tested ex vivo against the intracellular amastigotes of *Leishmania infantum* but showed no selectivity (Grigoletto Diana et al. 2020). Moreover, curing of leishmaniasis applying peptaibols, like anti-amoebin (AAM) and suzukacillin A (SZA), has created a new window for synergistic use in medical treatment with zero risk (Ramachander Turaga 2020).

7 *Trichoderma* as Anti-*Trypanosoma*

Recently, Iwatsuki et al. (2010) have isolated two new peptaibiotics, trichosporin analogs, designated as trichosporins B-VIIa (Fig. 13) and B-VIIb, together with five known trichosporins from the culture broth of *Trichoderma* sp. FKI-4452. All trichosporins showed antitrypanosomal activities against *Trypanosoma brucei*

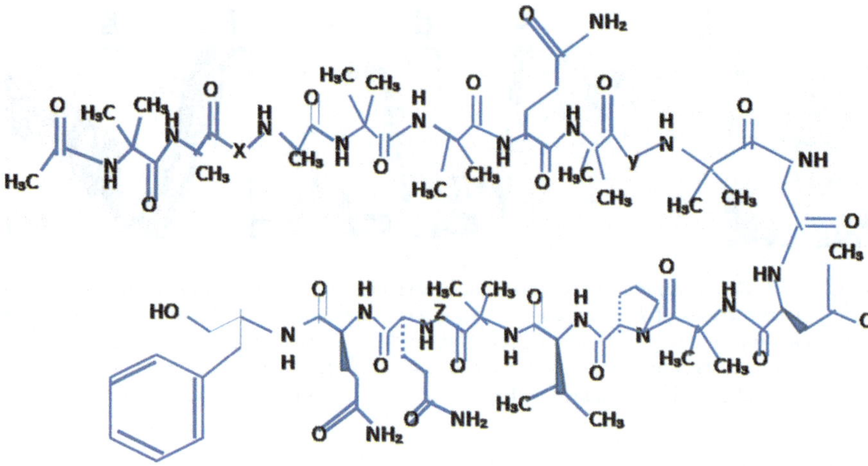

Fig. 13 Antitrypanosomal peptaibiotics, trichosporins B-VIIa. (Drawn on the basis of Iwatsuki et al. 2010)

brucei strain GUT. Among them, five showed the most potent activity, with an IC50 value of 0.16 mg /ml and the highest selectivity (cytotoxicity against MRC-5 cells/antitrypanosomal activity) of 60 times. Compounds trichosporins B-VIIa (Fig. 13) and B-VIIb showed moderate inhibitory activities, with IC50 values of 0.92 and 6.1 mg/ ml, respectively. This research team has previously reported that some peptaibiotics (leucinostatins and alamethicin I) exhibited antitrypanosomal activities (Ishiyama et al. 2009). Trichosporins might have the inhibitory mechanism, due to their membrane-interacting properties.

8 *Trichoderma* as Anticholesterol and Antiaging Drug Development

R-mevalonolactone (derivative of harzialactones) (Fig. 14) was isolated and characterized from a strain of *T. harzianum* (Amagata et al. 1998). This compound has the potentiality to activate the metabolism of cholesterol in aged skin. Later on, Yamashita (2000) exhibited that this antiaging compound, R-mevalonolactone must be a promising component of cosmetic drug for aged humans. Furthermore, the compound compactin, which later was renamed as mevastatin (Endo et al. 1985), also has been established as a good cholesterol-lowering agent. This compound hinders cholesterol biosynthesis (Goldstein et al. 1979). Compactin (Fig. 14a) also has been isolated from different sources including *T. longibrachiatum* and *T. pseudokoningii*. It is also a statin group drug. The key structural feature of compactin is the chiral b-hydroxy-d-lactone moiety, which closely mimics mevalonic acid, a crucial intermediate in the biosynthesis of cholesterol (Stokker et al. 1985). Lovastatin

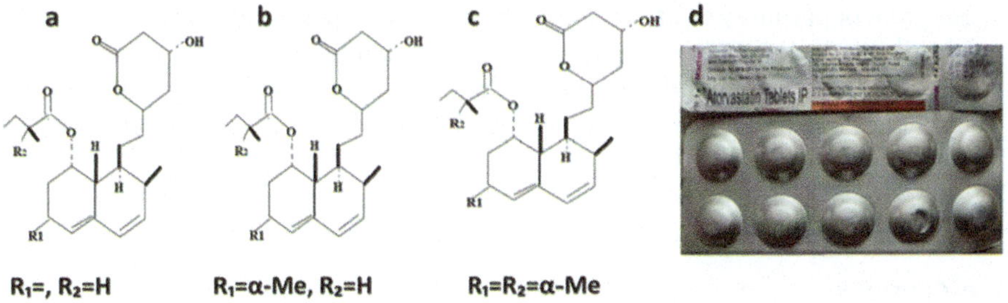

Fig. 14 Statin group of compounds. (**a**) Compactin. (**b**) Lovastatin. (**c**) Simvastatin (drawn as per the information of Reino et al. 2008). (**d**) Atorvastatin tablet in the market (purchased)

(Monacolin K) 1 (Fig. 14b), which is produced from some species of fungi like *Monascus, Aspergillus*, and simvastatin (Fig. 14c), a synthetic derivative of monacolin K, are major anticholesterol drug (Jones 1990) available in market as tablet (Fig. 14d) or other forms. The mechanism of action of these drugs lies on the ability to inhibit the function of the key enzyme, HMG CoA reductase, and as a result, the formation of mevalonate from acetoacetyl CoA becomes limited in cholesterol biosynthesis, and consequently the pathway responsible for the production of farnesyl pyrophosphate and geranylgeranyl pyrophosphate is inhibited. As these two pyrophosphates are very essential for many signal transduction pathways that influence a series of events directing to endothelial dysfunction, proliferation, apoptosis, inflammation, and other events important for atherogenesis (Jakobisiak and Golab 2003), their inhibition of the production indicates the effectiveness of the statin group drugs as anticholesterol biosynthesis.

9 Conclusions

After foregoing discussion, we found that the *Trichoderma* genus of fungi having many species is an important source of bioactive compounds of chemically diverse groups. Many bioactive compounds have been isolated, characterized, and tested of their bio-efficacy against many human pathogens, including multidrug-resistant strains, by several scientists from time to time. Some exhibited excellent efficacy and deserve for clinical application, and few have already been placed as essential drugs in clinical practices. The worth-mentioned compound is cyclosporine A (CsA), which is an essential drug or antibiotics used during organ transplantation and also to manage autoimmune disease. Similarly, anticholesterol drugs statin group (lovastatin, mevalostatin, etc.) which are produced from some species of this genus, has acquired a place in essential drug list in medical science. In addition, Trichoderma-derived compounds are tested against several human cancer cell lines; some are very promising for further trials. At the same time, many species of *Trichoderma* are still untouched for deciphering their bioactive compounds, and it is our great opportunity to harness more bioactive compounds from these members for medical uses. Another important thing is that the case report of species of *Trichoderma* as an opportunistic human pathogen is increasing by time course. Therefore, the genus *Trichoderma* has been recognized as a very important aspect in medical science.

References

Abdel-Lateff A, Fisch K, Wright AD (2009) Trichopyrone and other constituents from the marine sponge-derived fungus *Trichoderma* sp. Z Naturforsch 64c:186–192

Abdissa D, Geleta G, Bacha K, Abdissa N (2017) Phytochemical investigation of *Aloe pulcherrima* roots and evaluation for its antibacterial and antiplasmodial activities. PLoS One 12:e0173882

Abe N, Murata T, Hirota A (1998a) Novel DPPH radical scavengers, bisorbicillinol and demethyltrichodimerol, from a fungus. Biosci Biotechnol Biochem 62:661–666

Abe N, Murata T, Hirota A (1998b) Novel oxidized sorbicillin dimers with 1,1-diphenyl-2-picrylhydrazyl radical scavenging activity from a fungus. Biosci Biotechnol Biochem 62:2120–2126

Abe N, Sugimoto O, Tanji K, Hirota A (2000a) Identification of the quinol metabolite "Sorbicillinol", a key intermediate postulated in bisorbicillinoid biosynthesis. J Am Chem Soc 122:12606–12607

Abe N, Yamamoto K, Hirota A (2000b) Novel fungal metabolites, demethylsorbicillin and oxosorbicillinol, isolated from *Trichoderma* sp. USF-2690. Biosci Biotechnol Biochem 64:620

Abe N, Arakawa T, Yamamoto K, Hirota A (2002) Biosynthesis of bisorbicillinoid in *Trichoderma* sp. USF-2690; evidence for the biosynthetic pathway, via sorbicillinol, of sorbicillin, bisorbicillinol, bisorbibutenolide, and bisorbicillinolide. Bisosci Biotechnol Biochem 66:2090–2099

Adachi T, Aoki H, Osawa T, Namiki M, Yamane T, Ashida T (1983) Structure of trichodermaol, antibacterial substance produced in combined culture of Trichoderma sp. with Fusarium oxysporum or Fusarium solani. Chem Lett 6:923–926

Adams PM, Hanson JR (1972) Sesquiterpenoid metabolites of Trichoderma polysporum and T. sporulosum. Phytochemistry 11:423

Agarwal SK, Singh SS, Verma S, Kumar S (2000) Antifungal activity of anthraquinone derivatives from Rheum emodi. J Ethnopharmacol 72:43–46

Ali S, Watson MS, Osborne RH (2004) The stimulant cathartic, emodin, contracts the rat isolated ileum by triggering release of endogenous acetylcholine. Auton Autacoid Pharmacol 24:103–105

Amagata T, Usami Y, Minoura K, Ito T, Numata A (1998) Cytotoxic substances produced by a fungal strain from a sponge: physico-chemical properties and structures. J Antibiot 51:33–40

Ande SR, Fussi H, Knauer H, Murkovic M, Ghisla S, Frohlich KU, Macheroux P (2008) Induction of apoptosis in yeast by L-amino acid oxidase from the Malayan pit viper *Calloselasma rhodostoma*. Yeast 25:349–357

Andrade R, Ayer WA, Mebe PP (1992) The metabolites of *Trichoderma longibrachiatum*. Part 1. Isolation of the metabolites and the structure of trichodimerol. Can J Chem 70:2526–2535

Andrade R, Ayer WA, Trifonov LS (1997) The metabolites of *Trichoderma longibrachiatum* III. Two new tetronic acids: 5-hydroxyvertinolide and bislongiquinolide. Aust J Chem 50:255–257

Auvin-Guette C, Rebuffat S, Prigent Y, Bodo B (1992) Trichogin A IV, an 11-residue Lipopeptaibol from *Trichoderma Longibrachiatum*. J Am Chem Soc 114:2170–2174

Auvin-Guette C, Rebuffat S, Vuidepot I, Massias M, Bodo B (1993) Structural elucidation of trikoningins KA and KB, peptaibols from Trichoderma koningii. J Chem Soc Perkin Trans 1(2):249–255

Awan AR, Blount BA, Bell DJ, Shaw WM, Jch H, Mckiernan RM and Ellis T (2017) Biosynthesis of the antibiotic non-ribosomal peptide penicillin in baker's yeast. Nat. Commun, 8:15202

Azam A, Anjum T, Irum W (2012) *Trichoderma harzianum*: a new fungal source for the production of cyclosporin A. Bangladesh J Pharmacol 7:33–35

Balde ES, Andolfi A, Bruyère C, Cimmino A, Lamoral-Theys D, Vurro DMV, Altomare C, Mathieu V, Kiss R et al (2010) Investigations of fungal secondary metabolites with potential anticancer activity. J Nat Prod 73:969–971

Baldwin JE, Harjinder SB, Chondrogianni J, Field L, Taha AA, Thaller V, Brewer D, Taylor A (1985) Biosynthesis of 3-(3′-isocyanocyclopent-2-enylidene) propionic acid by *Trichoderma Hamatum* (Bon.) Bain Aggr. Tetrahedron 41:1931–1938

Baldwin JE, Harjinder SB, Chondrogianni J, Field L, Taha AA, Thaller V, Brewer D, Taylor A (1985) Biosynthesis of 3-(3′-isocyanocyclopent-2-enylidene) propionic acid by *Trichoderma Hamatum* (Bon.) Bain Aggr. Tetrahedron 41:1931–1938

Bauman G, Mueller P (1974) A model for excitability in membranes. J Supramol Struct 2:56

Beheim G, Hanke W, Jung G (1983) Alamethicin pore formation: voltage-dependent flipflop of α-helix dipoles. Biophys Struct Mech 9:181–191

Berg A, Grigoriev PA, Degenkolb T, Neuhof T, Härtl A, Schlegel BU, Gräfe E (2003) Isolation, structure elucidation and biological activities of trichofumins A, B, C and D, new 11 and 13mer peptaibols from *Trichoderma* sp. HKI 0276. J Pept Sci 9:810–816

Bergeron RJ, Bharti N, Singh S, McManis JS, Wiegand J, Green LG (2009) Vibriobactin antibodies: a vaccine strategy. J Med Chem 52:3801–3813

Bergeron RJ, Wiegand J, McManis JS, Bharti N (2014) Desferrithiocin: a search for clinically effective iron chelators. J Med Chem 57:9259–9291

Béven L, Duval D, Rebuffat S, Riddell FG, Bodo B, Wróblewski H (1998) Membrane permeabilisation and antimycoplasmic activity of the 18-residue peptaibols, trichorzins PA. Biochem Biophysi Acta 1372: 78-90

Bissett J, Gams W, Jaklitsch W, Samuels GJ (2015) Accepted *Trichoderma* names in the year 2015. IMA Fungus 6:263

Boheim G, Kolb H-A (1978) Analysis of the multi-pore system of alamethicin in a lipid membrane. I. Voltage –jump current—relaxation measurements. J Membr Biol 38:99–150

Boheim G, Hanke W, Jung G (1983) Alamethicin pore formation: voltage-dependent flipflop of a-helix dipoles. Biophys Struct Mechan 9:181-191

Brewer D, Mason FG, Taylor A (1987) The production of alamethicins by *Trichoderma* spp. Can J Microbiol 33:619–625

Brückner H, Konig WA, Aydin M, Jung G (1985) Trichotoxin A 40. Purification by counter-current distribution and sequencing of isolated fragments. Biochim Biophys Acta 827:51–62

Bruckner H, Graf H, Bokel M (1984) Paracelsin; characterization by NMR spectroscopy and circular dichroism, and hemolytic properties of a peptaibol antibiotic from the cellulolytically active mold *Trichoderma reesei*. Part B. Experientia 40:1189–1197

Brückner, H., Kirschbaum, J., and Jaworski, A. (2002). "Sequences of peptaibol antibiotic strichoaureocins from *Trichoderma aureoviride*," in Proceedings of the 27th European Peptide Symposium (Sorrento), 362–363

Buss JL, Torti FM, Torti SV (2003) The role of iron chelation in cancer therapy. Curr Med Chem 10:1021–1034

Cai F, Druzhinina IS (2021) In honor of John Bissett: authoritative guidelines on molecular identification of *Trichoderma*. Fungal Divers. https://doi.org/10.1007/s13225-020-00464-4

Cariaco Y, Lima WR, Sousa R, Nascimento LAC, Briceño MP, Fotoran WL, Wunderlich G, dos Santos JL, Silva NM (2018) Ethanolic extract of the fungus *Trichoderma stromaticum* decreases inflammation and ameliorates experimental cerebral malaria in C57BL/6 mice. Sci Rep 8:1547. https://doi.org/10.1038/s41598-018-19840-x

Carrouxa AI, Van Bohemena A-I, Roulliera C, du Ponta TR, Vansteelandta M, Bondonb A, Zalouk-Vergnouxa A, Pouchus YF, Ruiza N (2013) Unprecedented 17-residue Peptaibiotics produced by marine-derived *Trichoderma atroviride*. Chem Biodivers 10(5):772–786

CDCP (2013) Antibiotic resistance threats in the United States, CDC webpage https://www.cdc.gov/drugresistance/threat-report-2013/index.html.

Chambers HF, Deleo FR (2009) Waves of resistance: Staphylococcus aureus in the antibiotic era. Nat Rev Microbiol 7(9):629–41

Chang SJ, Huang SH, Lin YJ, Tsou YY, Lin CW (2014) Antiviral activity of *Rheum palmatum* methanol extract and chrysophanol against Japanese encephalitis virus. Arch Pharm Res 37:1117–1123

Chen L, Zhong P, Pan J-R, Zhou KL, Huang K, Fang ZX, Zhang Q-Q (2013) Asperelines G and H, two New Peptaibols from the marine-derived fungus *Trichoderma asperellum*. Heterocycles 87(3):645–655

Cheng CH, Yang CA, Liu SY, Lo CT, Huang HC, Liao FC, Peng KC (2011) Cloning of a novel L-amino acid oxidase from *Trichoderma harzianum* ETS 323 and bioactivity analysis of overexpressed L-amino acid oxidase. J Agric Food Chem 59:9142–9149

Choi JS (2016) Chrysophanic acid induces necrosis but not necroptosis in human renal cell carcinoma Caki-2 cells. J Cancer Prev 21:81–87

Chugh JK, Wallace BA (2001) Peptaibols: models for ion channels. Biochem Soc Trans 29:565–570

Chukwujekwu JC, Coombes PH, Mulholland DA, van Staden J (2006). Emodin, an antibacterial anthraquinone from the roots of Cassia occidentalis. J Bot 72:295–297

Chutrakul C, Alcocer M, Bailey K, Peberdy JF (2008) The production and characterization of trichotoxin peptaibols of *Trichoderma asperellum*. Chem Biodivers 5:1694–1706

Ciscotto P, Machado RA, de Avila EA, Coelho J, Oliveira CG, Farais LM, de Carvalho MA, Maria WS, Sanchez EF, Borges A, Chavez-Olortegui C (2009) Antigenic microbial and parasitic properties of an L-amino acid oxidase isolated from *Bothrops jararaca* snake venom. Toxicon 53:330–341

Coats JH, Myer CE, Pyke TR (1971) Antibiotic dermadin. USA Patent 3,627,882 (Chem Abstr 1972, 76:84524)

Corley DG, Miller-Wideman M, Durley RC (1994) Isolation and structure of harzianum A: a new trichothecene from *Trichoderma harzianum*. J Nat Prod 57:422–425

Corona A and Cattaneo D (2017) Dosing Colistin Properly: Let's Save "Our Last Resort Old Drug!" Clin Infect Dis 65: 870

Cutler HG, Cox RH, Crumley FG, Cole PD (1986) 6-pentyl-α-pyrone from Trichoderma harzianum: its plant growth inhibitory and antimicrobial properties. Agric Biol Chem 50:2943–294

Daniel JE, de S, Filho ER (2007) Peptiabols of *Trichoderma*. Nat Prod Rep 24:1128–1141

De Serrano LO (2017) Biotechnology of siderophores in high-impact scientific field. Bio Mol Concepts 8(3–4):169–178

De Stefano S, Nicoletti R (1999) Pachybasin and chrysophanol, two anthraquinones produced by the fungus Trichoderma aureoviride. Il Tabacco 7:21–24

De Lucca AJ and Walsh TJ (1999) Antifungal peptides: Novel therapeutic compounds against emerging pathogens. Antimicrob Agents Chemotherap 43: 1–11

De Zotti M, Biondi B, Formaggio F, Toniolo C, Stella L, Park Y, Hahm KS (2009) Trichogin GAIV: an antibacterial and protease resistant peptide. J Pept Sci 15:615–619

Demirezer LOO, Mine U, Yukselen O, Guvenalp Z, Sezerman OU (2016) Molecular docking of Anthranoids on some targeted human proteins. FABAD J Pharm Sci 41:1–16

Devireddy LR, Hart DO, Goetz DH, Green MR (2010) A mammalian siderophore synthesized by an enzyme with a bacterial homolog involved in enterobactin production. Cell 141:1006–1017

Dhungana S, Harrington JM, Gebhardt P, Möllmann U, Crumbliss AL (2007) Iron chelation equilibria, redox, and siderophore activity of a saccharide platform ferrichrome analogue. Inorg Chem 46:8362–8837

Diaz-Muñoz G, Miranda IL, Sartori SK, de Rezende DC, Diaz MAN (2018) Chapter 11-Anthraquinones: an overview. Stud Nat Prod Chem 58:313–338

Doble DMJ, Melchior M, Osullivan B, Siering JC, PierreXVC RKN (2003) A highly stable gadolinium complex with a fast, associative mechanism of water exchange. Inorg Chem 42:4930–4937

Dodge JA, Sato M, Vlahos CJ (1995) Inhibition of phosphatidylinositol 3-kinase with viridin and analogs thereof. Eur. Patent Appl. 648492, 19 Apr 1995

Donnelly DMX, Sheridan MH (1986) Anthraquinones from *Trichoderma polysporum*. Phytochemistry 25:2303–2304

Duclohier H, Alder G M, Bashford CL, Bruckner H, Chugh JK and Wallace BA (2004) Conductance studies on Trichotoxin_A50E and implications for channel structure. Biophys J 87: 1705-1710

Du L, Zhu T, LiLY CS, Zhao B, Gu Q (2009) Cytotoxic sorbicillinoids and bisorbicillinoids from a marine-derived fungus *Trichoderma* sp. Chem Pharm Bull 57:220–223

El-Hasan A F, Walker H Buchenauer (2008) Trichoderma harzianum and its Metabolite 6-Pentyl-alpha-pyrone Suppress Fusaric Acid Produced by Fusarium moniliforme. J Phytopath 156 (2) : 79–87

El-Shanawany A, El-Fakharany EM, Afifi MM, El-Shenawy FSH (2018) Purification, characterization and antitumor activity of L-lysine alpha-oxidase from *Trichoderma harzianum* Rifai AUMC No 848. J Sci Eng Res 5(9):350–367

Emri T, Toth V, Nagy CT, Nagy G, Pocsi I, Gyemant G, Antal K, Balla J, Balla G, Román G, Kovács I, Pócsi I (2013) Towards high-sidero-phore-content foods: optimisation of coprogen production in submerged cultures of *Penicillium nalgiovense*. J Sci Food and Agricult 93:2221–2228

Endo A, Hasumi K, Sakai K, Kanbe T (1985) Specific inhibition of glyceraldehyde-3-phosphate dehydrogenase by koningic acid (heptelidic acid). J Antibiot 38:920–925

EPAR (Environmental Protection Agency Report) (www.epa.gov) (1995, 2004, 2009)

Evidente A, Andolfi A, Cimmino A, Ganassi S, Altomare C, Favilla M, Cristofaro AD, Vitagliano S, Sabatini MA (2009) Bisorbicillinoids produced by the fungus *Trichoderma citrinoviride* affect feeding preference of the aphid *Schizaphis graminum*. J Chem Ecol 35:533–541

Fahad A, Abood A, Fisch KM, Osipow A, Davison J, Avramovic M, Butts CP, Piel J, Simpson TJ, Cox RJ (2014) Oxidative dearomatisation: the key step of sorbicillinoid biosynthesis. Chem Sci 5:523–527

Fleming A (1929) On the Antibacterial Action of Cultures of a Penicillium, with Special Reference to their Use in the Isolation of B. influenzæ. Br J Exp Pathol 10: 226–236

Fox RO, Richards FM (1982) A voltage-gated ion channel model inferred from the crystal structure of alamethicin at 1.5 Å resolution. Nature 300:325–330

Fujimoto H, Satoh Y, Yamaguchi K, Yamazaki M (1998) Monoamine oxidase inhibitory constituents from Anixiella micropertusa. Chem Pharm Bull 46:1506–1510

Fujita T, Takaishi Y, Okamura A, Fujita E, Fuji K, Hratsuka N, Komatsu M, Arita I (1981) New peptide antibiotics, Trichopolyns I and II, from *Trichoderma polysporum*. J Chem Soc Chem Commun 585–587. https://doi.org/10.1039/C39810000585

Fujita T, Wada S, Iida A, Nishimura T, Kanai M, Toyama N (1994) Fungal metabolites. XIII. Isolation and structural elucidation of new peptaibols, trichodecenins-I and -II, from Trichoderma viride. Chem Pharm Bull 42:489–494

Gardiner D M, Howlett B J (2005) Bioinformatic and expression analysis of the putative gliotoxin biosynthetic gene cluster of Aspergillus fumigatus. FEMS Microbiol Lett 248: 241–248

Ge Y, Xu Y, Sun W, Man Z, Zhu L, Xia X, et al. (2012) The molecular mechanisms of the effect of Dexamethasone and Cyclosporin A on TLR4/NFκB signaling pathway activation in oral lichen planus. Gene 508:157–164

Ghosh SK, Panja A (2020a) Chapter 13: Signatures of signaling pathways underlying plant growth promotion by fungi. In: Jogaiah S, Singh HB, Fraceto L, Lima R (eds) Biocontrol agents and secondary metabolites: applications in plant growth and Protection. Elsevier, pp 331–327. https://doi.org/10.1016/B978-0-12-822919-4.00013-2, pp 728

Ghosh SK, Panja A (2020b) Chapter 26: Molecular mechanisms/cell signaling pathways for plant protection from diseases by fungi. In: Jogaiah S, Singh HB, Fraceto L, Lima (eds) Biocontrol agents and secondary metabolites: applications in plant growth and protection. Elsevier, pp 591–612. https://doi.org/10.1016/B978-0-12-822919-4.00013-2, pp 728

Ghosh SK, Banerjee S, Sengupta C (2017) Bioassay, characterization and estimation of siderophores from some important antagonistic fungi. J Biopest 10:105–112

Ghosh SK, Banerjee S, Pal S, Chakrabarty N (2018) Encountering epidemic effects of leaf spot disease (*Alternaria brassicae*) on *Aloe vera* by fungal biocontrol agents in agrifields – an eco-friendly approach. PLoS One 13(3):e0193720

Ghosh SK, Bera T, Chakrabarty AM (2020) Microbial siderophore – a boon to agricultural sciences. Biol Control 144:104214

Goldstein JL, Helgeson JAS, Brown MS (1979) Inhibition of cholesterol synthesis with compactin renders growth of cultured cells dependent on the low density lipoprotein receptor. J Biol Chem 254:5403–5540

Grigoletto Diana F, Trivella Daniela BB, Tempone André G, André R, Correia Ana Maria L, Lira Simone P (2020) Antifungal compounds with anticancer potential from *Trichoderma* sp. P8BDA1F1, an endophytic fungus from *Begonia venosa*. Braz. J Microbiol 51(3):989–997

Hall JE (1975) Toward a molecular understanding of excitability. Alamethicin in black lipid films. Biophys J 15:934–936

Harned AM, Volp KA (2011) The sorbicillinoid family of natural products: isolation, biosynthesis and synthetic studies. Nat Prod Rep 28:1790–1810

Hermosa R, Viterbo A, Chet I, Monte E (2012) Plant-beneficial effects of *Trichoderma* and of its genes. Microbiol 158:17–25

Hlimi S, Rebuffat S, Gouldard C, Duchamp S, Bodo B (1995) Trichorzins HA and MA, antibiotic peptides from *Trichoderma harzianum*. II Sequence determination. J Antibiot (Tokyo) 48(11):1254–1261

Holden VI, Bachman MA (2015) Diverging roles of bacterial siderophores during infection. Metallomics 7:986–995

Huang Q, Tezuka Y, Hatanaka Y, Kikuchi T, Nishi A, Tubaki K (1995) Studies on metabolites of mycoparasitic fungi. Iv. Minor peptaibols of *Trichoderma koningii*. Chem Pharm Bull 43:1663–1667

Huang Q, Shen HM, Shui G, Wenk M, Ong CN (2006) Emodin inhibits tumor cell adhesion through disruption of the membrane lipid raft-associated integrin signaling pathway. Cancer Res 66:5807–5815

Huang Y, Jiang Y, Wang H, Wang J, Shin MC, Byun Y, He H, Liang Y, Yang VC (2013) Curb challenges of the "Trojan Horse" approach: smart strategies in achieving effective yet safe cell-penetrating peptide-based drug delivery. Adv Drug Deliv Rev 65(10):1299–1315

Hussein KA, Joo JH (2012) Potential of siderophore production by bacteria isolated from heavy metal: polluted and rhizosphere soils. Korean J Soil Sci Fert 45:798–804

Hyo Y, Yamada S, Ishimatsu M, Fukutsuji K, Harada T (2012) Antimicrobial effects of Burow's solution on Staphylococcus aureus and Pseudomonas aeruginosa. Med Mol Morphol 45:66–7

Iida A, Uesato S, Shingu T, Nagaoka Y, Kuroda Y, Fujita T (1993) Fungal metabolites. Part 7. Solution structure of an antibiotic peptide, trichosporin B-V, from Trichoderma polysporum. J Chem Soc Perkin Trans 1(3):375–379

Iida A, Sanekata M, Fujita T, Tanaka H, Enoki A, Fuse G, Kanai M, Rudewicz PJ, Tachikawa E (1994) Fungal metabolites. XVI. Structures of new peptaibols, trichokindins I-VII, from the fungus Trichoderma harzianum. Chem Pharm Bull 42:1070–1075

Iida A, Sanekata M, Wada S, Fujita T, Tanaka H, Enoki A, Fuse G, Kanai M, Asami K (1995) Fungal metabolites. Xviii. New membrane-modifying peptides, Trichorozins I–IV, from the fungus *Trichoderma harzianum*. Chem Pharm Bull 43:392–397

Isaka M, Prathumpai W, Wongsa P, Tanticharoen M (2006) Hirsutellone F, a dimer of antitubercular alkaloids from the seed fungus *Trichoderma* species BCC 7579. Org Lett 8:2815–2817

Ishiyama A, Otoguro K, Iwatsuki M, Namatame M, Nishihara A, Nonaka K, Kinoshita Y, Takahashi Y, Masuma R, Shiomi K, Yamada H, Omura S (2009) *In vitro* and *in vivo* antitrypanosomal activities of three peptide antibiotics: leucinostatin A and B, alamethicin I and tsuchimycin. J Antibiot 62:303–308

Ismaiel A, Ali DMI (2017). Antimicrobial properties of 6-pentyl-α-pyrone produced by endophytic strains of Trichoderma koningii and its effect on aflatoxin B1 production. Biologia, 72(12) :1403–1415

Iwatsuki M, Kinoshita Y, Niitsuma M, Hashida J, Mori M, Ishiyama A, Namatame M, Nishihara-Tsukashima A, Nonaka K, Masuma R, Otoguro K, Yamada H, Shiomi K, Omura S (2010) Antitrypanosomal peptaibiotics, trichosporins BVIIa and B-VIIb, produced by *Trichoderma polysporum* FKI-4452. J Antibiot (Tokyo) 63:331–333

Jaklitsch WM, Voglmayr H (2015) Biodiversity of *Trichoderma* (Hypocreaceae) in Southern Europe and Macaronesia. Stud Mycol 80:1–87

Jakobisiak M, Golab J (2003) Potential antitumor effects of statins (review). Int J Oncol 23:1055

Jaworski A, Brückner H (1999) Detection of new sequences of peptaibol antibiotics trichotoxins A-40 by on-line liquid chromatography–electrospray ionization mass spectrometry. J Chromatogr A 862:179–189

Jaworski A, Kirschbaum J, Bruckner H (1999) Structures of trichovirins II, peptaibol antibiotics from the mold Trichoderma viride NRRL 5243. J Pept Sci 5:341–351

Ji C, Miller PA, Miller MJ (2012) Iron transport-mediated drug delivery: practical syntheses and in vitro antibacterial studies of tris-catecholate siderophore–aminopenicillin conjugates reveals selectively potent anti-pseudomonal activity. J Am Chem Soc 134:9898

Jomova K, Valko M (2011) Advances in metal-induced oxidative stress and human disease. Toxicology 283:65–87

Jones PH (1990) Lovastatin and simvastatin prevention studies. Am J Cardiol 66:39B–43B

Juhász T, Kozma K, Szengyel Z, Réczey K (2003) Production of β-glucosidase by mixed culture of *Aspergillus niger* BKMF 1305 and *Trichoderma reesei* RUT C30. Food Technol Biotechnol 41(1):49–53

Jung ME, Node M, Pfluger RW, Lyster MA, Lowe JA (1982) Regio specific synthesis of bicyclic 6-alkoxy-2-pyrones and their use in the production of tetracyclic intermediates for 11-deoxyanthracycline synthesis. J Org Chem 47:1150–1152

Kalra S, Midha K, Monika KA (2016) Studies on anti- cancer activity of lysyl oxidase from *Trichoderma viride* mtcc 167. Int J Appl Sci Biotechnol 4(1):57–63

Kang SC, Lee CM, Choung ES, Bak JP, Bae JJ, Yoo HS, Kwak JH, Zee OP (2008) Anti-proliferative effects of estrogen receptor-modulating compounds isolated from Rheum palmatum. Arch Pharm Res 31(722–726):87

Katz E, Aydin M, Lucht N, Konig WA, Ooka T, Jung G (1985) Sequence and conformation of Suzukallicin. Liebigs Ann Chem:1041–1062

Klaiklay S, Rukachaisirikul V, Saithong S, Phongpaichit S, Sakayaroj J (2019) Trichothecenes from a soil-derived *Trichoderma brevicompactum*. J Nat Prod 82:687–693

Kohira N, West J, Ito A, Ito-Horiyama T, Nakamura R, Sato T, Rittenhouse S, Tsuji M, Yamano Y (2015) Antimicrobial agents. Chemothe 60:729

Konstantinovas C, de Oliveira Mendes TA, Vannier-Santos MA, Lima-Santos J (2017) Modulation of Human Immune Response by Fungal Biocontrol Agents. Front Microbiol 8:39 https://doi.org/10.3389/fmicb.2017.00039

Krause C, Kirschbaum J, Jung G, Brueckner H (2006) Sequence diversity of the peptaibol antibiotic suzukacillin-A from the mold Trichoderma viride. J Pept Sci 12:321–327

Kupferschmidt K (2016) Resistance fighters. Science 352: 758–761

Kusakabe H, Kodama K, Kuninaka A (2014) A new antitumor enzyme, L-lysine α-oxidase from *Trichoderma viride*: purification and enzymological properties. J Biol Chem 255:976–981

Lan W-J, Zhao Y, Xie Z-L, Liang L-Z, Shao W-Y, Zhu L-P, Yang D-P, Zhu X-F, Li H-J (2012) Novel sorbicillin analogues from the marine fungus *Trichoderma* sp. associated with the seastar *Acanthaster planci*. Nat Prod Commun 7:1337–1340

Latorre R, Alvarez O (1981) Voltage-dependent channels in planar lipid bilayer membranes. Physiol Rev 61:77–150

Lebeau J, Venkatachalam M, Fouillaud M, Petit T, Vinale F, Dufossé L, Caro Y (2017) Production and new extraction method of polyketide red pigments produced by ascomycetous fungi from terrestrial and marine habitats. J Fungi 3:34. https://doi.org/10.3390/jof3030034

Leclerc G, Rebuffat S, Goulard C, Bodo B (1998) Directed biosynthesis of peptaibol antibiotics in two *Trichoderma* strains. J Antibiot 51:170–177

Leclerc G, Goulard C, Prigent Y, Bodo B, Wróblewski H, Rebuffat S (2001) Sequences and antimycoplasmic properties of longibrachins LGB II and LGB III, two novel 20- residue peptaibols from *Trichoderma longibrachiatum*. J Nat Prod 64:164–170

Lee HB, Kim Y, Jin HZ, Lee JJ, Kim CJ, Park JY, Jung HS (2005a) A new *Hypocrea* strain producing harzianum a cytotoxic to tumour cell lines. Lett Appl Microbiol 40:497–503

Lee NJ, Choi JH, Koo BS, Ryu SY, Han YH, Lee SI, Lee DU (2005b) Antimutagenicity and cytotoxicity of the constituents from the aerial parts of *Rumex acetosa*. Biol Pharm Bull 28:2158–2161

Lee MS, Cha EY, Sul JY, Song IS, Kim JY (2011) Chrysophanic acid blocks proliferation of colon cancer cells by inhibiting EGFR/mTOR pathway. Phytother Res 25:833–837

Leistner E, Zenk MH (1969) Chrysophanol (1,8-dihydroxy-3-methylanthraquinone) biosynthesis in higher plants. J Chem Soc D Chem Commun 5:210–211

Liu W, Gu Q, Zhu W, Cui C, Fan G (2005) Dihydrotrichodimerol and tetrahydrotrichodimerol, two new bisorbicillinoids, from a marine-derived *Penicillium terrestre*. J Antibiot 58:621–624

Liu SY, Lo CT, Chen C, Liu MY, Chen JH, Peng KC (2007) Efficient isolation of anthraquinone-derivatives from *Trichoderma harzianum* ETS 323. J Biochem Biophys Methods 70:391–395

Liu SY, Lo CT, Shibu MA, Leu YL, Jen BY, Peng KC (2009) Study on the anthraquinones separated from the cultivation of *Trichoderma harzianum* strain Th-R16 and their biological activity. J Agric Food Chem 57:7288–7292

Liu K, Yang YB, Chen JL et al (2016) Koninginins N-Q, polyketides from the endophytic fungus *Trichoderma koningiopsis* harbored in *Panax notoginseng*. Nat Prod Bioprospect 6:49–55

Lopes DS, Santos UR, Dos Anjos DO, Silva LJC Jr, Paula VF, Vannier-Santos MA, Silva-Jardim I, Castro-Gomes T, Pirovani CP, Lima-Santos J (2020) Ethanolic extract of the fungus *Trichoderma asperelloides* induces ultrastructural effects and death on *Leishmania amazonensis*. Front Cell Infect Microbiol 10:306. https://doi.org/10.3389/fcimb.2020.00306

Lovejoy DB, Richardson DR (2003) Iron chelators as anti-neoplastic agents: current developments and promise of the PIH class of chelators. Cur Med Chem 10:1035–1049

Lu QM, Wei Q, Jin Y, Wei JF, Wang WY, Xiong YL (2002) L-amino acid oxidase from *Trimeresurus jerdonii* snake venom: purification, characterization, platelet aggregation-inducing and antibacterial effects. J Nat Toxins 11:345–352

Lu CC, Yang JS, Huang AC, Hsia TC, Chou ST, Kuo CL, Lu HF, Lee TH, Wood WG, Chung JG (2010) Chrysophanol induces necrosis through the production of ROS and alteration of ATP levels in J5 human liver cancer cells. Mol Nutr Food Res 54:967–976

Lucaciu M, Rebuffat S, Goulard C, Duclohier H,′ Molle G, Bodo B (1997) Interaction of the 14-residue peptaibols, harzianins HC, with lipid bilayers: permeability modifications and conductance properties. Biochim Biophys Acta 1323: 85–96

Ma L, Liu W, Huang Y, Rong X (2011) Two acid sorbicillin analogues from saline lands-derived fungus *Trichoderma* sp. J Antibiot 64:645–647

Malik EM, Muller CE (2016) Anthraquinones as pharmacological tools and drugs. Med Res Rev 36:705–748

Mandala SM, Thornton RA, Frommer BR, Dreikorn S, Kurtz MB (1997) Viridiofungins, novel inhibitors of sphingolipid synthesis. J Antibiot 50:339–343

Mathew MK, Balaram P (1983a) Alamethicin and related channel forming polypeptides. Mol Cell Biochem 50:47–64

Mathew MK, Balaram P (1983b) A helix dipole model for alamethicin and related transmembrane channels. FEBS Lett 157:1–5

Mazzucco CE, Warr G (1996) Trichodimerol (BMS- 182123) inhibits lipopolysaccharide-induced eicosanoid secretion in THP-1 human monocytic cells. J Leukoc Biol 60:271–277

McKenna M (2013) CDC Threat Report: We will soon be in a post-antibiotic era, 2013, https://www.wired.com/2013/09/cdc-amr-rpt1/.

Meng J, Wang X, Xu D, Fu X, Zhang X, Lai D, Zhou L, ZhangG. (2017) Sorbicillinoids from fungi and their bioactivities. Molecules 21:715. https://doi.org/10.3390/molecules21060715

Meyer CE, Reusser F (1967) A polypeptide antibacterial agent from *Trichoderma viride*. Experientia 23:85–86

Mihara T, Iida A, Akimoto N, Fujita T, Takaishi Y, Inoue K, Kushimoto S (1994) Structures of antibiotic peptides, trichopolyns, from the fungus *Trichoderma Polysporum*. Tennen Yuki Kagobutsu Toronkai Koen Yoshishu 36th, pp 713–720 (Chem. Abstr. 1995, 123:107378)

Mike LA, Smith SN, Sumner CA, Eaton KA, Mobley HLT (2016) Siderophore vaccine conjugates protect against uropathogenic *Escherichia coli* urinary tract infection. Proc Natl Acad Sci U S A 113:13468–13473

Milner SJ, Seve A, Snelling AM, Thomas GH, Kerr KG, Routledge A, Duhme-Klair AK (2013) Staphyloferrin A as siderophore component in fluoroquinolone-based Trojan horse antibiotics. Org Biomol Chem 11(21):3461–3468

Milov AD, Tsvetkov YD, Raap J, De Zotti M, Formaggio F, Toniolo C (2016) Conformation, self-aggregation, and membrane interaction of peptaibols as studied by pulsed electron double resonance spectroscopy. Biopolymers 106:6–24

Mislin GLA, Schalk IJ (2014) Siderophore-dependent iron uptake systems as gates for antibiotic Trojan horse strategies against *Pseudomonas aeruginosa*. Metallomics 6:408

Mohamed-Benkada M, Montagu M, Biard J, Mondeguer F, Verite P, Dalgalarrondo M, Bissett J, Pouchus YF (2006) New short peptaibols from a marine *Trichoderma* strain. Rapid Commun Mass Spectrom 20:1176–1180

Mohamed-Benkada M, Pouchus YF, Vérité P, Pagniez F, Caroff N, Ruiz N (2016) Identification and biological activities of long-chain peptaibols produced by a marine-derived strain of *Trichoderma longibrachiatum*. Chem Biodivers 13:521–530

Mootz HD, Schwarzer D, Marahiel MA (2002) Ways of assembling complex natural products on modular nonribosomal peptide synthetases. Chembiochem 3:490–504

Mukherjee PK, Horwitz BA, Herrera-Estrella A, Schmoll M, Kenerley CM (2013) *Trichoderma* research in the genome era. Annu Rev Phytopathol 51:105–129

Naár Z, Kecskés M (1995) A method for selecting *Trichoderma* strains antagonistic against *Sclerotinia minor*. Microbiol Res 150:239–246

Nagoba B, Vedpathak D (2011) Medical applications of siderophores. Eur J Gen Med 8:229–235

Nakouti I, Sihanonth P, Palaga T, Hobbs G (2013) Effect of a siderophore producer on animal cell apoptosis: a possible role as anti-cancer agent. Int J Pharma Med and Biol Sci 2(3):1–5

Naseby DC, Pascual JA, Lynch JM (2000) Effect of biocontrol strains of *Trichoderma* on plant growth, *Pythium ultimum* populations, soil microbial communities and soil enzyme activities. J Appl Microbiol 88:161–169

Neu MP (2000) Siderophore-mediated chemistry and microbial uptake of plutonium. Los Alamos Sci 26:416–417

Neumann K, Abdel-Lateff A, Wright AD, Kehraus S, Krick A, König GM (2007) Novel sorbicillin derivatives with an unprecedented carbon skeleton from the sponge-derived fungus *Trichoderma* species. Eur J Org Chem 2007:2268–2275

Neumann NKN, Stoppacher N, Zeilinger S, Degenkolb T, Brückner H, Schuhmacher R (2015) The peptaibiotics database—a comprehensive online resource. Chem Biodivers 12:743–751

New AP, Eckers C, Haskins NJ, Neville WA, Elson S, Hueso-Rodriguez JA, Rivera-Sagredo A (1996) Structures of polysporins A-D, four new peptaibols isolated from *Trichoderma polysporum*. Tetrahedron Lett 37:3039–3042

Nicolaou KC, Simonsen KB, Vassilikogiannakis G, Baran PS, Vidali VP, Pitsinos EN, Couladouros EA (1999) Biomimetic explorations towards the bisorbicillinoids: total synthesis of bisorbicillinol, bisorbibutenolide, and trichodimerol. Angew Chem Int Ed Engl 38:3555–3559

Oda S, Isshiki K, Ohashi S (2009) Production of 6-pentyl-α-pyrone with *Trichoderma atroviride* and its mutant in a novel extractive liquid-surface immobilization (Ext-LSI) system. Process Biochem 44:625–630

Oh SU, Lee SJ, Kim JH, Yoo ID (2000) Structural elucidation of new antibiotic peptides, atroviridins A, B and C from *Trichoderma atroviride*. Tetrahedron Lett 41:61–64

Oh SU, Yun BS, Lee SJ, Kim JH, Yoo ID (2002) Atroviridins A-C and neoatroviridins A-D, novel peptaibol antibiotics produced by *Trichoderma atroviride* F80317 I. taxonomy, fermentation, isolation and biological activities. J Antibiot (Tokyo) 55:557–564

Page MGP (2013) Siderophore conjugates. Ann N Y Acad Sci 1277:115

Page MPG (2019) The role of iron and Siderophores in infection, and the development of Siderophore antibiotics. Clin Infect Dis 69(S7):S529–S537

Pang G, Sun T, Yu Z, Yuan T, Liu W, Zhu H, Gao Q, Yang D, Kubicek CP, Zhang J, Shen Q (2020) Azaphilones biosynthesis complements the defence mechanism of *Trichoderma guizhouense* against oxidative stress. Environ Microbiol 22(11):4808–4824

Papavizas GC (1985) *Trichoderma* and *Gliocladium*: biology, ecology, and potential for biocontrol. Annu Rev Phytopathol 23:23–54

Patent No RU2465314C1 (2012) *Trichoderma harzianum* m 99/51 micromycete strain used as producer for preparing biologically active anticancer preparations; Inventors: ТатьянаИльиничнаГромовых (RU)ТатьянаИльиничнаГромовыхВераСергеевнаСады кова (RU) Вера Сергеевна СадыковаЕ ленаВикторовнаСвирщевская (RU) ЕленаВик торовнаСвирщевскаяЕкатеринаСергеевнаЗубарева (RU)ЕкатеринаСергеевнаЗубар ева2012-10-27 Publication of RU2465314C1

Peggion, Cristina, Formaggio, Fernando, Crisma, Marco, Toniolo, Claudio, Kamphuis, Johan, Kaptein, Bernard, Broxterman, Quirinus B., Vitale, Rosa Maria, Iacovino, Rosa, Saviano, Michele, Benedetti, Ettore (2001) Cα-Methyl, Cα-allylglycine (Mag) Homooligomers. Macromolecules 34: 4263–4269

Peltola J, Ritieni A, Mikkola R, Grigoriev PA, Pócsfalvi G, Adersson MA, Salkinoja-Salonen MS (2004) Biological effects of *Trichoderma harzianum* peptaibols on mammalian cells. Appl Environ Microbiol 70:996–5004

Petrik M, Zhai C, Haas H (2017) Decristoforo C Siderophores for molecular imaging applications. Clin Transl Imaging 5:15–27

Phupiewkham W, Sirithorn P, Weerasak Saksirirat W, Thammasirirak S (2015) Antibacterial agents from *Trichoderma harzianum* strain T9 against pathogenic bacteria. Chiang Mai J Sci 42(2):304–316

Pocsi I, Jeney V, Kertai P, Pocsi I, Emri T, Gyemant G, Fésüs L, Balla J, Balla G (2008) Fungal siderophores function as protective agents of LDL oxidation and are promising anti-atherosclerotic metabolites in functional food. Mol Nutrit Food Res 52:1434–1447

Podder D, Ghosh SK (2019) A new application of *Trichoderma asperellum* as anopheline larvicide for eco friendly management in medical science. Sci Rep 9:1108. https://doi.org/10.1038/s41598-018-37108-2-x

Prateeksha YMS, Singh BN, Sudheer S, Kharwar RN, Siddiqui S, Abdel-Azeem AM, Fraceto LN, Dashora K, Gupta VK (2019) Chrysophanol: a natural Anthraquinone with multifaceted biotherapeutic potential. Biomol Ther 9:68. https://doi.org/10.3390/biom9020068

Pruksakorn P, Arai M, Kotoku N, Vilchèze C, Baughn AD, Moodley P, Jacobs WR Jr, Kobayashi M (2010) Trichoderins, novel aminolipopeptides from a marine sponge-derived *Trichoderma* sp., are active against dormant mycobacteria. Bioorg Med Chem Lett 20:3658–3563

Pruksakorn P, Arai M, Liu L, Moodley P, Jacobs WR Jr, Kobayashi M (2011) Action mechanism of trichoderin A, an anti-dormant mycobacterial aminolipopeptide from marine sponge-derived *Trichoderma* sp. Biol Pharm Bull 34:1287–1290

Pyke TH, Dietz A (1966) A new antibiotic. I. Discovery and biological activity. Appl Microbiol 14:506–510

Qian-Cutrone J, Huang S, Chang LP, Pirnik DM, Klohr SE, Dalterio RA, Hugill R, Lowe S, Alam M, Kadow KF (1996) Harziphilone and fleephilone, two new HIV REV/RRE binding inhibitors produced by Trichoderma harzianum. J Antibiot 49: 990–997

Qin WT, Zhuang WY (2016) Seven wood-inhabiting new species of the genus *Trichoderma* (fungi, Ascomycota) in Viride clade. Sci Rep 6:27074

Ramachander Turaga VN (2020) Peptaibols: antimicrobial peptides from fungi. In: Singh J, Meshram V, Gupta M (eds) Bioactive natural products in drug discovery. Springer, Singapore, http://doi-org-443.webvpn.fjmu.edu.cn/10.1007/978-981-15-1394-7_26

Ramana LV, Apparao KC, Rao BN, Rao MS (2017) New phenolic constituents from bark of Walsura trifoliata. J Pharmacogn Phytochem 6:1314–1316

Raymond KN, Dertz EA, Kim SS (2003) Enterobactin: an archetype for microbial iron transport. Proc Natl Acad Sci 100:3584–3588

Rebuffat S, El Hajji M, Hennig P, Davoust D, Bodo B (1989) Isolation, sequence, and conformation of seven trichorzianines B from *Trichoderma harzianum*. Int J Pept Protein Res 34:200–210

Rebuffat S, Prigent Y, Auvin-Guette C, Bodo B (1991) Tricholongins B I and B II, 19-residue peptaibols from *Trichoderma longibrachiatum*. Solution structure from two-dimensional NMR spectroscopy. Eur J Biochem 201:661–674

Rebuffat S, Conraux L, Massias M, Auvin-Guette C, Bodo B (1993) Sequence and solution conformation of the 20-residue peptaibols, saturnisporins SA II and SAIV. Int J Pept Protein Res 41:74–84

Rebuffat S, Goulard C, Bodo B (1995) Antibiotic peptides from *Trichoderma harzianum*: harzianins HC, proline-rich 14-residue peptaibols. J Chem Soc Perkin Trans 1(14):1849–1855

Ritieni A, Fogliano V, Nanno D, Randazzo G, Altomare C, Perrone G, Bottalico A, Maddau L, Marras F (1995) Paracelsin E, a new peptaibol from Trichoderma saturnisporum. J Nat Prod 58:1745–1748

Reino JL, Guerrero RF, Hernández-Galán R, Collado IG (2008) Secondary metabolites from species of the biocontrol agent *Trichoderma*. Phytochem Rev 7:89–123

Ren J, Xue C, Tian L, Xu M, Chen J, Deng Z, Proksch P, Lin W (2009) Asperelines A–F, Peptaibols from the marine-derived fungus *Trichoderma asperellum*. J Nat Prod 72:1036–1090

Rizzo V, Stankowski S, Schwarz G (1987) Alamethicin incorporation in lipid bilayers: a thermodynamic study. Biochemistry 26:2751–2759

Rodríguez M A, Cabrera G, Godeas A (2006) Cyclosporine A from a nonpathogenic Fusarium oxysporum suppressing Sclerotinia sclerotiorum. J Appl Microbiol 100: 575–586

Salim N, Santhiagu A, Joji K (2020) Purification, characterization and anticancer evaluation of l-methioninase from *Trichoderma harzianum*. 3 Biotech 10:501

Sansom MSP (1991) The biophysics of peptide models of ion channels. Prog Biophys Mol Biol 55:139–235

Saravanakumar K, Vivek R, Boopathy NS, Yaqian L, Kathiresan K, Chen J (2015) Anticancer potential of bioactive 16-methylheptadecanoic acid methyl ester derived from marine *Trichoderma*. J Appl Biomed 13:199–212

Saravanakumar K, Mandava S, Chellia R, Jeevithan E, Yelamanchi RSB, Mandava D, Wen-Hui W, Lee J, Oh D-H, Kathiresan K, Wang M-H (2019) Novel metabolites from *Trichoderma atroviride* against human prostate cancer cells and their inhibitory effect on *Helicobacter pylori* and Shigella toxin producing *Escherichia coli*. Microb Pathog 126:19–26

Sassone-Corsi M, Chairatana P, Zheng T, Perez-Lopez A, Edwards RA, George MD, Nolan EM, Rafftellu M (2016) Siderophore-based immunization strategy to inhibit growth of enteric pathogens. Proc Natl Acad Sci USA113:13462–13467

Schalk IJ, Guillon L (2013) Fate of ferrisiderophores after import across bacterial outer membranes: different iron release strategies are observed in the cytoplasm or periplasm depending on the siderophore pathways. Amino Acids 44:1267–1277

Scheuer PJ (1992) Isocyanides and cyanides as natural products. Acc Chem Res 25:433–439

Schwarz G, Stankowski S, Rizzo G (1986) Thermodynamics analysis of incorporation and aggregation in a membrane: application to the pore-forming peptide alamethicin. Biochim Biophys Acta 861:141–151

Schwarzer D, Finking R, Marahiel MA (2003) Nonribosomal peptides: from genes to products. Nat Prod Rep 20:275–287

Serrano-Carreon L, Hathout Y, Bensoussan M, Belin J-M (1993) Metabolism of linoleic acid or mevalonate and 6-pentyl-α-pyrone biosynthesis by Trichoderma Species. Appl Environ Microbiol 59:2945–2950

Sheeba H, Ali MS, Anuradha V (2020) *In-vitro* anti-cancer activity of endophytic fungi isolated from *Ziziphus mauritiana* in cervical cancer cell line. Euro J Med Plants 31(4):38–48

Shi M, Wang H-N, Xie S-T, Luo Y, Sun A-Y, Chen X-L, Zhang Y-Z (2010) Antimicrobial peptaibols, novel suppressors of tumor cells, targeted calcium mediated apoptosis and autophagy in human hepatocellular carcinoma cells. Mol Cancer 9:26, http://www.molecular-cancer.com/content/9/1/26

Shi X-S, Wang D-J, Li X-M, Li H-L, Meng L-H, Li X, Pi Y, Zhou XW, Wang BG (2017) Antimicrobial polyketides from *Trichoderma koningiopsis* QA-3, an endophytic fungus obtained from the medicinal plant *Artemisia argyi*. RSC Adv 7:51335–51342

Shirota O, Pathak V, Hossain CF, Sekita S, Takatori K, Satake M (1997) Structural elucidation of trichotetronines: polyketides possessing a bicycle [2.2.2] octane skeleton with a tetronic acid moiety isolated from *Trichoderma* sp. J Chem Soc Perk Trans 1:2961–2964

Singh S, Dureja P, Tanwar RS, Singh A (2005) Production and antifungal activity of secondary metabolites of Trichoderma virens. Pestic Res J 17:26–29

Sinthujah S, Samarakoon SR, Tennakoon KH, Attanayake RN, Weerakoon G, Gunasegara DS, Paranagama PA (2017) Anticancer activity of *Trichoderma harzianum* extract against NCI-H292 lung cancer cells. In: International research symposium on pure and applied sciences, 2017 Faculty of Science, University of Kelaniya, Sri Lanka, p 57

Sivan A, Chet I (1986) Biological control of *fusarium* spp. in cotton, wheat and muskmelon by *Trichoderma harzianum*. J Phytopathol 116:39–47

Sivasithamparam K, Ghisalberti EI (2002) Secondary metabolism in *Trichoderma* and *Gliocladium*. In: Kubicek CP, Harman G (eds) *Trichoderma* and *Gliocladium* volume 1 basic biology, taxonomy and genetics. Taylor & Francis Ltd, London, pp 139–181, Pp 271

Snounou G, Viriyakosol S, Jarra W, Taithong S, Brown KN (1993) Identification of the four human malaria parasite species in field samples by the polymerase chain reaction and detection of a high prevalence of mixed infections. Mol Biochem Parasitol 58:283–292

Sperry S, Samuels GJ, Crews P (1998) Vertinoid polyketides from the saltwater culture of the fungus *Trichoderma longibrachiatum* separated from a Haliclona marine sponge. J Org Chem 63:10011–10014

Stokker GE, Hoffman WF, Alberts AW, Cragoe EJ Jr, Deana AA, Gilfillan JL, Huff JW, Novello FC, Prugh JD, Smith RL (1985) 3-Hydroxy-3-methylglutarylcoenzyme A reductase inhibitors. 1. Structural modification of 5-substituted 3,5-dihydroxypentanoic acids and their lactone derivatives. J Med Chem 28: 347–358

Stoppacher N, Neumann NK, Burgstaller L, Zeilinger S, Degenkolb T, Brückner H, Schuhmacher R (2013) The comprehensive peptaibiotics database. Chem Biodivers 10:734–743

Suga T, Asami Y, Hashimoto S, Nonaka K, Iwatsuki M, Nakashima T, Watanabe Y, Sugahara R, Shiotsuki T, Yamamoto T, Shinohara Y, Ichimaru N, Murai M, Miyoshi H, Ōmura S, Shiomi K (2015) Trichopolyn VI: a new peptaibol insecticidal compound discovered using a recombinant Saccharomyces cerevisiae screening system. J Gen Appl Microbiol 61:82–87

Sugaya K, Koshino H, Hongo Y, Yasunaga K, Onose J, Yoshikawa K, Abe N (2008) The biosynthesis of sorbicillinoids in *Trichoderma* sp. USF-2690: Prospect for the existence of a common precursor to sorbicillinol and 5-epihydroxyvertinolide, a new sorbicillinoid member. Tetrahedron Lett 49:654–657

Sullivan JL (1981) Iron and the sex difference in heart disease risk. Lancet 1:1293–1294

Sullivan JL (2009) Iron in arterial plaque: modifiable risk factor for atherosclerosis. Biochim Biophys Acta 1790:718–723

Sun Y, Tian L, Huang J, Ma HY, Zheng Z, Yasukawa K, Pei YH (2008) Novel polyketides from the marine-derived fungus *Trichoderma reesei*. Oncol Lett 10(3):393–396

Sun J, Pei Y, Li E, Li W, Hyde KD, Yin W-B, Liu X (2016) A new species of *Trichoderma hypoxylon* harbours abundant secondary metabolites. Sci Rep 6:37369. https://doi.org/10.1038/srep37369

Suwan S, Isobe M, Kanokmedhakul S, Lourit N, Kanokmedhakul K, Soytong K, Koga K (2000) Elucidation of high micro-heterogeneity of an acidic–neutral trichotoxin mixture from *Trichoderma harzianum* by electrospray ionization quadrupole time-of-flight mass spectrometry. J Mass Spectrom 35:1438–1451

Szekeres A, Leitgeb B, Kredics L, Antal Z, Hatvani L, Manczinger L, Vagvolgyi CS (2005) Peptaibols and related peptaibiotics of Trichoderma. Acta Microiol Immunol Hungarica 52: 137-168

Tacconelli E, Cataldo MA (2008) Vancomycin-resistant enterococci (VRE): transmission and control. Int J Antimicrob Agents 31:99–106

Takashima J, Wataya Y (1999) Trichothecene derivatives as antimalarial agents. Jpn Kokai Tokkyo Koho, JP Patent 11228408, 24 Aug

Tamandegani PR, Marik T, Zafari D, Balázs D, Vágvölgyi C, Szekeres A, Kredics L (2020) Changes in peptaibol production of *Trichoderma* species during in vitro antagonistic interactions with fungal plant pathogens. Biomol Ther 10:730. https://doi.org/10.3390/biom10050730

Tanaka Y, Shiomi K, Kamei K, Sugoh-Hagino M, Enomoto Y, Fang F, Yamaguchi Y, Masuma R, Zhang CG, Zhang XW, Omura S (1998) Antimalarial activity of radicicol, heptelidic acid and other fungal metabolites. J Antibiot 51:153–160

Taylor JT, Mukherjee PK, Puckhaber LS, Dixit K, Igumenova TI, Suh C, Horwitz BA, Kenerley CM (2020) Deletion of the *Trichoderma virens* NRPS, Tex7, induces accumulation of the anti-cancer compound heptelidic acid. Biochem Biophysi Res Communi 529(3):672–677

Thakur NL, Hentschel U, Krasko A, Pabel CT, Anr AC, Muller WEG (2003) Antibacterial activity of the sponge *Suberites domuncula* and its primmorphs: potential basis for epibacterial chemical defense. Aquat Microbial Ecol 31:77–83

Thell K, Hellinger R, Schabbauer G, Gruber CW (2014) Immunosuppressive peptides and their therapeutic applications. Drug Discovery Today 19(5):645–653

Touati I, Ruiz N, Thomas O, Druzhinina IS, Atanasova L, Tabbene O, Elkahoui S, Benzekri R, Bouslama L, Pouchus YF, Limam F (2018) Hyporientalin A, an anti-Candida peptaibol from a marine *Trichoderma orientale*. World J Microbiol Biotechno l34:98

Ueno Y, Umemori K, Niimi E, Tanuma S, Nagata S, Sugamata M, Ihara T, Sekijima M, Kawai K, Ueno I et al (1995) Induction of apoptosis by T-2 toxin and other natural toxins in HL-60 human promyelotic leukemia cells. Nat Toxins 3:129–137

Vaughn CB, Weinstein R, Bond B, Rice R, Vaughn RW, McKendrick A, Ayad G, Rockwell MA, Rocchio R (1987) Ferritin content in human cancerous and non-cancerous colonic tissue. Cancer Investig 5:7–10

Vinale F, Marra R, Scala F, Ghisalberti EL, Lorito M, Sivasithamparam K (2006) Major secondary metabolites produced by two commercial *Trichoderma* strains active against different phytopathogens. Lett Appl Microbiol 43:143–148

Walsh TR, Weeks J, Livermore DM, Toleman MA (2011) Dissemination of NDM-1 positive bacteria in the New Delhi environment and its implications for human health: an environmental point prevalence study. Lancet Infect Dis 11:355–362

Wandersman C, Delepelaire P (2004) Bacterial iron sources: from siderophores to hemophores. Annu Rev Microbiol 58:611–647

Wang X N, Zhang H M, Jian S, Liu YH, Feng YJ (2017) The MCR-1 colistin resistance: a new challenge to global public health. Chin Sci Bull 62:1018–1029

Weindling R (1934) Studies on a lethal principle effective in the parasitic action of *Trichoderma lignorum* on *Rhizoctonia solani* and other soil fungi. Phytopathology 24:1153–1179

WHO (World Health Organization) World Malaria Report (2010, 2011, 2012, 2013, 2014)

Wiest A, Grzegorski D, Xu BW, Goulard C, Rebuffat S, Ebbole DJ, Bodo B, Kenerley C (2002) Identification of peptaibols from *Trichoderma virens* and cloning of a peptaibol synthetase. J Biol Chem 277:20862–20868

Worasatit N, Sivasithamparam K, Ghisalberti E L, Rowland C Y (1994) Variation in pyrone production, lytic enzymes and control of Rhizoctonia root rot of wheat among single-spore isolates of Trichoderma koningii. Mycol Res 90:1357–1363

Wu SH, Zhao LX, Chen YW, Huang R, Miao CP, Wang J (2011) Sesquiterpenoids from the endophytic fungus *Trichoderma* sp. PR-35 of *Paeonia delavayi*. Chem Biodivers 8:1717–1723

Wu YW, Ouyang J, Xiao X, Gao WY, Liu Y (2006) Antimicrobial properties and toxicity of anthraquinones by microcalorimetric bioassay. Chin J Chem 24: 45–50

Yamashita H (2000) Commercial production of mevalonolactone by fermentation and the application to skin cosmetics with anti-aging effect. Frag J 28:62–65

Yao Y, Li J, Jiang C-S, Zhao X-X, Miao Z-H, Liu H-T, Zheng P, Yao W-X, Li W-Q (2015) Trichodimerol and sorbicillin induced apoptosis of HL-60 cells is mediated by reactive oxygen species. Pharmazie 70:394–398

Ying Y-M, Zhan Z-J, Ding Z-S, Shan W-G (2011) Bioactive metabolites from *Penicillium* sp. P-1, a fungal endophyte in *Huperzia serrata*. Chem Nat Compd 47:541–544

You J, Dai H, Chen Z, Liu G, He Z, Song F, Yang X, Fu H, Zhang L, Xiaoping (2010) Trichoderone, a novel cytotoxic cyclopentenone and cholesta-7,22-diene-3b, 5a, 6b-triol, with new activities from the marine-derived fungus *Trichoderma* sp. J Ind Microbiol Biotechnol 37:245–252

Zähner H, Keller-Schierlein W, Hütter R, Hess-Leisinger K, Deer A (1963) Stoffwechselprodukte von Mikroorganismen. Arch Microbiol 45:119–135

Zeilinger S, Gruber S, Bansal R, Mukherjee PK (2016) Secondary metabolism in *Trichoderma* chemistry meets genomics. Fungal Biol Rev 30:74–90

Zhai M-M, Qi F-M, Li J, Jiang C-X, Hou Y, Shi Y-P, Di D-L, Zhang J-W, Wu Q-X (2016) Isolation of secondary metabolites from the soil-derived fungus *Clonostachys rosea* YRS-06, a biological control agent, and evaluation of antibacterial activity. J Agric Food Chem 64:2298–2306

Zhang YB, Zhuang WY (2018) New species of *Trichoderma* in the Harzianum, Longibrachiatum and Viride clades. Phytotaxa 379:131–142

Zhao P, Xue Y, Li X, Li J, Zhao Z, Quan C, Gao W, Zu X, Bai X, Feng S (2019a) Fungi-derived lipopeptide antibiotics developed since 2000. Peptides 113:52–65

Zhao P, Xue Y, Li J, LiX ZX, Zhao Z, Quan C, Gao W, Feng S (2019b) Non-lipopeptide fungi-derived peptide antibiotics developed since 2000. Biotechnol Lett 41:651–673

Trichoderma Enzymes in the Wine and Beer Industry

Sukumar Debnath

Contents

1 Introduction

The discovery of the anamorphic hyphomycetes *Trichoderma* (Syn: *Hypocrea*) dates back to 1794 when the four species of the genus were described by Persoon in Germany. Teleomorph or sexual form of *Trichoderma*, i.e., *Hypocrea*, was discovered by Tulasne and Tulasne in 1865. The first taxonomic treatment and species diversity of *Trichoderma* was proposed based on colony growth rate and microscopic characteristics by Rifai in 1969. The genus was subdivided into nine species aggregates, which were distinguished from each other primarily on the basis of conidiophore branching patterns, phialide, and conidium morphology. The nine species aggregates proposed were (1) *T. piluliferum* Webster and Rifai, (2) *T. polysporum* (link ex Pers.), (3) *T. hamatum* (Bon.) Bain, (4) *T. koningii* Oudemans, (5)

S. Debnath (✉)
Mani Trust, Kalimpong, Relli Road, West Bengal, India

N. Amaresan et al. (eds.), *Advances in* Trichoderma *Biology for Agricultural Applications*, Fungal Biology, https://doi.org/10.1007/978-3-030-91650-3_20

T. aureoviride Rifai, (6) *T. harzianum* Rifai, (7) *T. longibrachiatum* Rifai, (8) *T. pseudokoningii* Rifai, and (9) *T. viride* (Pers. Ex. Fr.).

Taxonomy of *Trichoderma* is highly complex due to close morphological similarity among species and occurrence of cryptic species (Druzhinina et al. 2011). However, in the recent years, significant progress has been made in understanding *Trichoderma* species recognition, reidentification of already preserved species in microbial depository, ecology, and species diversity of *Trichoderma* (Cai and Druzhinina 2021). Molecular identification based on DNA bar codes (ITS, *tef1*, and *rpb2*) has been developed as a reliable protocol for the identification of *Trichoderma* species (Kubicek et al. 2003; Druzhinina et al. 2006, 2011; Samuels 1996; Harman et al. 2004). As of now, over 370 *Trichoderma* species have been described, and the majority of the descriptions have been based on the combined analyses of DNA bar code aided molecular and morphological data (www.trichoderma.info; Bissett et al. 2015; Cai and Druzhinina 2021).

2 *Trichoderma* Enzymes

A great diversity of secondary metabolites such as trichodermin, 6-pentyl-α-pyrone, koninginin, viridian, etc. (Sivasithampara and Ghisalberti 2002) and enzymes such as cellulases, β-glucanase, pectinase, chitinase, protease, lipase, and amylase (Kunamneni et al. 2014; Bhale and Rajkonda 2012; Tenkanen et al. 1992) are produced by various species of *Trichoderma*.

The groundbreaking discovery of *T. reesei* (Syn: *Hypocrea jecorina*) opened up several new avenues for research on *Trichoderma* cellulases (Bischof et al. 2016). *T. reesei* was isolated from Solomon Island during World War II in 1946 and was found to be associated with the degradation of canvas-based army materials. *T. reesei* wild-type strain QM6a and its mutants, RUT-C30, were identified to be a potent cellulose decomposer after the screening of 14,000 isolates of *Trichoderma* maintained at Quater Master Microbial Deposit Centre at Natick. RUT-C30, mutants of *T. reesei* QM6a, contains genes encoding proteins of industrial importance, and recent developments in fungal genomics have enhanced our understanding of the genetic aspects of *T. reesei* (Martinez et al. 2008; Gupta et al. 2016; Mukherjee 2017; Druzhinina et al. 2011). Genes isolated from *T. reesei* are used to engineer yeast cells to produce the desired metabolites (Bischof et al. 2016). Strains of *Trichoderma*, particularly *T. reesei* QM6a and its mutants, are good sources of extracellular enzymes (cellulases, xylanases, β-mannanase, α-L-arabinofuranosidase, α-galactosidase, pectin methyl esterases, acetylxylan esterases, and laccases) suitable for practical application in industrial sectors (Montenecourt and Eveleigh 1977; Gautam and Narayan 2020).

Many of the *Trichoderma* species are efficient producers of industrially important enzymes (Kunamneni et al. 2014; Bhale and Rajkonda 2012; Tenkanen et al. 1992). Genes responsible for the production of many enzymes have been studied, particularly with regard to *T. reesei* (Gautam and Narayan 2020). *Trichoderma*

Table 1 Enzymes produced by *Trichoderma* species

Enzymes	*Trichoderma* spp.	References
Pectinases	*T. reesei*	Haltmeier et al. (1983)
Cellulases	*T. reesei*	Xiong et al. (2013), Ohmiya et al. (1997), Reinikainen (1994), Kunamneni et al. (2014), Kumar and Ray (2014) and Montenecourt and Eveleigh (1977)
Xylans	*T. reesei* RUT C30	Tenkanen et al. (1992)
Chitinase	*T. asperellum*	Bech et al. (2014)
Lipases	*T. reesei*	Rajesh et al. (2010) and Kumar et al. (2014)
Protease	*T. harzianum*	Nirmal et al. (2011)
Amylase	*T. harzianum*	Bhale and Rajkonda (2012)
Ligninolytic enzymes	*Trichoderma* spp.	Saili et al. (2014)

enzymes (Table 1), namely, cellulase, hemicellulase (Wong and Saddler 1992), β-glucanase, pectinase, chitinase, protease, lipase, and amylase, have been explored for possible application in food, wine, bakery, brewery, biofuel, and textile industries (Galant et al. 1998). Fast growth rate, ability to produce diverse secondary metabolites, colonization of a wide variety of substrates, rich species diversity, and amenability to gene cloning have made *Trichoderma* an excellent candidate for enzyme production (Paloheimo et al. 2016; Merino and Cherry 2007).

3 *Trichoderma* Enzymes in the Wine Industry

The exogenous use of enzymes has been identified as one of the most cost-effective and attractive alternatives for enhancing the quality of the products in wine industries (Ottone et al. 2020; Canal-Llaubères 2010; Villettaz and Dubourdieu 1991). Traditionally, must fermentation has been performed by cultivating the natural yeasts present in grape skin, and the biotransformation has been facilitated by the enzymes produced by such yeasts (Pretorius 2000). This practice has long been abandoned owing to the advances in biotechnology and the better understanding of the source, isolation, and function of improved yeast strains for modern winemaking (Ottone et al. 2020). Nowadays, the use of exogenous enzymes in different stages of the process is a well-established practice in the large-scale production of wine (Rensburg and Pretorius 2000; Canal-Llaubères 2010).

There are four basic stages in winemaking: (1) pressing, maceration, and crushing, (2) fermentation, (3) clarification, and (4) stabilization, aging, and bottling (Table 2).

Grapes (*Vitis vinifera*) are harvested either by hand or by using a mechanical harvester (Canal-Llaubères 2010). The fruits are sorted at the winery, and only the good quality, ripe grapes are selected (Table 2). The crushing duration of the grapes

Table 2 Stages of winemaking and respective enzymes

Stage 1	Stage 2	Stage 3	Stage 4
Pressing and maceration ⇨	Fermentation ⇨	Clarification ⇨	Stabilization and aging
Enzymes	*Enzymes*	*Enzymes*	*Enzymes*
Cellulases *Hemicellulases* *Pectinases* *Glucose oxidases*	*Glucosidases* (aroma release)	*Pectinases*	*Lysozymes* *Protease* *Urease*

for white wine production is relatively short, while the grapes for red wine production are left in contact with the skin to enable the release of anthocyanins, thereby imparting color, flavor, and additional tannins during the fermentation process (Canal-Llaubères 2010).

Fermentation is initiated by the yeast naturally present in the grapes and also by commercially available improved yeast strains within 6–12 h of addition. The grape surface has a rich diversity of microorganisms, including naturally occurring yeast flora and lactic acid bacteria that grow along with the introduced yeast. The population of extraneous microbes drastically declines during fermentation owing to the harsh processing conditions. Fermentation continues until the sugar is entirely converted to alcohol. Sweet wine is produced when the fermentation is stopped before all the sugars are converted to alcohol. The clarification stage begins after completion of the fermentation. Filtration is accomplished using inert materials that retain the coarse solids. The clarified wine is then transferred to another vessel and is ready for bottling. Aging is carried out in bottles, stainless steel or ceramic tanks, large wooden ovals, or small barrels (Canal-Llaubères 2010).

4 Role of Cellulase and Pectinase Enzymes in the Wine Industry

Trichoderma spp. are known to produce hydrolytic enzymes such as chitinases, β-1,3- and β-1,6-glucanases, and proteases (β-1,3-glucans). Strains of *Trichoderma reesei* and its mutants are good sources of extracellular cellulases that are practically suitable for saccharification. The organisms secrete a complete cellulase complex containing endo- and exoglucanases as well as β-glucosidase (cellobiase), which act synergistically to totally degrade even the highly resistant crystalline cellulose to soluble sugars (Bischof et al. 2016). Generally, pectic enzymes are used in the clarification step to aid in the extraction process, maximize the juice yield, facilitate the filtration process, and intensify the flavor and color (Canal-Llaubères 2010). The addition of pectinase in winemaking reduces the viscosity and turbidity of the must. The turbidity is attributed to the electrostatic destabilization of the suspended, negatively charged pectin particles (Endo 1965).

Tannins and anthocyanins are the most important phenolic compounds involved in the production of red wine. Tannins add to the mouthfeel of the wine and also form pigmented polymers along with anthocyanins to provide the stable pigments that give red wine its long-term color stability. Tannins also exert mild antimicrobial activity. Grape anthocyanins are red pigments that are located mainly in the vacuoles (Barcelo et al. 1994) as well as in special structures called anthocyanoplasts (Pecket and Small 1980). The physical appearance of red wine, described by its color, brightness, and turbidity, is an important qualitative attribute. The use of pectolytic enzymes reduces the turbidity of the wine and contributes to its quality (Hagan 1996). Pectinases, β-glucanases, xylanases, and proteases are used to augment the clarification and processing of wine. Glycosidase is employed for the release of aroma from the precursor compounds (Chakraborty et al. 2016).

A combination of macerating enzymes such as cellulases and pectinases is added in different proportions during maceration and vinification (storage, aging) and/or before filtration to remove the pecto-cellulosic substances (Chakraborty et al. 2016). They catalyze the degradation of structural polysaccharides, thereby lowering the viscosity. This results in the release of fermentation compounds (tannins) and glycoside precursors from the cell wall and flesh, leading to better color intensity, stability, and improved overall mouthfeel and aroma (Villettaz et al. 1984).

Pectinases are able to degrade the cell wall components and release tannins and anthocyanins from the skin of red grapes. Color gain and improvement in sensory attributes, such as mouthfeel and aroma, after pectinase treatment are well documented (Main and Morris 2007).

Pectinases are added to the grapes in the pre-fermentation stage. The enzyme application improves the juice yield, which varies from 4% to 6% in red wine and from 6% to 11% in white wine, depending on the cultivars, maturity of the grapes, and extraction efficiency. Pectinase-treated grapes showed reduction in sediment volume and were therefore easier to clarify than the non-treated grapes (Canal-Llaubères 2010).

The wine becomes stable and attains the optimum level of aroma during the post-fermentation stage. Nonvolatile aroma compounds are released from the precursor molecules in addition to the metabolic products arising from the autolysis of yeast cells, thereby adding to the overall aroma of the wine. Urease (EC 3.5.1.5) was developed and tested to hydrolyze urea and lower the undesirable carcinogenic ethyl carbamate to acceptable levels (Canal-Llaubères 2010; Whitaker 1984).

Enzymes such as β-glucosidase, α-arabinosidase, α-rhamnosidase, and β-apiosidase are used for enhancing the levels of compounds contributing to the aroma of the wine (α-terpinol, geraniol, nerol, linalool, and citronellol) (Cabaroglu et al. 2003). *Botrytis cinerea* causes gray rot of grapes and acts as a source of pectinase enzymes which is responsible for the breakdown of grape pectin to a certain extent. However, this fungus is associated with earthy, musty, mushroom-like smell identified to be 2,4,6-trichloroanisole in wine bottle. It has been observed that β-glucanase from *T. harzianum* can be used successfully to process *Botrytis* infested in grapes in the wine (Dubourdieu et al. 1985). Villettaz et al. (1984) have documented the role of β-glucanase in the clarification and filtration of wine.

5 Regulatory Status

The use of enzymes in the wine industry is regulated by the International Organisation of Vine and Wine (OIV) (Hüfner and Haßelbeck 2017) and European Union. Enzymes from many GRAS (Generally Recognized as Safe) microorganisms, such as *Aspergillus niger*, *T. reesei*, *T. harzianum* (Almasy 2016), *Bacillus subtilis*, *B. amyloliquefaciens*, and *B. licheniformis*, are being investigated to achieve high yield and improved quality. *T. reesei* is excluded from the list of pathogens by the EU. It is also classified as one of the GRAS microbes by US Department of Public Health guidelines (Nevalainen et al. 1994; Pariza and Johnson 2001).

6 Uses of Enzymes in the Beer Industry

Beer is the third most popular beverage worldwide after water and tea. The global beer production amounted to 1.91 billion hectoliters in 2019. There are several steps in the brewing process, which include malting, mashing, lautering, boiling, fermentation, conditioning, filtering, and packaging. The brewing usually utilizes barley grains, although oats and wheat (*Triticum aestivum*) and sorghum (*Sorghum vulgare*) can also be used (Dennis et al. 2004).

The barley grains are harvested and processed by heating, drying, and cracking to stimulate the activity of the endogenous hydrolytic enzymes. Malt is added to the mash and left to steep in hot water, producing a sticky, sugary liquid known as wort (Dennis et al. 2004). The wort is transferred to a copper vessel and hops (*Humulus lupulus*) is added for bitterness and aroma, and the liquid is boiled. It is cooled and transferred to a fermentation vessel, after which yeast is added. Maintaining an optimum fermentation temperature is very important as it determines the quality of the beer. After the completion of fermentation, the beer is packed in bottles or cans for further processing and may also require carbonation depending on the type of beer. A detailed account of the science of beer production has been given by Dennis et al. (2004).

7 Role of Cellulase and Amylase Enzymes in the Beer Industry

Cellulases are typically added during the mashing step to reduce the viscosity of the wort and improve the separation of the wort from the spent grains (Chakraborty et al. 2016). The enzyme degrades the polymeric beta-glucans present in the endosperm cell wall of the grain into smaller, less-viscous molecules, thereby lessening the filtration time. The cellulase is typically denatured during the lautering (separating the mash into a clear liquid), mash filtration, or the pasteurization step after the

Table 3 Brewing enzymes and their functions

Enzymes	Process	Functions
α-Amylase	Malting, mashing	Starch hydrolysis; improve clarification
β-Amylase	Mashing, malting	Improve starch hydrolysis, malting, saccharification, fermentation yield
β-Glucanase	Malting, mashing, fermentation	Improve malting, lower viscosity, improve clarification
Fungal-β-amylase	Fermentation	Increase fermentation yield
Protease	Malting, mashing, storage	Improve malting, fermentation, clarification, chilling, and storage quality
α-Acetolactate-decarboxylase	Fermentation	Reduce fermentation time
Amyloglucosidase	Mashing	Increase glucose level in wort

fermentation. Besides, α- and β-amylases are used for starch degradation during the mashing and malting stages, while β-glucanase improves malting, aids clarification, and lowers the viscosity (Table 3).

Barley contains 6–11% of non-starch polysaccharides, and gluten is an important fraction. High gluten content is undesirable for the production of good quality beer; hence, it needs to be reduced (Ullrich 2011; Galante et al. 1998). Four different enzymes, i.e., α- and β-amylase, peptidase, and glucanase, are applied in different stages of beer production (Bamforth 1994, 2009). *Trichoderma* endoglucanase II and cellobiohydrolase II of the *Trichoderma* cellulose system causes a reduction in wort viscosity (Scheffler and Bamforth 2005). Exogenous application of cellulase causes a 90% decrease in the β-glucan content and a 30% improvement in the filtration rate. Bamforth and Kannauchi (2001) observed that the mixed application of β-endoglucanase with endoxylanase increases the solubility of glucan. β-Glucanases used in the beer industry are obtained from *Penicillium emersonii*, *A. niger*, *B. subtilis*, or *T. reesei*.

8 Prospects of Genetic Improvement of the Beer Yeast

The use of genetically modified organisms (GMOs) has been permitted in the food and beverage industry globally. In 1990, the United Kingdom became the first country to permit the use of GMO in food (Hammond 1995). A new yeast species, *S. pastorianus*, which is a hybrid of *S. cerevisiae* and *S. eubayanus*, is being utilized to enhance beer production in the United Kingdom (Gorter de Vries et al. 2019). Efforts are now being made to improve the strain and to genetically engineer and modify the ethanogenic yeast *S. cerevisiae* by incorporating the cellulose-degrading genes from *T. reesei* (Fujita et al. 2002; Katahira et al. 2004; Yamada et al. 2013; Nakatani et al. 2013; Matano et al. 2013; Liu et al. 2015). However, consumer acceptance of the beer manufactured using genetically modified yeast will largely depend on an extensive and reliable risk assessment process.

9 Conclusion

Naturally occurring filamentous fungi belonging to the genus *Trichoderma*, including mutants of *T. reesei*, have been recognized as a rich source of genes for enzymes for practical application in agricultural, environmental, and industrial setup. In the wine and beer industry, enzymes help in the extraction, filtration, and clarification of the juice and in the release of aroma from the sugar-bound precursors. Furthermore, they inhibit the formation of carcinogenic ethyl carbamates, help in the removal of haze from proteins, and prevent the oxidative burning of wine. Future research on the preservation of naturally occurring or improved strains of *Trichoderma* as well as its genes encoding useful enzymes and their production process with GRAS approval and benefit sharing should be a priority.

References

Almasy RW (2016) Federal Register, vol 81(225), 1–8 pp

Bamforth CW (1994) β-Glucan and β-glucanases in malting and brewing: practical aspects. Brew Dig 69:12–16

Bamforth CW (2009) Current perspectives on the role of enzymes in brewing. J Cereal Sci 50:353–357

Bamforth CW, Kannauchi M (2001) A simple model for the cell wall of the starchy endosperm in barley. J Inst Brew 107:235–240

Barcelo AR, Calderon AA, Zapata JM, Munoz R (1994) The histochemical localization of anthocyanins in seeded and seedless grapes (*Vitis vinifera*). Sci Hortic 57(3):265–268

Bech L, Busk PK, Lange L (2014) Cell wall degrading enzymes in *Trichoderma asperellum* grown on wheat bran. Fungal Genome Biol 4(1):1–10

Bhale UN, Rajkonda JN (2012) Enzymatic activity of *Trichoderma* species. Novus Nat Sci Res 1:1–8

Bischof RH, Ramoni J, Seiboth B (2016) Cellulases and beyond: the first 70 years of the enzyme producer *Trichoderma reesei*. Microb Cell Factories 15(106):1–13

Bissett J, Gams W, Jaklitsch W, Samuel GJ (2015) Accepted *Trichoderma* names in the year. IMO Fungus 6(2):263–295

Cabaroglu T, Selli S, Canbas A, Lepoutre JP, Günata Z (2003) Wine flavor enhancement through the use of exogenous fungal glycosidases. Enzym Microb Technol 33(5):581–587

Cai F, Druzhinina IS (2021) In honor of John Bissett : authoritative guidelines on molecular identification of *Trichoderma*. Fungal Biodiver. https://doi.org/10.1007/s13225-020-00464-4

Canal-Llaubères RM (2010) Chapter 4: enzymes in wine quality. In: Rollands AG (ed) Vol 2. Oenology and wine quality. Woodhead Publishing, Cambridge, pp 93–132.g

Chakraborty S, Gupta R, Jain KK, Gautam HS, Kuhad RC (2016) Chapter 17: cellulases: application in wine and brewery industry. In: New and future developments in microbial biotechnology and bioengineering. Elsevier B.V, pp 193–200

Dennis EB, Chris AB, Boulton BPA, Stevens R (2004) Brewing science and practice. CRC Press/ Woodhead Publishing Limited, Boston/Abington Hall, pp 1–848

Druzhinina I, Kopchinskiy A, Kubicek C (2006) The first 100 *Trichoderma* species characterized by molecular data. Mycoscience 47(2):55–64

Druzihinina IS, Seiboth VS et al (2011) *Trichoderma*: the genomics of opportunistic success. Nat Rev Microbiol 9:749–759

Dubourdieu D, Desplanques C, Villettaz JC, Ribéreau-Gayon P (1985) Investigations of an industrial β-D-glucanase from *Trichoderma harzianum*. Carbohydr Res 144:277–287

Fujita Y, Takahashi S, Ueda M, Tanaka A, Okada H, Morikawa Y, Kawaguchi T, Arai M, Fukuda H, Kondo A (2002) Direct and efficient production of ethanol from cellulosic material with yeast strain displaying cellulolytic enzymes. Appl Environ Microbiol 68:5136–5141

Galante YM, De Conti A, Monteverdi R (1998) Application of *Trichoderma* enzymes in food and feed industries. In: Harman GF, Kubicek CP (eds) *Trichoderma* and *Gliocladium* – enzymes, biological control and commercial applications. Taylor & Francis, London, pp 327–342

Gautam RL, Narayan R (2020) Chapter 5: *Trichoderma*, a factory of multipurpose enzymes: cloning of enzymatic genes. In: Hesham AEL (ed) Fungal biotechnology and bioengineering, fungal biology. Springer Nature Switzerland AG, pp 137–162

Gorter de Vries AR, Pronk JT, Daran JMG (2019) Lager brewing yeast in the era of modern genetics. FEMS Yeast Res 19:1–17

Gupta et al (2016) The post-genomic era of *Trichoderma reesei*: what is next? Trends Biotechnol 34(12):970–982

Hagan P (1996) Enzyme revisiting the principle and future development. In: Allen M, Leslke P, Baldwin G (eds) Advances in juice clarification and yeast inoculation. Proceedings of ASVO oenology seminar, pp 3–5

Haltmeier T, Leisola M, Ulmer D, Waldner R, Fiechter A (1983) Pectinase from *Trichoderma reesei* QM 9414. Biotechnol Bioeng 25(6):1685–1690

Hammond JRM (1995) Genetically modified yeast for the 21st century, progress to date. Yeast 11:1613–1627

Harman GE, Howell CR, Viterbo A, Vhet I, Lorito M (2004) Opportunistic avirulent plant symbiont. Nat Rev Microbiol 2:43–56

Katahira S, Fujita Y, Mizuike A, Fukuda H, Kondo A (2004) Construction of a xylan fermenting yeast strain through codisplay of xylanolytic enzymes on the surface of xylose utilizing *Saccharomyces cerevisiae* cells. Appl Environ Microbiol 70:5407–5417

Kubieck CP, Bissett J, Ddinger C, Druzhinina I, Kullnig-Gradinger C, Szakacs G (2003) Genetic and metabolic diversity of *Trichoderma*: a case study of on South East Asian isolates. Fungal Genet Biol 38:310–319

Kumar DS, Ray S (2014) Fungal lipase production by solid state fermentation. An overview. J Anal Bioanal Technol 6:1–10

Kumar MR, Kumaran MDR, Balashanmugayan P, Rebecca AIN, Kumar DJM, Kalaiselvan PT (2014) Production of cellulase enzyme by *Trichoderma reesei* Cef 19 and its application in the production of bioethanol. Pak J Biol Sci 17:735–739

Kunamneni A, Plou FJ, Alcalde M, Ballesteros A (2014) *Trichoderma* enzymes for food industries. In: Gupta VK, Schmoll M, Herrera-Estrella A, Upadhyay RS, Druzhinina I, Tuohy MG (eds) Biotechnol biology of *Trichoderma*. Elsevier, pp 339–343

Liu Z, Inokuma K, Ho SH, Haan RD, Hasunuma T, Zyi WH, Kondo A (2015) Combined cell surface display and secretion based strategies for production of cellulosic ethanol with *Saccharomyces cerevisiae*. Biotechnol Biofuels 28:1–12

Main GL, Morris JR (2007) Effect of macerating enzymes and post fermentation grape seed tannin on the color of Cynthiana wines. Am J Enol Vitic 58(3):365–372

Martinez D et al (2008) Genome sequencing and analysis of the biomass-degrading fungus *Trichoderma reesei* (Syn. *Hypocrea jacorina*). Nat Biotechnol 26(5):553–561

Matano Y, Hasunuma T, Kondo A (2013) A. Simultaneous improvement of saccharification and ethanol production from crystalline cellulose by alleviation of irreversible adsorption of cellulase with a cell surface engineered yeast strain. Appl Microbiol Biotechnol 97:2231–2237

Merino ST, Cherry J (2007) Progress and challenges in enzyme development for biomass utilization. Adv Biochem Eng Biotechnol 108:95–120

Montenecourt BS, Eveleigh DE (1977) Preparation of mutants of *Trichoderma reesei* with enhanced cellulase production. Appl Environ Microbiol 34(6):777–782

Mukherjee PK (2017) *Trichoderma*: genes, genomics and applications. BARC News Letter 2017, Founders Day Special issue, 1–7 pp
Nakatani Y, Yamada R, Ogino C, Kondo A (2013) Synergistic effect of yeast cell-surface expression of cellulase and expansin like protein on direct ethanol production from cellulose. Microb Cell Factories 12(66):1–7
Nevalainen H, Pirkko S, Kaarina T (1994) On the safety of *Trichoderma reesei*. J Biotechnol 37(3):193–200
Nirmal NP, Shankar S, Laxman RS (2011) Fungal proteases: an overview. Int J Biotechnol Biosci 1:1–40
Ohmiya K, Sakka K, Karita S, Kimura T (1997) Structure of cellulose and their application. Biotechnol Genet Eng Rev 14:365–414
Ottone C, Romero O, Aburto C, Illanes A (2020) Biocatalysis in the winemaking industry: challenges and opportunities for immobilized enzymes. Compr Rev Food Sci Food Saf 9:595–621
Paloheimo M, Haarmann T, Makinen S, Vehmaanpera J (2016) Production of industrial enzymes in *Trichoderma reesei*. In: Dattenabock SM (ed) Gene expression systems in fungi: advancements and applications. Fungal biology. Springer, pp 1–61
Pariza MW, Johnson EA (2001) Evaluating the safety of microbial enzyme preparations used in food processing: update for a new century. Regul Toxicol Pharmacol 33:173–186
Pecket RC, Small CJ (1980) Occurrence, location and development of anthocyanoplasts. Phytochemistry 19(12):2571–2576
Persoon CH (1794) Neuer Versuch einer systematischen Einteilung der Schwämme. Racodium Römer's Neues Magazin der Botanik 1:123
Pretorius IS (2000) Tailoring wine yeasts for the new millennium novel approaches to the ancient art of wine making. Yeast 16:75–729
Rajesh EM, Arthe R, Rajendran R, Balakumar C (2010) Investigation of lipase production by *Trichoderma reesei* and optimization of production parameters. Electron J Environ Agric Food Chem 9:1177–1189
Reinikainen T (1994) *Trichoderma reesei* cellulases in processing of cotton. VTT Publications, pp 1–77
Rensburg PV, Pretorius IS (2000) Enzymes in wine making: harnessing natural catalysts for efficient biotransformations: a review. S Afr J Enol Vitic 21:52–73
Rifai MA (1969) A revision of the genus *Trichoderma*. Mycol Pap 116:1–56
Saili NS, Siddiquee S, Wong CM, Ling V, González M, Kumar SV (2014) Lignocellulolytic activities among *Trichoderma* isolates from Lahad Datu, Sabah and Deception Island, Antarctic. J Microb Biochem Technol 6(5):295–302
Samuels GJ (1996) *Trichoderma*: a review of biology and systematic of the genus. Mycol Res 100:923–935
Scheffler A, Bamforth CW (2005) Exogenous b-glucanases and pentosanases and their impact on mashing. Enzym Microb Technol 36:813–817
Sivasithampara K, Ghisalberti EL (2002) Secondary metabolism in *Trichoderma* and *Gliocladium*. In: Kubicek CP, Harman GE (eds) Trichoderma and Gliocladium basic biology, taxonomy and genetics, vol 1. Taylor & Francis Ltd, pp 139–191
Tenkanen M, Puls J, Poutanen K (1992) Two major xylanases of *Trichoderma reesei*. Enzym Microb Technol 14:566–574
Tulasne LR, Tulasne C (1865) Selecta fungorum carpologia. Jussu, Paris
Ullrich SE (ed) (2011) Barley: production, improvement, and uses. Wiley-Blackwell, West Sussex, p 656
Villettaz JC, Dubourdieu D (1991) Enzymes in wine-making. In: Fox PF (ed) Food enzymology. Elsevier Science Publishers Ltd
Villettaz JC, Steiner D, Trogus H (1984) The use of a beta-glucanase as an enzyme in wine clarification and filtration. Am J Enol Vitic 35(4):253–256
Whitaker JR (1984) Pectin substances, pectin enzymes and haze formation in fruit juices. Enzym Microb Technol 6:341–349

Wong KKY, Saddler JN (1992) *Trichoderma* xylances, their properties and application. Crit Rev Biotechnol 12(5–6):413–435

Xiong L, Huang C, Peng W, Tang L, Yang X, Chen X, Chen X, Ma L, Chen Y (2013) Efficient cellulase production from low-cost substrates by *Trichoderma reesei* and its application on the enzymatic hydrolysis of corncob. Afr J Microbiol Res 7:5018–5024

Yamada R, Hasunuma T, Kondo A (2013) Endowing non cellulolytic microorganisms with cellulolytic activity aiming for consolidated bioprocessing. Biotechnol Adv 31:754–763

Part VI
Risk Factors Related to *Trichoderma* Applications in Agriculture

Trichoderma Green Mould Disease of Cultivated Mushrooms

László Kredics, Lóránt Hatvani, Henrietta Allaga, Rita Büchner, Feng Cai, Csaba Vágvölgyi, Irina S. Druzhinina, and Shahram Naeimi

Contents

L. Kredics (✉) · L. Hatvani · H. Allaga · R. Büchner · C. Vágvölgyi
Department of Microbiology, Faculty of Science and Informatics, University of Szeged, Szeged, Hungary
e-mail: kredics@bio.u-szeged.hu; csaba@bio.u-szeged.hu

F. Cai · I. S. Druzhinina
Key Laboratory of Plant Immunity, Fungal Genomics Laboratory (FungiG), Nanjing Agricultural University, Nanjing, China
e-mail: irina.druzhinina@njau.edu.cn

S. Naeimi
Department of Biological Control Research, Iranian Research Institute of Plant Protection, Agricultural Research, Education and Extension Organization (AREEO), Tehran, Iran

N. Amaresan et al. (eds.), *Advances in* Trichoderma *Biology for Agricultural Applications*, Fungal Biology, https://doi.org/10.1007/978-3-030-91650-3_21

1 Introduction

Mushrooms are ideal sources of vitamins, proteins, minerals and polysaccharides with low fat content (Khan et al. 1981) and are known to have a broad range of uses both as food and medicine (Alice and Kustudia 2004). Due to their various nutritional and functional food properties, edible mushrooms form an important part of human diet. Commercial cultivation of edible mushrooms was first introduced in France during the eighteenth century. *Lentinula, Pleurotus* and *Agaricus* are among the five most commonly cultivated mushroom genera in the twenty-first century along with *Auricularia* and *Flammulina* (Royse et al. 2017). The global annual edible mushroom production is estimated to be around $42 billion (Prescott et al. 2018).

With a production of 4.43 million tons in 2013, white button mushroom (*Agaricus bisporus*) is the most commonly produced mushroom in Europe and North America, accounting for about 15% of worldwide total mushroom production (Royse et al. 2017). It is produced on mushroom compost, which is a cultivation substratum usually containing a mix of wheat straw, gypsum and chicken manure. This mixture is processed through the composting procedure consisting of fermentation (Phase I), pasteurisation (Phase II) and conditioning and spawning with cereal grain inoculated with *A. bisporus* mycelium (Phase III). The compost generally reaches the farmers inoculated with *A. bisporus* (Phase III).

Shiitake (*Lentinula edodes*), usually grown on wheat straw or bed logs, is widely cultivated in East Asia as a food source and is dried and exported to several countries due to its special aroma and flavour (Chen 2005; Luo 2004). Shiitake also has valuable medicinal functions, including immunostimulant properties (Yamamoto et al. 1997). In China, it is one of the most important edible mushrooms (Wang et al. 2016), while in Korea, its consumption has been increasing annually since 1999. There are now about 20 cultivars of shiitake.

The cultivation of oyster mushroom (*Pleurotus ostreatus*) has grown exponentially in recent years with China as the most important *Pleurotus*-growing country. In Europe, Italy, Hungary and Poland are the main producers of *P. ostreatus* (Błaszczyk et al. 2013). In addition to being grown for human consumption, it is also used for the bioconversion of agricultural and industrial lignocellulose (Ballero et al. 1990; Puniya et al. 1996) and is a source of enzymes and metabolites important for industry and medicine (Marzullo et al. 1995; Gunde-Cimerman 1999; Gregori et al. 2007). Oyster mushrooms possess good nutritional value as well as anti-inflammatory and immunomodulatory properties (Lavi et al. 2010). They are rich in proteins, vitamins and minerals (Mattila et al. 2000). Oyster mushroom can be grown on a wide range of industrial wastes and agricultural by-products, including wheat straw, paddy straw, vegetable plant residues, maize stalks/cobs, bagasse, etc. (Hassan et al. 2011).

The production of the aforementioned mushrooms can be seriously affected by the so-called green mould diseases caused by certain members of the filamentous fungal genus *Trichoderma* (Hypocreales, Ascomycota). Some *Trichoderma* species are industrially significant, while others have long been known to antagonise various plant pathogenic fungi (Gupta et al. 2014). Among the most important background mechanisms of antagonism, the competition for nutrients and living space, antibiosis by the production of antibiotics, mycoparasitism aided by the activity of cell wall-degrading enzymes as well as the stimulation of plant germination, growth and defence responses need to be mentioned (Olmedo Monfil and Casas-Flores 2014). During the recent decades, it has also become clear that members of the genus may be opportunistic human pathogens (see Chap. 22), while some species were identified as pathogens of green mould diseases causing significant yield losses in the production of white button mushroom, oyster mushroom, shiitake and other cultivated mushrooms.

This chapter aims to provide a comprehensive overview about the epidemiology, biology and control options of mushroom green mould diseases caused by *Trichoderma* species.

2 Green Mould of White Button Mushroom (*Agaricus bisporus*)

In the case of *A. bisporus*, it has long been known that compost infected with *Trichoderma* may result in yield reduction (Sinden and Hauser 1953). Initially, the species *T. viride* and *T. koningii* were reported to be responsible for intermittent crop losses (Sinden and Hauser 1953). Sinden (1971) characterised representatives of the genus *Trichoderma* as competitors of cultivated mushrooms indicating the poor quality of compost, which appear under acidic conditions or in the case of the presence of soluble sugar residues. The green mould disease of *A. bisporus* mould was considered an insignificant problem until the 1980s, related to weak compost quality or inadequate hygienic conditions, that could be treated by improving the quality of the compost, ensuring adequate hygiene or applying chemical intervention (Geels et al. 1988). This view changed fundamentally due to the green mould outbreaks in Ireland, Northern Ireland (1985–1986) and later in the British Isles (1990–1991), causing around £3–4 million losses (Doyle 1991; Fletcher 1990; Morris et al. 1995a, b; Seaby 1987, 1989, 1996a, b, 1998). In 1994, the problem also occurred in the Netherlands resulting in severe losses (Geels 1997). In the early 1990s, a similar disease has emerged in the United States and Canada (Alberta, British Columbia, Ontario and Pennsylvania) and caused more than $30 million losses in *Agaricus* production (Castle et al. 1998; Ospina-Giraldo et al. 1998, 1999; Rinker 1993, 1994; Romaine et al. 1996; Spillmann 2002). In France, green mould became a known problem in 1997 (Mamoun et al. 2000a, b), while in Spain, *Trichoderma* strains much more aggressive than the ones previously known were first noticed in the winter of 1996–1997 (García-Morrás and Oliván 1997; Hermosa et al. 1999). *Trichoderma* green mould of *A. bisporus* appeared later also in Central Europe, from Poland (Szczech et al. 2008; Sobieralski et al. 2009a, b) through Hungary (Hatvani et al. 2007; Kredics et al. 2010) to Croatia (Hatvani et al. 2012) and Serbia (Kosanović et al. 2013), as well as in Turkey (Aydoğdu et al. 2020), Iran (Vahabi et al. 2009; Zargarzadeh et al. 2011), Mexico (Romero-Arenas et al. 2009) and Australia (Clift and Shamshad 2009). Due to the spread of the problem, intensive research programmes started to identify and study the green mould pathogens.

2.1 Epidemiology

Although a series of *Trichoderma* species including *T. citrinoviride*, *T. crassum*, *T. hamatum*, *T. koningii*, *T. longibrachiatum* and *T. spirale* (Fig. 1) were previously described from *Agaricus* compost (Castle et al. 1998), the aggressive colonisation leading to outbreaks and epidemics was initially attributed exclusively to strains of *T. harzianum* sensu *lato* (Doyle 1991; Seaby 1987, 1989). *T. harzianum* sensu *lato* isolates from compost in the British Isles were classified into three biotypes, Th1, Th2 and Th3, differing in their growth rate, conidium formation and aggressivity to *A. bisporus* (Doyle 1991; Seaby 1987). Inoculation experiments revealed that the

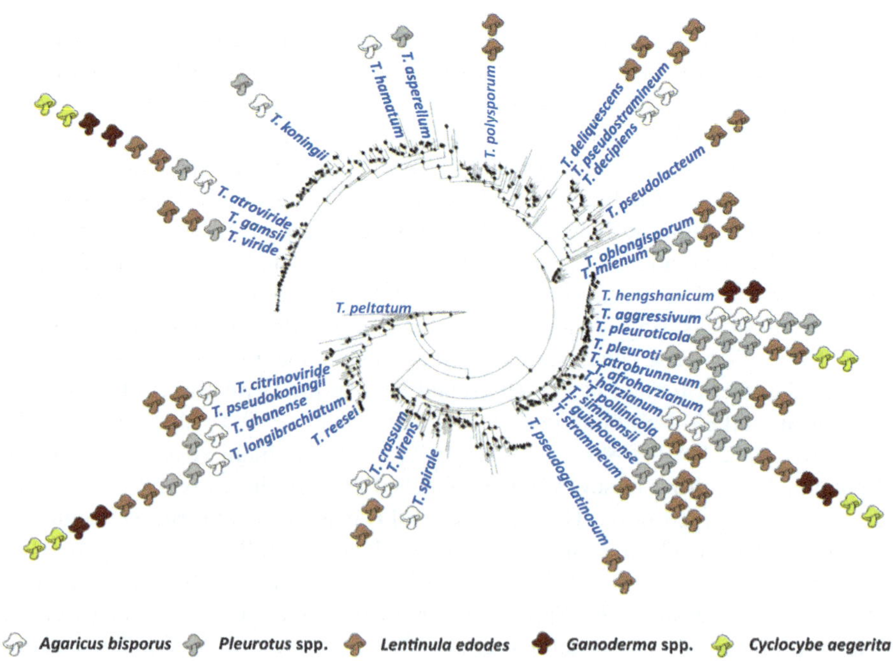

Fig. 1 Taxonomic position of *Trichoderma* species associated with cultivated mushrooms. Mushroom symbols of different colours indicate different host mushrooms. The number of symbols of the same colour reflects the significance of the particular mushroom (one symbol, detected in cultivation; two symbols, identified as pathogenic mould; three symbols, reported to cause epidemic outbreaks). The Bayesian tree was inferred from the alignment of 808 nucleotides of the *rpb2* gene for 349 sequences retrieved from NCBI GenBank (Kai et al. 2020). Nodes supported with posterior probabilities above 0.94 are marked by black diamonds

development of green mould epidemics can be attributed to the most aggressive biotype Th2 (Fletcher 1990; Seaby 1987, 1989; Staunton 1987).

The biotypes are different also in terms of their colony morphology and micromorphological characteristics (phialides and conidiospores) (Seaby 1996a). The Th1 biotype usually occurs in raw compost ingredients but rarely occurs in pasteurised compost (Seaby 1987). It has a high growth rate at 27 °C (1 mm/h) and produces conidia within 2 days under illumination. The sporulating culture is spinach-like green in colour and has a malt-like smell (García-Morrás and Oliván 1997). Biotype Th2 is rarely found in raw compost components; it occurs primarily in affected compost (Morris et al. 1995a, b). It shows rapid growth at 27 °C (1 mm/h) and forms a cotton-like layer of aerial mycelium. Conidiation begins after 4 days with the formation of concentric green bands. The Th3 biotype, like Th1, colonises raw compost components, but its conidia may occasionally also enter the pasteurised compost from the air (Seaby 1996a). Its growth rate is 0.5–1 mm/h, and the colony emits a coconut odour. The Th1-Th3 grouping was later supported by studies on 81 compost isolates of *Trichoderma* using molecular biological methods (restriction fragment length polymorphism (RFLP), random amplified polymorphic DNA (RAPD), sequence analysis of the internal transcribed spacer (ITS) region). Representatives of the Th2 biotype proved to be genetically homogeneous (Muthumeenakshi et al. 1994), supporting the assumption that the green mould epidemic in the British Isles may have proceeded from a single source, probably in Northern Ireland (Morris et al. 1995a, b), due to a mutation resulting in the development of aggressive compost colonisation ability (Seaby 1987). However, the British and Irish isolates can be distinguished from each other based on minor differences found in the mitochondrial DNA (Muthumeenakshi et al. 1994). This genetic diversity is hypothesised to be the result of several additional mutations following the first mutational event that enabled compost colonisation. The molecular techniques mentioned above were later also applied to *Trichoderma* isolates from mushroom farms of North America (Castle et al. 1998; Muthumeenakshi and Mills 1995; Muthumeenakshi et al. 1998; Qi et al. 1996). The aggressive Th2 biotype in the British Isles and the North American pathogen – named biotype Th4 – proved to be clearly distinguishable. Th4 strains appeared to be genetically uniform, suggesting that the Th4 biotype can also be derived from a single source. Based on sequence analysis of the ITS1 region, there was a 5 base pair difference between the Th2 and Th4 biotypes, and both showed a close phylogenetic relationship to the Th1 biotype of *T. harzianum* (Muthumeenakshi et al. 1998). The Th4 biotype has a growth rate of 0.8 mm/h, colonies with wavy edges forming aerial mycelium and conidiation occurring in bands (Seaby 1996a). Based on the above results, *Agaricus* green mould infection is not caused by a single strain; aggressive biotypes appear to have developed from at least two independent sources (British Isles and North America), which also explains the differences among British/Irish and North American green mould isolates of *Trichoderma*.

As members of the Harzianum clade of the genus *Trichoderma* are often used as biocontrol agents against plant pathogenic fungi (Naeimi et al. 2019), concerns have been raised that biocontrol strains may be able to cause green mould infections.

However, ITS sequence analysis and RAPD have shown that although the strains used in biological control and green mould-causing strains are close relatives, they can be clearly distinguished from each other (Hermosa et al. 2000; Ospina-Giraldo et al. 1999; Royse et al. 2001). The Th2 and Th4 biotypes and biocontrol strains thus presumably derive from a common ancestor. All this is supported by the phylogenetic analysis based on a fragment of the β-tubulin gene (Romaine et al. 1999), while the results of artificial infection experiments showed that in contrast to Th4 isolates but similarly to representatives of the Th1 biotype, *T. harzianum* strains used for biocontrol are unable to attack *Agaricus bisporus* (Rinker et al. 1997a; Romaine et al. 2001). Molecular differences between the Th1 and Th3 biotypes let Muthumeenakshi et al. (1994) to first hypothesise that these biotypes may represent three separate species. Later molecular phylogenetic studies have shown that Th3 is in fact *T. atroviride* (Castle et al. 1998; Ospina-Giraldo et al. 1998), while the Th1 biotype was identified as *T. harzianum* sensu stricto (Gams and Meyer 1998). These two species of *Trichoderma* occur most commonly in *Agaricus* production of Australia (Clift and Shamshad 2009). However, it has been shown in Croatia that members of the *T. harzianum* species complex (THSC) may also be able to cause green mould problems (Hatvani et al. 2012). A redescription of the two aggressive biotypes, Th2 and Th4 as *T. aggressivum* f. *europaeum* and *T. aggressivum* f. *aggressivum*, respectively (Fig. 2), was performed based on their morphological characteristics, as well as the phylogenetic analysis of their ITS1 region and a fragment of the translation elongation factor 1-alpha (*tef1*) gene (Samuels et al. 2002). The biotypes previously known as Th2 and Th4 are therefore referred to hereinafter as Ta2 and Ta4, respectively. Although *T. aggressivum* f. *europaeum* (Ta2) and *T. aggressivum* f. *aggressivum* (Ta4) show statistically significant micromorphological differences and diverse growth rates on synthetic low nutrient agar (SNA) at 25 °C, their

Fig. 2 Adult of the sciarid mushroom fly *Lycoriella ingenua*, a potential vector of *Agaricus* green mould disease under Zeiss STEMI-305 stereomicroscope. (Photo: Rita Büchner)

differentiation based on purely morphological basis is practically impossible. Ta2 was described to be responsible for most green mould problems in Europe, while Ta4 to cause problems in Mexican, US and Canadian mushroom farms. *T. aggressivum* causes severe economic losses in *A. bisporus* cultivation (O'Brien et al. 2014); the estimated worldwide losses in mushroom yield caused by this green mould species are in the tens of millions of dollars (Kredics et al. 2010). Green mould losses in the United States were estimated to be 14 million USD in 2011 (Pecchia 2012). To date, Ta2 has been prevalent and caused serious yield losses between 60% and 100% in European *A. bisporus* production (O'Brien et al. 2017; Kredics et al. 2010). Although a 6-month survey in England from December 2007 involving 15 *Agaricus* farms could not identify the presence of Ta4 (Lane 2008), the North American biotype was later detected in Europe and caused substantial losses in Hungarian *Agaricus* production (Hatvani et al. 2017). Similar problems occurred also in Australia and Iran, where the North American biotype Ta4 also appeared and caused severe crop losses (Khan et al. 2008; Zargarzadeh et al. 2011). More recently, Aydoğdu et al. (2020) isolated *T. aggressivum* f. *aggressivum* from an *Agaricus* farm in Turkey. Thus, the recent distribution of *T. aggressivum* f. *aggressivum* increases the genetic diversity of the species *T. aggressivum* worldwide, which will likely cause new, unexpected problems for mushroom growers.

T. aggressivum has not yet been isolated from the natural environment. To investigate whether the natural substratum of wild-grown *Agaricus* species and the surface of their fruiting bodies may be natural reservoirs of *T. aggressivum*, *Trichoderma* strains were isolated at three Hungarian locations (Kecskemét, Nagykőrös, Szeged) from the environment of wild-growing *Agaricus* species (Kredics et al. 2010). A total of 65 *Trichoderma* strains isolated from the environment of *Agaricus* species were analysed by PCR techniques specific for *T. aggressivum* and by analysing the sequences of the ITS region and a fragment of the *tef1* gene. Seven different species, *T. atroviride*, *T. hamatum*, *T. harzianum*, *T. koningii*, *T. koningiopsis*, *T. tomentosum* and *T. virens*, were identified based on the sequences. The presence of *T. aggressivum* in samples from the wild mushroom environment has not been detected so far, meaning that there are still no data on the natural occurrence of this species (Kredics et al. 2010).

T. aggressivum f. *europaeum* may initially have spread between composting plants in the British Isles, presumably through joint customers. Transport vehicles can carry large amounts of dust, conidia, mycelial debris, mites and mosquitoes, making them a risk factor for spreading between facilities. In addition, growers threw out or scattered thousands of bags infected with *Trichoderma* during the original epidemic, which may also have contributed to the initial spread of the infection (Seaby 1996b). Affected farms often face prolonged green mould infections, and infection of freshly spawned compost with green mould can only be avoided through strict hygienic interventions (Rinker et al. 1997b; Seaby 1987).

Another important question is the infection route. According to the record of a grower, green mould infestation is 60% higher when compost is bagged in dry, windy weather, and compost from heat treatment tunnels most exposed to wind is the most contaminated, suggesting that airborne dust particles can contribute greatly to contamination (Seaby 1996b). Besides mushroom compost, the casing material may be another inoculum source of *T. aggressivum* in *Agaricus* cultivation (Szczech et al. 2008; Aydoğdu et al. 2020).

Trichoderma species other than *T. aggressivum* are generally known to cause less and/or occasional damage in mushroom crops and supposed to be less adapted than *T. aggressivum* to grow in compost colonised by *A. bisporus*. In Iran, Vahabi et al. (2009) obtained 423 isolates of *Trichoderma* from compost, pili, spawn and casing soil from industrial mushroom production of *A. bisporus*. *T. atroviride*, *T. citrino-viride*, *T. ghanense*, *T. harzianum*, *T. longibrachiatum* and *T. virens* (Fig. 1) could be identified using morphological characters and ITS sequence analysis, but no isolates of *T. aggressivum* were obtained in this study. *Trichoderma* species with lower ability to cause disease symptoms, including *T. harzianum* sensu *lato*, *T. virens*, *T. atroviride*, *T. koningii*, etc., have also been isolated from Serbian *Agaricus* mushroom beds (Kosanović et al. 2013). Kosanović et al. (2013) collected 20 *Trichoderma* isolates from *A. bisporus* farms in Serbia as well as Bosnia and Herzegovina. Twelve isolates were identified (based on morphology and ITS sequence analyses) as members of *T. atroviride*, *T. koningii*, *T. virens*, *T. aggressivum* f. *europaeum* and the THSC. Hatvani et al. (2012) reported that at Croatian *Agaricus* farms, the green mould pathogens were exclusively from the THSC. This result is different from earlier reports, as previous studies from other countries identified only *T. aggressivum* as the agent of *Agaricus* green mould and indicated a widening spectrum of *Agaricus* green mould pathogens, also suggesting a continuous evolution of green mould disease in cultivated mushrooms and underlining the importance of monitoring these infections. Furthermore, a recent 'white mould' outbreak at a Hungarian *Agaricus* farm revealed *T. decipiens* as the causal agent (Fig. 1; Geösel and Hatvani, personal communications).

Sobieralski et al. (2012a) studied the effect of inoculation of the cultivation substratum with *T. harzianum*, *T. aggressivum* (Ta2), *T. atroviride* and *T. hamatum* isolates on the yield of *A. bisporus*. Although the greatest yield reduction was caused by an isolate of Ta2, other *Trichoderma* isolates also inhibited the mycelial growth of *A. bisporus*. Górski et al. (2014) investigated the effect of inoculation of cultivation substratum with different *Trichoderma* species including *T. aggressivum* f. *europaeum*, *T. atroviride*, *T. hamatum*, *T. harzianum*, *T. inhamatum*, *T. koningii* and *T. longibrachiatum* on the yield of four *A. bisporus* strains. Except for *T. atroviride*, all the examined *Trichoderma* isolates reduced the yield of the button mushroom strains.

Many factors contribute to green mould disease development in *Agaricus* farms, and no single factor acts alone in green mould epidemiology (Mazin et al. 2019). Regarding the nature of the raw materials used for the mushroom growing substratum and composting procedures (Anderson et al. 2001), composts with high carbohydrate but low nitrogen content proved to be suitable for green mould development (Sharma et al. 2007). Compost features such as temperature, moisture, pH, conductivity, C/N ratio as well as macro- and micronutrients influence the mycelium development of *A. bisporus* and green moulds. Higher temperature, humidity and the presence of organic matter in the growing house during the cultivation of *Agaricus* favour the development of *Trichoderma* spp. aggressive to mushrooms (Grogan 2005). This is particularly important in the cultivation of *A. bitorquis*, a button mushroom species that prefers higher incubation temperature than *A. bisporus* (Guler et al. 2006; Hasselbach and Mutsers 1971), which, due to the higher thermal requirements of *T. aggressivum* f. *europaeum*, can pose a serious problem. Hatvani et al. (2012) stated that the temperature profiles of pathogenic *Trichoderma* isolates from Croatian *Agaricus* (and *Pleurotus*) farms and their hosts overlapped, with optima between 25 and 30 °C; thus, no range was found that would allow optimal growth of the mushrooms without mould contamination. Kosanović et al. (2015) evaluated the effects of light and pH on the growth of *Trichoderma* isolates originating from farms in Serbia as well as Bosnia and Herzegovina and reported that the majority of isolates showed an optimal growth at pH 5.0, while the rest at pH 6.0, with a few isolates also growing well at pH 7. The weakest mycelial growth was detected in the range of pH 8.0–9.0. The growth of the isolates proved to be inhibited by light.

Seaby (1996b) conducted detailed epidemiological studies on green mould caused by the Ta2 biotype in the Republic of Ireland and Northern Ireland. In these areas, the so-called satellite system is used for mushroom cultivation, the essence of which is that compost production and cultivation are separated: composting plants supply many small satellite plants with bagged compost (Győrfi 2002). The study included both the examination of samples taken both from compost and mushroom farms as well as the results of artificial infection experiments. Ta2 strains could not be isolated from the wastewater of the farms, but almost all surfaces (packaging machines, growing unit floors, handrails, rungs, spawn dispensers, wooden pallets, trailers and tarpaulin road covers) proved to be positive for the Ta2 biotype. *Trichoderma* strains, most commonly the Ta2 biotype, were also recovered from samples taken from workers' clothing (jackets, picker's gloves, woollen jerseys) by electrostatic and vacuum methods. However, Ta2 strains could not be detected in the peak-heated, contaminated compost, suggesting that Ta2 does not survive heat treatment. The number of strains isolated from cotton bags was reduced to zero after 30 min of treatment with a tumbler dryer at 60 °C. The study also identified possible biological vectors of green mould infection: groups of *Trichoderma* conidia were observed on the body surface of pepper mites by microscopic examination, and Ta2 strains were also isolated. Red pepper mites were found to be able to reproduce on all *Trichoderma* species present in compost, so their presence is not an indicator of a specific green mould species (Clift and Shamshad 2009). Ta2 strains have also been isolated from sciarid mushroom flies, and when a mite-covered mouse cadaver came into contact with the spawn, green mould also appeared (Seaby 1996b). Sciarid mushroom flies, especially *Lycoriella ingenua* (Fig. 2), are considered as vectors for *T. aggressivum* f. *aggressivum* confirmed by scanning electron microscopy and molecular analysis (Mazin et al. 2019); therefore, they may be important parts of green mould disease epidemiology in *Agaricus* farms. The flies may acquire *Trichoderma* conidia upon landing on *T. aggressivum*-infested compost patches and then spread the conidia to other compost bed areas via further movement. Mazin et al. (2019) have demonstrated that the larvae of sciarid mushroom flies reared on spawned compost infested with *Trichoderma* green mould develop faster into adults than the larvae that reared on spawned compost uninfected with green mould.

Based on artificial infection experiments, early inoculation (on day 7) resulted in intensive colonisation of the compost by the Ta2 biotype, whereas late inoculation (after 14 days) showed no significant colonisation (Seaby 1996b). The use of experimentally infected CACing additive (CACing: the addition of a small amount of compost colonised by *Agaricus* mycelium to the casing material to accelerate spawn run) did not lead to the appearance of Ta2, although in some cases the green mould-infested compost used as CACing additive resulted in Ta2 conidiation on the surface of the casing layer. Spawned compost bags became highly infected if they had been touched with hands artificially contaminated with the Ta2 biotype. The use of metals essential for mushroom growth (Cu, Fe, Mn and Zn) did not reduce the colonisation of compost by the Ta2 biotype, but the use of some bacteria isolated from compost was able to inhibit it (Seaby 1996b).

The spatial distribution of the Ta4 biotype was investigated by Royse et al. (1999) in North American mushroom farms. Based on the results, the appearance pattern of Ta4 is not random: green mould foci appeared in an aggregated pattern at the edges of the shelves, suggesting non-airborne infection. Based on Rinker's (1996) experiments in Canada, air carries only small amounts of Ta4 conidia. Thus, in the North American cultivation system, infections originate presumably from workers and contaminated equipment, so the most likely factors influencing the appearance of green mould are activities involving movement along shelves (filling shelves, spawning, spreading the casing layer, nutrient supplementation, etc.) (Royse et al. 1999).

O'Brien et al. (2017) studied the effect of artificial green mould inoculation of fully colonised, bulk Phase III compost on mushroom yield. Phase III compost proved to be vulnerable to *T. aggressivum* infection when the fully colonised substratum was broken up and mixed during bulk handling operations, with a higher degree of mixing leading to significantly increased crop losses. The authors suggested that mixing results in the rupture of the fragile *A. bisporus* mycelial network, which enables the release of new nutrition sources and stimulatory metabolites for *T. aggressivum*, that are not accessible in intact, fully colonised substratum.

2.2 Biology

Pathogenic green moulds can colonise the compost used for the cultivation of *A. bisporus* but may even grow on the surface of developing fruiting bodies (Largeteau and Savoie 2010). There is a lack of symptoms till 10–35 days after the apparently normal run of *Agaricus* spawn. The whitish mycelium of *Trichoderma* species cannot be distinguished from the mycelium of the mushroom during the phase of spawn run, so infection at this stage is difficult to detect (Largeteau-Mamoun et al. 2002). Later, as the mycelium of *Trichoderma* begins to conidiate, large green spots are appearing on the surface (Fig. 3) (Rinker 1996; Seaby 1996a).

Fig. 3 *Agaricus* compost infected with *T. aggressivum* in Hungary. (Photo: Rita Büchner)

The main symptoms of green mould are the formation of green conidiospores on the casing layer or in the mushroom compost during the second and fifth week of the growing cycle, which gives the condition its common names, green mould disease or *Trichoderma* compost mould (Morris et al. 1995a, b; Seaby 1987; Fletcher and Gaze 2007). In areas colonised by *T. aggressivum*, the *Agaricus* fruiting body formation is retarded, while the fruiting bodies that do develop are generally of poor quality due to discolouration or damage (Largeteau and Savoie 2010). Yield loss is proportional to the size of the infected area: mushrooms do not grow in the case of serious infections. Even if the fruiting bodies of the mushroom appear, the crop is not marketable, as there are often brown necrotic lesions and distortions on them (Seaby 1989; Largeteau and Savoie 2010), and the fragrances of *Trichoderma* also attract red pepper mites (*Pygmephorus mesembrinae*), which consume the conidia of *Trichoderma* (Morris et al. 1995a, b; Krupke et al. 2003).

Spawning could be identified as a critical point of time for green mould infection, as the carbohydrates readily available in the grains of spawn provide *T. aggressivum* a backup for substratum colonisation (Fletcher 1997; Seaby 1996a, b). Compost entirely colonised by *A. bisporus* is considered to be more resistant to green mould infection due to the limited access to the substratum, and the infections after spawning tend to be limited to the edges of the bag or compost block (Fletcher 1997; Rinker and Alm 2000). Based on the results of Sharma et al. (1999), when inoculated into compost during spawning, Ta2 settles and spreads rapidly in all directions. On the other hand, *T. harzianum* (Th1) and *T. atroviride* (Th3) are colonising slowly and can only reach short distances from the inoculation point.

Although these biotypes also reduce the crop quality, yields are generally not significantly affected.

Mushroom growing conditions (presence of carbon and nitrogen sources, high relative humidity, high temperature, the lack of light during spawn run) provide ideal conditions for *Trichoderma* species, which can thus easily settle on compost. The rate and extent of *Trichoderma* colonisation in the compost is influenced by its degree of fermentation and moisture level (Sharma et al. 1999). Thus, parameters that positively affect the development of *Agaricus* also favour *Trichoderma* colonisation. *T. aggressivum* f. *europaeum* isolates can cause very serious reduction of *A. bisporus* pinhead formation (Sobieralski et al. 2010a). Sobieralski et al. (2010b) also determined the effect of Ta2 infection on the yield of wild *A. bisporus* strains derived from natural habitats in Poland, and yield losses up to 75% were reported. Studying the influence of substratum inoculation with Ta2 on yielding of several *A. bitorquis* strains from natural sites of different regions of Poland revealed significant yield reductions of both the cultivated mushroom strain and strains obtained from the natural environment (Sobieralski et al. 2010c).

Several compounds are involved in the interaction between *T. aggressivum* and *A. bisporus*, including volatile compounds, small non-volatile molecules and extracellular enzymes of *T. aggressivum*, which are inhibitory to *A. bisporus* (Mumpuni et al. 1998; Krupke et al. 2003; Guthrie and Castle 2006), as well as *Agaricus* metabolites, e.g. fungistatic compounds produced to counteract the growth of moulds, or to resist to growth-limiting metabolites produced by *Trichoderma* strains (Foulongne-Oriol et al. 2011). The disease severity may be reduced if *Agaricus* is able to successfully colonise the compost before the development of *T. aggressivum*. At the contact point between *Agaricus* and *Trichoderma* during infection, *A. bisporus* strains with the ability to adapt or resist to the antifungal metabolites and lytic enzymes of *T. aggressivum* have the potential to be less affected than others (Foulongne-Oriol et al. 2011). Under favourable conditions, *Trichoderma* moulds grow rapidly and compete more effectively for habitat and nutrients than mushrooms, henceforth producing toxic secondary metabolites, extracellular enzymes and volatile organic compounds (Mumpuni et al. 1998; Williams et al. 2003a), which can lead to drastic crop losses.

The *Agaricus* compost contains large amounts of cellulose and lignin, so this medium is strongly selective for basidiomycetes which can use lignin as a carbon source through the action of their extracellular enzymes like cellulases and laccases (Matcham and Wood 1992). The presence of *Agaricus* mycelium is required by compost-colonising *Trichoderma* biotypes for proliferation in the compost (Seaby 1987, 1996a): the Ta2 mycelium does not develop to a detectable extent without the presence of *A. bisporus* (Seaby 1996a; Romaine et al. 1998). According to Rinker's (1996) research, however, the aggressive *Trichoderma* is able to colonise compost even in the absence of *A. bisporus* but forms conidia only on freshly inoculated compost. Further studies indicated that Ta2 is capable of colonising both *A. bisporus*-inoculated and *A. bisporus*-free composts (Mamoun et al. 2000b; Morris et al. 1995a, b). As an explanation for the different observations mentioned above, Mamoun et al. (2000b) suggested that the intense conidium formation of Ta2 requires the mycelium of the mushroom. Ta2 is able to colonise the compost even in the absence of mushrooms, but its thin hyphae cannot be detected without a microscope. The ability to form high mycelial mass and the delayed conidium formation may be key factors in the ability of Ta2 to colonise spawned compost and reduce crop yield and quality. According to Mumpuni et al. (1998), the Th1, Ta2 and Th3 biotypes produce volatile metabolites in vitro that inhibit *A. bisporus* growth. Th1 and Th3 showed higher levels of toxicity, in contrast to Ta2, from which the strongest toxicity was expected. Henceforth, Th1 and Th3 were inhibited much more strongly than Ta2 by mushroom compound(s); therefore, the appearance of these biotypes was generally limited to small compost areas free of mushrooms (Seaby 1987). Based on the above, mutual inhibition between the Th1 and Th3 biotypes and *A. bisporus* can be assumed. The mushroom mycelium exhibited no defence mechanism against Ta2: unlike during contact between *L. edodes* or other leaf litter-degrading basidiomycetes and non-Ta2 *Trichoderma* spp. (Savoie and Mata 1999; Savoie et al. 2001), there was no formation of stationary assemblage of brown aerial hyphae and higher laccase activity in the substratum; in contrast, the stimulatory effect of compound(s) produced by the mushroom on growth and conidia germination of Ta2 could be demonstrated, which was not seen in the case of other *Trichoderma* species (Mamoun et al. 2000b; Mumpuni et al. 1998). It is hypothesised that this stimulation and the relative tolerance of the mushroom to Ta2 toxins allow the two fungi to parallelly grow in compost. The simultaneous growth of Ta2 and *A. bisporus* can be observed before the mushroom mycelia stimulate the conidiation of Ta2. As conidiation begins, the growth of mushroom mycelium slows significantly, and the typical symptoms of green mould develop rapidly (Mamoun et al. 2000b).

The metabolite 3,4-dihydro-8-hydroxy-3-methylisocoumarin, identified by Krupke et al. (2003), was produced in vitro by *T. aggressivum* f. *aggressivum* strains, but not in the case of non-aggressive isolates. This compound inhibits *A. bisporus* growth and fruiting body formation during green mould development, thereby allowing Ta4 to spread and utilise the nutrients released from the compost components by the extracellular enzymes of the mushroom. Ta4 needs the extracellular enzymes of the mushroom to decompose the complex components of compost into simple, absorbable and utilisable carbon sources (Krupke et al. 2003).

Marik et al. (2017) studied the production of peptaibols – non-ribosomally synthesised bioactive secondary metabolites – by *T. aggressivum* f. *europaeum* using HPLC-MS-based methods and detected several hypomurocin-like compounds. Peptaibols proved to be potential growth inhibitors of mushroom mycelia in in vitro experiments, and the host was shown to have an influence on the peptaibol profiles of the green mould, suggesting that peptaibols may play a role in the antagonistic action between Ta2 and *A. bisporus*.

Trichoderma species can produce a range of extracellular hydrolytic enzymes that degrade different polymers, which can thus be used as nutrient sources. In the case of compost-inhabiting *Trichoderma* strains, extracellular β-1,3-glucanases have substrates in the cell wall of both *A. bisporus* and wheat straw (the main component of *Agaricus* compost), while chitinases and proteases may facilitate saprotrophic growth on the rich fungal and bacterial compost microflora. For Ta2 and non-aggressive *Trichoderma* strains, 17 extracellular enzyme activities were examined, but no significant differences could be found (Largeteau-Mamoun et al. 2002; Savoie et al. 2001). Based on the results of confrontation tests performed with 27 bacterial strains, it was also found in the above studies that Ta2 is much less affected by bacteria compared to non-aggressive strains. Based on these, it can be hypothesised that the better mushroom compost adaptability of Ta2 is due to the tolerance to the inhibitory effects of compost-inhabiting bacteria, rather than the ability to degrade the components of the compost. This allows Ta2 to colonise certain areas of compost before direct interaction with *A. bisporus* takes place. When the amount of nutrients in the compost is already limited, lysis of *A. bisporus* hyphae by Ta2 occurs (Mumpuni et al. 1998).

The saprotrophic and mycoparasitic behaviour of compost-dwelling *Trichoderma* isolates belonging to the Th1 (*T. harzianum*), Th3 (*T. atroviride*), Ta2 and Ta4 biotypes was studied by Williams et al. (2003a) to elucidate the mechanism of aggressivity against *A. bisporus*. Mycoparasitic structures have been rarely observed, suggesting that antagonism of *T. aggressivum* against *A. bisporus* is not primarily based on mycoparasitism. For all *Trichoderma* groups studied, the production of polymer-degrading extracellular enzymes capable of degrading both the cell wall of mushrooms and wheat straw has been demonstrated. Some extracellular enzymes, e.g. trypsin-like protease and chimoelastase, were produced only by the Ta2 and Ta4 biotypes, so they could be associated with aggressivity. The results suggested the essentiality of polymer-degrading extracellular enzymes for both parasitic and saprotrophic growth. However, some isolates belonging to the Th3 biotype (*T. atroviride*) were able to colonise sterile compost to a similar extent as *T. aggressivum* and produced the same amount and quality of polymer-degrading extracellular enzymes, indicating that aggressivity cannot be explained by these factors alone. It was concluded that aggressivity depends on extensive saprotrophic colonisation ability and possibly also on closely related competition. Competition and colonisation may be prerequisites for antagonism against *A. bisporus*, of which mycoparasitism is just one factor among many (Williams et al. 2003a).

Foulongne-Oriol et al. (2011) examined the genetic control of *A. bisporus* resistance to lytic enzymes and metabolites of *Trichoderma* by quantitative trait locus (QTL) analysis. Sequential cultures on media with or without the commercial product Lysing Enzyme® were set up in an in vitro experiment. The traits used for QTL detection were mycelial growth rate under control condition, tolerance level and adaptation capacity. Based on the results, the tolerance to lytic enzymes and metabolites of *Trichoderma* is tightly related to mycelial growth ability and is quantitatively inherited under oligogenic control, suggesting that the genetic factors involved in the ability to resist or adapt to metabolites and lytic enzymes of *Trichoderma* are linked to the fitness of the *A. bisporus* strains.

Abubaker et al. (2013) studied the structures of three *T. aggressivum* genes, *prb1* (proteinase), *ech42* (endochitinase) and a β-glucanase gene in order to find out the role of cell wall-degrading enzymes in growth inhibition of *A. bisporus*. Promoter elements in the *prb1* and *ech42* suggest that the transcription of these genes is regulated by stress as well as carbon and nitrogen levels. Both genes possess mycoparasitism-related elements indicating the potential roles of their protein products in competition. The promoter of the β-glucanase gene contains no mycoparasitism-linked elements, but CreA and AreA binding sites are present, suggesting catabolite regulation. Two *A. bisporus* varieties sensitive and resistant to green mould disease (off-white and brown strains, respectively) were co-cultivated with *T. aggressivum* to assess the possible roles of the genes in disease development and severity by transcript level measurements. Results showed that *prb1* and *ech42* were upregulated coordinately after 5 days, while the transcription of the β-glucanase gene was upregulated from day 0 in the presence of *Agaricus* strains. In co-cultures of *T. aggressivum* with the resistant *Agaricus* strain, the upregulation was much less pronounced than with the sensitive strain. The results suggest that the proteins encoded by the examined genes have roles in both nutrition and green mould severity.

O'Brien et al. (2014) performed in vitro proteomic analysis to study the response of *T. aggressivum* f. *europaeum* to compost and *A. bisporus*. A differential expression study performed on the intracellular fraction of Ta2 grown in media containing Phase III mushroom compost or *A. bisporus* in comparison with a control medium resulted in the functional identification of 31 proteins, with differential expression observed for seven of them. Three and two proteins were up- and downregulated in both treatments, respectively, while two showed qualitatively different regulation in the two treatments. Proteins directly related to the degradation of fungal cell wall could not be observed, which may be due to the secretion of such proteins into the extracellular space, resulting in their relatively low abundance in the intracellular protein fraction. The differentially produced intracellular proteins were functionally related to cytoskeletal structure, oxidative stress tolerance and cell longevity. The identified proteins could be divided into structural, informational, metabolic and stress response functions. Differential production of these proteins may play a role in Ta2 growth in *Agaricus* compost and in its virulence towards *A. bisporus*. Most of the identified proteins were metabolic that are involved in nutrient uptake and the energy production. Specific proteins involved in pentose degradation may be relevant for degradation of carbohydrates liberated from fibrous mushroom compost components. An actin-binding and depolymerising protein was upregulated in response to both experimental treatments, suggesting that there may be a cytoskeletal structure alteration in Ta2 exposed to mushroom compost and *A. bisporus*.

A recent proteomic study performed on *A. bisporus* exposed to *T. aggressivum* (Kosanović et al. 2020) revealed an increased abundance of proteins associated with oxidative stress response in the mushroom (zinc ion binding, peroxidase, carboxylic ester hydrolase, dipeptidase and cluster assembly as well as proteins with pyruvate kinase activity and hydrolase activity), while the relative abundance of proteins associated with growth (structural constituent of ribosome, translation, deadenylation-dependent decapping of nuclear-transcribed mRNA, small GTPase-mediated signal transduction, deoxyribonucleotide catabolic process, GTP binding, glycine cleavage system P protein and proteasome subunit beta as well as proteins involved in the polysaccharide catabolic process, formation of extracellular region and lyase activity) decreased. The results of this study indicate that – although it is frequently considered as a saprophytic compost contaminant rather than a mushroom pathogen (Williams et al. 2003a) – *T. aggressivum* should be considered as a true mycopathogen due to its direct effect on *A. bisporus* development and the induction of its oxidative stress response.

With the release of the full genome sequences of the industrially important *T. reesei* (Martinez et al. 2008) and the biocontrol agents *T. atroviride* and *T. virens* (Kubicek et al. 2011), *Trichoderma* research entered the genomic and transcriptomic era. Meanwhile, the full genome of *T. aggressivum* f. *europaeum* (Urbán et al. 2016a, b) has also been sequenced, which, along with the available genome sequence of *A. bisporus* (Morin et al. 2012), opened the way to proceed in understanding the molecular background of green mould disease development.

2.3 Diagnosis

The more and more frequently occurring epidemics resulted in an emerging need for reliable methods that allow the early detection of *Agaricus* green mould. As both non-aggressive and aggressive forms of *Trichoderma* are able to grow in the same cultivation area (Morris et al. 1995a) and *Trichoderma* isolates are difficult to distinguish on a morphological basis, it has become necessary to develop rapid, efficient, sensitive and inexpensive tools for the identification of aggressive *Trichoderma* biotypes. Such methods may also help reveal the pathway of *Trichoderma* to farms and the spreading mechanisms of already established green mould infections, identify the activities that may contribute to the development of the disease and evaluate the effectiveness of hygiene measures (Castle et al. 1998).

Williams et al. (2003b) developed a selective medium containing chloramphenicol, streptomycin, quintozene and propamocarb for the isolation of *Trichoderma* strains from *Agaricus* compost. The medium also provides an opportunity to compare aggressive and non-aggressive biotypes.

Based on a RAPD amplified product of *T. aggressivum* f. *aggressivum* DNA, PCR primers (Th-F and Th-R) allowing the identification of the aggressive Ta2 and Ta4 biotypes were developed to assess the potential risk of *Trichoderma* strains intended to be used for the biological control of plant pathogenic fungi to mushroom production (Chen et al. 1999a). The two primers target a 444 bp long DNA segment present in the genome of Ta4, but also generate the same product from the genome of Ta2 (*T. aggressivum* f. *europaeum*). This PCR assay based on *T. aggressivum*-specific primers has also been shown to be useful in disease management programmes. The method has been used in conjunction with the RAPD-PCR technique in the United States to compare *Trichoderma* strains isolated before and during an outbreak of green mould (Chen et al. 1999b). The results suggest that the highly virulent genotype may have emerged recently, as the Ta4 biotype could not be identified among the *Trichoderma* strains isolated before the outbreak. This method was also used to detect the presence of *T. aggressivum* in Hungarian and

Polish mushroom growing plants (Hatvani et al. 2007; Szczech et al. 2008). The results of specific PCR were also confirmed by mitochondrial DNA-RFLP technique and ITS sequence analysis in the work of Hatvani et al. (2007) to clearly demonstrate the appearance of *T. aggressivum* f. *europaeum* in Central Europe.

O'Brien et al. (2017) used a quantitative polymerase chain reaction (qPCR) method with fluorescence detection and primers targeting the *tef1* gene for the detection of *T. aggressivum*. The qPCR method gave consistent and less variable results and proved to be more sensitive than microbiological counting methods. On the other hand, unlike culture-dependent methods, PCR tools do not distinguish between non-viable and viable cells, which should be considered during the interpretation of the results.

Green mould infected and uninfected *Agaricus* composts can be distinguished from each other also based on their volatile blend. In order to develop a sophisticated non-invasive detection tool of *T. aggressivum* in the process air of tunnels without the need to sample inside them during spawn run, Baars et al. (2011) tested the possibility to detect *T. aggressivum* based on the emitted volatiles. The authors sampled and analysed process air from both artificially infected and non-infected compost cultures by gas chromatography coupled with mass spectrometry (GC-MS). Volatile blends produced during normal compost colonisation proved to be significantly different from those appeared during colonisation of compost infected with *T. aggressivum*, and *T. aggressivum*-specific volatiles could be identified. Specific terpenoid volatiles that are present in the process air of *T. aggressivum*-infected compost could not be identified in the uninfected compost (Baars et al. 2011).

Radványi et al. (2020) identified medium-dependent and medium-independent biomarkers of Ta2 on different media. They detected emitted microbial volatile organic compounds (MVOCs) from the air by headspace solid-phase microextraction gas chromatography-mass spectrometry (HS SPME GC-MS) and examined the changes in their intensity values and linked them to the fungal growth phase. Such biomarkers have the potential to be used in quality control systems aiming to identify the presence of green mould disease in an early phase, thereby providing the producers more time to prevent yield losses.

2.4 Prevention and Control

Green moulds form billions of conidia that can be easily distributed by contaminated equipment and substratum materials that have not been properly pasteurised, or by the clothing of mushroom-growing workers, as well as by insects. Therefore, the infection spreads rapidly, and the treatment of the disease is extremely difficult (Anderson et al. 2001; Rinker and Alm 2000).

Pasteurisation of compost (Peil et al. 1996) and wood used to build growing rooms (Catlin et al. 2004) resulted in minimal green mould infestation, high yields and the appearance of a larger flush number. Catlin et al. (2004) proposed a 6 h treatment at 60 °C for post-harvest pasteurisation. However, this method is not always effective: green mould has already been isolated from freshly pasteurised compost (Morris et al. 2000), as pathogens are able to survive 60 °C for some time. Furthermore, sterilisation of mushroom substrata may favour *Trichoderma* growth due to the reduction of the natural microbiota in the substratum, which in turn increases the opportunity of colonisation by *Trichoderma* as a consequence of the reduced abundance of competitive microbiota (Velázquez-Cedeño et al. 2006; Colavolpe et al. 2014). Controlling the pH of the casing material is also a possible method of green mould management (Rinker and Alm 2008). Attention should be paid also to the quality of the building surfaces used in mushroom cultivation. Green mould contamination is more likely to persist on rougher surfaces like concrete or wood than on smoother, glazed surfaces (Abosriwil and Clancy 2002).

Green mould infections can be prevented by following strict hygiene procedures and treatments with common disinfectants (e.g. chlorine, ethanol, iodine, formaldehyde, phenol or quaternary ammonium compounds), some of which, however, are harmful to mushrooms and humans (Geels et al. 1988). In industrial-scale *Agaricus* cultivation, disinfectants are often used to supplement the general hygiene procedure (Lelley 1987). Such agents are also used to clean growing containers, shelves, machines, work surfaces, corridors, walls and foot dips (Fletcher et al. 1989; Lelley and Straetman 1986). Improper disinfection of growing equipment, the reduced attention to sanitation, the influx of contaminated air into spawning rooms and poor post-crop steam-off programmes can facilitate the entry of the pathogen. Sciarid mushroom flies need to be managed as well, as they are potential vectors of *T. aggressivum* (Mazin et al. 2019). Failure to control the infection at an early stage can have serious financial consequences, as conidia develop in poorly cleaned locations and spread the disease to other areas of the farm (Grogan 2008).

When *T. aggressivum* infects a mushroom crop, it has long been a common practice to spread salt onto the green mould spots. If the crop is seriously affected, it is generally treated with steam, then the infected casing material is removed, and new casing is applied (Győrfi 2002; Fletcher and Gaze 2007). However, chemical procedures proved to be the most effective means of treatment generally. Effective prevention of fungal mushroom pathogens can be achieved by the application of azoles, which inhibit the demethylation step within the biosynthesis of ergosterol, an essential compound responsible for lipoprotein membrane stability and function in many fungi, or by benzimidazole compounds (carbendazim, benomyl) binding to the fungal microtubules and stopping hyphal growth.

The use of several fungicides, including prochloraz, the combination of prochloraz and carbendazim, and thiabendazole, to control *Agaricus* compost colonisation by *Trichoderma* strains has been investigated (Abosriwil and Clancy 2003; Grogan and Jukes 2003). The application of fungicides to the spawn was found to be more economic and efficient in controlling the colonisation of green mould than the treatment of the compost (Rinker et al. 1997a, b; Abosriwil and Clancy 2003; Potočnik et al. 2015). However, registered fungicides are applied to the casing soil at a later stage of *Agaricus* cultivation, when disease outbreaks are more likely. The concentration of thiabendazole in casing soil was found to remain high throughout the cropping period, while prochloraz-manganese and carbendazim levels dropped considerably by the end of the second flush (Grogan and Jukes 2003). A high selectivity between green moulds and *A. bisporus* has been reported for prochloraz and benzimidazole fungicides; they proved to be highly toxic to pathogenic fungi without affecting the host mushrooms (Chrysayi-Tokousbalides et al. 2007; Hatvani et al. 2012; Potočnik et al. 2015). At the same time, among the commercial demethylation inhibitors (DMIs) tested, tebuconazole was more inhibitory to mushrooms than to the *Trichoderma* isolates (Hatvani et al. 2012). Among five commercial fungicides tested against the *Trichoderma* isolates originating from *Agaricus* farms in Serbia as well as Bosnia and Herzegovina, the highest susceptibility of the isolates was found to carbendazim and chlorothalonil, while they were less sensitive to iprodione, weakly resistant to thiophanate-methyl and resistant to trifloxystrobin (Kosanović et al. 2015). Considering the toxicity of fungicides to *A. bisporus*, carbendazim showed the best, iprodione and chlorothalonil a moderate, while thiophanate-methyl the lowest selective toxicity. Luković et al. (2021) examined the susceptibility of Serbian isolates of *T. aggressivum* f. *europaeum* and THSC recovered from *Agaricus* compost to the commercial fungicides prochloraz and metrafenone. Based on the observed ED_{50} (effective dose: fungicide concentrations inhibiting radial mycelial growth by 50%) values, both groups of isolates were found to be susceptible to the tested fungicides (*T. aggressivum*, ED_{50} = 0.04–1.34 μg mL^{-1} for both substances; THSC, ED_{50} = 0.03–3.64 and 0.04–3.64 μg mL^{-1} for metrafenone and prochloraz, respectively).

The frequent application of fungicides in mushroom cultivation results in the evolution of resistance in green mould pathogens, which is a common phenomenon and a serious problem. Benzimidazole fungicides initially provided a sufficient level of disease control, but in the early 2000s, resistance of Ta4 to benomyl and thiophanate-methyl was detected at *Agaricus* farms in North America (Romaine et al. 2005). An imidazole compound, imazalil sulphate, has been proposed as a solution against benzimidazole-resistant strains (Romaine et al. 2008).

Fungicides can also inhibit the mycelial growth of the mushroom and change the microbiota of the casing, so a balance must be struck between the benefits of inhibiting *Trichoderma* strains and the potential harmful effects on the crop. Prochloraz-manganese, the most effective fungicide in mushroom disease control (Grogan 2008), maintains a balance between the benefit of green mould control and the reduction in mushroom yield (Kosanović et al. 2015). It has also been found that certain fungicides (e.g. benomyl, carbendazim) are degraded by microorganisms (Fletcher et al. 1980; Yarden et al. 1990), thus reducing their efficacy against pathogens. Prochloraz has also been shown to be very susceptible to degradation by microorganisms present in the casing soil (Grogan et al. 2000; Papadopoulos 2006). Biodegradation of pesticides is a desirable trait, so that toxic chemicals do not accumulate in the environment, but it may lead to reduced control of pathogens which are still sensitive to the fungicide. Thus, there must be a balance between the time frame within which a chemical is effective against its target pathogen and its ultimate breakdown to nontoxic components (Grogan 2008).

In the expanding commercial cultivation of mushrooms, only a small number of fungicides have been proposed for compost, casing material and spawn treatment (Abosriwil and Clancy 2003). Fungicides officially recommended in *Agaricus* industry are prochloraz-manganese in Europe and worldwide, as well as thiabendazole, chlorothalonil and thiophanate-methyl in North America (Potočnik et al. 2018). EU pesticide reviews resulted in the withdrawal of approval for many chemicals, mainly in the group of benzimidazoles due to their mutagenicity (Fletcher et al. 1989; Grogan 2008). As the use of several chemicals is no longer permitted due to the concerns for their impact on the environment and human health, and there is a growing need to reduce pesticide use, growers must increasingly focus on prevention and the use of alternative, environmentally friendly control methods (Grogan 2008).

A variety of natural compounds (plant extracts, essential oils and their components) and beneficial bacteria have been tested against *Trichoderma* isolates causing green mould in *Agaricus* cultivation. Among plant essential oils tested for antimicrobial activity against *Trichoderma* green mould, thyme (*Thymus vulgaris*) oil and its major component, thymol, as well as oregano (*Origanum vulgare*) oil and its major component, carvacrol, exhibited very strong activity against Ta2, Th1 and *T. atroviride* (Soković and van Griensven 2006). On the other hand, it has to be considered that the application of these essential oils to compost may result in the inhibition of the bacterial microbiota, as *B. cereus* was found to be strongly inhibited by thymol and carvacrol (Gallucci et al. 2009), while the carvacrol-rich essential oil of *Thymus pubescens* strongly inhibited *B. subtilis* (Rasooli and Mirmostafa 2002). Đurović-Pejčev et al. (2014) evaluated the inhibitory and fungicidal activity of six essential oils to *Trichoderma aggressivum* f. *europaeum*, and the strongest activity was demonstrated by the oils of basil (*Ocimum basilicum*) and peppermint (*Mentha piperita*). Menthol, the major essential oil of peppermint, showed strong inhibitory activity against *Trichoderma* (Soković and van Griensven 2006); therefore, it was suggested as a potential biofungicide in mushroom compost. However, although another study also reported that thymol, (+)-menthol, (−)-menthol and ferulic acid inhibited the growth of green mould isolates in vitro at concentrations as low as 0.08 mg mL^{-1} to 1.25 mg mL^{-1} (Hatvani et al. 2012), they also blocked the growth of the host mushroom, suggesting that their use for green mould control may not be possible. Another opportunity might be the application of tea tree (*Melaleuca alternifolia*) oil to *Agaricus* casing layer, which was found to considerably inhibit *T. harzianum* (Kosanović et al. 2013).

Biological control based on the use of microorganisms may be an alternative to the chemical treatment of *Agaricus* green mould disease caused by *Trichoderma*. Certain bacteria naturally occurring in the casing material (e.g. *Bacillus* species) are potent antagonists of aggressive *Trichoderma* strains and are therefore potentially useful in the treatment of green mould disease. Savoie et al. (2001) demonstrated the inhibitory effect of *Bacillus* species to *T. aggressivum* growth. Bhatt and Singh (2002) investigated the utility of antagonistic bacteria naturally occurring in the casing material against mushroom pathogenic organisms, including *Trichoderma*, both in vitro and on *Agaricus* beds. Among the bacteria, isolate BI III significantly

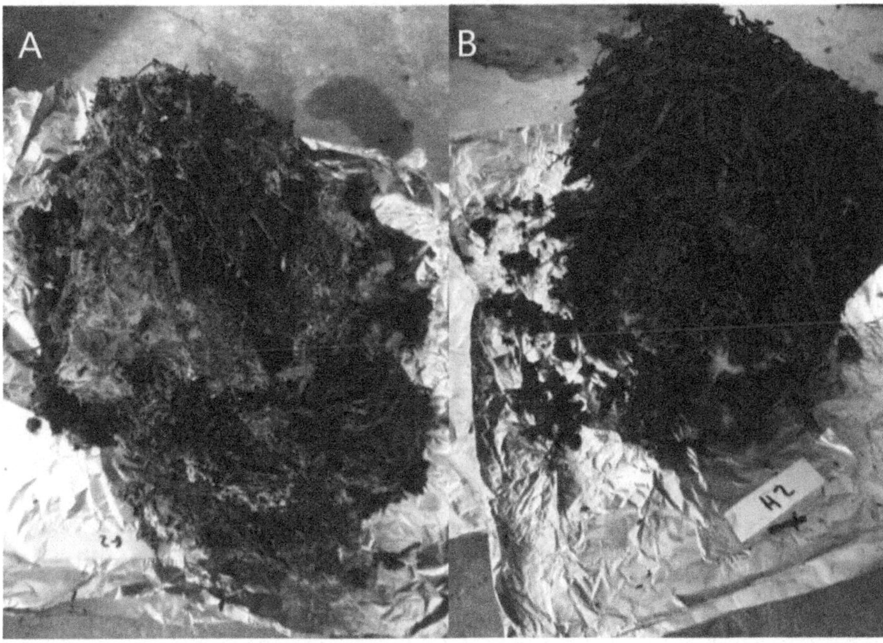

Fig. 4 Spawned mushroom compost at the end of an experimental cultivation cycle in pots. (**a**) Compost artificially infected with *Trichoderma aggressivum* f. *aggressivum* at the beginning of the cultivation cycle; (**b**) artificially infected with *T. aggressivum* f. *aggressivum* but also treated with a bacterial suspension of *Bacillus velezensis* at the beginning of the cultivation cycle. (Photo: Rita Büchner)

inhibited green mould and increased yield. Győrfi and Geösel (2008) investigated the protective effect of certain antagonistic bacteria (*Bacillus* species) against *T. aggressivum* infections. Two strains were shown to be effective in controlling *Trichoderma* strains under cultivation conditions, and the bacteria have also increased the yields.

Representatives of the species *Bacillus velezensis* are especially promising for the biological control of green mould in *Agaricus* production (Fig. 4). A biocontrol strategy based on the application of the commercial product Serenade MAX® (Bayer CropScience) containing *B. velezensis* QST713 (formerly known as *B. subtilis* QST713) was introduced in France to prevent crop losses (Pandin et al. 2018a, b), which now represents approximately 80% of the control measures in French *Agaricus* cultivation (Pandin et al. 2018a). Kosanović et al. (2013) evaluated the antagonistic activities of Serenade® WP (AgraQuest, Davis, Canada), also based on *B. velezensis* QST713 against green mould isolates of *Trichoderma* collected from Serbian *Agaricus* farms, and it was highly effective to all tested *Trichoderma* isolates in vitro and against aggressive *T. harzianum* isolates in a mushroom growing room. Similarly, Potočnik et al. (2018) applied the same biofungicide by coating *Agaricus* spawn and reported that *B. velezensis* QST713 effectively controlled green mould without inhibition of mycelial growth of *A. bisporus* and its efficacy was not significantly different from that of prochloraz-manganese. The complete genome sequence of this biocontrol strain has revealed that it harbours several antimicrobial clusters and an important arsenal enabling 3D biofilm formation (Pandin et al. 2018b). The impact of *B. velezensis* QST713 on the natural microbiota of *Agaricus* compost both in the presence and absence of Ta2 was also evaluated, and it was found that while Ta2 profoundly increased the fungal community and bacterial populations in the compost, the biocontrol strain decreased *Pseudomonas* populations and did not influence the naturally occurring fungal populations in the compost uninfected with green mould, while it strongly decreased the fungal population (mostly *T. aggressivum*) in Ta2-infected compost.

Potočnik et al. (2019b) evaluated the efficacy of a similar commercial biofungicide based on *B. subtilis* (Ekstrasol F SC) in comparison with Serenade® WP and prochloraz-manganese (Octave® WP) in a mushroom growing room. Although *B. subtilis* enhanced mushroom yield more than Serenade®, its bioefficacy was less than both prochloraz-manganese and *B. velezensis* QST713. Composting material represented a valuable source of antagonistic microorganisms with a potential for use in biological control of green mould in button mushroom production (Milijašević-Marčić et al. 2017). To find indigenous biocontrol agents against *T. aggressivum* f. *europaeum* and *T. harzianum*, Milijašević-Marčić et al. (2017) identified a *B. subtilis* isolate to inhibit the growth of the pathogens in vitro. Also, the bacterial isolate significantly lowered the green mould incidence in mushroom growing rooms, while it did not affect the mycelial growth of *A. bisporus*. There was no statistically significant difference found between the indigenous *B. subtilis*, prochloraz-manganese and the commercial isolate *B. velezensis* QST713 in terms of mushroom yield.

Stanojević et al. (2016) isolated bacteria from straw and chicken manure, compost and casing soil used for growing *A. bisporus*. They screened 108 bacterial isolates for antagonistic activity against green mould pathogens, and 23 isolates – representing *Bacillus subtilis*, *B. amyloliquefaciens*, *B. licheniformis* and *B. pumilus* – inhibited the growth of *T. aggressivum* f. *europaeum*, *T. harzianum* and *T. koningii*, while 13 isolates inhibited the growth of *T. atroviride*. Ta2 proved to be the most sensitive. Two *B. amyloliquefaciens* strains inhibited the growth of all four pathogenic fungi, thus exhibiting a wider spectrum of activity than other *Bacillus* spp. strains. In a subsequent study (Stanojević et al. 2019), the bioefficacy of the *Bacillus* strains was evaluated against Ta2 and *T. harzianum* in comparison with *B. velezensis* QST713 and a prochloraz-manganese-based fungicide. Strain B-241 of *B. amyloliquefaciens* had a performance in the suppression of both green mould and dry bubble disease like *B. velezensis* QST713 in all in vivo experiments, suggesting its potential applicability at a commercial scale. Another *B. amyloliquefaciens* strain, D747, is commercially available in Italy against *T. aggressivum* (Amylo-X®, Biogard, CBC (Europe) S.r.l., Nova Milanese, MB, Italy).

Šantrić et al. (2018) introduced *Streptomyces flavovirens* as a promising biocontrol agent of *T. aggressivum* f. *europaeum* with a bioefficacy similar to that of prochloraz-Mn (prochloraz-manganese complex) and no negative influence on the mycelial growth of *A. bisporus* in compost, while having a positive effect on mushroom yield. The application of *S. flavovirens* resulted in improved *Agaricus* production and better competitiveness of the mushroom with Ta2.

Kosanović et al. (2019) studied the interaction between *Pseudomonas putida* (which stimulates mushroom pinning), *Ps. tolaasii* (the brown blotch pathogen of *A. bisporus*) and *T. aggressivum*. *Ps. tolaasii* inhibited the growth of the green mould pathogen, increased conidiation and induced oxidative stress response and catabolic enzyme activation. On the other hand, *Ps. putida* stimulated the growth of *T. aggressivum* and increased the abundance of proteins associated with growth and development.

A reverse aspect of *T. aggressivum* f. *europaeum* and biological control was studied by Sánchez-Montesinos et al. (2020), who demonstrated the biostimulant capacity of Ta2 in tomato and pepper plants under commercial plant nursery and greenhouse conditions. Ta2 has been described as a growth promoter of melon seedlings under saline stress; in addition to its capacity to control *Pythium ultimum*, it was able to decrease the severity of the disease in seedlings (Sánchez-Montesinos et al. 2019). However, due to the substantial risk to mushroom growers, the agricultural use of any causal agent of green mould disease for biocontrol purposes is not recommended.

Another way to prevent economic damage caused by *Trichoderma* green mould in *Agaricus* production may be the cultivation of mushroom varieties resistant to aggressive *Trichoderma* strains. In the case of *T. aggressivum* infection, no defence reactions are observed in many *Agaricus* varieties (Mamoun et al. 2000b; Mumpuni et al. 1998). Savoie and Mata (2003) used extracellular metabolites from *T. harzianum* to increase the resistance of *A. bisporus* to *T. aggressivum* (induced resistance), but the mushroom was unable to adapt to these metabolites and showed a high degree of sensitivity. Anderson et al. (2001) compared the resistance of three *Agaricus* cultivars to the Ta4 *Trichoderma* biotype. White hybrid strains were extremely sensitive, off-white hybrid strains were moderately susceptible, while brown strains were found to be resistant to infection. Resistance of brown *Agaricus* strains has also been reported by Chen et al. (2003) and Sobieralski et al. (2009a). Anderson et al. (2001) found that fungal cell wall-degrading enzymes might be involved in the defence of *A. bisporus*. In dual cultures with Ta4, a 96 kDa N-acetylglucosaminidase was found to be produced by the brown strains Sylvan SB65 and Amycel 2400 earlier and at higher specific activity than by the off-white (Sylvan 130) and white (U1) strains, suggesting a role for this enzyme in the green mould resistance of commercial brown strains by its action on the cell walls of *T. aggressivum*. Three N-acetylglucosaminidases are also produced by Ta4, one of which may be an important indicator of antifungal activity, and they proved to be identically induced by both brown and white *A. bisporus* strains, suggesting that the resistance of brown strains is not due to a reduced induction of chitinase production in *T. aggressivum* (Guthrie and Castle 2006).

The grains present in the spawn may serve as a nutrient source for *Trichoderma*. Speer (2010) performed artificial infection experiments with *T. aggressivum* f. *aggressivum* on compost inoculated with different types of spawn. While the use of spawn with reduced grain content did not reduce the rate of *Trichoderma* infection, compost inoculated with grain-free spawn and artificially infected with *Trichoderma* did not show symptoms of green mould, suggesting that the use of grain-free spawn may prevent the development of *Trichoderma* green mould.

3 Green Mould of Oyster Mushroom (*Pleurotus* spp.)

Many pests and pathogens are known in the cultivation of oyster mushrooms (e.g. *Pseudomonas* species, insects), but in recent years, the most significant crop losses have been attributed to green mould infections caused by *Trichoderma*. Oyster mushroom growers have detected green mould infections in, among others, North America (Sharma and Vijay 1996), South Korea (Park et al. 2004, 2005), Italy (Woo et al. 2004), Romania (Kredics et al. 2006), Hungary (Kredics et al. 2006; Hatvani et al. 2007), Spain (Gea 2009), Poland (Siwulski et al. 2011), Croatia (Hatvani et al. 2012), Serbia and North Macedonia (Luković et al. 2021), Iraq (Al-Rubaiey and Al-Juboory 2020), Egypt (Ayman Daba, personal communication) as well as Sri Lanka (Jayalal and Adikaram 2007), which may indicate that *Pleurotus* green mould is becoming a global problem.

3.1 Epidemiology

Sharma and Vijay (1996) reported green mould caused by '*T. viride*' on oyster mushrooms in North America, but the identification of the pathogen was not verified by molecular taxonomic methods. The first epidemic causing significant crop losses was described in South Korea. Yu (2001) examined 110 samples from oyster mushroom cultivation and showed the presence of *T. viride* (13.6%), *T. harzianum* (8.2%) and *T. koningii* (5.5%) (Fig. 1); however, the majority of the isolates (65.5%) belonged to an unidentified *Trichoderma* species. Two *Hypocrea* species also occurred in cultivation, one of which (*Hypocrea* sp. 1) formed a brown and the other (*Hypocrea* sp. 2) a white fruiting body. *Hypocrea* sp. 1 also appeared in *Gliocladium*-like, asexual form, while no asexual form was found in the case of *Hypocrea* sp. 2. *Trichoderma* isolates were found to be dominant over oyster mushroom in both potato dextrose agar medium (in vitro) and under growing conditions (in situ).

Park et al. (2005) divided 26 *Trichoderma* strains isolated from South Korean oyster mushroom farms into the following four groups based on their cultural and morphological characteristics: *Trichoderma* sp. K1 (K1), *Trichoderma* sp. K2 (K2), *T. harzianum* and *T. atroviride*. The most frequently isolated species was K2, followed by K1 and *T. atroviride*. Based on colony morphology, growth rate and the morphology of phialides and conidia, K1 and K2 proved to be different from *T. harzianum*, *T. atroviride* and each other. Phylogenetic analysis of the ITS region revealed that K1 and K2 were clearly distinguishable from *T. harzianum*, *T.*

atroviride as well as *T. aggressivum* f. *aggressivum* and f. *europaeum*. It has been confirmed that the two groups differ from each other in a single A/C conversion: adenine is found in species K1 and cytosine in K2 in position 447 of the ITS2 region, while their sequences are identical in all other positions. According to the authors, species K1 and K2 could be distinguished from each other and *T. harzianum*, *T. atroviride*, Ta2 and Ta4 also by the phylogenetic analysis of the fourth intron of the *tef1* gene and motif 6 and 7 of the RNA polymerase II gene (*rpb2*) (Park et al. 2004, 2005). The two new species corresponding to groups K1 and K2 were finally described as *T. pleurotum* (currently accepted name: *T. pleuroti*) and *T. pleuroticola* (Fig. 1) (Park et al. 2006), but the type strains of the new species were not deposited in publicly available culture collections, and most of the sequences deriving from the molecular characterisation of the species were not submitted to public sequence databases.

At the same time, severe green mould infection of oyster mushrooms in Italy led to a crisis in the sector (Woo et al. 2004). Data from preliminary morphological and genetic characterisation suggested that the infectious agent belonged to the species *T. harzianum*, but it was later shown that representatives of the species *T. pleuroticola* and *T. pleuroti* caused the problem. Green mould infection also appeared in oyster mushroom cultivation in Hungary. Hatvani et al. (2007) isolated several *Trichoderma* strains from samples taken from oyster mushroom cultivation substratum infected with green mould. Sequence analysis of the ITS region revealed that the strains most aggressive to oyster mushroom were genetically closely related to *Trichoderma* sp. DAOM 175924, an isolate from a rotten poplar trunk found in Ontario, Canada (Kullnig-Gradinger et al. 2002). Isolates were found to be genetically heterogeneous based on their mitochondrial DNA-RFLP patterns, and the presence of a 2.2 kb mitochondrial plasmid was also detected in some strains (Hatvani et al. 2007). The isolates could be separated into two groups corresponding with the species *T. pleuroti* and *T. pleuroticola*. Interestingly, while *T. pleuroticola* proved to be more common on Italian oyster mushroom farms, the vast majority of isolates from Hungary were classified as *T. pleuroti*.

Subsequently, Komoń-Zelazowska et al. (2007) performed the detailed, comprehensive, scientifically documented characterisation of the two new species based on several *T. pleuroti* isolates from Hungary, Italy and Romania as well as *T. pleuroticola* isolates from Canada, Iran, the Netherlands, Germany and New Zealand. Similar to *T. aggressivum*, both new species are classified into the Harzianum clade of the genus *Trichoderma*. Morphological studies on strains of the new species have revealed that *T. pleuroticola* shows pachybasium-like morphology characteristic of the Harzianum clade, while *T. pleuroti* exhibits *Gliocladium*-like properties. Carbon source utilisation profiles of the isolates examined by the BIOLOG phenotype microarray method revealed a clear difference between the two species: the growth of *T. pleuroti* was slower on most carbon sources compared to *T. pleuroticola* showing a similar growth to *T. aggressivum* (Komoń-Zelazowska et al. 2007). These results suggest that the evolution of *T. pleuroti* may have been associated with a loss of utilisation ability of certain carbon sources. Phylogenetic analysis of the ITS, *tef1* and *chi18-5* loci has confirmed that the causal agents of oyster mushroom green

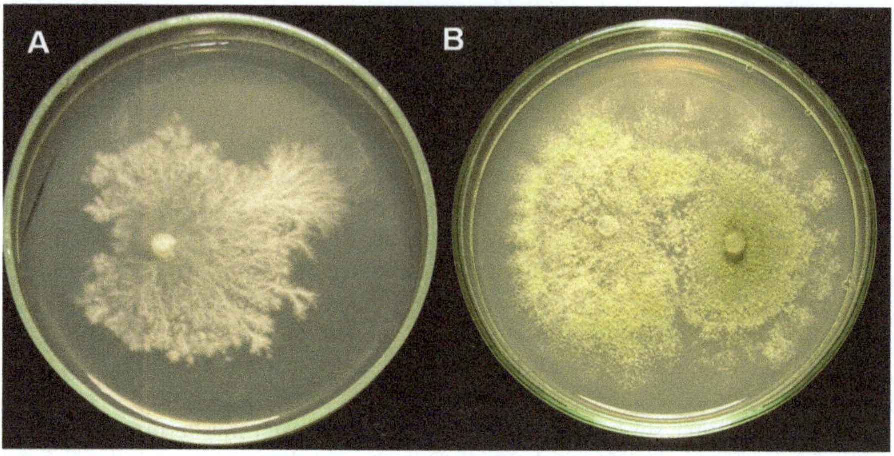

Fig. 5 In vitro confrontation between *Pleurotus ostreatus* and *Trichoderma pleuroti* on yeast extract, xylose medium. (**a**) *P. ostreatus*, (**b**) *P. ostreatus* + *T. pleuroti*. 5 mm mycelial disks from the actively growing edge of a *P. ostreatus* colony were inoculated onto the plates (**a**, **b**, left side), and after reaching a colony radius of approximately 1 cm, *T. pleuroti* was inoculated at a distance of 3 cm in the same way (**b**, right side). (Photo: Lóránt Hatvani)

mould, *T. pleuroticola* and *T. pleuroti*, are indeed two clearly distinct species. DNA barcodes for the identification of the two new species based on ITS1 and ITS2 sequences have also been identified and incorporated into the *Trichoderma* identification programme previously developed by Druzhinina et al. (2005). In vitro antagonism tests revealed that the two oyster mushroom pathogenic *Trichoderma* species were able to efficiently overgrow *P. ostreatus* colonies (Fig. 5); furthermore, both species were able to antagonise *A. bisporus* to a similar extent as *T. aggressivum* (Komoń-Zelazowska et al. 2007). However, no data are available on the possible damage caused by the two new species in *Agaricus* cultivation: oyster mushroom pathogens have not been detected in *Agaricus* compost even on farms where oyster mushrooms are grown in addition to *Agaricus*, and pathogens can be found only in the oyster mushroom substratum (Hatvani et al. 2007).

Among *Trichoderma* strains isolated from oyster mushroom cultivation substratum in Hungary, Croatia and Romania, Hatvani et al. (2007, 2008, 2012) identified *T. pleuroti* as the most prevalent, while *T. pleuroticola*, *T. atroviride*, *T. asperellum* and *T. longibrachiatum* (Fig. 1) were also detected. Woo et al. (2009) identified the majority of isolates pathogenic to *P. ostreatus* from Italian mushroom farms as *T. pleuroticola* and *T. harzianum* and less commonly as *T. pleuroti*. Innocenti and Montanari (2014) isolated *T. pleuroti* and *T. pleuroticola* from symptomatic areas of the cultivation substratum, while *T. harzianum* could be isolated only from areas without disease symptoms. Recently, Lee et al. (2020) isolated *T. pleuroticola* for the first time from a substratum of *P. eryngii*.

Błaszczyk et al. (2013) identified *T. pleuroti* and *T. pleuroticola* isolates obtained from Polish oyster mushroom farms based on morphological characteristics and by using ITS and *tef1* sequences and found the latter to be more frequent. The predominance of *T. pleuroti* in both Hungarian and Polish oyster mushroom farms may be due to the similar, wheat straw-based production technology, which is different from the methods used in Italy, where *T. pleuroticola* is the major agent of *Pleurotus* green mould (Komon-Zelazowska et al. 2007; Błaszczyk et al. 2013). Comparative ITS sequence analysis has shown that all Polish *T. pleuroti* isolates represent a single haplotype identical to that of Hungarian and Romanian *T. pleuroti* strains, while besides the type known from the Carpathian region, sequence analysis of the *tef1* locus also revealed a unique *tef1* allele from Poland (Błaszczyk et al. 2013). Detailed analysis of ITS and *tef1* sequences of two Polish *T. pleuroticola* isolates showed their identity with an Italian strain. As the substratum is supposed to be the source of green mould infection in oyster mushroom cultivation (Komon-Zelazowska et al. 2007), the composition of two *T. pleuroti* haplotypes in Poland most likely depends on the source of the wheat straw used for *Pleurotus* cultivation, which may also spread certain haplotypes between countries via trading (import-export) activities.

In the recent study of Luković et al. (2021), green mould-affected *P. ostreatus* samples were shown to harbour *T. pleuroticola* and *T. pleuroti* in North Macedonia. In addition to *T. pleuroticola*, members of the THSC were detected in Serbia (Luković et al. 2021), which was later specified as *T. afroharzianum* (Allaga et al. 2021).

Kredics et al. (2009) detected *T. pleuroticola*, *T. harzianum*, *T. atroviride*, *T. longibrachiatum* and *T. asperellum* – but not *T. pleuroti* – in the natural vicinity and on the fruiting bodies of wild-grown *P. ostreatus* in Hungary. The presence of *T. pleuroticola* in natural habitats suggests that these might be potential reservoirs of the pathogen and possible sources of contamination at mushroom farms.

While *T. pleuroti* has been isolated so far only from oyster mushroom cultivation, sets of data are available about the occurrence of *T. pleuroticola* in environmental samples (soil, tree) from different parts of the world, including soil samples from wheat fields in Hungary (Kredics et al. 2012) and Austrian soils, suggesting that these two species may occupy different ecological and trophic niches in nature (Komoń-Zelazowska et al. 2007). Interestingly, *T. pleuroticola* is used in New Zealand for biological control of the root rot pathogens *Armillaria novae-zelandiae* and *A. limonea* in the protection of kiwi and pine (Dodd et al. 2000). The 'HEND' strain used was initially identified as '*T. harzianum*', and it turned out only later that it was in fact a representative of *T. pleuroticola*. Using such strains for plant protection purposes may have catastrophic consequences if oyster mushrooms are grown near the area of their application. Therefore, a comprehensive risk assessment is essential during the development of biological control methods based on *Trichoderma* strains, the basis of which should be the accurate, species-level identification of the biocontrol candidates by molecular methods.

Other species isolated from green mould-infected *Pleurotus* cultivation materials include *T. harzianum* and *T. atroviride* in Poland (Błaszczyk et al. 2013) and Hungary (Hatvani et al. 2007). Additionally, Hatvani et al. (2007) also found individual isolates of *T. asperellum*, *T. ghanense* and *T. longibrachiatum* in Hungary. Al-Rubaiey and Al-Juboory (2020) reported *T. longibrachiatum* as the green mould pathogen of *P. eryngii* in Iraq. Sobieralski et al. (2010d) found that *T. aggressivum* f. *europaeum* isolates caused a significant crop reduction of *P. eryngii* in Poland without any significant impact on morphological features of the mushroom. *T. aggressivum* f. *aggressivum* was also found to occur on *P. ostreatus* at a Hungarian mushroom farm during a period of a Ta4 outbreak in the adjacent *A. bisporus* growing houses (Hatvani and Allaga, personal communication). Besides *T. harzianum* detected in oyster mushroom substratum in North Macedonia, other members of the Harzianum clade diagnosed in green mould-affected *P. ostreatus* cultivation are *T. guizhouense* (Serbia, Croatia), *T. atrobrunneum* (North Macedonia), *T. simmonsii* (North Macedonia, Serbia) and *T. afroharzianum* (North Macedonia, Serbia, Spain) (Fig. 1) (Allaga et al. 2021).

3.2 Biology

The appearance of *T. pleuroti* and *T. pleuroticola* in oyster mushroom cultivation cannot be related to a certain kind of substratum; the infection was detected from cultivation on rice straw, cotton, sawdust (Park et al. 2004; Yu 2001) and wheat straw (Hatvani et al. 2007) as well. In its advanced state, oyster mushroom green mould infection can be easily identified by the symptoms, i.e. thick green areas of conidiation on the surface of cultivation substratum, which is mostly exposed to *Trichoderma* infection during the spawn run phase. The pathogenic *Trichoderma* may also be able to grow on the surface of developing fruiting bodies, which can therefore often become distorted and severely spotted. In case of severe infections, fruiting body formation may not occur at all.

Both *T. pleuroti* and *T. pleuroticola* cause significant losses of *P. ostreatus* yields (Sobieralski et al. 2012b, c); however, *T. pleuroticola* results in greater yield drops up to 84%. The degree of yield decline depends on the pathogenic strain and the mushroom variety. Sobieralski et al. (2012d) determined the interactions between different *T. pleuroticola* and *T. pleuroti* isolates and six species of *Pleurotus* and found that *P. cornucopiae* was the most sensitive when interacted with both pathogenic species. Wild strains of *P. ostreatus* exhibit a relatively low yield drop compared to cultivated strains.

Woo et al. (2004) have observed that *Trichoderma* species are present in the initial phase of preparation of the substratum used to grow oyster mushrooms and then disappear upon pasteurisation but can be found again after spawning (inoculation with oyster mushroom), during the spawn run and the harvest cycles in the special shelf-growing system prevalent in South Korea. The peculiarity of this system is that the cultivation substratum is placed in bulk on the shelves after heat treatment and only spawned later, after the substratum has cooled down (Choi 2004). The bulk cultivation substratum can come into contact with pathogens on a large surface after heat treatment; therefore, the risk of re-infection is higher than in the case of bagged or blocked cultivation substratum, which is widespread in Hungary and Poland.

Chen and Moy (2004) have stated that the parameters of oyster mushroom cultivation, such as the nitrogen and carbon source, high relative humidity, elevated temperatures, the fluctuation of these factors as well as the absence of light during spawn run, are ideal environmental conditions for moulds, which may easily lead to contamination. Moulds rapidly grow under preferred conditions and more efficiently compete for nutrients and space than the oyster mushroom. Furthermore, they produce extracellular enzymes (e.g. glucanases), toxic secondary metabolites and volatile organic compounds, which may result in a drastic yield reduction or even the elimination of the entire crop.

The effect of osmotic and matrix potential on mycelial growth of three cultivated *Pleurotus* species (*P. florida*, *P. ostreatus*, *P. sajor-caju*) and two *Trichoderma* species (*T. atroviride* and *T. pleuroti*) was investigated by Lee et al. (2000). The growth optimum for both *Pleurotus* and *Trichoderma* strains ranged from −0.2 to −0.5 MPa, but the growth rate of *Trichoderma* strains was much higher than that of the *Pleurotus* strains. *Trichoderma* strains were also able to grow at the lowest water potential value tested, −4.0 MPa, which had already inhibited the growth of *Pleurotus* strains. The effect of moisture content of the cultivation substratum on the growth of oyster mushroom and *Trichoderma* was studied by Yu (2001). The optimum of *P. ostreatus* was between 60% and 70% moisture, while growth was inhibited above 80%. In contrast, the maximum mycelial growth of *Trichoderma* was observed at 80% moisture content. Too high moisture content has an adverse effect on the growth of the mycelium of oyster mushroom as it prevents the aeration of the substratum, but in turn favours the appearance of green mould and the growth of the pathogenic *Trichoderma* strains.

Trichoderma species were active against the mycelia of *P. ostreatus* by competing for space and nutrients, and neither hyphal interaction nor effect by volatile or non-volatile metabolites occurred (Innocenti et al. 2019). The results of studies on extracellular enzyme production (Kredics et al. 2008a, b) suggest that the two closely related oyster mushroom pathogenic species, *T. pleuroti* and *T. pleuroticola*, use different enzymatic strategies to adapt to oyster mushroom growing conditions. Isolation of mutants damaged in each enzyme system and comparison of their properties with wild-type oyster mushroom pathogenic strains (e.g. in in vitro antagonism experiments with oyster mushrooms or provoked infection experiments) may contribute to the identification of extracellular enzymes involved as virulence factors in green mould infection. Hatvani (2008) reported that *T. pleuroti* mutants deficient in their protease, chitinase and lipase enzyme systems have significantly reduced in vitro antagonistic ability against oyster mushrooms, suggesting that these enzyme systems may play an important role in the process of oyster mushroom infection by *Trichoderma*.

The production of peptaibols by *T. pleuroti* was studied by Marik et al. (2017) using HPLC-MS-based methods, and tripleurins – representing a new group of 18-residue peptaibols – were discovered. Tripleurins had an inhibitory effect on the mycelial growth of *P. ostreatus*, suggesting a possible role of these bioactive peptides in oyster mushroom green mould development. The gene sequence of the non-ribosomal peptide synthetase responsible for the production of tripleurins was mined from the full genome sequence of *T. pleuroti* (Marik et al. 2017), which has been made available to assist future studies on the biology of oyster mushroom green mould disease (Urbán et al. 2016a, b).

3.3 Diagnosis

Due to the rapid spread of green mould infection in oyster mushroom farms worldwide, the development of effective diagnostic methods for the detection of the causal agents had become urgent. A polymerase chain reaction (PCR)-based technique for the rapid identification of *T. pleuroti* and *T. pleuroticola* has been developed by Kredics et al. (2009). Based on the sequences of the variable introns of the *tef1* gene, three primers were designed, two of which are specific for both oyster mushroom pathogenic *Trichoderma* species, while the third one can bind only to the *tef1* gene of *T. pleuroti*. Accordingly, when used together in one reaction, the three primers amplify a single fragment from the genomic DNA of *T. pleuroticola* and two fragments from *T. pleuroti*. In addition to *T. pleuroti* and *T. pleuroticola*, 28 other *Trichoderma* species and several other fungi were tested with the multiplex PCR method outlined above, and no cross-reactivity was observed in any of the cases. Based on the results, *T. pleuroti* and *T. pleuroticola* can be clearly distinguished from each other and from other fungal species using this triple primer set. The method also allows the quick detection of the two oyster mushroom pathogenic *Trichoderma* species directly from substratum samples used for oyster mushroom cultivation without strain isolation and ITS sequence analysis. In this way, the method can help detect green mould infestation of oyster mushrooms at an early stage, paving the way for the application of appropriate control procedures. This PCR technique was also used to detect both species on the surface of insects present in the growing houses, suggesting their possible role as vectors of *Pleurotus* green mould (Hatvani et al. unpublished). The application of this method also revealed the presence of *T. pleuroticola* – but not that of *T. pleuroti* – in the natural substratum and on the fruiting body surface of wild-grown oyster mushrooms (Kredics et al. 2009).

Lee et al. (2000) developed a rapid and accurate detection method that involves a single *Trichoderma*-specific primer set designed based on the DNA sequence alignment of the ITS1 and ITS2 regions of 11 *Trichoderma* species occurring in mushroom cultivation substrata. The method can detect the *Trichoderma* mycelium both independently and in a mixture with *P. eryngii*, even at a very low amount during the early stage of spawn run.

3.4 Prevention and Control

For prevention and control of oyster mushroom green mould disease, the influence of temperature and pH was studied and optimised for oyster mushroom cultivation. Although the temperature optimum for oyster mushroom growth may vary between cultivated strains, approximately 25 °C is required for spawn run, 13–15 °C for induction of fruiting body development and 12–18 °C for fruiting (Choi 2004). The cultivation substratum is the most susceptible to green mould infection during spawn run. Its temperature can then rise up to 30 °C due to the metabolic heat generated by the oyster mushroom mycelia, and the *Pleurotus* pathogenic *Trichoderma* strains show maximal mycelial growth exactly in the temperature range of 25–30 °C. Woo et al. (2004) found that *Trichoderma* was able to grow well over a wider temperature range than oyster mushroom (20–28 °C) and its growth rate was three times higher than that of *P. ostreatus* at 25 °C. Based on this study (Woo et al. 2004), it is recommended to maintain a temperature between 15 and 18 °C for the post-spawn run phase to prevent the development and spread of green mould infection.

According to Woo et al. (2004), while the pH optimum for oyster mushroom growth is in the alkaline range (pH = 8–9), *Trichoderma* prefers acidic and neutral conditions (pH = 5–7). According to the authors, adjusting the cultivation stock to a pH between 8 and 9 may slow the growth of *Trichoderma*, resulting in a reduction in the spread of infection. Chang and Miles (2004), on the other hand, characterised the vegetative mycelial growth of *P. ostreatus* with an optimum of pH 5.4–6.0. As the mycelium of the oyster mushroom grows, it acidifies the straw-based substratum, and the pH decrease (from 8–9 to 4.5–5) occurs in 5–6 days. Thus, a higher pH can only provide protection in the beginning, but later, with the growth of its mycelium, the oyster mushroom itself creates favourable circumstances for the pathogens.

Yu (2001) also studied the effect of pasteurisation on the development of oyster mushroom green mould infections as a function of sterilisation temperature, time and moisture content of the cultivation substratum. The results showed that the growth of *Trichoderma* mycelium could be completely inhibited with pasteurisation for 10 h or more at 60 °C at both 50% and 70% moisture content of the growing substratum, while teleomorphic stages could not even survive a heat treatment at 50 °C for up to 5 h. When determining the duration of pasteurisation, the thermal conductivity of the cultivation substratum, which depends on its type, volume and moisture content, must also be considered. The immersion of the substratum in hot water at 60 °C for 30 min or in alkalinised water for 36 h is a treatment capable of reducing the incidence of contamination with *Trichoderma* sp. during the spawning phase of oyster mushroom cultivation (Colavolpe et al. 2014).

Yu (2001) tested the effect of fungicides (prochloraz, thiabendazole, benomyl, propineb, chlorothalonil) on conidial germination and mycelial growth of oyster mushroom pathogenic *Trichoderma* isolates. Several strains of *Trichoderma* were shown to be resistant to benomyl and thiabendazole. Prochloraz was found to be the most effective fungicide in inhibiting the growth of green moulds, and no resistance of *Trichoderma* strains has appeared to this fungicide. Prochloraz, benomyl and propineb had an inhibitory effect on the germination of conidia in benomyl-sensitive strains, while chlorothalonil was also able to inhibit conidia germination in benomyl-resistant strains (Yu 2001). Another seven fungicides (including captan) inhibited the mycelial growth of *P. ostreatus* better than that of *Trichoderma* species, meaning that these chemicals could not be used to control the infection. Fungicide treatment before sterilisation of the cultivation substratum can effectively prevent infection with green mould strains throughout the entire oyster mushroom cultivation process, but fungicide treatment after spawning can be risky from the aspect of food safety. However, none of these fungicides are authorised for oyster mushroom cultivation within the European Union, so their use is not possible even before heat treatment. The inhibitory efficacy of prochloraz and thiabendazole on oyster mushroom beds contaminated with green mould was 78.5% and 70.9%, respectively, whereas benomyl treatment had no inhibitory effect on *Trichoderma* (Yu 2001). Although at high concentrations prochloraz also had an inhibitory effect on the mycelial growth and fruiting body development of oyster mushroom cultivars, it was still considered to be the most effective fungicide against green mould infection in fungal beds. When cotton as a possible cultivation substratum was treated with 250 ppm prochloraz, fruiting body formation started in a shorter time and led to higher yields. After harvest, fungicide residues were analysed in the fruiting bodies of oyster mushrooms grown on chemically treated cultivation substratum. The levels of prochloraz, thiabendazole and benomyl residues in the fruiting bodies were well below the maximum residue limits for fungicides in cultivated mushrooms (Yu 2001). In another study, the inhibitory effect of several fungicides commonly used in agriculture (prochloraz, thiabendazole, dichloran, benomyl, propiconazole, thiofanatomethyl) was tested by Woo et al. (2004), and both prochloraz and thiabendazole were found to inhibit the growth of the aggressive *Trichoderma* isolates without having a negative effect on *Pleurotus*.

The benomyl-containing Chinoin Fundazol 50 WP had been used in Hungarian oyster mushroom cultivation for more than two decades (Szili 2008); however, the use of this fungicide alone, without adhering to hygienic and cultivation technology standards, did not inhibit the appearance and spread of green moulds. In the meantime, these drugs were permanently withdrawn from plant protection in the EU, as they were shown to be carcinogenic, teratogenic and endocrine disruptors. Adherence to the appropriate hygienic and cultivation technology regulations as well as the use of well heat-treated cultivation substratum can prevent the infection of *Trichoderma* green mould and its spread within oyster mushroom farms.

The oyster mushroom pathogenic *Trichoderma* strains examined in Serbia and North Macedonia showed high susceptibility to metrafenone and prochloraz: the ED_{50} values for *T. pleuroti*, *T. pleuroticola* and THSC isolates ranged between 0.02–0.14 and 0.001–0.01, 0.02–0.17 and 0.001–0.01, as well as 0.01–0.06 and 0.01.0.02 $\mu g\ mL^{-1}$, respectively (Luković et al. 2021).

High concentrations of prochloraz had harmful effects on mycelial growth and fruiting body development of *Pleurotus* spp. (Hatvani et al. 2008). Prochloraz was found to be effective against both *T. pleuroti* and *T. pleuroticola*, completely inhibiting the colony growth rate and spore germination both in vitro and in vivo; however, the treatment at the spawning phase was not sufficient to ensure protection during the entire cycle of oyster mushroom cultivation, which may be due to the reduction of fungicide activity under cultivation conditions, enabling secondary infections by airborne spores of *Trichoderma* and the lack of natural antagonists in the cultivation substratum (Innocenti et al. 2019).

As chemical control by pesticides is not an available option in oyster mushroom production in most parts of the EU (Nagy et al. 2012), similar to the case of *Agaricus* green mould, a promising, alternative solution for the control of *Trichoderma* green mould is the application of natural compounds or microorganisms as biocontrol agents. Angelini et al. (2008) reported that tea tree essential oil inhibited the mycelial growth of *T. harzianum* in vitro, while the growth of *Pleurotus* spp. (*P. eryngii*, *P. nebrodensis* and *P. hadamardii*) was stimulated. Application of the essential oil to the substratum of *Pleurotus* cultivation resulted in strong to total inhibition of *T. harzianum*; its pathogenicity proved to be weak or non-existent. During a subsequent study, the in vitro effects of methanol extract from *Ferula assa-foetida* oleo-gum-resin on *T. harzianum* and *Pleurotus* spp. were investigated in dual culture experiments (Angelini et al. 2009). The methanol extract showed fungistatic and fungicidal properties against the strains of both *T. harzianum* and *Pleurotus* spp. at higher concentrations. When methanol extracts were added to the cultivation substratum, the antagonistic activity of *T. harzianum* against *Pleurotus* spp. was only moderate or weak. Shah et al. (2011) evaluated the antifungal activities of eight botanicals against mycelia of both *Pleurotus* and the green mould pathogen *T. harzianum* in vitro and in vivo. Among them, *Azadirachta indica* showed a maximum increase in yield and exhibited a minimum disease incidence. Talavera-Ortiz et al. (2020) evaluated an extract of the fruiting body of *Pycnoporus* sp. (Polyporaceae, Agaricomycetes, Basidiomycota), an edible and medicinal mushroom, against *T. pleuroti* and *T. atrobrunneum* isolated from infected substratum of *P. ostreatus* farms. The results showed a decrease in mycelial growth rate up to 72% in vitro and delay of both mycelial growth and sporulation of the pathogens on lignocellulosic substratum up to 10 days.

A solution that can be followed in oyster mushroom green mould control may be making mushroom cultivation substrata resistant to green mould pathogens (Yu 2001). For this purpose, the use of beneficial bacteria seems to be appropriate, which, when mixed to the cultivation substratum, are capable of selectively killing or at least vigorously suppressing the aggressive moulds which develop during cultivation. In South Korea, studies have been conducted with a bacterial strain (CNU LI-1) that inhibits mycelial growth of *Trichoderma* species (Yu 2001). Inoculation of the pre-sterilised cultivation substratum with strain CNU LI-1 was shown to be effective in preventing the emergence of *Trichoderma* species. However, if green mould had already appeared on the fungal beds, the treatment was not able to eliminate the infection. Hatvani (2008) tested the effect of isolates of several bacterial species on the growth of *T. pleuroti* and *P. ostreatus* in in vitro antagonism tests and found bacterial strains that inhibited the *Trichoderma* strain without having a significant effect on *Pleurotus*. Nagy et al. (2012) reported that strains belonging to *Bacillus subtilis*, *B. amyloliquefaciens* and *B. licheniformis* were very effective in antagonising the oyster mushroom pathogenic *T. pleuroti* without any negative effect on *P. ostreatus*. The *B. amyloliquefaciens* strain is a potential biocontrol candidate, as in addition to the lack of antagonistic activity towards *P. ostreatus*, it also increased crop yield. Potočnik et al. (2019a) evaluated the antagonistic potential of *B. subtilis* strains isolated from oyster mushroom substratum against *T. pleuroti* and *T. pleuroticola* in vitro, and growth inhibition up to 62.22% and 69.62% could be achieved, respectively. Roberti et al. (2019) evaluated two yeast strains of *Aureobasidium pullulans* against *T. pleuroti* and *T. pleuroticola*. Both strains were effective in reducing colony growth of the two *Trichoderma* species in vitro without any negative effect on *P. ostreatus* growth. Furthermore, the yeast strains were more efficient than *Trichoderma* in substratum colonisation and produced volatile and non-volatile metabolites, which reduced *Trichoderma* growth. Under controlled conditions like those of a mushroom farm, only one of the yeasts, *A. pullulans* L8, was effective in controlling the disease, with control effects comparable with those of prochloraz.

4 Green Mould of Other Mushrooms

4.1 Lentinula edodes

The most important pathogens in shiitake cultivation are also from the genus *Trichoderma*, mainly attacking the mycelia of *L. edodes* in bed logs and sawdust cultures and causing serious damage negatively affecting the mushroom yield during production. Nine *Trichoderma* species (*T. atroviride*, *T. citrinoviride*, *T. harzianum*, *T. longibrachiatum*, *T. pseudokoningii*, *T. polysporum*, *T.* cf. *stramineum*, *T. virens* and *Trichoderma* sp.; (Fig. 1) were reported by Miyazaki et al. (2009) to cause economic damage in Japanese shiitake production. In addition, Kim et al. (2012a, 2013) introduced and described *T. mienum* and *T. pseudolacteum* (previously recognised as *H. lactea*) (Fig. 1) as new green mould pathogens from the bed logs of both shiitake and oyster mushroom farms in Japan. *Trichoderma* species reported from shiitake farms in Korea include *T. atroviride*, *T. citrinoviride*, *T. harzianum*, *T. polysporum*, *T. longibrachiatum*, *T. viride* (Fig. 1) as well as two unidentified species strongly invading the mycelial blocks of shiitake, which were re-described based on morphology, culture characteristics as well as ITS, *tef1* and *rpb2* sequences as *Hypocrea pseudogelatinosa* (current name: *T. pseudogelatinosum*) and *H. pseudostraminea* (current name: *T. pseudostramineum*) (Fig. 1; Kim et al. 2012b). Furthermore, Kim et al. (2010) reported *Gliocladium viride* (syn. *Hypocrea lutea*) as a new shiitake green mould agent in Korea. Although this species was morphologically similar to *Gliocladium*, its phylogenetic position located it within the genus *Trichoderma*; therefore, the scientific name of this species was changed to *T. deliquescens* (Fig. 1) by Jaklitsch (2011). The species were described based on cultural characteristics, holomorph morphology and the phylogenetic markers ITS, *rpb2*, *tef1*, endochitinase and actin. In China, Cao et al. (2014) described *T. oblongisporum* (Fig. 1) as a new causal agent of shiitake green mould, while Wang et al. (2016) confirmed six *Trichoderma* species, i.e. *T. harzianum*, *T. atroviride*, *T. viride*, *T. pleuroticola*, *T. longibrachiatum* and *T. oblongisporum* (Fig. 1), based on morphology characteristics as well as ITS and *tef1* sequence analysis, among which *T. harzianum* and *T. atroviride* proved to be the most prevalent. Luković et al. (2021) reported the isolation of THSC members in Serbia. The identity of these strains was refined later as *T. guizhouense*, *T. atrobrunneum* and

T. simmonsii, while additional species diagnosed in green mould-affected *L. edodes* cultivation in Hungary include *T. simmonsii* and *T. pollinicola* (Fig. 1; Allaga et al. 2021).

Infected shiitake mycelia in cultivated bags became rotten, wilted, yellow and finally died, and the surface of the cultivation bags became covered with dark green fungal colonies. Above 20 °C, a disease incidence of nearly 100% was experienced at some mushroom farms (Cao et al. 2014). Wang et al. (2016) reported that the metabolites of different *Trichoderma* species inhibited the mycelial growth of *L. edodes* and caused distortion and swelling of its hyphae in vitro. However, the inhibition rate considerably varied among different species. Also, *T. harzianum* hyphae overgrew shiitake mycelia and coiled around them, resulting in gradual withering of the mushroom.

A PCR-based diagnostic tool was developed by Miyazaki et al. (2009) for *T. harzianum* causing green mould in shiitake cultivation fields and facilities. Three forward and three reverse primers were designed based on the ITS sequences of *T. harzianum* strains and several other species, among which the primer pair THITS-F2 and THITS-R3 distinguished most *T. harzianum* strains from other *Trichoderma* species and successfully detected *T. harzianum* in infected *L. edodes* cultures.

According to Luković et al. (2021), shiitake pathogenic THSC strains were found to be highly sensitive to commercial fungicides, with ED_{50} values between 0.16–3.63 and 0.01–0.07 μg mL^{-1} for metrafenone and prochloraz, respectively. Strains of *B. licheniformis* and *B. subtilis* were able to efficiently inhibit the growth of *T. harzianum*, *T. pseudokoningii* and *T. viride* and in the cultivation of *L. edodes* and *P. sajor-caju* (Chittihunsa et al. 2007).

4.2 Ganoderma *spp.*

Ganoderma spp., including *G. lucidum* and *G. lingzhi*, mostly cultivated in China, Japan and South Korea have been extensively used as traditional medicine and functional food. Their 'reishi' extract significantly inhibits allergic reactions. It is effectively used in treating bronchitis, rheumatism, nephritis and hypertension and has antitumour properties; therefore, it is applied as a complementary treatment in patients undergoing chemotherapy (Lu et al. 2016). *Ganoderma* species are also known to be affected by *Trichoderma* green moulds, both in cultivation and in their natural environment. Lu et al. (2016) isolated and identified *T. harzianum* (Fig. 1) based on the sequences of ITS, *tef1*, actin and calmodulin genes as the causal agent of *G. lucidum* green mould in China, where the disease can cause great losses to local farmers. Based on morphological characters and ITS and *tef1* sequences, Zhang et al. (2019) identified the causal agent of *G. lingzhi* green mould as *T. longibrachiatum*, while Yan et al. (2019) detected *T. atroviride* (Fig. 1) as a new green mould pathogen of *G. lingzhi*. Cai et al. (2020) isolated a pathogenic *Trichoderma* from diseased *G. lingzhi* in China and identified it as *T. hengshanicum* (Fig. 1) based on morphology as well as *rpb2* and *tef1* sequence analysis. In *Ganoderma*, *Trichoderma* green mould can infect both mycelium tubes and fruiting bodies (stipe and cap). The allochroic patches appearing on the cap lead to gradual rot and the production of a pale green, mildew-like layer. Infected mediostrata turn golden orange into pale yellow and necrotic. The infected mycelium tubes initially produce white, villous mycelia and then green conidia, and are finally covered by dark green, thick mycelium. The severely infected mycelium tubes cannot produce fruiting bodies (Lu et al. 2016). The main symptoms on *G. lingzhi* fruiting bodies are spider-reticulated white mycelium under the cap and green rot on the stipe (Zhang et al. 2019). The infected fruiting bodies rot and wither away. The disease may result in drastic crop reductions due to deformation and deterioration of the mushrooms (Yan et al. 2019).

4.3 Cyclocybe aegerita

Cyclocybe aegerita (formerly *Agrocybe aegerita*), the poplar mushroom, is an important mushroom cultivated in Korea and distributed also in Japan, Europe and Africa. It is very fibrous and has a high antioxidant effect and free-radical scavenging ability, which is correlated with total phenolic content. Choi et al. (2010) identified and characterised 26 *Trichoderma* isolates belonging to four species, i.e. *T. harzianum*, *T. pleuroticola*, *T. longibrachiatum* and *T. atroviride* (Fig. 1), from the fruiting bodies and the substratum of commercially produced *C. aegerita*, with *T. harzianum* being the most prevalent (55.2%). *Trichoderma* spp. initially developed a dense pure white mycelium difficult to distinguish from the mycelium of *C. aegerita*, which later turned green in colour due to intense conidium production. Newly developing primordia infected by *Trichoderma* spp. produced brownish spots and lesions, which later joined and completely covered the fruiting bodies, resulting in badly spotted, brownish mushrooms with reduced growth and crop yield (Choi et al. 2010).

5 Conclusions

The substantial crop losses caused by green mould disease in mushroom production worldwide increased the need for efficient solution strategies. Biofungicides may be combined with chemicals in an integrated management of green mould or even provide entirely harmless, fully environment-friendly alternatives to synthetic fungicides in mushroom production. The development of such control measures can be expected to be substantially supported in the recent 'omics' era by the modern tools of molecular biology, which enable to gain genome-level knowledge about the molecular background of the pathogenic activities of *Trichoderma* green mould species and their infection process, as well as the composition and function of the microbiota associated with them in the mushroom cultivation substrata.

Taken the dimensions and economical importance of the green mould problem in mushroom cultivation, the risks of agricultural *Trichoderma* application to mushroom production should always be kept in mind. Besides harbouring most of the high priority green mould pathogens, the Harzianum clade of the genus also includes several *Trichoderma* strains that are recently being developed or are already registered as biocontrol agents for field applications. This increases the risk of the eventual registration of green mould agents as biocontrol product components. The complete genome sequence of a biocontrol strain designated as Tr1 and referred to as '*T. harzianum*' is available in the GenBank database (assembly accession number: GCA_002894145), but the genomic sequence clearly indicates that the strain actually belongs to *T. pleuroticola*. This example demonstrates that biocontrol agents previously identified and registered as '*T. harzianum*' may actually belong to closely related species with the potential to cause green mould diseases in mushroom cultivation. Although *Trichoderma* strains causing mushroom green mould disease may also possess good biocontrol properties, we strongly recommend that the biocontrol application of strains belonging to species known as high priority green mould pathogens of cultivated mushrooms should be abandoned. Furthermore, due to the frequent problem of misidentifications within the genus *Trichoderma*, we suggest that the active strains included in *Trichoderma*-based commercial biocontrol products available on the market should be subjected to species-level identification by sequence-based molecular methods, preferably targeting the *tef1* gene sequence, in order to minimise the risks to mushroom production by the inadvertent agricultural application of eventually misidentified *Trichoderma* strains that may prove to be green mould pathogens of cultivated mushrooms.

Conflicts of Interest/Competing Interests The authors declare no conflict of interest.

Ethics Approval Not applicable.

Consent to Participate Not applicable.

Consent for Publication All authors have seen and approved the final version of the manuscript.

Availability of Data and Material Not applicable.

Code Availability Not applicable.

Funding The preparation of this review was supported by projects K-116475 (Hungarian Scientific Research Fund) and GINOP-2.2.1-15-2016-00006 (Széchenyi 2020 Programme).

References

Abosriwil SO, Clancy KJ (2002) A protocol for evaluation of the role of disinfectants in limiting pathogens and weed moulds in commercial mushroom production. Pest Manag Sci 58:282–289

Abosriwil SO, Clancy KJ (2003) A mini-bag technique for evaluation of fungicide effects on *Trichoderma* spp. in mushroom compost. Pest Manag Sci 60:350–358

Abubaker KS, Sjaarda C, Castle AJ (2013) Regulation of three genes encoding cell-wall-degrading enzymes of *Trichoderma aggressivum* during interaction with *Agaricus bisporus*. Can J Microbiol 59:417–424

Alice B, Kustudia M (2004) Mushroom cultivation and marketing. NCAT ATTRA Publication No. IP087

Allaga H, Zhumakayev A, Büchner R, Kocsubé S, Szűcs A, Vágvölgyi C, Kredics L, Hatvani L (2021) Members of the Trichoderma harzianum species complex with mushroom pathogenic potential. Agronomy 11:2434.

Al-Rubaiey WL, Al-Juboory HH (2020) Molecular identification of *Trichoderma longibrachiatum* causing green mold in *Pleurotus eryngii* culture media. Plant Arch 20:181–184

Anderson MG, Beyer DM, Wuest PJ (2001) Yield comparison of hybrid *Agaricus* mushroom strains as a measure of resistance to *Trichoderma* green mold. Plant Dis 85:731–734

Angelini P, Pagiotti R, Granetti B (2008) Effect of antimicrobial activity of *Melaleuca alternifolia* essential oil on antagonistic potential of *Pleurotus* spp. against *Trichoderma harzianum* in dual culture. World J Microbiol Biotechnol 24:197–202

Angelini P, Pagiotti R, Venanzoni R, Granetti B (2009) Antifungal and allelopathic effects of asafoetida against *Trichoderma harzianum* and *Pleurotus* spp. Allelopath J 23:357–368

Aydoğdu M, Kurbetli I, Kitapçı A, Sülü G (2020) Aggressiveness of green mould on cultivated mushroom (*Agaricus bisporus*) in Turkey. J Plant Dis Protect 127:695–708

Baars J, Rutjens J, Mumm R (2011) Can volatiles emitted by compost during spawn run be used to detect green mould infection early? In: Proceedings of the 7th International Conference on Mushroom Biology and Mushroom Products (ICMBMP7), Arcachon, France, 4–7 October 2011, pp 469–478

Ballero M, Mascia E, Rescigno A, Teulada ESD (1990) Use of *Pleurotus* for transformation of polyphenols in waste waters from olive presses into proteins. Micol Ital 19:39–41

Bhatt N, Singh RP (2002) Casing soil bacteria as biocontrol agents against the mycoparasitic fungi of *Agaricus bisporus*. In: Proceedings of the 4th International Conference on Mushroom Biology and Mushroom Products, Cuernavaca, Morelos, Mexico, 20–23 February 2002, pp 1–9

Błaszczyk L, Siwulski M, Sobieralski K, Frużyńska-Jóźwiak D (2013) Diversity of *Trichoderma* spp. causing *Pleurotus* green mould diseases in Central Europe. Folia Microbiol 58:325–333

Cai M, Idrees M, Zhou Y, Zhang C, Xu J (2020) First report of green mold disease caused by *Trichoderma hengshanicum* on *Ganoderma lingzhi*. Mycobiology 48:427–430

Cao XT, Bian YB, Xu ZY (2014) First report of *Trichoderma oblongisporum* causing green mold disease on *Lentinula edodes* (shiitake) in China. Plant Dis 98:1440

Castle A, Speranzini D, Rghei N, Alm G, Rinker D, Bissett J (1998) Morphological and molecular identification of *Trichoderma* isolates on North American mushroom farms. Appl Environ Microbiol 64:133–137

Catlin NJ, Wuest PJ, Beyer DM (2004) Green mold harbored by wood: post-crop steaming and preservatives. Mushroom Sci 16:449–458

Chang ST, Miles PG (2004) Mushrooms: cultivation, nutritional value, medicinal effect and environmental impact. CRC Press, Boca Raton

Chen AW (2005) Chapter 1: what is shiitake. In: MushroomWorld (ed) Mushroom growers' handbook, 2nd edn. MushroomWorld, Seoul, pp 1–11

Chen AW, Moy M (2004) Mushroom cultivation: building mold contamination. Int Soc Mushroom Sci 16:82

Chen X, Romaine CP, Ospina-Giraldo MD, Royse DJ (1999a) A polymerase chain reaction-based test for the identification of *Trichoderma harzianum* biotypes 2 and 4, responsible for the worldwide green mold epidemic in cultivated *Agaricus bisporus*. Appl Microbiol Biotechnol 52:246–250

Chen X, Romaine CP, Tan Q, Schlagnhaufer B, Ospina-Giraldo MD, Royse DJ, Huff DR (1999b) PCR-based genotyping of epidemic and preepidemic *Trichoderma* isolates associated with green mold of *Agaricus bisporus*. Appl Environ Microbiol 65:2674–2678

Chen X, Ospina-Giraldo MD, Wilkinson V, Royse DJ, Romaine CP (2003) Resistance of pre- and post-epidemic strains of *Agaricus bisporus* to *Trichoderma aggressivum* f. *aggressivum*. Plant Dis 87:1457–1461

Chittihunsa T, Bangeekhan E, Wongsamitkul N, Subsomboon T (2007) Screening of *Bacillus* spp. suppressing the infection of *Trichoderma* sp. in mushroom cultivation. KMITL Sci Technol J 7:S19–S27

Choi KW (2004) Shelf cultivation of oyster mushroom with emphasis on substrate fermentation. In: Mushroom growers' handbook 1. Oyster mushroom cultivation. Mushworld-Heineart Inc, Seoul

Choi I, Choi J, Sharma PK, Lee W (2010) Isolation and identification of mushroom pathogens from *Agrocybe aegerita*. Mycobiology 38:310–315

Chrysayi-Tokousbalides M, Kastanias MA, Philippoussis A, Diamantopoulou P (2007) Selective fungitoxicity of famoxadone, tebuconazole and trifloxystrobin between *Verticillium fungicola* and *Agaricus bisporus*. Crop Prot 26:469–475

Clift AD, Shamshad A (2009) Modelling mites, moulds and mushroom yields in the Australian Mushroom Industry. In: Proceedings of the 18th World IMACS/MODSIM Congress, Cairns, Australia, 13–17 July 2009, pp 491–497

Colavolpe MB, Mejía SJ, Albertó E (2014) Efficiency of treatments for controlling *Trichoderma* spp. during spawning in cultivation of lignicolous mushrooms. Braz J Microbiol 45:1263–1270

Dodd SL, Crowhurst RN, Rodrigo AG, Samuels GJ, Hills RA, Stewart A (2000) Examination of *Trichoderma* phylogenies derived from ribosomal DNA sequence data. Mycol Res 104:23–34

Doyle O (1991) *Trichoderma* green mould update. Irish Mushroom Rev 3:13–17

Druzhinina IS, Kopchinskiy AG, Komon M, Bissett J, Szakács G, Kubicek CP (2005) An oligonucleotide barcode for species identification in *Trichoderma* and *Hypocrea*. Fungal Genet Biol 42:813–828

Đurović-Pejčev R, Potočnik I, Milijašević-Marčić S, Todorović B, Rekanović E, Stepanović M (2014) Antifungal activity of six plant essential oils from Serbia against *Trichoderma aggressivum* f. *europaeum*. Pestic Phytomed 29:291–297

Fletcher JT (1990) *Trichoderma* and *Penicillium* diseases of *Agaricus bisporus*. A literature review for the Horticultural Development Council. ADAS, London

Fletcher JT (1997) Mushroom spawn and the development of *Trichoderma harzianum* compost mold. Mushroom News 45(8):6–8

Fletcher JT, Gaze RH (2007) Mushroom pest and disease control a color handbook. CRC Press, Boca Raton

Fletcher JT, Connolly G, Mountfield EX, Jacobs L (1980) The disappearance of benomyl from mushroom casing. Ann Appl Biol 95:73–82

Fletcher JT, White PF, Gaze RH (1989) Mushroom pest and disease control, 2nd edn. Intercept, Andover

Foulongne-Oriol M, Minvielle N, Savoie J (2011) QTL for resistance to *Trichoderma* lytic enzymes and metabolites in *Agaricus bisporus*. In: Proceedings of the 7th International Conference on Mushroom Biology and Mushroom Products (ICMBMP7), Arcachon, France, 4–7 October 2011, pp 17–25

Gallucci MN, Oliva M, Casero C, Dambolena J, Luna A, Zygadlo J, Demo M (2009) Antimicrobial combined action of terpenes against the food-borne microorganisms *Escherichia coli*, *Staphylococcus aureus* and *Bacillus cereus*. Flavour Fragr J 24:348–354

Gams W, Meyer W (1998) What exactly is *Trichoderma harzianum*? Mycologia 90:904–915

Garcia-Morras JA, Olivan R (1997) Problemática actual de *Trichoderma* Pers. In: Proceedings of 2nd Technical Conference on Mushroom and Other Edible Fungi in Castilla-La Mancha, Casasimarro, Cuenca, Spain 4–5 November 1997, pp 131–140
Gea FJ (2009) First report of *Trichoderma pleurotum* on oyster mushroom crops in Spain. J Plant Pathol 91:504
Geels FP (1997) Rondetafel-bijeenkomst over *Trichoderma*. Champignoncultuur 41:13
Geels FJ, van de Geijin LJLD, Rutjens A (1988) Pests and diseases. In: van Griensven LJLD (ed) The cultivation of mushrooms. Md: Interlingua, East Grinstead, Sussex, England, pp 361–422
Górski R, Sobieralski K, Siwulski M, Frąszczak B, Sas-Golak I (2014) The effect of *Trichoderma* isolates, from family mushroom growing farms, on the yield of four *Agaricus bisporus* (Lange) Imbach strains. J Plant Prot Res 54:24–27
Gregori A, Svagelj M, Pohleven J (2007) Cultivation techniques and medicinal properties of *Pleurotus* spp. Food Technol Biotechnol 45:238–249
Grogan HM, Jukes AA (2003) Persistence of the fungicides thiabendazole, carbendazim and prochloraz-Mn in mushroom casing soil. Pest Manag Sci 59:1225–1231.
Grogan HM (2005) Zielone pleśnie – wczoraj i dziś. In: Materiały Ogolnopolskiej Konferencji Naukowej „Ochrona i uprawa pieczarki oraz innych grzybow uprawnych", Wyd. Instytutu Warzywnictwa, Skierniewice, pp 19–25
Grogan H (2008) Challenges facing mushroom disease control in the 21st century. In: Proceedings of the 6th International Conference on Mushroom Biology and Mushroom Products, Bonn, Germany, 29 September – 3 October 2008, pp 120–127
Grogan HM, Keeling C, Jukes AA (2000) In vivo response of the mushroom pathogen *Verticillium fungicola* (dry bubble) to prochloraz-manganese. In: Proceedings of Brighton Crop Protection Conference: Pests & Diseases, BCPC, Farnham, Surrey, UK, pp 273–278
Guler P, Ergene A, Tan S (2006) Production of high temperature-resistant strains of *Agaricus bitorquis*. Afr J Biotechnol 5:615–619
Gunde-Cimerman N (1999) Medicinal value of the genus *Pleurotus* (Fr.) P. Karst (*Agaricales* s.l. *Basidiomycetes*). Int J Med Mushrooms 1:69–80
Gupta VK, Schmoll M, Herrera-Estrella A, Upadhyay RS, Druzhinina I, Tuohy MG (eds) (2014) Biotechnology and biology of *Trichoderma*. Elsevier, Amsterdam
Guthrie JL, Castle AJ (2006) Chitinase production during interaction of *Trichoderma aggressivum* and *Agaricus bisporus*. Can J Microbiol 52:961–967
Győrfi J (2002) Zöldpenészek, *Trichoderma* fajok. Magyar Gomba 18:28–29
Győrfi J, Geösel A (2008) Biological control against *Trichoderma* species in *Agaricus* cultivation. In: Proceedings of the 6th International Conference on Mushroom Biology and Mushroom Products. Bonn, Germany, 29 September – 3 October 2008, pp 158–164
Hassan S, Mohammad AZ, Kiramat K (2011) Cultivation of the oyster mushroom (*Pleurotus ostreatus* (Jacq.) P. Kumm.) in two different agroecological zones of Pakistan. Afr J Biotechnol 10:183–188
Hasselbach OE, Mutsers P (1971) *Agaricus bitorquis* (Quel.) Sacc. ein warmeliebendes Familienmitglied der Champignon. Champignon 130:20–22
Hatvani L (2008) Mushroom pathogenic *Trichoderma* species: occurrence, biodiversity, diagnosis and extracellular enzyme production. PhD-dissertation, University of Szeged, Hungary
Hatvani L, Antal Z, Manczinger L, Szekeres A, Druzhinina IS, Kubicek CP, Nagy A, Nagy E, Vágvölgyi C, Kredics L (2007) Green mould diseases of *Agaricus* and *Pleurotus* spp. are caused by related but phylogenetically different *Trichoderma* species. Phytopathology 97:532–537
Hatvani L, Kocsubé S, Manczinger L, Antal Z, Szekeres A, Druzhinina IS, Komoń-Zelazowska M, Kubicek CP, Nagy A, Vágvölgyi C, Kredics L (2008) The green mould disease global threat to the cultivation of oyster mushroom (*Pleurotus ostreatus*): a review. Mushroom Sci 17:485–495
Hatvani L, Sabolic P, Kocsubé S, Kredics L, Czifra D, Vágvölgyi C, Kaliterna J, Ivić D, Đermić E, Kosalec I (2012) First report of mushroom green mould disease in Croatia. Arh Hig Rada Toksikol 63:481–487

Hatvani L, Kredics L, Allaga H, Manczinger L, Vágvölgyi C, Kuti K, Geösel A (2017) First report of *Trichoderma aggressivum* f. *aggressivum* green mold on *Agaricus bisporus* in Europe. Plant Dis 101(1052)

Hermosa MR, Grondona I, Monte E (1999) Isolation of *Trichoderma harzianum* Th2 from commercial mushroom compost in Spain. Plant Dis 83:591

Hermosa MR, Grondona I, Iturriaga EA, Diaz-Minguez JM, Castro C, Monte E, Garcia-Acha I (2000) Molecular characterization and identification of biocontrol isolates of *Trichoderma* spp. Appl Environ Microbiol 66:1890–1898

Innocenti G, Montanari M (2014) *Trichoderma* green mould disease: a case study in a *Pleurotus ostreatus* farm. Preliminary data. Micol Ital 43:49–57

Innocenti G, Montanari M, Righini H, Roberti R (2019) *Trichoderma* species associated with green mould disease of *Pleurotus ostreatus* and their sensitivity to prochloraz. Plant Pathol 68:392–398

Jaklitsch WM (2011) European species of *Hypocrea* part II: species with hyaline ascospores. Fungal Divers 48:1–250

Jayalal RGU, Adikaram NKB (2007) Influence of *Trichoderma harzianum* metabolites on the development of green mould disease in the oyster mushroom. Ceylon J Sci 36:53–60

Kai D, Zhixiang L, Qiong W, Mi N, Chuanjin Y, Meng W et al (2020) MIST: a multilocus identification system for *Trichoderma*. Appl Environ Microbiol 86:e01532–e01520

Khan SM, Kausar AG, Ali MA (1981) Yield performance of different strains of oyster mushroom (*Pleurotus* spp.) on paddy straw in Pakistan. Mushroom Sci 11:675–678

Khan I, Shah F, Bulman S, Scott I (2008) Molecular diagnostic tools for improved mushroom production. In: Abstracts of the XVII International Congress on the Science and Cultivation of Edible and Medicinal Fungi, Cape Town, South Africa

Kim JY, Yun YH, Hyun MW, Kim MH, Kim SH (2010) Identification and characterization of *Gliocladium viride* isolated from mushroom fly infested oak log used for shiitake cultivation. Mycobiology 38:7–12

Kim CS, Shirouzu T, Nakagiri A, Sotome K, Nagasawa E, Maekawa N (2012a) *Trichoderma mienum* sp. nov., isolated from mushroom farms in Japan. Anton Leeuw Int J G 102:629–641

Kim CS, Yu SH, Nakagiri A, Shirouzu T, Sotome K, Kim SC, Maekawa N (2012b) Re-evaluation of *Hypocrea pseudogelatinosa* and *H. pseudostraminea* isolated from shiitake mushroom (*Lentinula edodes*) cultivation in Korea and Japan. Plant Pathol J 28:341–356

Kim CS, Shirouzu T, Nakagiri A, Sotome K, Maekawa N (2013) *Trichoderma eijii* and *T. pseudolacteum*, two new species from Japan. Mycol Prog 12:739–753

Komoń-Zelazowska M, Bissett J, Zafari D, Hatvani L, Manczinger L, Woo S, Lorito M, Kredics L, Kubicek CP, Druzhinina IS (2007) Genetically closely related but phenotypically divergent *Trichoderma* species cause worldwide green mould disease in oyster mushroom farms. Appl Environ Microbiol 73:7415–7426

Kosanović D, Potočnik I, Duduk B, Vukojević J, Stajić M, Rekanović E, Milijašević-Marčić S (2013) *Trichoderma* species on *Agaricus bisporus* farms in Serbia and their biocontrol. Ann Appl Biol 163:218–230

Kosanović D, Potočnik I, Vukojević J, Stajić M, Rekanović E, Stepanović M, Todorović B (2015) Fungicide sensitivity of *Trichoderma* spp. from *Agaricus bisporus* farms in Serbia. J Environ Sci Health B 50:607–613

Kosanović D, Sheehan G, Grogan H, Kavanagh K (2019) Characterisation of the interaction of *Pseudomonas putida* and *Pseudomonas tolaasii* with *Trichoderma aggressivum*. Eur J Plant Pathol 156:111–121

Kosanović D, Grogan H, Kavanagh K (2020) Exposure of *Agaricus bisporus* to *Trichoderma aggressivum* f. *europaeum* leads to growth inhibition and induction of an oxidative stress response. Fungal Biol 124:814–820

Kredics L, Hatvani L, Antal Z, Manczinger L, Druzhinina IS, Kubicek CP, Szekeres A, Nagy A, Vágvölgyi C, Nagy E (2006) Green mould disease of oyster mushroom in Hungary and Transylvania. Acta Microbiol Immunol Hung 53:306–307

Kredics L, Cseh T, Körmöczi P, Hatvani L, Manczinger L, Vágvölgyi C (2008a) Extracellular enzyme production of the two causative agents of oyster mushroom green mould under inductive and non-inductive conditions. In: Proceedings of the 6th International Conference on Mushroom Biology and Mushroom Products, Bonn, Germany, 29 September – 3 October 2008, p 48

Kredics L, Cseh T, Körmöczi P, Hatvani L, Antal Z, Manczinger L, Vágvölgyi C (2008b) Proteolytic enzyme production of the causative agents of oyster mushroom green mould under inductive and non-inductive conditions. Acta Microbiol Immunol Hung 55:211–212

Kredics L, Kocsubé S, Nagy L, Komon-Zelazowska M, Manczinger L, Sajben E, Nagy A, Vágvölgyi C, Kubicek CP, Druzhinina IS, Hatvani L (2009) Molecular identification of *Trichoderma* species associated with *Pleurotus ostreatus* and natural substrates of the oyster mushroom. FEMS Microbiol Lett 300:58–67

Kredics L, Jimenez LG, Naeimi S, Czifra D, Urban P, Manczinger L, Vágvölgyi C, Hatvani L (2010) A challenge to mushroom growers: the green mould disease of cultivated champignons. In: Méndez-Vilas A (ed) Current research, technology and education topics in applied microbiology and microbial biotechnology. Formatex, Badajoz, pp 295–305

Kredics L, Láday M, Körmöczi P, Manczinger L, Rákhely G, Vágvölgyi C, Szekeres A (2012) Genetic and biochemical diversity among *Trichoderma* isolates in soil samples from winter wheat fields of the Great Hungarian Plain. Acta Biol Szeged 56:141–149

Krupke O, Castle A, Rinker D (2003) The North American mushroom competitor, *Trichoderma aggressivum* f. *aggressivum*, produces antifungal compounds in mushroom compost that inhibit mycelial growth of the commercial mushroom *Agaricus bisporus*. Mycol Res 107:1467–1475

Kubicek CP, Herrera-Estrella A, Seidl-Seiboth V, Martinez DA, Druzhinina IS, Thon M et al (2011) Comparative genome sequence analysis underscores mycoparasitism as the ancestral life style of *Trichoderma*. Genome Biol 12:R40

Kullnig-Gradinger CM, Szakács G, Kubicek CP (2002) Phylogeny and evolution of the fungal genus *Trichoderma* – a multigene approach. Mycol Res 106:757–767

Lane C (2008) *Trichoderma* green mould – determining diversity and highlighting risks. HDC Project Report M46, 17 pp

Largeteau ML, Savoie JM (2010) Microbially induced diseases of *Agaricus bisporus*: biochemical mechanisms and impact on commercial mushroom production. Appl Microbiol Biotechnol 86:63–73

Largeteau-Mamoun ML, Mata G, Savoie JM (2002) Green mold disease: adaptation of *Trichoderma harzianum* Th2 to mushroom compost. In: Proceedings of the 4th International Conference on Mushroom Biology and Mushroom Products, Cuernavaca, Mexico, 20–22 February 2002, pp 179–187

Lavi ID, Levinson D, Peri I, Hadar Y, Schwartz B (2010) Orally administered glucans from the edible mushroom (*Pleurotus pulmonarius*) reduce acute inflammation in dextran sulfate sodium induced experimental colitis. Br J Nutr 103:393–402

Lee HB, Magan N, Yu SH (2000) Comparison of water relations of three cultivated *Pleurotus* species and *Trichoderma* green moulds. Plant Pathol J 16:25–28

Lee SH, Jung HJ, Hong S-B, Choi JI, Ryua J-S. (2020) Molecular markers for detecting a wide range of Trichoderma spp. that might potentially cause green mold in Pleurotus eryngii. Mycobiology 48:313–320.

Lelley J (1987) Disinfection in mushroom farming-possibilities and limits. Mushroom J 14:181–187

Lelley J, Straetman U (1986) Hygiene in mushroom growing units-disinfection, disinfectants and their suitability for mushroom farms. Dev Crop Sci 10:621–636

Lu BH, Zuo B, Liu XL, Feng J, Wang ZM, Gao J (2016) *Trichoderma harzianum* causing green mold disease on cultivated *Ganoderma lucidum* in Jilin Province, China. Plant Dis 100:2524

Luković J, Milijašević-Marčić S, Hatvani L, Kredics L, Szűcs A, Vágvölgyi C, Duduk N, Vico I, Potočnik I (2021) Sensitivity of *Trichoderma* strains from edible mushrooms to the fungicides prochloraz and metrafenone. J Environ Sci Health B 56:54–63

Luo XC (2004) Progress of xiang-gu (shiitake) cultivation in China. Mushroom Sci 16:317–322

Mamoun ML, Iapicco R, Savoie J-M, Olivier JM (2000a) Green mould disease in France: *Trichoderma harzianum* Th2 and other species causing damages on mushroom farms. Mushroom Sci 15:625–632

Mamoun ML, Savoie J-M, Olivier JM (2000b) Interactions between the pathogen *Trichoderma harzianum* Th2 and *Agaricus bisporus* in mushroom compost. Mycologia 92:233–240

Marik T, Urbán P, Tyagi C, Szekeres A, Leitgeb B, Vágvölgyi M, Manczinger L, Druzhinina IS, Vágvölgyi C, Kredics L (2017) Diversity profile and dynamics of peptaibols produced by green mould *Trichoderma* species in interactions with their hosts *Agaricus bisporus* and *Pleurotus ostreatus*. Chem Biodivers 14:e1700033

Martinez D, Berka RM, Henrissat B, Saloheimo M, Arvas M, Baker SE et al (2008) Genome sequencing and analysis of the biomass-degrading fungus *Trichoderma reesei* (syn. *Hypocrea jecorina*). Nat Biotechnol 26:553–560

Marzullo L, Cannio R, Giardina P, Santini MT, Sannia G (1995) Veratryl alcohol oxidase from *Pleurotus ostreatus* participates in lignin biodegradation and prevents polymerization of laccase-oxidized substrates. J Biol Chem 270:3823–3827

Matcham SE, Wood DA (1992) Purification of *Agaricus bisporus* extracellular laccase from mushroom compost. Biotechnol Lett 14:297–300

Mattila PK, Suonpaa K, Piironen V (2000) Functional properties of edible mushrooms. Nutrition 16:694–696

Mazin M, Harvey R, Andreadis S, Pecchia J, Cloonan K, Rajotte EG (2019) Mushroom sciarid fly, *Lycoriella ingenua* (Diptera: Sciaridae) adults and larvae vector mushroom green mold (*Trichoderma aggressivum* f. *aggressivum*) spores. Appl Entomol Zool 54:369–376

Milijašević-Marčić S, Stepanović M, Todorović B, Duduk B, Stepanović J, Rekanović E, Potočnik I (2017) Biological control of green mould on *Agaricus bisporus* by a native *Bacillus subtilis* strain from mushroom compost. Eur J Plant Pathol 148:509–519

Miyazaki K, Tsuchiya Y, Okuda T (2009) Specific PCR assays for the detection of *Trichoderma harzianum* causing green mold disease during mushroom cultivation. Mycoscience 50:94–99

Morin E, Kohler A, Baker AR, Foulongne-Oriol M, Lombard V, Nagy LG et al (2012) Genome sequence of the button mushroom *Agaricus bisporus* reveals mechanisms governing adaptation to a humic-rich ecological niche. Proc Natl Acad Sci U S A 109:17501–17506

Morris E, Doyle O, Clancy KJ (1995a) A profile of *Trichoderma* species. I – mushroom compost production. Mushroom Sci 14:611–618

Morris E, Doyle O, Clancy KJ (1995b) A profile of *Trichoderma* species. II – mushroom growing units. Mushroom Sci 14:619–625

Morris E, Harrington O, Doyle ORE (2000) Green mould disease – the study of survival and dispersal characteristics of the weed mould *Trichoderma*, in the Irish mushroom industry. Mushroom Sci 15:645–652

Mumpuni A, Sharma HSS, Brown AE (1998) Effect of metabolites produced by *Trichoderma harzianum* biotypes and *Agaricus bisporus* on their respective growth radii in culture. Appl Environ Microbiol 64:5053–5056

Muthumeenakshi S, Mills PR (1995) Detection and differentiation of fungal pathogens of *Agaricus bisporus*. Mushroom Sci 14:603–610

Muthumeenakshi S, Mills PR, Brown-Averil E, Seaby DA (1994) Intraspecific molecular variation among *Trichoderma harzianum* isolates colonizing mushroom compost in the British Isles. Microbiology (UK) 140:769–777

Muthumeenakshi S, Brown AE, Mills PR (1998) Genetic comparison of the aggressive weed mould strains of *Trichoderma harzianum* from mushroom compost in North America and the British Isles. Mycol Res 102:385–390

Naeimi S, Khosravi V, Nouri M-Z, Hoda H, Vágvölgyi C, Kredics L (2019) Biological control of rice sheath blight disease with formulation of indigenous *Trichoderma* strains under paddy field conditions. Acta Biol Szeged 63:37–43

Nagy A, Manczinger L, Tombácz D, Hatvani L, Győrfi J, Antal Z, Sajben E, Vágvölgyi C, Kredics L (2012) Biological control of oyster mushroom green mould disease by antagonistic *Bacillus* species. IOBC-WPRS Bull 78:289–293

O'Brien M, Grogan H, Kavanagh K (2014) Proteomic response of *Trichoderma aggressivum* f. *europaeum* to *Agaricus bisporus* tissue and mushroom compost. Fungal Biol 118:785–791

O'Brien M, Kavanagh K, Grogan H (2017) Detection of *Trichoderma aggressivum* in bulk phase III substrate and the effect of *T. aggressivum* inoculum, supplementation and substrate-mixing on *Agaricus bisporus* yields. Eur J Plant Pathol 147:199–209

Olmedo Monfil V, Casas-Flores S (2014) Molecular mechanisms of biocontrol in *Trichoderma* spp. and their applications in agriculture. In: Gupta VK, Schmoll M, Herrera-Estrella A, Upadhyay RS, Druzhinina I, Tuohy MG (eds) Biotechnology and biology of *Trichoderma*. Elsevier, Amsterdam, pp 429–476

Ospina-Giraldo MD, Royse DJ, Thon MR, Chen X, Romaine CP (1998) Phylogenetic relationships of *Trichoderma harzianum* causing mushroom green mold in Europe and North America to other species of *Trichoderma* from world-wide sources. Mycologia 90:76–81

Ospina-Giraldo MD, Royse DJ, Chen X, Romaine CP (1999) Molecular phylogenetic analyses of biological controls strains of *Trichoderma harzianum* and other biotypes of *Trichoderma* ssp. associated with mushrooms green mold. Phytopatology 89:308–313

Pandin C, Vedie R, Rousseau T, Le Coq D, Aymericha S, Briandeta R (2018a) Dynamics of compost microbiota during the cultivation of *Agaricus bisporus* in the presence of *Bacillus velezensis* QST713 as biocontrol agent against *Trichoderma aggressivum*. Biol Control 127:39–54

Pandin C, Le Coq D, Deschamps J, Vedic R, Rousseau T, Aymerich S, Briandet R (2018b) Complete genome sequence of *Bacillus velezensis* QST713: a biocontrol agent that protects *Agaricus bisporus* crops against green mould disease. J Biotechnol 278:10–19

Papadopoulos G (2006) The fate of prochloraz in mushroom casing. PhD dissertation, University of Reading, UK

Park MS, Bae KS, Yu SH (2004) Molecular and morphological analysis of *Trichoderma* isolates associated with green mold epidemic of oyster mushroom in Korea. J Huazhong Agric Univ 23:157–164

Park MS, Seo GS, Lee KH, Bae KS, Yu SH (2005) Characterization of *Trichoderma* spp. associated with green mold of oyster mushroom by PCR-RFLP and sequence analysis of ITS regions of rDNA. Plant Pathol J 21:229–236

Park MS, Bae KS, Yu SH (2006) Two new species of *Trichoderma* associated with green mold of oyster mushroom cultivation in Korea. Mycobiology 34:111–113

Pecchia JA (2012) *Trichoderma* green mold survey results. In: Proceedings from the Penn State Mushroom Short Course. State College, Pennsylvania

Peil RM, Rosseto EA, Pierobom CR, Rocha MT (1996) Desinfestacao de composto para cultivo de cogumelo *Agaricus bisporus* (Lange) Imbach. Revista Bras Agrocienc 2:159–164

Potočnik I, Stepanović M, Rekanović E, Todorović B, Milijašević-Marčić S (2015) Disease control by chemical and biological fungicides in cultivated mushrooms: button mushroom, oyster mushroom and shiitake. Pestic Phytomed 30:201–208

Potočnik I, Todorović B, Rekanović E, Luković J, Paunović D, Milijašević-Marčić S (2018) Impact of *Bacillus subtilis* QST713 mushroom grain spawn treatment on yield and green mould control. Pestic Phytomed 33:205–211

Potočnik I, Rekanović E, Todorović B, Luković J, Paunović D, Stanojević O, Milijašević-Marčić S (2019a) The effects of casing soil treatment with *Bacillus subtilis* Ch-13 biofungicide on green mould control and mushroom yield. Pestic Phytomed 34:53–60

Potočnik I, Milijašević-Marčić S, Stanojević O, Berić T, Stanković S, Kredics L, Hatvani L (2019b) The activity of native *Bacillus subtilis* strains in control of green mould disease of oyster mushroom (*Pleurotus* spp.). Pestic Phytomed 34:97–102

Prescott TJ, Wong B, Panaretou E, Boa A, Bond A, Chowdhury S et al (2018) Useful fungi. In: Willis K (ed) State of the world's fungi, report. Royal Botanic Gardens, Kew/Richmond, pp 24–31

Puniya AK, Shah KG, Hire SA, Ahire RN, Rathod MP, Mali RS (1996) Bioreactor for solid-state fermentation of agro-industrial wastes. Indian J Microbiol 36:177–178

Qi T, Ospina-Giraldo MD, Romaine CP, Schlagnhaufer B, Xi C, Huff D, Royse DJ (1996) Genetic analysis of the *Trichoderma* spp. associated with the green mold epidemic in mushrooms. Phytopathology 86(S89)

Radványi D, Geösel A, Jókai Z, Fodor P, Gere A (2020) Detection and identification of microbial volatile organic compounds of the green mold disease: MVOC profile on different media. Int J Agric Environ Inf Syst 11:14–28

Rasooli I, Mirmostafa SA (2002) Antibacterial properties of *Thymus pubescens* and *Thymus serpyllum* essential oils. Fitoterapia 73:244–250

Rinker DL (1993) Disease management strategies for *Trichoderma* mould. Mushroom World 4:3–5

Rinker DL (1994) *Trichoderma* green mold: a seminar by Dr. Donald Betterley. Monterey Labs Mushroom News 42:28–32

Rinker DL (1996) *Trichoderma* disease: progress toward solutions. Mushroom World 7:46–53

Rinker DL, Alm G (2000) Management of green mould disease in Canada. Mushroom Sci 15:617–623

Rinker DL, Alm G (2008) Management of casing *Trichoderma* using fungicides. Mushroom Sci 17:496–509

Rinker DL, Alm G, Castle AJ, Rghei N (1997a) Not all green is *Trichoderma* green mould. Mushroom World 8:47–50

Rinker DL, Alm G, Castle AJ, Rghei N (1997b) Distribution of green mould on infected mushroom farms. Mushroom World 8:71–75

Roberti R, Di Francesco A, Innocenti G, Mari M (2019) Potential for biocontrol of *Pleurotus ostreatus* green mould disease by *Aureobasidium pullulans* De Bary (Arnaud). Biol Control 135:9–15

Romaine CP, Royse DJ, Wuest PJ, Beyer DM (1996) Mushroom green mold: cause, edaphic factors and control. Mushroom News 44:20–23

Romaine CP, Royse DJ, Wuest PJ, Beyer DM (1998) Mushroom green mold: cause, edaphic factors and control. Mushroom News 46:12–17

Romaine CP, Chen X, Ospina-Giraldo MD, Royse DJ (1999) Phylogenetic analysis of *Trichoderma* spp. associated with green mould on *Agaricus bisporus* using a sequence in β-tubulin gene 1. In: Proceedings of the 3rd International Conference on Mushroom Biology and Mushroom Products, 12–16 October 1999, Sydney, Australia, pp 116–124

Romaine CP, Chen X, Ospina-Giraldo MD, Royse DJ (2001) Molecular genetics and pathogenicity of biocontrol and mushroom *Trichoderma*. IOBC WPRS Bull 24(3):333–336

Romaine CPD, Royse DJ, Schlagnhaufer C (2005) Superpathogenic *Trichoderma* resistant to TopsinM found in Pennsylvania and Delaware. Mushroom News 53:6–9

Romaine CP, Royse DJ, Schlagnhaufer C (2008) Emergence of benzimidazole-resistant green mould, *Trichoderma aggressivum*, on cultivated *Agaricus bisporus* in North America. Mushroom Sci 17:510–523

Romero-Arenas O, Lara MH, Huato MAD, Hernandez FD, Victoria DAA (2009) The characteristics of *Trichoderma harzianum* as a limiting agent in edible mushrooms. Rev Colomb Biotechnol 11:143–151

Royse DJ, Boomer K, Du Y, Handcock M (1999) Spatial distribution of green mold foci in 30 commercial mushroom crops. Plant Dis 83:71–76

Royse DJ, Ospina-Giraldo MD, Chen X, Romaine CP (2001) Phylogenetic analyses of *Trichoderma harzianum* associated with mushroom culture or used for biological control of plant pathogens. IOBC WPRS Bull 24:341–344

Royse DJ, Baars J, Tan Q (2017) Current overview of mushroom production in the world. In: Zied DC, Pardo-Giménez A (eds) Edible and medicinal mushrooms: technology and applications, 1st edn. Wiley-Blackwell, Hoboken, pp 2–13

Samuels GJ, Dodd SL, Gams W, Castlebury LA, Petrini O (2002) *Trichoderma* species associated with the green mold epidemic of commercially grown *Agaricus bisporus*. Mycologia 94:146–170

Sánchez-Montesinos B, Diánez F, Moreno-Gavira A, Gea FJ, Santos M (2019) Plant growth promotion and biocontrol of *Pythium ultimum* by saline tolerant *Trichoderma* isolates under salinity stress. Int J Environ Res Public Health 16:2053

Sánchez-Montesinos B, Diánez F, Moreno-Gavíra A, Gea FJ, Santos M (2020) Role of *Trichoderma aggressivum* f. *europaeum* as plant-growth promoter in horticulture. Agronomy 10(7):1004

Šantrić L, Potočnik I, Radivojević L, Umiljendic JG, Rekanović E, Duduk B, Milijašević-Marčić S (2018) Impact of a native *Streptomyces flavovirens* from mushroom compost on green mold control and yield of *Agaricus bisporus*. J Environ Sci Health B 53:677–684

Savoie J-M, Mata G (1999) The antagonistic action of *Trichoderma* sp. hyphae to *Lentinula edodes* hyphae changes lignocellulolytic activities during cultivation in wheat straw. World J Microbiol Biotechnol 15:369–373

Savoie J-M, Mata G (2003) *Trichoderma harzianum* metabolites pre-adapt mushrooms to *Trichoderma aggressivum* antagonism. Mycologia 95:191–199

Savoie J-M, Iapicco R, Largeteau-Mamoun M (2001) Factors influencing the competitive saprophytic ability of *Trichoderma harzianum* Th2 in mushroom (*Agaricus bisporus*) compost. Mycol Res 105:1348–1356

Seaby DA (1987) Infection of mushroom compost by *Trichoderma* species. Mushroom J 179:355–361

Seaby DA (1989) Further observations on *Trichoderma*. Mushroom 197:147–151

Seaby DA (1996a) Differentiation of *Trichoderma* taxa associated with mushroom production. Plant Pathol 45:905–912

Seaby DA (1996b) Investigation of the epidemiology of green mould of mushroom (*Agaricus bisporus*) compost caused by *Trichoderma harzianum*. Plant Pathol 45:913–923

Seaby DA (1998) *Trichoderma* as weed mould or pathogen in mushroom cultivation. In: Harman GE, Kubicek CP (eds) *Trichoderma* and *Gliocladium* Vol. 2. Enzymes, biological control and commercial applications. Taylor and Francis, London, pp 267–287

Shah S, Nasreen S, Munshi NA (2011) Evaluation of some botanicals in controlling green mold (*Trichoderma harzianum*) disease in oyster mushroom cultivation. Int J Bot 7:209–215

Sharma SR, Vijay B (1996) Yield loss in *Pleurotus ostreatus* spp. caused by *Trichoderma viride*. Mushroom Res 5:19–22

Sharma HSS, Kilpatrick M, Ward F, Lyons G, Burns L (1999) Colonisation of phase II compost by biotypes of *Trichoderma harzianum* and their effect on mushroom yield and quality. Appl Microbiol Biotechnol 51:572–578

Sharma SR, Kumar S, Sharma VP (2007) Diseases and competitor moulds of mushrooms and their management. Technical Bulletin, National Research Centre for Mushroom (Indian Council of Agricultural Research), Chambaghat, India, pp 1–81

Sinden JW (1971) Ecological control of pathogens and weed moulds in mushroom culture. Annu Rev Phytophatol 9:411–432

Sinden J, Hauser E (1953) Nature and control of three mildew diseases of mushrooms in America. Mushroom Sci 2:177–180

Siwulski M, Sobieralski K, Błaszczyk L, Frąszczak B, Frużyńska-Jóźwiak D, Sas-Golak I (2011) Evaluation of mycelium growth of several *T. pleurotum* and *T. pleuroticola* isolates and their biotic interaction with *Pleurotus florida*. Phytopathologia 59:43–48

Sobieralski K, Siwulski M, Fruzynska-Jozwiak D, Gorski R (2009a) Impact of *Trichoderma aggressivum* f. *europaeum* Th2 on the yielding of *Agaricus bisporus*. Phytopathologia 53:5–10

Sobieralski K, Siwulski M, Fruzynska-Jozwiak D (2009b) Growth of aggressive isolates of *Trichoderma aggressivum* f. *europaeum* in dependence on temperature and medium. Phytopathologia 53:5–10

Sobieralski K, Siwulski M, Frużyńska-Jóźwiak D, Górski R (2010a) Impact of infection with *Trichoderma aggressivum* f. *europaeum* isolates on carpophore setting and yielding of *Agaricus bisporus*. Phytopathologia 55:35–41

Sobieralski K, Siwulski M, Fruyska-Józwiak D, Błaszczyk L, Sas-Golak I, Jasińska A (2010b) Impact of infections with two *Trichoderma aggressivum* f. *europaeum* isolates on the yielding of some wild strains of *Agaricus bisporus* (Lange) Imbach. J Plant Prot Res 50:501–504

Sobieralski K, Siwulski M, Jasińska A, Frużyńska-Joźwiak D, Sas-Golak I, Szymański J (2010c) Impact of infection with *Trichoderma aggressivum* f. *europaeum* isolates on the yielding of some wild strains of *Agaricus bitorquis* from different regions of Poland. Phytopathologia 58:5–11

Sobieralski K, Siwulski M, Górski R, Fruyska-Jówiak D, Nowak-Sowiska M (2010d) Impact of *Trichoderma aggressivum* f. *europaeum* isolates on yielding and morphological features of *Pleurotus eryngii*. Phytopatol Pol 56:17–25

Sobieralski K, Siwulski M, Błaszczyk L, Fruyska-Józwiak D, Lisiecka J (2012a) The effect of infestation with isolates of *Trichoderma* sp. on mycelium growth and yielding in single-spore heterokaryotic cultures of *Agaricus bisporus* (Lange) Imbach. Acta Sci Pol Hortic Cult 11:47–57

Sobieralski K, Siwulski M, Błaszczyk L, Fraszczak B, Sas-Golak I (2012b) Impact of infections with *Trichoderma pleurotum* and *Trichoderma pleuroticola* isolates on yielding of wild strains of *Pleurotus ostreatus* (fr.) Kumm. obtained from natural sites. Acta Sci Pol Hortic Cult 11:239–249

Sobieralski K, Siwulski M, Kommon-Żelazowska M, Błaszczyk L, Sas-Golak I, Frużyńska-Jóźwiak D (2012c) Impact of *Trichoderma pleurotum* and *T. pleuroticola* isolates on yielding of *Pleurotus ostreatus* (FR.) Kumm. J Plant Protect Res 52:165–168

Sobieralski K, Siwulski M, Komon-Żelazowska M, Błaszczyk L, Górski R, Spiżewski T, Sas-Golak I (2012d) Evaluation of the growth of *Trichoderma pleurotum* and *Trichoderma pleuroticola* isolates and their biotic interaction with *Pleurotus* sp. J Plant Protect Res 52:235–239

Soković M, van Griensven JLD (2006) Antimicrobial activity of essential oils and their components against the three major pathogens of the cultivated button mushroom, Agaricus bisporus. Eur J Plant Pathol 116: 211–224

Speer M (2010) Effect of spawn type on *Trichoderma* disease. Mushroom News 4:4–6

Spillmann A (2002) What's killing the mushrooms of Pennsylvania? (a mushroom mystery). Agric Res 50:12

Stanojević O, Milijašević-Marčić S, Potočnik I, Stepanović M, Dimkić I, Stanković S, Berić T (2016) Isolation and identification of *Bacillus* spp. from compost material, compost and mushroom casing soil active against *Trichoderma* spp. Arch Biol Sci 68:845–852

Stanojević O, Berić T, Potočnik I, Rekanović E, Stanković S, Milijašević-Marčić S (2019) Biological control of green mould and dry bubble diseases of cultivated mushroom (*Agaricus bisporus* L.) by *Bacillus* spp. Crop Prot 126:104944

Staunton L (1987) *Trichoderma* green mould in mushroom compost. Mushroom J 179:362–363

Szczech M, Staniaszek M, Habdas H, Ulinski Z, Szymanski J (2008) *Trichoderma* spp. – the cause of green mould on Polish mushroom farms. Veg Crop Res Bull 69:105–114

Szili I (2008) Gombatermesztők könyve. Mezőgazda Kiadó, Budapest

Talavera-Ortiz A, Chaverri P, Díaz-Godínez G, Acosta-Urdapilleta ML, Villegas E, Téllez-Téllez M (2020) Mycelial inhibition of *Trichoderma* spp. (*Hypocreaceae*) isolated from the cultivation of *Pleurotus ostreatus* (*Pleurotaceae*) with an extract of *Pycnoporus* sp. (*Polyporaceae*). Acta Bot Mex 127:e1537

Urbán P, Miao Y, Fekete C, Schönhardt K, Valasek A, Hatvani L et al (2016a) Complete genome sequence of *Trichoderma aggressivum* f. *europaeum*, the causal agent of green mould disease in the cultivation of champignon (*Agaricus bisporus*). In: Márialigeti K (ed) Book of abstracts, Annual Meeting of the Hungarian Society of Microbiology, Eötvös Loránd University, Budapest, Hungary, p 62

Urbán P, Miao Y, Fekete C, Hatvani L, Büchner R, Vágvölgyi C et al (2016b) Complete genome sequence of the green mould pathogen *Trichoderma pleuroti*. In: Škrbić B (ed) 18th Danube-

Kris-Mures-Tisa (DKMT) Euroregional Conference on Environment and Health: Book of abstracts. University of Novi Sad, Novi Sad, pp 51–52

Vahabi K, Sharifnabi B, Zafari D (2009) Genetic diversity of *Trichoderma* spp. associated with button mushroom, *Agaricus bisporus*, inferred from AFLP markers and ITS sequencing. Acta Phytopathol Entomol Hung 44:239–253

Velázquez-Cedeño MA, Farnet AM, Mata G, Savoie JM (2006) Estudio preliminar de la microflora bacteriana termo tolerante de la pulpa de café y la paja de trigo con potencial de inhibición contra *Trichoderma sp viride* en el cultivo de *Pleurotus*. Rev Mex Micol 22:33–39

Wang G, Cao X, Ma X, Guo M, Liu C, Yan L, Bian Y (2016) Diversity and effect of *Trichoderma* spp. associated with green mold disease on *Lentinula edodes* in China. Microbiol Open 5:709–718

Williams J, Clarkson JM, Mills PR, Cooper RM (2003a) Saprotrophic and mycoparasitic components of aggressiveness of *Trichoderma harzianum* groups toward the commercial mushroom *Agaricus bisporus*. Appl Environ Microbiol 69:4192–4199

Williams J, Clarkson JM, Mills PR, Cooper RM (2003b) A selective medium for quantitative reisolation of *Trichoderma harzianum* from *Agaricus bisporus* compost. Appl Environ Microbiol 69:4190–4191

Woo SL, Di Benedetto P, Senatore M, Abadi K, Gigante S, Soriente I, Ferraioli S, Scala F, Lorito M (2004) Identification and characterization of *Trichoderma* species aggressive to *Pleurotus* in Italy. J Zhejiang Univ Agric Life Sci 30:469–470

Woo SL, Kubicek CP, Druzhinina IS, Vinale F, Cavallo P, Lorito M (2009) Characterization of *Trichoderma* species associated with the production of *Pleurotus ostreatus* in Italy. J Plant Pathol 91:S94

Yamamoto Y, Shirono H, Kono K, Ohashi Y (1997) Immunopotentiating activity of the water-soluble lignin rich fraction prepared from LEM – the extract of the solid culture medium of *Lentinus edodes* mycelia. Biosci Biotechnol Biochem 61:1909–1912

Yan Y, Zhang C, Moodley O, Zhang L, Xu J (2019) Green mold caused by *Trichoderma atroviride* on the lingzhi medicinal mushroom, *Ganoderma lingzhi* (*Agaricomycetes*). Int J Med Mushrooms 21:515–521

Yarden O, Salomon R, Katan J, Aharonson N (1990) Involvement of fungi and bacteria in enhanced and nonenhanced biodegradation of carbendazim and other benzimidazole compounds in soil. Can J Microbiol 36:15–23

Yu SH (2001) Integrated control of oyster mushroom green mould. Ministry of Agriculture and Forestry, Korea, 157 pp

Zargarzadeh Z, Mohammadi Goltapeh E, Danesh YR, Mehrparvar M (2011) Identification of aggressive species of *Trichoderma* from button mushroom farms (*Agaricus bisporus*) using morphological and molecular methods. Rostaniha 12:83–90

Zhang T, Lu MZ, Zhang CL, Xu JZ (2019) First report of *Trichoderma longibrachiatum* causing green mould disease on *Ganoderma lingzhi*. Plant Dis 103:156

Trichodermosis: Human Infections Caused by *Trichoderma* Species

Shahram Naeimi, Lóránt Hatvani, Tamás Marik, Dóra Balázs, Ilona Dóczi, Feng Cai, Csaba Vágvölgyi, Irina S. Druzhinina, and László Kredics

Contents

S. Naeimi
Department of Biological Control Research, Iranian Research Institute of Plant Protection, Agricultural Research, Education and Extension Organization (AREEO), Tehran, Iran
e-mail: sh.naeimi@areeo.ac.ir

L. Hatvani · T. Marik · D. Balázs · C. Vágvölgyi · L. Kredics (✉)
Department of Microbiology, Faculty of Science and Informatics, University of Szeged, Szeged, Hungary
e-mail: csaba@bio.u-szeged.hu; kredics@bio.u-szeged.hu

I. Dóczi
Institute of Clinical Microbiology, Faculty of Medicine, University of Szeged, Szeged, Hungary
e-mail: doczi.ilona@med.u-szeged.hu

F. Cai · I. S. Druzhinina
Key Laboratory of Plant Immunity, Fungal Genomics Laboratory (FungiG), Nanjing Agricultural University, Nanjing, China
e-mail: irina.druzhinina@njau.edu.cn

N. Amaresan et al. (eds.), *Advances in* Trichoderma *Biology for Agricultural Applications*, Fungal Biology, https://doi.org/10.1007/978-3-030-91650-3_22

1 Introduction

So far, three English and one Chinese language reviews have been published about the clinical relevance of *Trichoderma* species, together summarizing the data of 59 reports, including 38 detailed case descriptions (Kredics et al. 2003, 2011; Hatvani et al. 2013; Zhang and Li 2019). However, as the appearance of *Trichoderma* in clinical samples is usually considered as the result of contamination, *Trichoderma* species as the causal agents of human fungal infections are often disregarded, suggesting that the incidence of human mycoses due to *Trichoderma* is likely to have remained underestimated. Nevertheless, along with the rising number of immunocompromised patients and the widening knowledge about *Trichoderma* species, the incidence of confirmed human infections caused by members of this genus is growing permanently.

Zhang and Li (2019) used the term "trichodermasis" for *Trichoderma* infections; however, the linguistically correct term is "trichodermosis." To the best of our knowledge, this term has only been mentioned so far a single time in the medical literature, in relation with fungemia of a cystic fibrosis patient (Khan et al. 2001). Here, we propose the usage of the term "trichodermosis" (plural: "trichodermoses") for all types of *Trichoderma* infections in humans. Trichodermoses may be transmitted by air, water, food, dust, soil, building materials, and medical devices, particularly catheters (Lübeck et al. 2000; Colakoğlu 2003; Hageskal et al. 2006; Hatvani et al. 2013). Trichodermosis is widely distributed in the world, and all age groups are affected (Fig. 1). In the decreasing order of frequency, *Trichoderma* may cause peritonitis; pulmonary infections; disseminated infections; heart infection; fungemia; sinusitis; skin, CNS, liver, and corneal infections; otitis; and stomatitis, while the predisposing conditions include peritoneal dialysis, hematological malignancies, organ transplantations, cardiac surgery, HIV infection, carcinoma, asthma, as well as intravenous transfusion, parenteral nutrition, soft contact lens wear, cataract, eye surgery, brain surgery, and pulmonary fibrosis (Fig. 2). The presumed virulence factors of *Trichoderma* species with the potential to colonize human tissues include the capability of growing at elevated temperatures (Fig. 3) and pH of the human body, the production of different extracellular proteolytic enzymes and different secondary metabolites – such as peptaibols – that disrupt mammalian cells (Antal et al. 2005; Kredics et al. 2004; Hatvani et al. 2013; Marik et al. 2019), and also their frequent resistance to multiple, routinely used antifungal substances

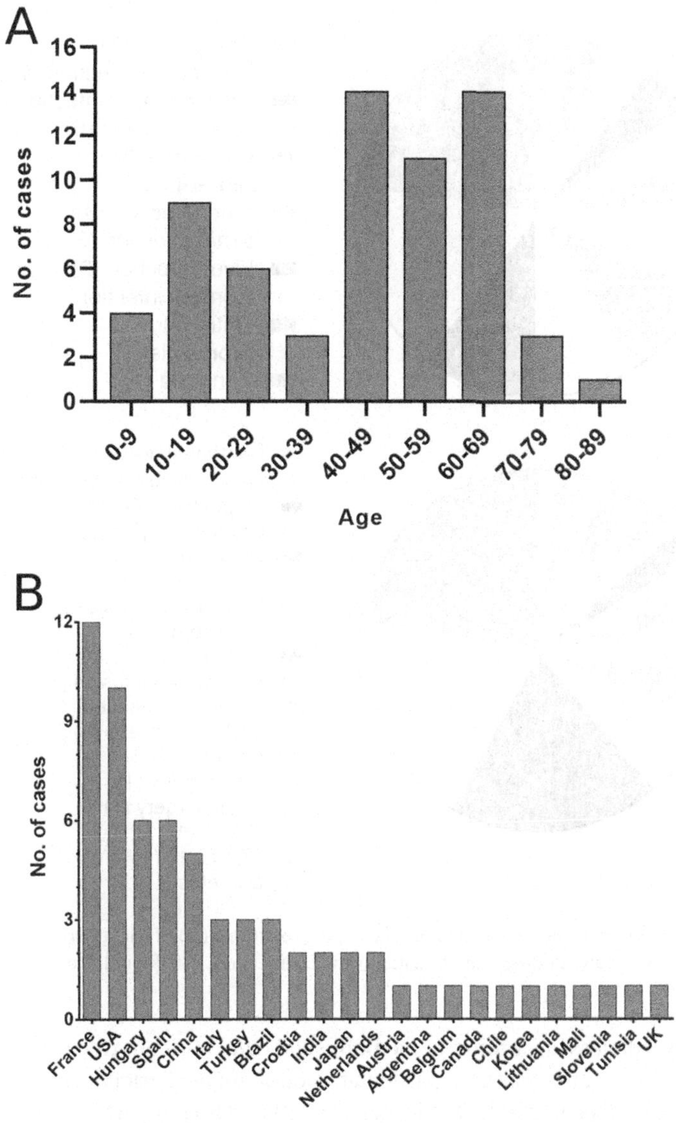

Fig. 1 Distribution of 65 trichodermosis cases among age groups (**A**) and countries (**B**). (Diagrams were corrected and updated from Zhang and Li 2019 published for 38 cases)

(Kredics et al. 2011). The resistance to fluconazole, 5-fluorocytosine, and amphotericin B is widespread, but certain clinical *Trichoderma* isolates could also be characterized with high minimum inhibitory concentration (MIC) values of ketoconazole (Druzhinina et al. 2007; Guarro et al. 1999; Kviliute et al. 2008), itraconazole (Antal et al. 2005; Guarro et al. 1999; Hatvani et al. 2012; Hennequin et al. 2000; Myoken et al. 2002), posaconazole (Hatvani et al. 2012), and voriconazole (Ranque et al.

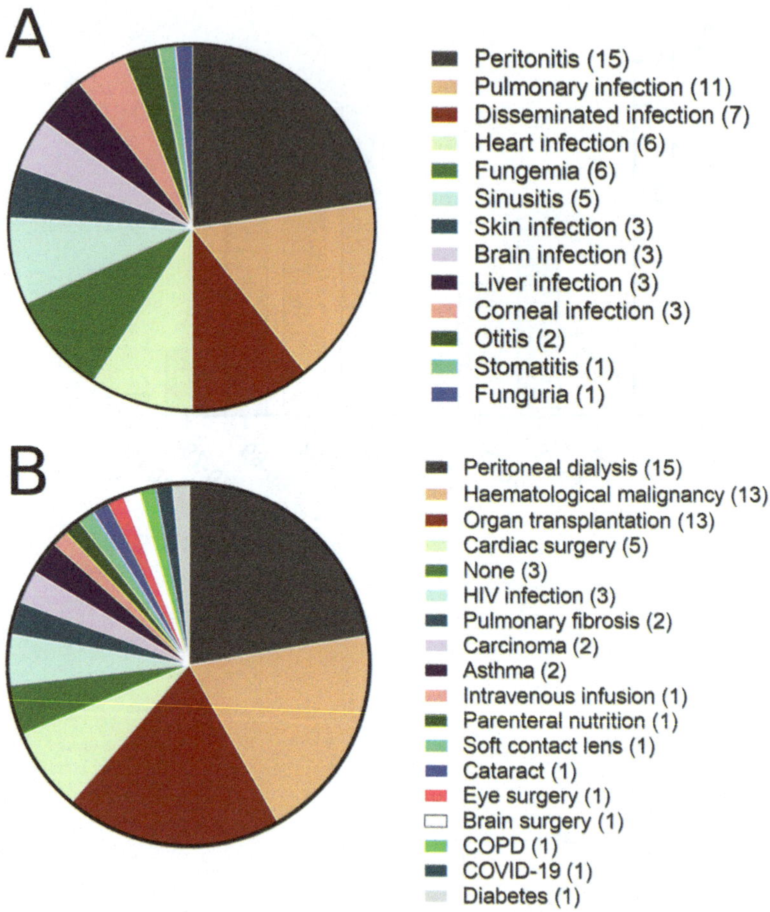

Fig. 2 Clinical manifestation (**A**) and predisposing conditions (**B**) of trichodermosis based on 65 cases. (Diagrams were modified and updated from Zhang and Li 2019 published for 38 cases)

2008). Nevertheless, according to the published data, voriconazole can still be used efficiently for treating severe, particularly disseminated trichodermoses (Alanio et al. 2008; Antal et al. 2002; De Miguel et al. 2005; Druzhinina et al. 2007; Espinel-Ingroff 2001; Espinel-Ingroff et al. 2002; Hatvani et al. 2012; Kantarcioğlu et al. 2009; Kratzer et al. 2006; Lagrange-Xélot et al. 2008; Marco et al. 1998; Myoken et al. 2002). At the same time, *Trichoderma* strains are exposed to various antifungal substances in agricultural environments, which – due to the similarities in the chemical structure of the compounds – may lead to the development of resistance to drugs used in the clinical practice (Hatvani et al. 2019). In the case of *T. longibrachiatum*, Paredes et al. (2016) found low virulence but high resistance to amphotericin B, micafungin, and voriconazole in an immunosuppressed mouse model: only

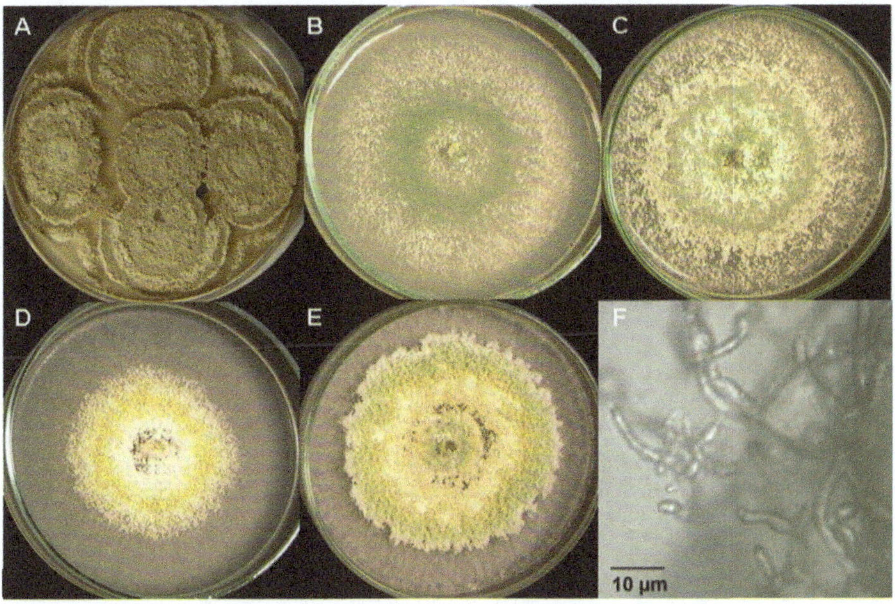

Fig. 3 Morphology of the *Trichoderma longibrachiatum* strain isolated from sinusitis sphenoidalis in an immunocompetent patient with headache (Molnár-Gábor et al. 2013). (**A**) Sabouraud agar at 25 °C; (**B**) minimal medium at 25 °C; (**C**) minimal medium at 37 °C; (**D**) yeast extract - glucose medium at 25 °C; (**E**) yeast extract - glucose medium at 27 °C; (**F**) hyphae of the strain with phialides and conidia

the highest inoculum concentration (10^7 CFU per animal) used for intravenous injection was able to kill all mice after 15 days, with the liver and spleen as the most affected organs.

This chapter is intended to present an inventory about *Trichoderma* species of clinical importance.

2 The Siblings *T. longibrachiatum* and *T. orientale*

Within the genus *Trichoderma*, the sibling species *T. longibrachiatum* and *T. orientale* (originally described as *Hypocrea orientalis*) (Table 1 and Fig. 4) are the most frequent to cause fatal trichodermosis in immunocompromised patients; their reliable and rapid diagnosis is therefore vital. However, the differential diagnosis of these two species is not possible based on their internal transcribed spacer (ITS) sequences alone; this should be complemented at least with sequence analysis of a fragment of the translation elongation factor 1α (*tef1*) gene.

Table 1 Clinical cases of trichodermosis confirmed by sequence-based molecular identification of the causal agent

Causal agent	Patients' age/sex, country	Underlying condition	Clinical diagnosis	Source(s) of isolation	Therapeutic interventions	Therapeutic outcome	References
T. longibrachiatum	45/F, France	HM, TX	Lung and skin dissemination	Lung, brain, heart, stomach, bronchoalveolar lavage, skin biopsy	5FC, ABLC, AMB, FCZ	Death	Gautheret et al. (1995) and Druzhinina et al. (2008)
	41/M, Spain	CAPD	Peritonitis	Peritoneal fluid	5FC, AMB, FCZ	Death	Campos-Herrero et al. (1996) and Druzhinina et al. (2008)
	29/M, USA	TX	Disseminated infection	Intestinal wall, liver, lung, stool, autopsy	ABLC, AMB, ICZ	Death	Richter et al. (1999) and Druzhinina et al. (2008)
	27/F, Hungary	None	Rhinosinusitis	Sinus lavage	AMB, surgery	Recovery	Molnár-Gábor et al. (2013) and Druzhinina et al. (2008)
	21/M, Croatia	ND	Gastrointestinal symptoms	Stool	Diet	Recovery	Hatvani et al. (2012)
	69/M, France	HM	Invasive pulmonary infection	Bronchoalveolar lavage	VCZ, CSP	Recovery	Sautour et al. (2018)
	56/F, Croatia	None	Otitis externa	Ear discharge	TRB	Recovery	Hatvani et al. (2019)
	65/M, India	Diabetes, cataract	Keratitis in the right eye	Corneal infiltrate	VCZ, NTM, therapeutic keratoplasty	Recovery	Hatvani et al. (2019)

	71/M, Hungary	Aortic valve implantation	Aorta inflammation	Aortic valve, wall of aorta	VCZ, aortic valve removal	Death	Hatvani et al. (2019)
	75/F, Hungary	Pacemaker implant	Pacemaker sac infection	Fluid from implant	Pacemaker removal	Recovery	Hatvani et al. (2019)
	12/F, Spain	TX	Lung colonization	Skin biopsy	VCZ, CSP	Death	Román-Soto et al. (2019)
	76/M, Hungary	COVID-19, hepatic cirrhosis	Fungemia	Blood culture	ND	Death	Dóczi et al. (unpublished)
T. longibrachiatum/T. orientale	63/F, France	CAPD	Peritonitis	Peritoneal fluid	MCZ, catheter removal	Survival	Ragnaud et al. (1984) and Kuhls et al. (1999)
	17/F, France	HM	Brain abscess	Cerebral pus, brain biopsy	5FC, AMB, ICZ, KCZ, surgery	Recovery	Seguin et al. (1995)
	12/M, France	None	Otitis externa	Ear discharge	Nystatin	Recovery	Hennequin et al. (2000)
	63/F, France	TX	Invasive infection	Subcapsular hepatic collection	Surgical debridement	Recovery	Chouaki et al. (2002)
	11/M, USA	HM	Skin infection	Skin biopsy	ABLC, AMB	Recovery	Munoz et al. (1997)
	11/M, Mali	TX	Invasive infection	Pleural drains, bronchoalveolar lavage	ABLC	Death	Chouaki et al. (2002)
	66/F, Japan	HM	Necrotizing stomatitis	Ulcerative mucogingiva	AMB, ICZ	Death	Myoken et al. (2002)

(continued)

Table 1 (continued)

Causal agent	Patients' age/sex, country	Underlying condition	Clinical diagnosis	Source(s) of isolation	Therapeutic interventions	Therapeutic outcome	References
	52/F, Canada	Asthma	Allergic fungal sinusitis	Bilateral endoscopic antral lavage	ICZ, sinus lavage	Recovery	Tang et al. (2003)
	58/M, France	HIV, bronchopulmonary adenocarcinoma	Fungemia	Blood	AMB, VCZ, catheter removal	Recovery	Lagrange-Xélot et al. (2008)
	16/M, France	HM	Invasive pulmonary infection	Bronchoalveolar lavage, bronchoaspiration, sputum	CSP, VCZ	Recovery	Alanio et al. (2008)
	46/M, Tunisia	TX	Suprapubic abscess next to the old intertrigo lesion	Fluid puncture, skin biopsy	FCZ, VCZ	Recovery	Trabelsi et al. (2010)
	3/F, USA	Complex congenital cardiac disease	Peritonitis	Pericardium, sternum tissue, pulmonary infiltrates, peritoneal fluid	CSP, FCZ, VCZ, amphotericin deoxycholate instillations, catheter removal	Death	Santillan Salas et al. (2011)
	51/M, Spain	Short bowel	Endocarditis over catheter	Surgical specimen	Catheter removal, CSP	Recovery	Rodríguez Peralta et al. (2013)
	30/M, Italy	Cardioverter defibrillator implantation	Endocarditis	Catheter tip	VCZ, liposomal AMB	Recovery	Tascini et al. (2016)
	64/F, China	CAPD	Peritonitis	Peritoneal fluid	VCZ + AMB	Recovery	Guo et al. (2017)

	29/M, Japan	TX	Invasive pulmonary infection	Sputum, bronchoaspiration, bronchoalveolar lavage	Liposomal AMB	Recovery	Akagi et al. (2017)
	59/M, Spain	TX	Pericarditis	Pericardial tissue	AND, IVZ	Death	Recio et al. (2019)
	57/M, China	Pulmonary spindle cell carcinoma	Invasive pulmonary infection	Lung biopsy	VCZ	Recovery	Zhou et al. (2020)
T. orientale	3/F, Hungary	Acute lymphoid leukemia	Fungemia	Blood	Nystatin	Recovery	Kredics et al. (2006) and Druzhinina et al. (2008)
	16/F, Hungary	Non-Hodgkin lymphoma		Stool	ECZ	Recovery	Kredics et al. (2006) and Druzhinina et al. (2008)
T. harzianum species complex	68/M, Spain	TX	Disseminated infection	Lung and brain abscesses, autopsy	-	Death	Guarro et al. (1999)
T. harzianum species complex	9/M, Turkey	HM	Invasive fungal infection	Serum, skin lesions, sputum	ABLC	Death	Kantarcioğlu et al. (2009)
T. atroviride	49/M, France	TX	Liver infection	Liver biopsy	FCZ	Death	Ranque et al. (2008)
T. peltatum	19/F, USA	Asthma, interstitial lung disease	Pulmonary fibrosis	Lung biopsy	Bronchodilators	Recovery	Druzhinina et al. (2007)

CAPD chronic ambulatory peritoneal dialysis, *TX* transplant, *HM* hematological malignancy, *ND* no data available, *5FC* 5-fluorocytosine, *ABLC* amphotericin B lipid complex, *AMB* amphotericin B, *AND* anidulafungin, *CSP* caspofungin, *ECZ* econazole, *FCZ* fluconazole, *ICZ* itraconazole, *IVZ* isavuconazole, *KCZ* ketoconazole, *MCZ* miconazole, *NTM* natamycin, *TRB* terbinafine, *VCZ* voriconazole

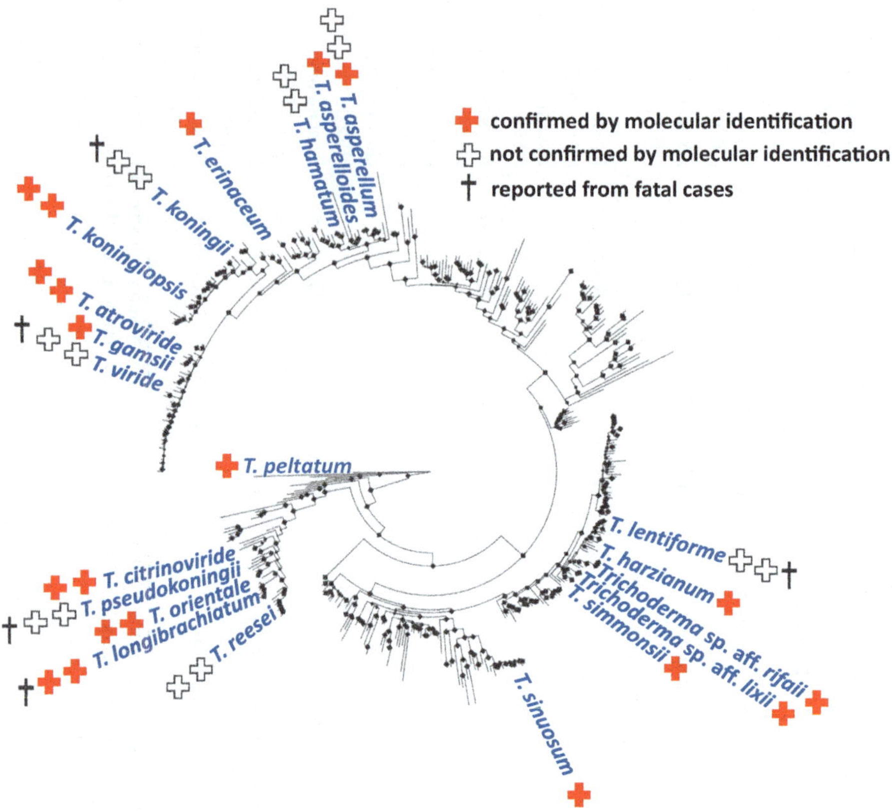

Fig. 4 Taxonomic position of *Trichoderma* species isolated from clinical specimens. The number of symbols reflects the clinical significance of the particular species (one symbol, detected in clinical specimens; two symbols, identified as the causal agent of infection). The Bayesian tree was inferred from the alignment of 808 nucleotides of the *rpb2* gene for 349 sequences retrieved from NCBI GenBank (Dou et al. 2020). Nodes supported with posterior probabilities above 0.94 are marked by black diamonds

2.1 Cases Caused by T. longibrachiatum/T. orientale *Diagnosed by ITS Sequence Analysis*

A series of case reports were published in the literature about the ITS sequence-based diagnosis of *T. longibrachiatum* (meaning actually *T. longibrachiatum* or *T. orientale*) in opportunistic infections of mostly immunocompromised patients (Table 1). Brain abscess developed in a patient with leukemia after the invasive progression of ethmoidal sinusitis, which was successfully medicated with extremely prolonged amphotericin B, itraconazole, ketoconazole, and 5-fluorocytosine therapy along with neurosurgery performed for the resection of the abscess (Seguin et al. 1995). An invasive skin infection occurred in a pediatric patient with severe aplastic anemia and neutropenia, which could be healed by the intravenous administration of amphotericin B (Munoz et al. 1997). The ear discharge of another pediatric patient with inflammation of the right eardrum revealed *Trichoderma* in culture; the infection could be mended with a combination of local nystatin, polymyxin B, and oxytetracycline (Hennequin et al. 2000). A case of stomatitis in a neutropenic patient with malignant lymphoma has rapidly disseminated from the oral mucosa to the lungs; the patient died despite the intensive antifungal therapy with amphotericin B and itraconazole (Myoken et al. 2002). In a liver transplant patient with hepatitis C virus-induced cirrhosis, a sample collected from the subcapsular area related to biopsy specimens of perilesional tissues and suture threads yielded pure fungal culture (Chouaki et al. 2002). Full recovery could be achieved with concomitant surgical debridement and local povidone iodine treatment. The authors have also reported a fatal case of trichodermosis in a patient who underwent pulmonary transplantation because of terminal respiratory failure due to cystic fibrosis (Chouaki et al. 2002). Transcutaneous tracheal puncture and bronchoalveolar lavage revealed cultures of *T. longibrachiatum*/*T. orientale*. A case of allergic fungal sinusitis in an asthmatic patient could be managed with a combination of oral corticosteroids, itraconazole, sinus lavage, and immunotherapy (Tang et al. 2003). The invasive lung infection in a B cell acute lymphoblastic leukemia patient confirmed by culture-positive sputum, bronchoaspiration, and bronchoalveolar lavage fluid samples could be remedied with a caspofungin-voriconazole combination (Alanio et al. 2008). Blood culture from peripheral vein of an HIV patient with bronchopulmonary adenocarcinoma also yielded *T. longibrachiatum*/*T. orientale* (Lagrange-Xélot et al. 2008). Trabelsi et al. (2010) described a case of cutaneous trichodermosis in a renal transplant recipient where the fungus could be isolated from skin biopsy and the fluid punctured of inguinal abscess. The patient could be healed with voriconazole. The treatment with caspofungin alone, followed by a combination of caspofungin, systemic and topical voriconazole, and intraperitoneal amphotericin B, remained unsuccessful in the case of a postoperative mediastinitis and peritonitis in a child with functional asplenia and complex congenital cardiac disease (Santillan Salas et al. 2011). Trichodermosis caused by *T. longibrachiatum*/*T. orientale* was diagnosed in a man with endocarditis acquired via indwelling device (Rodríguez Peralta et al. 2013). He received parenteral nutrition at home because of short bowel syndrome. Antibiotic treatment combined with catheter removal by conventional surgery resulted in full recovery. The fungus could be isolated from the surgical specimen. A case of cardiac implantable electronic device (CIED)-associated endocarditis successfully treated with voriconazole and liposomal amphotericin B was reported in a non-immunocompromised patient, who underwent the implantation of a cardioverter defibrillator (Tascini et al. 2016). A case of peritonitis reported by Guo et al. (2017) from a CAPD patient with chronic nephritis could be successfully cured with voriconazole and amphotericin B. Invasive pulmonary infection was diagnosed by Akagi et al. (2017) in an allogeneic stem cell transplant recipient with severe aplastic anemia. Direct microscopic examination of sputum, bronchoaspiration, and bronchoalveolar lavage samples revealed septate fungal hyphae. The infection was effectively managed with 1 mg/kg/day liposomal amphotericin B. The culture of the biopsy after pericardiectomy of an immunocompromised lung transplant recipient also yielded *T. longibrachiatum*/*T. orientale* (Recio et al. 2019). Matrix-assisted laser desorption/ionization time of flight mass spectrometry (MALDI-TOF) and subsequent ITS sequence analysis were used for the identification of the causal agent. Despite the initial anidulafungin and subsequent isavuconazole therapy, the patient died after progressive clinical deterioration. Laser capture microdissection (LCM) and polymerase chain reaction (PCR) were used by Zhou et al. (2020) to identify *T. longibrachiatum*/*T. orientale* as the causal agent of invasive pulmonary trichodermosis in a pulmonary spindle cell carcinoma patient with a mass at the left hilum revealed by computed tomography (CT) scan. The patient was successfully cured with oral voriconazole for 4 months. The authors suggested that the LCM-based technology might be a promising diagnostic tool for fungal infection.

Further known clinical isolates of *T. longibrachiatum*/*T. orientale* are IP-92 0647 from an acute lymphoblastic leukemia patient, IP-94 0958 from a lung transplant recipient, IP-96 0086 from a hematic wound, IP-93 1282 from the bronchoalveolar lavage of a bone marrow recipient, IP-94 1510 from maxillary sinusitis, and IP-97 0711 from the liquid of a chylothorax (Kuhls et al. 1999).

2.2 *Cases Caused by* T. longibrachiatum *Confirmed by* tef1 *Sequence Analysis*

Among the *T. longibrachiatum* cases retrospectively confirmed by *tef1* sequence analysis (Druzhinina et al. 2008) (Table 1 and Fig. 4), a strain had caused fatal trichodermosis disseminated to the brain, heart, lungs, pretracheal abscesses, and stomach of an allogeneic bone marrow transplant recipient with erythroleukemia (Gautheret et al. 1995). In another fatal case in a patient who received allogeneic bone marrow transplant for acute lymphoblastic leukemia, the fungus could be isolated from stool surveillance cultures and a perirectal ulcer biopsy specimen, suggesting a possible entry through the gastrointestinal tract (Richter et al. 1999). The species could also be associated with a fatal case of peritonitis in a patient with continuous ambulatory peritoneal dialysis (CAPD) (Campos-Herrero et al. 1996; Druzhinina et al. 2008). Molnár-Gábor et al. (2013) isolated *T. longibrachiatum* (Fig. 3) from the secretion obtained from the sphenoidal sinus of a non-immunocompromised rhinosinusitis patient, who was successfully cured by amphotericin B and the surgical removal of the fungal mass. A case of invasive pulmonary trichodermosis was reported by Sautour et al. (2018) in a leukemia patient, who fully recovered after antifungal therapy with voriconazole and caspofungin. The fungus could also be isolated from the ear discharge of a non-immunocompromised patient with otitis externa cured with terbinafine, and from the corneal infiltrate of a diabetic patient with keratitis in the left eye medicated with voriconazole, natamycin, and therapeutic keratoplasty (Hatvani et al. 2019). The same study also reported a fatal case of aortic valve infection by *T. longibrachiatum* treated with voriconazole and a pacemaker sac infection remedied by the extraction of the device (Hatvani et al. 2019). A lung colonization and skin infection by *T. longibrachiatum* was reported in a ventilatory-supported pediatric patient who underwent allogeneic hematopoietic stem cell transplantation due to bone marrow failure, and died to multiple complications of her underlying disease (Román-Soto et al. 2019). The authors suggested the direct contact with the ventilation tube continuously disconnected by the patient as a possible route of skin entry.

Further occurrence of *T. longibrachiatum* in clinical specimens confirmed by *tef1* sequence analysis includes the isolation of strains ATCC 208859 from an HIV-positive host, UAMH 9515 from the peritoneal effluent of a female, CBS 446.95 from the lung of a patient, who died, CNM-CM 1798 from the blood culture of a liver transplant recipient, CNM-CM 2171 from the subcutaneous foot skin lesions of a premature infant, CNM-CM 2277 from the sputum of a tuberculosis patient (Druzhinina et al. 2008), most recently a strain from the blood culture of a COVID-19 patient (Dóczi et al. unpublished), as well as further isolates from human blood, bronchoalveolar lavage, plural fluid, cerebrospinal fluid, peritoneal fluid, sputum, lung tissue, sinuses, nails, ear, foot, bone, mediastinal mass, and vertebral body (Sandoval-Denis et al. 2014).

2.3 *Cases Caused by* **T. orientale** *Confirmed by* **tef1** *Sequence Analysis*

The species *T. orientale* (Table 1 and Fig. 4) could be isolated from the stool of a 15-year-old child with non-Hodgkin lymphoma and the hemoculture of a 3-year-old child with acute lymphoblastic leukemia (Kredics et al. 2006). Both were patients of the same pediatrics clinic suggesting a possible epidemiological connection. The occurrence of *T. orientale* has also been reported from further blood samples as well as human sputum, sinus, arm, bronchoalveolar lavage, peritoneal fluid, and vascular prosthesis (Sandoval-Denis et al. 2014).

2.4 *Cases Putatively Caused by* **T. longibrachiatum** *Unconfirmed by Molecular Identification*

The species *T. longibrachiatum* has also been reported from further trichodermoses, but without molecular confirmation (Table 2). Tanis et al. (1995) and Aroca et al. (2004) described fatal cases of peritonitis in peritoneal dialysis patients. The peritonitis and intra-abdominal abscess of another patient was successfully treated with antifungal agents, catheter removal, and appropriate drainage (Lee et al. 2007), while along with *Candida tropicalis*, *T. longibrachiatum* was diagnosed by MALDI-TOF MS as one of the causal agents in a mixed case of CAPD peritonitis successfully treated with voriconazole and amphotericin B (Yang et al. 2019). Surgical debridement and amphotericin B followed by oral itraconazole proved successful

Table 2 Clinical cases of trichodermosis without sequence-based molecular identification of the causal agent

Causal agent	Patients' age/sex, country	Underlying condition	Clinical diagnosis	Source(s) of isolation	Therapeutic interventions	Therapeutic outcome	References
T. longibrachiatum	48/M, Netherlands	CAPD	Peritonitis	Peritoneal fluid, autopsy	AMB	Death	Tanis et al. (1995)
	29/F, USA	TX	Acute invasive sinusitis	Sinus debridement	AMB, ICZ, surgery	Survival	Furukawa et al. (1998)
	13/F, Chile	CAPD	Peritonitis	Peritoneal fluid	AMB, FCZ, catheter removal	Death	Aroca et al. (2004)
	67/M, Korea	CAPD	Peritonitis, intra-abdominal abscess	Aspirated fluid, peritoneal fluid	AMB, catheter removal	Recovery	Lee et al. (2007)
	49/F, China	CAPD	Peritonitis	Peritoneal fluid	VCZ, AMB	Recovery	Yang et al. (2019)
T. citrinoviride	49/F, Lithuania	HM	Pneumonia	Bronchoalveolar lavage	AMB	Recovery	Kviliute et al. (2008)
T. pseudokoningii	33/M, Italy	APD	Peritonitis	Peritoneal fluid	Catheter removal	Recovery	Rota et al. (2000)
T. reesei	61/M, France	Cerebrospinal fluid shunt placement	Infection of cerebrospinal shunt device	Shunt device, cerebrospinal fluid	ABLC, CSP, VCZ	Recovery	Piens et al. (2004)
T. harzianum	82/M, France	CAPD	Peritonitis	Peritoneal fluid	5FC, KCZ	Death	Guiserix et al. (1996)
	8/F, USA	Cystic fibrosis	Fungemia	Blood	ICZ	Recovery	Khan et al. (2001)
T. asperellum	44/F, Brazil	Asthma, chronic allergic rhinitis, sinonasal polyps	Rhinosinusitis	Secretion of ethmoid and sphenoid sinuses	5FC, AMB, ICZ, polypectomy, ethmoidectomy, sphenoidectomy	Recovery	Cardoso et al. (2015)

T. viride	26/F, UK	Intravenous infusion	Fungemia by contaminated saline	Blood	AMB	Recovery	Robertson (1970)
	46/M, ND	ND	Pulmonary mycetoma	Lung biopsy, sputum	Surgery	Recovery	Escudero et al. (1976)
	47/M, USA	CAPD	Peritonitis	Peritoneal fluid, autopsy	AMB	Death	Loeppky et al. (1983)
	44/M, France	CAPD	Peritonitis	Peritoneal fluid	AMB	Death	Warnock and Johnson (1991)
	44/F, Belgium	TX	Abdominal dissemination	Peritoneal fluid, hematoma	AMB, FCZ, surgery	Death	Jacobs et al. (1992)
	54/F, Spain	HM	Pulmonary infection	Fine-needle aspiration from pulmonary consolidation	ABLC, CSP, VCZ	Recovery	De Miguel et al. (2005)
	38/M, Brazil	Kaposi's sarcoma, HIV and CMV infection	Rhinosinusitis	Maxillary sinus secretion	AMB, ICZ	Death	Cardoso et al. (2015)
Trichoderma sp.	60/M, Slovenia	CAPD	Peritonitis	Peritoneal fluid	KCZ, catheter removal	Recovery	Bren (1998)
	66/M, Argentina	Ascending aortic replacement	Endocarditis	Aortic conduit	Antifungal drugs, surgery	Recovery	Bustamante-Labarta et al. (2000)
	50/ND, Brazil	AIDS	Brain infection	Subdural collection	AMB	Recovery	Paula-Amato et al. (2002)
	40/M, Turkey	CAPD	Peritonitis	Peritoneal fluid	AMB, ICZ, catheter removal	Death	Esel et al. (2003)

(continued)

Table 2 (continued)

Causal agent	Patients' age/ sex, country	Underlying condition	Clinical diagnosis	Source(s) of isolation	Therapeutic interventions	Therapeutic outcome	References
	62/M, Austria	TX	Disseminated infection	Gastrointestinal tract, heart, kidneys, liver, lung, skin	VCZ	Death	Stelzmueller et al. (2008)
	64/F, Netherlands	Laparotomy and a hyperthermic intraperitoneal chemotherapy	Pneumonia	Bronchoalveolar lavage	AMB, CSP, VCZ	Death	Ariese et al. (2013)
	45/M, Italy	Multiple myeloma	Fungemia	Blood	VCZ	Recovery	Festuccia et al. (2014)
	50/M, India	Diabetes, chronic alcoholic liver disease, hepatorenal syndrome type 2	Abdominal pain, fever, vomiting	Urine	ND	ND	Chakraborty et al. (2015)
	50/F, USA	COPD, pneumonia with sepsis, antibiotic allergy	Pneumonia	Lung	VCZ	Recovery	Morrell (2017)
	50/F, USA	Soft contact lens	Keratitis	Corneal scraping	NTM	Recovery	Hodkin and Gustus (2018)
	41/M, Turkey	Laser in situ keratomileusis	Keratitis	Corneal scraping	AMB, NTM	Recovery	Mergen et al. (2019)
	51/M, China	HM	Pulmonary infection	Lung	Lobectomy, CSP	Recovery	Dong et al. (2019)
	67/F, China	CAPD	Peritonitis	Peritoneal fluid	VCZ	Recovery	Ning and Yang (2020)
	65/M, USA	PD	Peritonitis	Peritoneal fluid	Catheter removal, AND, VCZ, AMB	Death	Bachu et al. (2020)

APD automated peritoneal dialysis, *CAPD* chronic ambulatory peritoneal dialysis, *PD* peritoneal dialysis, *TX* transplant, *HM* hematological malignancy, *ND* no data available, *5FC* 5-fluorocytosine, *ABLC* amphotericin B lipid complex, *AMB* amphotericin B, *CSP* caspofungin, *FCZ* fluconazole, *ICZ* itraconazole, *KCZ* ketoconazole, *MCZ* miconazole, *NTM* natamycin, *VCZ* voriconazole for the treatment of acute invasive sinusitis in a patient who underwent small bowel and liver transplantation (Furukawa et al. 1998). An additional keratitis case caused by *T. longibrachiatum* is mentioned by He et al. (2016), while the FungiScope database (Seidel et al. 2017) contains further information about the isolation of *T. longibrachiatum* from the blood of a patient with neuroblastoma and neutropenia, who was efficiently treated with liposomal amphotericin B, from the left ear of a chronic otitis patient treated with terbinafine, from the central nervous system of a patient who died, as well as from the lungs of an acute lymphoblastic leukemia patient who died in spite of voriconazole, amphotericin B lipid complex, and caspofungin therapy (http://www.fungiquest.net).

2.5 *Characterization of Clinical* T. longibrachiatum *and* T. orientale *Isolates*

The species *T. longibrachiatum* from the clade *Longibrachiatum* of the genus was suggested to be the possible anamorph of *Hypocrea orientalis* (recently accepted name: *T. orientale*) by Samuels et al. (1998) based on isoenzyme data and ITS1 sequence analysis. A comparative population genetics study was performed by Druzhinina et al. (2008) on *T. longibrachiatum* and *T. orientale* isolates derived from clinical specimens as well as fungal cultivation substrata and soil samples. In addition to the *tef1* marker, fragments of the calmodulin (*cal1*) and endochitinase (*chit18-5*) genes were also examined. The analyses separated *T. longibrachiatum* and *T. orientale* from each other in terms of reproduction, revealing that in contrast to previous views (Samuels et al. 1998), the two species are not in a teleomorph-anamorph relation with each other. The species *T. longibrachiatum* is a widespread but less frequent component of soil *Trichoderma* communities, in contrast to closed habitats, e.g., water-damaged buildings (Thrane et al. 2001) or mushroom-growing houses (see Chapter 21). As a result of population genetic studies, it has been concluded that while *T. longibrachiatum* is a strictly clonal imperfect species, the genetically closely related *T. orientale* is characterized by a reproductive strategy based on sexual recombination (Druzhinina et al. 2008). Sandoval-Denis et al. (2014) separated *T. bissettii* from *T. longibrachiatum* as a new species with frequent clinical implications; however, recently, Hatvani et al. (2019) proposed its placement to the rank of "phylotype" as *T. longibrachiatum* f. sp. *bissettii* nom. prov.

Isoenzyme analysis based on cellulose acetate electrophoresis used to identify strains isolated from the wheat rhizosphere may also be suitable for the rapid identification of the species *T. longibrachiatum* and *T. orientale*. Seven enzyme activities analyzed by cellulose acetate electrophoresis – glucose-6-phosphate dehydrogenase, glucose-6-phosphate isomerase, 6-phosphogluconate dehydrogenase, peptidases A, B, and C, and phosphoglucomutase – proved to be applicable for the rapid differentiation between clinical *T. longibrachiatum* and *T. orientale* strains (Szekeres et al. 2006). The enzymes used were found to be polymorphic in the study population. Using enzyme patterns, the authors were able to identify ten electrophoretic types, which divided the examined *T. longibrachiatum* isolates into four distinct groups on the prepared dendrogram and clearly separated them from *T. orientale* strains (Szekeres et al. 2006).

Based on the differences in the restriction fragment length polymorphism (RFLP) patterns of the mitochondrial DNA (mtDNA) generated from *T. longibrachiatum* and *T. orientale* strains isolated from both clinical and soil samples, clinical strains were classified into five, while soil strains into four mtDNA types (Antal et al. 2006). Although fragments of the same size were observed in samples from clinical and soil-derived strains, identical patterns were not found. Based on these results, the isolates can be characterized by a high degree of polymorphism also within *T. longibrachiatum* and *T. orientale* at the mtDNA level. Based on the sizes of the obtained fragments, the mtDNA of clinical and saprophytic *T. longibrachiatum* and *T. orientale* strains ranges from 34.9 to 39.5 kbp (Antal et al. 2006). RFLP analysis of mtDNA resulted in better resolution than ITS sequence analysis, and the observed patterns allowed the separation of three *T. longibrachiatum* and one *T. orientale* groups on the dendrogram.

Sequence analysis of the ITS, *tef1*, *chit18-5*, and *cal1* genes, mtDNA RFLP, the examination of carbon source utilization profiles, and isoenzyme analysis did not separate clinical *T. longibrachiatum* isolates from those deriving from environmental samples. Therefore, while certain fungi of clinical significance, e.g., *Exophiala dermatitidis*, can be divided into pathogenic and nonpathogenic subpopulations (Matos et al. 2003), the existence of such subpopulations in *T. longibrachiatum* has not been shown, suggesting that among preferred conditions, each *T. longibrachiatum* strain might be able to cause trichodermosis.

3 *Trichoderma citrinoviride*

Molecularly confirmed strains of the species *T. citrinoviride* (Fig. 4 and Table 1) were reported from the stool of a patient with gastrointestinal symptoms (Hatvani et al. 2012), from a peritoneal catheter tip (UAMH 9573; Antal et al. 2006), from the blood culture of a patient with lymphoma-associated aplasia (IP-95 1151; Kuhls et al. 1999), from a cerebrospinal derivative catheter (IP-93 1792; Kuhls et al., 1999), from a chronic bronchitis patient with fever and respiratory infection (CNM-CM 1792; Kredics et al., unpublished), as well as from further blood samples, human bronchoalveolar lavage, sputum, ascitic fluid, pleural fluid, lung, eye, toenail, and abdominal wound (Sandoval-Denis et al., 2014). In contrast with the above examples, no details of identification were provided in a case of pneumonia of a patient with acute myeloid leukemia, where *T. citrinoviride* was reported from bronchoalveolar lavage and the infection was successfully treated with amphotericin B (Kviliute et al. 2008; Table 2), as well as in a case of the isolation of this species from the liver and lungs of a transplant recipient with diabetes and chronic liver disease (http://www.fungiquest.net).

4 Species from Other Clades of the Genus *Trichoderma* Confirmed by Molecular Identification

4.1 The Harzianum *Clade*

A systemic *T. harzianum* infection was diagnosed by Guarro et al. (1999); the fungus was isolated postmortem during the necropsy study from mycotic brain lesions and lung tissue microabscesses, initially identified based on morphological features, and subsequently confirmed by ITS sequence analysis as a member of the *Harzianum* clade (Kredics et al. 2003; Szekeres et al. 2006), while based on *tef1* the strain seems to be closely related to *T. lentiforme*, which, however, needs further confirmation (Kredics et al. unpublished). Another fatal infection due to a *Harzianum* clade member was reported in an acute lymphoblastic leukemia patient (Kantarcioğlu et al. 2009); the fungus was recovered from serum, skin lesions, sputum, and the throat of the patient and identified by ITS sequence analysis. Further confirmed clinical isolates from the *Harzianum* clade include a strain from the lungs of an acute lymphoblastic leukemia and neutropenia patient who was successfully treated with amphotericin B lipid complex and voriconazole (http://www.fungiquest.net), as well as *T. harzianum* from human blood (UTHSC 07-2109) and stool (UTHSC 11-3234), *T. simmonsii* from human maxillary sinus (UTHSC 02-2663) and human sputum (UTHSC 05-2749), and the representatives of two putative yet undescribed new species closely related to *T. lixii* (UTHSC 10-1527 from human bronchoalveolar lavage) and *T. rifaii* (UTHSC 09-3558 from human cornea) (Fig. 4) (Sandoval-Denis et al. 2014).

Additional cases attributed to *T. harzianum* – but without any sequence-based molecular confirmation – were a fungal peritonitis in a peritoneal dialysis patient, who died in spite of oral ketoconazole and intraperitoneal 5-fluorocytosine treatment (Guiserix et al. 1996), as well as the detection of the fungus in the blood culture of an 8-year-old female cystic fibrosis patient with allergic bronchopulmonary infection (Khan et al. 2001) (Table 2).

4.2 Further Confirmed But Very Rarely Occurring Trichoderma *Species*

The postmortem isolation of *T. atroviride* (Table 1 and Fig. 4) was reported from a liver biopsy specimen of a patient who underwent liver transplantation for the treatment of hepatocellular carcinoma in the context of alcoholic cirrhosis (Ranque et al. 2008). The initial morphology-based identification has been confirmed by ITS sequence analysis. This species was also detected in a clinical sample from human lung mass (Sandoval-Denis et al. 2014) and in the paranasal sinuses of a patient who underwent functional endoscopic sinus surgery (http://www.fungiquest.net).

Cardoso et al. (2015) reported a case of rhinosinusitis by *T. asperellum* (Table 1 and Fig. 4) in a 44-year-old female patient from Brazil with asthma, chronic allergic rhinitis, and sinonasal polyps. The fungus was isolated from the secretion of ethmoid and sphenoid sinuses. The patient was treated with polypectomy, ethmoidectomy, and sphenoidectomy; however, identification details were not provided. *T. asperellum* is also known from human sputum, while the closely related *T. asperelloides* (Fig. 4) has been detected in a clinical specimen from human nails (Sandoval-Denis et al. 2014).

A *Trichoderma* species with *Hypocrea*-like teleomorph morphology was isolated from the lung of a nonfatal pulmonary fibrosis case (Druzhinina et al. 2007) and later identified as *T. peltatum* (originally described as *Hypocrea peltata*) (Table 1 and Fig. 4) (Samuels and Ismaiel 2011). Clinical and mycological studies could not reveal whether the isolate was the causal agent of – or just contributed to – the disease development.

Further *Trichoderma* species confirmed to occur in human clinical specimens by sequence-based molecular identification are *T. erinaceum* from human nails, *T. gamsii* from human sputum, *T. koningiopsis* from human nails and bronchoalveolar lavage, as well as *T. sinuosum* from human skin (Sandoval-Denis et al. 2014; Fig. 4).

5 Species Reported as Causal Agents of Human Infections Without Molecular Confirmation

5.1 T. viride

The diagnosis of *T. viride* (Table 2 and Fig. 4) in the case of several trichodermoses may be due to the fact that the name *T. viride* had been applied for a long time to all *Trichoderma* strains with round-shaped, roughened conidia, which, however, may belong to phylogenetically diverse species. The species *T. viride* was later redefined and separated from *T. asperellum* (Lieckfeldt et al. 1999). The first documented case of trichodermosis was the accidental infection of a 26-year-old immunocompetent patient with *T. viride* by contaminated intravenous infusion due to a cracked bottle (Robertson, 1970), which was successfully treated with amphotericin B. The infection of a patient with pulmonary mycetoma due to *T. viride* cured with surgical resection was reported by Escudero et al. (1976); sputum and lung biopsy samples revealed cultures of the fungus. Two patients on CAPD had peritonitis due to *T. viride* and died in spite of amphotericin B therapy (Loeppky et al. 1983; Warnock and Johnson 1991). An immunocompromised liver transplant recipient was reported to suffer from the *T. viride* infection of a perihepatic hematoma (Jacobs et al. 1992); despite surgical removal of the infected hematoma and amphotericin B treatment, the fungus could persist in the patient, who died of unrelated complications. Summerbell (2003) has already pointed out that the morphological descriptions in the case reports of Escudero et al. (1976), Loeppky et al. (1983), and Jacobs et al. (1992) had not supported the diagnosis of the case isolates as *T. viride*. De Miguel et al. (2005) isolated *T. viride* from the pulmonary aspirate of an adult patient with acute myeloid leukemia suffering from pulmonary infection. The initially administered liposomal amphotericin B proved unsuccessful; therefore, the therapy was switched to voriconazole and caspofungin, which resulted in recovery. The isolation of *T. viride* was also reported from the maxillary sinus secretion of a rhinosinusitis patient with Kaposi's sarcoma, HIV and CMV infection, as well as pansinusitis, who died in spite of itraconazole and amphotericin B therapy (Cardoso et al. 2015). A *T. viride* keratitis case is mentioned in Chouaki et al. (2002), while further two *T. viride* strains were isolated from the nasal mucus of chronic rhinosinusitis patients; however, along with a large number of other fungi (Braun et al. 2003), thus, in these cases, the responsibility of *Trichoderma* for the eosinophilic reaction could not be proven.

5.2 *Other Species*

Cases of human infections due to *T. pseudokoningii* (Table 2 and Fig. 4) reported in the literature but not confirmed by sequence-based molecular identification include an isolate deriving from peritonitis (Rota et al. 2000; Table 2) and a fatal infection in a bone marrow transplant recipient (Gautheret et al. 1995). The causal agent of the latter case has later been reidentified as *T. longibrachiatum* by ITS and *tef1* sequence analyses (Kuhls et al., 1999; Druzhinina et al., 2008). The isolation of a further unconfirmed strain of this species (CCFC 007754) is known from a liver and bowel transplant recipient based on culture collection data (Kredics et al., 2003). The clinical relevance of *T. koningii* (Fig. 4) – a species neotypified by Lieckfeldt et al. (1998) – could not be confirmed yet, as the two fungal peritonitis isolates originally identified as *T. koningii* based on their morphological characteristics (Ragnaud et al., 1984; Campos-Herrero et al., 1996) were later reidentified as *T. longibrachiatum* (Kuhls et al., 1999; Szekeres et al., 2006). *T. hamatum* is mentioned in a study as the causal agent of fungal keratitis (Gharamah et al. 2021). The cerebrospinal fluid and shunt device of a 61-year-old non-immunocompromised male patient who had received two cerebrospinal fluid shunt placements for congenital hydrocephalus revealed *T. reesei* (Table 2 and Fig. 4) as an infectious complication, which was treated with amphotericin B, caspofungin, and voriconazole (Piens et al. 2004). Identification details were not provided.

6 Cases Reported at the Genus Level Only

A series of case reports are also available where the causal agent was reported only as *Trichoderma* sp. (Table 2). Further patients on CAPD with fungal peritonitis were successfully (Bren 1998; Ning and Yang 2020) or unsuccessfully (Esel et al. 2003; Bachu et al. 2020) medicated with antifungals and catheter removal. Fungal endocarditis due to *Trichoderma* sp. was reported in a 66-year-old man with hypertension and ascending aortic replacement (Bustamante-Labarta et al. 2000). Paula-Amato et al. (2002) isolated *Trichoderma* sp. from samples taken from subdural area of a 50-year-old AIDS patient, who was cured with liposomal amphotericin B. The *Trichoderma* fungemia with pulmonary involvement diagnosed in a severely immunocompromised multiple myeloma patient, who underwent autologous hematopoietic cell transplantation, could be cured with voriconazole (Festuccia et al. 2014). *Trichoderma* sp. has also been isolated from the urine of a diabetic male patient with underlying chronic alcoholic liver disease and hepatorenal syndrome type 2 (Chakraborty et al. 2015). In a 51-year-old man with acute myeloid leukemia, Dong et al. (2019) reported a case of invasive pulmonary fungal infection, which was successfully treated with caspofungin and surgery.

Mixed infections involving *Trichoderma* are also known from the literature. A *Trichoderma* sp. and *Absidia corymbifera* infection disseminated to the gastrointestinal tract, liver, kidneys, lung, heart, and skin was diagnosed postmortem in a liver transplant recipient with generalized exanthema, esophageal ulcers, and diarrhea (Stelzmueller et al. 2008). In another fatal case of mixed infection, the co-occurrence of *Trichoderma* sp. with *Aspergillus* sp. was detected in the bronchoalveolar lavage of a 64-year-old female patient with an abdominal sepsis 6 days after a laparotomy and a hyperthermic intraperitoneal chemotherapy (Ariese et al. 2013). *Trichoderma* fungal pneumonia in addition to *Acinetobacter* pneumonia was diagnosed in a 50-year-old female with past medical history of COPD, pneumonia with sepsis, drug withdrawal with seizures, drug overdose requiring intubation, and an allergy history to multiple antibiotics (vancomycin, penicillin, loracarbef, erythromycin, cefixime, clindamycin, and amoxicillin) (Morrell 2017). The treatment with meropenem and voriconazole resulted in recovery. A simultaneous keratitis by *Trichoderma* sp. and *Staphylococcus aureus* was described from the left eye of a 41-year-old man after laser in situ keratomileusis and healed with topical antibacterial and antifungal agents (Mergen et al. 2019), while Hodkin and Gustus (2018) reported about mixed keratitis caused by *Trichoderma* and *Penicillium* in a 50-year-old soft contact lens user, which was effectively treated with natamycin. Further keratitis cases caused by *Trichoderma* sp. mentioned but not detailed in the literature include 14 cases from China (Wang et al. 2009), two cases from India (Venugopal et al. 1989; Sharma et al. 2015), as well as single cases from the USA (Ritterband et al. 2006) and Malaysia (Mohd-Tahir et al. 2012).

7 Conclusions

A high proportion of strains isolated from clinical specimens belong to the *Longibrachiatum* clade of the genus, primarily to *T. longibrachiatum*, while trichodermosis cases have not been attributed to the vast majority of the *Trichoderma* species described so far. Several reports about the occurrence of species belonging to *Trichoderma* sections other than *Longibrachiatum* were inaccurate in many cases due to difficulties of morphological identification. Accordingly, *T. longibrachiatum* deserves an increasing attention in the clinical practice as a potential opportunistic pathogen.

Zhang et al. (2019) documented a *T. longibrachiatum* strain to promote the growth of wheat and act as a biocontrol agent under conditions of salinity stress. In addition, attempts are being made towards the application of this species in agricultural systems in areas with tropical climate, taking the advantage of its tolerance to high temperature values. As *T. longibrachiatum* is the primary causal agent of trichodermoses within the genus, planning the agricultural use of *T. longibrachiatum* strains for biocontrol purposes (Migheli et al. 1998; Sánchez et al. 2007) or their biotechnological exploitation (Sidhu and Sandhu 1980) should be rather abandoned or at least performed with special precaution.

It must also be mentioned that although they are not common, thermotolerant strains can be found also in *Trichoderma* clades other than *Longibrachiatum*. To have the potential to colonize human tissues and cause deep infection, a fungus needs to be able to grow at 37 °C. Thus, to prevent jeopardizing human health, the agricultural application of *Trichoderma* isolates showing considerable growth at 37 °C – particularly those belonging to the *Longibrachiatum* clade – is contraindicated. However, the use of certain bioactive metabolites such as peptaibols (Marik et al. 2019) or enzymes (Urbina-Salazar et al. 2019) of the desired strains may represent an alternative solution.

References

Akagi T, Kawamura C, Terasawa N, Yamaguchi K, Kubo K (2017) Suspected pulmonary infection with *Trichoderma longibrachiatum* after allogeneic stem cell transplantation. Intern Med 56:215–219

Alanio A, Brethon B, Feuilhade de Chauvin M, de Kerviler E, Leblanc T, Lacroix C, Baruchel A, Menotti J (2008) Invasive pulmonary infection due to *Trichoderma longibrachiatum* mimicking invasive aspergillosis in a neutropenic patient successfully treated with voriconazole combined with caspofungin. Clin Infect Dis 46:116–118

Antal Z, Kredics L, Dóczi I, Manczinger L, Kevei F, Nagy E (2002) The physiological features of opportunistic *Trichoderma* strains. Acta Microbiol Immunol Hung 49:393

Antal Z, Kredics L, Pakarinen J, Dóczi I, Andersson M, Salkinoja-Salonen M, Manczinger L, Szekeres A, Hatvani L, Vágvölgyi C, Nagy E (2005) Comparative study of potential virulence factors in human pathogenic and saprophytic *Trichoderma longibrachiatum* strains. Acta Microbiol Immunol Hung 52:341–350

Antal Z, Varga J, Kredics L, Szekeres A, Hatvani L, Manczinger L, Vágvölgyi C, Nagy E (2006) Intraspecific mitochondrial DNA polymorphism within the emerging filamentous fungal pathogen *Trichoderma longibrachiatum*. J Med Microbiol 55:31–35

Ariese K, Hulshoff L, Jansen R, van der Voort PHJ (2013) *Trichoderma*: an unusual bystander in invasive pulmonary aspergillosis. Neth J Crit Care 17:21–24

Aroca TS, Piontelli LE, Cruz CR (2004) Case report: *Trichoderma longibrachiatum* infections in a pediatric patient with peritoneal dialysis. Bol Micol 19:13–17

Bachu R, Hemmings SC, Siddamreddy S. (2020) A rare case of *Trichoderma*-related peritonitis in a patient on peritoneal dialysis. In: Posters, Kidney Week 2020 Reimagined, American Nephrological Society. https://www.asn-online.org/education/kidneyweek/2020/program-abstract.aspx?controlId=3434479

Braun H, Stammberger H, Buzina W, Freudenschuss K, Lackner A, Beham A (2003) Incidence and detection of fungi and eosinophilic granulocytes in chronic rhinosinusitis. Laryngol Rhinol Otol 82:330–340

Bren A (1998) Fungal peritonitis in patients on continuous ambulatory peritoneal dialysis. Eur J Clin Microbiol 17:839–843

Bustamante-Labarta MH, Caramutti V, Allende GN, Weinschelbaum E, Torino AF (2000) Unsuspected embolic fungal endocarditis of an aortic conduit diagnosed by transesophageal echocardiography. J Am Soc Echocardiogr 13:953–954

Campos-Herrero MI, Bordes A, Perera A, Ruiz MC, Fernandez A (1996) *Trichoderma koningii* peritonitis in a patient undergoing peritoneal dialysis. Clin Microbiol Newsl 18:150–151

Cardoso ICE, Severo CB, de Mattos OF, Araújo E, Hochhegger B, Irion KL, Severo LC (2015) Non-*Aspergillus* fungal rhinosinusitis at a tertiary care hospital and the first report of human infection by *Trichoderma asperellum*. Rev Patol Trop 44:395–408

Chakraborty M, Ghosh SD, Bhattacharyya I, Paul D, Gupta MK, Hoque S (2015) *Trichoderma* species in urine: an emerging pathogen or a contaminant. Int J Curr Res 7:15793–15795

Chouaki T, Lavarde V, Lachaud L, Raccurt CP, Hennequin C (2002) Invasive infections due to *Trichoderma* species: report of 2 cases, findings of in vitro susceptibility testing and review of the literature. Clin Infect Dis 35:1360–1367

Colakoğlu G (2003) Airborne fungal spores at the Belgrad forest near the city of Istanbul (Turkey) in the year 2001 and their relation to allergic diseases. J Basic Microbiol 43:376–384

De Miguel D, Gómez P, González R, García-Suárez J, Cuadros JA, Bañas MH, Romanyk J, Burgaleta C (2005) Nonfatal pulmonary *Trichoderma viride* infection in an adult patient with acute myeloid leukemia. Report of one case and review of the literature. Diagn Microbiol Infect Dis 53:33–37

Dong M, Li X, Liu J, Song Z, Zhao H, Wei S, Chen G, Chen J (2019) Successful surgical management of invasive pulmonary fungal infection in patients with leukemia. Infect Drug Resist 12:1675–1681

Dou K, Lu Z, Wu Q, Ni M, Yu C, Wang M, Li Y, Wang X, Xie H, Chen J, Zhang C (2020) MIST: a Multilocus Identification System for *Trichoderma*. Appl Environ Microbiol 86:e01532–e01520

Druzhinina IS, LaFe K, Willinger B, Komoń-Zelazowska M, Ammirati J, Kubicek CP, Rogers JD (2007) An unknown *Hypocreaceae* species isolated from human lung tissue of a patient with non-fatal pulmonary fibrosis. Clin Microbiol Newsl 29:180–184

Druzhinina IS, Komoń-Zelazowska M, Kredics L, Hatvani L, Antal Z, Belayneh T, Kubicek CP (2008) Alternative reproductive strategies of *Hypocrea orientalis* and genetically close but clonal *Trichoderma longibrachiatum*, both capable to cause invasive mycoses of humans. Microbiol-SGM 154:3447–3459

Escudero CMR, Pino Corral E, Munoz R (1976) Pulmonary mycoma cause by *Trichoderma viride*. Actas Dermosifiliogr 67:673–680

Esel D, Koç AN, Utaş C, Karaca N, Bozdemir N (2003) Fatal peritonitis due to *Trichoderma* sp. in a patient undergoing continuous ambulatory peritoneal dialysis. Mycoses 46:71–73

Espinel-Ingroff A (2001) In vitro fungicidal activities of voriconazole, itraconazole, and amphotericin B against opportunistic moniliaceous and dematiaceous fungi. J Clin Microbiol 39:954–958

Espinel-Ingroff A, Chaturvedi V, Fothergill A, Rinaldi MG (2002) Optimal testing conditions for determining MICs and minimum fungicidal concentrations of new and established antifungal agents for uncommon molds: NCCLS collaborative study. J Clin Microbiol 40:3776–3781

Festuccia M, Giaccone L, Gay F, Brunello L, Maffini E, Ferrando F, Talamo E, Boccadoro M, Serra R, Barbui A, Bruno B (2014) *Trichoderma* species fungemia after high dose chemotherapy and autologous stem cell transplantation: a case report. Transpl Infect Dis 16:653–657

Furukawa H, Kusne S, Sutton DA, Manez R, Carrau R, Nichols L, Abu-Elmagd K, Skedros D, Todo S, Rinaldi MG (1998) Acute invasive sinusitis due to *Trichoderma longibrachiatum* in a liver and small bowel transplant recipient. Clin Infect Dis 26:487–489

Gautheret A, Dromer F, Bourhis J-H, Andremont A (1995) *Trichoderma pseudokoningii* as a cause of fatal infection in a bone marrow transplant recipient. Clin Infect Dis 20:1063–1064

Gharamah AA, Moharram AM, Ismail MA, Al-Hussaini AK (2021) Bacterial and fungal keratitis in Upper Egypt: in vitro screening of enzymes, toxins and antifungal activity. Indian J Ophthalmol 62:196–203

Guarro J, Antolin-Ayala MI, Gene J, Gutierrez-Calzada J, Nieves-Diez C, Ortoneda M (1999) Fatal case of *Trichoderma harzianum* infection in a renal transplant recipient. J Clin Microbiol 37:3751–3755

Guiserix J, Ramdane M, Finielz P, Michault A, Rajaonarivelo P (1996) *Trichoderma harzianum* peritonitis in peritoneal dialysis. Nephron 74:473–474

Guo J, He LH, Ni LJ, Lv L, Deng SW, Wu WJ (2017) Peritonitis caused by *Trichoderma longibrachiatum* infection: a case report. Lab Med 12:1176–1178. (in Chinese)

Hageskal G, Knutsen AK, Gaustad P, de Hoog GS, Skaar I (2006) Diversity and significance of mold species in Norwegian drinking water. Appl Environ Microbiol 72:7586–7593

Hatvani L, Kocsubé S, Kredics L, Vágvölgyi C, Šegvic Klaric M, Mlinaric-Missoni E, Druzhinina IS, Atanasova L, Kosalec I. (2012) *Trichoderma* on the dark side: report on mushroom green mould disease and clinical cases in Croatia. In: 14th DKMT Euroregional conference on environment and health, Szeged, Hungary, Proceedings CD.

Hatvani L, Manczinger L, Vágvölgyi C, Kredics L (2013) *Trichoderma* as a human pathogen. In: Mukherjee PK, Horwitz BA, Singh US, Mukherjee M, Schmoll M (eds) *Trichoderma* – biology and applications. CABI, Wallingford, pp 292–313

Hatvani L, Homa M, Chenthamara K, Cai F, Kocsubé S, Büchner R, Atanasova L, Mlinaric-Missoni E, Manikandan P, Revathi R, Dóczi I, Bogáts G, Narendran V, Vágvölgyi C, Kredics L (2019) Agricultural systems as potential sources of emerging human mycoses caused by *Trichoderma*: a successful, common phylotype of *Trichoderma longibrachiatum* in the frontline. FEMS Microbiol Lett 366:fnz246

He D, Hao J, Gao S, Wan X, Wang W, Shan Q, Wang L (2016) Etiological analysis of fungal keratitis and rapid identification of predominant fungal pathogens. Mycopathologia 181:75–82

Hennequin C, Chouaki T, Pichon JC, Strunski V, Raccurt C (2000) Otitis externa due to *Trichoderma longibrachiatum*. Eur J Clin Microbiol 19:641–642

Hodkin MJ, Gustus RC (2018) Fungal keratitis associated with airborne organic debris and soft contacts lenses: case reports and review of the literature. Eye Contact Lens 44:16–21

Jacobs F, Byl B, Bourgeois N, Coremans-Pelseneer J, Florquin S, Depré G, Van de Stadt J, Adler M, Gelin M, Thys JP (1992) *Trichoderma viride* infection in a liver transplant recipient. Mycoses 35:301–303

Kantarcioğlu AS, Celkan T, Yücel A, Mikami Y, Kurugoglu S, Mitani H, Altas K (2009) Fatal *Trichoderma harzianum* infection in a leukemic pediatric patient. Med Mycol 47:207–215

Khan A, Giusti R, Silverman B, Schneider A (2001) An unusual etiology of persistent cough in an 8-year-old girl with cystic fibrosis. Ann Allerg Asthma Immunol 87:191–195

Kratzer C, Tobudic S, Schmoll M, Graninger W, Georgopoulos A (2006) In vitro activity and synergism of amphotericin B, azoles and cationic antimicrobials against the emerging pathogen *Trichoderma* spp. J Antimicrob Chemother 58:1058–1061

Kredics L, Antal Z, Dóczi I, Manczinger L, Kevei F, Nagy E (2003) Clinical importance of the genus *Trichoderma*. Acta Microbiol Immunol Hung 50:105–117

Kredics L, Antal Z, Szekeres A, Manczinger L, Dóczi I, Kevei F, Nagy E (2004) Production of extracellular proteases by human pathogenic *Trichoderma longibrachiatum* strains. Acta Microbiol Immunol Hung 51:283–295

Kredics L, Dóczi I, Antal Z, Bartyik K, Molnár EG, Manczinger L, Hatvani L, Vágvölgyi C, Nagy E (2006) Emergence of the filamentous fungal opportunist *Trichoderma longibrachiatum* in Hungary. Acta Microbiol Immunol Hung 53:305

Kredics L, Hatvani L, Manczinger L, Vágvölgyi C, Antal Z (2011) Chapter 62. *Trichoderma*. In: Liu D (ed) Molecular detection of human fungal pathogens. Taylor & Francis Group, London, pp 509–526

Kuhls K, Lieckfeldt E, Börner T, Guého E (1999) Molecular reidentification of human pathogenic *Trichoderma* isolates as *Trichoderma longibrachiatum* and *Trichoderma citrinoviride*. Med Mycol 37:25–33

Kviliute R, Paskevicius A, Gulbinovic J, Stulpinas R, Griskevicius L (2008) Nonfatal *Trichoderma citrinoviride* pneumonia in an acute myeloid leukemia patient. Ann Hematol 87:501–502

Lagrange-Xélot M, Schlemmer F, Gallien S, Lacroix C, Molina JM (2008) *Trichoderma* fungaemia in a neutropenic patient with pulmonary cancer and human immunodeficiency virus infection. Clin Microbiol Infect 14:1190–1192

Lee HJ, Kim DW, Cho HS, Lim MH, Jung EY, Lee DW, Park DJ, Chang SH (2007) Peritonitis and intra-abdominal abscess by *Trichoderma longibrachiatum* in a patient undergoing continuous ambulatory peritoneal dialysis (CAPD). Korean J Nephrol 26:254–257

Lieckfeldt E, Samuels GJ, Gams W (1998) Neotypification of *Trichoderma koningii* and its *Hypocrea koningii* teleomorph. Can J Bot 76:1507–1522

Lieckfeldt E, Samules GJ, Nirenberg HI, Petrini O (1999) A morphological and molecular perspective of *Trichoderma viride*: is it one or two species? Appl Environ Microbiol 65:2418–2428

Loeppky CB, Sprouse RF, Carlson JV, Everett ED (1983) *Trichoderma viride* peritonitis. South Med J 76:798–799

Lübeck M, Poulsen SK, Lübeck PS, Jensen DF, Thrane U (2000) Identification of *Trichoderma* strains from building materials by ITS1 ribotyping, UP-PCR fingerprinting and UP-PCR cross hybridization. FEMS Microbiol Lett 185:129–134

Marco F, Pfaller MA, Messer SA, Jones RN (1998) Antifungal activity of a new triazole, voriconazole (UK-109,496), compared with three other antifungal agents tested against clinical isolates of filamentous fungi. Med Mycol 36:433–436

Marik T, Tyagi C, Balázs D, Urbán P, Szepesi Á, Bakacsy L, Endre G, Rakk D, Szekeres A, Andersson MA, Salonen H, Vágvölgyi C, Kredics L, Druzhinina IS (2019) Structural diversity and bioactivities of peptaibol compounds from the *Longibrachiatum* clade of the filamentous fungal genus *Trichoderma*. Front Microbiol 10:1434

Matos T, Haase G, Gerrits van den Ende AHG, de Hoog GS (2003) Molecular diversity of oligotrophic and neurotropic members of the black yeast genus *Exophiala*, with accent on *E. dermatitidis*. Anton Leeuw Int J G 83:293–303

Mergen B, Sarici AM, Baltu F, Bahar-Tokman H, Iskeleli G (2019) Mixed keratitis caused by *Trichoderma* and methicillin-resistant *Staphylococcus aureus* after uneventful laser *in situ* keratomileusis. JCRS Online Case Rep 7:14–16

Migheli Q, González-Candelas L, Dealessi L, Camponogara A, Ramón-Vidal D (1998) Transformants of *Trichoderma longibrachiatum* overexpressing the beta-1,4-endoglucanase gene *egl1* show enhanced biocontrol of *Pythium ultimum* on cucumber. Phytopathology 88:673–677

Mohd-Tahir F, Norhayati A, Siti-Raihan I, Ibrahim M (2012) A 5-year retrospective review of fungal keratitis at Hospital Universiti Sains Malaysia. Interdisc Persp Infect Dis 2012:851563

Molnár-Gábor E, Dóczi I, Hatvani L, Vágvölgyi C, Kredics L (2013) Isolated sinusitis sphenoidalis caused by *Trichoderma longibrachiatum* in an immunocompetent patient with headache. J Med Microbiol 62:1249–1252

Morrell K (2017) Successful treatment of *Trichoderma* and superimposed *Acinetobacter* pneumonia with voriconazole and meropenem. In: Resident posters, ASHP Midyear 2017 Clinical Meeting & Exhibition, Orlando, Florida, December 3–7, 2017. https://www.eventscribe.com/2017/ASHP-Meeting/ajaxcalls/PosterInfo.asp?efp=WUFORk1KUU8yMDk5&PosterID=123481&rnd=0.9524601

Munoz FM, Demmler GJ, Travis WR, Ogden AK, Rossmann SN, Rinaldi MG (1997) *Trichoderma longibrachiatum* infection in a pediatric patient with aplastic anemia. J Clin Microbiol 35:499–503

Myoken Y, Sugata T, Fujita Y, Asaoku H, Fujihara M, Mikami Y (2002) Fatal necrotizing stomatitis due to *Trichoderma longibrachiatum* in a neutropenic patient with malignant lymphoma: a case report. Int J Oral Maxillofac Surg 31:688–691

Ning N, Yang K (2020) Experience of clinical pharmacists participating in the selection of antimicrobial agents in a patient with peritoneal dialysis associated peritonitis caused by *Trichoderma*. Strait Pharm J 32:192–194. (in Chinese)

Paredes K, Capilla J, Mayayo E, Guarro J (2016) Virulence and experimental treatment of *Trichoderma longibrachiatum*, a fungus refractory to treatment. Antimicrob Agents Chemother 60:5029–5032

Paula-Amato JG, Amato VS, Amato Neto V, Heins-Vaccari EM (2002) An AIDS patient with *Trichoderma*. Rev Soc Bras Med Trop 35:269

Piens M-A, Celard M, de Monbrison F, Grando J, Vandenesch F, Mottolese C, Picot S (2004) *Trichoderma* infection of cerebrospinal fluid shunt device in a non immunocompromised patient. J Mycol Med 14:49–51

Ragnaud JM, Marceau C, Roche-Bezian MC, Wone C (1984) *Trichoderma koningii* peritoneal infection in continuous ambulatory peritoneal dialysis. Méd Maladies Infect 7:402–405

Ranque S, Garcia-Hermoso D, Michel-Nguyen A, Dumon H (2008) Isolation of *Trichoderma atroviride* from a liver transplant. J Mycol Med 18:234–236

Recio R, Melendez-Carmona MA, Martín-Higuera MC, Perez V, Lopez E, Lopez-Medrano F, Perez-Ayala A (2019) *Trichoderma longibrachiatum*: an unusual pathogen of fungal Pericarditis. Clin Microbiol Infect 25:586–587

Richter S, Cormican MG, Pfaller MA, Lee CK, Gingrich R, Rinaldi MG, Sutton DA (1999) Fatal disseminated *Trichoderma longibrachiatum* infection in an adult bone marrow transplant patient: species identification and review of the literature. J Clin Microbiol 37:1154–1160

Ritterband DC, Seedor JA, Shah MK, Koplin RS, McCormick SA (2006) Fungal keratitis at the New York Eye and Ear Infirmary. Cornea 25:264–267

Robertson MH (1970) Fungi in fluids – a hazard of intravenous therapy. J Med Microbiol 3:99–102

Rodríguez Peralta LI, Manas Vera MR, Garcia Delgado MJ, Perez de la Cruz AJ (2013) Endocarditis caused by *Trichoderma longibrachiatum* in a patient receiving home parenteral nutrition. Nutr Hosp 28:961–964

Román-Soto S, Álvarez-Rojas E, García-Rodríguez J (2019) Skin infection due to *Trichoderma longibrachiatum* in a haematological paediatric patient. Clin Microbiol Infect 25:1383–1384

Rota S, Marchesi C, Farina C, de Biévre C (2000) *Trichoderma pseudokoningii* peritonitis in automated peritoneal dialysis patient successfully treated by early catheter removal. Perit Dial Int 20:91–94

Samuels GJ, Ismaiel A (2011) *Hypocrea peltata*: a mycological Dr Jekyll and Mr Hyde? Mycologia 103:616–630

Samuels GJ, Petrini O, Kuhls K, Lieckfeldt E, Kubicek CP (1998) The *Hypocrea schweinitzii* complex and *Trichoderma* sect. Longibrachiatum. Stud Mycol 41:1–54

Sánchez V, Rebolledo O, Picaso RM, Cárdenas E, Córdova J, González O, Samuels GJ (2007) In vitro antagonism of *Thielaviopsis paradoxa* by *Trichoderma longibrachiatum*. Mycopathologia 163:49–58

Sandoval-Denis M, Sutton DA, Cano-Lira JF, Gene J, Fothergill AW, Wiederhold NP, Guarro J (2014) Phylogeny of the clinically relevant species of the emerging fungus *Trichoderma* and their antifungal susceptibilities. J Clin Microbiol 52:2112–2125

Santillan Salas CF, Joshi AY, Dhiman N, Banerjee R, Huskins WC, Wengenack NL, Henry NK (2011) Fatal post-operative *Trichoderma longibrachiatum* mediastinitis and peritonitis in a paediatric patient with complex congenital cardiac disease on peritoneal dialysis. J Med Microbiol 60:1869–1871

Sautour M, Chrétien ML, Valot S, Basmaciyan L, Legouge C, Verrier T, Gonssaud B, Abou-Hanna H, Dalle F, Caillot D (2018) First case of proven invasive pulmonary infection due to *Trichoderma longibrachiatum* in a neutropenic patient with acute leukemia. J Mycol Med 28:659–662

Seguin P, Degeilh B, Grulois I, Gacouin A, Maugendre S, Dufour T, Dupont B, Camus C (1995) Successful treatment of a brain abscess due to *Trichoderma longibrachiatum* after surgical resection. Eur J Clin Microbiol 14:445–448

Seidel D, Graeff LAD, Vehreschild MJGT, Wisplinghoff H, Ziegler M, Vehreschild JJ, Liss B, Hamprecht A, Köhler P, Racil Z, Klimko N, Sheppard DC, Herbrecht R, Chowdhary A, Cornely OA, FungiScope Group (2017) FungiScope™ – Global emerging fungal infection registry. Mycoses 60:508–516

Sharma S, Das S, Virdi A, Fernandes M, Sahu SK, Koday NK, Ali H, Garg P, Motukupally SR (2015) Re-appraisal of topical 1% voriconazole and 5% natamycin in the treatment of fungal keratitis in a randomised trial. Br J Ophthalmol 99:1190–1195

Sidhu M, Sandhu D (1980) Single-cell protein production by *Trichoderma longibrachiatum* on treated cane bagasse. Biotechnol Bioeng 22:689–692

Stelzmueller I, Lass-Floerl C, Geltner C, Graziadei I, Schneeberger S, Antretter H, Mueller L, Zelger B, Singh N, Pruett TL, Margreiter R, Bonatti H (2008) Zygomycosis and other rare filamentous fungal infections in solid organ transplant recipients. Transpl Int 21:534–546

Summerbell RC (2003) Ascomycetes: *Aspergillus*, *Fusarium*, *Sporothrix*, *Piedra*, and relatives. In: Howard DH (ed) Pathogenic fungi in humans and animals. Marcel Dekker Inc., New York, pp 437–449

Szekeres A, Láday M, Kredics L, Varga J, Antal Z, Hatvani L, Manczinger L, Vágvölgyi C, Nagy E (2006) Rapid identification of clinical *Trichoderma longibrachiatum* isolates by cellulose-acetate electrophoresis mediated isoenzyme analysis. Clin Microbiol Infect 12:369–375

Tang P, Mohan S, Sigler L, Witterick I, Summerbell R, Campbell I, Mazzulli T (2003) Allergic fungal sinusitis associated with *Trichoderma longibrachiatum*. J Clin Microbiol 41:5333–5336

Tanis BC, van der Pijl H, van Ogtrop ML, Kibbelaar RE, Chang PC (1995) Fatal fungal peritonitis by *Trichoderma longibrachiatum* complicating peritoneal dialysis. Nephrol Dial Transpl 10:114–116

Tascini C, Cardinali G, Barletta V, Di Paolo A, Leonildi A, Zucchelli G, Corte L, Colabella C, Roscini L, Consorte A, Pasticci MB, Menichetti F, Bongiorni MG (2016) First case of *Trichoderma longibrachiatum* CIED (Cardiac Implantable Electronic Device)-associated endocarditis in a non-immunocompromised host: biofilm removal and diagnostic problems in the light of the current literature. Mycopathologia 181:297–303

Thrane U, Poulsen SB, Nirenberg HI, Lieckfeldt E (2001) Identification of *Trichoderma* strains by image analysis of HPLC chromatograms. FEMS Microbiol Lett 203:249–255

Trabelsi S, Hariga D, Khaled S (2010) First case of *Trichoderma longibrachiatum* infection in a renal transplant recipient in Tunisia and review of the literature. Tunis Med 88:52–57

Urbina-Salazar A, Inca-Torres AR, Falcón-García G, Carbonero-Aguilar P, Rodríguez-Morgado B, del Campo JA, Parrado J, Bautista J (2019) Chitinase production by *Trichoderma harzianum* grown on a chitin-rich mushroom byproduct formulated medium. Waste Biomass Valor 10:2915–2923

Venugopal PL, Venugopal TL, Gomathi A, Ramakrishna ES, Ilavarasi S (1989) Mycotic keratitis in Madras. Indian J Pathol Microbiol 32:190–197

Wang L, Sun S, Jing Y, Han L, Zhang H, Yue J (2009) Spectrum of fungal keratitis in central China. Clin Exp Ophthalmol 37:763771

Warnock DW, Johnson EM (1991) Clinical manifestation and management of hyalohyphomycosis, phaehyphomycosis and other uncommon forms of fungal infection in the compromised patient. In: Warnock DW, Richardson MD (eds) Fungal infection in the compromised patient. Wiley, Chichester, pp 248–310

Yang SH, Ma SH, Cao ZS, Ma JX (2019) A case report of peritoneal dialysis infection caused by *Candida tropicalis* combined with *Trichoderma longibrachiatum*. Chin J Lab Diag 23:2002–2003. (in Chinese)

Zhang GJ, Li DM (2019) Trichodermasis: a review of the literature. Mycosystema 38:1287–1297

Zhang S, Xu B, Gan Y (2019) Seed treatment with *Trichoderma longibrachiatum* T6 promotes wheat seedling growth under NaCl stress through activating the enzymatic and nonenzymatic antioxidant defense systems. Int J Mol Sci 20:3729

Zhou YB, Zhang GJ, Song YG, Sun LN, Chen YH, Sun TT, Li RY, Liu W, Li DM (2020) Application of laser capture microdissection and polymerase chain reaction in the diagnosis of *Trichoderma longibrachiatum* infection: a promising diagnostic tool for 'fungal contaminants' infection. Med Mycol 58:315–321

Index

N. Amaresan et al. (eds.), *Advances in Trichoderma Biology for Agricultural Applications*, Fungal Biology, https://doi.org/10.1007/978-3-030-91650-3

C

D

R

S

T

U

V

The manufacturer's authorised representative in the EU is Springer Nature Customer Service Centre GmbH, Europaplatz 3, 69115 Heidelberg, Germany. If you have any concerns regarding our products, please contact ProductSafety@springernature.com

Printed and bound by CPI Group (UK) Ltd, Croydon, CR0 4YY
30/09/2026
02225329-0019